U0343765

马业英汉双解词典

English–Chinese Equine Dictionary

陆泉枝　编　著

中国农业出版社
北　京

图书在版编目（CIP）数据

马业英汉双解词典 / 陆泉枝编著. —北京：中国农业出版社，2022.12
ISBN 978-7-109-23607-3

Ⅰ. ①马… Ⅱ. ①陆… Ⅲ. ①马-双解词典-英、汉
Ⅳ. ①S821-61

中国版本图书馆 CIP 数据核字（2017）第 289479 号

马业英汉双解词典
MAYE YINGHAN SHUANGJIE CIDIAN

中国农业出版社出版
地址：北京市朝阳区麦子店街 18 号楼
邮编：100125
责任编辑：郭永立 张艳晶
版式设计：王 晨 责任校对：吴丽婷 责任印制：王 宏
印刷：北京通州皇家印刷厂
版次：2022 年 12 月第 1 版
印次：2022 年 12 月北京第 1 次印刷
发行：新华书店北京发行所
开本：880mm × 1230mm 1/32
印张：18.5
字数：982 千字
定价：198.00 元

词典特邀审校人员

（以姓氏笔画为序）

丁伯良　王　煜　王振山
文　立　李　桢　汪雅丽
陈卫忠　郭永立

For my parents who gave me life,

&

the chestnut horse who worked along with us.

谨以本词典献给养育我的父母

和

那匹曾与我们一道劳作的栗马。

Key to Dictionary Usage
凡　例

一部完整的词典，在体例上基本由词目、注音、词类、复数、门类、释义、例证等部分构成，以下分别简单予以说明：

1　词目

词目是词条说明的对象，词目可以是单词、短语、复合词、缩略语等。

1.1　词目均用黑体排印，并依照英文字母顺序排列。对于拼写相同而词义毫不相关的词，分别设立，并在词目的右上方标注 1、2 等数码。例如，**bay**[1] 与 **bay**[2]、**draft**[1] 与 **draft**[2]。

1.2　部分源于外来语的词目，采用鱼尾号【　】进行标识。例如，【Latin】、【French】。

1.3　对于词目的同义词采用 syn 标识，反义词采用 opp 标识；相关词目之间的比较采用 compare 标识；普通短语用 PHR 标识，动词短语则用 PHR V 标识。

1.4　对于词目的派生词，大多不单独以词条形式列出，而是用“ | ”与单词释义部分隔开。例如，**trauma** /ˈtrɔːmə; *AmE* ˈtraʊmə/ n. a serious physical injury resulting from an accident（意外事故给身体造成的严重伤害）创伤，外伤 | **traumatic** /trɔːˈmætɪk/ adj.

2　注音

注音是词目发音的基础，本词典采用最新国际音标，基本只对单词进行了标注。

2.1　单词注音紧接词目之后，放在两斜杠之间。例如，**stress** /stres/、**thorax** /ˈθɔːræks/。

2.2　对于美式与英式音标存在差异的单词，美式注音单独采用 *AmE* 标出。例如，**horse** /hɔːs; *AmE* hɔːrs/。

2.3　对有些法语外来词，注音采用法语体系。例如，**Mèrens** /mɛrã/。

2.4　对于汉语外来词，发音不予标注。例如，**Sanhe**、**Yili**。

3　词类

本词典中收录单词绝大多数为名词、形容词和动词，名词用“n.”标识，形容词用“adj.”标识；动词分为及物动词和不及物动词，前者用“vt.”标识，后者用“vi.”标识，两种兼用则以“v.”标识。

4　复数

英语中绝大多数名词复数直接加“s”，这里仅对少数特殊名词复数进行标示，同时给出音标。

4.1　名词复数用“*pl.*”标识，例如，**meninx** /ˈmiːnɪŋks/n.（*pl.* **meninges** /mɪˈnɪndʒiːz/）。

4.2　个别名词仅以复数形式存在，在区分词类时采用“n. pl.”标识。例如，**breeches** /ˈbrɪtʃɪz/ n. pl.。

5　门类

对于属于特定门类的词目，采用鱼尾号【　】对其进行了标识。

5.1　本词典大多数词目没有进行门类区分。

5.2　对于词目的具体门类划分，请参见词典标识页。

6　释义

词典在词条释义上，力求言简意赅、准确得体。

6.1　词目释义为英中双语对照，英语释义在前，汉语释义随后置于括号（　）之中，然后是相应的汉语词目，如下所示：

jaw /dʒɔː/ n.【anatomy】either of the two bony structures in vertebrates that form the framework of the mouth and contain the teeth（构成脊椎动物口腔的两块骨质结构，其上附生牙齿）颌骨

6.2　单个词目有多个释义，采用数字 1、2 等分别标识，如下所示：

carriage /ˈkærɪdʒ/ n. **1** a horse - drawn vehicle 马车，舆 **2** the manner or way in which a horse carries his head or moves（马举头或运步的方式）姿态，姿势

7　例证

词目例证分为短语和例句，大多出自国外教科书和相关网站，在提供惯用表达的同时，也提供了许多基本信息和马业的相关知识。

7.1　例证部分与释义部分之间用“:”隔开，两个例证之间则用“◇”相隔。如下所示：

salivary /səˈlaɪvəri; *AmE* ˈsæləveri/ adj. of or relating to saliva 唾液的：salivary gland 唾液腺 ◇ salivary amylase 唾液淀粉酶

7.2　对于词目中的例句，提供相应的汉语译文。例如，Horses' heart rates vary from about 25 beats/min at rest to 250 beats/min during maximal exercise. 马的心率可以从静息时的 25 次/分到最大运动强度时的 250 次/分之间变动。

Labels Used in the Dictionary

词典标识

Subject Labels 门类标识

【anatomy】解剖学
【biochemistry】生物化学
【color】毛色
【cutting】截牛
【dressage】盛装舞步
【driving】驾车
【farriery】钉蹄
【genetics】遗传学
【harness racing】轻驾车赛
【hunting】狩猎
【jumping】障碍赛
【mythology】神话
【nutrition】营养学
【physiology】生理学
【polo】马球
【racing】赛马
【rodeo】牛仔竞技
【roping】套牛
【vaulting】马上体操
【vet】兽医

Speech Labels 语言标识

abbr. 缩写
AD 公元后
adj. 形容词
also 亦作
AmE 美式英语
Arabic 阿拉伯语
BC 公元前
BrE 英式英语
Chinese 汉语
colloq 口语
compare 比较
dated 旧用
French 法语
IDM 习语
interj. 叹词
Irish 爱尔兰语
Latin 拉丁语
n. 名词
n. pl. 复数名词
opp 反义词
PHR 短语
PHR V 动词短语
pl. 复数
pt 过去式
pp 过去分词
pref. 前缀
saying 谚语
sb 某人
Scots 苏格兰语
see also 参看
slang 俚语
Spanish 西班牙语

sth 某事
syn 同义词
usu. 常作
v. 动词
vi. 不及物动词
vt. 及物动词

前　言

《马业英汉双解词典》是一部专业性双语辞书，共收录词目一万一千余条，内容涵盖马术、赛马、牛仔竞技、马球以及马进化、解剖、生理、遗传、繁育、营养等各个领域的词汇，可供广大马业人士和英语研习者参考使用。

本词典编排体例以国内外权威词典为范本，释意力求言简意赅。在充分占有资料和广泛收集的基础上，内容丰富而详实。词条编排以字母为序并进行了学科项目分类，并提供了不少例句，以佐读者更好地使用。词典附录分多个专项，主要包括世界马种、奥运马术奖牌统计、世界经典赛事记录、国内外机构和相关网络资源等版块，具有重要的参考价值。

十几年来，纵然编者竭尽所能，力求本词典在广度、布局、语言和体例等诸多方面精益求精，以适应新时代读者的需要。然而，直到付梓之时，仍深感词典存在许多尚需改进之处。现在我们把这本《马业英汉双解词典》奉献给广大读者和同行专家，请大家在使用过程中批评、指正，以便今后逐步完善本词典，更好地为读者服务。

陆泉枝
上海理工大学
2022 年 9 月

Foreword

The *English-Chinese Equine Dictionary* (ECED), listing more than 11,000 horse words and terms, is a bilingual reference book for both professionals in horse industry and learners of English in general. The equine terms included here cover a wide range of areas and disciplines, roaming from horse racing, equestrian sports, rodeo, and polo to the evolution, anatomy, physiology, genetics, breeding, and nutrition of horses.

Modeling upon the format of several most-acclaimed dictionaries, the ECED endeavors to interpret horse words in the most concise yet comprehensive manner through extensive reference work and detailed explanation. Arranged alphabetically, this dictionary groups most of the words into a dozen categories, providing sample phrases and sentences wherever necessary. In addition, the ECED carries cartloads of valuable data in its appendices, the horse breeds of the world, the classical horse races, and the winners of the Olympic equestrian events, to name only a few.

During the past fifteen years, the author has been trying to improve the dictionary in terms of its scope, diction, layout as well as its exemplification so as to meet the readers' rising needs in the new age. However, I find there is still much space for its perfection in the future. Here, I present this dictionary to the readership, and I welcome any comments and suggestions from the public for revision in its following edition.

Frank Lu
University of Shanghai for Science and Technology
September, 2022

Foreword

致　谢

从构思到完工，一晃二十年已逝。经过不懈的努力与坚持，《马业英汉双解词典》终于要付梓出版了。以一人之力编纂这本词典，所费心力难以计量。唯有践行之后，我方才彻悟塞缪尔·约翰逊（1709—1784）缘何将这项繁琐的工作讽喻为“无害的苦工”。

诚然，《马业英汉双解词典》的顺利完工，无疑也离不开众多人士和机构的大力协助和鼎力支持。首先，要感谢中国农业大学昔日同窗岳高峰（中国马业协会秘书长），是他多方筹措资金，促成了本词典的问世。在本词典的编辑过程中，承蒙马业诸多前辈的悉心指导和教诲，这其中有王铁权博士（中国农业科学院畜牧所研究员）、王绍松副教授（宁夏农学院）、王振山博士（中国纯血马登记管理委员会）、韩国才副教授（中国农业大学）、李伟教练（上海马术运动场）等。此外，业界许多马友也提供了非常有用的参考意见。

能持之以恒地完成这项工作，同样离不开家人的鼓励和亲友的支持，是他们在我每欲放弃之时，又让我鼓起勇气，抖擞精神走到了终点。此外，好友刘玉方在词典附录整理过程中也分担了不少任务，这里也表示由衷的谢意。在词典竣工之时，本人研究生期间的导师金衡山教授（华东师范大学）还就词典修缮提出了非常中肯的建议，美国友人雪莉玛·鲍尔博士还对个别生僻单词的发音进行了核对，这里一并予以感谢！

限于篇幅和记忆所限，对于那些曾帮助过我但未在此列出的诸多友人，这里也要衷心地说一声谢谢，并祝愿大家一切顺利！

Acknowledgements

After two decades of hard work since its conception in 2001, the *English-Chinese Equine Dicitonary* (ECED) is soon to be published. To finish a dictionary of equine terms of this scope, I spent far more time and energy than I had thought at first. It is through the process of compiling the ECED that I came to understand fully why Dr. Samuel Johnson (1709 – 1784) would explain laconically the term "lexicographer" as "a harmless drudge."

Fortunately, I am not alone in this journey. First of all, my sincere thanks to my college roommate, Mr. Yue Gaofeng (CHIA Secretary-General), for raising funds and facilitating the publication of this dictionary, without his help this dictionary could never have taken its material form in the first place. I am also indebted to many friends and experts for their unfailing guidance and direction, in particular to late Dr. Wang Tiequan (Research Fellow of the Institute of Animal Sciences, CAAS), late Associate Professor Wang Shaosong (Ningxia Agricultural College), Dr. Wang Zhenshan (China Stud Book Authority), Associate Professor Han Guocai (China Agricultural University), and Mr. Li Wei (Shanghai Equestrian Association). Besides, I also feel deeply grateful to many horse lovers for their kind advice and timely help.

Here, I would like to take this opportunity to express my heart-felt gratitude to my family who have encouraged me and cheered me up throughout this odyssey whenever I call it quits. In addition, I also thank my old friend, Mr. Liu Yufang, for his assistance in sorting out the appendices. Near the finish line of this marathon, my PhD advisor Professor Jin Hengshan (East China Normal University) gave me very constructive comment concerning the layout and revision of the dictionary; while my American friend, Dr. Sherrema Bower, also provided unfailing assistance in helping me nail down the pronunciation of several rare words.

Here, I conclude my acknowledgements with my thanks to those who are not listed here yet have offered me much support during the past years.

Acknowledgements

Contents
目　　录

The Oictionary

词 典 正 文

A

A /eɪ/abbr. artery 动脉

AAA abbr. American Albino Association 美国白马协会

AAEP abbr. American Association of Equine Practitioners 美国马兽医协会

AAEP Lameness Scale n. a system developed recently by the American Association of Equine Practitioners to evaluate degrees of lameness of horses, ranging from 0 to 5 according to situations (由美国马兽医协会制定的等级,用来评估马匹的跛行级别,根据严重程度划分为0~5六个等级)美国马兽医协会跛行级别 compare Obel Lameness Grades

AAHC abbr. American Albino Horse Club 美国白马俱乐部

AAOBPPH abbr. American Association of Owners and Breeders of Peruvian Paso Horses 美国秘鲁帕索马马主与育马者协会

abandon /əˈbændən/ vt.【racing】to withdraw or cancel a horse race 取消(比赛): an abandoned horse race 一场取消了的赛马比赛

abasia /əˈbeɪʒə/ n. inability to walk due to impaired muscular coordination (肌肉协调性紊乱导致个体无法行走)步行不能,步行障碍

abattoir /ˈæbətwɑː(r)/ n. a place where animals are butchered; slaughterhouse 屠宰场

Abbot buggy n. a small four-wheeled carriage with semi-elliptical side springs popular in the early 19th century (19世纪早期流行的一种小型四轮马车,装有半椭圆形减震弹簧)艾博特马车

abdomen /ˈæbdəmən/ n. the part of the body between the chest and the pelvis (躯体介于胸腔与骨盆之间的部分)腹部 syn belly

abdominal /æbˈdɒmɪnl; *AmE* -ˈdɑːm-/ adj.【anatomy】of or relating to the abdomen 腹部的,腹腔的: abdominal cavity 腹腔 ◇ abdominal gestation 腹腔妊娠 ◇ abdominal pains 腹痛 ◇ abdominal worm 腹腔寄生虫

n. pl. **abdominals** (abbr. **abs**) the muscles of the abdomen 腹肌

abdominocentesis /æbˌdɒmɪnəsenˈtiːsɪs/ n. the medical practice of removing accumulated fluid in the abdominal cavity through a needle (通过针头抽吸腹腔内积液的医疗方法)腹腔穿刺术

abduct /æbˈdʌkt/ vi.【physiology】to pull or draw away from the midline of the body or limb (偏离身体或四肢中轴)外展 compare adduct | **abduction** /æbˈdʌkʃn/ n.

abductor /æbˈdʌktə(r)/ n.【anatomy】a muscle that draws a body part away from the midline of the body or limb (使部位偏离身体或四肢中轴的肌肉)外展肌

abnormal /æbˈnɔːml; *AmE* -ˈnɔːrml/ adj. not typical or regular; not normal; aberrant 反常的,异常的: abnormal estrus 异常发情 ◇ abnormal sexual behavior 异常性行为 | **abnormality** /ˌæbnɔːˈmæləti; *AmE* -nɔːrˈm-/ n.

abort /əˈbɔːt; *AmE* əˈbɔːrt/ v. **1** to end a pregnancy early (提前终止妊娠)使流产,引产 **2** to give birth to a foal too early for it to survive (产驹过早使其难以存活)流产: The virus can cause pregnant mare to abort. 这种病毒可导致怀孕母马流产。syn slip, miscarry

aborticide /əˈbɔːtɪsaɪd; *AmE* əˈbɔːrt-/ n. an agent that causes abortion 引产药,堕胎药

abortifacient /əˌbɔːtɪˈfeɪʃənt; *AmE* əˌbɔːrt-/ adj. causing abortion 引产的,堕胎的

n. any substance or device used to induce abortion 堕胎药;堕胎器械

abortion /əˈbɔːʃn; *AmE* əˈbɔːrʃn/ n. the premature expulsion of the fetus due to injury or infection (由于受伤或感染造成胎儿提早产出)流产,早产: abortion storm 暴发性流产 ◇ early abortion 早期流产 ◇ equine abortion 马匹流产 ◇ infectious abortion 感染性流产 ◇ late abortion 晚期流产

abrade /əˈbreɪd/ vt. to rub or damage the skin by friction; chafe (由于摩擦造成皮肤损伤)蹭破,擦伤

abrasion /əˈbreɪʒn/ n. any injury caused by rubbing against a hard and rough surface; bruise (皮肤因摩擦而损伤)擦伤

abrupt transition n. a change from one gait to another performed unexpectedly 突然换步,转换过急

abscess /ˈæbses/ n. a localized collection of pus on the skin caused by bacterial infection (由细菌感染引起皮肤局部淤积脓液)脓肿,脓疮: acute abscess 急性脓肿 ◇ traumatic abscess 创伤性脓肿 [syn] pus pocket

vi. to form an abscess 化脓

abscessation /ˌæbsəˈseɪʃn/ n. the formation of an abscess 脓疮形成,化脓

absolute ensurer n.【slang】a whip 鞭子

absorbent /əbˈsɔːbənt,-zɔːb-; *AmE* əbˈsɔːrbənt,-zɔːrb-/ adj. capable of absorbing water or fluid 吸水的: absorbent cotton 吸水棉

acari /ˈækəraɪ/n. the plural form of acarus [复]螨

acariasis /ˌækəˈraɪəsɪs/ n. a disease caused by the infestation of mites (由于大量寄生螨虫导致的疾病)螨病,疥癣 [syn] mange

acarus /ˈækərəs/ n. (*pl.* **acari** /-raɪ/) a mite or tick 螨

accessory /əkˈsesəri; *AmE* ækˈs-/ adj. having a secondary function 次要的,副的: accessory yellow body 副黄体,次级黄体 ◇ accessory glands 副性腺 ◇ accessory carpal bone 副腕骨

Acchetta /əˈtʃetə/ n. (also **Acchetta pony**) a pony breed from Sardinia, Italy (源于意大利撒丁岛的矮马品种)阿凯特矮马: The Acchetta pony stands about 120 to 130 cm. 阿凯特矮马体高在120~130厘米。

Acchetta pony n. *see* Acchetta 阿凯特矮马

accouterments /əˈkuːtərmənts/ n. pl. (BrE **accoutrements** /əˈkuːtrəmənts/) the ancillary items or tack worn by a horse, such as the saddle, bridle, and bit (给马配备的鞍、勒和口衔等必要物件)装备;马具

accumulator bet n. *see* parlay bet 累积投注

acepromazine /ˌæsəˈprəʊməzɪː n; *AmE* -ˈproʊ-/ n.【vet】(abbr. **ACP**; also **acetyl promazine**) a chemical commonly used as tranquilizer for horses (马常用的一种化学镇静剂)乙酰丙嗪: Acepromazine is usually administered to calm frightened or aggressive animals. 乙酰丙嗪常用来安抚受到惊吓或脾性乖戾的动物。

acetyl promazine n. *see* acepromazine 乙酰丙嗪

acey-deucy /ˈeɪsiːˈdjuːsi; *AmE* -ˈduːsi/ n.【racing】a riding style with uneven stirrups in which the left is placed lower than the right to enable the jockey to keep better balance on the turns (两马镫左侧低、右侧高的骑乘方式,可使骑手在转弯处更好地保持平衡)高低镫骑乘: A relatively unknown American jockey named Jackie Westrope developed acey-deucey, although it was popularized by a far more famous rider, Eddie Arcaro. 一位不太出名的美国骑师杰克·韦斯特罗普发展了高低镫骑乘,但其推广要归功于著名骑手埃迪·阿克洛。

adv. riding with uneven stirrups 高低镫: to ride acey-deucy 以高低镫骑乘

ACHA abbr. American Cutting Horse Association 美国截牛协会

Achaean /əˈkiːən/ n. (also **Achean**) an extinct pony breed of ancient Greece (古希腊矮马品种,现已灭绝)阿凯亚矮马

Achilles tendon n.【anatomy】the tendon of gaskin; hamstring 跟腱

acid brand n. (also **acid mark**) a brand burned on the skin of a horse by using acid (通过强酸在马体上烫的标记)酸蚀烙印

vt. to mark a registration brand with an acid 酸烙

acid mark n. *see* acid brand 酸蚀烙印

acidemia /ˌæsɪˈdiːmiə/ n. *see* acidosis 酸血症

acidosis /ˌæsɪˈdəʊsis; *AmE* -ˈdoʊ-/ n. (also **acidemia**) an abnormal increase of the acidity of the body's fluids (机体体液酸度异常增高) 酸血症: respiratory acidosis 呼吸性酸血症 ◇ metabolic acidosis 代谢性酸血症

acme sporting cart n. a light horse-drawn gig popular in Australia during the late 19th century (19 世纪后期流行于澳大利亚的一种轻型马车)极品运动马车

ACP abbr. acepromazine 乙酰丙嗪

acquired /əˈkwaɪəd/ adj. gained or obtained after birth 获得的;后天的: acquired character 获得性状 ◇ acquired defect 后天缺陷

acquired leukoderma n. the whitening of hair or skin due to skin trauma (皮肤受伤后皮肤或毛发变白的现象)获得性白斑病

acquired markings n. (also **acquired marks**) any permanent marks produced on the skin of a horse after birth, such as brands, scars, and tattoos (马匹出生后在皮肤上刻下的永久性标记,如烙印、伤疤和刺纹) 获得性别征: Acquired markings could be used for identification purposes. 获得性别征可用于马匹鉴定。

acquired marks n. *see* acquired markings 获得性别征

acrosome /ˈækrəˌsəʊm; *AmE* -ˌsoʊm/ n. 【physiology】 a caplike structure at the anterior end of a sperm (精子前端的帽状结构)顶体 | **acrosomal** /ˈækrəˌsəʊml; *AmE* -ˌsoʊml/ adj.: acrosomal reaction 顶体反应

ACTH abbr. adrenocorticotropic hormone 促肾上腺皮质激素

actin /ˈæktɪn/ n. 【physiology】 a protein present in muscle fiber that together with myosin functions in muscle contraction (见于肌纤维中的一种蛋白,与肌球蛋白一起负责肌肉收缩) 肌动蛋白

acting hand n. *see* active hand 控缰手

acting master n. 【hunting】 one appointed temporarily to organize a hunt during the absence of the Hunt Master (在猎主缺职下临时任命组织打猎的人)委任猎主,代理猎主

Actinobacillus equuli n. 【Latin】 a species of bacteria causing foal abortion or death shortly after birth (一种导致马驹流产或产后死亡的细菌)马放线杆菌

action /ˈækʃn/ n. the manner or way that a horse moves (马匹运行的方式)运步,步态: free action 运步自如 ◇ a horse with good action 一匹步态优美的马

active hand n. (also **acting hand**) the rider's hand used to work the reins 控缰手

active immunity n. a long-lasting immunity that is acquired when the body is stimulated to produce its own antibodies (机体受到刺激后产生抗体而获得的持久免疫力)主动免疫 [compare] passive immunity

acuity /əˈkjuːəti/ n. the ability to see and perceive clearly (视觉的)锐利,敏锐

acupression /ˈækjʊpreʃn/ n. (also **acupressure**) a form of therapeutic massage in which pressure is applied with the thumbs and palms to those acupuncture points in the body (一种用手指和手掌对身体针灸部位进行的治疗性按摩) 针灸指压

acupressure /ˈækjupreʃə(r)/ n. *see* acupression 针灸指压: equine acupressure 马匹针灸指压

acupuncture /ˈækjupʌŋktʃə(r)/ n. a traditional Chinese clinical method for relieving pain or inducing regional anesthesia by inserting needles into precise nerve points (中医采用细针扎入身体穴位以减轻疼痛或进行局部麻醉的临床疗法)针灸: acupuncture points 针灸穴位

acute /əˈkjuːt/ adj. (of disease) severe and occurring over a short period of time (指疾病)剧烈的;急性的: acute poison 剧毒 ◇ acute arthritis 急性关节炎 ◇ acute laminitis 急性蹄叶炎

A

acute equine respiratory syndrome n. (abbr. **AERS**) a lethal viral disease of horses, symptomized by shallow breathing and respiratory distress (马患的一种病毒致死性疾病,症状表现为呼吸急促而困难)马急性呼吸综合征

ad libitum /æd ˈlɪbɪtəm/ adv.【Latin】freely 自由地:The horses should best be left to feed ad libitum on pasture. 马匹最好放在草场上自由采食。

added money n.【racing】extra money added to the winner's purse in addition to entry fees paid by the owners and breeders (除出场费以外由马主和育马者多余支付给获胜者的金额)追加奖金,增加奖金

added weight n.【racing】the extra weight carried by a racehorse over the stated weight limit (赛马规定负重之外额外担负的重量)额外负重

Addison's disease n. a degenerative condition caused by deficiency in the secretion of adrenocortical hormones (由于肾上腺皮质激素分泌不足导致的机能衰退性疾病)阿狄森氏症

additive /ˈædətɪv/ n.【nutrition】any substance added in small account in horse's ration to improve its quality (为提高马匹体能而在日粮中少量添加的成分)添加剂

adduct /ˈædʌkt/ vt.【physiology】to draw toward the middle axis of the body (趋向身体中轴)内收 [compare] abduct | **adduction** /əˈdʌkʃn/ n.

adductor /əˈdʌktə(r)/ n.【anatomy】a muscle that draws body part toward the median axis of the body (将机体部位向身体中轴内收的肌肉)内收肌:adductor muscle 内收肌 [compare] abductor

Adeav /ˈædiːv/ n. an ancient horse breed originating in Kazakhstan (源于哈萨克斯坦的古老马种)阿迪乌马

adenitis /ˌædəˈnaɪtɪs/ n. inflammation of the lymph node or gland (淋巴结或腺体发炎的症状)腺炎:equine adenitis 马腺疫

adenoma /ˌædəˈnəʊmə/ n. a benign epithelial tumor found in the glandular structures (源于腺体组织的良性上皮细胞瘤)腺瘤

adenosine triphosphate n.【biochemistry】(abbr. **ATP**) an organic compound consisting of an adenosine molecule bonded to three phosphate groups; present in all living organisms as the universal energy-transfer molecule (由一个腺苷分子结合三个磷酸基构成的有机复合物,是所有生物的通用能量传递分子)三磷酸腺苷

Adequan /ædɪˈkwæn; *AmE* -kwən/ n.【vet】a brand name for polysulfate glycosaminoglycan injected to treat some arthritic conditions in horses (聚硫葡胺聚糖的商品名,可注射治疗马的关节炎)安迪快

adipose /ˈædɪpəʊs; *AmE* -poʊs/ adj. of or relating to animal fat 脂肪的:adipose tissue 脂肪组织

adjusted score n.【rodeo】a competitor's score reviewed and changed under the rules of the monitor system (根据监控系统对竞赛者的得分予以调整)调整后得分

ADMS abbr. American Donkey and Mule Society 美国驴骡协会

adrenal /əˈdriːnl/ adj. of or relating to the adrenal glands 肾上腺的: adrenal cortex 肾上腺皮质 ◇ adrenal medulla 肾上腺髓质

adrenal gland n.【anatomy】either of a pair of endocrine glands located over each kidney (位于肾上方的一对内分泌腺体)肾上腺 [syn] suprarenal gland

adrenalin /əˈdrenəlɪn/ n. (also **adrenaline**) a hormone secreted by the adrenal gland and acting primarily as a stimulant (肾上腺分泌的一种激素,有刺激兴奋的作用)肾上腺素

adrenaline /əˈdrenəlɪn/ n. *see* adrenalin 肾上腺素

adrenocorticotropic /əˈdriːnəʊˌkɔːtɪkəʊˈtrɒpɪk/ adj. having a stimulating effect on the adrenal cortex 促肾上腺皮质的: adrenocorticotropic hormone 促肾上腺皮质激素

A

ADS abbr. **1** American Driving Society 美国驾车协会 **2** Australian Driving Society 澳大利亚驾车协会

ADT abbr. American Discovery Trail 美国探索大道

adult /ˈædʌlt, əˈdʌlt/ n. a horse who has attained maturity 成年马

adj. fully developed and mature 成年的: an adult stallion 一匹成年公马

advanced /ədˈvɑːnst; *AmE* -ˈvænst/ n. (also **advanced class**) the highest level of equestrian competition (马术比赛最高的赛事级别)高级— *see also* **intro**, **pre-novice**, **novice**, **intermediate**

adventitial marks n. (also **adventitious marks**) any mark of white coat hairs resulting from destruction of the pigment cells of the underlying skin due to trauma (皮肤受伤导致色素细胞受损后被毛变白而形成的标记)创伤性别征

adventitious marks n. *see* adventitial marks 创伤性别征

AEF abbr. Asian Equestrian Federation 亚洲马术联合会(简称亚马联)

AEI abbr. Average Earnings Index 平均收益指数

a-equi-1 n. a strain of the equine influenza virus Ⅰ型马流感病毒

a-equi-2 n. a strain of the equine influenza virus Ⅱ型马流感病毒

aerobic /eəˈrəʊbɪk; *AmE* eˈroʊ-/ adj. needing oxygen to live 需氧的,有氧的: aerobic bacteria 需氧细菌 ◇ aerobic exercise 有氧运动 [opp] anaerobic

aerophagia /eərəˈfeɪdʒɪə; *AmE* ˌerəˈfeɪdʒə/ n. *see* wind sucking 咽气癖,吞气症

AERS abbr. acute equine respiratory syndrome 马急性呼吸综合征

AFA abbr. American Farriers Association 美国蹄铁匠协会,美国蹄师协会

afebrile /eɪˈfiːbrəl/ adj. having no fever 无热的,不发热的

African dwarf donkey n. a rare donkey breed living in a few remote parts of northeast Africa, is small in stature and regarded as the ancestor of other miniature donkeys (生活在非洲东北偏远地区的驴品种,体型小,被认为是其他微型驴的先祖)非洲侏儒驴

African horse sickness n. a highly infectious, fatal viral disease of horses, symptomized by high fever, labored breathing, and a dry cough (马患的一种致死性病毒传染病,其症状表现为高热、呼吸困难和干咳)非洲马瘟

African Wild Ass n. a wild ass of the genus *Equus* living in Africa, generally thought to be the progenitor of the domestic donkey (栖息于非洲的野驴,普遍被认为是家驴的祖先)非洲野驴 [syn] *Equus africanus*

afterbirth /ˈɑːftəbɜːθ; *AmE* ˈæftərbɜːrθ/ n. the placenta and amnion expelled from the uterus following birth (分娩后从子宫排出的胎盘和羊膜)胎衣: Some individuals, usually those foaling for the first time, gallop away from their foals, frightened by the afterbirth hanging behind them. 有的马匹,尤其是那些初次分娩的母马,被挂在身后的胎衣所惊吓,不顾幼驹便惊惶失逃。

against the clock adv.【jumping】*see* jump-off (障碍赛第二轮)争时赛,附加赛: to jump against the clock 进入争时赛

agalactia /ægəˈlækʃə; *AmE* -ˈlæktiə/ n. *see* agalactosis 泌乳缺乏,无乳症

agalactosis /ægəˈlæktəʊsɪs/ n. (also **agalactia**) lack of milk in the mare's udder after foaling 泌乳缺乏;无乳症

age /eɪdʒ/ n. the period of time that a horse has lived 年龄: The age of a horse could be judged by the condition of his teeth. 马匹年龄可以根据牙齿情况进行判断。

vi. to grow or become old 长老,衰老

vt. to judge the age of a horse by evaluating its teeth (通过马匹牙齿)判定年龄

aged /eɪdʒd/ adj. (of a horse) over the age of 12 years (指马年龄在 12 岁以上)年老的: an aged horse 一匹老马

aggressive /əˈgresɪv/ adj. (of a horse) inclined to behave in a savage or domineering manner 悍性强的，烈性的

agile /ˈædʒaɪl; *AmE* ˈædʒl/ adj. (of a horse) quick and light in movement 敏捷的，迅捷的 | **agility** /əˈdʒɪləti/ n. 迅速：The speed and agility of horse enable him to escape from predators with a flight, and he is ever vigilant. 马的速度快且动作迅捷，这使它能急速逃脱食肉动物的追捕，此外它的警觉性也非常高。

aging /ˈeɪdʒɪŋ/ n. **1** the process of determining the age of a horse by checking its teeth 根据马匹牙齿判定年龄 **2** the process of growing old or mature 衰老；成熟

agist /əˈdʒɪst/ vt. to care for and feed horses or cattle for payment 代人放牧；代人饲养 | **agistment** /-mənt/ n.

agonistic /ˌægəˈnɪstɪk/ adj. likely to fight; combative 打斗的；好斗的：agonistic behavior 打斗行为

agouti /əˈguːtɪ/ n.【genetics】a gene considered to control the distribution pattern of eumelanin (控制真黑色素分布模式的基因) A 基因

ah /ɑː/ interj. used to command a horse to turn to the left (用来命令马匹向左转的吆喝声) 哎！
vi. to command a horse to turn to the left (命令马匹) 左转

AHC abbr. American Horse Council 美国马匹理事会

AHS abbr. African Horse Sickness 非洲马瘟

AHSA abbr. American Horse Shows Associations 美国马术展示协会

AI abbr. artificial insemination 人工授精

aid /eɪd/ n. (usu. **aids**) a signal used by a rider to give instructions or directions to the horse (骑手给马传达指令时所采用的信号) 扶助：artificial aids 人工扶助 ◇ lower aids 下身扶助 ◇ natural aids 自然扶助 ◇ rein aids 缰绳扶助 ◇ upper aids 上身扶助 ◇ voice aids 声音扶助 ◇ weight aids 体重扶助 PHR **on the aids** (of a horse) fully responsive to the aids of rider 听从扶助的 syn between hand and leg

Aintree /ˈeɪntrɪ/ n. a 7.22 km long racecourse consisting of 30 fences and a water jump 4.5 m wide, located near Liverpool in Lancashire, England, where the Grand National Steeplechase has been run since 1839 (位于英国兰开夏郡利物浦市的赛马跑道，长 7.22 千米，设有 30 道障碍和一道 4.5 米宽的水障，自 1839 年起在此举行全英越野障碍大奖赛) 安翠赛道

Aintree breast girth n. (also **Aintree girth**) a canvass breastplate used to prevent a racing saddle from slipping backwards (帆布制成的胸带，用来防止赛马用鞍向后滑动) 安翠胸带 syn Newmarket girth, Newmarket breast girth, racing girth, racing breastplate

Aintree girth n. *see* Aintree breast girth 安翠胸带

air /eə(r); *AmE* er/ n. the bearing of a horse in its movements and gaits (马在运步中的) 姿态，步态
vi.【racing】(of a horse) to run a race as if he were only out for exercise 轻松获胜

air-dry /ˈeə(r)ˌdraɪ; *AmE* ˈerˌ-/ vt. to dry in the open air 风干，晒干：air-dried feed 风干饲草

airer saddle n. a wooden stand on which a saddle is placed to air out or dry (用来晾晒马鞍的木架) 晾鞍架

airs /eəz; *AmE* erz/ n. pl. any movements and postures trained of a horse in classical equitation (古典骑术中马匹受训的) 动作，姿势：low airs 地面动作 ◇ high airs 空中动作

airs above the ground n. any of the movements performed with the forelegs or all four feet off the ground in classical equitation (古典骑术中马的前肢或四肢同时离地时表演的高难动作) 空中动作 syn high airs, movements off the ground, schools above the ground

airway /ˈeəweɪ; *AmE* erweɪ/ n.【anatomy】the passage way of lung in which air ventilates and

A

exchanges (肺内进行气体交换的通道)管道,气管:airway resistance 气道阻力

AIT abbr. Area International Trial 国际场地障碍赛

Akhal-Teké /ˌækəlˈtɛk,-ˈtɛki/ n. a warmblood horse breed descended from the Turkmene more than 3,000 years ago, bred around the oases of the Turkmenistan desert; stands between 14.2 and 16 hands and usually has a dun coat with golden luster, although black and gray also occur; has a long head and neck, slender legs, and sparse mane and tail; is noted for its speed and endurance on long marches (3 000 多年前由土库曼马繁衍而来的一个古老的温血马品种,育成于土库曼斯坦沙漠中的绿洲,体高 14.2～16 掌,毛色以兔褐色为主且色泽金亮,但黑和青亦有之,头颈长,四肢修长,鬃尾稀疏,在长途赛中以速力著称)阿哈捷金马,汗血马

albata /ælˈbɑːtə/ n. an alloy of copper and nickel used for decorating tack and bits (用于装饰马具和衔铁的一种铜镍合金)镍黄铜 [syn] German silver

ALBC abbr. American Livestock Breeds Conservancy 美国畜种保护委员会

albinism /ˈælbɪnɪzəm/ n. **1** congenital absence of normal pigmentation in the skin of a horse (马皮肤中先天缺乏色素沉着的病症)白化 **2** the condition of being an albino 白化现象

albino/ælˈbiːnəʊ/ n. **1** a horse with white hairs and pink skin due to lack of normal pigmentation (由于色素无法正常沉着,毛色为白而皮肤粉红的马匹)白化马:albino gene 白化基因 **2** (usu. **Albino**) the American White Horse 美国白马: Old King, a white stallion with pink skin and brown eyes, is said to be the foundation horse for the American Albino. "老国王"据说是美国白马的奠基公马,这匹马毛色为白,皮肤粉红,眼珠褐色。

al-Buraq /æl bʊˈrɑːk/ n. *see* Buraq 布拉卡

Alcock Arabian n. (also **Alcock's Arabian**) a gray Arabian stallion imported into England in 1720s, from which all gray Thoroughbreds are descended in direct sire line (18 世纪 20 年代后引入英国的一匹青毛阿拉伯公马,所有的青毛纯血马都以其为父本繁育而来)安考科·阿拉伯 [syn] Pelham Arabian, Ancaster Arabian

Alcock's Arabian n. *see* Alcock Arabian 安考科·阿拉伯

alcohol block n. an illegal practice of injecting alcohol into the tail nerves of a horse to numb them in order to achieve a quiet tail carriage (在马匹尾部非法注射乙醇来麻醉尾神经达到稳定尾姿的做法)乙醇阻断

Alexandra dogcart n. an English two-wheeled horse-drawn cart popular in the late 19th century (19 世纪末在英国使用的一种双轮马车)亚历山大马车

alfalfa /ælˈfælfə/ n. a perennial legume plant widely cultivated as feed for farm animals (一种多年生豆科植物,被用作牧草而广泛种植)【紫花】苜蓿: alfalfa cubes 苜蓿块 ◇ alfalfa hay 苜蓿干草 ◇ alfalfa meal 苜蓿粉 ◇ alfalfa pellets 苜蓿颗粒料 [syn] lucerne

alight /əˈlaɪt/ vi. to get off from a horse; dismount 下马:to alight from one's mount 从坐骑上下来

alimentary /ælɪˈmentəri/ adj. of or relating to nutrition or digestion 营养的;消化的:alimentary canal 消化道 ◇ alimentary disturbance 消化机能障碍 ◇ alimentary system 消化系统 ◇ alimentary tract 消化道

alkalosis /ˌælkəˈləʊsɪs; *AmE* -ˈloʊsɪs/ n. an abnormal increase of pH value of the blood and body fluids (血液和体液中 pH 异常增高的症状)碱中毒:metabolic alkalosis 代谢性碱中毒 ◇ respiratory alkalosis 呼吸性碱中毒 | **alkalotic** /-ˈlɔtik/ adj.

All England Jumping Course n. a jumping course with permanent obstacles built by Douglas Bunn on his estate, Hickstead Place, in

1960 for British horses and riders (1960 年由杜格拉斯·波恩在自家庄园为英国骑手搭建的障碍赛场)全英障碍赛场

all-age race n.【racing】a horse race for horses over two years old (2 岁以上马匹皆可参加的比赛)全龄赛

allanto-chorion /æˌlæntəʊˈkɔːrɪɒn/ n.【anatomy】the fused allantois and chorion 尿囊绒毛膜

all-around /ˌɔləˈraʊnd/ adj. (also **all-round**) versatile, comprehensive 全能的, 兼用的; 综合的: all-around competition 综合全能比赛; 三日赛

allelomemtic /əˌliːləˈmemtɪk/ adj. imitating others 模仿的: allelomemtic behavior 模仿行为

allergen /ˈælədʒən; *AmE* ˈælərdʒən/ n. any substance that causes an allergic reaction (能引起过敏反应的任何物质)过敏原

allergic /əˈlɜːdʒɪk; *AmE* əˈlɜːrdʒ-/ adj. **1** ~ (to sth) having an allergy to sth 过敏的 **2** caused by an allergy 过敏性的: an allergic reaction 过敏反应

allergy /ˈælədʒi; *AmE* ˈælərdʒi/ n. an abnormally high sensitivity to certain substances (对某些物质的高度敏感性)过敏

alleyway /ˈælɪwei/ n. a narrow passage in front of stalls; aisleway (马舍前的狭窄通道)过道, 走廊: The alleyway should be wide enough to allow the horse to be turned around with ease. 厩舍走廊要足够宽, 以便马匹比较容易转身。

alligator /ˈælɪgeɪtə(r)/ n.【cutting】a calf who is difficult to be seperated from the herd 难截的牛

allowance /əˈlaʊəns/ n.【racing】a weight reduction granted to a horse, usually on account of its being inferior to other horses (比赛中由于马匹处于劣势而减少的负重)让磅数: weight allowance 负重让磅数 ◇ sex allowance 性别让磅数 ◇ apprentice allowance 见习让磅数

allowance race n.【racing】a race in which weight allowances are given according to the competitors' previous performance (根据赛马以往赛绩而减少负重的比赛)让磅赛 compare welter race

all-purpose saddle n. (also **general purpose saddle**) an English saddle used for multiple purposes (可作多种用途的英式马鞍)兼用鞍, 综合鞍

all-round /ˌɔːlˈraʊnd/ adj. *see* all-around 全能的: an all-round athlete 全能运动员 ◇ all-round horse 兼用型马, 全能马

all-round cow horse n. a horse capable of performing the tasks required of it by a cowboy (牛仔日常工作中使用的马匹)全能牧牛马

all-rounder /ˌɔːlˈraʊndə(r)/ n.【polo】a versatile polo player who can take both the positions of defense and offense (马球比赛中攻守兼备的选手)全能选手

allures /əˈlʊərz; *AmE* əˈlʊrz/ n.【French】the gait of a horse (马的)步法: natural allures 自然步法 ◇ artificial allures 人工步法

all-weather /ˌɔːlˈweðə/ adj.【racing】suitable for all kinds of weather condition 全天候的: an all-weather racecourse 全天候赛场 ◇ all-weather track 全天候跑道

alopecia /ˌæləˈpiːʃɪə/ n. a local or general loss of hair without visible skin disease (无显著皮肤病但局部或周身被毛脱落的情况)秃毛症; 脱毛症

Alpha /ˈælfə/ n. *see* alpha horse 头马

Alpha horse n. (also **Alpha**) the leader of a herd of horses (马群的)头马: The Alpha horse provides leadership and security to all others in their herd. 头马在马群中有统率作用, 并保证其他个体的安全。

also-ran /ˈɔːlsəʊˌræn; *AmE* -soʊˌ-/ n. (also **also-ranner**) **1**【racing】a racehorse who did not finish in the first three 未入三甲的赛马 **2** a loser in a race or contest 败将; 输家

Altai /ˈæltaɪ/ n. (also **Altai horse**) a horse breed indigenous to Russia; stands around 13.2 hands and generally has a chestnut, bay, black, or gray coat; used mainly for riding (俄国本土的马品种,体高约13.2掌,毛色以栗、骝、黑或青为主,主要用于骑乘)阿尔泰马

alter /ˈɔːltə(r)/ vt. to castrate or spay an animal 阉割,去势;卵巢摘除

Alter /ˈɔːltə(r)/ n. a Spanish warmblood with similar lineage to the Andalusian, mainly used for riding and jumping (西班牙温血马品种,与安达卢西亚马的培育谱系相似,主要用于骑乘和障碍赛)阿特马

Alter-Real /ˌɔːltəˈriːəl/ n. a light horse breed indigenous to Portugal; stands 15.1 to 16.1 hands and may have a bay, brown, chestnut, or gray coat; is intelligent, agile and athletic (葡萄牙轻型马品种,体高15.1~16.1掌,毛色为骝、褐、栗或青,反应敏快,动作迅捷,擅长竞技)阿特瑞尔马

alveolar /ælˈviːələ(r)/ adj. of or relating to alveoli 肺泡的:alveolar duct 肺泡小管 ◇ alveolar-capillary membrane 肺泡毛细血管膜

alveolar emphysema n. another term for chronic obstructive pulmonary disease (慢性阻塞性肺炎的别名)肺气肿

alveolus /ælˈviələs/ n.【anatomy】(*pl.* **alveoli** /-laɪ/) **1** a tiny, thin-walled, capillary-rich air sac in the lungs where the exchange of oxygen and carbon dioxide takes place (肺中壁薄且富有毛细血管的微小气囊,是氧气和二氧化碳交换的场所)肺泡 [syn] air sac **2** a small cavity or pit, such as the tooth socket in the jawbone (小窝或小槽,如牙床中的牙槽)窝;齿槽

Amalgamated Society of Farriers and Blacksmiths n. a trade union founded in 1805 in England to regulate the employment relations for farriers and blacksmiths (1805年在英国成立的组织机构,旨在协调蹄铁匠的劳资关系)英国蹄铁匠联合会

amateur /ˈæmətə(r)/ n. a person who practices sth for pleasure rather than as a profession (出于兴趣开展某种活动而非以此为业的人)业余者,爱好者:amateur huntsman 业余猎手

amateur rider n. **1** one who rides a horse solely for pleasure 骑马爱好者 [syn] gentleman rider **2** a rider who lacks the talent of a professional 业余骑手

amble /ˈæmbl/ n. a four-beat artificial gait of horse derived from the walk (由慢步演变而来的四蹄音人工步法)溜花蹄

vi. to move or go with the gait of amble 以溜花蹄行进

ambler /ˈæmblə(r)/ n. a horse naturally gaited or bred to amble 善走溜花蹄的马

Americaine /əˈmerɪkeɪn/ n. *see* American buggy 美式轻便马车

American Albino n. (also **Albino**) *see* American White Horse 美国白马

American Albino Association n. (abbr. **AAA**) an organization founded in the United States in 1970 to register pure white horses and ponies (1970年在美国成立的组织机构,负责纯种白马和矮马的登记)美国白马协会

American Albino Horse Club n. (abbr. **AAHC**) an organization founded in Nebraska, USA in 1936 to promote and preserve the breeding records of the American White Horse; was incorporated in 1970 into the American Albino Association (1936年在美国内布拉斯加州成立的组织机构,旨在促进和保护美国白马的育种登记,于1970年并入美国白马协会)美国白马俱乐部

American Association of Equine Practitioners n. (abbr. **AAEP**) an organization founded in 1954 in the United States to supervise the activities of equine veterinarians (1954年在美国成立的一个组织,是监管马兽医行业规范的权威机构)美国马兽医协会

American Association of Owners and Breeders of Peruvian Paso Horses n. (abbr. **AAOBPPH**) an organization founded in the United States in 1962 to register purebred Peruvian Pasos, promote the breed, and educate the public

(1962 年在美国成立的组织机构，主要负责纯种秘鲁帕索马的登记、马种推广以及公众教育)美国秘鲁帕索马马主与育马者协会

American Bashkir n. a horse breed used by the American Indians for riding, packing and light draft in the early 1800s; stands 13.1 to 14 hands and usually has a bay, chestnut or palomino coat which grows distinctively thicker and curly in winter against harsh weather; has a short neck, low withers, deep chest, short and strong legs, and small and hard hoofs; is hardy, docile and quiet (19 世纪初北美印第安人用于骑乘、驮运和轻挽的马品种，体高 13.1 ~ 14 掌，毛色以骝、栗和银鬃为主，为抵抗严寒冬季被毛变得浓密而蜷曲，颈短，鬐甲低，胸身宽广，四肢粗短，蹄小而坚实，抗逆性强，性情温驯而安静)美国巴什基尔马 [syn] American Bashkir Curly

American Bashkir Curly n. *see* American Bashkir 美国巴什基尔卷毛马

American break n. a horse-drawn passenger carriage used in the Western United States for long-distance travel in the 19th century (19 世纪美国西部使用的一种长途客用马车)美式客用马车

American Buckskin n. an American light breed developed recently by crossing Norwegian dun with Spanish Sorraia (美国最近采用挪威峡湾马和西班牙索拉亚矮马杂交培育而成的轻型马品种)美国鹿皮马

American buggy n. (also **buggy**) a light, four-wheeled wagon popular in the United States from the 1830s to the 1920s, usually drawn by a single horse (在 19 世纪 30 年代至 20 世纪 20 年代流行于美国的轻型四轮马车，多由单马拉行)美式轻型马车 [syn] Americaine

American cab n. a two-wheeled horse-drawn vehicle first used in New York, United States around the early 1830s (19 世纪 30 年代初出现于美国纽约的双轮出租马车)美式出租马车

American cabriolet n. a four-wheeled, horse-drawn carriage popular in the United States during the second half of the 19th century (19 世纪下半叶流行于美国的四轮马车)美式小型出租马车

American Cream n. *see* American Cream Horse 美国乳色马

American Cream Horse n. (also **American Cream**) an American breed of horse with pink skin and cream-colored coat (肤色粉红、被毛乳白的美国马品种)美国乳色马

American Cutting Horse Association n. (abbr. **ACHA**) an organization founded by a group of amateur cutters in the United States in 1946 to promote and regulate the cutting contests as a sport (由截牛者于 1946 年在美国成立的组织机构，旨在促进并监管截牛运动)美国截牛协会

American Discovery Trail n. (abbr. **ADT**) a non-motorized, coast-to-coast trail developed in the late 20th century across the United States; covering more than 6,000 miles from Henelopen State Park in Delaware to Point Reyes National Seashore in California (美国在 20 世纪末建成的连接两大洋海岸的非机动车道，全长超过 9 600 千米，东起特拉华州国家公园，西至加利福尼亚州雷耶斯角国家海滩)美国探索大道

American Donkey and Mule Society n. (abbr. **ADMS**) an organization founded in the United States in 1968 to register all types of asses and mules foaled (1968 年在美国成立的组织机构，主要对全美所有的驴及骡进行登记)美国驴骡协会

American Farriers Association n. (abbr. **AFA**) an association of farriers founded in the United States in 1971 to provide certification tests for its members since 1979 (1971 年在美国成立的蹄铁匠协会，自 1979 年起组织开展蹄铁匠资格考试)美国蹄铁匠协会，美国蹄师协会

A

American Foxhound Club n. a non-profit organization founded in the United States in 1912 to promote all pack hounds in North America (1912 年在美国成立的非营利机构，负责整个北美猎狐犬的事务)美国猎狐犬俱乐部

American Horse n. an all-purpose, generic riding horse developed in the Eastern America in the early 1700s (18 世纪初在美国东部培育而成的兼用型乘用马种)美国马

American Horse Council n. (abbr. **AHC**) an organization founded in Washington DC, USA in 1969 to promote and protect the horse industry (1969 年在美国华盛顿特区成立的组织机构，旨在促进和保护美国马业的发展)美国马匹理事会

American Horse Shows Association n. (abbr. **AHSA**) an organization founded in the United States in 1917 to govern equestrian sports in the United States, merged in 2003 with the United States Equestrian Team to form the United States Equestrian Federation (1917 年在美国成立的组织机构，负责监管全美的马术运动，2003 年与美国马术队合并成立美国马术联合会)美国马匹展览协会

American Livestock Breeds Conservancy n. (abbr. **ALBC**) a non-profit organization founded in 1977 to preserve some rare breeds of livestock from extinction, with its name shortened to the Livestock Breeds Conservancy in 2013 (1977 年成立的非营利组织机构，旨在保护濒临灭绝的稀有畜禽品种，2013 年名称简缩为畜种保护委员会)美国畜种保护委员会

American mail coach n. another term for the Concord coach (协和四轮马车的别名)美式邮递马车

American Paint Horse Association n. (abbr. **APHA**) an organization founded in the United States in 1962 to keep registration of Paint Horses (1962 年在美国成立的组织机构，主要负责美国花马品种登记)美国花马协会

American Pelham bit n. one kind of Pelham bits 美式佩勒姆衔铁

American Quarter Horse Association n. (abbr. **AQHA**) an organization founded in the United States in 1940 to establish breed specifications and maintain the registry for Quarter horses (1940 年在美国成立的组织机构，主要负责夸特马的品种鉴定和登记)美国夸特马协会

American Saddle Horse n. *see* American Saddlebred 美国骑乘马

American Saddlebred n. a horse breed developed in Kentucky by putting a Thoroughbred foundation sire to local mares; stands 15 to 16 hands and may have a coat of any color; has a high-set tail carriage, a long and arched neck, and a sloped shoulder; is smooth and animated in action, and used for riding, show competition, and harness (美国肯塔基州采用一匹雄性纯血马与当地母马育成的马种，体高 15 ~ 16 掌，各种毛色兼有，尾高翘，颈长且呈弓形，斜肩，步态流畅，举蹄较高，主要用于骑乘、展览和挽车)美国骑乘马 [syn] Saddlebred, Kentucky Saddler, American Saddle Horse

American Saddlebred Horse Association n. (abbr. **ASHA**) an association founded in 1891 in Louisville, Kentucky, USA to keep the breed registry for the American Saddlebred (1891 年在美国肯塔基州路易斯维尔市成立的组织机构，负责美国乘用马的登记)美国骑乘马协会

American Shetland n. an American pony breed descended from the Shetland pony (由设特兰矮马繁育而来的美国矮马品种)美国设特兰矮马

American Stage Coach n. *see* stagecoach 美式驿站马车

American Standardbred n. *see* Standardbred 美国标准马

American Stud Book n. the American registry for all Thoroughbred horses bred in the United States, Puerto Rico, and Canada (负责美国、波

多黎各和加拿大纯血马登记工作的组织机构)美国纯血马登记册

American Team Penning Championships n. (abbr. **ATPC**) one of three national events established in the United States since 1980, which is devoted solely to the sport of team penning (自1980年起在全美专门为团体截牛比赛而举行的三项比赛之一)美国团体截牛锦标赛

American Triple Crown n. *see* Triple Crown 美国三冠王

American Trotter n. another term for Standardbred (美国标准马的别名)美国快步马

American Vaulting Association n. (abbr. **AVA**) an organization founded in California in 1968 to promote the equestrian sport of vaulting in the United States (1968年在美国加利福尼亚州成立的组织机构,旨在促进美国马上杂技表演的发展)美国马上体操协会

American Veterinary Medical Association n. (abbr. **AVMA**) a non-profit organization founded in 1863 in the United States to represent the interests of veterinarians and regulate the profession (1863年在美国成立的非营利组织机构,旨在维护兽医的利益并对该行业进行监管)美国兽医协会

American Warmblood n. a relatively new breed of light horse developed in the United States; stands 15 to 17 hands and usually has a coat of solid color; used mainly for dressage, jumping, and eventing (美国最近培育的马品种,体高15~17掌,被毛多为单色,主要用于盛装舞步、障碍赛和三日赛)美国温血马

American White n. *see* American White Horse 美国白马

American White Horse n. (also **American White**) an American color breed developed in the early 20th century; stands 15 to 17 hands on average and usually has white hair, pink skin, and brown or black eyes (美国在20世纪初育成的毛色品种,体高15~17掌,被毛为白,皮肤粉红,眼珠褐或黑)美国白马 [syn] Albino, American Albino

AMHA abbr. American Miniature Horse Association 美国微型马协会

amnion /ˈæmnɪən/ n.【anatomy】(*pl.* **amnia** /-nɪə/) a tough, membranous sac in the uterus of the pregnant mare that contains fluid and the fetus (妊娠母马子宫内包裹胎儿的膜囊)羊膜: The amnion forms a protective cloak around the fetus, preventing any contact with the allantoic fluid and also, at birth, acting as a means to reduce friction between the foal and the birth canal. 羊膜作为胎儿外围的保护屏障,可防止胎儿与尿囊液的接触,且在分娩时有润滑胎儿和产道的作用。[syn] amnionic sac, water bag

amnionic /ˌæmniˈɒnɪk/ adj. of or having amnia 羊膜的,有羊膜的: amnionic sac 羊膜囊 ◇ amnionic fluid 羊水

Ampeton /ˈæmptən/ n. a light, horse-drawn open carriage first used in London, England during the 1850s (19世纪50年代出现于英国伦敦的轻型无篷马车)安普顿马车

ampoule /ˈæmpuːl/ n. (also **ampule**) a small glass container holding a drug solution used for hypodermic injection (贮存皮下注射液的小玻璃瓶)安瓿

ampulla /æmˈpulə/ n.【anatomy】(*pl.* **ampullae** /-liː/) a small dilatation in a canal or duct (管道的)壶腹部: ampulla of uterine tube 输卵管壶腹部 ◇ ampulla of vas deferens 输精管壶腹部

amulet /ˈæmjʊlət/ n. anything worn to act as a charm against evil or mischief (用来驱魔避灾的饰物)护身符

anabolic /ˌænəˈbɒlɪk/ adj.【biochemistry】of or relating to anabolism 合成代谢的: anabolic steroids 合成类固醇 [opp] catabolic

anabolism /əˈnæbəˌlɪzəm/ n.【biochemistry】the phase of metabolism in which simple substances are synthesized into the complex materials of living tissue (代谢中由简单物质合成复杂有

A

机物的过程)合成代谢 opp catabolism

anaemia /ə'niːmiə/ n.【BrE】*see* anemia 贫血

anaerobic /ˌænεə'rəʊbɪk/ adj. needing no oxygen 厌氧的;无氧的:anaerobic respiration 无氧呼吸 ◇ anaerobic exercise 无氧运动 opp aerobic

anal /'eɪnl/ adj. of or relating to the anus 肛门的:anal temperature 肛温

anal atresia n. a congenital condition of foals born without an anal opening 肛门闭锁

analeptic /ˌænə'lεptɪk/ adj. (of an agent) restorative or stimulating (指药剂)复壮的;刺激兴奋的

n. an agent used as a central nervous system stimulant (可使中枢神经系统兴奋的制剂)强壮剂,兴奋剂

analgesia /ˌænəl'dʒiːziə; *AmE* -dʒiːʒə/ n. a deadening or absence of the sense of pain without loss of consciousness (在未失去知觉的情况下痛感减弱或消失)痛觉消失;止痛,镇痛

analgesic /ˌænəl'dʒiːzɪk/ adj. of or causing analgesia 止痛的:analgesic drugs/effects 止痛药/效果

n. a medication that relieves or eliminates pain 止痛药,镇痛剂

anaphylactic /ˌænəfɪ'læktɪk/ adj. of or causing allergic reaction 过敏性的:anaphylactic shock 过敏性休克

anaphylaxis /ˌænəfɪ'læksɪs/ n. an allergic reaction or hypersensitivity to foreign substances (对异物的)过敏性反应

anatomy /ə'nætəmi/ n. the scientific study of the shape and structure of organisms by dissection (通过解剖研究有机体组成结构的学科)解剖学 | **anatomical** /ˌænə'tɒmɪkl/ adj.

Ancaster Arabian n. another name for Alcock Arabian (安考科·阿拉伯马的别名)安卡斯特·阿拉伯马

ancestor /'ænsestə(r)/ n. a primitive form or species from which later kinds evolved; origin (物种进化的)祖先;先祖

ancestral /æn'sestrəl/ adj. of, relating to, or evolved from ancestors 祖先的;古代的

Ancient English Pacer n. a horse breed used throughout the British Isles during the Middle Ages (中世纪在不列颠岛上广为使用的马种)古英吉利对侧步马

Andalusian /ˌændə'luːʃn/ n. a warmblood breed originating in Spain around the 8th century AD, thought to have either descended from Barb and Arab stock crossed with native breeds or descended from the Iberian horse; stands around 15 hands and generally has a gray coat, although bay, black, chestnut and roan do occur; is particularly good at jumping and dressage (公元8世纪源于西班牙的温血马品种,普遍认为由柏布马和阿拉伯马与当地品种杂交或由伊比利亚马繁育而来,体高约15掌,毛色以青为主,但骝、黑、栗或沙亦有之,擅长障碍赛和盛装舞步)安达卢西亚马 syn Spanish Horse

androgen /'ændrədʒən/ n.【physiology】a steroid hormone that controls the development and maintenance of masculine characteristics (控制雄性发育并维持其性征的类固醇激素)雄激素

anemia /ə'niːmiə/ n. an abnormal decrease of erythrocyte in the blood caused by excessive bleeding, dietary iron deficiency or infection of viruses, symptoms include fatigue, lack of performance and listlessness (由于大量失血、日粮铁缺乏或病毒感染引起红细胞减少的症状,表现为乏力、体力下降和精神不振)贫血:equine infectious anemia 马传染性贫血 ◇ hemorrhagic anemia 出血性贫血 ◇ hemolytic anemia 溶血性贫血 ◇ nutritional anemia 营养性贫血

anesthesia /ˌænəs'θiːziə; *AmE* ˌænɪs'θiʒə/ n. a local or general insensibility to pain without loss of consciousness (在清醒状态下身体局部或全身对疼痛无感)麻醉:general anesthesia 全身麻醉 ◇ local anesthesia 局部麻醉 ◇ gaseous anesthesia 吸入麻醉

anesthetic /ˌænəsˈθetɪk/ adj. of or causing anesthesia 麻醉的;引起麻醉的

n. an agent injected or inhaled to induce desensitization without loss of consciousness(一种注射或吸入式药剂,可使机体在清醒状态下丧失痛感)麻醉剂,麻药

anesthetize /əˈniːsθətaɪz/ vt. to induce anesthesia 麻醉

anestrous /eɪˈnestrəs/ adj. of or relating to anestrus 乏情期的,乏情的

anestrum /əˈnestrʌm/ n. the anestrus 乏情期,休情期

anestrus / æˈniːstrəs,-ˈnes-/ n. (*pl.* **anestra** /-trə/) an interval of sexual inactivity in the mare during which there is no observable heat or acceptance of the stallion (母马性活动消极阶段,期间无发情迹象,且不接受公马交配)乏情期,休情期— *see also* **diestrus**, **estrus**

Anglo-Arab /ˌæŋgləʊˈærəb/ n. a cross-breed developed from crossing Thoroughbred with Arab (由纯血马和阿拉伯马杂交培育而成的马品种)盎格鲁阿拉伯马: An Anglo-Arab should have no less than 25%, but no more than 75% Arabian blood. 盎格鲁阿拉伯马所含阿拉伯马的血统既不少于1/4,也不多于3/4。

Anglo-Argentine /ˌæŋgləʊ ˈɑːdʒəntaɪn/ n. a horse breed developed in South America by crossing the Thoroughbred with the Criollo (由纯血马与克里奥尔马杂交在南美育成的马品种)盎格鲁阿根廷马

Anglo-Norman /ˌæŋgləʊˈnɔː(r)mən/ n. *see* Norman 盎格鲁诺曼马

Anglo-Persian /ˌæŋgləʊˈpɜːʃn; *AmE* -ˈpɜːrʒn/ n. a horse breed developed in Iran by crossing the Arab with Persian(在伊朗由阿拉伯马和波斯马杂交培育而成的马品种)盎格鲁波斯马

Angular Landau n. *see* Shelburne Landau 谢尔本兰道马车

anhydrosis /æhaɪˈdrəʊsɪs/ n. a failure of the sweating mechanism of the body (身体排汗机制失常)缺汗症,无汗症

Animalia /ˌænəˈmeɪliə/ n. the kingdom of animals 动物界

animated /ˈænɪmeɪtɪd/ adj. (of horse's gait) lively and spiritedly (马运步)活泼的,轻快的

ankle /ˈæŋkl/ n. **1** *see* fetlock joint 踝关节 **2** a leg marking extending from the coronary band to just above the fetlock (马四肢别征之一,白色区域从蹄冠至球节)踝白

ankle boots n. a light-duty boot used to protect a horse's fetlock joint and tendons from brushing (保护马匹踝关节及筋腱的)护踝 [syn] sesamoid boots

ankle cutter n.【racing】a horse who knocks his fetlock with the opposite foot while running (行进中)撞踝的马,踝交突的马

ankylosis /ˌæŋkɪˈləʊsɪs/ n. the stiffening and immobility of a joint as the result of disease, trauma, surgery, or abnormal fusion of bones (由于疾病、外伤、手术或骨异常愈合导致的关节僵硬)硬化;愈合: ankylosing arthritis 愈合性关节炎

annular /ˈænjələ(r)/ adj. shaped like a ring 环形的

anoestrus /əˈnəʊstrəs/ n.【BrE】anestrus 乏情期,休情期

Anoplocephala manga n.【Latin】a tapeworm varying in length from 8 to 25 cm and found in the small intestine and stomach, which might cause digestive disturbances and anemia in heavy infestations (体长在8~25厘米的一种绦虫,主要寄生于动物的小肠和胃内,当大量繁殖时,患畜表现消化机能紊乱和贫血)大裸头绦虫

Anoplocephala perforliata n.【Latin】a tapeworm varying in length from 8 to 25 cm and found in the cecum, which might cause digestive disturbances, unthriftiness, and anemia in heavy infestations (体长在8~25厘米的一种绦虫,主要寄生于动物的盲肠,大量繁殖时,患畜表现消化紊乱、体况下降,还可能导致贫血)叶

A

状裸头绦虫

anorexia /ˌænəˈreksiə/ n. loss of appetite, especially as a result of disease or stress (由疾病或应激造成的食欲消退)厌食

anovulatory /ænˈɒvjulətəri; *AmE* -ˈɑːvjələˌtɔri/ adj.【physiology】having no ovulation 不排卵的:anovulatory estrus 无卵发情

anoxia /æˈnɒksɪə; *AmE* -ˈnɑːk-/ n. a pathological absence of oxygen (病理性)缺氧[症]:cerebral anoxia 脑缺氧

antebrachial /ˌæntəˈbreɪkɪəl/ adj. of or relating to the forearm 前臂的:antebrachial region 前臂部

antebrachium /ˌæntəˈbreɪkɪəm/ n.【anatomy】(*pl.* **antebrachia** /-kiə/) the forearm of a vertebrate 前臂

ante-post betting n.【racing】(also **antipost bet**) a betting option in which wagering occurs on a day prior to the race 赛前赌马,赛前投注

anterior /ænˈtɪəriə(r); *AmE* -tɪr-/ adj. located on or near the front of an organ or the body (器官或身体)前部的,前面的:the anterior pituitary gland 垂体前叶

anterior enteritis n. an acute inflammation of the small intestine resulting in abdominal distress such as colic and diarrhea (小肠的急性炎症,表现为腹痛和腹泻)小肠炎

anthelmintic /ænθelˈmɪntɪk; *AmE* anθ(ə)lˈ-/ adj. acting to get rid of intestinal worms 驱虫的:anthelmintic drugs 驱虫药

n. (usu. **anthelmintics**) an agent used to get rid of intestinal parasites; wormer (用来驱除肠道寄生虫的药剂)驱虫药

anthrax /ˈænθræks/ n. (*pl.* **anthraces** /-siːz/) an acute, infectious, usually fatal disease of mammals caused by the bacterium *Bacillus anthracis*, and symphomized by high fever, abnormal swelling in the throat and neck, ulcerative skin lesions, and sudden death (由炭疽杆菌引起的急性致死性传染病,其症状表现为高热、咽喉肿胀、皮肤溃烂损伤和猝死)炭疽

antibacterial /ˌæntibækˈtɪəriəl/ adj. (of agents) inhibiting the growth and reproduction of bacteria 杀菌的;抗菌的:antibacterial treatment 抗菌处理

n. an agent or drug that destroys or inhibits the growth and reproduction of bacteria(可抑制病菌生长的制剂)抗菌药;杀菌剂

antibiotic /ˌæntibaɪˈɒtɪk/ n. any of various drugs used to destroy or inhibit the growth of bacteria and other disease-causing organisms (用来抑制病菌生长的药物)抗生素:Antibiotics are widely used in the prevention and treatment of infectious diseases, and may be administered orally, topically, or by injection depending on the drug used and the nature of the infection. 抗生素广泛用于传染病的预防和治疗,可根据使用的药物和所患传染病的情况选择口服、局部用药或注射。

antibody /ˈæntibɒdi/ n. a blood protein produced in immune response to an alien substance introduced into the body (异物进入机体后通过免疫应答产生的血液蛋白)抗体:antibody test 抗体检测

anticipate /ænˈtɪsɪpeɪt/ vt. (of a horse) to realize the aids or requests of its rider or trainer and act accordingly before actually receiving it (在未接到骑手或练马师的扶助或指令之前)预料,预判:A well-trained horse can anticipate its rider's aids accurately. 训练有素的马匹可以准确预判骑手的扶助。

anti-cribbing device n. a device fitted around the upper neck of a horse to apply pressure to the horse's gullet to prevent cribbing (套在马颈前段以防止咬槽恶习形成的装置)防咬槽装置

antidiuretic /ˌæntɪˌdaɪjʊˈrɛtɪk; *AmE* -daɪəˈretɪk/ adj. (of a hormone or drug) acting on the kidneys to control water excretion (指激素或药物有利于肾脏控制水分排泄)抗利尿的 [opp] diuretic

antidiuretic hormone n. another term for vasopressin (血管升压素的别名)抗利尿激素

antidote /ˈæntidəʊt; *AmE* -doʊt/ n. a remedy used to neutralize or counteract the effects of a poison（用来中和或对抗中毒物质的药剂）解毒药，解毒剂

antifebrile /ˌæntɪˈfiːbraɪl/ adj. capable of reducing fever; antipyretic 退热的
n. an agent that reduces fever 退热药

antifungal /ˌæntɪˈfʌŋgəl/ n. any substance capable of destroying or inhibiting the growth of fungi 抗真菌剂，杀真菌剂

antigen /ˈæntɪdʒən/ n. an alien substance that stimulates the production of an antibody when introduced into the body（进入体内后可刺激机体产生抗体的物质）抗原

antihistamine /ˌæntiˈhɪstəmiːn/ n.【biochemistry】a drug used to counteract the physiological effects of histamine（用来对抗组胺生理效应的药物）抗组胺药

anti-inflammatory /ˌæntiɪnˈflæməˌtɔri/ adj. preventing or reducing inflammation 抗炎的；抗感染的
n. an agent that reduces inflammation 消炎药

antimere /ˈæntɪˌmɪə/ n.【anatomy】a body part corresponding to the opposite side of the body 体辐，对称部

antiparasitic /ˌæntɪˌpærəˈsɪtɪk/ adj.（of drugs）destroying or inhibiting the growth and reproduction of parasites 抗寄生虫的，驱虫的
n. an antiparasitic agent; anthelmintic 抗寄生虫药，驱虫剂

antiphlogistic /ˌæntɪfləʊˈdʒɪstɪk/ adj. reducing inflammation and fever; anti-inflammatory 消炎的
n. an agent or drug that reduces inflammation 消炎药

antipost bet n.【racing】*see* ante-post betting 赛前押注

antipyretic /ˌæntɪpaɪˈrɛtɪk/ adj.（of a medicine）reducing or tending to reduce fever 退热的，解热的
n. a medication effective in reducing fever 退热药，解热剂

anti-roll bar n. *see* anti-sway bar 防摆杆，抗侧倾杆

antiseptic /ˌæntɪˈsɛptɪk/ adj. capable of preventing infection by inhibiting the growth of microbes（通过抑制微生物生长而防止感染）防腐的，杀菌的
n. a substance that prevents the growth and reproduction of bacteria 杀菌剂；抑菌剂

antiserum /ˌæntɪˈsɪərəm; *AmE* -ˈsɪr-/ n. a serum containing antibodies that are specific for one or more antigens and used for injection into others to provide immunity against certain diseases（一种内含与抗原特异性结合抗体的血清，注射后可增强机体对某些疾病的免疫力）抗血清

anti-sway bar n. a metal bar fixed to stabilize the chassis against sway（防止马车底盘摆动的）防摆杆 [syn] anti-roll bar, stabilizer bar, sway bar

antisweat sheet n. a cloth covering for horses that enables the horse to cool down slowly following exercise（运动后可降低马体温的外套）防汗马衣

antitoxic /ˌæntɪˈtɒksɪk/ adj. **1** counteracting a toxin or poison 抗毒素的 **2** of, relating to, or containing an antitoxin 抗毒的；含抗毒素的：an antitoxic serum 抗毒血清

antitoxin /ˌæntɪˈtɒksɪn; *AmE* -ˈtɑːk-/ n. a substance such as the antibody produced in response to and capable of counteracting a specific toxin（机体对某种毒素作出免疫应答后产生的物质）抗毒素

Antoine de Pluvinel n. *see* de Pluvinel, Antoine 安托万·德·普鲁维奈尔

anus /ˈeɪnəs/ n.【anatomy】the opening at a horse's bottom through which solid waste is eliminated from its body（马臀端排泄固体废物的开口）肛门

anvil /ˈænvɪl/ n. **1**【farriery】a heavy iron block with a flat surface upon which horseshoes are

A

shaped (锤制蹄铁的) 铁砧 **2**【anatomy】the middle bone of the ear; incus 砧骨

aorta /eɪˈɔːtə; *AmE* eɪˈɔːrtə/ n.【anatomy】(*pl.* **aortas**, or **aortae** /-tiː/) the main artery that carries blood from the left side of the heart to all limbs and organs except the lungs (将血液由左心室运送至四肢和除肺脏之外的器官的主要动脉) 主动脉: abdominal aorta 腹主动脉 ◇ ascending aorta 升主动脉 ◇ descending aorta 降主动脉 ◇ fetal aorta 胎儿主动脉 ◇ thoracic aorta 胸主动脉

aortic /eɪˈɔːtɪk/ adj. of or relating to aorta 主动脉的

aperture /ˈæpətʃə(r)/ n. an opening or slit 开裂;缝隙:vulval aperture 阴门裂

apex /ˈeɪpeks/ n.【anatomy】(*pl.* **apexes**, or **apices** /ˈeipisiːz/) the pointed end of an organ 尖:cecal apex 盲肠尖 ◇ apex of tongue 舌尖 ◇ apex of frog 蹄叉尖

APHA abbr. American Paint Horse Association 美国花马协会

ApHC abbr. Appaloosa Horse Club 阿帕卢萨马俱乐部

Appaloosa /ˌæpəˈluːsə/ n. (abbr. **Appy**) an American warmblood breed descended from the Spanish stock brought by the conquistadors in the 16th century and crossed with native mares; is noted for its spotted coat pattern, striped hoofs and white sclera encircling the eye; used extensively for rodeo events and Western disciplines (由16世纪殖民者引入的西班牙马与当地母马杂交繁育而来的美国温血马品种,被毛带花斑,蹄上有纹,眼周巩膜泛白,多用于牛仔竞技和西部马上项目) 阿帕卢萨马: The Appaloosa breed was developed by the Nez Percé Indians and named for the River Palouse. 阿帕卢萨马由一个名为内兹·佩尔塞的印第安部落育成,并以帕卢斯河命名。 syn spotted horse, Palouse Pony, Palousy

Appaloosa Horse Club n. (abbr. **ApHC**) an organization founded in Idaho, USA in 1938 to preserve and maintain the registry of the Appaloosa breed (1938年在美国爱达荷州成立的组织机构,主要负责阿帕卢萨马的保种和登记工作) 阿帕卢萨马俱乐部

appearance /əˈpɪərəns; *AmE* əˈpɪr-/ n. the outward aspects of a horse (马的) 外貌,体貌

appetite /ˈæpɪtaɪt/ n. an instinctive desire for food or drink 食欲:loss of appetite 食欲减退

appetizer /ˈæpɪtaɪzə(r)/ n.【nutrition】a feed used to stimulate the appetite of horses (用于增进食欲的) 开胃料

appetizing /ˈæpɪtaɪzɪŋ/ adj. appealing to or stimulating the appetite 开胃的;促进食欲的:The first feed after foaling may generally be the wheat bran mash, as this is appetizing and easy to eat. 母马分娩后要先喂麸皮粥,这主要因为麸皮粥既增加食欲又便于进食。

apple-picker n.【slang】a fork used to clean a horse's stall (清理马厩的) 粪叉

appoint /əˈpɔɪnt/ vt. to furnish; equip 装饰;布置: a coach with elegantly appointed interiors 内部装饰华丽的马车

appointments /əˈpɔɪntmənts/ n. the equipments such as the saddlery and harness worn by a horse at work or show (马作业或表演时所用的) 装备;鞍具

apprentice /əˈprentɪs/ n. **1** one bound by a contract to learn a profession (订立合同学习某种职业的人) 学徒;见习生:apprentice jockey 见习骑师 **2** a beginner; apprentice jockey 新手;见习骑师

apprentice allowance n.【racing】a weight reduction granted to an apprentice jockey when competing against jockeys (赛马中给见习骑师减少的负重) 见习让磅:The amount of the apprentice allowance as well as the conditions relating to such varies from area to area, usually 5 pounds until the jockey has five wins, or 7 pounds until the 35th win, but is not granted apprentice jockeys when competing in stakes races. 不同地方的见习让磅以及让磅附则都

有所不同。一般来讲，骑师获胜5次，让5磅；获胜35次，让7磅。但是，奖金赛中则不给予见习让磅。

approach /əˈprəʊtʃ; *AmE* əˈproʊtʃ/ vi. (of a horse) to come or draw near to an obstacle in jumping (指马在跨栏时)接近，靠近

n. the act of drawing near to an obstacle, as in a horse to jump a fence (马匹跨栏时)接近

appuyer /ɑːˈpjʊje/ n.【French】two track 斜横步，双蹄迹：appuyer en renvers (haunches-out) 臀朝外横斜步 ◇ appuyer en travers (haunches-in) 臀朝里横斜步 ◇ appuyer épaule en dedans (shoulder-in) 肩向内横斜步 ◇ appuyer épaule en dehors (shoulder-out) 肩向外横斜步

vi. to walk on two track 斜横步行进

Appy /ˈæpi/ n.【colloq】Appaloosa 阿帕卢萨马

apricot dun n.【color】the lightest shade of red dun (红兔褐色中最淡的毛色)杏兔褐

apron /ˈeɪprən/ n. **1** a leather or canvas covering worn by a farrier to protect his thigh while shoeing a horse (蹄师穿的皮革或帆布套兜，钉马掌时可保护大腿)围裙，护裙 [syn] shoeing chaps **2** a cloth or leather covering used by the driver of a horse-drawn vehicle to keep him warm or clean (车夫驾车时用来保暖或护身的织物或皮革)护裙

AQHA abbr. **1** American Quarter Horse Association 美国夸特马协会 **2** Australian Quarter Horse Association 澳大利亚夸特马协会

aquapuncture /ˌækwəˈpʌŋktʃər/ n. a type of acupuncture practiced by injecting fluids or solutions into the acupuncture points (向穴位注射针剂或溶液的针灸方法)注水针灸

aqueduct /ˈækwɪdʌkt/ n.【anatomy】a channel in an organ or a body part for conveying fluid (机体器官内用于输送液体的管道)导管：aqueduct of cochlea 耳蜗管 ◇ aqueduct of mesencephalon 中脑导水管 ◇ aqueduct of vestibule 前庭导水管

aqueductus /ˌækwɪˈdʌktəs/ n.【Latin】aqueduct 导水管

AR abbr. Arabian 阿拉伯马

Arab /ˈærəb/ n. (also **Arabian**, or **Arabian horse**) a horse breed originating from Saudi Arabia and Middle East more than 3,000 years ago; stands 14.3 to 16 hands and generally has a bay, brown, chestnut, gray, or black coat; has a dished profile, small ears and possess excellent endurance and speed; used mainly for riding and light draft (3 000 多年前源于中东和沙特阿拉伯地区的马品种，体高 14.3～16 掌，常见毛色有骝、褐、栗、青和黑，面部凹陷，耳小，耐力好，速度快，主要用于骑乘和轻挽)阿拉伯马：Arab has 17 ribs, 5 lumbar bones, and 16 tail vertebrae in comparison with the 18-6-18 formation of other breeds. 阿拉伯马有17根肋骨、5个腰椎和16个尾椎，而其他品种相应为18-6-18。

Arab Horse Society n. an organization founded in 1918 in England to govern and promote the breeding of Arab breeds (1918年在英国成立的组织机构，旨在监管和促进阿拉伯马的培育工作)阿拉伯马协会

Araba /əˈrɑːbə/ n. a horse-drawn wagon with a canopy top and crosswise seating, used in the 18th and 19th centuries by the Turkish court (18—19世纪土耳其宫廷使用的一种四轮马车，带有顶篷和横排座位)阿拉巴马车

arabesque /ˌærəˈbesk/ n.【vaulting】a position in which the vaulter stands on the back of a moving horse on one leg with the other stretching out backward (马上杂技姿势，表演者以单足站立在疾驰的马背上，另外一条腿则向后伸展)燕式平衡

Arabian /əˈreɪbɪən/ n. *see* Arab 阿拉伯马

Arabian horse n. *see* Arab 阿拉伯马：The Arabian horse is generally known for its elegant, delicate appearance, graceful gaits and dished face. 阿拉伯马外貌俊秀、端庄，步伐优美，面颊凹陷。

Arabian Race n. a part-bred Arab bred at the

Bablona stud, Hungary in the early 19th century, developed by crossing purebred Arab stallions with mares who carried strains of Spanish, Hungarian, and Thoroughbred blood; is usually considered as the foundation horse of the Shagya Arabian (19 世纪初在匈牙利巴布罗纳马场育成的一个半血阿拉伯马品种,通过纯种阿拉伯公马与带有西班牙、匈牙利以及纯血马血统的母马杂交培育而成,普遍认为是沙迦阿拉伯马的奠基马)阿拉伯马亚种

arachnoid /əˈræknɔɪd/ n.【anatomy】a delicate membrane that encloses the spinal cord and brain (包裹脊髓和大脑的一层软膜)蛛网膜
adj. of or relating to a delicate membrane enclosing the spinal cord and brain 蛛网膜的

Arappaloosa /əˌræpəˈluːsə/ n. a cross-breed developed by crossing Arabian with Appaloosa (由阿拉伯马和阿帕卢萨马杂交培育而来的马种)阿拉帕卢萨马

arc /ɑːk; *AmE* ɑːrk/ n.【rodeo】the bend formed by the horse when turning around a barrel (马在绕桶时形成的曲线)弧度,弧线

Arcaro, Eddie n. (1916 – 1997) the professional name of the American jockey George Edward Arcaro who was widely regarded as the greatest jockey in the history of American horse racing (美国职业骑手乔治·爱德华·阿克洛的艺名,普遍认为是美国赛马史上最伟大的骑师)埃迪·阿克洛

arch /ɑːtʃ; *AmE* ɑːrtʃ/ n. the curve of a horse's spine (脊柱的)曲弓,曲弧
vi. to form a curved line or shape 成弧状,使弯曲

arch Pelham n. a bit having a mouthpiece that curves in the center to accommodate the tongue (衔杆中段呈弧形的衔铁,以便容纳马的舌头)弧形佩勒姆衔铁

arched /ɑːtʃt; *AmE* ɑːrtʃt/ adj. bent or curved like an arch 弓形的,拱状的: arched back 背部拱起

archer /ˈɑːtʃə(r)/ n. one who shoots with a bow and arrows (采用弓箭射击的人)射手,弓箭手: The Mongolian archers were legendary marksman and dextrous on foot or horseback. 蒙古弓箭手曾是传奇的神射手,步行或骑马都身手矫健。

archery /ˈɑːrtʃəri/ n. the sport or skill or shooting with a bow and arrows at a target (采用弓箭向目标射击的运动或技艺)射箭: mounted archery 骑射

Ardennais /ɑːrˈdeneɪs/ n.【French】(also **Ardennes**) a heavy draft breed originating in the Ardennes region of France from which the name derived; stands 15 to 16 hands and may have a bay, roan, gray, palomino, or chestnut coat; has a heavy head, small ears, large eyes, short neck and back, and short feathered legs with broad hoofs; used mainly for heavy draft and farm work (源于法国阿尔登省并由此得名的重挽马品种。体高 15 ~ 16 掌,毛色可为骝、沙、青、银鬃或栗,头重,耳小,眼大,颈背短,四肢粗短且附生距毛,蹄大,主要用于重挽和农田作业)阿[尔]登马

Ardennes /ɑːrˈden/ n. *see* Ardennes 阿[尔]登马

arena /əˈriːnə/ n. an area or center where equestrian events and contests are held (举行马术项目和比赛的地点)竞技场,赛场: competition arena 比赛场 ◇ indoor arena 室内赛场 ◇ outdoor arena 室外赛场 ◇ warm-up arena 热身场 ◇ arena director 场内指导 ◇ arena footing 赛场地况

arena polo n. (also **indoor polo**) a polo game played indoors 室内马球: The field of the arena polo is approximately 91 m × 46 m in size, and rules and principles are of the same with the polo except that each team consists of only three player. 室内马球场地一般为 91 米 × 46 米,其比赛规则与马球相同,但每队仅由 3 名球员组成。

Argentina /ˌɑːdʒənˈtiːnə/ n. a breed of hot-blood horse 阿根廷马

Argentine Criollo n. *see* Criollo 阿根廷克里奥罗马

Argentine polo pony n. (also **Argentine pony**) a polo pony developed in Argentina by crossing imported Thoroughbred stallions with native Criollo mares; stands around 15 hands and has strong joints and sturdy hoofs; is quick and agile in movements and possess good stamina (阿根廷采用纯血马与本地克里奥罗马杂交培育而成的马品种。体高约 15 掌，关节强健，蹄质结实，反应迅捷，体力出色）阿根廷马球马

Argentine pony n. *see* Argentine polo pony 阿根廷马球马

Argentine snaffle n. a snaffle bit mainly used in Argentina 阿根廷衔铁

arginine /ˈɑːdʒəˌniːn/ n.【nutrition】an essential amino acid of vertebrates (脊椎动物的一种必需氨基酸）精氨酸

Ariégeois /arjeʒwa/ n.【French】*see* Ariégeois pony 阿列日马

Ariégeois pony n. (also **Ariégeois**) the original name for the Mérens (梅隆马的旧称）阿列日马

armchair ride n.【racing】an easily won victory of a horse race 轻松获胜的比赛

armor /ˈɑːmə; *AmE* ˈɑːrmər/ n. (also **armour**) a metal covering formerly worn by soldiers for protection (过去战士所穿起防护作用的金属外套）铠甲，盔甲

armored /ˈɑːməd; *AmE* ˈɑːrmərd/ adj. covered or protected with armour 身披盔甲的：armoured knights 身披盔甲的骑士

armour /ˈɑːmə/ n.【BrE】*see* armor 铠甲，盔甲

aroma /əˈrəʊmə; *AmE* əˈroʊmə/ n. a pleasant, noticeable smell 香味；芳香

aromatherapy /əˌrəʊməˈθerəpi; *AmE* əˌroʊmə-/ n. the use of essential oils to treat emotional and physical ailments through the sense of smell (采用精油气味来治愈病痛的疗法）芳香疗法

arouse /əˈraʊz/ vt. to stimulate or excite sexually 性唤起，性刺激

arrhythmia /əˈrɪθmiə/ n. an irregularity of the heart rate and rhythm (心率和节奏紊乱的症状）心律不齐，心律失常：cardiac arrhythmia 心律不齐 [syn] heartbeat irregularity

art. abbr. articulus 关节

arterial /ɑːˈtɪərɪəl/ adj. of, like, or in an artery or arteries 动脉的：arterial pressure 动脉压

arterial bleeding n. the loss of blood from an artery 动脉出血

arteriole /ɑːˈtɪərɪəʊl/ n.【anatomy】one of the small terminal branches of an artery, especially one that connects with a capillary (动脉末端的细小分支，连接毛细血管）微动脉

arteritis /ˈɑːtəraɪtɪs/ n. inflammation of the arteries 动脉炎：equine viral arteritis 马病毒性动脉炎

artery /ˈɑːtəri/ n.【anatomy】any of the muscular-walled tubes of the circulation system by which blood is carried from the heart to all parts of the body (血液循环体系中的肌性管道，负责将血液从心脏输送到全身各处）动脉 [compare] vein

arthritis /ɑːˈθraɪtɪs; *AmE* ɑːrˈθ-/ n. inflammation of a joint or joints resulting in pain and swelling (关节所患的炎症，常导致关节疼痛和肿大）关节炎：serous arthritis 浆性关节炎 ◇ osteoarthritis 骨关节炎 ◇ infectious arthritis 传染性关节炎 ◇ ankylosing arthritis 僵直性关节炎

Arthrobotrys oligospora n.【Latin】a species of predatory fungus existing widely in the world, often used to control small strongyle populations by destroying them (一种分布广泛的捕食性真菌，常用来控制小圆线虫的繁殖）寡孢节丛孢菌：*Arthrobotrys oligospora* is considered an effective alternative to chemical antiparasitics. 寡孢节丛孢菌被认为是化学抗虫剂有效的替代物。

arthroscope /ˈɑːθrəskəʊp/ n. a medical instru-

A

ment consisting of an optic fiber through which the interior of a joint may be inspected or operated on (带有光纤细管的医用设备,借此对关节内部进行观察或实施手术)关节[内窥]镜

arthroscopic surgery n. a micro-surgery operated on a joint by using an arthroscope (借助关节内窥镜对关节实施的一种显微手术)关节微创手术

arthroscopy /ɑːˈθrɒskəpi/ n. the examination of the interior of a joint by using an endoscope that is inserted into the joint through a small incision (从小切口插入内窥镜而对关节内部进行检查的方法)关节内窥镜检查,关节镜检查

articular /ɑːˈtɪkjʊlə; *AmE* ɑːrˈtɪkjələr/ adj. of or relating to a joint or joints 关节的:articular capsule 关节囊 ◇ articular cavity 关节腔 ◇articular disc 关节盘 ◇ articular surface 关节面

articular cartilage n. one type of cartilage that is firmly attached to and covers the ends of the bones where they meet in a joint (紧贴于骨末端关节面处的软骨)关节软骨 syn joint cartilage

articular fracture n. a break in a bone which extends into a joint 关节骨折

articular windgall n. (also **articular windpuff**) a soft, painless swelling that develops between the cannon bone and the suspensory ligament due to excessive accumulation of synovial fluid (管骨和悬韧带间滑液淤积导致的关节水肿)关节软肿 compare tendinous windgall

articular windpuff n. *see* articular windgall 关节软肿

articulate /ɑːˈtɪkjuleɪt/ vi.【anatomy】to form a joint; be jointed 连接,连结

articulating ring bone n. a ring bone condition where the bony growth has attached to the joints between the long and short pastern or the short pastern and the pedal bone (长系骨与短系骨或短系骨与趾骨关节间所附生的赘骨)关节赘骨

articulation /ɑːˌtɪkjuˈleɪʃn/ n. a joint between bones 关节

articulus /ɑːˈtɪkjuləs/ n.【Latin】(*pl.* **articuli**/-laɪ/) joint 关节

artificial /ˌɑːtɪˈfɪʃl/ adj. made rather than natural 人工的: artificial lighting 人工光照 ◇ artificial coloring 人工着色

artificial aids n. any mechanical means by which a rider conveys his requests to the horse, such as the use of the spurs or whips (骑手借用马刺或鞭子等用具传达命令的方式)人工辅助 compare natural aids

artificial airs n. any artificial gait of a horse 人工步法

artificial allures n. any trained gait or pace of a horse 人工步法 syn artificial airs

artificial breeding n. the breeding of horses by means of artificial insemination, embryo transfer or cloning rather than natural cover (采用人工授精、胚胎移植和克隆而非自然交配进行的繁育方式)人工繁育: Artificial breeding is still not approved by Jockey Clubs in many countries now. 目前,许多国家的赛马会仍然反对采用人工繁育。

artificial drag n.【hunting】a trail of scent used in a drag hunt, often left artificially by dragging an object soaked in certain strong-smelling liquid similar to the scent of a fox for drag hunt (寻踪猎狐中人工设计的气味踪迹,通过将物件在带有类似狐味的液体中浸泡然后拖于地面产生)人造踪迹

artificial insemination n. (abbr. **AI**) an artificial method of impregnating the mare (使母马受孕的人工方法)人工授精

artificial scent n.【hunting】a scent similar to that of a fox or other quarry laid in a drag hunt (在寻踪狩猎中撒布的类似于狐狸或其他猎物的气味)人工气味

artificial vagina n. (abbr. **AV**) a device used to collect semen from stallions for artificial insemination or laboratory analysis (人工授精或试验

分析时采集公马精液的用具)假阴道

ascarid /ˈəskəraɪd/ n. a type of parasite hosting the bowels, liver and lungs of certain animals(见于某些动物肠道、肝和肺的一类寄生虫)蛔虫:equine ascarid 马蛔虫 syn roundworm

ascending oxer n.【jumping】an oxer with the front rail lower than the back rail(前杆比后杆低的双横木跨栏)梯形双横木

ascorbic acid n. *see* vitamin C 抗坏血酸,维生素 C

Ascot /ˈæskət/ n. a village southwest of London, where the Royal Ascot horse races, initiated by Queen Anne in 1711, are held annually in June(位于英国伦敦西南的小镇,皇家阿斯科特赛马比赛自 1711 年由安妮女王首次发起,此后每年 6 月在此举行)阿斯科特

Ascot Landau n. a horse-drawn English vehicle of the Landau type(一种英式马车)阿斯科特兰道马车

aseptic /ˌeɪˈseptɪk/ adj. free from pathogenic germs 无菌的

ASH abbr. Australian Stock Horse 澳大利亚牧牛马

ASHA abbr. American Saddlebred Horse Association 美国骑乘马协会

Asian Racing Federation n. (abbr. **ARF**) the governing body of horse racing in Asia eatablished in 1960 as the Asian Racing Conference and changed to its present name in 2001(亚洲赛马行业的监管机构,1960 年成立时名为亚洲赛马委员会,2001 年易为今名)亚洲赛马联合会

Asian wild horse n. *see* Asiatic wild horse 亚洲野马

Asiatic wild ass n. a wild ass of the genus *Equus* living in herds through western Asia, having a rust-colored dun coat and a light dorsal stripe(栖息在西亚地区的马属动物,被毛兔褐色,背线较浅)亚洲野驴 syn *Equus hemionus*

Asiatic wild horse n. (also **Asian wild horse**) a species of the genus *Equus* once lived in Gobi desert of Mongolia, having a heavy head, yellow dun coat and short, upright mane(曾经栖息于蒙古戈壁沙漠的马属动物,头重,被毛黄兔褐色,鬃毛短而直立)亚洲野马,蒙古野马 syn Mongolian wild horse

Asil /ˈæsɪl/ n.【Arabic】(also **Asil hosre**) a purebred Arabian horse indigenous to Khuzestan, Iran; stands about 14.3 hands and usually has a chestnut, grey, or bay coat; has a dished head with small muzzle, large expressive eyes, short back and strong legs; is intelligent and spirited and used mainly for riding(伊朗胡齐斯坦省的纯种阿拉伯马,体高约 14.3 掌,毛色以栗、青和骝为主,脸面内凹,口鼻小,眼大,背短,四肢强健,反应敏捷,性格活泼,主要用于骑乘)纯种阿拉伯马

asinine /ˈæsɪnaɪn/ adj. of, relating to, or resembling an ass 驴的;似驴的

Asinus /ˈæsɪnəs/ n.【Latin】ass 驴:*Equus asinus*[学名]驴

Asinus iasbellinus n.【Latin】the scientific name for the Isabella quagga[学名]伊莎贝拉野驴

asparagine /əˈspærəˌdʒiːn/ n. a crystalline non-essential amino acid(一种非必需氨基酸)天冬酰胺

aspartic acid n. one kind of non-essential amino acids(一种非必需氨基酸)天冬氨酸

aspermia /əˈspɜːmiə/ n. an absence of sperm 无精子发生;不排精

asphyxia /æsˈfɪksɪə; əsˈf-/ n. a state of being unable to breathe, causing loss of consciousness or even death(呼吸困难的症状,可引发晕厥甚至死亡)窒息 | **asphyxial** adj.

ass /æs/ n. a species of the genus *Equus* with small build and long ears(一种马属动物,体小而耳长)驴:jack ass 公驴 ◇ jenny ass 母驴 ◇ Asiatic wild ass 亚洲野驴 ◇ Tibetan wild ass 藏野驴 ◇ Persian wild ass 波斯野驴 ◇ Indian wild ass 印度野驴 ◇ Mongolian wild ass 蒙古野驴 syn donkey

Assateague /ˈæsəˌtiːg/ n. *see* Assateague pony 安

萨狄格矮马

Assateague pony n. (also **Assateague**) a pony breed existing on the Assateague island of the United States (见于美国安萨狄格岛上的矮马品种)安萨狄格矮马

assess /ə'ses/ vt. to estimate the quality of sth 评定 | **assessment** /-mənt/ n.: conformation assessment 体貌鉴定

Assil /'æsɪl/ n. one type of Arabian horse bred by Bedouins (阿拉伯马的类型之一,由贝都因人培育)阿西利阿拉伯马 [syn] Bedouin Arab

Associate Farriers Company of London n. (abbr. **AFCL**) a recognition of advanced farrier skill formerly awarded by the Worshipful Company of Farriers, London, England, which later was replaced by the Associate of the Worshipful Company of Farriers (英国伦敦蹄铁匠公会对蹄铁匠技能的资格认证,其后由蹄铁匠公会准会员取代)伦敦公会准蹄铁匠

Associate of the Worshipful Company of Farriers n. (abbr. **AWCF**) a certification of advanced farrier skill awarded by the Worshipful Company of Farriers, London, England (英国伦敦蹄铁匠公会向蹄铁匠授予的资格证书)蹄铁匠公会准会员 [compare] Fellow of the Worshipful Company of Farriers, Registered Shoeing Smith

asterisk /'æstərɪsk/ n. the figure " * " used to denote an imported horse when placed in front of a horse's name, or to denote an apprentice jockey when placed in front of a jockey's name (星形符号" * ",置于马名前用来表示进口马匹,在骑手名字前则表示见习骑师)星号

asternum /ə'stɜːnəm; *AmE* -'stɜːrn-/ n.【anatomy】a piece of tissue in the region of, but not attached to, the sternum (胸骨附近的软骨组织)胸软骨

Astley, Philip n. (1742 – 1814) an English circus owner who was regarded as the father of the modern circus (英国马戏团主,被誉为现代马戏之父)菲利普·阿斯特利

astragalus /æ'strægələs/ n.【anatomy】(*pl.* **astragali** /-gəli/) talus 距骨

astride /ə'straɪd/ adv. with legs on each side or wide apart 跨着,叉开: riding astride 跨骑

astringent /ə'strɪndʒənt/ adj. tending to constrict or tighten body tissues (可紧缩机体组织的)收敛的

n. any agent applied to constrict body tissues 收敛剂

asymmetry /eɪ'sɪmətri/ n. a lack of balance or symmetry 不对称;不匀称: Asymmetry of leg markings is very common in horses. 四肢别征的不对称性在马中是比较常见的。| **asymmetrical** /ˌeɪsɪ'metrɪkl/ adj.

asymptomatic /ˌeɪsɪmptə'mætɪk/ adj. exhibiting no symptoms of disease 无症状的: Many heterozygous animals carrying defective gene appear to be asymptomatic to their owners. 对所有者来说,携带缺陷基因的杂合子动物并不表现明显症状。

ataxia /ə'tæksiə/ n. inability to coordinate muscular movement caused by injury or infection of the spinal cord (脊髓受伤或感染后导致的肌肉运动协调能力丧失)运动失调,共济失调 [syn] wobbles, wobbler disease, wobbler syndrome

ATCP abbr. American Team Penning Championships 美国团体截牛锦标赛

athletic /æθ'letɪk/ adj. **1** of or relating to athletics or athletes 运动的: The event horses need a fair amount of natural athletic ability. 综合全能比赛马匹需要有相当好的运动天赋。**2** physically strong; muscular 体格健壮的;肌肉发达的

atlas /'ætləs/ n.【anatomy】the first cervical vertebra of the spinal column (脊柱的第一颈椎)寰椎 [compare] axis

atony /'ætəni/ n. a lack of normal muscle tone (缺乏正常的肌肉收缩节律)弛缓

atresia /ə'trɪʒə/ n. the absence or closure of a normal body orifice or tubular passage, such as the anus, intestine, or external ear canal (身体开口如肛门、肠道或外耳道等管道的缺失或闭合)闭锁

atresic /əˈtreɪsɪk/ adj. of or relating to atresia 闭锁的:atresic follicle 闭锁卵泡

atrophy /ˈætrəfi/ n. a wasting away or degeneration of bodily tissue as caused by inadequate nutrition or disease (由于营养不良或疾病导致机体组织的衰退)萎缩,退化:muscular atrophy 肌肉萎缩 | **atrophic** /əˈtrɒfik/ adj.

vi. to degenerate or decline due to illness or underuse (由于疾病或弃用导致衰退)萎缩,退化: Without exercises, the muscles will atrophy. 不运动的话,肌肉就会萎缩。

atropine /ˈætrəpiːn/ n. an alkaloid compound used as an antispasmodic (一种抗痉挛药物)阿托品

attentive /əˈtentɪv/ adj. (of a horse) giving attention to; watchful 留神的,专心的

attire /əˈtaɪə(r); *AmE* ˈtaɪr/ n. clothing or apparel worn in formal occasions 服装,装束:riding attire 骑马装束,骑装

auburn /ˈɔːbən/ n. a reddish-brown color 赤褐色

adj. of a reddish-brown color 赤褐色的

auction /ˈɔːkʃn/ n. a public sale in which properties and real estate are sold to the highest bidder (财产的公开竞价出售)拍卖[会]:Yearling Auction 岁驹拍卖会

vt. to sell at or by a auction 拍卖

auctioneer /ˌɔːkʃəˈnɪə(r); *AmE* -ˈnɪr/ n. one that presides an auction 拍卖商;拍卖人

Australian bow wagon n. a horse-drawn farm wagon with an arched top, used in Australia during the late 19th century and pulled by a single horse in shafts (19 世纪末在澳大利亚使用的一种农用四轮马车,顶篷为拱形,由单马驾辕拉行)澳洲拱篷马车

Australian pony n. a pony breed descended from Welsh ponies brought to Australia in the early 19th century; stands 11 to 14 hands and usually has a gray coat; has a light head, arched neck, full mane, prominent withers, straight back, and short and strong legs; used mainly as mounts for children (由 19 世纪初引入澳大利亚的威尔士矮马繁育而来的矮马品种。体高 11 ~ 14 掌,多为青毛,头轻而颈弯,鬃毛浓密,鬐甲突出,背平直,四肢粗短,主要供儿童骑乘)澳洲矮马

Australian Pony Stud Book Society n. an organization founded in Australia in 1929 to preserve the breeding and registration of the Australian Pony (1929 年在澳大利亚成立的组织机构,监管澳大利亚矮马的育种、登记工作)澳洲矮马登记委员会

Australian road wagon n. a horse-drawn wagon used on the rough roads of the Australian outback (澳大利亚内地崎岖道路上使用的一种马车)澳洲马车

Australian simplex safety iron n. a safety stirrup designed to ensure the escape of the rider's foot in case of a fall (一种安全马镫,骑手落马时脚可轻易脱出)澳洲简易安全马镫

Australian spring dray n. a horse-drawn, two-wheeled heavy cart historically used in Australian cities to transport dead or injured horses (过去澳大利亚用来拉运受伤或死亡马匹的重型双轮马车)澳洲弹簧板车

Australian Stock Horse n. (abbr. **ASH**) a warmblood breed developed by putting Arab, Thoroughbred, and Anglo-Arab stock to native Australian mares; stands 14.2 to 16 hands and may have a coat of any solid color; has a long neck, prominent withers, deep chest, broad back, and powerful hindquarters; is hardy, agile, docile and possesses excellent endurance; used in competitive sports and as a working horse on farms (由澳大利亚当地母马与阿拉伯马、纯血马、盎格鲁阿拉伯马公马繁育而来的温血马品种。体高 14.2 ~ 16 掌,各种单毛色均可见,颈长,鬐甲明显,胸深,背宽,后躯发达,抗逆性强,反应敏捷,性情温驯,

A

耐力好，用于竞技比赛和牧场工作）澳洲牧牛马

Australian Stock Horse Society n. an organization founded in New South Wales, Australia in 1971 to preserve and promote bloodlines of the Australian Stock Horse（1971 年在澳大利亚新南威尔士成立的组织机构，旨在促进澳洲种马的育种）澳洲牧牛马协会

Australian Veterinary Association n.（abbr. **AVA**）a non-profit organization founded in 1921 to represent the interests of veterinarians across Australia and regulate the profession（1921 年成立的非营利机构，代表澳洲兽医的权益，并对该行业进行监管）澳大利亚兽医协会，澳洲兽医协会

Australian Waler n. *see* Waler 澳洲威尔士马

automatic /ˌɔːtəˈmætɪk/ adj. acting or done by machine; mechanical 机械的，自动的；自主的：automatic drinker 自动饮水器 ◇ automatic nervous system 自主神经系统

automatic timer n. an electronic device used to record the time of equestrian competitions（竞技马术项目中所使用的电子计时装置）自动计时器 [syn] electronic eye

autopsy /ˈɔːtɒpsi; *AmE* -tɑːpsi/ n. a postmortem examination of a cadaver to determine or confirm the cause of death（为确定死因而对尸体进行的检验）尸检，验尸 [syn] postmortem examination

autosome /ˈɔːtəsəʊm; *AmE* -soʊm/ n.【genetics】any chromosome that does not determine the sex of a horse（不决定马匹性别的染色体）常染色体：Horse has 31 pairs of autosomes. 马有 31 对常染色体。[compare] sex chromosome

Auto-top buggy n. an American buggy with a folding top, popular during the early 1900s（一种带折叠式车篷的轻便美式马车，流行于 20 世纪初）翻盖马车，折顶马车

Autumn Double n. a racing event consisting of the Cesarewitch Stakes and the Cambridgeshire Stakes held annually each autumn in Newmarket, England（每年秋天在英国纽马基特市举行的恺撒维奇奖金赛和剑桥郡奖金赛）秋季赛

auxiliary /ɔːɡˈzɪliəri/ adj. giving help or support; additional 辅助的；备用的：auxiliary reins 辅助缰

auxiliary starting gate n.【racing】a spare starting gate used when the number of entrants exceeds the capacity of the primary starting gate（参赛马匹数目超额时使用的）备用起跑闸，备用马闸

Auxois /ɔːksˈwɑː/ n.【French】an old horse breed indigenous to France; stands 15 to 16 hands and may have a bay, roan, or chestnut coat; has a light head, heavy body, and slender legs; is quiet, docile and used mainly for heavy draft and farm work（公元 6 世纪时法国本土马品种。体高 15～16 掌，毛色多为骝、沙或栗，头轻体重，四肢修长，性情安静而温驯，主要用于重挽和农田作业）奥斯瓦马

AVA abbr. **1** American Vaulting Association 美国马上体操协会 **2** Australian Veterinary Association 澳大利亚兽医协会

Avelignese /ˈævliɡˌniːz/ n.（also **Avelignese pony**）an ancient coldblood pony breed originating in Avelengo, Italy, from which the name derived（源于意大利艾维林格并由此得名的矮马品种）艾维林格矮马

Avelignese pony n. *see* Avelignese 艾维林格矮马

average earnings index n.【racing】（abbr. **AEI**）an index that shows the statistics of the racing earnings of a stallion or mare's foals（对公马或母马后代所获奖金的统计表）平均收益指数 [compare] standard starts index

avirulent /əˈvɪəruːlənt/ adj. not virulent 无毒的

avitaminosis /əˌvɪtəmɪˈnəʊsɪs/ n. a condition that results from deficiency of one or more essential vitamins（必需维生素缺乏导致的病症）维生素缺乏症

AVMA abbr. American Veterinary Medical Association 美国兽医协会

avulsion /əˈvʌlʃn/ n. the tearing away of a body part 撕裂,开裂:heel avulsion 蹄踵开裂

AWCF abbr. Associate of the Worshipful Company of Farriers [伦敦]蹄铁匠公会会员

awn /ɒn/ n. a slender, bristlelike appendage found on the spikelets of many grasses (禾本科植物穗上的刺状附属结构)芒

awned /ɒnd/ adj. having awns on spikelets (穗)有芒的

awnless /ˈɒnlɪs/ adj. having no awns 无芒的:awnless brome 无芒雀麦

axis /ˈæksɪs/ n. 【anatomy】 **1** an imaginary line through the center of the body (身体的)中轴 **2** the second cervical vertebra of the horse (脊柱的第二颈椎)枢椎 [compare] atlas

axle /æksl/ n. *see* axle-tree 车轴

axle-tree /ˈæksltriː/ n. (also **axle**) a crossbar supporting a vehicle that has terminal spindles on which the wheels revolve (支撑车身的横轴,车轮绕其上的轴承旋转)车轴

axle-tree maker n. a blacksmith who makes axle-trees for horse-drawn vehicles (为马车制造车轴的铁匠)车轴制造者,车轴匠

Aylesbury wagon n. a box-type English farm or road wagon usually drawn by a single horse (一种英国农用或乘用厢式马车,多为单马套驾)爱斯勃雷四轮马车

Ayrshire harvest cart n. a low-sided cart mainly used in the western Lowlands of Scotland (苏格兰西部低地使用的矮帮马车)艾尔夏农用马车

azoospermia /ˌeɪzəʊəˈspɜːmiə/ n. an absence of sperm in the semen 精子缺乏症 | **azoospermic** /-mɪk/ adj.

azotemia /ˌæzəˈtiːmiə/ n. a condition in which large amounts of nitrogenous waste accumulate abnormally in the blood (血液中含氮化合物异常积累的症状)氮血症

azoturia /ˌæzəˈtjʊəriə; *AmE* ˈtʊər-/ n. an exercise-related myopathy symptomized by stiffness and cramping of muscles, and discharge of red or brown urine (一种运动相关性肌病,症状表现为肌肉强直和痉挛,尿液呈红褐色)氮尿症 [syn] Monday morning disease

Azteca /æzˈtekə/ n. (also **Azteca horse**) a horse breed developed in the 1970s in Mexico by crossing Andalusian with Quarter Horse (墨西哥在20世纪70年代采用安达卢西亚马与夸特马杂交培育而成的马品种)阿兹特克矮马

Azteca horse n. *see* Azteca 阿兹特克矮马

B

B

b /biː/ abbr.【color】bay 骝毛(马)

babble /ˈbæbl/ vi.【hunting】(of hound) to utter a low, murmuring sound; bark 低吠,叫哮

babbler /ˈbæblə/ n.【hunting】a hound who barks on the hunt 吠叫的猎犬

babesiosis /beɪˈbiːsiːəʊsɪs/ n. *see* equine piroplasmosis 巴贝斯虫病

Babolna State Stud n. the Hungarian state stud established in 1789 and famous for its Arab breeding program (建于1789年的匈牙利国家种马场,以培育阿拉伯马而闻名)巴伯黎纳国家种马场

baby /ˈbeɪbi/ n.【racing】a two-year old horse 两岁驹: baby race 两岁驹赛,幼驹赛

baby teeth n.【anatomy】the deciduous teeth of horses 乳齿;脱落齿

babysitter /ˈbeɪbisɪtə(r)/ n. **1** a gentle horse used to give lessons to novice riders (新手骑乘教学用马)保姆马 **2** a cow or horse who is easy to control 易驾驭的牛或马

back /bæk/ n. **1** the dorsal part of horse's body between the withers and the loins (马背从鬐甲至后腰的部分)背部: hollow back 凹背 ◇ long back 长背 ◇ short back 短背 **2**【polo】a polo player who assumes the position of defense (马球中负责防守的球员)后卫

vt. **1** to teach a horse to accept a rider on his back 驯马接受骑乘: to back a horse 驯马受骑 **2**【driving】(of a horse) to move backwards 后退,倒退 PHR V **back up** (of a horse) to slow down noticeably (指马)减速 **3**(of a rider) to command a horse to move backwards 后撤,后退: to back off one's mount 骑马后退 **4**【racing】to bet or wager on a horse 赌马,下注

back band n.【driving】the leather strap that passes over the back of the horse in harness tack (挽具中搭在马背的皮带)背带,搭腰

back breeding n.【genetics】the practice of breeding progeny back to a certain sire to preserve a particular desirable trait (采用后代与父本交配以固定某些性状的育种手段)回交繁育

back cinch n. *see* flank cinch 肋带

back fence n.【cutting】the fence located directly behind the herd 后方围栏: In competition, the cutting horse is penalized three points each time the cow being worked stops or turns within 3 feet of the back fence. 截牛比赛中,所拦截的牛如果在后方围栏3英尺(约91厘米)以内停下或转向,将给予马匹3分的罚分。

back hander n. *see* backhand shot 反手击球,反向击球

back jockey n. the top skirt of a Western saddle (西部牛仔鞍顶端的鞍裙)鞍背

back pad n.【vaulting】a protective pad buckled around the back of horse in vaulting (系在体操用马脊背上的护垫)背垫

back racking n. the removal of feces from the horse's rectum in case of constipation (马便秘时从直肠掏粪的措施)直肠掏粪

back shot n.【polo】*see* backhand shot 反手击球

back strap n. a leather strap that holds the harness crupper in position around the dock of the horse's tail (挽具中用来固定尻兜的一条皮带)背带

backbone /ˈbækbəʊn/ n.【colloq】the vertebral column of a horse; spine 脊柱,脊梁

back-comb /ˈbækkəʊm/ vt. to comb against the way the hair grows 反梳,倒梳

Backdoor Cab n. a horse-drawn wagon popular in the late 19th century and entered from the rear (流行于19世纪末的小型四轮马车,由后门上车)后门马车 syn Backdoor Omnibus

Backdoor Omnibus n. *see* Backdoor Cab 后门马车

backer /ˈbækə(r)/ n. **1**【racing】one who bets on a horse; bettor 赌马者，投注者 **2** one who rides a horse for the first time（初次骑马的人）新手

backhand /ˈbækhæd/ n.【polo】a way of hitting the ball in the reverse direction to the travel（朝运行相反的方向击球）反向[击球]，反手[击球] compare forehand

backhand cut n.【polo】a stroke of the ball hit backwards at an angle away from the horse（反向朝偏离坐骑的角度击球）反手削球，反手侧击

backhand shot n.【polo】a stroke of the ball in the reverse direction to the travel（朝运动相反的方向击球）反手击球，反向击球 syn back hander, back shot

backline /ˈbæklaɪn/ n.（also **back line**）**1**【polo】a line marking the back of a polo field（标示马球赛场底部的边界）底线：A ball passing beyond backline is considered out of play. 球滚出底线即视为犯规。**2** *see* top line 背部轮廓线

backover saddle fall n. a maneuver performed by a rider who rolls backwards off the croup of his mount（骑马特技动作，骑手后滚翻身下马）后翻式下马

backside /ˈbæksaɪd/ n.【racing】*see* backstretch 远侧直道

backstretch /bækstretʃ/ n.【racing】（also **backside**）the straight part of the track on the far side 远侧直道 compare homestretch

backward deviation of the carpal joints n. *see* calf knees 凹膝

backwards shoe n. a horseshoe attached to the hoof in reverse position 反向蹄铁 syn reverse shoe, Napoleon shoe

bacteria /bækˈtɪəriə; *AmE* -ˈtɪr-/ n. the plural form of bacterium [复]细菌

bacteria-free /bækˈtɪərɪəˈfriː/ adj. free of bacteria; aseptic 无菌的

bacterial /bækˈtɪərɪəl; *AmE* -ˈtɪr-/ adj. of or relating to bacteria 细菌的：bacterial diseases 细菌病 ◇ bacterial infection 细菌感染

bactericide /bækˈtɪərɪsaɪd; *AmE* -ˈtɪr-/ n. an agent used to kill bacteria（用来杀灭细菌的制剂）杀菌剂

bacterin /ˈbæktərɪn/ n. a suspension of killed or attenuated bacteria used as a vaccine（细菌灭活或弱化后作疫苗用的悬液）菌苗：polyvalent bacterin 多价菌苗

bacteriostatic /bækˌtɪrɪəsˈtætɪk/ adj. capable of preventing the reproduction of bacteria 抑菌的 n. an agent used to prevent the reproduction of bacteria 抑菌剂

bacterium /bækˈtɪəriəm; *AmE* -ˈtɪr-/ n.（*pl.* **bacteria** /-riə/）any of the unicellular microorganisms that cause diseases（可导致疾病的单细胞微生物）细菌

bad actor n. a horse who is difficult to ride 难骑的马

bad cow n.【cutting】a cow who is difficult to be separated from the herd; alligator 难截的牛

bad doer n. a horse who lacks appetite; an unthrifty horse 食欲不振的马；难喂的马

bad hands n. the rider's hands that are heavy and rough in using reins（骑者操控缰绳）手法过重，手法不当 compare good hands

bad keeper n. *see* bad doer 食欲不振的马，难喂的马

bad mouth n. any of various jaw or tooth misalignments（马的）口齿不正— *see also* **overshot mouth, undershot mouth**

Badge of Honor n. an award given by the Fédération Equestre Internationale to riders competing in Prix des Nations events（国际马术联合会授予参加全国大奖赛骑手的奖项）荣誉奖章

Badminton /ˈbædmɪntən/ n. the location of the first three-day event trials held in Great Britain in 1949（英国 1949 年首次举办三日赛的地

点)巴德明顿赛场 syn Badminton House

Badminton Horse Trials n. (also **Badminton Three Day Horse Trials**) a three-day equestrian event held annually since 1949 in the village of Badminton, England from which the name is derived(自 1949 年起每年在英国巴德明顿举行的三日赛)巴德明顿三日赛:The first Badminton Horse Trials was sponsored by the Duke of Beaufort in 1949 on his estate. 首届巴德明顿三日赛于 1949 年由英国博福特公爵在他的官邸举办。

Badminton House n. *see* Badminton 巴德明顿赛场

Badminton Three Day Horse Trials n. *see* Badminton Horse Trials 巴德明顿三日赛

bag fox n.【hunting】a fox held in captivity and turned loose until needed for the hunt(装到袋里带到猎场放出的狐狸)袋狐

Baggage Cart n. a two-wheeled, horse-drawn vehicle used from the late Middle Ages to the mid-17th century to transport military supplies (从中世纪末到 17 世纪中叶用来运输军备物资的一种双轮马车)双轮辎车

Baggage Wagon n. a larger, four-wheeled wagon used formerly to transport military supplies(过去用来运输军备物资的大型四轮马车)四轮辎车

Baiga /ˈbaɪgə/ n. a long-distance horse race held cross country in Russia(俄国的)长途赛马

Baise n.【Chinese】(also **Baise Pony**) a pony breed originating in Guangxi, China; stands about 11 hands and usually has a bay coat, although gray, chestnut and black also occur; has a heavy head, small pricking ears, straight and broad back, strong legs with hard hoofs; mainly used for riding and as pack animal(源于中国广西的矮马品种。体高约 11 掌,毛色以骝色常见,单青、栗和黑亦有之,头较重,耳小前竖,背部平直,四肢健壮,蹄质结实,主要用于骑乘并作驮畜使用)百色矮马

balance /ˈbæləns/ n. a state of equilibrium or parity 平衡,均匀:balance not maintained 步调失衡 ◇ balance rein 平衡缰

balanced /ˈbælənst/ adj. (of a horse and rider's weight) distributed equally over the four feet of the horse(马匹与骑手体重均匀分配于马匹四肢上的)负重均衡的:A balanced horse will move freely and correctly. 四肢负重均衡的马运步轻快自如。

balanced seat n. a balanced position of the mounted rider(骑手的)平衡骑姿,平衡坐姿

balancing strap n. a leather strap that serves as a handhold for the sidesaddler(供侧鞍骑者手抓的皮革)平衡革

bald /bɔːld/ adj. **1** without hair 秃的 **2** having white markings on the head 有白斑的

bald face n. (also **white face**) a facial marking that covers most of the face of horse(一种面部别征,白色区域覆盖大部分脸面)白脸 —*see also* **blaze, snip, star, stripe, white lip**

bald-faced /ˈbɔːldfeɪst/ adj. (of a horse) having the marking of white face 白脸的:a bald-faced horse 一匹白脸马

baldy /ˈbɔːldi/ n.【slang】a white-faced cow 白脸牛

bale /beɪl/ n. a large square-shaped bundle of hay pressed and bound tightly with twine or wire(用绳或铁丝打包的方形干草捆)草块,草捆

vt. to pack hay into bales 捆扎,打包

Balearic /ˌbælɪˈærɪk; *AmE* bəˈlɪərɪk/ n. (also **Balearic pony**) an ancient pony breed found on the island of Majorca off the coast of Spain (西班牙近海马略卡岛上生活的一个古老的矮马品种)巴利阿里矮马

Balearic pony n. *see* Balearic 巴利阿里矮马

baler /ˈbeɪlə/ n. a machine used to pack hay into bales 打包机;压捆机

Bali /ˈbɑːli/ n. (also **Bali pony**) a pony breed originating on the island of Bali, Indonesia; stands 12 to 13 hands and generally has a bay coat; has low withers and a short, straight back; is strong, frugal, and docile; used for riding and packing (源于印度尼西亚巴厘岛的矮马品种。体高12~13掌,毛色以骝为主,鬐甲低平,背短而直,体格强健,耐粗饲,性情温驯,用于骑乘和驮运)巴厘矮马

Balikun n.【Chinese】(also **Balikun horse**) a breed of horse originating in the area of Baliqun, Sinkiang, from which the name derived; stands around 15 hands and generally has a bay, chestnut, or gray coat; has a strong build, muscular neck, full mane and forelock, broad chest, strong legs with hard hoofs; is frugal, hardy and used mainly as a riding horse in the pastoral area (产于中国新疆巴里坤地区的马品种。体高约15掌,毛色主要为骝、栗或青,体质结实,颈粗壮,鬃毛和额发浓密,胸身宽广,四肢强健,蹄质结实,耐粗饲,抗逆性强,是牧区主要的骑乘工具)巴里坤马

Balius /ˈbeɪliəs/ n.【mythology】one of two immortal horses who pulled the chariot of Achilles (传说中为阿基里斯挽战车的神马之一)贝利斯 — *see also* **Xanthus**

balk /bɔːk/ vi. (BrE also **baulk**) (of a horse) to stop and refuse to move forward (指马)止步不前,逡巡不前: The horse balked at the jump, and just won't listen to the commands of the rider. 马在跨栏前突然止步不前,就是不听骑手的命令。 syn jib

balking /ˈbɔːkɪŋ/ n. a condition in which a horse stops and refuses to move forward 止步不前,逡巡

balky /ˈbɔːki/ adj. (of a horse) stopping and refusing to move on 逡巡不前的: a balky horse 一匹逡巡不前的马

ball[1] /bɔːl/ n. **1**【polo】the ball used in the sport of polo 马球 **2** (usu. **balls**) testes 睾丸 **3** a medicine ball 药丸

vt. to administer medicine ball to a horse 给马服药丸: to ball a horse 给马服药丸 | **balling** /ˈbɔːlɪŋ/ n.

ball[2] /bɔːl/ vi. (of snow) to form lump on the sole of a horse's hoof (雪在蹄底)结块: Balling may be prevented by applying grease to soles of the hoof or use of a balling pad. 在马蹄蹄底涂上油脂或使用蹄垫可以防止马蹄上结雪块。

ball and bucket race n. a mounted event in which the competitors ride from one end of an arena to the other carrying one ball at a time and placing that ball in a bucket and the one who moves the balls in the shortest amount of time wins the race (一种马上竞技项目,其中参赛者从场地一端取球放到另一端的桶中,在最短时间内转移完所有球的人获胜)球桶比赛

ball and socket joint n.【anatomy】a joint in which the end of one bone rests in a socket or cup of another, as in the hip joint (骨的一端嵌入到另一骨末端陷窝的关节,如髋关节)球窝关节

ballet on horseback n. another term for dressage (盛装舞步的别名)马上芭蕾

balling gun n. a wooden or steel tube formerly used to administer medicine balls to a horse (过去给马匹服药丸的木制或铁制的)服药管,灌药枪

balling pad n. a rubber pad fitted between the horseshoe and a horse's hoof to prevent snow from packing (装在蹄铁和马蹄之间的橡胶垫,可防止雪块冻结)防抱垫

balloon bit n. a curb bit with a fancy cheek in the shape of a balloon (口衔似球形的)球形口衔

ballot /ˈbælət/ n. **1** the system of voting 投票 **2** a piece of paper used for voting 选票

vi. to vote about sth 投票 IDM **ballot out** to reject or withdraw a nominated horse from a race (将报名参赛马匹从比赛中除名)撤回,取消资格

ballotade /bæləˈteɪd/ n. an air above the ground in which the horse half rears and then jumps forward as though he is about to kick (古典骑术空中动作之一,其中马半直立后向前跳起后踢)直立后踢

B

band /bænd/ n. a group of animals; herd 畜群

bandage /ˈbændɪdʒ/ n. **1** a leg wrapping used to provide support or protection against injury to the horse's lower leg（缠在马腿上起保护作用或防止受伤的布带）绑腿：standing bandage 站立绑腿 ◇ exercise bandage 训练绑腿 ◇ stable bandage 马房绑腿 ◇ stocking bandage 配种绑腿 ◇ tail bandage 扎尾绷带 **2** a strip of cloth or gauze used for tying around a wound or injured body part（用来包扎伤口或身体受伤部位的布带或纱布）绷带

vt. to apply a bandage to a body part or dress a wound 用绷带包扎：In order to keep a pulled tail neat and tidy, the tail should be bandaged whilst the horse is stabled during the day. 为了保持尾巴梳理后的整洁，马匹白天厩养时就应当包扎起来。

bandage bow n. a swelling of limb tendon caused by too tight or uneven wrapping of bandages（由于四肢绷带缠裹过紧造成的屈腱肿胀）绷带压迫性屈腱肿胀 [syn] compression bow

bandana /bænˈdænə/ n.（also **bandanna**）a large, colorfully printed handkerchief formerly worn by the American cowboys around their neck（美国西部牛仔围在脖子上的印花手巾）花色围巾，印花围巾

bandy /ˈbændi/ adj. bowed or bent in an outward curve 弓形的；O 形的

bandy legs n. *see* bow legs 膝弓形，膝 O 形；飞节弓形，O 形飞节

bandy-legged /ˈbændiːˌlegɪd/ adj. **1** *see* bow-legged（前膝或飞节）弓形的 **2**（of a rider）having legs set wider apart at the knees than at the ankles when riding a horse（指骑手策马时两膝开张比脚踝大）腿弓形的

bang /bæŋ/ n. **1** a fringe of hair cut short and hanging over the forehead of a horse（马前额剪齐的毛发）额发，刘海 **2** the act of strapping and striking 捶打

vt. **1** to cut hair off level 剪齐：banged tail 剪齐的尾巴 **2** to tie or bind together 捆，绑：The tail of harness horse is usually banged up so the tail hair does not catch in the harness and reins. 挽马的尾巴通常绑起来，以免尾毛与挽具或缰绳缠结。**3** to strike the body of horse with a pad to stimulate the muscular circulation（用软垫捶打马体以促进血液循环）捶打（按摩）；拍打 | **banging** /ˈbæŋgɪŋ/ n.

bang tail n. **1**【slang】a racehorse 赛马 **2** *see* bangtail 捆扎的尾巴

banged tail n. the horse's tail that has been cut short or trimmed level at the bottom（剪短且末梢修整过的马尾）剪齐的尾巴：The banged tails are often seen in jumpers and dressage. The long hair of the tail is chopped straight across the bottom in a blunt cut so that the end falls even with the hock. 将尾巴剪齐在障碍马和舞步马中较为常见，修剪时采用钝头剪刀，将马尾剪齐至飞节处。

banger /ˈbæŋə(r)/ n. a pad used for stroking and massaging the body of a horse（用来捶打按摩马体的软垫）捶打带，捶打垫

bangtail /ˈbæŋteɪl/ n. **1**（also **bang tail**）a horse's tail that has been tied up short for safety reasons（出于安全考虑而捆扎起来的马尾）捆扎的尾巴 **2**【slang】a racehorse 赛马

bank /bæŋk/ n. a cross-country obstacle consisting of a mound of dirt with a flat top（越野赛中由泥土构筑的平台障碍）跨堤：There is often a slope on the take-off side of the bank and a steep on the downside. 跨堤障碍的起跳侧多为斜坡，但落地侧则为垂直峭壁。

bank and ditch n. a cross-country jumping obstacle consisting of a bank and a ditch located on the take-off side（跨堤起跳侧带水沟的越野障碍）跨堤-水沟障碍

bant /bænt/ vi.【racing】(of a jockey) to reduce his body weight（骑师）减重，减肥

bar /bɑː(r)/ n. **1** the gap between a horse's incisors and molars on its lower jaw（马下颌切齿与臼齿间的空缺）缺口：bars of the mouth 马嘴缺口 **2**（also **bars of the saddle**）a metal bar located under the saddle flap to which the stirrup leathers are attached (位于鞍翼下方的金属杆,用来拴系马镫)系镫杆 **3** (also **hoof bar**) part of the hoof wall which turns inward at the heel (马蹄位于蹄踵的部分)蹄支:bar of the hoof 蹄支 **4**【driving】the singletree to which the harness traces are attached 横杠,挂杆 PHR V **hang up one's bars** (of a driver) to retire from driving as a profession (指车夫)退休

bar bit n. a driving bit with a straight mouthpiece 直杆衔铁

barb /bɑ:b; *AmE* bɑ:rb/ n. a sharp point projecting on an arrow or fishhook (箭或鱼钩上的)刺,倒刺

vt. to furnish with barbs 装倒刺:barbed wire 带刺铁丝网

Barb /bɑ:b; *AmE* bɑ:rb/ n. an ancient oriental horse breed originating from North Africa and introduced later into Europe in the 7th century; has a long head, sloping quarters, and rather low-set tail; stands 14 to 15 hands and may have a brown, black, bay, chestnut, or gray coat; is hardy, extremely fast over short distances and possesses great stamina over longer ones; used mainly for riding (源于北非的古老的马品种,于公元 7 世纪引入欧洲。头长,斜尻,尾础较低,体高 14 ~ 15 掌,毛色多为褐、黑、骝、栗或青,抗逆性强,短途速力快且长途耐力好,主要用于骑乘)柏布马 syn Barbary horse

Barbary horse n. *see* Barb 柏布马

bard /bɑ:d; *AmE* bɑ:rd/ n. (also **barde**) a piece of armor historically worn by a war horse for protection (古代战马穿的铠甲)铁甲,甲衣

vt. to equip a horse with bards 给马穿铁甲

barde /bɑ:d; *AmE* bɑ:rd/ n. *see* bard [马穿]铁甲

bardella /bɑ:ˈdelə; *AmE* bɑ:rˈ-/ n. a rough wooden saddle with hooded stirrups used by the Italian buttero (意大利牛仔使用的一种木制马鞍,其马镫带护罩)意式牛仔鞍

Bardi horse n. *see* Bardigiano 巴迪奇诺马,巴迪马

Bardigiano /bɑ:dɪˈdʒɪɑ:nəʊ/ n. (also **Bardi horse**) an ancient mountain pony originating in the northern Appenine region of Bardi, Italy; stands 13.1 to 14.1 hands and may have a coat of bay, brown, or black; has a small head, broad forehead, small ears, arched neck, wide withers, short loins, deep girth, wide chest, and well-formed legs with clean joints; is strong, hardy, docile, and quick moving; used for riding, light draft, and farm work (源于意大利巴迪亚平宁北部地区的古老矮马品种。体高 13.1 ~ 14.1 掌,毛色多为骝、褐或黑,头小,额宽,耳小,颈弧形,鬐甲平宽,腰短,胸身宽广,四肢端正,关节发达,体格强健,性情温驯,运步较快,用于骑乘、轻挽和农田作业)巴迪奇诺马,巴迪马

barding /ˈbɑ:dɪŋ; *AmE* ˈbɑ:r-/ n. **1** the process of equipping a horse with bards 给马穿甲衣 **2** the armor worn by a horse; bard 马穿甲衣

bareback /ˈbeəbæk/ adv. & adj. riding a horse without using a saddle 无鞍的,光背的: a bareback rider 无鞍骑手 ◇ to ride bareback 无鞍骑马,骣骑

bareback riding n. the style of riding a horse without the aid of saddle 无鞍骑乘,骣骑 syn Indian style

barefoot /ˈbeəfʊt/ adj. (of a horse) not shoed; unshod 未钉掌的,未装蹄的,跣蹄的:a barefoot horse 未钉掌的马 syn smooth

Barenger, James n. (1780 – 1831) a noted British animal painter (英国动物绘画大师)詹姆士 · 巴瑞格

barge /bɑ:dʒ; *AmE* bɑ:rdʒ/ n. a flat-bottomed boat used chiefly for transportation on inland waterways (一种内陆河流运载货物的平底船)驳船:barge horse 拉驳船的马,拖驳船的马

B

bark /bɑːk; *AmE* bɑːrk/ n.【hunting】the loud cry of a hound or fox; babble（猎犬或狐狸的）嚎叫，吠

vi.（of a hound）to utter harsh, abrupt cry 吠，嗥叫

barker /ˈbɑːkə; *AmE* bɑːrkə/ n. **1** a dog who barks 吠叫的犬 **2**（also **barker foal**）a foal suffering from neonatal maladjustment syndrome（患新生适应不良综合征的幼驹）吠驹症 syn dummy foal, wanderer, convulsive foal

barker foal n. *see* barker 吠驹症

barley /ˈbɑːli; *AmE* ˈbɑːrli/ n. a grass of the genus *Hordeum* that is grown widely as grain crop around the world（大麦属草本植物，是世界上种植较为广泛的谷类作物）大麦：barley grain 大麦粒

barley hay n. cut and dried grass hay made from barley grass（大麦收割晒干后制成的干草）大麦干草：Barley hay contains approximately 5 percent digestible protein and 27 percent fiber. 大麦干草约含有5%的可消化蛋白和27%的粗纤维。

Barlow, Francis n.（1626 – 1704）a renowned British painter noted for his animal and hunting scenes（英国著名画家，擅长动物与狩猎场景写生）弗朗希斯·巴罗

barn /bɑːn; *AmE* bɑːrn/ n. **1** a large farm building used for storing farm products or fodder（储存农产品或饲料的牧场建筑物）仓库 **2** a large building for sheltering livestock 畜舍，牲口棚

barn crazy adj.（of a horse）tired of or resenting being kept in an enclosed area, such as a barn（指马）厌恶圈养的

barn rat n. a horse who is reluctant to leave the barn 不愿出厩的马，恋厩的马

barn sour adj.（of a horse）unwilling to be ridden away from the barn（指马）恋厩的 compare buddy sour

barn-dried adj.（of hay）dried in a barn 舍内干燥的，阴干的

barn-dry vt. to dry fresh grass in a barn 舍内阴干

barnyard /ˈbɑːnjɑːd; *AmE* ˈbɑːrnjɑːrd/ n. the area surrounding a barn 谷仓场地，谷仓院落

barnyard grass n. an annual grass of the genus *Echinochloa* used sometimes for forage（稗属一年生草本植物，可作饲草使用）稗草

baroque movement n. the way in which a horse moves as characterized by high and elastic strides with much suspension（马运步中举蹄高且富有弹性的步姿）巴洛克步法：The baroque movement derived its name because it was characteristic of the horses in the 17th century. 巴洛克步法由于在17世纪的马匹比较常见，故而得名。

Barouche /bəˈruːʃ/ n. a four-wheeled horse-drawn carriage with a collapsible top, two double seats inside opposite each other and a box seat in front for the driver（一种四轮马车，带折叠式顶篷，内设双人对座，车夫驾驶座位于前面）巴洛齐马车：Barouche Landau 巴洛齐兰道马车 ◇ Barouche Sociable 巴洛齐社交马车

barrage /ˈbærɑːʒ; *AmE* bəˈrɑːʒ/ n.【jumping】*see* jump-off（障碍赛）附加赛，争时赛

Barraud, Henry n.（1811 – 1874）a British equestrian painter of French descent who worked with his brother William Barraud and created many famous works of hunting scenes（法裔英国马艺术画家，与其兄威廉·巴洛德合作创作了许多描写狩猎场景的名作）亨利·巴洛德

Barraud, William n.（1810 – 1850）a British equestrian painter of French descent who collaborated with his brother Henry Barraud and created many famous works of hunting scenes（英国法裔马艺术家，与弟亨利·巴洛德合作创作了许多描写狩猎场景的名作）威廉·巴洛德

barrel /ˈbærəl/ n. **1**【anatomy】the part of a horse's body which is enclosed by the ribs; ribs

(马体躯由肋骨构成的部分)体侧;肋部 **2**【rodeo】a metal cylindrical vessel used as a turning point in barrel racing (绕桶赛使用的金属圆桶)铁桶

barrel horse n.【rodeo】a horse used for competition in a barrel race 绕桶马

barrel race n.【rodeo】(also **barrel racing**) a rodeo event in which riders race around three barrels set in a cloverleaf pattern, and the one who rides around all three barrels in shortest time wins (牛仔竞技项目之一,骑手策马绕过3个三叶草形摆放的铁桶,顺利绕过铁桶且用时最短者获胜)绕桶赛

barrel racer n.【rodeo】one who competes in barrel races 绕桶赛骑手

barren /ˈbærən/ adj. (of a mare) not in foal or incapable of producing young; infertile 空怀的;不育的,不孕的:a barren mare 空怀母马 syn eild

barrenness /ˈbærənnəs/ n. the state or condition of being barren (母马)不育;空怀

barrier /ˈbæriə(r)/ n. **1**【jumping】a structure, such as a fence, built for jumping; obstacle (在场地障碍赛中设置的跨栏等结构)障碍 **2**【racing】a movable gate that keeps racehorses in line before the start of a race (开赛前将赛马拦在起跑闸内的可活动门栅)闸门,栅栏 PHR V **to have barrier trial** 试闸:It is quite natural that most racehorses having a barrier trial for the first time are to certain degrees nervous or a little bit excited. 大多数赛马在第一次试闸时,多少都会有些紧张或兴奋,这是非常自然的事。

barrier draw n.【racing】the position a horse assumes in the starting gate at the start of a race (马出赛起跑时的位置)闸位

barstock /ˈbɑːstɒk; *AmE* ˈbɑːrs-/ n. (also **stock**) the metal block from which horseshoes are forged (用来锻制蹄铁的铁条)蹄铁条

Barthais /bɑːrˈθeɪ/ n.【French】(also **Barthais pony**) a French pony breed indigenous to the plains of Challosse near the Adour River; stands 11.2 to 13 hands and usually has a bay, chestnut, black, or brown coat; has a short, wide back and a sloping croup; used commonly as a children's pony (源于法国阿杜尔河附近卡洛斯平原的矮马品种。体高11.2～13掌,毛色以骝、栗、黑或褐为主,背短而宽,斜尻,常用于儿童骑乘)巴赛矮马:Once considered to be a separate breed, the Barthais pony is now often treated as a heavier and taller type of Landais pony. 巴赛矮马以前被视为一个独立的马种,现在则经常作为郎德矮马中体型较大的个体对待。

Barthais Pony n. *see* Barthais 巴赛矮马

basal /ˈbeɪsl/ adj.【anatomy】of, relating to, located at, or forming a base 基底的

basal cell n. any one of the innermost cells of the deeper epidermis of the skin (表皮最内层的细胞)基底细胞

basal crack n. a crack beginning at the ground surface of the hoof which splits towards the coronet (由马蹄着地面开始并延伸至蹄冠的裂缝)蹄基底开裂

basal metabolic rate n.【physiology】(abbr. **BMR**) the rate at which heat is given off by a horse at complete rest (马在完全静息时的产热比率)基础代谢率

bascule /ˈbæskjuːl/ n.【jumping】the arc a horse makes when jumping over a fence (马跨越障碍时身体形成的)弧线

base color n.【color】the dominant color of a horse (马的)基础毛色,基色

base narrow n. (also **stand narrow**) a conformation defect in which the horse's front or hind feet stand closer than the limbs at their origin (马左右两蹄间距比两肢起始处小的体格缺陷)狭踏 opp base wide

base wide n. (also **stand wide**) a conformation defect in which the horse's front or hind feet stand further apart than the limbs at their origin (马左右两蹄间距比两肢起始处大的体格缺

陷）广踏 opp base narrow

Bashkir /ˈbæʃkɜː(r)/ n. *see* Bashkir Curly 巴什基尔卷毛马

Bashkir Curly n.（also **Bashkir**, or **Bashkirsky**）an old horse breed originating in Bashkiria, Russia; stands 13 to 14 hands and usually has a bay, chestnut or palomino curly coat with thick mane, tail and forelock; has a short neck, low withers, wide and deep chest, short and strong legs with small and hard feet; is docile, strong, quite, and hardy and used mainly for packing, riding and light draft（源于俄国巴什基尔的马品种。体高 13 ~ 14 掌，毛色以骝、栗和银鬃为主，被毛卷曲，鬃尾和额发浓密，颈短，鬐甲低，胸身宽广，四肢粗短，蹄小而结实，性情温驯而安静，抗逆性强，主要用于驮运、轻挽和骑乘）巴什基尔卷毛马 syn Curly horse

Bashkirsky /ˈbæʃˌkɜːski/ n. *see* Bashkir 巴什基尔卷毛马

basic seat n.【vaulting】a compulsory exercise in which the vaulter sits in the deepest part of the horse's back with both arms stretched outwards（马上技巧中参赛者坐立马背两臂外展的规定动作）基本坐姿 — *see also* **flag**, **flank**, **mill**, **scissors**, **stand**

basophil /ˈbeɪsəfɪl/ adj.（of cells）stained strongly or definitely by basic dyes（指细胞可被碱性染料着色的）嗜碱性的

n. a basophil white blood cell 嗜碱性粒细胞 | **basophilic** /ˌbeɪsəˈfɪlɪk/ adj.

Basque /bɑːsk/ n.（also **Basque pony**）a semi-wild pony breed indigenous to the mountainous Basque regions of Spain and Southwest France; stands 11 to 14.2 hands and generally has a brown, black coat, although bay, chestnut, piebald and skewbald also occur; has a well-proportioned head, long ears, a short neck, long back, full, but shaggy mane and tail and small hoofs; is very hardy and tough; used for riding, light draft, farming, and jumping（源于西班牙巴斯克山部地区和法国西南地区的半野生矮马品种。体高 11 ~ 14.2 掌，毛色多为褐或黑，但骝、栗、花毛亦有之，头型适中，耳长，颈短，背长，鬃、尾浓密而杂乱，蹄小，抗逆性极强，耐粗饲，主要用于骑乘、轻役、农田作业和障碍赛）巴斯克矮马，又名波特克矮马 syn Pottok, Pottoka

Basseri /ˈbæsəri/ n. a horse breed indigenous to Iran（伊朗本土马种）白瑟力马

basset /ˈbæsɪt/ n.（also **basset hound**）a breed of hound with a long body, short legs and big ears（一种体长、腿短、耳大的猎犬）短腿猎犬，巴吉度犬

bastard strangles n. a condition occurring when the swollen glands typical of strangles burst（马腺疫病灶胀破的情况）恶性马腺疫

Basuto /bəˈsuːtəʊ/ n.（also **Basuto pony**）a pony breed originating in Basutoland, South Africa; stands about 14 hands and may have a bay, brown, gray, or chestnut coat; has a strong build, long, muscular neck, prominent withers, long back, low-set tail, and slender, strong legs; is reliable, hard-working and tolerant and used for polo and flat racing as well as general riding（源于非洲巴苏陀兰的矮马品种。体高约 14 掌，毛色多为骝、褐、青或栗，体格粗壮，颈长而肌肉发达，鬐甲明显，背较长，尾础低，四肢健壮，习性稳重，工作力强，耐粗饲，多用于马球、平地赛及骑乘）巴苏陀矮马

Basuto Pony n. *see* Basuto 巴苏陀矮马

bat /bæt/ n. a crop or whip around two feet long with a flattened end〔一种长约 2 英尺（约 61 厘米）且末端扁平的鞭子〕短马鞭 syn cosh

vt.【racing】to whip a horse to gee up or gain attention 鞭打，赶马

Batak /ˈbætək/ n.（also **Batak pony**）a pony breed descended from Arab stallions put to selected native mares on the island of Sumatra, Indonesia; stands 12 to 13 hands and may have a coat of any color, although brown is most common; has a light head, arched, well-shaped neck, short back, and sturdy legs; is gentle, lively, and frugal, and used mainly for riding and packing (印度尼西亚苏门答腊岛采用阿拉伯马公马与当地精选母马繁育而来的矮马品种。体高12～13掌,各种毛色兼有,但以褐最为常见,头轻,颈弓优美,背短,四肢粗壮,性情温驯、活泼,耐粗饲,主要用于骑乘和驮运)巴特克矮马

Batak pony n. *see* Batak 巴特克矮马

battery /ˈbætri; *AmE* ˈbætəri/ n.【racing】an illegal electrical device used by a jockey to stimulate the horse to run faster (一种禁用电子设备,骑手以此刺激马匹跑得更快)电击棍

Battlesden cart n. a small, two-wheeled, horse-drawn pleasure cart popular in England during the second half of the 19th century (19世纪下半叶流行于英国的小型双轮马车)巴特斯腾马车

batwing chaps n. *see* batwings 蝠翼套裤

batwings /ˈbætwɪŋz/ n. pl. (also **batwing chaps**) one type of the chaps that cowboys wear over their legs (牛仔穿在腿上的一种套裤)蝠翼套裤

baudet de Poitou n.【French】*see* Poitou ass 普瓦图驴

baulk /bɔːk/ vi.【BrE】to balk (指马)止步不前,逡巡不前

Bavarian Warmblood n. a warmblood breed originating in southern Germany; stands 15.2-16.2 hands and may have any of the dark solid colors; is used mainly as sport horses in the Olympic equestrian disciplines and combined driving (源于德国南部地区的温血马品种。体高15.2～16.2掌,多为单毛色,作为体育竞技用马,主要用于奥运会马术项目和驾车赛)巴伐利亚温血马: All registered Bavarian Warmbloods carry the brand of a crowned shield outside the letter B on the left thigh. 所有登记的巴伐利亚温血马,都在左侧大腿上留有外围带皇冠盾形的"B"字母。

bay[1] /beɪ/ n. **1**【color】a reddish brown coat with black points (被毛红褐而末梢为黑)骝毛: bay brown 褐骝 ◇ bay roan 沙骝 ◇ blood bay 血骝 ◇ bright bay 亮骝 ◇ dark bay 深骝 ◇ red bay 红骝 ◇ mahogany bay 褐骝 ◇ sand bay 黄骝 ◇ light bay 淡骝 ◇ Bay is perhaps the most popular and attractive color of all. 骝毛大概是所有毛色中最受青睐和最具吸引力的毛色。**2** a horse of reddish brown body color with black points (体色红褐而末梢为黑的马匹)骝马

bay[2] /beɪ/ n.【hunting】the deep bark made by the hounds while hunting (打猎时猎犬发出的低沉的叫声)低沉的嗥叫声,低沉的吠叫声 vi. (of hounds) bark with a deep note 低沉的嗥叫

bay roan n. the color of a horse who has a bay coat with white hairs interspersed (骝毛中散生有白毛的毛色)沙骝毛 [syn] red roan

bayard /ˈbeɪəd/ n. a bay horse 栗毛马

BBF abbr. bronze buckle finals 铜扣带决赛

BBQ abbr. bronze buckle qualifiers 铜扣带资格赛

BCN abbr. Breeding Clearance Notification 繁育许可报告,繁育许可证

BDS abbr. British Driving Society 英国马车协会

Beach Carriage n. *see* Basket Phaeton 海滨马车

Beach Phaeton n. a four-wheeled, horse-drawn American carriage of the Phaeton type, used for summer excursions at seaside resorts (一种美式四轮费顿马车,主要用于夏季海滨短途度假旅行)海滨费顿马车

Beach Wagon n. a horse-drawn American pleasure wagon used formerly for holiday excursions around the coastal regions of New England (过去美国新英格兰沿海地区用于假日短途旅行使用的美式四轮马车)海滨四轮马车

beagle /ˈbiːgl/ n. a small short-legged hound having a smooth coat and drooping ears, used for hunting hares and other small game on foot (一种四肢较短的小型猎犬,被毛光滑,耳朵下垂,主要用来徒步追捕野兔或其他小型猎

物）猎兔犬，比格犬

beagling /ˈbiːglɪŋ/ n.【hunting】the sport of hunting hares on foot with beagles 猎野兔

beak /biːk/ n.【farriery】(also **bick**) the round, pointed end of a farrier's anvil（蹄匠所用铁砧的尖端）砧尖

beaked shoe n. *see* extended-toe shoe 尖嘴蹄铁

beam /biːm/ n.【hunting】**1** one of the main stems of a stag's antlers 鹿角主干 **2** the distance between the main antlers of a stag 鹿角幅

bear /beə(r); *AmE* ber/ v. **1** (of a horse) to move or behave in a certain way or manner（指马）运步，表现 **2** (of a horse) to move or turn in a certain direction 转向，走向 PHR V **bear in**【racing】(of a horse) to run with its body turned to the inside rail（指马）侧向内栏跑 syn lug in **bear out**【racing】(of a horse) to run with its body turned away from the inside rail（指马）侧离内栏跑 syn lug out

bear foot n. *see* club foot 狭蹄，立系

bearing /ˈbeərɪŋ; *AmE* ˈber-/ n. the general balance and carriage of the horse（马的）姿态

bearing rein n. a rein attached to each side of the bit to achieve a high head carriage of the horse（安在衔铁两侧用来提高马头姿的缰绳）姿态缰 syn check rein

bearing surface n. the ground-side surface of the hoof upon which the horse's weight is borne（马蹄的）负重面

bear-skin cape n.【driving】a cape made out of bear hide and worn by coachmen during cold weather（寒冬车夫穿的熊皮制披风）熊皮斗篷

beast /biːst/ n. **1** any large, four-footed animal 兽，兽类 **2** a large domestic animal, especially a horse or bull（指马或牛等大家畜）家畜，牲畜

beat[1] /biːt/ n. (also **heartbeat**) a pulsation or throb of heart（心脏的）跳动，搏动：The heart rates are usually expressed as beats per minute. 心率通常用每分钟心跳数表示。

vi. to pulsate or throb 跳动，搏动

beat[2] /biːt/ vt. to flog a horse with a whip 鞭打

vi. to hunt through woods or underbrush in search of game（穿越丛林灌木搜寻猎物）搜猎

beat[3] /biːt/ n.【racing】an unfortunate defeat（比赛）失利，溃败

vt. to defeat the opponents in a race 击败；获胜 IDM **couldn't beat a fat man**【racing】(of a racehorse) to run too slowly to win（赛马）跑不过肥佬，跑得太慢

Beaufort Phaeton n. a four-wheeled, horse-drawn carriage designed by the Duke of Beaufort after whom it was named（由博福特公爵设计并因此得名的一种四轮马车）博福特费顿马车 syn Hunting Phaeton

Beberbeck /ˈbiːbəbek; *AmE* -bərbek/ n. (also **Beberbeck horse**) a light horse breed developed by putting Arab and Thoroughbred stallions to local mares at the Beberbeck Stud in western Germany throughout the 18th and 19th centuries; stands about 16 hands and generally has a bay or chestnut coat; used mainly for riding and light draft（德国西部地区的波波贝克种马场自18～19世纪采用阿拉伯马和纯血马与当地母马交配育成的马品种。体高约16掌，毛色多为骝或栗，主要用于骑乘和轻挽）波波贝克马：Very few Beberbeck horses exist today. 现今，波波贝克马仅存无几。

bed /bed/ n. the bedding for a horse in a stable 垫草，垫料

vt. to lay beddings for a horse 铺垫草：to bed down 铺垫草

bedding /ˈbedɪŋ/ n. (also **bed**) the material such as straw or sawdust placed on the floor of a stall or stable as a bed for horses（铺在马房或厩舍地板上的稻草或锯末）垫草，垫料：Different bedding materials can be used depending on availability and cost factors. 根据原料获取方式和价格因素，可选用不同的垫料。 syn litter

Bedford Cart n. a two-wheeled, horse-drawn English farm cart popular in the 19th century; had fixed sideboards above high wheels and an unsprung or dead axle (19世纪在英国使用较为普遍的一种农用双轮马车,车轮大且上方带侧板,车身无减震弹簧或为死轴)贝德福德马车

Bedford cord n. a hard fabric used to make cushion and cover on horse-drawn carriages and for driving aprons and breeches (一种耐磨织物,主要用于制作马车坐垫、马夫围裙及马裤)贝德福德条绒

Bedouin Arab n. *see* Assil 贝都因·阿拉伯马;阿西利马

beefy /ˈbiːfi/ adj. (of the horse's hocks) thick and meaty (指马飞节)粗壮的:beefy hocks 飞节粗壮

beet /biːt/ n. **1** *see* sugar beet 甜菜 **2**【AmE】beetroot 甜菜根

beet pulp n. (also **sugar beet pulp**) a by-product of the sugar production process, made by drying the residual beet chips after the sugar extraction (蔗糖生产工艺中的副产品,由甜菜提炼蔗糖后的残渣干燥加工而成)甜菜渣:Beet pulp is a good source of fibre with a high energy, mainly used for supplementing an inadequate hay supply and as a dust-free roughage substitute. 甜菜渣是一种很好的高能纤维饲料,主要用来补充干草供给不足或作为粗饲料代用料使用。

Beetewk /ˈbiːtjuːk/ n. a heavy Russian breed developed along the banks of Beetewk river by putting Dutch stallions to local mares; is extremely obedient and has high spirits, good action, and great strength; used mainly for farm work in the past (育成于俄国彼秋格河畔的重挽马品种,主要通过选用荷兰公马与当地母马交配繁育而来。性情温驯而活泼,步态优雅,挽力惊人,过去主要用于农田作业)彼秋格马

beetroot /ˈbiːtruːt/ n. the root of beet from which sugar is extracted 甜菜根

BEF abbr. British Equestrian Federation 英国马术联合会

behavior /bɪˈheɪvjə/ n. the actions or reactions of a horse in response to external or internal stimuli (马对内外界刺激所作出的行动或反应)行为:abnormal behavior 异常行为 ◇ agonistic behavior 争斗行为 ◇ eliminative behavior 排泄行为 ◇ ingestive behavior 摄食行为 ◇ investigative behavior 探究行为 ◇ mimicry behavior 模仿行为 ◇ sexual behavior 性行为 ◇ sleep and rest behavior 休眠行为 | **behavioral** /-vjərəl/ adj.

belch /beltʃ/ vi. to expel gas from the stomach noisily through the mouth 打嗝:Belch is uncommon in horses and a sign of a more serious condition. 马很少打嗝,如若发生则情况比较严重。

Belgian /ˈbeldʒən/ n. a breed of heavy draft horse originated in Belgium and is often used for heavy draft work (源于比利时的重挽马品种,常用于重役作业)比利时挽马:The Belgian draft horse is commonly known for it's chestnut coat, flaxen mane and tail, and it's strong, bulky muscled quarters. 比利时挽马体色通常为栗色,鬃、尾为亚麻黄,体格强健,后躯肌肉发达。

Belgian Ardennais n. *see* Belgian Ardennes 比利时阿登马

Belgian Ardennes n. (also **Belgian Ardennais**) a division of French Ardennais originating from the Ardennes plateau in France from which the name derived; stands about 15.3 hands and generally has a bay, roan, or chestnut coat; is compact and heavily built with a wide, deep chest, a big, broad head, short, heavily feathered, massive legs, and a well-crested neck; is hardy, frugal and has tremendous endurance and gentle temperament; used for heavy draft and farm work and is particularly well suited to work in hilly country (源于法国阿登高原的

冷血马品种并由此得名。体高约15.3掌，毛色以骝、沙或栗为主，体格结实，胸身宽广，头重，前额较宽，四肢粗短且距毛浓密，颈峰曲线优美，抗逆性强，耐粗饲，耐力惊人，性情温驯，主要用于重挽和农田作业，尤其适于山地耕作）比利时阿登马

Belgian Brabant n. *see* Brabant 布拉邦特马

Belgian Draft n. *see* Brabant 比利时重挽马，布拉邦特马

Belgian Draft Horse n. *see* Brabant 比利时重挽马，布拉邦特马

Belgian Heavy Draft n. *see* Brabant 比利时重挽马，布拉邦特马

Belgian road cart n. a light, two-wheeled, horse-drawn cart popular in Belgium during the 1900s（20世纪初流行于比利时的一种轻型双轮马车）比利时公路马车

bell /bel/ n. **1** a hollow metallic device which makes a ringing sound when struck（敲击时可发出响声的）铃铛，响铃：Small bells may be attached on show harness for decoration purposes. 出于装饰目的，展览挽具上可系上小铃铛。**2** a signal sounded to start a performance or to mark the close of betting at a racetrack（宣布比赛开始或赛马投注停止时的信号）铃声 **3** the bellowing cry made by a stag during mating season（发情期的雄鹿发出的）吼叫，鸣叫：the bell of a stag 牡鹿的叫声
vi. to utter long, deep, resonant sounds; bellow 吼叫，鸣叫 | **belling** /ˈbelɪŋ/ n.

bell boots n.【AmE】the over-reach boots 护蹄碗

bell mare n. a mare with a bell tied around her neck to lead a herd of horses（颈上系有铃铛的母马，是马群的头马）系铃母马 [syn] madrina

Bellerophon /bəˈlɛrəfən/ n.【mythology】the prince of Corinth in Greek mythology who tamed the winged horse Pegasus and slew the monster Chimera（希腊神话中科林斯国的王子，在驯服了飞马佩加索斯后，骑马杀死了蛇怪喀迈拉）柏勒罗丰

belligerent /bəˈlɪdʒərənt/ adj.（of a horse）inclined or eager to fight; aggressive 好斗的：a belligerent horse 一匹好斗的马

belly /ˈbeli/ n. *see* abdomen 腹部

belly band n. **1** a wide leather strap that passes under the belly of the horse to keep the harness tugs in place and to prevent the collar from choking the horse（束在马腹部的宽皮带，用来固定挽具并可防止颈圈勒住马颈）腰带，肚带 **2** *see* girth 肚带

belly clip n.（also **neck clip**）a pattern of clipping the coat of a horse from under the belly upwards between the forelegs, and up along the lower line of the neck to the jaw（一种马匹被毛修剪方式，对腹部至前肢以及颈下侧的毛都进行修剪）腹修剪

Belmont Stakes n.【racing】a 1.5-mile race run annually every June by three-year-olds since 1905 at Belmont Park in New York（自1905年起每年6月在纽约贝尔蒙特公园举行的一项比赛，由3岁马参加，赛程2.4千米）贝尔蒙特大赛

bench knees n. *see* offset knees 管骨外偏，膝外偏

bend /bend/ n. **1**【racing】any turn in a race track（赛道上的）转弯处 **2** the flexion of the horse from nose to tail（马从头至尾的弯曲）背部弧线，曲线

bending line n.【jumping】an imaginary curved line from one obstacle to another where such obstacles are not placed in a straight line（非直线排列的障碍之间假想的曲线）弧线，曲线：The rider and horse must follow the bending line when jumping. 骑手跨越障碍时须沿衔接障碍间的曲线行进。

bending race n. *see* pole bending 绕杆，窜杆

bending tackle n. a piece of tack used to alter the carriage of the horse's head by encouraging flexion of the neck（一种通过鼓励马匹曲颈来改善头姿的用具）弯头索具

benign /bɪˈnaɪn/ adj.（of a disease）having no

danger to health 无害的;良性的:benign tumor 良性肿瘤 [opp] malignant

Benjamin /ˈbendʒəmɪn/ n.【driving】a heavy overcoat worn by coachmen(马车夫穿的)外套,大衣

Benna /ˈbenɑː/ n.【dated】a crude horse-drawn farm cart used in Ancient Rome(古罗马使用的农用马车,制作简陋)勃纳马车

Bennett, James Gordon n.(1841 – 1918) an American who introduced the game of polo into the United States in 1876 and organized the first polo matches there; generally considered as the father of American polo(1876 年将马球运动引入本国并最早组办比赛的美国人,普遍认为是美国马球之父)詹姆士·高登·班尼特

bent-panel cart n. a light, horse-drawn, two-wheeled cart with outwardly curved sides(一种车帮呈弧形的轻型双轮马车)弧帮马车

berlin /bɜːˈlɪn/ n.(also **berlin coach, berline**) a four-wheeled closed carriage first appeared in Berlin in the mid-17th century and was usually drawn by a pair of horses in pole gear(17 世纪中期出现在柏林的四轮马车,常由双马套驾杆拉行)柏林马车 [syn] berline

berlin coach n. *see* berlin 柏林马车

berline /bɜːˈlɪn/ n. *see* berlin 柏林马车

Bermuda grass n. a perennial grass of the genus *Cynodon dactylon*, widespread in warm regions and grown as a lawn and pasture grass(一种多年生狗牙根属草本植物,分布于温暖地区,常作为草坪和牧场草本种植)百慕大草,狗牙根 [syn] devil grass

Bermuda hay n. the cut and dried hay made from Bermuda grass 百慕大干草,狗牙根干草

bestride /bɪˈstraɪd/ vt.(pt **bestrode**; pp **bestridden**) to sit or ride with the legs astride; straddle(两腿分开跨坐或骑)叉骑,跨骑:In the past, it was regarded inappropriate for ladies to bestride horses. 过去,女性跨骑马匹被认为有伤大雅。

bet /bet/ n. **1** an agreement to risk money on an event of which the result is uncertain(对非确定事件结果的押注)打赌,博彩 IDM **to make a bet** 打赌 [syn] wager, play **2**【racing】a sum of money laid on the result of a horse race(在赛马上所押的)赌金,赌注:bet declared off 投注取消

vt.【racing】to risk a sum of money or place a bet on a horse race 押注,赌马 PHR V **bet on sth** 打赌,押注

beta-carotene /ˌbiːtəˈkærətiːn/ n.【nutrition】a derivative of vitamin A or retinol(维生素 A 或视黄醇的衍生物)β-胡萝卜素

beta-oxidation /ˈbiːtəˌɒksɪdˈeɪʃn/ n.【biochemistry】the biochemical process during which fatty acids are converted to acetyl-CoA for entering the tricarboxylic acid cycle(脂肪酸转变为乙酰辅酶 A 进入三羧酸循环的生化过程)β-氧化

better /ˈbetə(r)/ n.【racing】a bettor 投注者,赌马者

betting /ˈbetɪŋ/ n.【racing】the act or practice of placing or making a bet in a horse race(在赛马中投注的行为)押注,赌马:ante-post betting 赛前押注 ◇ betting guide 投注指南 ◇ betting pool 投注池

betting board n.【racing】a board upon which the odds of any horse competing in a given race are displayed(赛马中用来出示赔率的屏板)赌金显示板

betting book n.【racing】a ledger maintained by a bookmaker in which all bets made and paid are logged(过去赌马经纪人用来记录投注的账簿)赌马册

betting favorite n.【racing】a racehorse upon which most bets are placed 投注热门马

betting ring n.【racing】**1** the area at a racetrack where authorized bookmakers conduct their business(赛马场上赌马经纪人开业的地方)赌马场,投注点 **2** an organized, yet illegal group of bookmakers(非法的)赌马团伙

betting shop n.【racing】a licensed, offtrack office

which takes bets on horse races（场外的）赌马处，投注处

betting ticket n.【racing】a ticket serving as an evidence of the bet placed on a racing horse（作为赌马凭证的票据）彩票，赌票

bettor /'betər/ n.【racing】(also **better**) one that bets or places a bet 赌马者，博彩者 syn player

BEVA abbr. British Equine Veterinary Association 英国马兽医协会

BFSS abbr. British Field Sports Society 英国野外运动协会

Bhotia /'bəʊtɪə/ n. (also **Bhutia**) a pony breed originating in the Himalayan mountains of India; stands around 13 hands and generally has a gray coat, short neck, shaggy mane, straight shoulder, and short strong legs; is hardy, frugal, and has good endurance; used mainly for riding and packing（源于印度喜马拉雅山区的矮马品种。体高约 13 掌，毛色为青，颈短，鬃毛杂乱，直肩，四肢粗短，抗逆性强，耐粗饲，耐力好，主要用于骑乘和驮运）普提亚矮马

BHS abbr. British Horse Society 英国马会

BHSAI abbr. British Horse Society Assistant Instructor 英国马会助理教练

BHSI abbr. British Horse Society Instructor 英国马会教练

Bhutia /'buːtɪə/ n. *see* Bhotia 普提亚矮马

Bian /'bɪən/ n. *see* Bianconi Car 彼安考尼马车

Bianconi Car n. (also **Bian**) a four-wheeled, horse-drawn passenger vehicle used in Ireland during the early 19th century（19 世纪初在爱尔兰出现的一种四轮客用马车）彼安考尼马车

bib /bɪb/ n. a piece of leather or plastic secured around a horse's muzzle to prevent him from biting his rugs, bandages, etc.（遮在嘴上用来防止马匹撕咬马衣或绷带的皮革或塑料挂件）围嘴，围兜：a leather bib 皮围兜

bib martingale n. a running martingale in which the area between the split straps is filled with a piece of hide; used to prevent a horse from entangling its fronts legs and other equipment in the martingale（分叉处缝有皮革的跑步鞅，可防止马前肢或其他部件与之缠结）围嘴鞅 syn web martingale

biceps /'baɪseps/ n.【anatomy】(*pl.* **biceps**) a muscle with two heads or points of origin 二头肌

bick /bɪk/ n. *see* beak 砧尖，砧嘴

biconcave /baɪ'kɒnkeɪv/ adj. concave on both sides or surfaces 双凹的：Red blood cells are biconcave discs with no nucleus. 红细胞呈双凹圆盘状，是无核细胞。

bicuspid valve n.【anatomy】the atrioventricular valve found between the left atrium and the left ventricle（左心房与左心室之间的房室瓣）二尖瓣 compare tricuspid valve

bid /bɪd/ n. a price offered or proposed to purchase a horse at an auction（拍卖会上马匹的报价）出价；要价

vt. to offer or propose a price for sth at an auction 出价；要价

bidder /'bɪdə/ n. one who offers a bid at an auction 竞价者，出价者

Bidet Breton n. an ancient type of Breton believed to have originated in Brittany, France; stood about 14 hands and was used mainly as a military mount during the Middle Ages（古代布雷顿马的类型之一，据信源于法国不列塔尼地区，体高约 14 掌，中世纪多用作战马）彼德布雷顿马

big head disease n. *see* osteodystrophia fibrosa 大头症，纤维性骨营养不良

big hitch n. a team of four, six, eight, or more horses harnessed together（由四、六、八或更多匹马联驾的车队）多马联驾

Big Jake n. a chestnut Belgian gelding foaled in 2000 and kept at Smokey Hollow Farm in Wisconsin, USA; is the world's tallest living horse standing 82.75 inches and weighing 2,600

pounds（生于2000年饲养在美国威斯康星州烟谷农场的一匹栗色比利时骟马，是世界上现存最高的马匹，体高2.1米，体重1 179千克）大杰克

big knee n. *see* carpitis 膝关节炎，膝部囊瘤

big leg n. *see* stock-up 四肢肿胀

big race n.【racing】the primary race of a race day 主赛

Big Red n.【racing】either of the two chestnut-colored racing champions, Man O' War or Secretariat（赛马史上的两匹栗色纯血马冠军，即"战神"和"秘书处"）大红马

biga /ˈbiːɡə; *AmE* ˈbaɪɡə/ n.（*pl.* **bigae**/-giː/）an ancient Roman chariot drawn by a pair of horses（古罗马时期的战车，由双马并驾拉行）双马战车 — *see also* **quadriga, triga**

bight /baɪt/ n. **1** a loop in a rope 绳环 **2** the middle or slack part of an extended rope 绳子松弛部分；绳子中段：bight of the reins 缰绳中段

Bigourdan /bɪˈɡʊədæn; *AmE* -ˈɡʊrd-/ n. a horse breed descended from the Iberian Horse improved with Arab and Thoroughbred blood（伊比利亚马经阿拉伯马和纯血马改良培育的马品种）彼哥丹马

bike /baɪk/ n.【AmE, slang】a harness racing sulky 轻驾赛车

bilateral /baɪˈlætərəl/ adj. of or pertaining to both sides 双侧的，双边的

bile /baɪl/ n. the yellow brownish fluid secreted by the liver to facilitate the digestion of fats（由肝脏分泌的黄褐色液体，有助于消化脂肪）胆汁 [syn] gall

biliary /ˈbɪliəri; *AmE* -eri/ adj. of or relating to bile 胆汁的

biliary fever n. *see* equine piroplamosis 马巴贝斯虫病，胆瘟

bilirubin /ˌbɪlɪˈruːbɪn/ n.【physiology】a reddish-yellow bile pigment derived from the degradation of heme（胆汁中血红素降解生成的红棕色色素）胆红素

Bill Daly n. an American racehorse trainer who promoted the racing style of taking the lead early on in the race and staying in the lead to the finish（美国著名练马师，提倡开跑抢先并保持优势至终点）比尔·达利 [PHR] **on the Bill Daly**（of a jockey）taking the lead at the start and staying there to the finish（指骑手）先发制人；采取领跑策略

billet /ˈbɪlɪt/ n. **1**（also **billet strap**）a piece of leather strap by which the girth is buckled under the saddle's flap（将肚带扣在鞍襟上的皮革）扣带：billet flap 护扣革 **2** a metal loop or hook by which the reins attach to the headstall of the bridle（用于将缰和马勒连接起来的金属搭环或搭扣）勒环；勒扣 [syn] bridle stud billet **3**（*pl.* billets）fox manure 狐狸粪便

binocular /bɪˈnɒkjələ(r); *AmE* bɪˈnɑːk-/ adj. with both eyes focusing on an object at the same time 双目聚焦的：binocular vision 双目共视；双目视觉

biological control n. a method of controlling insects around the stable by using their natural enemies（利用昆虫天敌来控制马厩及周围昆虫的防治手段）生物防控

biopsy /ˈbaɪɒpsi; *AmE* -ɑːpsi/ n. the examination of a sample of tissue taken from a living body for diagnosis（从活体采样后进行组织化验）活体组织检查；活检 | **biopsic** /baɪˈɒpsɪk/ adj.

biotin /ˈbaɪətɪn/ n.【biochemistry】an internally synthesized vitamin necessary for proper skin, hair, and hoof health of horses（一种体内合成的维生素，对马的皮肤、被毛和蹄质健康较为重要）生物素 [syn] vitamin H

bird-eyed /ˈbɜːdˌaɪd; *AmE* ˈbɜːrd-/ adj.（of a horse）shying at imaginary objects 易惊的

Birdsville disease n.【vet】a fatal disease of horses occurring in Australia caused by the ingestion of the poisonous plant *Birdsville indigo*; early symptoms include incoordination, drowsiness, and some lack of control of the hind limbs（澳大利亚马匹所患的致死性疾病，主要由

马采食有毒植物槐蓝所致,早期症状包括共济失调、呆滞和后肢失控等)槐蓝中毒症:Horses affected with Birdsville disease are prone to stand rather than move about. 马患槐蓝中毒症后,多呆立而不愿走动。

Birotum /baɪˈrɒtəm/ n.【Latin】a small, two-wheeled, horse-drawn chariot used in Ancient Rome(古罗马使用的小型双轮战车)双轮马车

birth /bɜːθ; *AmE* bɜːrθ/ n. the act or process of bearing young; parturition 出生;生产:the mare's second birth 母马第二胎

birth bag n. *see* amnionic sac 羊膜,胞衣

birth date n. the foal's birthday considered as January 1st in the Northern Hemisphere and August 1st in the Southern Hemisphere in racing or showing events, regardless of the actual month it was born(比赛中不论幼驹的实际出生日期,而在北半球统一视为1月1日,南半球统一视为8月1日)出生日期

birthweight /ˈbɜːθweɪt/ n.(also **birth weight**) the weight of a foal at birth 初生重:percentage of low birthweight foals 幼驹初生重过低的比例

bishop /ˈbɪʃəp/ n. an old, horse-drawn vehicle refurbished to appear new 翻新的马车
vt. **1** to refurbish old horse-drawn vehicle to appear new 翻新(旧马车) **2** to dishonestly alter the teeth of older horses to represent them young(窜改老马牙口使其显得年轻)镶齿

bishoping /ˈbɪʃəpɪŋ/ n. the dishonest practice of altering the teeth of older horses to represent them as young horses(通过人工改变马匹牙口使之显得年轻)镶齿 [syn] tampering

bit /bɪt/ n. a metal or plastic bar attached to the bridle and serving to help the rider control and direct the horse(连在马勒上的金属或塑料杆,用来协助骑手驾驭马匹)衔铁,嚼子:Liverpool bit 利物浦衔铁 ◇ medicine bit 灌药口衔 ◇ Pelham bit 佩勒姆衔铁 [PHR] **above/over the bit**(of a horse) refusing the bit contact by carrying his head high(指马举头过高)抗衔的 **behind the bit**(of a horse) evading the bit contact(指马)避衔 **accept the bit**(of a horse) to respond to rider's rein aids without resistance(指马)受衔 **bend below the bit**(of a horse) to draw his chin into his chest to evade the pressure applied to the bit by rein contact(指马为逃避衔铁施力而)低头过度 **on the bit**(of a horse) willingly accepting the rider's gentle contact on the reins(指马)处于受衔状态 **up to their bits**【driving】(of the horses) pulling a vehicle without too much exertion and moving freely(指马拉车)不费力的,轻松的
vt. to place a bit in the mouth of a horse 戴嚼子,上口衔 [PHR V] **to bit up** 上嚼子

bit burr n.(also **burr**) a round piece of plastic or rubber with bristles on one side, fitted around the mouthpiece just inside the cheekpiece with the bristled side against the corner of horse's mouth to prevent the horse from leaning to the side(单面带刺的塑料或橡胶片,穿在衔杆靠近颊革处,带刺面朝向嘴角,可防止马靠向单侧)衔铁刺环

bit cover n. a tubular rubber piece fitted over the mouthpiece of bit 衔杆套管

bit guard n.(also **bit shield**) a round piece of leather or rubber fitted over the bit to prevent the bit from catching on the horse's lips(套在衔铁上的革环,可防止衔铁夹住马的口角)衔铁护套

bit keys n. *see* keys 衔铁缀饰

bit roller n.(also **roller**) a round, bead-shaped, stainless ball which rotates freely about the mouthpiece(衔铁杆上可自由转动的钢珠)衔铁滚珠:The bit roller may encourage salivation of a horse. 衔铁滚珠可以增加马的唾液分泌。

bit shield n. *see* bit guard 衔铁护套

bitch /bɪtʃ/ n. **1** a female hound 雌猎犬 :bitch hound 雌猎犬 **2** a female fox 雌狐:bitch fox 雌狐

bitch pack n.【hunting】a pack consisting entirely of bitch hounds 雌猎犬队

bite /baɪt/ vt.（of a horse）to cut with teeth 啃,咬

biter /ˈbaɪtə/ n. a horse who bites 有咬癖的马;咬人的马

biting /ˈbaɪtɪŋ/ n. a vice of horse who bites people 咬人恶习,咬癖

biting louse n. an external, blood-sucking parasite which feeds on the dander and hair of the horse（一种吸血性外部寄生虫,多附生在马皮屑和被毛中）咬虱 [syn] *Damalinia equi*

bitless /ˈbɪtləs/ adj.（of a bridle）having no bit（马勒）无衔铁的: bitless bridle 无衔铁的马勒

bitted /ˈbɪtɪd/ adj.（of a horse or mule）wearing bridle and bit（马或骡）戴口衔的: a bitted mule 一匹戴口衔的骡子

bl abbr.【color】black 黑毛

black /blæk/ n. **1**【color】a coat color in which all hair of the body are black except white markings（除白色别征外,被毛全黑的毛色类型）黑毛,骊毛: jet black 乌黑 ◇ smoky black 烟黑 ◇ The black may fade to a rusty color in the sun. 黑色被毛在阳光暴晒下可能会褪色变为朽栗。**2** a horse with black coat 骊马,黑马

Black Beauty n. a novel written by a British woman named Anna Sewell and published in 1877 that tells the story of horse of the same name（由安娜·斯维尔撰写并于1877出版的一部以马为题材的小说）《黑骏马》: In English literature, *Black Beauty* is perhaps the best known fiction about the life of a horse. 在英语文学中,《黑骏马》或许是迄今与马有关的最杰出的小说。

black brigade n. the black horses used to pull hearses in the past（过去用来拉灵车的黑色马匹）出殡黑马

black brown n.【color】a black body with the muzzle, and sometimes the flanks, brown or tan（体色为黑而口鼻为褐色的毛色类型）褐黑色

black bullfinch n.【jumping】a type of jumping obstacle made of thick, live hedge which is jumped over, not through（用活树篱搭建的障碍,需要从上面跨越而非从中穿越）黑棘篱障碍

black cell tumor n. *see* melanoma 黑[色]素瘤

black fly n. any of various small, dark-colored flies of the genus *Simulium* that feeds on blood of animals by biting hairless body areas（一种蚋属黑色蝇,靠吸食动物无毛区的血液为生）黑蝇

Black Maria n. a four-wheeled, horse-drawn vehicle used formerly to transport prisoners in Britain and North America（过去在英国和北美用来运载囚犯的四轮马车）囚车

black marks n. small areas of black hairs that appear anywhere on the body of a horse（马被毛上的）黑斑

Black Master n. a funeral service owner who used black brigade to pull hearses prior to the advent of the automobiles（汽车出现前用黑色马队运送灵柩提供服务的人）殡仪业主

black points n.【color】the black ear tips, mane, tail, and lower legs of a horse（包括耳梢、鬃、尾和四肢下部在内的黑色部位）黑色末梢

black saddler n. a saddle maker specializing in cart and carriage harnesses（马车）挽具匠 [compare] brown saddler

black tobiano n. *see* piebald 黑白花马

blacksmith /ˈblæksmɪθ/ n. a craftsman who shapes metal into tacks with anvil and hammer（用砧和锤制作马具的匠人）铁匠

blackthorn /ˈblækθɔːn; *AmE* -θɔːrn/ n. a thorny, deciduous Eurasian shrub often used to make whips owing to its elastic quality（产于欧亚大陆的带刺落叶灌木,因木质弹性较好而常用来制作马鞭）黑刺李

bladder /ˈblædə(r)/ n.【anatomy】(also **urinary bladder**) a distensible, muscular sac which

B

serves as a receptacle for the urine secreted by the kidneys (体内的一个可伸缩性肌囊,用于储存肾脏分泌的尿液)膀胱

blade /bleɪd/ n. the flat-edged cutting part of a tool, such as a clipper or plow (剃刀或犁等工具的锋刃)剃刀;犁铧

Blagden /ˈblædən/ n. (also **Blagdon**) a breed of spotted horse formerly bred at the Royal Stud in Britain (过去育成于英国皇家种马场的斑毛马品种)布兰登马

Blagdon /ˈblædən/ n. *see* Blagden 布兰登马

blank /blæŋk/ adj.【hunting】(of a den or hiding place) having no fox in it (指洞内或藏身处)无狐的;空的

blank day n.【hunting】a day in which the hounds fail to find a fox (猎狐无果而归的一天)空猎日

blanket /ˈblæŋkɪt/ n. **1**【color】a color pattern in the Appaloosa breed with white hair on hips and loins with or without spots (阿帕卢萨马的毛色类型之一,其中马后躯白色区域上的花斑时有时无)花毯型 compare frost, leopard, marble, snowflake **2** the horse blanket 马衣 **3** the saddle blanket 鞍毯,鞯

vt. to put a blanket over a horse 穿马衣:Horses should best be blanketed in winter. 马在冬季最好穿上马衣。

blanket clip n. a type of coat clip in which the body hair of the horse from the neck and belly is cut (马匹被毛修剪类型,其中仅对颈和腹部被毛予以修剪)披毯型修剪

blanket finish n.【racing】an extremely close race finish in which the horses could be, figuratively, covered with one blanket (指比赛中马匹过于接近以至于可同披一件马衣)胜负难分的冲刺

blanket riding n. a riding style using only a blanket held in position by strap without a saddle (仅用皮带固定鞍毯而不用马鞍的骑乘方式)鞍毯骑乘

blastula /ˈblæstʃələ/ n.【biology】(*pl.* **blastulas** or **blastulae** /-iː/; also **blastosphere**) an embryo at the early stage of development when it is a hollow ball of cells (胚胎发育的早期阶段,由细胞组成空心球状结构)囊胚 compare morula

blaze /bleɪz/ n.【color】a facial marking consisting of a white band that runs from the forehead to the muzzle; is wider than stripe (马的面部别征之一,为前额延伸至口鼻部的白色条纹,比流星要宽)宽流星,广流星— *see also* **bald face, snip, star, stripe, white lip**

Blazer /ˈbleɪzə(r)/ n. a horse breed developed in Idaho, USA around the 1960s; is known for their easy maintenance and versatility (20 世纪 60 年代左右育成于美国爱达荷州的马品种,以易于驾驭和用途广泛而知名)布莱泽马

blazing /ˈbleɪzɪŋ/ adj.【hunting】(of a scent) very strong; burning (猎物气味)浓烈的:a blazing scent 浓烈的气味

bleed /bliːd/ vi. to emit or lose blood 失血,流血 vt. to take or draw blood from a horse using a syringe(用注射器进行)放血,采血

bleeder /ˈbliːdə(r)/ n. a horse who suffers from bleeders 肺出血的马

bleeders /ˈbliːdəz; *AmE* -dərz/ n.【vet】a respiratory disease of racehorses in which the capillaries of the lung raptured due to strenuous exercise during a workout or race, thus causing pulmonary hemorrhage and bleeding from nostrils (赛马多发的呼吸性疾病,多由比赛或训练中运动强度过大造成肺部毛细血管破裂,进而导致肺部出血并由鼻孔流出)肺出血 syn exercised-induced pulmonary hemorrhage

blemish /ˈblemɪʃ/ n. a minor conformation fault or permanent mark of a horse (马的次要体型缺陷或永久斑点)瑕疵,斑点:Common examples of blemishes would be curbs, girth galls or scars. 马体常见的缺陷有关节硬瘤、肚带勒迹和伤疤。

blind /blaɪnd/ adj. losing or lack of the sense of

sight due to inheritance or injury (由于遗传或受伤引起的视力丧失)失明的,瞎的:blind spot 盲区 ◇ A blind horse will rely more heavily on sound as a sense and will be very hesitant in its gait. 瞎马主要依靠其听觉进行辨别,而在运步中犹豫不前。

n. (usu. **blinds**) the blinkers 眼罩

blind boil n.【vet】a painful inflammatory sore on the skin caused by microbic infection (由于微生物感染引起的皮炎疮)瞎疖子

blind bucker n. a horse who bucks violently without any sense of direction 狂奔乱跳的马

blind country n.【hunting】countryside with exuberant growth of weeds and brush that obscures the obstacles and terrain ahead (杂草丛生的乡野地带,其中障碍和地貌难以辨认)盲猎地带

blind obstacle n.【jumping】any obstacle of which the landing side is not visible at the take-off side (起跳时落地侧看不清的障碍)盲点障碍

blind spavin n.【vet】arthritis of the lower joints of the hock where the bone degenerates without visible bone growth (马飞节下侧关节发炎的症状,患处骨逐渐退化而不增生)隐性跗关节炎

blind splint n.【vet】inflammation of the interosseous ligament in the leg (骨间韧带发炎的症状)内生骨刺

blinder /ˈblaɪndə(r)/ n. a horse who performs unexpectedly good 表现出乎意料的马;黑马 IDM **play the blinder** (of a horse) to perform excellently (赛马)表现极佳

blinders /ˈblaɪndəz/ n. *pl.* *see* blinkers 眼罩

blindness /ˈblaɪndnəs/ n. the condition of being blind 瞎眼,失明:moon blindness 夜盲症

blinkers /ˈblɪŋkəz; *AmE* -kərz/ n. pl. a pair of leather shields fastened to the bridle behind a horse's eye to curtail side vision (遮在赛马眼睛上的皮罩,可遮挡侧面视野)眼罩:racing blinkers 赛用眼罩 syn blinders, blinds, winkers

blister /ˈblɪstə(r)/ n. **1** a local swelling of the skin caused by burning or irritation (皮肤因烫伤或瘙痒而形成的局部肿胀)水疱 **2** a counterirritant applied to the skin of the leg to cause blistering and inflammation (涂在体表引发炎症和水疱的药剂)发疱剂 syn vesicant, resicant

vi. (of skin) to break out in blisters or form blisters 起疱

blister beetle n. any of various beetles capable of blistering the skin 斑蝥

blistering /ˈblɪstərɪŋ/ n. the practice of applying caustic agent to the skin to promote internal healing in some cases (为促进治愈而在皮肤上涂抹发疱剂)发疱疗法

bloat /bləʊt/ n. a swelling of the stomach or intestines caused by excessive gas following fermentation of ingested forage (家畜采食饲草后发酵造成肠胃臌胀)胀气

v. (to cause) to swell up or inflate, as with gas (指肠胃)胀气

block /blɒk; *AmE* blɑːk/ vt. to stop the passage or movement; obstruct 阻碍,阻塞:The ascarids hosting in foal will cause inflammation of the lining of the gut and in more severe cases may block the small intestine, causing rupture, peritonitis and death. 寄生在幼驹体内的马蛔虫常造成肠壁发炎,严重时可阻塞小肠并引发肠道破裂、腹膜炎,甚至引起死亡。

blond sorrel n.【color】a sandy-red coat with paler areas around the eyes and on the muzzle, flanks, and the inside of the legs (体色为沙红色,而眼睛周围、口鼻、两胁及大腿内侧稍浅的毛色)金栗毛

blood /blʌd/ n. the red, viscid fluid that circulates through the vascular system of the body (在体内脉管系统中循环的红色黏液)血液:blood calcium 血钙含量 ◇ blood cell count 血细胞计数 ◇ blood clot 血凝块 ◇ blood glucose level 血糖水平 ◇ blood group 血型 ◇

blood plasma 血浆 ◇ blood platelet 血小板 ◇ blood poisoning 毒血症 ◇ blood profile 血象 ◇ blood pressure 血压 ◇ blood testing 血液化验 ◇ blood type 血型 PHR **out/short of blood**【hunting】(of hounds) not having killed prey for some time (指猎犬)没杀生的:The hounds have gone out of blood for one year. 这些猎犬已有一年都没有杀生了。

vt.【hunting】**1** to give (a hound) its first taste of blood from killing (让猎犬首次)尝腥,杀生 **2** to initiate one in a particular activity 使入行:This will be her first time to perform on the stage, she hasn't yet been blooded. 这将是她首次登台演出,之前她还尚未入行。

blood bay n.【color】a dark shade of bay 血骝毛

blood blister n. *see* hematoma 血泡

blood cell n. (also **blood corpuscle**) any of the cells existing in the blood 血细胞

blood corpuscle n. *see* blood cell 血细胞

blood count n. the counting of the number and proportion of red and white cells in a specific volume of blood sample (对定量血样内血细胞的数量和红/白细胞比例予以计量)血细胞计数

blood heat n. the normal temperature of a horse's blood (马体的)血液温度

blood horse n. **1** a pedigree horse 纯种马 **2** a Thoroughbred horse 纯血马

blood marks n.【color】patches of red colored hairs growing into a gray coat (青色被毛上着生的红色斑点)血斑,红斑

blood plasma n. the clear fluid of the blood when separated from blood corpuscles by centrifuging (血液经离心富集血细胞后所得的清亮液体)血浆

blood poisoning n. *see* septicemia 败血症

blood pressure n. the pressure exerted by the blood against the inner walls of the blood vessels (血液对血管内壁所施加的压力)血压

blood sports n. any pastime or sports in which animals or birds are killed, such as fox-hunting (涉及杀生的体育娱乐项目,如猎狐)血腥运动,杀生项目

blood test n. examination of a blood sample to obtain information required for specific purposes (出于特定目的对血样进行检查)血液检测,血检

blood tit n. *see* blood weed 体格瘦弱的马

blood type n. (also **blood group**) n. any of various types of horse blood that are seperated into for testing purposes (用于检测目的而将马匹血液所分的类型)血型

blood urea nitrogen n. (abbr. **BUN**) the urea nitrogen level in the blood 血尿素氮:The blood urea nitrogen level could be used as an indicator of the kidney's ability to excrete urea from the body. 血液中的尿素氮含量可作为肾脏排泄尿素氮性能的指标。

blood vessel n. any of the tubes through which blood circulates in the body (体内血液循环流经的管道)血管

blood weed n.【dated】a lightly built Thoroughbred of poor quality lacking bone and substance 体格瘦弱的纯血马 syn blood tit

blood worm n. *see* large strongyles 大圆线虫

blooded /ˈblʌdɪd/ adj. **1**【hunting】(of a young hound) have been given its first taste of blood from the kill (指猎犬)尝过血腥的 **2** (of an animal) of good blood or breed (指动物)血统优良的,纯种的:blooded breeding stock 优良种畜

blood-engorged /ˌblʌdənˈgɔːdʒd; *AmE* -ˈgɔːrdʒd/ adj. (of body tissue) engorged with blood (身体组织)充血的:a blood-engorged tick 一只吸足血的扁虱

blood-letting /ˈblʌdletɪŋ/ n. the act of bleeding a horse, as a remedial measure (作为治疗手段的)放血

bloodline /ˈblʌdlaɪn/ n. (also **blood line**) the lineage of descent; pedigree 谱系,血统

bloodstock /ˈblʌdstɒk; *AmE* -tɑːk/ n. (also **blood-stock**) **1** the stock or blood lines used to preserve a breed（稳固品种时使用的品系）纯种，种畜 **2** the Thoroughbred horses considered collectively（总体）纯血马：blood-stock agent 纯血马买卖经纪人 ◇ New Zealand bloodstock 新西兰纯血马

bloodstock sale n. a sale or auction of Thoroughbred horses used in racing 纯血马交易会

blood-typing /ˌblʌdˈtaɪpɪŋ/ n. a method formerly used to determine the true identity and parentage of a horse by examining blood factors unique to each horse（过去通过检测马匹血液中特异的蛋白因子来确定其身份和血统的方法）血型鉴定：Now, blood-typing has been replaced by DNA profiling as a more accurate method to identify parentage. 目前，DNA 检测已经取代血型鉴定成为亲子鉴定中更为准确的方法。

bloom /bluːm/ n. the gloss or luster on the coat of a well-fed or groomed horse（马匹被毛上的）光泽，光亮

blow /bləʊ; *AmE* bloʊ/ v. **1** to exhale with heavy, loud, and accelerated breathing following strenuous exercise（指训练后沉重而急促地呼气）喘气，喘息：to blow wind 喘气 **2** to overtrain and sour a young horse 训练过度；使厌烦：to blow a horse 练马过度 **3**【hunting】to call by sounding a horn 吹号角 PHR V **blow (hounds) home** to declare the hunt is over by blowing a horn 吹号返回 **blow out** (of the huntsman) to call the hounds out of an empty covert 将猎犬唤出树丛 **blow a stirrup** (of foot) to fall out of the stirrup（指脚）脱蹬 **blow up** (of a horse) to break from its gait or buck off（指马）受惊；弓背跳

blowfly /ˈbləʊflaɪ; *AmE* ˈbloʊ-/ n. (also **blow fly**) a species of fly that lays its eggs in animal carrion or in open sores and wounds（在动物腐尸或伤口上产卵的蝇类）丽蝇

blowout /ˈbləʊaʊt/ n.【racing】a brief workout designed to sharpen or maintain the condition of the horse shortly prior to a race（为保持并提高马匹竞技水平而在赛前进行的短时热身）适应性训练，短跑训练

blue dun n.【color】*see* grullo 青兔褐

blue riband n. *see* blue ribbon 蓝带

blue ribbon n. **1** (also **blue riband**) a narrow strip of blue fabric that is awarded as the first prize in a competition（比赛中奖给获胜者的一窄条蓝色织物）蓝带 **2** a horse who wins the blue ribbon in a show or competition 获胜马，头马

adj. first-rate; excellent 一流的，出色的：blue-ribbon horse 骏马；蓝带马，头马

blue roan n. **1**【color】a coat color with red, black, and white hairs（被毛中混生红、黑、白的毛色）蓝沙毛 **2** any roan horse with a bluish color 蓝沙毛马

bluegrass /ˈbluːgrɑːs; *AmE* ˈ-græs/ n. any of various grasses of the genus *Poa* used as forage for livestock and rich in phosphates and lime（早熟禾属牧草，其钙磷含量丰富）蓝草：Kentucky bluegrass 肯塔基蓝草

bluegrass country n. a district of Kentucky, USA where the bluegrass is grown extensively（美国肯塔基州的辖区，那里蓝草种植极其广泛）蓝草之乡：The bluegrass country is an ideal place for breeding horses. 美国蓝草之乡是育马的理想场所。

bluestem /ˈbluːstem/ n. (also **bluestem grass**) any tufted grass grown widely for hay and forage（一种簇生草本植物，因作牧草而广泛种植）蓝茎冰草

bluestem grass n. *see* bluestem 蓝茎冰草

bluestem grass hay n. the cut and dried hay made from bluestem grass 蓝茎冰草干草

bluff /blʌf/ n. a headpiece having eye sockets put over the head of an excitable horse to keep it quiet and calm（一种带眼孔的头罩，可让悍性高的马保持安静）头罩

Bo Le n.【Chinese】*see* Sun Yang 伯乐：In Chinese culture, the term Bo Le is often used as synonym for an excellent judge of talents. 在中

B

国文化中,“伯乐”已经成为慧眼识才者的代名词。

board /bɔːd; *AmE* bɔːrd/ n. **1** a long, flat piece of wood; plank 木板 **2**【racing】the totalizator board 赌金显示板 PHR **across the board** a combination betting option in which the bettor wins if the selected horse either wins, places or shows (赌马中博彩者所选的赛马只要进入前三名即可赢得奖金的组合投注)统压;全投:A $2 bet across the board would be a $6 wager. 一注2美元的统压其投注总额为6美元。**on the board** (of a racehorse) finishing in one of the first three positions in a race (指赛马)跑入三甲:to run on the board 跑入三甲 **off the board** (of a racehorse) not finishing in the money (指赛马) 未获奖金,未跑入三甲:Lucky Star ran off the board in the 130th Kentucky Derby. “幸运之星”没能在第130届肯塔基德比赛中跑入三甲。

board fence n.【jumping】a jumping obstacle consisting of long, narrow, flat, and often painted board supported by the standards (由窄长的油漆木板制作而成的跨栏,其两侧多用立柱固定)木板跨栏 syn plank jump

boarder /ˈbɔːdə(r); *AmE* ˈbɔːrdə(r)/ n. a horse kept and tended in a commercial stable (寄养在商业厩舍的马匹)寄养马

boarding stable n. a facility where horse owners may keep their horse for a monthly fee (马主付费寄养马匹的场所)寄养马厩:Boarding stables usually feed, water and turn out boarded horses for their owners. 寄养马厩通常替马主给马喂料、饮水并将马赶到草场吃草。

boards /bɔːdz; *AmE* bɔːrdz/ n.【polo】wooden plank used to define the side lines of the polo field (马球场上用来标明边界的木板)边线木板 syn sideboards, polo boards

boat race n. *see* fixed race 名次事先确定的比赛;黑哨赛

bob /bɒb; *AmE* bɑːb/ vi. (of a horse) to raise and lower its head restlessly (马头)上下晃动,不断点头 | **bobbing** n.

bobbery pack n.【hunting】a mixed pack of hounds used for hunting 混合犬群

bobble /ˈbɒbl; *AmE* ˈbɑːbl/ vi.【racing】(of a horse) take a bad step out of the starting gate (指赛马)出闸不利,出闸失利

bobby back n. *see* sway back 凹背

bobby backed adj. *see* sway backed 凹背的

bobtail /ˈbɒbteɪl; *AmE* ˈbɑːb-/ n. **1** a short or shortened tail 截短的尾巴;短尾巴 **2** an animal having a short or shortened tail 截短尾巴的动物:Bells on bobtail ring. 马尾上的铃铛响叮当。

bobtailed /ˈbɒbteɪld/ adj. **1** (of a horse) having a short or docked tail (指马)尾巴被截短的 **2**【hunting】(of a fox) having little or no tail (指狐狸)无尾巴的

body /ˈbɒdi; *AmE* ˈbɑːdi/ n. the main part or physical structure of a horse 身体;躯体:body parts of horse 马体部位

body brush n. a soft-bristled brush used to remove grease, dirt, and sweat from a horse's coat (用来去除马体上油污和汗渍的软毛刷)体刷

body clip n. *see* full clip 全修剪

body horse n.【driving】the middle horse of a lead team pulling a horse-drawn carriage (驾车马队的)中马,主马

body length n. the length from the point of a horse's shoulder to its buttock (马肩端至臀端的斜线长度)体长

body temperature n. (abbr. **BT**) *see* temperature 体温

body wash n. *see* brace 爽身液,爽身油

body weight n. the net weight of a horse (马的)体重

bog rider n. a cowboy responsible for rescuing cattle mired in the mud (负责营救陷入泥潭牛的牛仔)沼地牛仔

bog spavin n.【vet】a condition in horses in which lymph accumulates in the hock joint and results in a puffy swelling (跗关节内淋巴液淤积造成关节水肿的症状)跗关节软肿;飞节内肿 syn tarsal hydrarthrosis

bogey /ˈbəʊgi; *AmE* ˈboʊgi/ n. **1**【jumping】a problem fence 有问题的跨栏 **2** a muddy ground 泥地

Bohai n.【Chinese】(also **Bohai horse**) a breed of horse developed in certain areas of Shandong province, China by crossing native horses with light Soviet Thoroughbred and later with Don in the 1960s; stands about 14 hands and usually have a bay or chestnut coat with black and gray occurring rarely; is frugal and possesses great strength; used mainly for draft and farm work (中国山东省一些地区在20世纪60年代通过采用当地马与轻型苏纯血马杂交后经顿河马改良育成的马品种。体高约14掌,毛色以骝、栗为主,黑和青较少,耐粗饲,挽力大,主要用于拉挽和农田作业)渤海马

boil /bɔɪl/ n. a painful pus-filled swelling on the skin caused by microbic infection (由微生物感染引起的皮肤化脓性炎症)疖;疔疮

vi. (of a horse) to act in an excited, agitated manner 激怒,狂躁 PHR V **boil over 1** (of a horse) to start bucking (指马)狂跳 **2**【racing】(of a horse) to win by a decisive margin when not expected to do so (指马)意外险胜

Bokara /bəˈkɑːrə/ n. (also **Bokara pony**) a pony breed descended from the Persian horse (由波斯马演化而来的矮马品种)波卡拉矮马

Bokara pony n. *see* Bokara 波卡拉矮马

bold /bəʊld; *AmE* boʊld/ adj. (of a horse's eyes) large, expressive and clear (马眼睛)大而亮的:a horse with bold eyes 一匹眼大而明亮的马

bold front n. a proud head carriage of the horse 头姿昂扬:to put up a bold front 头姿昂扬

bolet /ˈbəʊlɪt/ n. *see* bolus 食糜,食团

bollard /ˈbɒlɑːd; *AmE* ˈbɑːlərd/ n. *see* roller bolt 挂杆闩,挂杆桩

bolster /ˈbəʊlstə(r); *AmE* ˈboʊ-/ n. the timber running parallel to the front and rear axles on a horse-drawn vehicle (马车上与前后车轴平行的木杆)枕杆

Bolster Wagon n. a four-wheeled, horse-drawn buggy developed in the United States in 1814 (1814年出现在美国的一种四轮马车)枕杆四轮马车

bolt[1] /bəʊlt; *AmE* boʊlt/ vi. (of a horse) to break out of control and run away suddenly (指马)脱缰,惊逃 PHR V **bolt away** to run away suddenly 惊逃

vt. (of a hound) to drive or force fox out of a covert, hole or other shelter (指猎犬将狐狸从隐藏处)赶出 PHR V **bolt a fox** 将狐狸赶出藏身处

bolt[2] /bəʊlt; *AmE* boʊlt/ vt. to eat feed hurriedly with little chewing; gulp (不加细嚼)吞食:to bolt feed 吞食 | **bolting** n.

bolter /ˈbəʊltə(r); *AmE* ˈboʊ-/ n. **1** a horse who breaks out of control and run away suddenly 脱缰马 **2** a horse who eats too fast with little chewing 吞食的马

bolus /ˈbəʊləs; *AmE* ˈboʊləs/ n. a soft mass of chewed forage swallowed by the horse (马匹咀嚼吞咽的饲草)食糜,食团 syn bolet

bomb-proof /ˈbɒmpruːf/ adj.【BrE】(of a horse) safe, reliable, and quiet (指马)炸弹惊不动的;性情安静的

bone /bəʊn; *AmE* boʊn/ n. **1** the dense, calcified connective tissue forming the skeleton of a horse (构成马体骨架的致密结缔组织)骨,骨骼:compact bone 致密骨 ◇ flat bone 扁骨 ◇ irregular bone 不规则骨 ◇ long bone 长骨 ◇ shaft of bone 骨干 ◇ bone and meat meal 肉骨粉 **2** the measurement of circumference around the cannon bone halfway between the knee and fetlock (马前膝与球节间管骨中段的周长)管围:bone measurement 管围测量 ◇ The size of a horse's bone is a recognized indication of the weight-bearing capacity. 马匹管围大小是

B

其载重能力的一个公认指标。PHR **have good bone** 管围较粗 ◇ **short of bone** 管骨过细

bone chip n. a bone fragment resulting from injury that does not result in fracture of the bone (受伤后导致的)碎骨,骨裂片

bone meal n.【nutrition】crushed and coarsely ground bones fed as a calcium-phosphorous feed supplement in animal feed (粉碎后作动物饲料钙-磷添加剂使用的骨头)骨粉

bone shaker n.【slang】any rickety, uncomfortable, horse-drawn carriage 颠簸的马车

bone spavin n. an ossification or new bone growth caused by degenerative arthritis in the lower joints of the hock and characterized by a bony enlargement on the front and inside of the hock joint (由于跗关节患关节炎诱发的关节骨化或骨质增生,症状表现为飞节前内侧肿大突起)飞节内生赘骨,跗关节骨疣

bonnet /ˈbɒnɪt; *AmE* ˈbɑːnət/ n. **1** a facial marking of horse with white head if only the ears and eyes remain pigmented (马面部别征之一,其中除耳和眼周围有颜色外头部其他地方均为白色)白头 **2**【dated】one who praises the merits of a horse to assist the owner to sell it while pretending that he does not know the owner (在马匹买卖中假装与马主不相识但却鼓吹马匹优点的人)诱拐贩,卖马托 syn shill

bony growth n. *see* exostosis 外生骨疣

book /bʊk/ n. **1**【racing】a jockey's schedule of riding assignments (赛事)日程表 **2**【racing】(also **betting book**) a bookmaker's record of all bets placed on a horse race (赛马经纪人用来记录投注的账簿)赌马记录册 **3** the stud book 育种登记册,良马簿 PHR **have a full book** (of a stallion) be bred to the maximum number of mares allowed by his manager (指公马)配种满额 **in the book** (of a horse) registered in the General Stud Book for the breed (指马)登记在册

Book of Horses n. a book written on the management and health of horses by Yang Shiqiao (1531－1609), a scholar and court official of the Ming dynasty in ancient China (中国古代明朝官员杨时乔撰写的有关马匹管理和医药方面的著作)《马书》

bookie /ˈbʊki/ n.【racing, slang】a bookmaker 赌马经纪人,庄家

bookmaker /ˈbʊkmeɪkə(r)/ n. (also **bookie**) a professional who accepts and pays off bets placed by others on horses competing in a race (赛马中接受他人投注并按特定赔率返还奖金的职业经纪人)赌马经纪人;庄家

booster /ˈbuːstə(r)/ n. an extra dose of vaccine given to a horse to enhance protection against a particular disease (为提高马对某疾病的抵抗力而额外注射的疫苗)免疫增强剂

boot /buːt/ n. **1** a protective or supportive covering for the legs of horse (缠在马四肢上起保护作用的)绑腿,护腿:brushing boot 防蹭护腿 ◇ polo boot 马球护腿 **2** the riding boot 马靴:a pair of top boots 一双高帮马靴 **3** a container on a coach or carriage used for storing luggage (马车上存放行礼的)行李箱:front boot 车前行李箱 ◇ rear boot 车后行李箱
vt. **1** to place boots on the legs of a horse 给马戴护腿 **2**【racing】(of a jockey) to kick his horse hard (骑手)踢马 PHR V **boot home a winner** to ride a winning horse in a race 骑马获胜

boot hook n. (also **boot pull**) a J-shaped metal hook used to pull on riding boots (用来提马靴的铁钩)靴钩

boot jack n. (also **bootjack**) a forked device used to assist the rider in removing boots 脱靴器

boot pull n. *see* boot hook 靴钩,提靴钩

boot tred n.【driving】the step attached to the front boot of a horse-drawn coach and used by travelers to get into the box (安装在马车前行李箱上的踩脚,乘客借此上车)上车踩脚,上

车踏板

boot tree n.（also **boottree**）a wooden or plastic form inserted into a tall boot to preserve its shape when not being worn（马靴不穿时塞在内部以保持形状的木质或塑料器件）靴楦

boots and saddles n.【racing】a bugle call sounded when jockeys mount and enter the racetrack for the post parade（骑手上马举行入场仪式时响起的号角声）入场号角

borer /ˈbɔːrə; *AmE* ˈboʊrə/ n. a horse who constantly pulls on the bit by thrusting his head up and down（上下摆头而扯动衔铁的马）烦躁不安的马

boring /ˈbɔːrɪŋ/ adj.（of a horse）pulling the bit by thrusting its head up and down constantly（指马不停地扯动衔铁的）烦躁的

bosal /ˈbɔːsl/ n. a braided, leather, or rope noseband used in Western equitation（西部骑术中采用的鼻革，由皮革或绳子编制而成）鼻索：Western bosal bridle 西部鼻索马勒

Bosnian /ˈbɒznɪən; *AmE* ˈbɑːz-/ n. an ancient pony breed descended from the Tarpan and originating in Bosnia-Herzegovina; stands 12.1 to 14 hands and may have a bay, brown, black, gray, chestnut, or palomino coat; has a rather heavy head and short neck, deep and wide chest, and short, muscled legs; is docile, steady, and hardy; used for riding, packing, light farm work and light draft（源于南斯拉夫波黑地区的古老矮马品种，主要由塔畔马演化而来。体高12.1～14掌，毛色多为骝、褐、黑、青、栗或银鬃，头重，颈短，胸身宽广，四肢粗短，性情温驯，步法稳健，主要用于骑乘、驮运、农田作业和轻役）波斯尼亚马

bosomy /ˈbʊzəmi/ adj.（of a horse）having an over-wide chest（指马）胸腔过宽的

bot /bɒt; *AmE* bɑːt/ n. the parasitic yellow larvae of botfly living in the stomach of horse（寄生在马胃中的马胃蝇幼虫）马胃蝇蛆 [syn] horse bot, stomach bot

bot larvae n. the larvae of botfly; bot 马胃蝇蛆

botfly /ˈbɒtflaɪ/ n.（also **horse botfly**）any of various beelike flies of the genera *Gasterophilus* and *Oestrus*, having larvae that are parasitic on various animals, especially horses and sheep（胃蝇属和狂蝇属的多种形似蜜蜂的蝇类，其幼虫寄生于马和羊等动物体内）马胃蝇

bottom[1] /ˈbɒtəm; *AmE* ˈbɑːt-/ n. **1** the buttocks of an animal（动物的）臀部，后躯 **2**【racing】the foundation of a race track（赛场的）地基 **3** a racehorse listed last in the race program and assigned an outside post position（列在比赛栏最后且排到闸位最外侧的赛马）置后赛马，垫底赛马

bottom[2] /ˈbɒtəm; *AmE* ˈbɑːt-/ n. **1** a ditch or brook running at the bottom of a valley or ravine 谷底溪流 **2**【hunting】a fence with a deep and jumpable ditch 带水沟的跨栏

botulin /ˈbɒtjulɪn/ n. any of several potent neurotoxins produced by botulinum（由肉毒杆菌产生的烈性神经毒素）肉毒杆菌毒素

botulism /ˈbɒtjulɪzəm; *AmE* ˈbɑːtʃə-/ n. a fatal food poisoning caused by ingestion of feed containing botulin（马误食含有肉毒梭菌的饲草所致的致死性中毒症状）肉毒杆菌中毒 [syn] forage poisoning, shaker foal syndrome

bought sight unseen adj.（of a horse）bought on the basis of verbal or written description rather than visual evaluation（指根据口头或书面描述而非外貌鉴定进行马匹交易）未看马购买

Boulnois Cab n. a four-wheeled, enclosed, horse-drawn carriage with a small, elongated body resembling a small Omnibus（一种封闭式四轮马车，长方形车身形似小型公共马车）布勒诺出租马车

Boulonnais /buːˈlɔneɪ/ n.【French】a breed of draft horse originating in Northern France; stands 15.3 to 16.2 hands and usually has a dappled gray coat, although chestnut and bay also occur occasionally; has a distinctive profile and large eyes, a thick, but graceful neck, prominent withers, a compact body, broad and

B

straight back, short, strong limbs with wide joints, thick cannons and a lack of feathering; the mane and tail are full and the tail is set high; has straight, relatively long, swift, and energetic action; used mainly for draft and farm work (源于法国北部的挽马品种。体高15.3~16.2掌,毛色多为斑点青,但栗和骝亦有之,面貌特征明显,眼大,颈粗但线条优美,鬐甲突出,体格结实,背部平直,四肢粗短,关节粗大且无距毛,鬃尾浓密,尾础较高,步幅较大,动作灵活而有力,主要用于重挽和农田作业)布洛涅马

Boulster Wagon n. a four-wheeled, horse-drawn vehicle 布斯特四轮马车

bounce fence n.【jumping】a jumping obstacle consisting of two parallel fences between which the horse touches down and takes off without taking a stride (由两道平行跨栏组成的障碍,马跳入其中间后不迈步又从中跳出)弹跳式跨栏

bouncing Bet n. *see* soapwort 肥皂草

boundary rider n. an Australian ranch worker who rides around a range to check and repair holes in the fence (在澳洲牧场上骑马巡视围栏并修补破损处的人)牧场巡骑

bout /baʊt/ n. **1** a sudden attack or breaking out of a disease (病症突然)发作: recurrent bouts 复发 **2** a period of time during which intense activity is conducted (进行剧烈运动的片刻)一阵: exercise bout 一阵剧烈运动

bovine /ˈbəʊvaɪn; *AmE* ˈboʊ-/ adj. of or relating to cattle 牛的: bovine diseases 牛患的疾病
n. an animal of the cattle group 牛

bow /bəʊ; *AmE* boʊ/ n. **1** a weapon used for shooting arrows (用来射箭的武器)弓 **2** the condition of having bowed tendon 腱鞘炎: high bow 上位腱鞘炎 ◇ middle bow 中位腱鞘炎 ◇ low bow 下位腱鞘炎
v. to bend into the shape of a bow 弯曲

bow hocks n. a conformation defect in which the horse's hock joints turn outwards (马飞节朝外的体格缺陷)飞节弓形, O 形飞节 [compare] cow hocks

bow legs n. a conformation defect in which the horse's knees or hock joints turn outwards (马前膝或飞节朝外的体格缺陷)膝弓形,膝 O 形;飞节弓形, O 形飞节 [syn] bandy legs

bowed /bəʊd; *AmE* boʊd/ adj. **1** curved or bent into the shape of a bow; bandy 弯曲的,弓形的 **2** (of a horse) having or suffering from a bowed tendon (指马)患腱鞘炎的

bowed knees n. a conformation defect in which the horse's knees turn outwards when viewed from the front (前观时马前膝朝外偏离的体格缺陷)膝弓形,膝O 形 [syn] carpus varus, lateral deviation of the carpal joints

bowed tendon n. an injury to the superficial digital flexor tendon or tendon sheath symptomized by diffuse swelling over the tendon area, heat, and pain (指浅屈肌腱受伤的情况,症状表现为患处肿胀、发热和疼痛)腱鞘炎: Bowed tendons usually take a minimum of one year to heal, and the horse usually needs to stay stalled or in a small paddock to prevent it from running too hard and making the condition worse. 马患腱鞘炎后至少需要一年以上的休养才能痊愈,而且患病马匹应当圈养或放养在小型围场中,以免剧烈奔跑而使病情恶化。 [syn] tendosynovitis

bow-hocked /ˌbəʊˈhɔkt/ adj. (of the hocks) turning outwards 飞节弓形的,飞节 O 形的 [compare] cow-hocked

bow-legged /ˌbəʊˈlegɪd; *AmE* ˌboʊ-/ adj. (of the knees or hocks) bent or curved outwards (前膝或飞节)弓形的, O 形的: A horse bow-legged in front tends to be very rare, so this term is usually referred to the hind legs. 前肢弓形的马比较少见,因此该词通常是指马后肢。

bowler /ˈbəʊlə(r); *AmE* ˈboʊ-/ n. (also **bowler hat**) a derby hat 常礼帽

bowline /ˈbəʊlɪn/ n. *see* bowline knot 单套结,死结

bowline knot n. (also **bowline**) a knot forming a loop that does not slip and get smaller (不能滑脱缩小的绳结)单套结,死结 [compare] release knot

box /bɒks; *AmE* bɑːks/ n. **1**【BrE】(also **loose box**) a stall for keeping horse in a stable 厩舍 **2** (also **box seat**) the seat a coachman sits to drive (车夫的)驾驶座 [syn] coachman's seat **3**【BrE】a horse trailer (运马的)拖车 **4**【eventing】an area where the veterinary inspection is conducted (三日赛中进行兽医检查的区域)检测室

vt. to put a horse in a box or enclosure 圈养;围困 PHR V **box in**【racing】to be trapped behind other horses in the field (指赛马在场内)被挡,受困

box cloth n. a fabric formerly used for gaiters and outdoor wear (过去用来制造绑腿和外套的织物)厚呢料

box foot n. *see* club foot 狭蹄

box seat n. *see* box (车夫的)驾驶座

box spur n. a spur consisting of a long, straight shank terminating in a thin square (马刺的一种,直杆末端为方形)方块马刺

box stall n. *see* stall 马厩,厩舍

box trifecta n.【racing】*see* boxed trifecta 头三组合彩

box wagon n. a light, four-wheeled, pneumatic-tired wagon having a shallow, box-shaped body (一种轻型充气轮胎四轮马车,带厢式车身)厢式四轮马车 [syn] show wagon, Hackney show wagon

box walker n. a horse who has the vice to walk around the stable restlessly 患绕舍癖的马: Boxwalkers are difficult to keep weight on and therefore not easy to train. 有绕舍癖的马不易保持膘情,因此训练也比较困难。

box walking n. (also **stall walking**) a vice in which the horse walks around the box restlessly (马不停地绕厩舍走圈的恶习)厩舍绕行

boxed /bɒkst; *AmE* bɑːkst/ adj.【farriery】(of a horseshoe) with the outer edge rounded or beveled (指蹄铁)圆边的,磨边的: boxed horseshoe 圆边蹄铁

boxed trifecta n.【racing】(also **box trifecta**) a betting option in which the bettor selects the first three horses in the exact or any order (赌马者选出前三名而不考虑其获胜次序的押注方式)头三组合彩 [compare] straight trifecta

boxing /ˈbɒksɪŋ; *AmE* ˈbɑːk-/ n. **1** the transport of horses in a trailer or truck (拖车)马匹运输 **2**【racing】a technique in which the bettor selects three key horses in a trifecta and bets them in all possible combinations (一种头三彩的押注方法,其中赌马者选出 3 匹潜在获胜马匹并对所有组合进行投注)厢式押注,组合押注: Boxing is an expensive option that may yield high monetary rewards. 组合投注所需赌金较高,但却有可能赢得高额奖金。

boxy /ˈbɒksi; *AmE* ˈbɑːk-/ adj. *see* club-footed 狭蹄的;立系的: boxy hoofs 狭蹄,立系蹄

boxy foot n. (also **boxy hoof**) a conformation defect in which the hoof has a a small frog with high heel (马蹄叉过小而蹄踵过高的肢蹄缺陷)狭蹄: Boxy feet tend to occur with club feet in the same horse. 马狭蹄和立系的缺陷倾向于同时出现。— *see also* **club foot**, **coon foot**

boxy hoof n. *see* boxy foot 狭蹄

boy /bɔɪ/ n.【racing, slang】a jockey 骑师

br abbr.【color】brown 褐毛

Brabant /brəˈbænt/ n. (also **Belgian Brabant**, ***Belgian Draft***, ***Belgian Draft Horse***) an ancient horse breed developed from the Ardennais through centuries of selective breeding in Brabant, Belgium from which the name derived; stands 16 to 17 hands and generally has a red-roan, sorrel, and chestnut coat, but bay, dun, and gray also occur; has a relatively small head, short, thick neck, thick back, short and extremely strong limbs with feathering, and huge and powerful quarters; is docile and obedient, strong,

B

willing, matures early, long lived and has a slow but vigorous action; used for heavy draft and farm work(比利时布拉邦特地区经过几个世纪对阿尔登马的选育形成的古老马品种并由此得名。体高16~17掌,毛色多为红沙、红栗或栗,但骝、兔褐和青亦有之,头偏小,颈粗短,背短宽,四肢粗短且附生距毛,后躯发达,性情温驯而刚毅,成熟早且寿命长,动作缓慢而有力,主要用于重挽和农田作业)布拉邦特马:The Brabant is instrumental in the development of many European draft breeds. 布拉邦特马在欧洲诸多挽马的育成中都起到了很大的作用。syn Belgian Draft, Belgian Draft horse, Belgian Heavy Draft, Flanders horse

brace /breɪs/ n. **1** any liniment or lotion applied to the horse's body to create mild stimulation and increase circulation (涂在马身上通过产生刺激以促进血液循环的护理液)爽身液,爽身油 syn body wash **2**【hunting】a pair of foxes 一对狐狸 **3** a pair of geldings used to pull a vehicle (挽车的)一对骟马 **4** a leather strip from which the body of some horse-drawn vehicles were hung on the chassis (马车底盘上用来悬挂车身的皮带)悬带

vi. to assume an extremely stiff, erect posture (姿势)僵硬,僵直

brace bandage n. *see* exercise bandage 训练绷带

brachial /ˈbreɪkɪəl/ adj. of, relating to, or resembling the upper arm 臂的,上臂的: brachial muscle 臂肌 ◇ brachial region 上臂部

brachi(o)- /ˈbreɪkɪə-/ pref. of or relating to arm 臂

brachiocephalic /breɪtʃiːəʊˈkefælɪk; *AmE* -tʃiːoˈke-/ adj. of or relating to the arm and head 臂头的: brachiocephalic muscle 臂头肌

brachium /ˈbreɪkɪəm/ n.【anatomy】(*pl.* **brachia**/-kɪə/) the part of the upper forelimb extending from the shoulder to the elbow (前肢从肩至肘的部分)肱,上臂

brachygnathia /ˌbrækɪgˈnæθiə/ n. (also **brachygnathism**) *see* over-shot jaw 下颌突出

brachygnathism /ˌbrækɪgˈnæθɪzəm/ n. a variant of brachygnathia 下颌突出

bracken /ˈbrækən/ n. a widespread, poisonous fern having large, triangular, pinnately compound fronds and often forming dense thickets (一种分布广泛的蕨类有毒植物,三角形羽状复叶,常形成茂密的灌木丛)欧洲蕨,羊齿

bradoon /ˈbræduːn; *AmE* brəˈduːn/ n. *see* bridoon 轻勒,小勒

braid /breɪd/ n. a length of hair made up of interlaced strands (由几缕毛发结成的)辫子

vt. to interweave the hair of a horse's mane or tail in braids 辫鬃,辫尾

braid aid n. a grooming tool having a row of teeth, used in braiding the mane of a horse (用作辫鬃的带齿用具)辫鬃梳

brain fever n. *see* encephalomyelitis 脑脊髓炎

brain staggers n. *see* encephalomyelitis 脑脊髓炎

braincase /ˈbreɪnkeɪs/ n.【anatomy】the part of the skull that encloses the brain; the cranium (容纳大脑的头骨)脑壳;颅骨

brake[1] /breɪk/ n. a mechanical device used to decelerate or stop the motion of a wheel or vehicle by means of contact friction (通过接触摩擦使车辆减速或停止的机械装置)闸,刹车

Brake[2] /breɪk/ n. an open, four-wheeled, horse-drawn passenger vehicle with an elevated driving seat, used for sporting and general purposes in the 19th century (19世纪使用的敞篷客用四轮马车,驾驶座较高,主要用于户外运动等用途)布雷克马车

brake lever n. a long-handled metal bar used to decelerate or stop the motion of a wheel or vehicle (使马车减速或制动的带柄铁杆)刹车杆,制动杆

bran /bræn/ n. (also **broad bran**) the outer covering of grain such as wheat, removed during the process of milling and used as a source of dietary fiber (小麦磨面时去掉的种皮,可作日粮粗纤维使用)麸皮;糠: wheat bran 麸皮

◇ Bran contains less digestible energy, but more protein, fiber, and phosphorus than whole grain. 麸皮的消化能含量较低,但蛋白质、纤维和磷的含量要比全麦高。

bran disease n. *see* osteodystrophia fibrosa 纤维性骨营养不良,麦麸病

bran mash n. a boiled mixture of bran, wheat and water (用麸皮、小麦和水煮成的饲料)麸皮粥:In preparing of bran mash, salt can be added to make it more palatable. 在煮麸皮粥时,加盐可以使之适口性更佳。

branch /brɑːntʃ; *AmE* bræntʃ/ n. **1** the part of a horseshoe from toe to heel on each side (蹄铁两侧的部分)蹄铁支 **2** an arm-like part diverging from the main stem of a stag's antler (从主干发出的)鹿角分支

brand /brænd/ n. a mark burned on farm animals to show ownership (烫在家畜身上表示所有权的标识)烙印,印号:freeze brand 冷冻烙印 ◇ hot brand 火烫烙印

vt. to mark a horse or cattle with a brand 打烙印

brander /ˈbrændər/ n. **1** one who marks farm animals with a brand 打号工 **2** *see* branding iron 烙铁

branding /ˈbrændɪŋ/ n. the act or process of marking with a hot iron 烙印

branding iron n. (also **brander**) an iron rod that is heated to brand a horse or cattle (烧热给马或牛打烙印的铁棍)烙铁,印铁

brass /brɑːs; *AmE* bræs/ n. *see* horse brass 马用铜饰:crescent brass 半月形铜饰 ◇ heart brass 心形铜饰 ◇ horseshoe brass 蹄铁形铜饰 ◇ sunflash brass 日形铜饰

bray /breɪ/ n. the loud, harsh cry of a donkey 驴叫(声)

vi. to utter the loud, harsh cry of a donkey 驴叫

braze /breɪz/ vt. to join two pieces of metal together by using brass or copper as solder 铜焊,铜接

breach /briːtʃ/ n. an opening or tear; rupture 裂口,裂缝;疝

vi. break or tear 破裂

break /breɪk/ vt. **1** to train a young horse to obey commands and accept disciplines 调教(马匹),驯服 **2** to fracture or crack 破裂,折断:a bone break 骨裂,骨折 PHR V **break out** (of a horse) to sweat again after being cooled out following exercise (指马凉下来后)发冷汗:Cooled horses would break out due to illness, excitement, stress or electrolyte imbalance. 凉下来的马发冷汗可能由于患病、体况不佳、兴奋、应激或体内电解质失衡所致。**break up**【hunting】(of the hounds) to tear a fox into pieces (猎犬)撕碎,咬碎 **break water** (of a mare) to expel allantoic fluid during foaling (指母马)羊水破裂 **3** (of a horse) to change gait suddenly at the command of its rider (指马)突然换步:to break into a gallop 突然换为袭步 **4** to run off or away 跑掉,跑脱 PHR V **break covert**【hunting】(of a fox) to leave or escape from its hiding place (指狐狸)逃离,出窝 **5**【racing】(of a horse) to leave the starting gate (指赛马)出闸,破闸 PHR V **break maiden** (of a horse or rider) to win the first race (指马或骑手)首次获胜 **break in the air** (of a horse) to leap into the air at the start of a race rather than running forward (指赛马)腾空跳起

breakage /ˈbreɪkɪdʒ/ n.【racing】the difference between true mutuel odds and the lesser, rounded amounts given to the winning bettor(s) (实际赔率与赌马者获胜后所得赔率之间的差额)赔率差额:The breakage is usually divided between the track and the state. 赔率差额在不同赛场和州都不尽相同。

breakdown /ˈbreɪkdaʊn/ n. **1** breaking into smaller parts or elements; disintegration 分解:The anaerobic breakdown of glycogen is more rapid but less efficient than aerobic breakdown. 糖原无氧分解比有氧分解速度快,但效率低。**2** a condition in which the body fails to perform due to illness or injury（由于患病或受伤致使机能丧失的情况）崩溃，衰弱

break-in cart n. *see* breaking cart 调教马车

breaking /ˈbreɪkɪŋ/ n. the early schooling and training of a young horse（马驹早期的）调教：breaking yard 驯马圈

breaking bit n. a straight bar bit used to break a horse（调教马匹使用的一种直杆口衔）调教口衔 [syn] mouthing bit

Breaking Cart n.（also **break-in cart**）a low, two-wheeled, training or exercise vehicle used to break a single horse to draft（调教马匹挽车的双轮马车，底盘较低）调教马车

breaking cavesson n. *see* lungeing cavesson 调教笼头

breaking head collar n. *see* lungeing cavesson 调教笼头

breaking tackle n. any tack used for breaking a horse 调教用具：Lungeing cavesson, side reins, lungeing reins, bridle, and crupper are the most common breaking tackles used. 常用的调教用具有调教索、侧僵、长缰、笼头和尻兜等。

breast /brest/ n. the part of horse extending from the base of neck to the shoulder（马从颈础到肩胛的部分）前胸 [syn] brisket

breast collar n. a wide leather strap fitted around the front of the horse's chest and held in place by a narrow wither strap, used to prevent the saddle from slipping backwards（戴在马前胸上的皮带，由鬐甲处的皮带固定，可防止马鞍向后滑动）攀胸：The breast collar could be easily adjusted to fit various horses of similar size, but with different shaped necks. 攀胸可根据体型相当而颈形不同的各种马匹随意进行调整。[syn] Dutch collar

breast piece n. *see* breastplate 攀胸

breast strap n. *see* breastplate 攀胸

breastbone /ˈbrestbəʊn; *AmE* -boʊn/ n.（also **breast bone**）*see* sternum 胸骨

breastgirth /ˈbrestgɜːθ/ n. *see* breastplate 攀胸

breastplate /ˈbrestpleɪt/ n.（also **breast plate**）a wide leather strap that encircles around the horse's chest and attaches to the girths between the forelegs（绕在马前胸上的皮带，其下穿过前肢与肚带相连）攀胸：Breastplates are generally used to prevent the saddle from slipping back or for ornamental and show purposes. 攀胸主要用来防止马鞍后滑或起装饰美观作用。[syn] breast girth, breast piece, breast strap

breath /breθ/ n. **1** the air taken into or expelled from the lungs（吸入或排出肺部的空气）气息 **2** the act of inhaling or exhaling of air from the lungs 呼吸

breathe /briːð/ vi. to inhale and exhale air naturally 呼吸 | **breathing** n. 呼吸：abdominal breathing 腹式呼吸 ◇ thoracic breathing 胸式呼吸

breather /ˈbriːðə(r)/ n. a slowing of pace to allow a horse to rest or renew his strength for a short spell during exercises（训练中让马慢行歇息片刻使其恢复体力）喘气时间，短暂歇息 PHR V **take a breather** to take a short rest 短暂歇息

bred /bred/ adj.（of a horse）developed from a specific geographic area（指马）育成的：a Kentucky-bred horse 育成于肯塔基州的马

breech birth n. the delivery of a foal with the hindquarters presented first in the birth canal（幼驹产出时后躯先从产道出来的分娩方式）臀位分娩 [syn] breech delivery

breech delivery n. *see* breech birth 臀位分娩

breeches /ˈbrɪtʃɪz/ n. pl.（also **britches**）a close-fitting riding pant made of natural or synthetic fabric（由天然或合成纤维材料制成的紧身骑马长裤）马裤：Breeches usually have suede patches on the insides of the knees for better grip and less slippage while riding. 马裤靠近膝盖的内侧常衬有一块绒面革，在骑乘中有防滑作用。[syn] jodhpurs

breeching /ˈbrɪtʃɪŋ/ n.【driving】a broad leather strap passing around the haunches of a horse and fastened to the traces of driving harness to hold back the weight of brakeless, two-wheeled vehicles when descending a hill or slope (驾车挽具中绕过马后臀且与挽绳相连的一条宽皮带,无制动的两轮马车下坡时可借此抵抗载重的冲力)坐鞧,后鞧

breed /bri:d/ n. a group or line of horses having certain consistent, distinguishable characteristics which are passed on through successive generations (带有特定性状且能稳定遗传给后代的群体)品种;马种: Horse breeds are mainly developed by artificial selection over a period of time. 马匹品种主要通过长期的人工选育而获得。

vt. to cause to reproduce by controlled mating and selection; cover (有选择地进行)育种;配种 PHR V **color breed** to develop desired coat colors by selective breeding 毛色选育

breed type n. (also **type**) the ideal or standard form for any breed as detailed in the related stud book or in registration papers (良种簿或登记册中所描述的马种标准体型)良马模型,良马式

breeder /ˈbriːdə(r)/ n. **1** a person who selects the sire to which a dam is mated 配种员 **2** one who owns a stud or farm where mares are bred (拥有马场进行马匹繁育的人)育马者 **3** a mare kept to produce offspring; broodmare 繁育母马,种用母马

Breeders' Cup n. an annual series of American Thoroughbred races consisting of eight races conducted on one day at a different racetrack each year (美国每年举行纯血马系列赛,其中八项赛事同一天在不同赛马场举行)育马者杯

Breeders' Cup Classic n. a 2-km American race for three-year old horses and older run annually since 1984 as part of the Breeders' Cup (美国自1984年起每年为3岁以上马匹举行的比赛,赛程2千米,属育马者杯八项锦标赛之一)育马者杯经典大赛

Breeders' Cup Day n. the single day per year on which the eight Breeders' Cup races are conducted in the United States (美国每年举行八项育马者杯系列赛的日期)育马者杯比赛日

Breeders' Cup Distaff n. a 1.8-km American race for mares and fillies over three years old run annually since 1984 as part of the Breeders' Cup (美国自1984年起每年为3岁以上雌驹和母马举行的比赛,赛程1.8千米,属育马者杯锦标赛之一)育马者杯雌驹赛

Breeders' Cup Filly and Mare Turf n. a weight-for-age American race held annually for fillies and mares three years old and up since 1999 at a different racetrack on grass as part of the Breeders' Cup (美国自1999年起每年在不同赛马场举行的草地赛,由3岁以上母马参加,实行按龄负重,属育马者杯锦标赛之一)育马者杯雌驹草地赛: The Breeders' Cup Filly & Mare Turf is run at either 1-1/4 mile or 1-3/8 mile, depending on the turf course configuration at the Breeders' Cup host track. 育马者杯雌驹草地赛依据承办方赛场情况,其赛程或为2千米或2.2千米。

Breeders' Cup Grand National Steeplechase n. (also **Breeders' Cup Steeplechase**) a 2-5/8 mile G-1 steeplechase race for three-year old and up Thoroughbreds conducted annually since 1986 over a turf course in the United States (美国自1986年起每年举办的一级草地越野障碍赛,由3岁以上纯血马参加,赛程4.2千米)育马者杯全美越野障碍赛,育马者杯越野障碍赛: The Breeders' Cup Grand National Steeplechase may not be conducted on Breeders' Cup Day or necessarily on the same track. 育马者杯全美越野障碍赛或许不会在育马者杯比赛日举行,甚至赛场也会有所变动。

Breeders' Cup Juvenile n. a 1-1/16 mile American race for two-year old colts and geldings run annually since 1984 as part of the Breeders' Cup (美国自1984年起每年由2岁雄驹和骟马参加的比赛,赛程1.7千米,属育马者杯八项锦标赛之一)育马者杯两岁雄驹赛

B

Breeders' Cup Juvenile Fillies n. a 1-1/16 mile American race for two-year old fillies run annually since 1984 as part of the Breeders' Cup（美国自1984年起每年为2岁雌驹举行的比赛，赛程1.7千米，属育马者杯八项锦标赛之一）育马者杯两岁雌驹赛

Breeders' Cup Mile n. a 1-mile American turf race held annually for three-year old horses and older since 1984 as part of the Breeders' Cup（美国自1984年起每年为3岁以上马匹举行的草地赛，赛程1.6千米，是育马者杯八项锦标赛之一）育马者杯一哩赛

Breeders' Cup Sprint n. a 6-furlong, weight-for-age G-1 stakes race held annually for three-year old horses and older since 1984 as part of the Breeders' Cup in the United States（美国自1984年起每年为3岁以上马匹举行的一级奖金赛，实施按龄负重，赛程1.2千米，为育马者杯八项锦标赛之一）育马者杯短途赛

Breeders' Cup Steeplechase n. *see* Breeders' Cup Grand National Steeplechase 育马者杯越野障碍赛

Breeders' Cup Turf n. a 1.5-mile American turf race for three-year old horses and older run annually since 1984 as part of Breeders' Cup（美国自1984年起每年为3岁以上马匹举行的草地赛，赛程2.4千米，为育马者杯八项锦标赛之一）育马者杯草地赛

Breeders' Stakes n. a 1.5-mile Canadian stakes race held annually for three-year-old Thoroughbred horses at Woodbine racetrack in Toronto since 1889（加拿大自1889年起每年为3岁纯血马举行的奖金赛，赛程2.4千米，地点在多伦多活拜赛马场）育马者大赛：At a distance of one-and-a-half miles, the Breeders' Stakes is the longest of the three Canadian Triple Crown races and is the only one raced on turf. 在加拿大三冠王比赛中，2.4千米的育马者大赛赛程最长，而且也是唯一在草地上进行的比赛。 compare Queen's Plate, Prince of Wales Stakes

breeding /ˈbriːdɪŋ/ n. **1** the act or process of reproducing offspring 繁育，配种：breeding certificate 配种证明 ◇ breeding fund 配种基金 ◇ breeding soundness evaluation 繁育性能评定 **2** the careful and controlled selection and mating of the dam and sire to achieve desired quality or characteristics in offspring（经过对母本和父本严格选育使后代获得期望性状）育种：breeding efficiency 育种效率 ◇ breeding establishments 育种机构 ◇ selective breeding 选育

breeding boots n.（also **covering boots**）a padded boot strapped onto the hind feet of a mare during mating to protect the stallion from being kicked（配种时绑在母马后腿上的衬垫绑腿，可防止公马被踢伤）配种靴，配种护腿

Breeding Clearance Notification n.（abbr. **BCN**）an official document issued by a Stud Book to qualify a Thoroughbred mare to be covered and foal（纯血马登记委员会出具的文件，母马凭此方可参加繁育）繁育许可报告，繁育许可证

breeding contract n. an agreement between the owners of the stallion and mare that states the service fees and the obligations of each party（公马马主与所配母马马主之间拟定的书面协议，其中包括配种费用及双方应有的义务）配种合同

breeding hobbles n. a hindleg restraint used on a mare to prevent her from kicking the stallion during mating（系在母马后肢的绊腿，可防止在配种时踢伤公马）配种绊腿，配种蹄绊 syn serving hobbles, mating hobbles

breeding season n. the period during which the male and female mate 交配季节，配种季节

breeze[1] /briːz/ vt.【racing】to exercise the horse at a controlled and relaxed speed（对马进行适当的放松训练）遛马

n. a rider hired to exercise or breeze racehorses 遛马员

breeze[2] /briːz/ vi.【racing】（of a horse）win a

race easily 轻松获胜 IDM **to breeze in race** 比赛中轻松获胜

Breton /ˈbretən/ n. a coldblood breed originating in Brittany, France more than 4,000 years ago; stands 15 to 16 hands and generally has a chestnut coat but bay, gray, roan and red roan also occur; has a well-proportioned head with a heavy jaw, broad forehead, short ears, flared nostrils, a short, broad, muscular and arched neck, short and straight back, broad loins, sloping croup and short and powerful legs with heavy joints; is energetic, lively, and has great endurance; has been used to improve other heavy draft breeds as well as heavy draft and farm work (4 000 多年前源于英国和法国的冷血马品种。体高 15 ~ 16 掌，毛色以栗为主，但骝、青、沙毛和沙栗亦有之，头型适中，颌骨较大，前额宽，耳短，鼻孔大，颈粗短且呈弧形，背短而平直，后腰宽阔，斜尻，四肢粗短且关节粗大，性情活泼，耐力极强，主要用于改良其他挽马品种、重挽和农田作业) 布雷顿马: The registered Breton foals are branded on the left side of the neck with a cross surmounting a splayed, upturned V. 登记过的布雷顿马幼驹在颈左侧都烙有十字符号，下面是倒写的 V 字。

Brett /bret/ n. a horse-drawn passenger vehicle with a shallow body, usually drawn by two or more horses in pole gear (一种车身较浅的客用马车，多由两匹以上的马套驾杆拉行) 布雷特马车

brewers' dried grains n. (also **brewers' grains**) the dried extracted residue of barley malt and other grains (麦芽与其他谷物发酵酿酒后的残渣) 干酒糟 syn distillers' grains, spent grains

brewers' grains n. *see* brewers' dried grains 酒糟

brewer's horse n. a heavy draft horse used by the breweries to pull vans prior to the 1900s (20 世纪前酿酒厂用来拉酒桶马车的重挽马) 酿酒厂用马

Breyer horse n. a model horse made of plastic or porcelain manufactured by Breyer Molding Company since 1954 (布雷叶模具公司自 1954 年起生产的塑料或瓷制马模型) 布雷叶模型马: The Breyer horses are often modeled after live horses and noted for their realistic appearance. 布雷叶模型马的制作多源于真实马匹，常以其外形逼真而著称。

Breyerfest /ˈbraɪəfest/ n. (also **Breyer Fest**) an annual model horse exposition held at the Kentucky Horse Park in July in Lexington, Kentucky, USA (每年 7 月在美国肯塔基州莱克星顿的马公园举行的模型马展览会) 布雷叶模型马展览会

briar /ˈbraɪə/ n. *see* brier 荆棘；野蔷薇

brick wall n.【jumping】(also **puissance wall**) a solid jumping obstacle consisting of false brick blocks (由假砖块堆砌的障碍) 砖墙

bride hunt n. a mounted game popular in Central Asia in which young men ride after the girls and attempt to kiss them, and the girls retaliate fiercely with their whips (盛行于中亚等地的马上娱乐项目，其中年青小伙们骑马追逐姑娘并试图亲吻她们，姑娘则以皮鞭回敬) 追姑娘

bridle /ˈbraɪdl/ n. a piece of tack consisting of a head collar, bit and reins, fitted on the horse's head to help a rider control the horse (由笼头、衔铁和缰绳构成的马具，戴在马头上以协助骑手驾驭马匹) 马勒: bridle cheekpiece 马勒颊革 PHR **in the bridle** (of a horse) wearing bridle and bits (指马) 戴勒的

vt. to put or fit bridle on 戴勒，上勒

bridle gate n.【hunting】*see* hunting gate (狩猎场上的) 小门

bridle hand n.【dated】a rider's left hand that holds the reins, thus freeing the right to hold a weapon (骑手持缰的左手，由此右手可持武器) 持缰手，左手

bridle head n. *see* headstall 顶革

B

bridle path n. **1** (also **bridle trail**) a trail or path used only for horseback riding (仅供骑马的小路)马道 **2** the area near the poll just behind the ears where the headstall of bridle passes (项顶供马勒的顶革绕过的部位)勒迹: The mane in bridle path is commonly clipped to prevent hairs from tangling in the bridle. 为了防止鬃毛与马勒缠结,勒迹处的鬃常被剪掉。

bridle teeth n. *see* canine teeth 犬齿

bridle trail n. *see* bridle path 马道

bridled /ˈbraɪdld/ adj. (of a horse) fitted with a bridle (指马)戴勒的

bridle-hook /ˈbraɪdlhʊk/ n. (also **bridle hook**) a hook fixed in the wall for hanging bridles (固定在墙上用来放马勒的挂钩)马勒挂钩

bridlewise /ˈbraɪdlwaɪz/ adj. (said of a horse) trained to obey the aids of the bridle and bit (指马)听勒控制的

bridoon /ˈbraɪduːn; *AmE* brɪˈduːn/ n. (also **bradoon**) a light, jointed snaffle bit used in conjunction with a curb bit in a double bridle (双勒中的绞式轻型衔铁,与链式衔铁共用)小勒

bridoon rein n. the rein attached to the bridoon bit 小勒缰

brier /ˈbraɪə/ n. (also **briar**) thorny bush; wild rose 荆棘;野蔷薇

bright /braɪt/ adj. **1** (of a horse) quick-witted or intelligent (指马)敏快的 **2**【cutting】(of cutting horse) alert in action (指截牛马)警觉的: The horse is bright on the cow. 截牛马对牛警觉性很高。

brisket /ˈbrɪskɪt/ n. the breast of an animal (动物的)胸部

britchin /ˈbrɪtʃɪn/ n. *see* breeching [坐]鞧

British Driving Society n. (abbr. **BDS**) an association founded in Great Britain 1957 to encourage and assist those involved with driving horses and ponies (1957 年在英国成立的组织机构,旨在鼓励驾车爱好者并提供支持)英国马车协会

British Equestrian Federation n. (abbr. **BEF**) an organization founded in 1972 which represents both the British Horse Society and the British Show Jumping Association in policy issues pertaining to the FEI (1972 年在英国建立的组织机构,负责代表英国马会和英国障碍赛协会办理国际马术联合会的相关事务)英国马术联合会

British Equine Veterinary Association n. (abbr. **BEVA**) the governing body representing most veterinarians licensed in Great Britain (代表英国注册兽医的监管机构)英国马兽医协会

British Field Sports Society n. an organization founded in 1930 in Great Britain to encourage interest in all equine field sports (1930 年在英国成立的组织机构,旨在增进人们对场地马术运动的兴趣)英国场地运动协会

British Horse Society n. (abbr. **BHS**) an organization founded in England in 1947 by uniting the National Horse Association of Great Britain and the Institute of the Horse and Pony Club to organize and promote all horse-related activities in Great Britain as well as to regulate the riding instructor standards (英国在 1947 年由英国马匹协会和马与矮马俱乐部研究所合并成立的组织机构,负责组织英国各种与马相关的活动并制定骑术教练标准)英国马会

British Horse Society Assistant Instructor n. (abbr. **BHSAI**) an assistant riding instructor certified by the British Horse Society (由英国马匹协会授予的资格证书)英国马会助理教练

British Horse Society Instructor n. (abbr. **BHSI**) a riding instructor certified by the British Horse Society (由英国马匹协会授予资格证书的骑术教练)英国马会教练

British Percheron Horse Society n. an organization founded in Great Britain in 1919 to establish and maintain the purity of the Percheron in Great Britain (1919 年在英国成立的组织机

构,旨在促进英国佩尔什马的品种保护工作)英国佩尔什马协会

British Riding Club n. an organization founded in Great Britain in 1936 to improve training of the riding horses (1936 年在英国成立的组织机构,旨在提高骑乘用马的训练水平)英国骑乘俱乐部

British Show Hack and Cob Association n. an organization founded in 1938 in Great Britain to promote interest in riding hacks and cobs (1938 年在英国成立的组织机构,旨在促进公众对骑乘马和柯柏马的兴趣)英国乘用马和柯柏马协会

British Show Jumping Association n. an organization founded in 1923 in Great Britain to regulate all show jumping activities conducted there (1923 年在英国成立的组织机构,主要负责监管国内的所有障碍比赛)英国场地障碍赛协会

British Show Pony Society n. an organization founded in Great Britain in 1949 to regulate showing of children's riding ponies (1949 年在英国成立的组织机构,旨在监管儿童骑乘用矮马的展览比赛)英国矮马展览协会

British Spotted Horse n. a British-bred horse type of mixed lineage who carries the spotting gene common in such breeds as the Appaloosa and Knabstrup (英国育成的马匹类型,系谱混杂,携带有阿帕卢萨马和纳普斯特鲁马等马种的斑毛基因)英国斑点马

British Spotted Horse Society n. an organization founded in Great Britain in 1947 to promote the interests of the riding-type British Spotted Horse (1947 年在英国成立的组织机构,旨在促进英国花斑马的相关育种福利工作)英国花斑马协会

British Thoroughbred n. *see* Thoroughbred 英纯血马

British Warmblood n. a breed of warm-blood horse developed in England 英国温血马

brittle feet n. *see* brittle hoofs 干裂蹄

brittle hoofs n. (also **brittle feet**) abnormally dry hoofs which are prone to breaking due to a dry condition of the horn (由于角质干燥而易于开裂的马蹄)干裂蹄

britzka /ˈbrɪtskə/ n. an open, four-wheeled, horse-drawn traveling wagon popular in Poland and Eastern Europe, generally drawn by four or six horses (流行于波兰和东欧国家的敞篷四轮旅行马车,多由四匹或六匹马拉行)敞篷马车

broad bean n. an annual plant of the genus *Vicia faba* in the pea family (一年生野豌豆属豆科植物)蚕豆 [syn] fababean, horse bean

broad bran n. *see* bran 麦麸,稻糠

brocket /ˈbrɒkɪt/ n. a two-year-old male deer 两岁雄鹿 [syn] knobbler

broke /brəʊk; *AmE* broʊk/ adj. (of a horse) tamed and trained (指马)调教好的:broke to ride 调教好可骑乘的 [compare] green, green-broke

broken amble n. *see* rack 单蹄快步

broken canter n. *see* disunited canter 失调跑步,分裂跑步

broken crest n. a heavy neck which falls to one side 颈峰歪斜

broken down adj. (of a horse) in poor physical condition due to injury, trauma or overuse (指马由于受伤或使用过度而体况较差)衰败的,差劣的

broken knee n. the knee of the horse when injured 膝部受伤

broken kneed adj. (of a horse) with knees injured 膝部受伤的

broken wind n. *see* chronic obstructive pulmonary disease 慢性阻塞性肺炎;呼吸阻气

brokk /brɒk/ n. a unique trot of the Icelandic Horse when traversing rough country (冰岛马穿越山地时特有的一种快步)冰岛快步

brome /brəʊm/ n. *see* bromegrass 雀麦[草]

bromegrass /ˈbrəʊmgrɑːs; *AmE* ˈbroʊ-/ n. any of various grasses of the genus *Bromus* native to temperate regions and including several weeds and some species important for forage（雀麦属多种草本，原产于温带地区，包括几种杂草和多种饲草）雀麦［草］：bromegrass hay 雀麦干草

bronc /brɒŋk/ n.【rodeo】（also **bronco**, or **broncho**）a wild or semi-wild horse of western North America entered into the rodeo（源于美国西部的半野生马匹，多用于牛仔竞技比赛）野马

bronc buster n.【rodeo】（also **buster**）one who breaks or trains broncs 驯野马者，驯马师 syn bronc twister, bronc peeler, bronc snapper

bronc busting n.【rodeo】the handling and breaking of broncs so they may be ridden 驯野马

bronc peeler n.【rodeo】*see* bronc buster 驯野马者，驯马师

bronc rider n.【rodeo】one who rides wild, unbroken horses in a rodeo event（牛仔竞技中骑野马的人）骑野马者

bronc riding n.【rodeo】a rodeo event in which a competitor rides a bareback bronc（西部牛仔竞技项目，选手光背骑野马比赛）骑野马比赛：In bronc riding, the rider tests how long he can stay on the bucking horse. 骑野马比赛中，根据骑手在狂跳的马背上停留的时间长短来决定胜负。syn buck jumping

bronc saddle n.【rodeo】a saddle used in breaking horses 调马鞍

bronc snapper n.【rodeo】*see* bronc buster 驯野马者，驯马师

bronc twister n.【rodeo】*see* bronc buster 驯野马者，驯马师

bronchial /ˈbrɒŋkiəl; *AmE* ˈbrɑːŋ-/ adj. of or relating to the bronchi 支气管的

bronchitis /brɒŋˈkaɪtɪs; *AmE* brɑːŋ-/ n. a highly contagious disease marked by acute inflammation of the mucous membrane of the bronchial tubes（一种高度传染性疾病，症状表现为支气管黏膜发炎）支气管炎：equine contagious bronchitis 马传染性支气管炎

broncho /ˈbrɒntʃəʊ; *AmE* -tʃoʊ/ n.（*pl.* **broncos**）*see* bronc 野马

bronch(o)- /brɒntʃəʊ/ pref. bronchus 支气管

bronchoconstrictor /ˌbrɒntʃəʊkənˈstrɪktər/ n. a medication causing the air passages of lung to narrow（促进肺部通气管道收缩的药物）支气管收缩药

bronchodilator /ˌbrɒŋkəʊdaɪˈleɪtə/ n. a medication that causes the air passages of the lung to wide（促进肺部通气管道扩张的药物）支气管扩张药

bronchospasm /ˈbrɒŋkəˌspæzəm/ n. abnormal involuntary contraction of the smooth muscles surrounding the bronchioles（细支气管周围环形肌的异常自主性收缩）支气管痉挛

bronchus /ˈbrɒŋkəs/ n.【anatomy】（*pl.* **bronchi** /-kaɪ/）either of two main branches of the trachea, leading directly to the lungs（气管通向肺的两个主要分支）支气管：primary bronchus 主支气管 ◇ secondary bronchus 次级支气管 ◇ tracheal bronchus 气管支气管

broncobuster /ˈbrɒŋkəʊˌbʌstə; *AmE* -ˈkoʊ-/ n. *see* bronc buster 驯野马者，驯马师

bronk /ˈbrɒŋk/ n. a bronc 野马

broodmare /ˈbruːdmɛə(r)/ n.（also **brood mare**）a mare used specifically for breeding 种用母马，繁育母马

broodmare sire n. *see* dam sire 外祖父马

broom /bruːm/ n. a long-handled tool used for sweeping or cleaning stable and yard（一种长把工具，用来清扫马厩和院落）扫帚
vt. to sweep with a broom 打扫，清扫

broomstock /ˈbruːmstɒk/ n. *see* broomtail 矮野马

broomtail /ˈbruːmteɪl/ n. **1** *see* swith tail 扫帚尾 **2**（also **broomstock**）any small horse, such as the mustangs, not considered worth the effort to break（个头较小而不值调教的马）矬马，矮野马

brother /ˈbrʌðə; *AmE* -ðər/ n. a relationship of male horses by the same sire and out of the same dam（公马之间亲本相同的亲缘关系）兄弟关系— *see also* **full brother**, **half brother**

Brotherhood of Working Farriers Association n.（abbr. **BWFA**）an organization founded in the United States in 1989 of promote farrier education and certification（1989年在美国成立的组织机构，旨在促进蹄铁匠的教育和资格认证）蹄铁匠兄弟会，钉蹄师兄弟会

Brougham /ˈbruːəm/ n. a four-wheeled, horse-drawn carriage designed around 1839 by the British politician Lord Brougham for whom it was named（由英国政客布鲁厄姆在1839年前后设计的四轮马车，故而得名）布鲁厄姆马车

browband /ˈbraʊbænd/ n.（also **brow band**）a leather strap of bridle that lies across the forehead below the ears and above the eyes of the horse, combined with the throat lash to prevent the headpiece from slipping backwards（马勒中处于前额的一条皮革，位于马匹耳朵下方、眼睛上方，与咽革一起可防止顶革后滑）额革

brown /braʊn/ n.【color】a coffee or chocolate coat with black points（体色为咖啡色而末梢为黑色的马匹毛色类型）褐色，褐色：dark brown 黑褐 ◇ light brown 淡褐 ◇ seal brown 海狸褐；深褐

brown saddler n. a saddle maker specializing in saddles, bridles, etc.（专门制作鞍和勒等马具的工匠）鞍具匠 [compare] black saddler

browse /braʊz/ vi. to feed on grass; graze 吃草 [PHR V] **to browse on sth** 吃，采食

brucellosis /ˌbruːsəˈləʊsɪs; *AmE* -ˈloʊ-/ n. an infectious disease of domestic animals caused by brucellae, often resulting in spontaneous abortions in infected animals（由布鲁氏菌引起的家畜传染病，常造成患病母畜胎儿流产）布鲁氏菌病 [syn] Malta fever, undulant fever

bruise /bruːz/ n. an injury to underlying tissues without breaking the skin, often characterized by ruptured blood vessels and discoloration（皮下组织受伤而表皮未破裂的情况，常伴有血管破裂和瘀青）挫伤，青肿

vt. to injure the underlying tissue by a blow or abrasion without laceration（皮下组织受伤而未撕裂）挫伤，青肿

bruised sole n. a condition in which the sole is injured by striking with hard object or poor shoeing（马蹄与硬物相撞或装蹄不当所致的）蹄底挫伤

Brumby /ˈbrʌbi/ n.（also **Brumbie**）a wild horse indigenous to Australia descended from domestic horses turned loose on the ranges during the mid-19th century; stands about 15 hands and may have a coat of any color; is hardy, unrefined, and has a rather heavy head, short back, and sturdy legs（澳大利亚本土的野马种群，于19世纪中叶由家马放生繁衍而来。体高约15掌，各种毛色兼有，抗逆性强，外貌粗糙，头较重，背短，四肢粗壮）澳洲野马：Brumbies were once brought close to extinction due to organized culling of the wild herds in the 1960s. 20世纪60年代由于人类对野生种群的杀戮，澳洲野马几近陷入灭绝的境地。

Brumby runner n. an Australian horseman who captures and tames Brumbies（澳大利亚追捕并调教野马的骑手）捕野马者，驯野马者

brush /brʌʃ/ n. **1** a hand-held grooming tool consisting of bristles set in a block of wood or plastic, used for cleaning or polishing horses（木质或塑料块上嵌有硬毛的刷拭工具，主要用来清洁马体）刷子：body brush 体刷 ◇ dandy brush 硬毛刷 **2**【hunting】the tail of a fox 狐尾 **3** a dense growth of bushes or shrubs; undergrowth 灌木丛，树篱：brush fence 树篱障碍 **4**【racing】the brief peak of speed reached by a horse in a race or training（马在比赛或训练中短期的高速奔驰）疾驰

vt. **1** to clean or sweep (a horse) with a brush; groom 刷拭 **2**【racing】(of two horses) touch

B

or collide slightly while running（两匹马）轻撞，轻碰 3【jumping】（of a horse）to touch but does not drop a rail when jumping over a fence（马在跨越障碍时）触击跨栏 4 to run a horse at top speed over a short distance 策马疾驰

brush boots n. *see* boots 防蹭护腿

brush fence n.【jumping】a natural or artificial jumping obstacle consisting of brush wood（由灌木构成的自然或人工障碍）灌木跨栏

brushing /ˈbrʌʃɪŋ/ n. a gait fault in which the forelegs or hindlegs scrape or strike slightly（一种运步缺陷，其中马的前肢或后肢间发生蹭撞）蹭蹄：brushing bandages 防蹭绷带

brushing boots n.（also **brush boots**）special boots used to protect the horse's legs from injury due to brushing（用来保护马四肢以防蹭蹄造成损伤的护腿）防蹭护腿 syn splint boots

brushing ring boot n. *see* fetlock ring boot 护蹄圈

BSJA abbr. British Show Jumping Association 英国障碍赛协会

bucca /ˈbʌkə/ n.【Latin】cheek 颊

buccal /ˈbʌkəl/ adj.【anatomy】of or relating to the cheeks or the mouth cavity 颊的，口腔的：buccal infections 口腔感染

buccinator /bʌksɪˈneɪtə/ n.【anatomy】the thin, flat muscle forming the wall of the cheek（构成口腔壁的薄层肌肉）颊肌：buccinator muscle 颊肌

Bucephalus /bjuːˈsefələs/ n. the war horse ridden by Alexander the Great（亚历山大大帝所骑的战马）布赛弗勒斯：It was recorded that Bucephalus died at the age about twenty-one in 326 BC. 据历史记载，布赛弗勒斯于公元前326年离世，时年21岁。

buck[1] /bʌk/ vi. **1**（of a horse）jump in the air with its back arched and its head dropped（指马弓背低头跳起）弓背跳，尥蹶子：Wild horse usually bucks as a natural response to fend off an intruder which the horse believes is dangerous. 野马弓背跳是对其所认为的危险入侵者作出的自然反应。IDM **buck itself out**（of a horse）to stop bucking due to exhaustion or because it cannot unseat the rider（指马）弓背跳累 **2 buck sb（off）** throw the rider to the ground by jumping like this（将骑手）撂下马：to buck off a rider 把骑手撂下马
n. the act or instance of bucking 弓背跳，尥蹶子：Buck is an unwanted behavior in riding horses, but is desired in some rodeo events. 弓背跳在乘用马中视为恶习，但在有些牛仔竞技项目中却希望发生。

buck[2] /bʌk/ vi.（of a horse）to suffer from bucked shin（马）患胫凸：the young Thoroughbred bucked 这匹年青的纯血马患了胫凸

buck[3] /bʌk/ n. **1** an adult male deer or hare; stag 雄鹿；公兔 **2** the lower part of the body of a horse-drawn carriage 马车底盘

buck eye n. *see* toad eye 眼球外突，蛤蟆眼

buckaroo /ˈbʌkəruː/ n.（*pl.* **buckaroos**）a cowboy or bronc buster 牛仔；驯马师 syn vaquero

buckboard /ˈbʌkbɔːd; *AmE* -bɔːrd/ n. a four-wheeled open carriage popular in the United States throughout the second half of the 19th century; was generally drawn by a single horse or pair（19世纪下半叶流行于美国的四轮客运马车，多由单马或两匹马拉行）平板马车，桌式马车

bucked knees n. a conformation defect in which the knees bend forward in front of the vertical line of the leg（马前膝外凸超出前肢垂线的体格缺陷）凸膝 syn forward deviation of the carpal joints, goat knees, over at the knees

bucked shins n. an unsoundness of the horse's forelimbs in which the shins buck as a result of concussion or injury（马前肢的一种缺陷，多由重撞或受伤致使胫骨外凸）胫凸，胫痛 syn shinbuck, sore shins

buckeroo /ˈbʌkəruː/ n.（*pl.* **buckeroos**）buckaroo 牛仔

bucket[1] /ˈbʌkɪt/ n. a cylindrical vessel used for

watering horses; pail(用来饮马的圆柱形容器)水桶:water buckets 水桶

bucket² /ˈbʌkɪt/ v.【racing】to ride a horse hard 策马飞奔;疾驰:The jockeys bucketed their mounts along the homestretch towards the finish. 在迈向终点直道时,骑手们策马疾驰。

bucketful /ˈbʌkɪtfʊl/ n. the volume that can be contained by a bucket 一桶(的量):The horse drank three bucketfuls of water. 这匹马喝了三桶水。

buckhound /ˈbʌkhaʊnd/ n.【hunting】a hound used to hunt deer 猎鹿犬

buckjump /ˈbʌkˌdʒʌmp/ vi. (of a horse) to buck 弓背跳

buckjumper /ˈbʌkˌdʒʌmpə(r)/ n. a horse who bucks 弓背跳的马

buckjumping /ˈbʌkˌdʒʌmpɪŋ/ n. **1** the act of a horse who arches his back and jumps into the air of all four legs at once repeatedly(马弓背不断四肢腾空跳起的动作)弓背跳 **2**【rodeo】the bronc riding 骑野马比赛

buck-knee /ˈbʌkniː/ n. the knee of a horse's foreleg that bends forward when looked at from the side(侧观时马的前膝向前弯曲)凸膝 syn calf-knee

buck-kneed /ˈbʌkniːd/ adj. (of a horse) having buck-knees 凸膝的

buckle /ˈbʌkl/ n. a clasp used to fasten two loose ends together(用来系两皮带末梢的搭扣)带扣

vt. to fasten or join together with a buckle 扣系,扣紧

buckle guard n. (also **buckle safe**) a piece of leather fitted over a buckle to prevent it from rubbing on the horse's hide(遮在带扣上的皮革,可防止马被蹭伤)带扣盖,带扣护罩

buckle safe n. *see* buckle guard 带扣盖

buckskin /ˈbʌkskɪn/ n.【color】a type of dun having a grayish-brown yellow coat with black points(一种兔褐毛类型,马被毛为淡褐黄色,而末梢为黑色)鹿皮色,沙黄:dusty buckskin 土沙黄 ◇ silver buckskin 银沙黄

buddy sour adj. (of a horse) objecting to be ridden from or led away from its pasture mates(指马抗拒牵行而离开厩伴)恋伴的:a buddy sour horse 一匹恋伴的马 compare barn sour

Budenny /ˈbuːdəni/ n. *see* Budyonny 布琼尼马

budget /ˈbʌdʒɪt/ n. a leather storage box attached to the fore-carriage of a horse-drawn vehicle(马车前身安放的皮革储物箱)物品箱,储物箱

Budyonny /buːˈdjɔːni/ n. (also **Budenny**) a Russian warm-blood developed at a military stud in Rostov in the 1930s by putting Thoroughbred stallions to Don mares; stands 15.1 to 16 hands and generally has a chestnut coat, although bay and gray also occur; has a close-coupled body with deep chest, strong legs, an elegant head, and long neck; is calm, sensible, and has great stamina; used for steeplechasing, jumping, and riding(苏联在20世纪30年代采用纯血马与顿河马育成的军用马品种。体高15.1~16掌,毛色以栗为主,骝和青亦有之,体格紧凑,胸身宽广,四肢结实,面目清秀,颈长,性情安静,反应敏捷,耐力好,主要用于越野障碍赛、障碍赛和骑乘)布琼尼马

buffer /ˈbʌfə(r)/ n.【farriery】*see* clinch cutter 钉节刀

bug boy n.【racing】*see* apprentice jockey 见习骑师

buggie /ˈbʌgi/ n.【colloq】*see* buggy 轻型马车

buggy /ˈbʌgi/ n. **1** (also **buggie**) a light, two-wheeled, English horse-drawn vehicle popular in the early 19th century, having curved shafts and a folding head(流行于19世纪初的一种英式双轮轻型马车,辕弧形,带折叠顶篷)轻型马车 syn hooded gig **2** the American buggy 美式轻型马车

build /bɪld/ n. the physical makeup of a horse (马的)体格:a horse of solid build 一匹体格健壮的马

bulb /bʌlb/ n.【anatomy】(also **heel bulb**) the rounded projection of the heel 蹄球

bulbus /ˈbʌlbʌs/ n.【Latin】(*pl.* **bulbi** /-bai/) a bulb-like structure of an body organ (身体器官上的球状结构)球: bulbus oculi 眼球 ◇ bulbus olfactorius 嗅球

bulk /bʌlk/ n. the main or large part of sth; stem 主干;茎

bulky /ˈbʌlki/ adj. having huge bulks; coarse 粗大的;粗的: bulky feed 粗饲料;块状饲料

bull /bʊl/ n. **1** an adult male cattle 公牛 **2** an adult male horse; stallion 公马,雄马

bull dogger n.【rodeo】*see* steer wrestler 摔牛者

bull neck n. a short, thick neck of horse that is difficult to flex (马粗而短的颈)牛颈

bull pen n. **1** an enclosed area where a horse can be worked or trained (用来训练马匹的围栏)调马圈 [syn] round pen **2** an auction ring 拍卖场 **3**【AmE, racing】a small racetrack usually less than one mile in length with sharp turns (全长不足1.6千米且弯度很急的小型赛马场)小赛场 [syn] bull ring

bull riding n. a rodeo event in which the contestant rides a bull with only one hand holding a rope wrapped around its chest (西部牛仔竞技项目,其中参赛者骑着公牛,仅用单手抓着系在牛肩处的皮带)骑公牛

bulldog /ˈbʊldɒg/ n. a sturdy dog with a large head, a wrinkled face, and a thick neck (一种体格结实的犬,头大,面部皱褶,颈粗)斗牛犬

vi.【rodeo】to throw a calf or steer to the ground by seizing its horns and twisting its neck (抓住牛角,扭住牛颈,将其撂倒在地)摔牛,斗牛

bull-dog bite n. *see* under-shot jaw 下颌突出

bulldogging /bʊlˈdɒgɪŋ/ n.【rodeo】*see* steer wrestling 摔牛

bulldogging horse n.【rodeo】a horse used for steer wrestling 摔牛用马

bullfight /ˈbʊlfaɪt/ n. a public performance popular in Spain in which bulls are fought and usually killed (流行于西班牙的一种公开表演,其中一头公牛被挑逗并最终被杀)斗牛 | **bullfighting** n.

bullfighter /ˈbʊlfaɪtə(r)/ n. one who performs in a bullfight 斗牛士 — *see also* **matador, picador**

bullfinch /ˈbʊlfɪntʃ/ n.【jumping】a jumping obstacle consisting of a thick, overgrown live hedge (一道由浓密天然树篱构成的障碍)树篱障碍: black bullfinch 黑树篱障碍 ◇ hairy bullfinch 毛树篱障碍 ◇ Due to its height, the bullfinch could be jumped through by horses. 由于高度较高,树篱障碍可以从中间跨越。

bullock /ˈbʊlək/ n. **1** a castrated bull; a steer 阉牛;肉牛 **2** a young bull 青年公牛

Bullock dray n. a heavy, two-wheeled, flat vehicle pulled by a team of 8 to 12 bullocks (一种重型双轮平板车,多由8~12头阉牛拉行)牛拉平板车: A large log usually would be dragged along behind the brakeless Bullock dray when descending a steep hill. 不带刹车的牛拉平板车下山时,后面常拖一根巨大的圆木。

bullring /ˈbʊlrɪŋ/ n.【racing】*see* bull pen 小型赛马场

bump /bʌmp/ n.【polo】the collision between players in a polo match (马球运动中球员相互碰撞)相撞,冲撞: The angle of bump should not exceed 45 degrees in polo. 马球比赛中球员撞击角度不得超过45°。

vi. to strike or collide with each other 撞击,冲撞

bumper /ˈbʌmpə(r)/ n.【racing】an amateur rider 业余骑手

bunch /bʌntʃ/ n. a wisp of hair tied together; bundle, cluster 束;绺

vt. to tie or bind together 扎成束

Buraq /bʊˈrɑːk/ n.【mythology】(also **Al-Buraq**) a winged horse that the Prophet Mohammad rode to Heaven; is said to be white in color with a human head and possessed of incredible speed (伊斯兰神话中先知穆罕默德升天所骑的一匹飞马,据说体色为白,头如人形,且速力非凡)先知坐骑;布拉卡

Burchell's zebra n. *see* common zebra 普通斑马

Burford saddle n. a military saddle used in the early 17th century (17 世纪初使用的一种军鞍)伯弗德鞍 syn Morocco saddle

Burghley Horse Trials n. a three-day event held annually since 1961 in Burghley, UK (自 1961 年起每年在英国伯利举行的一项三日赛)伯利三日赛

Burmese pony n. a mountain pony bred primarily in the Shan state of Eastern Burma by the local tribes; stands about 13 hands and generally has a bay or chestnut coat; has a light head, fairly straight back, sloping croup, strong legs with hard feet; used for packing and riding (缅甸东部掸州山区培育的矮马品种。体高约 13 掌,毛色多为骝或栗,头轻,背直,斜尻,四肢健壮,蹄质结实,主要用于驮运和骑乘)缅甸矮马 syn Shan pony

burning /ˈbɜːnɪŋ/ adj.【hunting】(of a smell or scent) strong or intense; blazing 浓郁的,浓烈的:Chasing after the burning scent, the greyhound found the burrow of the hare. 追随着猎物浓烈的气味,灰狗找到了野兔的巢穴。

burro /ˈbʌrəʊ/ n.【AmE】(*pl.* **burros**) an ass or donkey used as a pack animal (作为驮畜使用的驴)驮驴

burrow /ˈbʌrəʊ; *AmE* ˈbɜːroʊ/ n. a hole dug in the ground by a small animal for habitation or refuge (小动物在地下挖的洞)地洞,巢穴

vt. to dig a hole or tunnel in or through sth. 挖洞,钻洞:Some parasites burrow into the wall of intestines and cause severe damage to the guts. 有些寄生虫钻到动物的肠壁内,使肠道严重受损。

bursa /ˈbɜːsə; *AmE* ˈbɜːrsə/ n.【anatomy】(*pl.* **bursae**/-siː/) a sac or saclike bodily cavity 囊,黏液囊

bursitis /ˌbɜːˈsaɪtɪs; *AmE* ˌbɜːrˈ-/ n. inflammation of a bursa, characterized by swelling, local heat, and pain (关节黏液囊发炎的症状,临床表现为肿胀、局部发热和疼痛)黏液囊炎,滑膜囊炎

burst /bɜːst; *AmE* bɜːrst/ n.【hunting】a fast, quick chase of a fox 追赶,猛追

vi. (of the hounds) to run after or chase a quarry suddenly(指猎犬)追赶,追击

bush track n.【racing】an unofficial or illegal race meeting 非法赛马,非正式赛马

buster /ˈbʌstə(r)/ n. *see* bronc buster 骑野马者,驯野马者

Butazolidin /ˌbjuːtəˈzɔːlɪdɪn/ n. a brand name for phenylbutazone, used to reduce pain and inflammation (苯基丁氮酮的商品名,可减轻疼痛并消除炎症)布他酮,保泰松

Butazone /ˈbjuːtəˌzəʊn; *AmE* -ˌzoʊn/ n. another brand name for phenylbutazone (苯基丁氮酮的商品名)布他酮,保泰松

butcher boots n.【hunting】black boots without tops worn in hunting (狩猎时穿的黑色马靴)矮帮马靴

bute /bjuːt/ n. the commercial name for phenylbutazone (苯基丁氮酮的商品名)保泰松

butteris /ˈbʌtərɪs/ n.【farriery】a handled tool with a sharp blade, formerly used for trimming hoofs of horse (过去装蹄工用来削蹄的带柄工具)蹄铲

buttermilk horse n. *see* Palomino horse 帕洛米诺马,乳色马

buttermilk roan n. *see* Sabino 白化马

buttero /ˈbʌtərəʊ/ n. a shepherd or cowboy in Tuscany, Italy (意大利托斯卡纳地区的牧人或牛仔)意大利牛仔:The buttero habitually rides a Maremmana to tend cattle. 意大利牛仔通常骑一匹摩雷曼纳马放牛。

buttock /ˈbʌtək/ n. either of the two rounded prominences of the horse's rear body that are posterior to the hips (马体后躯位于腰角后的圆形突起)臀部:point of buttock 臀端

B

buttress foot n. *see* pyramidal disease 锥蹄病

buy-back /ˈbaɪˌbæk/n.【racing】a condition in which a horse is retained in public auction because it did not reach the reserve set by the owner（指马在拍卖会上由于竞拍价格低于底价而未出售的情况）竞拍失败；回购 [syn] charge-back

Buzkashi /ˌbuːzˈkɑːʃiː/ n.（also **Buz Kashi**）a mounted game popular in Central Asia in which riders struggle to grab a stuffed goat skin from others on galloping horses without dismounting（盛行于中亚地区的马上运动，骑手们策马争抢填满沙土的羊皮）马背叼羊，叼羊大赛 [syn] goat grabbing, goat snatching

buzzer /ˈbʌzə(r)/ n. an electric device that gives a buzzing sound or produces electric shock, mainly used to frighten a horse into running faster or jumping in training（可发出嗡鸣声或产生电脉冲的电子装置，训练马匹时用来刺激其跑得更快或起跳）蜂鸣器

BVMS abbr. Bachelor of Veterinary Medicine and Surgery 兽医内外科学士

BVSc abbr. Bachelor of Veterinary Science 兽医学学士

BWFA abbr. Brotherhood of Working Farriers Association 蹄铁匠兄弟会，钉蹄师兄弟会

Byerley Turk n.（c. 1680 – c. 1706）one of three foundation sires of the English Thoroughbred; was a dark bay Arabian horse captured from the Turks at the Battle of Buda（1686）and brought back to England by Captain Robert Byerley for whom it was named（育成纯血马的三大奠基公马之一，在1686年的布达佩斯战争中，这匹骝色的阿拉伯马由统帅罗伯特·比艾而力从土耳其人手中夺得并带回英国故而得名）比艾而力·土耳其 — *see also* **Darley Arabian**, **Godolphin Arabian**

C

C /siː/ abbr.【racing】colt 雄驹

CAA abbr. Carriage Association of America 美国马车协会

cab /kæb/ n. a two-wheeled, horse-drawn vehicle used widely in Europe for public hire during the 19th century (19 世纪在欧洲广泛使用的一种双轮公用马车)出租马车

caballero /ˌkæbəlˈjerəʊ/ n.【Spanish】(*pl.* **caballeros**) a skilled horseman or cowboy (西班牙的)骑手,牛仔

cable link n. *see* cable-link snaffle 铰链衔铁

cable-link snaffle n. (also **cable link**) a snaffle bit with its mouthpiece consisting of a rough-link chain (口衔由铰链构成的一种衔铁)铰链衔铁

cabriole /ˈkæbriəʊl/ n. *see* cabriolet 轻型出租马车

cabriolet /ˈkæbriəʊleɪ; *AmE* -oʊleɪ/ n. (also **cabriole**) a two-wheeled, open, horse-drawn carriage with a shell-shaped body and a folding hood, used widely in 19th-century England by the upper class (一种敞篷双轮马车,车体呈贝壳状,带折叠式车篷,19 世纪英国上流阶层使用较为广泛)轻型双轮马车

Cachar Polo Club n. a polo club founded in Cachar, India in 1859 by British officers (于 1859 年在印度喀查成立的马球俱乐部)喀查马球俱乐部

cactus cloth n. a coarse fabric used to rub out dirt and sweat from the horse's coat (用来擦去马身上污渍的织物)粗布,麻布

cad /kæd/ n. **1**【BrE, slang】a driver of horse buses in London during the 19th century (19 世纪英国伦敦驾驶公共马车的人)车夫 **2**【dated】one who did miscellaneous tasks in a stable (马厩的)杂役,勤杂工

cade /keɪd/ adj.【BrE】(of a foal) reared by hand; bottle-raised (马驹)人工喂养的: a cade foal 人工喂养的幼驹

cadence /ˈkeɪdns/ n. the rhythm in the gait of a horse (马运步中的)节奏

Cadre Noir /ˈkɑːdə(r) nwɑː(r)/ n.【French】an equestrian team of the French Cavalry School founded at Saumur in 1828, which gets its name from the black uniform the horsemen wear (1828 年在法国索米尔建立的骑术学校,操练古典骑术的骑兵团,名称源于骑手们的黑色制服)黑骑士

Caesarean /sɪˈzeəriən; *AmE* -ˈzer-/ adj. of caesarean section 剖宫产的,剖腹产的

Caesarean section n. (also **Cesarean section**) a surgical incision through the abdominal wall and uterus performed to deliver a fetus (切开腹腔和子宫以利胎儿分娩的外科手术)剖宫产,剖腹产

CAF abbr. Certified Apprentice Farrier 见习蹄铁匠资格证书

Calabrese /ˈkæləbriːs/ n. a warmblood indigenous to Calabria, Italy; originally descended from Arab and was later crossed with Andalusian and Thoroughbred blood; has a well-formed head, prominent withers, slanted croup, short loins, and is short-coupled; stands 16 to 16. 2 hands and may have a brown, black, bay, chestnut, or gray coat; used mainly for riding (源于意大利卡拉布里亚地区的温血马品种,最初由阿拉伯马培育而来,后引入安达卢西亚马和纯血马进行过杂交改良。体格良好,鬐甲突出,斜尻,腰短,体高 16 ~ 16. 2 掌,毛色可为褐、黑、骝、栗或青,主要用于骑乘)卡拉布里亚马

Calash /kəˈlæʃ/ n. (also **Calèche**) a two- or four-wheeled, hooded, horse-drawn vehicle (一种两轮或四轮带篷马车)带篷马车

C

calcaneal /kælˈkeɪnɪəl/ adj.【anatomy】of or relating to the heel bone 跟骨的: The common calcaneal tendon is the aggregate of tendons attached to the point of hock, which includes Achilles' tendon, superficial digital flexor tendon and the accessory or tarsal tendon of the hamstring muscle. 总跟腱由连接在飞节端的多个筋腱融合构成,其中包括跟腱、趾浅屈肌腱和胭肌跗腱。

calciferol /kælˈsɪfərəʊl/ n.【nutrition】a kind of fat-soluble vitamin that functions in the absorption, uptake and transportation of calcium (一种脂溶性维生素,参与钙的摄入、吸收和运输)钙化醇,骨化醇: Calciferol has two forms, ergocalciferol and cholecalciferol. 钙化醇有角钙化醇和胆钙化醇两种形式。 syn vitamin D

calcification /ˌkælsɪfɪˈkeɪʃn/ n. the process of deposition of calcium salts 钙化作用

calcify /ˈkælsɪfaɪ/ vt. to make or become bony or stiff by deposition of calcium salts (骨由于钙质沉淀而变硬)钙化

calcitonin /ˌkælsɪˈtəʊnɪn/ n. a polypeptide hormone produced by the thyroid gland that lowers plasma calcium and phosphate (由甲状腺分泌的多肽类激素,可降低血液中的钙、磷水平)降钙素 syn thyrocalcitonin

calcium /ˈkælsɪəm/ n. a mineral element essential in the feed of horses (马匹日粮中的必需矿物质元素)钙

calcium blood level n. the amount of the mineral calcium carried in the blood (血液中的钙质含量)血钙水平

calcium-phosphorus ratio n.【nutrition】the amount of calcium compared to phosphorus in the horse's diet (马匹日粮中钙与磷的含量比例)钙磷比: An improper calcium-phosphorus ratio may result in serious diseases or physical problems such as osteomalacia, osteoporosis, osteofibrosis, crooked or enlarged joints or weakening of the bones, shifting lameness, especially connected with the hocks, and rickets. 钙磷比例搭配不当就可能导致严重疾病或体况下降,常见的如骨软化、骨质疏松、骨纤维化、关节变形增大、骨纤弱、飞节颠跛和佝偻病。

calculator /ˈkælkjuleɪtə(r)/ n.【racing】a clerk responsible for computing pari-mutuel odds in the past (过去赛马场负责计算投注赔率的雇员)赌金计算师

Calèche /kɑːˈleʃ/ n. *see* Calash 带篷马车

Calesa /kəˈleɪsɑː/ n. a two-wheeled, horse-drawn passenger cart or heavy gig used in the Philippines from the mid-1800s through the 1940s (19世纪中叶至20世纪40年代在菲律宾使用的双轮客用马车)两轮马车

calesh /kəˈleʃ/ n. the folding top on a horse-drawn carriage (马车的)折叠车篷

Calesso /kəˈlesəʊ/ n. a two-wheeled, horse-drawn, hooded gig popular in Italy from the late 1700s through the early 1900s (从18世纪末至20世纪初流行于意大利的双轮带篷轻便马车)带篷马车

calf /kɑːf/ n. **1** a young cow in its first year (不满1岁的小牛)犊牛 IDM **in/with calf** (of a cow) pregnant (母牛)怀犊的 **2** *see* gaskin 小腿

calf horse n.【roping】a horse used for calf-roping 套牛马

calf knees n. a conformation defect in which the knees bend backward into a concave shape when viewed from the side (侧观时马前膝向内凹陷的肢体缺陷)凹膝 syn back at the knee, backward/palmar deviation of the carpal joints, sheep knees

calf-kneed /ˈkɑːfniːd/ adj. (of the knees) bending backward into a concave shape 凹膝的

calf-rope /ˈkɑːfrəʊp/ vt.【roping】to rope a cow using a lariat 套牛

calf-roping n.【rodeo】(also **tie-down roping**) a rodeo event in which a mounted rider gallops to rope a calf using a lariat and dismounts to tie the calf by three legs with a string (牛仔竞技

项目之一,其中骑手策马掷出套索捕捉小牛并下马用绳捆绑牛犊的三肢)套小牛,套牛

calk /kɔːk/ n.【farriery】(also **caulk**, or **calkin**) a cleat or grip formed on the heels of horseshoes to increase friction (蹄铁踵的铁棱或抓手,主要用来增加摩擦力)铁棱,铁刺: Calk was used as early as in the 10th century and is largely replaced by studs now. 蹄铁刺早在10世纪就已开始使用,目前大多已经被防滑钉所取代。 syn sticker, cork, mud sticker

calkin /ˈkɔːkɪn/ n. *see* calk 铁棱,铁刺

calking /ˈkɔːkɪŋ/ n. an injury to the coronary band caused by the calks on horseshoes (由带铁刺的蹄铁在马蹄冠部形成的伤痕)铁刺伤

calking boots n. *see* stable boots 厩用绑腿

caloric /kəˈlɒrɪk/ adj. of or relating to heat 热的,产热的

calorie /ˈkæləri/ n.【nutrition】the unit of heat (热量单位)卡

Camargue /kəˈmɑːrg/ n. (also **Camargue pony**) a pony breed originating in the Camargue area of Southern France; has a large head, short ears, a short neck thick at the base, deep girth, straight back, long loins, and full yet shaggy mane; stands 13.1 to 14.2 hands and usually has a white or light gray coat, although bay and brown do occur rarely; is fiery, independent, and courageous; used for riding and packing (源于法国南部卡马格地区的矮马品种。头大,耳短,颈粗短,胸深,背平直,腰长,鬃浓密而杂乱,体高13.1~14.2掌,毛色多为白或淡青,骝和褐少见,悍性高,独立性强而勇猛,主要用于骑乘和驮运)卡马格矮马

Camargue pony n. *see* Camargue 卡马格矮马

camped /kæmpt/ adj. (of a horse) having the front limbs placed too far forward or the hind limbs too far backward when viewed from the side (侧观时马前肢姿势过于靠前或后肢姿势过于靠后)外踏

camped behind adj. (also **camped out**) a conformation defect in which the hind limbs are placed too far backward when viewed from the side (侧观时马后肢姿势过于外张的体格缺陷)后肢外踏 compare stand under behind

camped in the front adj. a conformation defect in which the forelegs are placed too far forward when viewed from the side (侧观时马前肢姿势过于外张的四肢缺陷)前肢外踏 compare stand under in the front

camped out adj. *see* camped behind 后肢外踏

Canadian Cutting Horse n. a Canadian warmblood descended from the American Quarter horse; stands 15.2 to 16.1 hands and may have a coat of any color; is intelligent, fast, and agile; used for riding and the rodeo discipline of cutting (由美国夸特马培育而来的温血马品种。体高15.2~16.1掌,各种毛色兼有,反应敏快,步法迅捷,用于骑乘和截牛比赛)加拿大截牛马

Canadian Pacer n. a pacing breed developed in Canada during the first half of the 19th century (19世纪上半叶在加拿大育成的对侧步马种)加拿大侧步马

canal /kəˈnæl/ n.【anatomy】a tube, duct, or passage 管,道:birth canal 产道 ◇ inguinal canal 腹股沟管

canine /ˈkeɪnaɪn/ adj. of or related to dogs 犬的 n. (also **canine tooth**) one of the pair of pointed teeth located between the incisors and the first premolars (位于切齿和第一前臼齿之间的牙齿)犬齿

canine tooth n. *see* canine 犬齿 syn wolf teeth

canker /ˈkæŋkə(r)/ n. a chronic deterioration of the horn-producing tissues of the hoof (马蹄角质层逐渐腐败的病症)蹄叉糜烂,蹄溃烂

cannon bone n. a bone between the knee and fetlock of horse (位于膝关节与球节之间的骨)管骨 syn shin bone

cannon bone circumference n. a measurement taken around the cannon bone (绕管骨一周的尺寸)管围

canopy /ˈkænəpi/ n. an ornamental covering for a

horse-drawn carriage (马车的)华盖,顶盖

canteen /kæn'tiːn/ n. a small watter vessel, used by soldiers and riders (士兵或骑者使用的)水壶

canter /'kæntə(r)/ n. a three-beat natural gait of the horse (马的三蹄音步法)跑步: left-lead canter 左起跑步,左领跑步 ◇ right-lead canter 右起跑步,右领跑步 ◇ The canter is known to be the most comfortable for both the horse and rider. 无论是对人还是对马,跑步都是最为舒适的步法。PHR V **win in a canter**【racing】(of a horse) to win a race with ease (指马)以跑步获胜;轻松取胜— *see also* **walk, trot, gallop**

cantle /'kæntl/ n. the raised rear end of a saddle 后鞍桥 compare pommel

cap fee n.【hunting】the fee paid by one to participate in a day of hunting (参加狩猎缴纳的费用)人头费

Cape Cart n. (also **Cape Wagon**) a large, two-wheeled, horse-drawn country cart used in South Africa (南非乡村使用的大型双轮马车)海角马车

Cape Horse n. a horse breed originating in South Africa, descended from Barb and Arab horses imported by the Dutch East India Company in the 1650s (源于南非的马品种,由17世纪中叶荷兰东印度公司进口的柏布马和阿拉伯马繁衍而来)海角马

Cape Wagon n. *see* Cape Cart 海角马车

capel /'kæpəl/ n. **1** a heavy farm horse used in the Middle Ages (中世纪的)役用马,农用马 **2** a horse; nag 马;驽马

capellet /'kæpəlɪt/ n.【vet】a swelling on the point of the elbow or point of the hock (马前肘或飞节肿胀的症状)前肘肿;飞节肿

caper /'keɪpə(r)/ n. a playful leap or hop 蹦跳 vi. (of a colt or filly) to leap or frisk about; frolic 欢蹦乱跳

capillary /kə'pɪləri; *AmE* 'kæpəleri/ n.【anatomy】(also **blood capillary**) one of the minute blood vessels that connect arterioles and venules (连接微动脉和微静脉的微血管)毛细血管: The blood capillaries form an intricate network throughout the body for the interchange of various substances, such as oxygen and carbon dioxide, between blood and tissue cells. 毛细血管在体内形成复杂的毛细血管网,使氧气、二氧化碳等各种物质可在血液和组织细胞间进行交换。

adj. of or relating to the capillaries 毛细血管的: capillary network 毛细血管网 ◇ capillary pressure 毛细血管压

capped elbow n. a swelling at the point of the elbow 肘关节肿胀

capped hock n. a swelling or puffiness at the point of the hock 跗关节肿胀,飞节肿胀: Capped hocks are often caused by a blow or injury, or from laying down on insufficient bedding. 飞节肿胀多由于撞击、受伤或垫料不足所致。

Caprilli, Federico n. (1868 – 1907) an Italian cavalry officer who introduced the forward seat jumping technique in 1907 (意大利骑兵部队军官,于1907年最先采用前倾式跨栏姿势)费德里克·卡普瑞利

capriole /'kæpriəʊl/ n. one of the high airs in which the horse leaps into the air with all four legs off the ground and kicks out with it's hind legs in mid-leap (高等骑术中马表演的空中高难动作之一,其中马四肢腾空跳起的同时后肢向后踢开)腾空后踢

cardi(o)- /'kɑːdɪəʊ/ pref. of heart 心脏的

cardiogram /'kɑːdɪəgræm/ n. *see* electrocardiogram 心电图

cardiograph /'kɑːdɪəgrɑːf/ n. *see* electrocardiograph 心电图仪

cardiovascular /ˌkɑːdiəʊ'væskjələ(r)/ adj. of, relating to, or involving the heart and the blood vessels 心血管的: cardiovascular disease 心血管疾病

cardiovascular system n. the system including

the heart and the blood vessels that circulates blood through the body (机体包括心脏和血管在内的系统,血液借此在周身循环)心血管系统 syn circulatory system

caries /ˈkeəriːz; *AmE* ˈker-/ n. (*pl.* **caries**) *see* tooth decay 龋齿,蛀牙

carnivore /ˈkɑːnɪvɔː(r); *AmE* ˈkɑːr-/ n. an animal that eats meat 食肉动物 compare herbivore, omnivore

carotene /ˈkærətiːn/ n.【nutrition】a pigment found in animal tissue and certain plants that provides the source for vitamin A synthesis (见于动植物中的一种色素,是合成维生素 A 的原料)胡萝卜素

carotid /kəˈrɒtɪd/ n.【anatomy】either of the two major arteries on each side of the horse's neck (位于马颈部两侧的主要动脉)颈动脉

adj. of or relating to either of these arteries 颈动脉的

carpal /kɑːpl; *AmE* kɑːrpl/ adj. of or relating to the carpus 腕骨的;腕的: carpal region 腕部 ◇ carpal joints 腕关节;膝关节

Carpathian /kɑːˈpeiθiən/ n. the Carpathian pony 喀尔巴阡矮马

Carpathian pony n. (also **Carpathian**) *see* Hucul (胡克尔矮马的别名)喀尔巴阡矮马

carpetgrass /kɑːpɪtˈgrɑːs/ n. (also **carpet grass**) any of several coarse, sod-forming grasses cultivated for turf and pasture in warm, humid regions (温带地区用于草坪和牧场种植的禾本类植物)地毯草

carpitis /kɑːˈpaɪtiːz/ n. inflammation of the knee joint of horse's foreleg, resulting from torn ligaments or fracture of any of the carpal bones (马匹前肢膝关节的炎症,多由韧带拉伤或某块掌骨骨折所致)膝关节炎 syn big knee, popped knee

carpus /ˈkɑːpəs/ n.【anatomy】the group of bones forming the knee joint of horse's foreleg (构成马前膝关节的多个骨)腕骨;膝关节: The carpus is composed of seven bones, including the radial carpal bone, intermediate carpal bone, ulna carpal bone and the accessory carpal bone; the 2nd carpal bone, 3rd carpal bone and the 4th carpal bone. 腕骨有七块骨头组成,包括桡侧腕骨、中间腕骨、尺侧腕骨和副腕骨,以及第二腕骨、第三腕骨和第四腕骨。

carpus varus n. *see* bowed knees 膝弓形,膝 O 形

carpus vulgus n. *see* knock knees 膝内向,X 形膝

carriage /ˈkærɪdʒ/ n. **1** a horse-drawn vehicle 马车,舆 **2** the manner or way in which a horse carries his head or moves (马举头或运步的方式)姿态,姿势: tail carriage 尾姿 ◇ head carriage 举头姿势

carriage dog n. *see* Dalmatian 斑点狗,跟车犬

carriage hood n. a fabric or leather top of a carriage (马车的)顶篷,顶盖 syn hood, top

carriage horse n. a light, elegant horse used for carriage driving 驾车用马,舆马

carrot /ˈkærət/ n. a biennial Eurasian plant widely cultivated for its edible taproot (一种产自欧亚大陆的二年生植物,由于主根可食而广泛种植)胡萝卜

carry /ˈkæri/ vt. **1** to hold in a certain way 持有,带有 PHR V **carry the ball**【polo】(of a player) to move the ball down the field towards the goal (球员)带球,运球 IDM **carry the target**【racing】(of a horse) to run in last position the entire race (赛马)始终落尾,跑在末尾 **2**【hunting】(of the ground) to hold the scent of the fox (指猎犬)带有气味 **3** (of a horse) to hold his head in certain manner (指马)举头: A horse should not carry its head high in jumping over a fence. 马在跨栏时不应该举头过高。

cart /kɑːt; *AmE* kɑːrt/ n. a two-wheeled vehicle drawn by a single horse for farm work (单马拉的农用马车)双轮马车: Scotch cart 苏格兰马车 ◇ water cart 运水马车 compare wagon IDM **put the cart before the horse** to reverse the logical order of things 本末倒置

vt. to convey or carry in a cart (用马车)运输

cart horse n. a draft-type horse used to pull a cart (用来拉车的)挽马,辕马

cart road n. *see* cart track 马车道

cart track n. (also **cart road**) a rough track or road for horse-drawn vehicles 马车道

carter /ˈkɑːtə/ n. one who drives a cart 驾车者,车夫

cartload /ˈkɑːtləʊd/ n. (also **cart-load**) the amount that a cart holds (马车的)货运量,装载量

cartwheel /ˈkɑːtˌwiːl; *AmE* ˈkɑːrt-/ n. (also **cart-wheel**) the wheel of a cart with thick wooden spokes and a metal rim (农用马车车轮,为木制辐条和金属轮缘)车轮

Caspian /ˈkæspɪən/ n. (also **Caspian pony**) a pony breed originating in present Iran around 3000 BC; has a small head, short ears, large eyes, flared nostrils, a slightly arched neck with full mane and tail, short back, and sturdy legs; stands 9.2 to 11.2 hands and may have a bay, chestnut, or gary coat; is quiet, docile and has good endurance; used mainly for riding (公元前3000多年源于现今伊朗地区的矮马品种。头小,耳短,眼大,鼻孔开张,颈弧形,鬃尾浓密,背短,四肢粗壮,体高9.2～11.2掌,毛色可为骝、栗或青,性情安静而温驯,耐力好,主要用于骑乘)里海矮马

cast /kɑːst/ v. **1** (of a horse) to throw a horseshoe (蹄铁)脱落 **2** to throw a horse onto the ground using hobbles (用绊腿将马)放倒:to cast a horse onto ground 将马放倒在地 **3** to become stuck or lodged 卡住

adj. (of a horse) being stuck or wedged against a corner, manger, etc. (指马卡在墙角或食槽)受困的,被卡的:When a cast horse is first discovered help should be summoned to extricate him immediately. 当发现受困马匹时,应立即找人帮助使之脱困。

cast horse n. **1** (also **principal horse**) a horse who has a primary role in a film (电影中的)主角马:The gray Arabian Alain Delon rode in *La Tulipe Noire* is undoubtedly the cast horse in the film.《黑郁金香》中阿兰·德龙所骑的那匹青毛阿拉伯马无疑是这部电影的主角马。 **2** any horse used in a film 电影用马

Castiglione, Giuseppe n. (1688－1766) an Italian missionary who came to China in 1715 and later became the court painter of several emperors of the Qing dynasty; noted for his paintings of portraits, hunting and equestrian scenes (意大利传教士,于1715年抵达中国,后成为清朝几代皇帝的宫廷画家,以绘制肖像、狩猎和马术场景见长)朱塞佩·伽斯底里奥内;郎世宁

castrate /kæˈstreɪt; *AmE* ˈkæstreɪt/ vt. to remove the testicles of a male horse; geld (摘除公马睾丸)阉割,去势:The castrated male horse is generally called gelding and easier to handle. 去势后的公马称为骟马,易于调教。

castration /kæˈstreɪʃn/ n. the practice of castrating a male horse 去势,阉割

cat hairs n. the long, untidy hairs growing in a horse's coat following clipping (马被毛修剪后长出的凌乱毛发)软毛

catabolic /kætəˈbɒlɪk/ adj. of or relating to catabolism 分解代谢的 opp anabolic

catabolism /kəˈtæbəlɪzəm/ n.【biochemistry】the phase of metabolism in which compounds are broken into small particles for the body to absorb (代谢中复杂有机物分解为简单分子以利机体吸收的过程)分解代谢 opp anabolism

cataract /ˈkætərækt/ n. a condition that affects the lens of the eye and causes a gradual loss of sight (影响眼晶状体的病症,可逐渐导致失明)白内障

catch /kætʃ/ vt. **1** to take hold of 抓;拽 PHR V **catch hold** (of a horse) to take hold of the bit and pull suddenly (指马)拽缰,扯缰 **catch up** to round up a herd of cattle (将牛群)赶拢

2 (of a stallion) to find a mare in season (公马)寻找(发情母马)

catch pole n. (also **catching pole**) a long pole with a loop at its end used by the Mongolians to catch horses (末端带绳套的长杆,蒙古人以此套马)套马杆

catch rope n.【roping】the lariat holding a roped animal 套绳

catch-driver n.【harness racing】a driver hired for his skill (轻驾车赛的)车手

caterpillar /ˈkætəpɪlə(r); *AmE* -tərp-/ n. *see* larva 幼虫,毛虫

cathartic /kəˈθɑːtɪk; *AmE* -θɑːr-/ n. *see* purgative 泻药

cattle /ˈkætl/ n. **1** cows, bulls, and steers that are kept as farm animals 牛;牛群 PHR V **to read cattle**【cutting】to anticipate the movements of a cow being worked 预判牛的动向 syn read a cow **2**【slang】a herd of horses 马群

cattle blocking n.【rodeo】a mounted South American rodeo game held in a round arena in which a steer is blocked by the rider against the wooden partition (南美举行的牛仔竞技项目,公牛被拦到圆形木围栏隔板边上)拦牛

Cattle Men's Carnivals n. *see* rodeo 牛仔竞技

cattle settler n.【cutting】a mounted rider who calms a herd of cattle when the cutting horse and the rider penetrate the herd (截牛项目中的骑手,当截牛者策马突入牛群时,协助安稳牛群)安定牛群者

cauda equina neuritis n.【vet】(abbr. **CEN**) a disease which affects the spinal cord and nerves at the horse's tail end, symptoms include paralysis of the tail and surrounding areas (马尾部脊椎神经所患得的疾病,症状表现为尾巴及周围区域的瘫痪)马尾神经炎

caudal /ˈkɔːdl/ adj.【anatomy】situated near the tail of a horse 尾部的:caudal vertebrae 尾椎

cavalier /ˌkævəˈlɪə(r); *AmE* -lɪr/ n. a mounted soldier or horseman 骑兵;骑士

cavalletti /ˌkævəˈletɪ/ n. (also **cavaletti**) *see* ground pole 地杆:The most common application of cavalletti is in the regulation of the horse's stride at the trot. 地杆的主要用途是调整马在快步中的步伐。

cavalry /ˈkævlri/ n. the part of an army composed of mounted soldiers 骑兵:a cavalry school 骑兵学校

cavalry hold n. a method of holding all the four reins of a double bridle in one hand with the other carrying a weapon (一种单手握双勒四根缰绳的方法,另一只手用来拿兵器)骑兵单手持缰

cavalry mount n. (also **cavalry mount**) any horse used for service in an army unit 军马,戎马

cavalry twill n. a type of fabric used for cavalry uniforms 骑兵斜纹布

cavalryman /ˈkævlrɪmən/ n. a soldier who fought on horseback 骑士,骑兵 syn horseman

Cavendish, William n. (1592 – 1676) a celebrated 17th-century cavalryman and horse trainer noted for his humane treatment of horses (17世纪著名的骑兵将领、驯马师,以其仁慈的驯马方式而出名)威廉·卡文迪什

cavesson /keɪˈvesʌn/ n. a simple noseband attached to the bridle and used for lungeing a horse (一种打圈用的简易鼻革)鼻勒;调教笼头,调教索:lungeing cavesson 打圈调教笼头

CB abbr. coldblood 冷血马

CBC abbr. complete blood count 全血细胞计数

CEA abbr. Chinese Equestrian Association 中国马术协会

cecum /ˈsiːkəm/ n.【anatomy】the large blind-ended sac situated at the end of the small intestine (小肠末端的憩室)盲肠

cee-spring n.【driving】(also **C-spring**) a C-shaped, metal spring used in vehicles to absorb shock (马车减震用的)C-形弹簧:Cee-spring was first introduced in 1790 as an improvement to the whip-spring. C-形弹簧于1790年首次使用,是由减震皮带改进而来。

C

cellulose /ˈseljuləʊs; *AmE* -loʊs/ n.【biochemistry】an insoluble complex carbohydrate that forms the main constituent of most plants (一种不溶性多糖,是构成植物的主要成分)纤维素

CEM abbr. contagious equine metritis 马传染性子宫炎

cement /sɪˈment/ n.【anatomy】the outer covering of the teeth (牙齿的最外层)垩质

CEN abbr. cauda equina neuritis 马尾神经炎

centaur /ˈsentɔː(r)/ n.【mythology】a mythical beast having the head, arms, and trunk of a man and the body and legs of a horse (古希腊神话中前半身为人、后半身为马的怪兽)半人马,人头马

center /ˈsentə/ n.【driving】one of the horse of the center pairs (多马联驾的)中马: near center 内侧中马,左中马 ◇ off center 外侧中马,右中马 [compare] leader, wheeler

center of gravity n. an imaginary point around which the mass of a body or limb is balanced (一个假想的中心,身体四肢在此处于平衡状态)重心

center pairs n. the extra pairs of horses harnessed between the leaders and the wheelers in a team (多马联驾中套在头马和辕马间的一对马)中间双马 [compare] wheelers, leaders

central incisors n. (also **centrals**) the four front middle incisors 中切齿 [syn] first incisors

centrals /ˈsentrəlz/ n. *see* central incisors 中切齿

cereal /ˈsɪəriəl; *AmE* ˈsɪr-/ n. **1** any of various grasses widely grown for its edible grain (禾本科植物,由于种子可食而广泛种植)谷类作物: cereal crops 谷类作物 **2** the grain produced by cereal crops 谷粒,谷子: cereal meal 谷类日粮

cerebral /ˈserəbrəl; *AmE* səˈriːbrəl/ adj. of or relating to the brain [大]脑的

certificate of registration n. a legal document recording the horse's age, pedigree, breeder, owner, and physical description; generally issued by the appropriate breed registry (由马匹育种协会颁发的证书,其上记录马匹的年龄、系谱、育马者、马主以及马匹外貌描述的凭证)马匹登记证明

certificate of veterinary inspection n. (abbr. **CVI**) a document issued by a veterinary hospital that certifies a horse is in good health and is vaccinated against specific diseases (由兽医院出具的文件,用于证明马匹健康状况良好并接种过特定疫苗)兽医检验证明: The certificate of veterinary inspection is usually required when a horse is transported across country borders. 马匹进出境时,通常都需要出示兽医检验证明。

Certified Apprentice Farrier, Level One n. (abbr. **CAF Ⅰ**) a certification awarded to novice farriers by the Brotherhood of Working Farriers Association on the basis of a practical examination (由蹄铁匠兄弟会经过专业技能考试颁发给蹄铁匠实习生的资格证书)一级见习蹄铁匠证书

Certified Apprentice Farrier, Level Two n. (abbr. **CAF Ⅱ**) a certification awarded to graduates and professional farriers just starting practice by the Brotherhood of Working Farriers Association on the basis of written and practical examinations (由蹄铁匠兄弟会经过专业技能考试颁发给毕业学员或职业蹄铁匠的资格证书)二级见习蹄铁匠证书

Certified Farrier n. (abbr. **CF**) a certification of knowledge and skill awarded to farriers practicing in the United States by the American Farrier's Association on the basis of written and practical examinations (由美国蹄铁匠协会经过书面和技能考试颁发给蹄铁匠的资格证书)蹄铁匠资格证书;注册蹄铁匠

Certified Journeyman Farrier n. (abbr. **CJF**) a certification awarded to a farrier by the Brotherhood of Working Farriers Association on the basis of written and practical examinations (由蹄铁匠兄弟会经过书面和技能考试颁发给

蹄铁匠的证书)职业蹄铁匠资格证书

Certified Journeyman Farrier, Level One n. (abbr. **CJF Ⅰ**) a certification awarded to farriers with more than one year of practical experience by the Brotherhood of Working Farriers Association on the basis of written and practical examinations (由蹄铁匠兄弟会经过书面和技能考试颁发给从业1年以上的蹄铁匠的一种资格证书)一级职业蹄铁匠资格证书

Certified Journeyman Farrier, Level Two n. (abbr. **CJF Ⅱ**) a certification awarded to farriers with more than three years of practical experience by the Brotherhood of Working Farriers Association on the basis of written and practical examinations (由蹄铁匠兄弟会经过书面和技能考试颁发给从业3年以上的蹄铁匠的一种资格证书)二级职业蹄铁匠资格证书

Certified Master Farrier n. (abbr. **CMF**) the highest level of achievement and certification awarded to a farrier by the Brotherhood of Working Farriers Association (由蹄铁匠兄弟会颁发给蹄铁匠的一种级别最高的资格证书)大师级蹄铁匠资格证书

cervical /ˈsɜːvɪkl; *AmE* ˈsɜːr-/ adj. of or relating to the neck or cervix 颈的;子宫颈的

cervix /ˈsɜːvɪks; *AmE* ˈsɜːr-/ n. **1**【anatomy】the neck 颈 **2** the narrow portion of the uterus 子宫颈

Cesarean section n. *see* Caesarean section 剖宫产,剖腹产

CF abbr. Certified Farrier 注册蹄铁匠

CG abbr. chorionic gonadotropin 绒毛膜促性腺激素

ch abbr.【color】chestnut 栗毛

chafe /tʃeɪf/ v. to wear away or cause irritation by rubbing 磨损;蹭伤

chaff /tʃæf/ n. **1** the dry husk of grains removed during threshing (谷类脱粒时除去的外壳)谷壳 **2** chopped hay or straw 铡草

Chaise /tʃeɪz/ n. any of various light, two-wheeled, open carriage with a folding hood over the driver's seat (一种双轮轻型马车,车夫驾驶座上方带折叠式顶篷)轻型马车: post chaise 驿站马车

chalk /tʃɔːk/ n. *see* chalk horse 投注热门马

chalk board n.【racing, dated】a board upon which the betting odds of racehorses were formerly notified by bookmakers (赌马经纪人过去发布赛马赔率的告示板)赔率公告板 compare totalizator board

chalk horse n.【racing】(also **chalk**) the betting favorite in a race 投注热门马

chalk player n.【racing】one who bets on favorites 在热门马上投注的人

Challon, Henry Bernard n. (1770 – 1849) a court painter of animals of Kings George Ⅳ and William Ⅳ (英国国王乔治四世和威廉四世的宫廷画家)亨利·伯纳德·查朗

champron /ˈtʃæmprən/ n. a piece of armor used to protect the horse's face (用来保护马匹脸面的盔甲)护头,面罩 — *see also* **crinet, croupiere, peytral**

change /tʃeindʒ/ vt. to replace one with another 更换;替换 PHR V **change lead** (of a horse) to change its leading foreleg in canter or gallop (指马)换领先肢 **change ponies**【polo】to exchange mount used in one chukker for a fresh horse (马球中)更换坐骑;换马

n. the instance of changing with another 更换;变换 PHR **change (of the leg) in the air** *see* flying lead change 空中换蹄 **change of lead** (also **lead change**) a change of horse's leading foreleg performed at a gallop or canter (指马)换领先肢,换起步肢

channel /ˈtʃænl/ n. *see* saddle channel 鞍沟

chaparejos /ʃæpəˈreɪəʊs/ n. (also **chaparajos**) *see* chaps 套裤

chape /tʃeɪp/ n.【hunting】the tip of a fox's tail (狐狸的)尾尖,尾梢

chaps /tʃæps/ n. pl. (also **chaparejos**, or **shaps**) a pair of heavy leather trousers without a seat, worn by cowboys over ordinary trousers to protect their legs（牛仔穿在长裤外用来保护大腿的皮裤）护腿；套裤：leg chaps 皮护腿 ◇ full chaps 皮套裤 ◇ half chaps 短护腿 ◇ Legs chaps, usually made of durable cowhide, are an informal substitute for boots or full chaps. 皮护腿通常由结实的生牛皮制成，是马靴和皮套裤的代用品。

charge /tʃɑːdʒ; *AmE* tʃɑːrdʒ/ vi. to rush or attack violently 冲锋 | **charging** /-dʒɪŋ/ n.

charge-back n.【racing】*see* buy-back 竞拍失败

charger /ˈtʃɑːdʒə(r); *AmE* tʃɑːrdʒ-/ n. a cavalry horse 战马，戎马

chariot /ˈtʃæriət/ n. an ancient, two-wheeled, horse-drawn vehicle used in battle, races, and processions（古时战争、竞赛或游行时用的双轮马车）战车：chariot horse 拉战车的马

chariot racing n. the national sport of ancient Greece in which horse-drawn chariots were raced on a straight or oval course（古希腊运动项目，马拉战车在直线或椭圆赛道上比赛）战车赛：In 408 BC, chariot racing became an Olympic sport. 公元前408年，双轮战车赛成为奥运会项目。

charioteer /ˌtʃæriəˈtɪə(r); *AmE* -ˈtɪr/ n. one who drove a chariot（驾驭战车的）车夫，御夫

Charley /ˈtʃɑːli/ n. *see* Charlie 狐狸

Charlie /ˈtʃɑːli/ n.（also **Charley**）a fox, so called in fables and literature（狐狸在寓言和文学中的称谓）狐狸，查理

Charollais /ˈʃærəˌleɪ/n. a French horse breed developed specifically in the Charolles region from which the name derived; has a compact body with a short neck and strong legs; stands 15 to 16.2 hands and usually has a solid color; formerly used as cavalry mount and is versatile in eventing, jumping, and dressage（法国夏洛莱地区培育的本土马品种并由此得名。体格结实，颈短，四肢结实，体高15～16.2掌，多为单毛色，过去用作军马，目前主要用于三日赛、障碍赛和盛装舞步）夏洛莱马：In 1958 the Charollais horses were officially merged into the French Saddle horse and as a distinct breed they no longer exist. 夏洛莱马从1958年正式并入法兰西乘用马，自此作为一个品种已不复存在。

charro /ˈtʃɑːrəʊ/ n.（*pl.* **charros**）a Mexican cowboy or horseman, especially one in elaborate traditional dress（墨西哥牛仔，经常身着传统服饰）墨西哥牛仔

chart /ˈtʃɑːt; *AmE* ˈtʃɑːrt/ n.【racing】*see* result chart 比赛结果表

chase /tʃeɪs/ vt. to follow in order to capture or hunt 追赶，追捕：to chase foxes 追捕狐狸
n. **1** the act or process of chasing or pursuit 追赶，追捕 **2**【colloq】*see* steeplechase 越野赛

check /tʃek/ vi.【hunting】(of hounds) to stop and sniff about for the scent（指猎犬）停下搜寻

check rein n. *see* bearing rein 姿态缰

cheek /tʃiːk/ n. **1** either of the two sides of a horse's face（马的）面颊 **2** the vertical portion of the bit outside of a horse's mouth（衔铁位于马口外的竖直部分）颊杆

cheek tooth n.（*pl.* **cheek teech**）*see* molar 臼齿

cheekbone /ˈtʃiːkbəʊn; *AmE* -boʊn/n. *see* zygoma 颧骨

cheekpiece /ˈtʃiːkpiːs/ n. either of the two straps of a bridle that connects to the bit rings（马勒上与衔环相连的皮革）颊革 [syn] sidepiece

cheer /tʃɪə(r); *AmE* tʃɪr/ n.【hunting】the call of the huntsman to urge the hounds forward（猎狐者激励猎犬前行的喊声）前进！

chef d'équipe /ʃef ˈdekɪp/ n.【French】the lead of a national equestrian team responsible for making all arrangements for team competition abroad（负责国家马术队所有国外比赛事务的带队）总领队：After his competitive career, Michael Page served as chef d'équipe of the U.S three-day team. 退役后，麦克尔·佩奇担任了美国三日赛马术队的总领队。

cheiloschisis /kaɪˈlɒskɪsɪs/ n. *see* cleft lip 唇裂

cherry bay n.【color】a reddish shade of bay 红

骝毛

chest /tʃest/ n. the part of a horse's body formed by the ribs and breastbone; thorax（马体位于胸骨和肋骨之间的部分）胸

chestnut /ˈtʃestnʌt/ n. **1**【color】a reddish brown coat with points nearly of the same color（被毛红褐且末梢也与之基本相同的毛色）栗毛: reddish chestnut 红栗 ◇ flaxen chestnut 金栗 ◇ light chestnut 淡栗 ◇ liver chestnut 朽栗 ◇ All chestnuts have shades of red in their coats, and the points (mane, tail, legs and ears) are the same color as the horse's body. 所有的栗毛马其被毛都有红色基调，且鬃、尾、四肢及耳朵的毛色与身体的毛色基本一致。**2** the horny growth on the insides of the horse's legs（马四肢内侧附生的角质结构）附蝉，夜眼 syn night-eye

chestnut roan n.【color】the color of a horse who has a chestnut coat color interspersed with white hairs（栗色被毛中散生有白毛的毛色）沙栗毛

cheval /ʃəˈval/ n.【French】horse 马

cheval de Mèrens n.【French】*see* Mèrens 梅隆马

Cheval du Poitou n.【French】*see* Poitevin 普瓦图马

chew /tʃuː/ v. to bite and grind with the teeth（用牙齿）啃咬，咀嚼: Application of creosote to the manger and wooden board will frustrate the horse from chewing. 在饲槽和木板表面涂上木焦油可阻碍马匹啃咬。

CHI abbr. Championnat Hippique International 国际马术锦标赛

CHIA abbr. China Horse Industry Association 中国马业协会

CHIA Cup n.【racing】a 1,650-meter race held annually for three-year old horses and older since 2015 at Xilinguole racetrack in Inner Mongolia, China（自2015年起每年在中国内蒙古锡林郭勒赛马场举行的比赛，由3岁以上马匹参加，赛程1 650米）中国马会杯

Chickasaw /ˈtʃɪkəˌsɔː/ n. a light horse breed developed by the native Americans of Tennessee and Oklahoma during the 18th century; has a very short neck and well-developed body; stands about 14 hands and may have a bay, black, chestnut, gray, roan, or palomino coat; used for packing, hunting and working fields（美国田纳西州和俄克拉荷马州的印第安人在18世纪培育的轻型马品种。颈短，体躯匀称，体高约14掌，毛色多为骝、黑、栗、青、沙或银鬃，主要用于驮运、狩猎和农场作业）契卡索马

chicken coop n.【jumping】an obstacle consisting of a two-sided panel with a peaked roof（由两块木板搭成的尖顶障碍）鸡舍障碍

chigger /ˈtʃɪgə(r)/ n. any of various small, six-legged larvae of mites of the family *Trombiculidae* that affects horses' feet and muzzles and causes severe itching（一种恙螨科六足寄生虫，常寄生在马蹄和口鼻等处，可导致患处瘙痒）恙螨 syn harvest mite

Chileno /tʃɪˈleɪnəʊ/ n. a Criollo horse bred in Chile 智利克里奥尔马

chill /tʃɪl/ n. sense of coldness, often accompanied by shivering of the skin 寒战；伤风: Never wash or bath a horse when he is even slightly unhealthy or he may catch a chill. 如果马有丝毫的病弱迹象，切勿进行洗浴，否则它可能患上伤风。

chime /tʃaɪm/ vi.【hunting】(of the hounds) to bark together when smelling the scent of the quarry（指猎犬嗅到猎物踪迹时一起出声）齐声吠叫

chin /tʃɪn/ n. the central forward portion of the lower jaw（下颌向前突出的部分）下颌，颐: prominence of chin 下颌突

chin groove n. a groove in the underside of the lower jaw（马下颌上的凹槽）颐沟，颐凹

chin strap n. the strap used to tie a helmet under the rider's chin（系住头盔的）颏带，颌革 PHR **on the chin strap**【racing】(of a horse)

C

winning by a wide margin (指赛马)显著胜出

China Breeders' Cup n.【racing】a 1,000-meter race held annually for two-year old horses at Yijinhuoluo racetrack since 2015 in Inner Mongolia, China (自 2015 年起每年在中国内蒙古伊金霍洛赛马场举行的比赛,由 2 岁驹参加,赛程 1 000 米)中国育马者杯

China Cup n.【racing】a 2,000-meter race held annually for three-year old horses and older since 2015 at Xiyu racetrack in Xinjiang, China (自 2015 年起每年在中国新疆西域赛马场举行的比赛,由 3 岁以上马匹参加,赛程 2 000米)中国杯

china eye n. *see* wall eye 玻璃眼,玉石眼

China Horse Industry Association n. (abbr. **CHIA**) an organization that originated in 1976, responsible for the management and service of horse breeding, studbooks, events, horse performance test, equine veterinarians and medicine, welfare, and horse culture promotion, etc., in China. It is a member of the ISBC, ARF, IAHA, WBFSH and WAHO. (源于 1976 年,是负责管理、服务马匹繁育、血统登记、赛事活动、性能测定、兽医兽药、马匹福利和马文化推广等的行业组织,是国际纯血马登记管理委员会、亚洲赛马联盟、世界阿哈尔捷金马(汗血马)协会、世界运动马繁育联盟和世界阿拉伯马协会的成员)中国马业协会

Chincoteague /ˌʃɪŋkəˈtiːg/ n. *see* Chincoteague pony 辛柯狄格矮马

Chincoteague pony n. (also **Chincoteague**) a pony breed indigenous to the Chincoteague Islands off the eastern coast of USA; has a long head and neck, slightly pronounced withers, straight back, and solid legs; stands about 12 hands and generally has a piebald or skewbald coat, although all coat colors occur; is rebellious, hardy and resistant to harsh weather conditions (源于美国东海岸辛柯狄格岛上的矮马品种。头颈较长,鬐甲突出,背平直,四肢粗壮,体高约 12 掌,各种毛色兼有,但以花毛最为常见,性情暴躁,抗逆性强)辛柯狄格矮马

Chinese Equestrian Association n. (abbr. **CEA**) an organization founded in 1979, to organize and regulate all kinds of national equestrian sports and affiliated to the FEI in 1982(成立于 1979 年,旨在组织管理全国马术运动项目的组织,于 1982 年加入国际马术联合会)中国马术协会

Chinese Pony n. a generic term for all the pony breeds indigenous to China; has a small head, with a straight profile, shaggy forelock, mane, and tail, small ears, short neck, and straight back; stands 12 to 13 hands and often has a yellow dun coat, although other colors also occur; primitive markings, specifically the dorsal strip and zebra markings on the legs, are common; is quite hardy, strong, sure-footed, and quite rebellious; is fast over short distances and used for riding, packing, and farm work (对中国本土矮马的泛称,头小,面貌平直,额发、鬃及尾浓密杂乱,耳小,颈短,腰背平直,体高 12 ~ 13 掌,毛色以兔褐为主,兼有其他毛色,背线和斑马纹等原始别征较为常见,抗逆性强,体格健壮,步法稳健,性情暴躁,短途步速快,主要用于骑乘、驮运和农田作业)中国矮马

chinked back n. *see* jinked back 脊柱扭曲

CHIO abbr. Championnat Hippique International Official 国际马术锦标赛

chip /tʃɪp/ vi. (of a horse) to add a short, unexpected stride just right before jumping a fence (指马在跨越障碍前突然多迈一短步)多跨半步,多迈半步

n. a short, unexpected stride added when jumping a fence; a stutter step 多迈的半步,多跨的半步

chlorhexidine /klɔːˈheksɪdaɪn/ n.【vet】an antiseptic agent used in surgery (外科手术中使用的消毒剂)氯己啶

chloride /ˈklɔːraɪd/ n. a chemical compound of chlorine 氯化物:sodium chloride 氯化钠

chlorpromazine /klɔːˈprəʊməzɪn/ n. *see* acetylpromazine 氯丙嗪,乙酰丙嗪

CHN abbr. Championnat Hippique National 全国马术锦标赛

chocolate chestnut n.【color】a shade of chestnut with a brown chocolate coat and brown points（被毛暗褐而末梢为褐色的一种栗色）褐栗色，深栗色

choke /tʃəʊk; *AmE* tʃoʊk/ n. a partial or complete blockage of the gullet（食管的）噎塞，梗塞

vi. to become blocked up or obstructed 噎塞，梗塞

choline /ˈkəʊliːn; *AmE* ˈkoʊ-/ n.【biochemistry】an intestinally synthesized B-complex vitamin essential in the metabolism of fat（肠道合成的B族维生素，为脂肪代谢所必需）胆碱

chondri-, chondro- pref. cartilage 软骨

chondric /ˈkɒndrɪk/ adj. of or relating to cartilage; cartilaginous 软骨的

chondrocyte /ˈkɒndrəsaɪt/ n. the cartilage cell 软骨细胞

chondroitin sulfates n.（abbr. **CS**）a popular nutraceutical used to reduce or relieve joint problems（治疗关节疼痛的常用药物）硫酸软骨素

chondroprotective /ˌkɒndrəˈprətektɪv/ n. any animal-derived protein used to heal cartilage damage of joints（用来治疗关节软骨损伤的一种动物提取蛋白）软骨保护剂：chondroprotective agents 软骨保护剂

chop[1] /tʃɒp; *AmE* tʃɑːp/ vt. to cut by striking with a heavy, sharp tool, such as hay cutter（用刀）铡碎；切碎：chopped straw 铡草

n. straw chopped as coarse fodder of livestock 铡草，铡料

chop[2] /tʃɒp; *AmE* tʃɑːp/ vt. **1**【hunting】（of hounds）to kill a fox in his covert before it had a chance to run（猎犬）杀死（狐狸）：a chopped fox 被杀死的狐狸 **2**【driving】to strike a horse with the whip on the thigh 用鞭抽马

chop[3] /tʃɒp; *AmE* tʃɑːp/ n. an exchange of horses between two sides（双方的）马匹交换

vt. to exchange horses 交换马匹

chop bedding n. chopped straw or hay used to bed a stall 铡草垫料

chorion /ˈkɔːrɪɒn; *AmE* ˈkoʊrɪˌɑːn/ n.【anatomy】the outermost membrane surrounding an embryo that contributes to the formation of the placenta（胚胎最外的膜组织，参与胎盘的形成）绒毛膜 | **chorionic** /ˌkəʊrɪˈɒnɪk; *AmE* ˌkoʊr-/ adj.

chorionic gonadotropin n.【physiology】（abbr. **CG**）a hormone produced by the chorion and excreted into the urine（由胎盘绒毛膜分泌且存在于尿液中的一种激素）绒毛膜促性腺激素

chorioptic mange n. a skin disease of horse's lower legs caused by a species of mite *Chorioptes equii*, characterized by skin lesions, severe itching, scales, crusts, and thickening of the skin（由足螨所致的马下肢皮肤病，表现为皮肤溃烂、痛痒、蜕皮、结痂和增厚等症）足螨病：Horses affected with chorioptic mange will stamp his feet and rub his lower legs in an attempt to relieve the pain. 患足螨病的马匹常顿足并蹭其下肢以减轻疼痛。 syn leg mange, itchy heel, itchy leg, foot mange, ear mange

chromosome /ˈkrəʊməsəʊm; *AmE* ˈkroʊməsoʊm/ n.【genetics】a thread-like linear strand of DNA in the cell that carries the genes and functions in the transmission of hereditary information（存在于细胞中的线状DNA链，携带基因并有传递遗传信息的功能）染色体：X chromosome X染色体 ◇ sex chromosome 性染色体

chronic /ˈkrɒnɪk; *AmE* ˈkrɑːn-/ adj.（of a disease）long lasting or recurring frequently over a long period of time（疾病持续周期长且不断复发）慢性的：chronic poison 慢性毒物，慢性中毒 compare acute

chronic arthritis n. a slowly developing, low-grade inflammation of the joint usually resulting in some degree of permanent damage（骨关节

的慢性发炎症状,常由于永久损伤而导致)慢性关节炎

chronic laminitis n. a persistent and long-term inflammation of the laminae 慢性蹄叶炎

chronic obstructive pulmonary disease n. (abbr. **COPD**) a chronic allergic airway disease caused by poor ventilation; symptoms include coughing, labored breathing, and a slight increase in respiratory rate (由于马厩通风差所致的过敏性呼吸道疾病,其症状表现为咳嗽、呼吸困难及呼吸急促等)慢性阻塞性肺炎 syn obstructive pulmonary disease, heaves, broken wind, alveolar emphysema, equine asthma

chronic osselet n. a permanent build-up of synovial fluid in a joint 慢性关节肿

chuck wagon n. a four-wheeled, horse-drawn vehicle used as a mobile kitchen on farms and ranches in the Western states of America (美国西部的农场和牧场用作活动厨房的一种四轮马车)野餐马车, 野营马车 syn camp wagon, round-up wagon, food wagon

chukka /ˈtʃʌkə/ n.【polo】*see* chukker (马球比赛的)局

chukker /ˈtʃʌkə(r)/ n.【polo】(also **chukka**) one period of play in polo (马球比赛的)局: A polo match usually consists of six chukkers. 一场马球比赛有 6 局。

chute /ʃuːt/ n. **1**【rodeo】a stall where wild horses and bulls are saddled, bridled, and mounted and from which they are released into the arena (牛仔竞技中野马或公牛备鞍和策骑的围栏)巷道,过道 **2**【racing】an extension of the backstretch or homestretch of a track (赛马场直道的延伸部分)甬道

Chyanta /ˈʃjæntə/ n. (also **Chyanta pony**) a pony breed originating in the Himalayan mountains of northern India; stands around 12 hands and generally has a bay coat, short neck, straight shoulder, and short strong leges; is frugal and has good endurance; used for riding and packing (源于印度北部喜马拉雅山区的矮马品种。体高约 12 掌,毛色以骝为主,颈短,直肩,四肢粗短,耐粗饲,耐力较好,用于骑乘和驮运)史洋塔矮马

chyme /kaɪm/ n.【nutrition】the thick semi-fluid mass of partially digested food that is passed from the stomach to the duodenum (由胃送入十二指肠的半消化食团)食糜

CID abbr. combined immuno-deficiency 综合免疫缺陷症

ciliary /ˈsɪliəri/ adj. of, relating to, or resembling cilia 睫毛的; 睫状的: ciliary body 睫状体 ◇ ciliary muscle 眼睫肌

cilium /ˈsɪliəm/ n. (*pl.* **cilia**/ˈsɪliə/) hair on the eyelid; eyelash 睫毛

cinch /sɪntʃ/ n.【AmE】a strap fastened around the horse's chest in order to secure the saddle on the horse's back (系在马前胸用以稳固马鞍的皮革)肚带 syn girth

v. to fasten the cinch 系肚带 PHR V **cinch up** to fasten or tighten the girth of a saddle 扣紧肚带,系紧肚带

cinch bound adj. (also **cinch-bound**) (of a horse) objecting or sensitive to the tightening of cinch (指马)紧肚带时不安的,紧肚带时反抗的 syn girth shy, cinchy, girthy

cinchy /ˈsɪntʃi/ adj. *see* cinch bound 紧肚带时不安的,紧肚带时反抗的

circulate /ˈsɜːkjəleɪt; *AmE* ˈsɜːrk-/ vi. to move or flow around in a circle 循环

circulation /ˌsɜːkjəˈleɪʃn; *AmE* ˌsɜːrk-/ n. the flow of blood through bodily vessels as a result of the heart's pumping action (血液通过心脏的泵压在体内的流动)血液循环: pulmonary circulation 肺循环 ◇ systemic circulation 体循环

circulatory /ˌsɜːkjəˈleɪtəri; *AmE* ˈsɜːrkjələtɔːri/ adj. relating to the circulation of blood 血液循环的: circulatory system 循环系统

circus /ˈsɜːkəs; *AmE* ˈsɜːrkəs/ n. a public show performed by a group of entertainers or trained animals in a large tent (由艺人和调教的动物

在帐篷内进行的公开表演)马戏:circus horses 马戏马 ◇ Philip Astley is generally considered the father of the modern circus. 菲利普·安斯特利是公认的现代马戏之父。

Cisium /ˈsɪzɪəm/ n.【Latin】a two-wheeled, horse-drawn vehicle used in ancient Rome(古罗马使用一种双轮马车)两轮马车

citrus pulp n. the residue of citrus fruit resulting from juice extraction(柑橘榨汁后的残余物)柑橘渣

CJF abbr. Certified Journeyman Farrier 职业蹄铁匠资格

CJF I abbr. Certified Journeyman Farrier, Level One 一级职业蹄铁匠资格证书

CJF II abbr. Certified Journeyman Farrier, Level Two 二级职业蹄铁匠资格证书

CK abbr. creatine phosphokinase 肌酸磷酸激酶

CL abbr. corpus luteum 黄体

claim /kleɪm/ n.【racing】the money paid by a licensed buyer of a horse to the owner of that horse following a claiming race(赛后买马者按议定价格付给马主的钱款)转让费,竞价 vt. to buy a horse in a claiming race 竞价买马

claimer /ˈkleɪmə(r)/ n.【racing】(also **claiming horse**) a horse who runs in a claiming race 参加竞价赛的马

claiming box n.【racing】a small storage box in which claims are placed before the start of a claiming race(竞价赛开始前为赛马出价的厅室)竞价厅,申购室

claiming horse n. *see* claimer 参加竞价赛的马

claiming race n.【racing】a class of horse race in which all entered horses are eligible for purchase at a set price(参赛马随后可按议定价格买卖的比赛)竞价赛,申购赛 [compare] allowance race, stakes race

Clarence /ˈklærəns/ n. a horse-drawn farm coach first introduced in England in the 1840s(19 世纪 40 年代在英国出现的四轮农用马车)克拉伦斯马车

class /klɑːs; *AmE* klæs/ n. **1** a taxonomic category ranking below a phylum and above an order(级别低于门而高于目的分类学单位)纲 [compare] kingdom, phylum, order, family, genus, species **2** a division of equestrian activities based on skills, age of horse or trophies(马术比赛项目的等级划分)级别

classic distance n.【racing】a specific distance of a horse race(赛马的)经典赛程: The American classic distance is 1-1/4 miles, while its European counterpart is 1-1/2 miles. 美国平地赛经典赛程是 2 千米,而欧洲则为 2.4 千米。

Classic Race n. **1**【BrE, racing】any of the five major English flat races held for three-year-old Thoroughbreds, including the 1,000 Guineas and the Oaks for fillies, and the 2,000 Guineas, the Derby, and the St. Leger for both colts and fillies(英国为 3 岁纯血马举行的五项重要赛事,其中包括雌驹参加的一千几尼赛和欧克斯大赛以及性别不限的两千几尼赛、德比赛和圣莱切大赛)经典大赛 **2**【AmE, racing】any of the three major American flat races held for three-year-old Thoroughbreds, including the Kentucky Derby, the Preakness Stakes, and the Belmont stakes(美国为 3 岁纯血马举行的三项重要赛事,包括肯塔基德比赛、普瑞克尼斯大赛和贝尔蒙特大赛)经典大赛

classical airs n.【dressage】any of the high school movements performed by a horse both on and above the ground(高等骑术学校训练的各种地面和空中高难动作)古典骑术动作

classical equitation n. a riding style developed in the Baroque period during the Renaissance based on the works of Xenophon(欧洲文艺复兴时期盛行的骑乘方式,大多基于色诺芬的著作)古典骑术

claybank /ˈkleɪbæŋk/ n. *see* claybank dun 泥兔褐

claybank dun n.【color】(also **claybank**) a yellow dun in which the coat hairs are mixed with red and the points red (灰色被毛中散生红色毛发的兔褐)泥兔褐

clean /kliːn/ adj. **1** (of a horse) without defect, blemishes or unsoundness (指马)无缺陷的 **2** free from dirt or filth 清洁的,干净的 **3**【jumping】without incurring any faults 无罚分的:clean round 零罚分跨越障碍 ◇ to jump clean 顺利跨越[栏杆]

vt. to make stable or the body of a horse clean or free of dirt, etc. 打扫,清洁:to clean out the stables 对马厩进行彻底清扫

clean ground n.【hunting】a piece of ground without the scent of a fox 地面无狐狸踪迹

clean-legged /ˈkliːnˈlegɪd/ adj. **1** (of horse's legs) having little or no feathering around the fetlock (马四肢)无距毛的 **2** (of horse's legs) free of blemishes or signs of prior injury (马四肢)无缺陷的;无伤痕的

clear /klɪə(r); *AmE* klɪr/ adj.【jumping】(of a horse) jumping without incurring any faults (指马)顺利跨越的;零罚分的 PHR **clear round** a show-jumping round completed without incurring any jumping or time faults 零罚分跨越障碍:When two or more entrants clear round in an event, another round called a "jump-off" results. If two or more entrants are stilled tied in jumping faults, the fastest time around the course will determine the winner. 当有两名以上参赛者在比赛中以零罚分跨过障碍时,比赛就进入附加赛。如果仍有两名以上骑手在罚分上再度打平,则用时少者获胜。 syn clean round

v.【jumping】to jump an obstacle without touching or dropping a rail; jump clean (在不触落栏杆的情况下)顺利跨过:to clear off the obstacles 顺利跨过障碍

cleat /kliːt/ n.【farriery】a projecting piece of metal attached to the underside of a horseshoe to provide traction (蹄铁底面上用来增加摩擦力的凸起)蹄铁棱

cleft /kleft/ n. a crack of the hoof wall 蹄裂

adj. partially split or divided half way 开裂的:cleft lip 唇裂,兔唇 ◇ cleft palate 腭裂

clemency /ˈklemənsi/ n. mildness, especially of weather (气候)温和

clement /ˈklemənt/ adj. (of weather) mild or favorable (气候)温和的;适宜的:Under natural conditions, the stock breed only in the spring and summer months, which determined that the birth of their foals a year later would be in the season of adequate nourishment and clement climatic condition. 在自然条件下,种畜只在春、夏两季繁育,这也就决定了幼畜次年正好出生在营养充沛、气候适宜的季节。

clench /klentʃ/ vt. (also **clinch**) to fix a nail firmly by hammering down or flattening the nail tip that sticks out (将突出的蹄钉尖敲弯或锤平加以固定)敲弯,铆牢

clencher /ˈklentʃə(r)/ n. *see* clincher 弯钉钳

clerk of scales n.【racing】a racing official whose duty is to weigh the riders before and after a race (负责在比赛前后称重的人员)司磅员,过磅员

Cleveland Bay n. a coldblood originating in the Cleveland district of Yorkshire, England in the Middle Ages; stands 16 to 16.2 hands and may have a brown or bay coat; is docile, strong, and clean-legged; originally used as a carriage horse, now mainly for eventing, hunting, and driving (源于中世纪英国约克郡克利福兰地区的冷血马品种。体高 16 ~ 16.2 掌,毛色为褐或骝,性情温驯,体格健壮,四肢无距毛,过去主要用于拉车,现在则用于三日赛、狩猎和驾车)克利福兰骝马

Clever Hans n. an Orlov stallion trained by the Russian Wilhelm von Osten (1838 – 1909) to perform arithmetic and other intellectual skills in the 1890s (由威廉·冯·奥斯汀调教的一匹奥尔洛夫公马,在 19 世纪末曾表演算术和其他智力项目)聪明马汉斯:The case of Clever Hans was taken to show an advanced level of number sense in animal. 聪明马汉斯

的个案被用来表明,动物对数的概念也有很高的认识。

click /klɪk/ vi. *see* forge 追突

clinch /klɪntʃ/ vt. *see* clench 敲弯

clinch cutter n.【farriery】a farrier tool used to cut off the tips of shoe nails (蹄铁匠用来切断蹄钉末梢的工具)钉节刀

clincher /ˈklɪntʃə(r)/ n. (also **clencher**) a tool used by farriers to pull down the protruding tip of horseshoe nails (蹄铁匠用来将蹄钉尖敲弯的工具)弯钉钳 syn clinching tong, alligator

clinching tong n. *see* clincher 弯钉钳

clinker /ˈklɪŋkə(r)/ n. a top-class horse 骏马,良驹

clip /klɪp/ vt. **1** to trim and cut the body hair or shave the whiskers of a horse 修剪,剪毛:a clipped horse 一匹修剪过的马 **2** (of the horse) to strike the leg or foot with the opposing or rear foot or horseshoe attached thereon (马的对侧或前后马蹄或蹄铁相撞)磕蹄

n. **1** the way or pattern of clipping the hair of a horse (马匹被毛的)修剪:apron clip 围裙型修剪 ◇ belly clip 腹修剪 ◇ bib clip 围嘴型修剪 ◇ blanket clip 披毯型修剪 ◇ chaser clip 越野型修剪 ◇ clip pattern 修剪类型 ◇ full clip 全修剪 ◇ hunter clip 狩猎型修剪 ◇ trace clip 挽缰型修剪 ◇ trace low clip 低缰型修剪 ◇ trace high clip 高缰型修剪 **2**【farriery】a thin, V-shaped extension of the horseshoe (薄的呈V形的蹄铁延伸部位)蹄铁唇

clip-clop /ˈklɪpklɒp/ n. the sound of a horse's hoofs beating on a hard surface (马蹄撞击地面发出的响声)嘚嘚声,铮铮响

vi. (of horse's hoofs) to make such a sound 嘚嘚响,铮铮响

clipper /ˈklɪpə/ n. **1** one who cuts or clips 修剪工 **2** (usu. **clippers**) a tool used for cutting or clipping the coat of horses (修剪马被毛的工具)剃刀,剪毛刀:rechargeable clippers 充电剃刀 ◇ electric clippers 电动剃刀 ◇ hand-cranked clippers 手动剃刀

clitoral /ˈklɪtərəl/ adj. of or relating to the clitoris 阴蒂的:clitoral fossa 阴蒂窝

clitoris /ˈklɪtərɪs/ n.【anatomy】a small, elongated erectile organ at the anterior part of the vulva (母马位于阴门前的勃起器官)阴蒂,阴核

clocker /ˈklɒkə(r); *AmE* ˈklɑːk-/ n.【racing】one responsible for timing races (赛马的)计时员

clop /klɒp/ n. a sharp, hollow sound, as of a horse's hoof striking hard ground (似马蹄敲击地面发出的响声)嘚嘚声,铮铮响

vi. to make such sound 嘚嘚响,铮铮响

close /kləʊz; *AmE* kloʊz/ vi. **1**【racing】to end in a final odds betted on a horse in a race 以……赔率结束 **2**【racing】(of a horse) to gain ground on the leader 紧追(头马)

close-coupled adj. *see* short-coupled 短背的

closed carriage n. a horse-drawn vehicle with a falling hood and raised sides 带篷马车,封闭式马车

closed top n.【driving】a falling hood on a horse-drawn vehicle 马车顶篷

closer /ˈkləʊzə; *AmE* ˈkloʊ-/ n.【racing】a horse who runs best in the latter part of the race 后程居上的赛马,后段发力的赛马

Clostridium tetani n.【Latin】the bacteria that cause tetanus of horses 破伤风梭菌

clot /klɒt/ n. a thick, viscous, or coagulated mass or lump, as of blood (血液等的)凝块

v. (to cause to) form into a clot or clots; coagulate 凝结,凝固

clothing /ˈkləʊðɪŋ; *AmE* ˈkloʊð-/ n. **1** rugs or blankets used as covering for horse 马衣 **2** the action of covering horses with rugs for warmth 穿马衣

clover /ˈkləʊvə(r); *AmE* ˈkloʊ-/ n. any of the several herbs of the legume family, having trifoliolate leaves and usually planted as forage (多种豆科植物,生三小叶,常用作饲料而广泛种植)三叶草:alsike clover 杂三叶 ◇ clover hay 三叶干草 ◇ scarlet clover 绛三叶 ◇ sweet clover 草木樨 ◇ white clover 白三叶

club foot n. (also **donkey foot**) a conformation defect in which the hoof is abnormally upright (马蹄角度过直的肢蹄缺陷)立系,狭蹄 syn bear foot, boxy foot, coon foot

club-footed /klʌbˈfutɪd/ adj. (of horse's hoof) narrow and upright in shape (蹄形)狭窄的,立系的 syn boxy

Clydesdale /ˈklaɪdzdeɪl/ n. a draft breed originating at the beginning of the 18th century in Scotland's Clyde valley from which the name derived; descended from hardy native breeds put to Flemish and Friesian stallions; has a strong build and prominent feathering around its fetlock on all four legs; stands 16 to 17 hands and generally has a brown, bay, or black coat with white markings on the face and legs; is calm, sociable, strong and hardy; used for farm work, heavy draft, and driving (18 世纪初源于苏格兰克莱德山谷的重挽马品种,主要用当地母马与佛兰德马和弗里斯马公马交配繁育而来。体格强健,四肢球节上附生浓密距毛,体高 16 ~ 17 掌,毛色多为褐、骝或黑,面部和四肢别征常见,性情安静、温驯,抗逆性强,主要用于农田作业、重挽和驾车)克莱兹代尔马

coach[1] /kəʊtʃ; *AmE* koʊtʃ/ n. a large, enclosed, four-wheeled carriage 四轮马车

v. to transport by or ride in a coach 马车载运;乘马车

coach[2] /kəʊtʃ; *AmE* koʊtʃ/ n. one who trains or directs athletes or athletic teams (训练或指导运动员的人)教练: an equestrian coach 马术教练

vt. to train, direct, or act as a trainer or coach 训练,指导

coach dog n. *see* Dalmatian 斑点狗,跟车犬

coach horn n. (also **coaching horn**) a funnel-shaped, brass horn used by coachman to announce the approach of a horse-drawn carriage (车夫用来开道的漏斗形铜号)车夫号角 syn yard of tin

coach horse n. a horse used for pulling a coach 舆马,挽马

coach pole n. *see* pole (马车的)驾杆,辀

coaching club n. **1** a club established to organize and regulate coaching events (马车)驾车俱乐部 **2** (esp. **Coaching Club**) an organization founded in England in 1870 to organize coaching events (1870 年在英国成立的组织机构,主要负责驾车运动比赛的组办)英国驾车俱乐部

coaching crop n. *see* coaching whip 驾车鞭

coaching horn n. *see* coach horn 车夫号角

coaching marathon n. a long-distance race for four-horse coaches first organized by the Royal International Horse Show in 1909 in England (首次于 1909 年由英国皇家国际马匹展览赛协会组织举办的驷马联驾长途赛)驾车马拉松,长途驾车赛

coaching whip n. a long-handled whip used by coachmen to drive harness and coach horses (车夫使用的长柄鞭杆)驾车鞭 syn driving whip, coaching crop

coachman /ˈkəʊtʃmən; *AmE* ˈkoʊ-/ n. one who drives a coach or carriage (驾驭马车的人)车夫 syn dragsman

coachman's seat n. the seat on which the coachman sits to drive a horse-drawn vehicle 驾驶座,车夫座

coach-step n. a step by which the passengers get into a coach 马车踏脚

coagulant /kəʊˈægjʊlənt; *AmE* koʊ-/ n.【physiology】a substance causing a liquid or blood to coagulate 凝血剂,凝血因子

coagulate /kəʊˈægjʊleɪt; *AmE* koʊ-/ vi. to change to a semisolid state 凝结,凝固: blood had coagulated around the edges of the cut 伤口边的血液已经凝固

vt. to cause to form a semisolid or solid mass; clot 使凝固 | **coagulation** /kəʊˌægjʊˈleɪʃn;

AmE koʊ-/ n.

coagulation vitamin n. one kind of fat-soluble vitamin acting as a coagulant in blood curdling; vitamin K (一种脂溶性维生素,参与血液凝固)凝血维生素,维生素 K

coarse /kɔːs; *AmE* kɔːrs/ adj. **1** (of a horse) lacking refinement and quality (指马)粗糙的,低劣的:a coarse horse 劣马,驽马 **2** (of feed or forage) rough (饲草)粗糙的

coat /kəʊt; *AmE* koʊt/ n. the natural outer covering of a horse (马的)被毛:coat color 毛色 ◇ coat color pattern 毛色类型 ◇ summer coat 夏季被毛 ◇ winter coat 冬季被毛

cob /kɒb; *AmE* kɑːb/ n. **1** *see* corncob 玉米棒子 **2** (usu. **Cob**) a type of short-legged, stocky horse 矮壮马;柯柏马: Welsh Cob 威尔士柯柏马

cobalt /ˈkəʊbɔːlt; *AmE* ˈkoʊ-/ n.【nutrition】a trace element vital in synthesis of vitamin B_{12} (一种微量元素,参与维生素 B_{12} 的合成)钴

Cobb & Co coach n. *see* New Zealand Cobb & Co coach 新西兰考伯四轮马车

cobby /ˈkɒbi/ adj. stout or thick like a cob 粗壮的:cobby neck 颈粗短

coccygeal /kɒkˈsɪdʒɪəl/ adj. of or relating to the coccyx 尾骨的:coccygeal vertebrae 尾椎骨

coccygeus /kɒksɪdˈʒɪəs/ n.【Latin】coccygeal 尾骨的:coccygeus muscle 尾骨肌

coccyx /ˈkɒksɪks; *AmE* ˈkɑːk-/ n.【anatomy】(*pl.* **coccyxes**/-ksiːz/) a small triangular bone at the base of the spinal column, consisting of several vertebrae (脊柱末梢的一块三角形骨,由多块脊椎组成)尾骨 [syn] tail bone

cock fence n.【jumping】an obstacle made from a natural hedge trimmed to a low level 矮树篱障碍

cockhorse /ˈkɒkˌhɔːs/ n. **1** a toy horse, such as a rocking horse 挽具马;摇摆木马 **2**【driving】a horse added to a team to assist in pulling a heavily loaded wagon uphill (重载马车上坡时加在马队前的马匹)外加马,外援马

cockhorse boy n. a young, light-weight boy trained to ride a cockhorse 外加马策骑者

Cocking cart n. a two-wheeled, horse-drawn British cart popular in the late 18th century, used for travel to sport events such as cock fights from which the name derived (18 世纪末流行于英国的双轮马车,用来参加斗鸡比赛而得名)斗鸡马车

cocktail /ˈkɒkteɪl; *AmE* ˈkɑːk-/ n.【slang】a horse with a docked tail 尾巴截短的马

coconut meal n.【nutrition】shredded coconut 椰子粕

coffin bone n.【anatomy】a small bone encapsulated by the hoof (马蹄内的一块小骨)蹄骨

coffin joint n. the lowermost joint of the horse's leg located between the short pastern bone and the coffin bone within the hoof (马四肢位于短系骨与蹄骨之间的关节)蹄骨关节

Coggins test n.【vet】a test performed by a veterinarian for detecting equine infectious anemia (兽医对马传染性贫血进行的检测)柯金斯检测,马传染性贫血检测:A Coggins test is usually required to attend horse shows, travel out of state, or begin boarding at a new stable. 马匹在参加展览、出境或入住新的厩舍时一般都要求进行马传染性贫血检测。

coitus /ˈkɔɪtəs, ˈkəʊɪtəs; *AmE* ˈkoʊ-/ n. the sexual mating between a stallion and a mare (公马与母马的)交配:natural coitus 自然交配

cold /kəʊld; *AmE* ˈkoʊld/ adj.【racing】(of a horse) unlikely to win a race (指马)无望获胜的,冷门的

cold back n. a condition in which a horse shows resistance to saddling and tightening of the girth (马备鞍和系肚带时有对抗情绪的情况)冷背

cold brand n. *see* freeze brand 冷冻烙印,冻号

cold fit vt.【farriery】to cold shoe a horse 冷装蹄 [compare] hot fit

cold hose vt. to run a stream of cold water over an injury to reduce inflammation (用冷水冲伤口

C

以消除炎症)冲水冷敷

cold line n.【hunting】*see* stale line 变淡的踪迹

cold scent n.【hunting】*see* stale line 变淡的气味

cold shoe n. a ready-made horseshoe 预制蹄铁 compare hot shoe

vt. to shoe a horse with ready-made shoes shaped to fit the hoof on the anvil without heating it in a forge; cold fit (不用熔炉煅烧而在铁砧上将预制蹄铁打造成形进行装钉)冷装蹄 | **cold shoeing** n. : Compared with hot shoeing, cold shoeing is a much faster shoeing process. 与热装蹄相比,冷装蹄要快很多。

cold shoer n. a farrier who shoes horses without heating the horseshoe in a forge to shape it to the hoof (不用熔炉煅烧蹄铁而直接进行装蹄的蹄匠)冷装蹄工 compare hot shoer

cold-backed /kəud'bækt/ adj. (of a horse) showing resistance to saddling and tightening of the girth 冷背的: The rider should avoid sitting down heavily into the saddle especially if the horse is cold-backed. 对于冷背的马,骑者在坐到马鞍上时切忌动作过重。

coldblood /'kəuldˌblʌd/ n. (also **cold blood**) a breed of heavy draft horses originating in Europe and descended from the prehistoric Forest Horse; generally has a heavy body, strong legs with large hoofs and feathering, short muscular neck with a heavy head; used mainly for farm work and carriage (源于欧洲的重挽马品种,主要由史前森林马演化而来。体型高大,四肢粗壮,蹄大且附生距毛,颈粗短,主要用于农田作业和驾车)冷血马: coldblood breeds 冷血马品种 compare hotblood, warmblood

coldblooded /ˌkəuld'blʌdɪd/ adj. of or relating to a coldblood horse 冷血马的

cold-jawed /'kəulddʒɔːd/ adj. (of a horse) having a tough mouth and resistant to the effects of the bit (指马)口角不灵的;抗衔的

colic /'kɒlɪk; *AmE* 'kɑːlɪk/ n.【vet】a sharp and severe pain of the abdomen 腹痛,绞痛: spasmodic colic 痉挛性腹痛 ◇ impactive colic 阻塞性腹痛 ◇ sand colic 积沙性腹痛 ◇ flatulent colic 胀气性腹痛 syn gripes, stomach staggers

colitis /kə'laɪtɪs/ n. inflammation of the colon 结肠炎 syn colonitis

collagen /'kɒlədʒən; *AmE* 'kɑːl-/ n. the fibrous protein in bone, cartilage and other connective tissue (存在于骨和软骨等结缔组织的纤维蛋白)胶原[蛋白]: collagen granuloma 胶原肉芽瘤

collar /'kɒlə(r); *AmE* 'kɑːlə(r)/ n. **1** a band of leather put around the neck of a dog or hound (套在猎犬颈上的皮圈)项圈 **2**【driving】the harness collar fitted around the the shoulder of a carriage horse (套在车马肩胛上的皮圈)臁脖,套包,颈圈: heavy transport collar 重载运输颈圈

collar work n.【driving】any work that requires the horse to strain against the collar (马在套包上使劲的)出力活,卖力活: To drive uphill certainly needs all the collar work of a carriage horse. 驾车上山时马需要出大力气。

collect /kə'lekt/ v. **1** to harvest semen from a stallion 采精 **2** (of a horse) to shorten the stride or pace (指马)收步,缩步: After receiving the cues from its owner, the horse collected his stride and came to a stop. 听到主人的讯号后,马收步并停了下来。**3** (of a rider) to restrain or shorten the pace of a horse by pulling the reins (骑手)收缰 **4**【racing】(of a bettor) to win a wager in a horse race (赛马中投注)赢利

collected /kə'lektɪd/ adj. (of gait) shortened and engaged in stride (步法)收缩的,缩短的: collected canter 收缩跑步 ◇ collected trot 收缩快步 ◇ collected walk 收缩慢步

collection /kə'lekʃn/ n. the shortening of a horse's stride in movement (马步幅)缩短,收缩 PHR V **lack collection** (of a horse) to move with insufficient shortening of the gait or pace (指马)运步收缩不足

collie /'kɒli; *AmE* 'kɑːli/ n. a sheepdog originating in Scotland who has shaggy hair and a long, pointed muzzle (源于苏格兰的牧羊犬，毛粗而口鼻尖)柯利牧羊犬

collier /ˈkɒliə(r); *AmE* ˈkɑːl-/ n. *see* pit pony 矿井矮马

collum /ˈkɒləm; *AmE* ˈkɑːl-/ n. 【Latin】(*pl.* **colla**/ˈkɒlə/) neck 颈：collum costae 肋颈

colon /ˈkəʊlən; *AmE* ˈkoʊ-/ n. 【anatomy】the section of the large intestine extending from the cecum to the rectum (大肠中从盲肠到直肠的部分)结肠：ascending colon 升结肠 ◇ descending colon 降结肠 ◇ sigmoid colon 乙状结肠

colon impaction n. difficult passage of feces resulting from inactivity, debility, senility, obstruction due to a foreign body, or poor feeding (由于活动量不足、体虚、异物阻塞或营养不良造成的排粪阻塞)结肠阻塞，便秘 syn constipation

colonitis /kɒləˈnaɪtɪs/ n. *see* colitis 结肠炎

colonize /ˈkɒlənaɪz; *AmE* ˈkɑː-/ vi. to settle in a colony or colonies 寄居

colony /ˈkɒləni; *AmE* ˈkɑː-/ n. a visible growth of microorganisms in a group (微生物)群体；菌落

color /ˈkʌlə(r)/ n.【AmE】the general pigmentation of hairs and skin of horses (马被毛和皮肤的颜色)毛色：base color 基色 ◇ body color 体色 ◇ color breed 毛色品种 ◇ compound color 复毛色 ◇ coat color 毛色 ◇ dark color 暗色 ◇ light color 浅色 ◇ solid color 单毛色 ◇ The horse's muzzle is referred to when identifying his true color. 在鉴定马真正的毛色时，要以其口鼻为准。

color gene n. a gene that influences the coat color of a horse (影响马匹毛色的基因)毛色基因

color settings n. the standards and system of distinguishing different coat colors of horse (区分马匹毛色的标准和体系)毛色体系：It is quite a normal thing that color settings in many cultures may vary and have different standard. 马匹毛色体系在不同文化下分类标准可能不尽相同，这是非常正常的事。

Colorado Ranger n. (also **Colorado Ranger Horse**) a breed of horse originated in Colorado, USA; heavily influenced by the Appaloosa horses; stands 14.2 to 16 hands and may have any of the coat colors except pinto; has a long, muscular neck, sloping shoulder, and deep chest; used mainly as ranch horses (源于美国科罗拉多州的马品种，育成过程受阿帕卢萨马影响较大。体高14.2～16掌，除花毛外各种毛色兼有，颈长，斜肩，胸深，主要用于牧场作业)科罗拉多牧用马

colored /ˈkʌləd; *AmE* -lərd/ adj. **1** (of an Appaloosa) showing any coat color combination (指阿帕卢萨马)带花毛色的 **2** (of a horse) showing more than one coat color (指马有两种以上毛色)花色的

colored horse n. a horse having more than one coat color 花色马，花马：A colored horse always stands out. 花马总是非常抢眼。

colors /ˈkʌləz; *AmE* -lərz/ n. **1**【racing】*see* racing colors 赛服颜色，骑师彩衣 **2**【hunting】the distinctively colored hunt coat worn by the hunt staff and such members of the field (猎狐成员所穿的彩色服装)猎服颜色

colostrum /kəˈlɒstrəm; *AmE* kəˈlɑː-/ n. the thick, yellowish milk secreted by a mare's mammary glands during the first three or four days after foaling (母马产驹后三四天内分泌的淡黄色黏稠乳汁)初乳：Colostrum is rich in antibodies and minerals that will provide the foal with temporary immunity against infectious diseases. 母马初乳富含大量的抗体和矿物质，可保证幼驹在短期内对传染病有足够的免疫力。syn foremilk, first milk

colour /ˈkʌlə(r)/ n.【BrE】color 颜色；毛色

colpitis /kɒlˈpaɪtɪs/ n. *see* vaginitis 阴道炎

colt /kəʊlt; *AmE* koʊlt/ n. a male horse under three years old (3岁以下的小公马)雄驹 compare filly

colt bit n. *see* yearling bit 周岁驹衔铁

colt foal n. the male sex of a foal 雄驹

coma /ˈkəʊmə; *AmE* ˈkoʊ-/ n. a state of deep unconsciousness as the result of injury, disease, or poison(由于受伤、患病或中毒所致的知觉丧失)昏迷

comb /kəʊm; *AmE* koʊm/ n. a toothed plastic or metal tool used for pulling or cleansing the mane and tail of a horse(用来梳理清洁马鬃尾的带齿工具)梳子: mane comb 鬃梳 ◇ tail comb 尾梳

vt. to dress the hair with a comb 梳理

combination /ˌkɒmbɪˈneɪʃn; *AmE* ˌkɑːm-/ n. **1** *see* combination fence 组合障碍 **2** *see* combination bet 组合彩

combination bet n.【racing】(also **combination**) any across-the-board bet for which a single mutuel ticket is issued 组合彩

combination fence n.【jumping】(also **combination**) any obstacle consisting of two or more separate jumps with each set apart by one or two strides(由两个以上的独立跨栏组成的障碍,其中两跨栏间隔1~2步)组合障碍: The maximum distance between any two jumps of combination fence may not exceed 39 feet 4 inches. 组合障碍中两障碍的间距最大不得超过39英尺4英寸(12米)。[syn] combination obstacle

combination horse n. a horse used for both riding and driving(用于骑乘和驾车的马匹)兼用马

combination obstacle n. *see* combination fence 组合障碍

combined driving trail n. a driving competition consisting of presentation, dressage test, the marathon, and the obstacle course(由表演赛、舞步赛、长途赛和障碍赛构成的驾车比赛)综合驾车赛

combined immunodeficiency n. (abbr. **CID**) a genetically based fatal disease resulting in a failure of the immune system(由于免疫机能缺陷所致的致死性遗传病)综合免疫缺陷

combined test n.【eventing】a special combined training event conducted in one day which consists of dressage and show jumping phases(在1天之内进行盛装舞步和障碍赛两个项目的考核)综合赛

combined training n. (also **combined training event**) an equestrian competition consisting of show jumping, cross country, and dressage phases conducted over a period of one to three days(在1~3天内进行盛装舞步、越野赛和障碍赛三个项目的比赛)综合训练赛,综合三项赛: Combined training differs from the Three-Day Event in its exclusion of the endurance phase. 综合三项赛与三日赛的区别在于,它不包括耐力赛。[syn] eventing, militaire

combined training event n. *see* combined training 综合训练赛,综合三项赛

come /kʌm/ vi. **1** to reach a certain place 到达 PHR V **come to a halt** (of a horse) to slow up and come to a standstill(指马)放慢停下 **come again** (of a horse) to regain its energy or strength 再次发力 **2** (of a horse) to reach a certain age(指马)到达某个年龄: The bay was coming three years of age. 这匹骝毛马快三岁了。

commensurate /kəˈmenʃərət/ adj. equal or corresponding in size or degree(在大小或程度上)等同的,对等的

comment card n. a card on which the judge writes comments to help him place the riders at the end of the show(展览赛中裁判对骑手做记录便于随后排名的卡片)评判卡,评分卡: Riders may view the comment card after the show has ended. 骑手可以在比赛结束后查看评分卡。

commingle /kəˈmɪŋgl/ vt.【racing】to combine the mutuel pools from off-track sites with the host track(主场和外场赌金)合并

comminute /ˈkɒmɪnjuːt/ vt. to reduce to particles or powder 粉碎

comminuted fracture n.【vet】a break of a bone consisting of more than two fragments 粉碎性骨折

commission bet n.【racing】a wager in which a percentage of the resulting winnings is paid to the party who placed the bet on behalf of someone（受委托替他人投注并按比例抽取奖金的投注方式）代理投注，委托投注

commissure /ˈkɒmɪsjʊə/ n.【anatomy】the point or surface where two parts of an organ join or form together（器官两部分的结合面）接合处，联合：dorsal commissure 背侧联合 ◇ ventral commissure 腹侧联合

commit /kəˈmɪt/ vt.【cutting】（of a horse or rider）to prepare to work on a cow（指马或骑手）准备截牛，开始拦截

common /ˈkɒmən; *AmE* ˈkɑːm-/ adj.（of a horse）having a coarse simple appearance（指马）体貌普通的：common breed 普通马种

common alum n. a mineral salt used in medicine as an astringent and styptic（医疗上作收敛剂和止血剂使用的矿物质）明矾 [syn] potash alum

common salt n. sodium chloride 食盐，氯化钠

common zebra n. a species of horse family that lives in West Africa，noted for the wide prominent stripes all over the body（一种生活在西非的马属动物，身上的黑白条纹宽而明显）普通斑马：Among the wild *Equidae* only the common zebra has survived in largest numbers. 在所有野生马属动物中，只有普通斑马现今存活数量最大。[syn] plains zebra，Burchell's zebra

common-bred adj.（of a horse）bred from mixed，non-pedigree or generic stock（指马）血统普通的

compact /kəmˈpækt; *AmE* ˈkɑːm-/ adj. closely and firmly united or packed together; dense 结实的；紧密的，致密的：a horse having a compact body 一匹体格结实的马

compact bone n.【anatomy】（also **dense bone**）the tissue on the outside of the bone shaft（骨干外围的组织）骨密质 [compare] spongy bone

companion /kəmˈpæniən/ n. a person or animal who accompanies sb; mate 同伴；伴侣：companion animal 伴侣动物

vt. to be a companion to sb; accompany 陪伴

compete /kəmˈpiːt/ vi. to strive or contest with others in a game or activity 比赛，竞争

competition /ˌkɒmpəˈtɪʃn; *AmE* ˌkɑːm-/ n. an athletic contest between riders or horses（马或骑手之间的）竞赛，比赛：competition horses 竞技用马

competitive /kəmˈpetətɪv/ adj. **1** of，involving，or determined by competition 竞争的 **2** liking competition or inclined to compete 爱竞争的，好斗的：competitive temperament 喜好争斗

complementary lameness n. lameness in a previously sound limb due to pain in another limb（病肢疼痛而使健康肢出现跛行的情况）代偿性跛行

complete /kəmˈpliːt/ adj. total or thorough; entire 完全的，完整的

vt. to make whole or bring to an end 完成，使完整

complete blood count n.【physiology】（abbr. **CBC**）the counting of each type of blood cell in a given sample of blood（对给定血样中各种类型的血细胞进行的计数）全血细胞计数

complete test n. *see* three-day event 三日赛

composition /ˌkɒmpəˈzɪʃn; *AmE* ˌkɑːm-/ n. the combining of distinct parts or elements to compose a whole（不同部分或元素构成整体）组成，构成：composition of mare's milk 马奶组成成分 ◇ composition of blood 血液构成

compound /ˈkɒmpaʊnd; *AmE* ˈkɑːm-/ adj. consisting of two or more ingredients or elements（由两种以上成分或元素组成）混合的，复合的：compound feeds 配合饲料

n.（usu. **compounds**）mixed feeds containing all the required nutrients 混合饲料

compound fracture n.【vet】a fracture in which

C

the broken bone lacerates soft tissue and penetrates through an open wound in the skin (骨折后断端刺伤软组织并穿出伤口)复合性骨折,开放性骨折 syn open fracture

compression bow n. *see* bandage bow 绷带压迫性屈腱肿胀

compulsory exercises n.【vaulting】a set of movements required to perform in a vaulting competition (马上体操比赛中的要求动作)规定动作 compare freestyle exercises

compulsory halt n.【eventing】the 10-minute break between the speed and endurance test phases in a three-day event (三日赛中速度与耐力赛中两个阶段之间的10分钟休息)规定暂息时间,例行休息时间

computerized tomography n. (abbr. **CT**) a diagnostic technique in which a large scanning instrument revolves around a body part in a 180-degree arc to form an image of the internal structure (采用大型仪器绕体周扫描以对内部结构成像的诊断方法)计算机断层造影术,CT扫描 syn CT scanning

concave /kɒnˈkeɪv; *AmE* kɑːnˈ-/ adj. having an outline or surface that curves inward (外形或表面向内凹陷)凹面的,凹陷的:concave face 面颊凹陷 opp convex

conceive /kənˈsiːv/ vi. to become pregnant 受孕,妊娠

concentrate /ˈkɒnsntreɪt; *AmE* ˈkɑːn-/ n.【nutrition】an animal feed low in fiber and high in total digestible nutrients (粗纤维含量低而可消化养分含量高的动物饲料)浓缩料,精料:concentrate mix 混合精料

conception /kənˈsepʃn/ n. the state or instance of conceiving a foal 怀孕,妊娠:conception rate 受胎率

conceptus /kənˈseptəs/ n.【anatomy】the fetus of conception at any point between fertilization and birth (从受精到出生期间的胎儿)孕体:The conceptus may include the embryo or the fetus as well as the extraembryonic membranes. 孕体包括胚胎、胎儿以及胚胎外膜。

Concord /ˈkɒŋkɔːd; *AmE* ˈkɑːŋkɔːrd/ n. *see* Concord coach 康科德四轮马车

Concord coach n. (also **Concord**) a horse-drawn carriage developed in the early 20th century by Downing and Company of Concord, New Hampshire, USA, for which the vehicle was named (20世纪初由美国新罕布什尔州的唐宁康科德公司制造的四轮马车,由此得名)康科德马车 syn American mail coach

concours /kʊŋˈkuːər/ n.【French】competition 比赛,竞赛:concours hippique amical 马术友谊赛

Concours Complet n.【French】*see* three-day event 全能赛,三日赛

Concours Complet d'Equitation n.【French】*see* three-day event 全能赛,三日赛

Concours Voltige d'Amitié n.【French】(abbr. **CVA**) an international vaulting event open to competitors from the host country and up to four foreign nations (由主办国和其他多国选手参加的国际马上杂技比赛)马上体操友谊赛

Concours Voltige Frontière n.【French】(abbr. **CVF**) an international vaulting event open to competitors from the host country and up to four foreign nations (由主办国和其他多国选手参加的国际马上杂技比赛)马上体操国际赛

Concours Voltige Internationale n.【French】(abbr. **CVI**) an international vaulting event with no limit to the number of foreign nations participating(一项国际马上体操比赛,对国外参赛选手不加限制)国际马上体操比赛

condition[1] /kənˈdɪʃn/ n. a horse's overall health and fitness 体况,膘情:in bad/good condition 体况差/好 ◇ loss of condition 体况下降 ◇ to put on condition 恢复体况 ◇ condition scoring 体况评分 ◇ The condition scoring is a method to assess the condition of horse subjectively, it is measured using a scoring system from 0 to 5, assessing manually and visually the horse's weight displacement along the neck, over the ribs and over the quarters. 体况评分是主观性地评定马匹体况的一种方法，通过触摸和目测马颈部、肋骨以及臀部的膘情分布，把体况划分为0～5六个等级的评分体系。

condition² /kən'dɪʃn/ vt. to make a horse fit for work by training 对马进行适应性训练

conditioned /kən'dɪʃnd/ adj. subject to or dependent on a condition or conditions; trained 条件性的；训练的：conditioned reflex 条件反射 opp unconditioned

conditioner /kən'dɪʃənə(r)/ n.【nutrition】an additive or supplement that improves the quality or palatability of feeds（能改进饲料品质或适口性的添加剂）调理剂

conditions race n.【racing】(also **terms race**) a horse race in which the weights carried by the horses according to their age, sex, and quality are laid down by the conditions attached to the race（根据马匹特定性别、年龄和级别条件规定其负重的比赛）规格赛，条款赛 compare handicap race

conduce /kən'djuːs/ vi. to contribute or lead to a specific result 导致，有利于

conducive /kən'djuːsɪv/ adj. contributive; favorable 有助于：If the weather is not conducive to shampooing the horse will have to be groomed and strapped. 如果天气不利于马匹洗浴，就只能对马匹进行刷拭和清洁了。

condylar /'kɒndɪlə(r)/ adj. of or relating to condyle 踝的

condylar fracture n. a break in the distal knobby end of a long bone（长骨末端开裂）踝关节骨折

condyle /'kɒndɪl/ n.【anatomy】a rounded prominence at the end of a bone, most often for articulation with another bone（骨末端的圆状突起，多起骨连接作用）踝：lateral condyle 外踝 ◇ medial condyle 内踝

condylus /'kɒndɪləs/ n.【Latin】condyle 踝：condylus occipitalis 枕踝 ◇ condylus temporalis 颞踝

Conestoga /ˌkɔnis'təʊgə; *AmE* ˌkɑːnə'stoʊgə/ n. *see* Conestoga Wagon 四轮篷车

Conestoga Wagon n. a large, horse-drawn, covered farm wagon popular in the United States during the 1850s（19世纪中叶在美国较为流行的一种大型农用四轮马车）四轮篷车 syn Conestoga

confidential /ˌkɒnfɪ'denʃl; *AmE* ˌkɑːnfɪ'-/ adj.【BrE, dated】(of a horse) submissive and suitable for a novice or elderly rider（指马）忠诚的，听话的

confine /kən'faɪn/ vt. **1** to keep within bounds 限制，限定 **2** to restrict the freedom and movement 束缚，禁锢

confinement /kən'faɪnmənt/ n. the act of confining or the state of being confined 限制；束缚，禁锢：The boredom of the solitary permanent confinement to the four walls of a stable is the major cause of most stable vices. 马大多数的恶习都是由于长期被禁锢于厩舍而产生厌倦所致。

conformation /ˌkɒnfɔː'meɪʃn; *AmE* ˌkɑːnfɔːr'm-/ n. the bodily structure and shape of a horse（马的）身体结构；体貌：conformation defects 体貌缺陷，失格特征 ◇ ideal conformation 正肢势，理想肢势 ◇ A horse with good conformation is stronger and more likely to stay sound than one with weak conformation. 体格优良的马比体格差的马更为强健，而且体况多半也更为健康。

congenital /kən'dʒenɪtl/ adj. **1** acquired before birth in the uterus; innate 先天的 **2** existing at or dating from birth 天生的：congenital defect 先天缺陷 ◇ congenital deformity 先天畸形

congenital abnormality n. any structural or functional defect existing at birth 先天异常，先天畸形 syn congenital defect, congenital deformity

congenital mark n. any mark on the coat of the horse present from birth（幼驹出生时身上的标记）先天别征

C

congenital trait n. any trait acquired during the development of the fetus in the uterus（胎儿在妊娠发育过程中获得的特征）先天性状

congested /kən'dʒestɪd/ adj.（of body parts or organ）accumulated with excessive blood（机体器官）充血的

congestion /kən'dʒestʃən/ n. the state of being congested 充血：congestion of the lungs 肺充血

congestive /kən'dʒestɪv/ adj. of or relating to congestion 充血的：congestive heart failure 充血性心力衰竭

conjunctiva /ˈkɒndʒʌŋk'taɪvə/ n.【anatomy】（*pl.* **conjunctivae** /-viː/）the mucous membrane that lines the inner surface of the eyelid（衬在眼睑内的黏膜）结膜

connection /kə'nekʃn/ n.【racing】the relation between the owner, trainer, and others involved with custodianship of a horse（马主、练马师等之间与马的联系）关系，关联

connective tissue n.【anatomy】tissue that forms the supporting and connecting structures of the body and includes collagen, elastic and reticular fibers, adipose tissue, cartilage, and bone（身体中起支撑和连接作用的组织，主要包括胶原、弹性和网状纤维、脂肪组织、软骨和骨）结缔组织：Various types of connective tissue together with the skeletal muscles make up the musculoskeletal system. 各种结缔组织和骨骼肌组成了肌肉骨骼系统。

Connemara /ˌkɒnɪ'mɑːrə/ n.（also **Connemara pony**）a breed of pony originating in Ireland; has a small head with a straight profile, a long, well-formed neck, prominent withers, full mane and tail, sturdy and well-muscled legs, and a slightly sloping croup; stands 13 to 14.2 hands and generally has a gray, dun, black, bay, brown, and occasionally roan or chestnut coat; is hardy, docile, intelligent, sound, and generally raised in the wild; historically used for farming and packing, but now used for riding and jumping（源于爱尔兰的矮马品种。头小，面貌平直，颈长且姿态优美，高鬐甲，鬃、尾浓密，体格健壮，四肢结实，尻稍斜，体高 13～14.2 掌，毛色多为青、兔褐、黑、骝或褐，沙毛和栗少见，抗逆性强，性情温驯，反应敏快，多在户外饲养，过去主要用于农田作业和驮运，现今则主要用于骑乘和障碍赛）康尼马拉矮马：The Connemara is the only breed of horse indigenous to Ireland. 康尼马拉矮马是爱尔兰唯一的本土马种。

Connemara Breeder's Society n. *see* Connemara Pony Breeder's Society 康尼马拉矮马育种者协会

Connemara Pony n. *see* Connemara 康尼马拉矮马

Connemara Pony Breeder's Society n.（also **Connemara Breeder's Society**）an organization founded in Ireland in 1923 to register and promote the breeding of the Connemara（1923 年在爱尔兰成立的组织机构，旨在促进康尼马拉矮马的育种工作并对其进行品种登记）康尼马拉矮马育种者协会

Connemara Pony Society n. *see* English Connemara Pony Society 英国康尼马拉矮马协会

conquistador /kɒn'kwɪstədɔː(r); *AmE* kɑːŋ'kist-/ n. one of the 16th-century Spanish explorers of Latin America who introduced the Spanish stock there（16 世纪将西班牙马种引入拉丁美洲的探险者）西班牙殖民者

consignor /kən'saɪnə/ n. one who offers a horse for sale through an auction（拍卖会上出售马匹的人）卖马者，移交者

conspicuous /kən'spɪkjuəs/ adj. easy to notice; obvious, noticeable 显见的；明显的：The presence of markings in domestic horses is one of the conspicuous features that distinguish them from their closest wild relatives, Przewalski's horses. 家马的白色别征是它们区别于最近的野生家系普氏野马的明显特征之一。

constipation /ˌkɒnstɪ'peɪʃn; *AmE* ˌkɑːn-/ n. *see* colon impaction 便秘

constrict /kən'strɪkt/ v. to make or become smaller or thinner by squeezing; compress; contract 压缩;收缩:The cervix relaxes or constricts according to the sexual state, being open in estrus and closed in diestrus and pregnancy. 子宫颈根据性周期有规律地松弛和收缩,在发情期开张,在间情期和妊娠时闭合。| **constriction** /-kʃn/ n.

contact /ˈkɒntækt; *AmE* ˈkɑːn-/ n. the connection between the rider's hands and the horse's mouth achieved through the reins (骑手通过缰绳与马嘴建立的联系)接触

contagion /kənˈteɪdʒən/ n. **1** the transmission of a disease by contact (疾病的)传染,传播 **2** a disease transmitted by direct or indirect contact 传染源,传染病

contagious /kənˈteɪdʒəs/ adj. of or relating to contagion; infectious 传染的,传染病的

contagious equine metritis n. (abbr. **CEM**) a highly contagious bacterial venereal disease transmitted during mating of horses; usually decrease the conception rate of mare after infection (通过马匹交配传播的一种高度传染性细菌病,常造成母马受孕率下降)马传染性子宫炎

content /ˈkɒntent; *AmE* ˈkɑːn-/ n. 【nutrition】 the proportion of a specified substance (某种物质的比例)含量:fibre/protein content 纤维/蛋白质含量

contraceptive /ˌkɒntrəˈseptɪv; *AmE* ˌkɑːn-/ adj. capable of preventing conception 阻止受孕的,节育的

contract n. /ˈkɒntrækt; *AmE* ˈkɑːn-/ an official written agreement 合同,合约: to make/ sign a contract with the owner 与所有者签订合同

v. /kənˈtrækt/ to become less or smaller or reduce in size; shrink 收缩:the heart muscles contract to pump the blood. 心肌通过收缩泵血。

contract rider n.【racing】a jockey obligated by contract to ride in races 签约骑师,签约骑手

contracted feet n. *see* contracted heels 蹄踵萎缩

contracted heels n. a narrowing of the heels and frog of the hoof; may be caused by improper shoeing or trimming of the feet; symptoms include a shortened stride and heat around the quarters and heels (马蹄踵和蹄叉变窄的情况,多由装蹄或削蹄不当所致,症状表现为步幅缩短和蹄踵肿痛)蹄踵萎缩 [syn] contracted feet

contractile /kənˈtræktaɪl/ adj. capable of contracting or causing contraction 可收缩的,收缩性的:contractile tissue 收缩性组织

contraction /kənˈtrækʃn/ n. the act of contracting or the state of being contracted 收缩

controlled rear n. *see* levade 静立,直立

contuse /kənˈtjuːz/ vt. to injure without breaking the skin; bruise 挫伤

contusion /kənˈtjuːʒn; *AmE* -ˈtuː-/ n. a blunt injury that damages the deep tissues without breaking the skin (表皮未破而深层组织受伤的情况)挫伤

convalesce /ˌkɒnvəˈles; *AmE* ˌkɑːn-/ vi. to return to health and strength after illness; recuperate (患病后恢复原有体力和健康)康复;复原

convalescence /ˌkɒnvəˈlesns; *AmE* ˌkɑːn-/ n. **1** the recovery to health from illness (患病后)康复,复原 **2** the period of time needed for recovery 康复期

conversion /kənˈvɜːʃn; *AmE* -ˈvɜːrʒn/ n. the act or instance of converting 转化,转变:feed conversion efficiency 饲料转化率

convert /kənˈvɜːt; *AmE* -ˈvɜːrt/ vt. to change sth into another form or state; transform 转化,转变

convex /ˈkɒnveks; *AmE* ˈkɑːn-/adj. having an outline or surface that curves outward (外形或表面向外突出)凸面的,突起的 [opp] concave

convolution /ˌkɒnvəˈluːʃn; *AmE* ˌkɑːn-/ n.【anatomy】one of the convex folds of the surface of the brain (大脑表面凸出的皱褶)脑回:convolution of cerebrum 大脑回 ◇ occipito-

C

temporal convolution 枕颞回 | **convolutionary** /-ʃnəri/ adj.

convulsion /kən'vʌlʃn/ n. a sudden, intense, involuntary muscular contraction; spasm (肌肉突然非自主性强烈收缩)抽搐,痉挛

convulsive /kən'vʌlsɪv/ adj. marked by or having convulsions 抽搐的,痉挛的

cook /kʊk/ vt.【AmE, slang】to overtrain a horse to the point of boredom; exhaust 使马力竭,训练过度

cooked /kʊkt/ adj. (of a young horse) overtrained to the point of boredom; overstressed 训练过度的;厌倦的

cool /kuːl/ v. to become or cause to become less hot or excited (使)降温;放松 PHR V **cool down/out** to restore one's normal temperature slowly after physical exertion (在训练后使体温缓慢降低)[使]放松,[使]降温: Horses need to cool down after strenuous exercises. 在大强度训练后,马需要放松下来。◇ The horse could take a bath after being cooled out. 马匹体温降下来后就可以洗浴。

cooler /'kuːlə(r)/ n. **1** a cotton or fabric blanket covered over the neck and back of a horse to allow him cool down slowly after a workout or race (马训练或赛后披在颈背上的薄单,用来使马缓慢降温)凉爽马衣 **2**【racing】a horse restrained by the rider to prevent it from running well 速度受遏制的赛马

coon foot n. a conformation defect in which the pastern slopes too much (马蹄系过于倾斜的四肢缺陷)卧系蹄 — *see also* **boxy foot**, **club foot**

coon-footed /kuːn'fuːtid/ adj. (of feet) with the pastern sloping more than the hoof angle (马蹄系比蹄壁更为倾斜)卧系的

COPD abbr. chronic obstructive pulmonary disease 慢性阻塞性肺炎,气喘症

coper /'kəu pə(r)/ n. *see* horse-coper 牙子,马贩子

coprophagy /kəprɒ'fædʒɪ/ n. (also **coprophilia**) a vice of horse having the habit of eating feces 食粪癖

coprophilia /ˌkɒprə'fɪliə/ n. *see* coprophagy 食粪癖

copulate /'kɒpjuleɪt; *AmE* 'kɑːp-/ vi. to mate or cover 交配

copulation /ˌkɒpjʊ'leɪʃn/ n. the act of copulating; intercourse 交配

copulation time n. period of time from the moment of insertion into the vagina until the stallion dismounts after ejaculation (从公马阴茎插入母马阴道至射精后下马所需的时间)交配时间

cord /kɔːd; *AmE* kɔːrd/ n. a rope made of twisted strands 绳索

cordage /'kɔːdɪdʒ/ n. cords or ropes called collectively [总称]绳索

cording up n. *see* tying-up syndrome 僵直症,强拘综合征

cordovan /'kɔːdəvən; *AmE* 'kɔːr-/ n. a fine leather made of split horse hides, pig skin, and goat hides (用马、猪或山羊皮鞣制成的革)科尔多瓦革

adj. made of cordovan leather 科尔多瓦革制的

Corinthian whip n. a driving whip used in England in the early 19th century (19 世纪初在英国使用的一种驾车用鞭)科林斯鞭

corium /'kɔːrɪəm/ n.【anatomy】*see* dermis 真皮

cork /kɔːk/ n. **1** blocked heel 折弯的蹄铁尾 **2** *see* calk (蹄铁上的)尖铁

Corlay /'kɔːleɪ; *AmE* 'kɔːrleɪ/ n. a coldblood originating in Brittany, France more than 4,000 years ago; descended from native Brittany mares crossed with Norfolk Roadsters; mainly used for riding and light draft work; is now quite rare if not extinct (4 000 多年前源于法国布列塔尼半岛的冷血马品种,由当地布列塔尼马母马与诺福克马公马杂交培育而来,主要用于骑乘和轻挽,现存数目很少,已几近消失)科尔莱马

corn /kɔːn; *AmE* kɔːrn/ n. **1** (AmE) a cereal

grain used as the feed of horses 玉米 [syn] maize **2** (usu. **corns**) bruises on the sole of the foot 蹄底硬伤;钉胼

corncob /ˈkɔːnkɒb; *AmE* ˈkɔːrnkɑːb/ n. (also **cob**) the long, hard, cylindrical core on which the yellow grains of corn grow (圆锥形的)玉米棒;玉米芯

corner incisors n.【anatomy】four of the incisor teeth of the horse located on either side of the laterals top and bottom (位于马上下颌两侧的4个切齿)隅齿,边齿:The corner incisors usually appear about 10 months after birth. 隅齿约在幼驹出生10个月后长出。[syn] corners, third incisors

corner man n.【cutting】a mounted rider responsible for keeping a herd of cattle to be worked by a cutter in the center of the pen (截牛中负责将牛群赶到赛场中心以待拦截的骑手)边角骑手

corners /ˈkɔːnəz; *AmE* ˈkɔːrnəz/ n. pl. *see* corner incisors 隅齿,边齿

coronary /ˈkɒrənri; *AmE* ˈkɔːrəneri/ adj. **1** of or relating to coronet 蹄冠的 **2** of or relating to the coronary vessels of the heart 冠状动/静脉的:coronary veins 冠状静脉 ◇coronary sinus 冠状窦

coronary artery n.【anatomy】either of two arteries that originate in the aorta and supply blood to the muscular tissue of the heart (主动脉的两条分支动脉,为心肌组织供应血液)冠状动脉 — *see also* **coronary vein**

coronary band n. *see* coronet band 蹄冠

coronary contraction of the foot n. a narrowing of the heels at the horn immediately below the coronary cushion (蹄冠下侧蹄踵角质萎缩的情况)蹄冠萎缩 [syn] local contraction of the foot

coronary crack n. a crack in the coronary band 蹄冠裂

coronary sinus n. *see* quittor 马蹄疽

coronary vein n.【anatomy】any of the veins that drain blood from the muscular tissue of the heart and empty into the coronary sinus (将血液从心脏肌肉组织汇入冠状窦的静脉)冠状静脉— *see also* **coronary artery**

coronation coach n. *see* state coach 加冕马车,仪仗马车

coronet /ˈkɒrənet; *AmE* ˌkɔːrəˈnet/ n. **1** *see* coronet band 蹄冠 **2** (also **coronet white**) a leg marking consisting of a narrow band of white hair which extends circumferentially around the coronet just above the hoof (马四肢别征之一,其中蹄冠周围毛发为白毛)蹄冠白

coronet band n. (also **coronet**) the part of the leg where the hoof meets the pastern (马蹄与系骨的结合部位)蹄冠 [syn] coronary band

coronet white n. *see* coronet 蹄冠白

corpus /ˈkɔːpəs; *AmE* ˈkɔːrpəs/ n.【anatomy】(*pl.* **corpora** /-rə/) the main part of a bodily structure or organ (体躯或器官的主要部分)体

corpus albicans n.【physiology】*see* white body 白体

corpus luteum n.【physiology】(*pl.* **corpora lutea**) a yellow, progesterone-secreting mass that forms from ovarian follicular cells after the release of a mature egg (成熟卵子排出后,由卵泡细胞形成黄色实体结构,可分泌孕酮)黄体:primary corpus luteum 初级黄体 ◇ secondary corpus luteum 次级黄体 [syn] yellow body

corpuscle /ˈkɔːpʌsl; *AmE* ˈkɔːrp-/ n. the blood cells, including the leucocytes and the erythrocyte 血细胞:red corpuscle 红细胞 ◇ white corpuscle 白细胞

corpuscular /kɔːˈpʌskjʊlə/ adj. of or relating to corpuscles 血细胞的;小体的

corral /kəˈrɑːl; *AmE* kəˈræl/ n. a small enclosure where a horse is turned out for exercise (牲畜)运动场;围场:A corral usually does not have grass, but rather sand, dirt or woodchips. 牲畜运动场通常没有草皮,而多为沙土或木屑铺垫。

C

corrected score n. a score awarded in a competition that is changed by the officials due to objection or a calculation error（比赛中由于异议或计分失误而对判分进行更改后的得分）更改后得分

corrective shoe n. **1** any horseshoe that corrects a defect in the stance or gait of a horse（可矫正马匹站姿或运步缺陷的蹄铁）矫正性蹄铁 **2** *see* therapeutic shoe 治疗性蹄铁

corrective shoeing n. the practice of shoeing a horse to change its balance or way of going as to modify gait or conformation faults（为了矫正马蹄着地平衡或运步缺陷而装钉蹄铁）矫正性装蹄

corrida /kɒˈriːdə/ n.【Spanish】a bullfighting performance 斗牛

corrugator /ˈkɒrʊˌgeɪtə(r)/ n.【anatomy】the muscle that wrinkles the brows 皱眉肌

cortex /ˈkɔːteks; *AmE* ˈkɔːr-/ n.【anatomy】(*pl.* **cortices**, or **cortexes**) the outer layer of an internal organ or body structure（体内器官或结构的外层）皮层，皮质：cerebral cortex 大脑皮层 ◇ renal cortex 肾皮质 compare medulla

cortical /ˈkɔːtɪkl/ adj. of, relating to, derived from, or consisting of cortex 皮质的：cortical hormone 皮质激素 ◇ cortical reaction 皮质层反应 ◇ cortical vesicles 皮质囊泡

corticosteroid /ˌkɔːtɪkəʊˈstɪərɔɪd/ n. any of the steroid hormones produced by the adrenal cortex, such as cortisol and aldosterone（由肾上腺皮质所生成的多种类固醇激素，如氢化皮质酮和醛固酮）皮质类固醇

cortisone /ˈkɔːtɪzəʊn/ n. a naturally occurring corticosteroid that functions primarily in carbohydrate metabolism and is used in the treatment of rheumatoid arthritis, adrenal insufficiency, certain allergies, and gout（一种天然类固醇皮质激素，能影响糖类代谢，常用于风湿性关节炎、肾上腺分泌不足、某些过敏症或痛风的治疗）皮质酮；可的松

cosh /kɒʃ; *AmE* kɑːʃ/ n. *see* bat 马鞭

costa /kɒstə/ n. a rib or rib-like part 肋骨

costal /ˈkɒstl/ adj. of or relating to the ribs 肋骨的：costal bone 肋骨

costal arch n. the arch of the chest cavity formed by the last nine ribs and the costal cartilage（由胸腔最后九根肋骨和肋软骨构成的弧弓）肋弓

costal cartilage n. the cartilage that joins the ribs to the sternum（连接肋骨和胸骨的软骨）肋软骨

cotton eye n. the eye of a horse with a lot of white showing around the iris（虹膜白色区域过多）白眼

cottonseed meal n.【nutrition】a high protein, high energy concentrate produced from the residue of the cottonseeds after oil extraction（棉籽榨油后产生的残渣，其能量和蛋白含量都较高）棉籽粕：The supplementation of cottonseed meal should not exceed 15 to 16 percent of the total ration due to the presence of the toxic gossypol. 由于棉籽粕含有有毒物质棉酚，因而在日粮中的含量不得超过15%~16%。

cough /kɒf; *AmE* kɔːf/ n. a sudden expulsion of air from the lungs marked by loud noise, usually resulting from inflammation or irritation to the airways（机体从肺内突然带声喷出气流，声音较大，多由呼吸道发炎或受刺激引起）咳嗽

vi. to force air suddenly from the lungs 咳嗽

counter /ˈkaʊntə(r)/ n. the rear strip which covers the seam on the heel of a riding boot（压在马靴后跟针脚上的窄条皮革）压条

adv.【hunting】(of a hound) running the line of a fox in the opposite direction to that the fox is traveling（猎犬与狐狸行进方向相反）反向，逆向：run counter 跑反，逆跑

counter canter n. a schooling dressage movement where the horse canters on the outside lead, instead of the correct inside lead（盛装舞步训练动作，马以外侧肢而非内侧肢起步）反跑步

[syn] false canter, counter lead

counter half-pass n.【dressage】a movement in which the halfpass is performed in a zigzag formation through the center of the arena（盛装舞步中马以横斜步 Z 形行进通过赛场中心）反向斜横步

counterirritant /ˌkauntərˈ ɪrɪtənt/ n.【vet】an agent applied to induce local inflammation to counteract irritation or relieve inflammation and pain elsewhere（通过诱发局部炎症来对抗别处痛痒和炎症的外用药物）抗刺激剂，抗刺激药

adj. of or producing the effect of such an agent 抗刺激作用的

Coupé /ˈkuːpeɪ; *AmE* kuːˈpeɪ/ n. an enclosed, four-wheeled, horse-drawn town carriage with a truncated body designed for greater compactness and improved appearance（一种城镇用的四轮马车，为了增加紧凑性和美观，其车顶平直）平顶四轮马车

couple /ˈkʌpl/ n.【hunting】two hounds linked or joined together in a pair（系在一起的两条猎犬）猎犬搭档

vt. to fasten or link together 拴或系在一起

coupling /ˈkʌplɪŋ/ n. the part of the loin area of horse between the last rib and the point of hip that connects the hindquarters to the forequarters（马最后肋骨与腰角间的部分，连接前体和后躯）后腰，侧腰

courbette /kʊəˈbet/ n.【dressage】a classical airs above the ground in which the horse rears to an angle about 45 degrees and leaps forward several times in series on his hind legs while maintaining that position（古典骑术空中动作之一，马以后肢直立约 45°时连续向前跳跃多次）直立跳跃 [syn] curvet

course /kɔːs; *AmE* kɔːrs/ n. **1**【racing】the track used for horse racing（赛马的）赛场，跑道：race course 赛道，赛马场 **2**【jumping】a route consisting of a series of obstacles which the rider and horse must jump in a specified order（障碍赛的路线，骑手须策马以特定顺序跨越内设的多道障碍）路线，线路

vt.【hunting】(of hounds) to hunt by sight rather than scent（猎犬凭视觉而非嗅觉）捕猎，追猎

course builder n.【jumping】one who builds show-jumping or cross-country courses（场地障碍或越野障碍的）场地搭建者

course designer n.【jumping】one that is responsible for designing show-jumping or cross-country courses（场地障碍或越野障碍的）路线设计师 [syn] jump designer

coursing /ˈkɔːsɪŋ/ n. the sport of chasing and hunting quarry with hounds by their keen sight（借助猎犬）捕猎，追猎：hare coursing 追猎野兔

cover /ˈkʌvə(r)/ vt. **1** (of a stallion) to mate with a mare（公马）配种 **2** to travel or pass over; traverse 运行，经过；穿过：A very good endurance horse is equivalent of the marathon runner, a lean, wiry individual who covers ground effortlessly. 出色的耐力赛马就好比优秀的马拉松运动员，体格要精瘦强健，行进步法要轻松。

covered /ˈkʌvəd/ adj. (of a carriage) having a cover or top; enclosed（马车）带顶篷的

covered wagon n. *see* prairie schooner 草原篷车

covering /ˈkʌvərɪŋ/ n. the action of a stallion to mate with a mare 交配；配种：covering yard 配种圈 ◇ covering season 配种季节

covering boots n. *see* breeding boots 配种靴，配种护腿

covert /ˈkʌvət; *AmE* -vərt/ n.【hunting】a small bush or thicket where foxes may hide or lay（狐狸藏身的灌木丛）藏身处，狐窝

cow /kaʊ/ n. **1** a large farm animal kept to produce milk or beef in general 奶牛；肉牛：a dairy cow 一头奶牛 **2**【cutting】a fully grown female of cattle 母牛 [PHR V] **lose a cow** (of a horse) to fail to prevent a cow from returning to the herd（指马）拦截失败：Five points will be penalized to a horse if he loses the cow being worked. 如果马在拦截牛时失败，将给予5个罚分。**pull off a cow** to stop working a cow 停止截牛 **read a cow** to prejudge the movement of a cow being worked 揣测牛的动向 **run short on a cow** (of a horse) to fall behind a cow when running parallel with it (指马)落在牛后，追不上牛 **step into the cow** (of a horse) to move toward the cow at the encouragement of the rider (指马)上前截牛

cow hand n. *see* cowboy 牛仔

cow hocks n. a conformation defect in which the horse's hocks turn inwards when viewed from behind (马匹飞节后观时偏向内侧的体格缺陷)飞节内偏；X形飞节 [syn] tarsus valgus, medial deviation of the hock joints

cow horn n.【hunting】an American hunting horn used exclusively when hunting with foxhounds (猎狐时使用的一种美式猎狐号角)牛角号

cow horse n. a horse ridden by a cowboy when working cattle (牛仔在牧场工作时所骑的马)牧牛马 [syn] cow pony, range horse

cow kick n. a type of kick in which the horse strikes forward and to the outside with one of his hind legs (马用某一后肢向前侧踢)牛弹腿

cow man n.【AmE】an owner of cattle; a rancher; a cowboy 牧场主，牧场工人；牛仔

cow poke n.【slang】a cowboy 牛仔

cow pony n. *see* cow horse 牧牛马

cow puncher n. one who punches or brands cattle in a ranch 牛场打号工

cow sense n. the innate ability for horse cutting 截牛天性: a horse who has good cow sense 有很好的截牛天性的马

cow smart adj.【cutting】(of a horse) having the ability to predict the movements of a cow 有截牛天性的 [syn] cowy, cow sense

cowbane /ˈkaʊbeɪn/ n. a perennial North American herb having pinnately compound leaves and umbels of small white flowers (一种生长于北美洲的多年生草本植物，羽状复叶，开伞状小白花)毒芹

cowboy /ˈkaʊbɔɪ/ n. **1** a mounted cattle-herder in western part of US (美国西部骑马的牧牛人)牛仔 [syn] cow poke, cow hand **2** a man who competes in rodeo events (参加竞技表演的人)牛仔

cowgirl /ˈkaʊɡɜːl; *AmE* -ɡɜːrl/ n. **1** a woman who drives cattle on horseback 牛仔女郎，女牛仔 **2** a woman who competes in rodeo events (参加竞技表演的女人)女牛仔

cowhide /ˈkaʊhaɪd/ n. the hide or skin of a cow 生牛皮: Almost all the saddlery is made from cowhide. 几乎所有的马鞍都由生牛皮制成。

cow-hocked /ˌkaʊˈhɒkt/ adj. (of horse's hocks) turning inwards when viewed from behind (马后观时飞节偏向内侧)内弧的，X形的 [compare] bow-hocked

cowlick /ˈkaulɪk/ n. a tuft of hair that grows in a different direction from the rest of the hair; whorl (与周围毛发生长方向相反的一缕毛发)旋毛

cowy /ˈkauɪ/ adj. *see* cow smart 有截牛天性的

coyote /kaɪˈəʊti; *AmE* -ˈoʊti/ n. a small, wolf-like carnivorous animal native to the plains of North America (生活在北美草原上的食肉动物，体型小，外表似狼)郊狼，草原狼 [syn] prairie wolf

crab/kræb/ n. an unfavorable feature of a horse such as a blemish, fault, or unsoundness (马匹体貌上的毛病)缺陷，毛病

vi. to give unfavorable remark about a horse in order to reduce its value (对马给出不利评价以贬损其价值)挑刺，找茬；贬低

crabber /ˈkræbə/ n. one who gives an unfavorable criticism of a horse 挑剔马匹毛病的人

crack /kræk/ vi. (of hoof) to break or split (马蹄)裂开，开裂

n. a partial split or breaking of hoof 蹄裂: heel crack 蹄踵裂 ◇ toe crack 蹄尖裂 [compare] cleft

cracked heels n. *see* scratches 蹄踵炎,乘踵裂

cracked hoof n. a vertical split in the hoof wall 裂蹄

cradle /ˈkreɪdl/ n. a light wooden framework fitted around the neck of a wounded horse to prevent him interfering with the bandages or blisters (马身体某处受伤后戴在其颈上木架栏,可防止马碰触绷带或水疱)护架,颈固定架

cramp /kræmp/ n. a sudden, involuntary muscular contraction causing severe pain (肌肉的非自主性突然收缩,可引起剧烈疼痛)痉挛;抽筋

vi. to suffer from cramps 痉挛;抽筋

cramped action n. the action of a horse who does not move freely 运步受限,运步狭促

cranial /ˈkreɪnɪəl/ adj. **1** of, near, or related to the skull or cranium 头骨的,颅(骨)的;颅侧的 **2** anterior 前面的:cranial angle of the scapula 肩胛骨前角

cranialis /ˈkreɪnɪəlɪs/ adj.【Latin】cranial 颅侧的;前面的

cranium /ˈkreɪnɪəm/ n.【anatomy】(*pl.* **crania** /-niə/) **1** the skull of a vertebrate (脊柱动物的)头骨 **2** the portion of the skull enclosing the brain 颅骨

crash skull n. the racing helmet (赛用)头盔

crawler /ˈkrɔːlə(r)/ n.【dated】a cab driver who wandered to pick up passengers 兜揽乘客的车夫

crazyweed /ˈkreɪzɪwiːd/n. *see* locoweed 疯草

creamy dun n. *see* golden dun 淡兔褐,金兔褐

crease /kriːs/ n. a groove cut into the ground-side surface of a horseshoe; fuller (蹄铁着地面上的凹槽)蹄铁槽

crease iron n. *see* fullering iron 开槽钢

creasing /ˈkriːzɪŋ/ n.【farriery】the act or process of cutting a crease into the ground-side surface of a horseshoe (在蹄铁的着地侧切出凹槽)开槽工艺 [syn] fullering

creep feeder n. a manger that allows a foal, but prevents his dam from feeding (允许幼驹但阻止母马采食的饲槽)补饲槽

creep feeding n. a method of providing the young foal a special concentrate feed that is not available to its mother (给幼驹添加精料而不让母马采食的做法)补饲

creep ration n. the concentrate fed to a foal (喂给幼驹的精料)补饲日粮

cremello /krɪˈmeləʊ; *AmE* -loʊ/ n.【color】a white to cream-colored coat and points with blue eyes (被毛颜色和末梢为乳白色而眼睛为蓝色的毛色)乳色马

crescent brass n. a brass decoration with the pattern of the crescent moon, placed on the harness or other tack of the horse to ward off the evil eye (形似半月的铜饰,常佩戴在马具上用来避邪)半月铜饰

crest /krest/ n. the topline of a horse's neck 颈峰,颈脊

crew /kruː/ n.【racing】a group looking after the horse and rider during a race or competition (比赛中负责料理马和骑手的人员)工作小组,工作人员:The skill of the crew can contribute significantly to the performance of horse and rider on the day of competition. 工作人员的技术水平对马和骑手在比赛那天的表现起着比较重要的作用。

crib /krɪb/ n. a rack or trough for fodder; manger 饲槽;食槽

cribber /ˈkrɪbə/ n. *see* crib-biter 咬槽的马

cribbing /ˈkrɪbɪŋ/ n. (also **crib-biting**) a vice of the horse who grasps hold of a fixed object, usually the tying post or manger, with his incisor teeth and swallows in air (马咬住木桩或饲槽等物体咽气的恶习)咬槽癖 [syn] stump sucking

crib-biter /krɪbˈbaɪtə/ n. (also **cribber**) a horse who has the vice of cribbing 咬槽马

crib-biting /krɪbˈbaɪtɪŋ/ n. *see* cribbing 咬槽癖:The main difference between wind-sucking and crib-biting is that the horse does not catch hold of anything with his teeth to do so in the former vice. 咽气癖和咬槽癖的主要区别是患有咽气恶习的马不用牙齿咬任何物体。

crinet /ˈkriːnɪt/ n. (also **criniere**) the medieval armor used to protect the neck and throat of the horse (中世纪马穿盔甲,可保护马的颈和喉)护颈 — *see also* **champron**, **peytral**, **croupiere**

Criollo /kriːˈəʊləʊ; *AmE* kriːˈoʊloʊ/ n. (also **Argentine Criollo**) a horse breed indigenous to Argentina which descended from Spanish stock brought to South America by the Conquistadors in the 16th century; has a compact body, broad head, straight profile, wide-set eyes, and short legs with hard feet; stands 14 to 15 hands and may have any of the coat colors; is willing, tough, agile, and possesses great endurance; used mainly as the mounts of cowboys of South America (由16世纪西班牙殖民者引入南美的西班牙马繁育而来的阿根廷马品种。体格结实强健,头宽短,面平直,瞳距大,四肢粗短,蹄质结实,体高14~15掌,各种毛色兼有,性情顽强,步法轻盈,耐力好,是南美牛仔主要的坐骑)克里奥罗马

crop[1] /krɒp; *AmE* krɑːp/ n. **1** a short whip used in hunting or jumping disciplines (狩猎或障碍赛中使用的一种短把鞭杆)短鞭,马鞭 [syn] hunting-crop **2** the stock of a whip 鞭杆

crop[2] /krɒp; *AmE* krɑːp/ vt. to clip the ears of an animal 剪耳

crop-eared adj. (of a horse) having the ears cropped (指马)耳尖被剪去的,被剪过耳的

cropper /ˈkrɒpə(r); *AmE* ˈkrɑːpə(r)/ n. 【racing】a heavy fall or tumble of the horse and rider in races (赛马中)跌倒,摔倒

cross /krɒs; *AmE* krɔːs/ n. **1** an animal produced by crossbreeding; hybrid 杂种 **2** the process of crossbreeding; hybridization 杂交 **3** a primitive marking often in donkeys, with the dorsal stripe crossed at the withers with another darker line (驴身上常见的原始别征,背线与鬐甲处暗色条纹呈十字交叉)十字叉

vt. **1** to form or make a cross 交叉,交错 [PHR V] **cross the stirrups** (of a rider) to put the right and left stirrups across over the withers of the horse (指骑手)对搭马镫,交错马镫 **2** to crossbreed or cross-fertilize 杂交(繁育)

cross noseband n. *see* Grackle noseband X形鼻革

cross tie vt. to secure the head of a horse by two straps both clipped at one end to the halter and to the posts on the other (用两根绳索扣在笼头两侧,另两端系在立柱上来固定马头)双侧拴系 [compare] tie

cross ties n. one of the two straps fitted with clips at one end and attached to posts on the other between which the horse is tied when grooming or tacking up (刷拭或备鞍时用来将马固定在两立杆之间的两根绳索)双侧拴马绳 [syn] pillar reins

crossbred /ˈkrɔːsbred/ adj. (of horses) produced by the mating of individuals of different breeds or species (由不同品种的个体交配育成)杂交的

n. crossbred animal 杂交品种

crossbreed /ˈkrɔːsbriːd/ n. crossbred animal 杂交品种,杂种

vi. to mate so as to produce a hybrid; interbreed 杂交

crossbreeding /ˈkrɔːsbriːdɪŋ/ n. the breeding between a mare and stallion that are of different breeds (不同品种的公马和母马间的配种)杂交

cross-country /ˌkrɔːsˈkʌntri/ adj. (abbr. **XC** or **X-C**) moving or directed across open country 越野的: a cross-country race 越野赛

cross-firing /ˈkrɔːsˌfaɪrɪŋ/ n. a gait defect in which the inside toe or wall of a hindfoot strikes the inner quarter of the diagonal forefoot (一种运步缺陷,马匹后蹄内侧撞到对角线前蹄的内侧)交叉追突: As a gait fault, cross-firing is quite common in pacers. 交叉追突这种运步缺陷,在对侧步马中较为常见。

[compare] underreaching

crossrail /ˈkrɔːsreɪl/ n.【jumping】a jumping obstacle consisting of two rails that cross like an X (由两根木杆交叉成 X 形的障碍)交叉障碍,X 形障碍

crotch seat n. *see* light seat 胯式骑姿,轻骑姿

crouch /krautʃ/ vi. to bend over knees with the upper body brought forward 屈膝,蹲伏:Most jockeys crouch over their mounts when racing. 大多骑师在赛马时都蹲骑于马背。

n. a crouching stance or posture 蹲伏: a crouch style 蹲骑姿势

croup /kruːp/ n. the upper part of a horse's hindquarters from loins to the root of the tail (马后躯从腰至尾根的部分)尻部:point of croup 尻端 [syn] rump,crupper

croupade /kruːˈpeɪd/ n. a movement above the ground in which the horse rears from a standstill and performs a single vertical jump (马直立后向前跳跃的空中高等骑术动作)直立跳跃

crouper mount n. a method of mounting the horse in which the rider runs and leaps onto the back of horse from behind (骑手从马尻部后面跳上马背的上马方式)尻后上马

croupiere /ˈkruːpiə/ n. a piece of armour used to protect the horse's hindquarters (马穿盔甲,可保护后躯)护尻 — *see also* **champron**, **peytral**, **crinet**

crow hop vi. (of a horse) to hop repeatedly into the air with all four feet off the ground at the same time (指马四肢离地)蹦跳,欢腾:Seeing his mother, the colt crow hopped straightly toward her. 雄驹看到母马后,蹦跳着径直朝它奔去。

crowd /kraud/ vi.【racing】(of a jockey) to ride too close to other horses (指骑师策马)扎堆,拥挤

crown /kraun/ n. the upper part of the tooth 齿冠

crown piece n. (also **crownpiece**) *see* head piece 顶革,项革

CRT abbr. capillary refill time 毛细血管回流时间

crude /kruːd/ adj. in a natural or raw state; rough 天然的;粗的: crude ash 粗灰分 ◇ crude fiber 粗纤维 ◇ crude fat 粗脂肪 ◇ crude protein 粗蛋白

crupper /ˈkrɒpə(r)/ n. **1** the rump or hindquarters of horse (马的)臀部 **2**【driving】(also **harness crupper**) a piece of leather strap looped under a horse's tail and fastened back to a harness or saddle to keep it from slipping forward (兜在马尾下的皮带,另一端扣于挽具或马鞍以防其向前滑动)兜带,尻带

crural /ˈkruərəl/ adj.【anatomy】of or relating to the leg or shank 腿的;小腿的:crural region 小腿部 ◇ crural fascia 小腿筋膜

crust of the hoof n. the hoof wall 蹄壁

cry /kraɪ/ n.【hunting】the sound made by a hound when in pursuit of a fox (猎犬追捕狐狸发出的叫声)叫哮,嚎叫 **IDM** **in full cry** (of hounds) barking fiercely (指猎犬)狂吠的

cryogenic branding n. *see* freeze branding 冷冻烙印,冻号

cryogenics /ˌkraɪəˈdʒenɪks/ n. the scientific study dealing with very low temperatures 低温学: Cryogenics has been used in many applications including branding and cryosurgery. 低温学已经应用于冻号和冷冻外科手术等方面。

cryosurgery /ˌkraɪəʊˈsɜːdʒəri/ n. the selective destruction or elimination of abnormal tissues by freezing with liquid nitrogen (用液氮冷冻去除病变组织的方法)冷冻手术

cryptorchid /ˌkrɪpˈtɔːkɪd; *AmE* -ˈtɔːrk-/ n. a stallion having one or both testicles undescended 隐睾马:bilateral cryptorchid 双侧隐睾 ◇ unilateral cryptorchid 单侧隐睾 [syn] rig

cryptorchidism /ˌkrɪpˈtɔːkɪdɪzəm; *AmE* -ˈtɔːrk-/n. a natal defect in which one or both testicles are undescended into the scrotum (单侧或两侧睾丸未落入阴囊)隐睾[症]

CS abbr. chondroitin sulfates 硫酸软骨素

CT abbr. Computerized Tomography 计算机断层

C

造影术:CT scanning CT 扫描

cub /kʌb/ n. the young of a fox, wolf, or lion 幼兽,幼崽

vt. to hunt the young of a fox 猎幼狐:to go cubbing 猎幼狐 | **cubbing** n.

cubbing season n. the cub hunting season 猎幼狐季节

cube /kju:b/ n. form of compound feeds made into small blocks 饲料块:alfalfa cubes 苜蓿块

cub-hunting /ˈkʌbhʌntɪŋ/ n.【hunting】(also **cubbing**) a sport in which the fox cubs are hunted at the end of August (8 月末进行的)猎幼狐:The cub hunting season usually ranges from the end of July to the end of September in America and the end of August until the end of October in England. 在美国,猎幼狐季节一般从 7 月底持续到 9 月底,而在英国则从 8 月底一直持续到10 月底。

cubitus /ˈkju:bɪtəs/ n.【anatomy】elbow of the foreleg (马前肢的)肘

cue /kju:/ n. a signal made by the rider to communicate his requests to the horse (传达骑手指令的信号)暗示,信号 compare aids

cull /kʌl/ vt. to get rid of inferior members from a herd 剔除;淘汰

n. an unwanted animal disposed of by the owner 被淘汰的牲畜

cup /kʌp/ n. **1** a funnel-like indentation in the central region of the crown of tooth; infundibulum (齿冠中央漏斗状的凹陷)齿坎,黑窝 **2**【jumping】a curved metal piece that holds the rail to the jump standards (障碍标杆上用来支撑跨栏的铁件)杯托:deep cup 深杯托 ◇ shallow cup 浅杯托 **3**【racing】(usu. **Cup**) a decorative cup-shaped vessel awarded as a prize or trophy to winners 奖杯:Hong Kong Cup 香港杯

cuppy /ˈkʌpi/ adj. (of a racetrack) marked by shallow depressions of hoofprints (指赛场)有浅坑的,坑坑洼洼的:The loose, cuppy track will greatly slows a horse. 松软且带坑洼的赛道会大大降低马的速度。

cur /kɜː(r)/ n.【hunting】(also **cur dog**) a dog of mixed breeds 杂种狗

cur dog n. *see* cur 杂种狗

curable /ˈkjʊərəbl/ adj. (of a disease) able to be treated or cured 可治愈的

curb /kɜːb; *AmE* kɜːrb/ n. **1** (*pl.* **curbs**) a thickening or bowing of the ligament located below the point of the hock due to strain (马飞节下侧跗韧带出现硬瘤的病症)关节硬瘤 **2** a curb chain or strap 颐沟链,颐沟革 **3** *see* curb bit 链式衔铁

curb bit n. (also **curb**) a bit consisting of two metal cheek pieces and a ported mouthpiece, usually used in conjunction with a curb strap or curb chain (由颊杆和带凸起铁杆构成的衔铁,多与颐沟链或颐沟革配套使用)链式衔铁,霸王衔铁

curb chain n. a metal chain that passes under the horse's chin groove and serves to restrain a horse in conjunction with the bit (位于马颐沟的铰链,与衔铁联合起来控制马匹)颐沟链

curb groove n. a furrow in the underside of the lower jaw just above the lower lip where the curb chain or strap rests (紧靠马下唇上方的凹沟,衔铁链绕过此处)颐沟,颐凹 syn chin groove

curb rein n. the rein attached to the lower rings of the cheeks of a curb bit (穿过克柏衔铁颊杆下端圆环的缰绳)缰

curb strap n. a strap that passes under a horse's lower jaw and serves in conjunction with the bit to restrain the horse (位于马颐沟的皮带,与衔铁联合起来控制马匹)颐沟革,颐沟带

curby /ˈkɜːbi; *AmE* ˈkɜːrbi/ adj. (of hocks) suffering from curbs 生硬瘤的:curby hocks 关节瘤

curby hocks n. *see* sickle hocks 镰形飞节

cure /kjʊə/ vt. to restore to health by medical practice; heal 治愈,治疗

curl /kɜːl; *AmE* kɜːrl/ v. to form or twist into

small coils and ringlets 使卷曲
n. a coil or ringlet of hair 卷发,卷曲

curly /ˈkɜːli; *AmE* ˈkɜːrli/ adj. (of horse's hair) forming or having curls (被毛)卷曲的

Curly horse n. *see* Bashkir curly 巴什基尔卷毛马

curricle /ˈkʌrɪkl/ n. a light, open two-wheeled carriage drawn by a pair of horses harnessed abreast and popular during the late 18th and early 19th centuries (18 世纪末至 19 世纪初在英国流行的一种轻型双轮马车,多由双马联驾拉行)轻型双轮马车

curry comb n. a plastic or metal grooming tool used to clean and tidy up the coat of the horse (用来清洁马匹被毛的塑料或金属制刷拭工具)刷梳:metal curry comb 铁篦子,毛刷刨 ◇ rubber curry comb 橡胶梳

curtal /ˈkɜːtl; *AmE* ˈkɜːrtl/ n.【dated】a horse whose tail has been docked 断过尾的马
adj. (of tail) cut short or docked (尾巴)剪断的,断尾的:a curtal horse 一匹截过尾的马

curvet /kɜːˈvɛt/ n.【BrE, dated】*see* courbette 直立跳跃

cushion /ˈkʊʃn/ n.【racing】the loose top surface of a race track (赛道表面的松软层)垫层,表层

cut /kʌt/ vt. **1** to geld or castrate 阉割,去势 **2** to sever or make an incision in the body tissue 切,割 PHR V **cut and set** to cut the tendons of tail and set it to achieve a high carriage when healed (割断尾部肌腱后将其垫高)割尾垫高:a cut and set tail 割后垫高的马尾 syn dock and set **cut corners**【dressage】(of a rider) to fail to ride the horse into the corners of the arena (盛装舞步中骑手没能策马步入场地拐角)折过拐角 **cut down**【racing】(of a horse) be injured by the horseshoe of another horse or by its own shoe in a race (指马被蹄铁)撞伤,蹭伤 **3** to separate a calf from a herd 拦牛,截牛 PHR V **cut off/out**【cutting】to select and separate a calf from a herd 挑牛拦截

cut shot n.【polo】any forehand or backhand stroke of the ball hit at an angle away from the horse (马球中用球杆正手或反手朝偏离马体方向击打)削击,侧击

cutaneous /kjuˈteɪniəs/ adj. of, relating to, or affecting the skin 皮肤的:cutaneous muscle 皮肌

cutaneous habronemiasis n. a skin disease of horses caused by the parasite of the genus *Habronema* that lays eggs on open wounds, characterized by painful granular lesions (由马胃线虫属寄生虫在伤口产卵造成的皮肤病,症状表现为皮肤粒状溃烂)皮肤线虫病 syn summer sores

cutis /kjuːtɪs/ n.【anatomy】(*pl.* **cutes** /-tiːz/) *see* dermis 真皮

cutter /ˈkʌtə/ n. **1**【cutting】one who rides to separate a cow from a herd 截牛者 **2**【AmE】a light horse-drawn sleigh with a high S-shaped dashboard that prevents snow thrown by the horse's hoofs (一种轻型马拉雪橇,前侧的 S 形挡板可挡住马蹄扬起的雪)轻型雪橇:Portland cutter 波特兰马拉雪橇 **3** a light, four-wheeled passenger vehicle drawn by a single horse 由单匹马拉动的轻型四轮马车

cutting /ˈkʌtɪŋ/ n.【rodeo】*see* horse cutting 截牛,卡停

cutting horse n. (also **cutting pony**) any horse used to work cattle selected from the herd 截牛用马 syn cow horse

cutting pony n. *see* cutting horse 截牛马

cutting saddle n. a western-type saddle used for horse cutting 截牛鞍

CVA abbr. Concours Voltige d' Amitié 马上杂技表演友谊赛

CVF abbr. Concours Voltige Frontière 马上杂技表演

CVI abbr. **1** Concours Voltige Internationale 国际马上杂技表演赛 **2** certificate of veterinary inspection 兽医检验证明

cyanocobalamin /ˌsaɪənəʊkəˈbæləmɪn/ n.【nutrition】one kind of water-soluble vitamin containing cobalt(含钴元素的水溶性维生素)钴胺素 [syn] vitamin B_{12}

cycle /ˈsaɪkl/ n. the estrus cycle 发情周期
vi. (of estrus) to occur in cycle 周期性发情 | **cycling** n. 发情

cyclic /ˈsaɪklɪk; *AmE* ˈsɪk-/ adj. of or relating to estrus cycle 发情周期的:cyclic corpus luteum 周期黄体

cyst /sɪst/ n. an abnormal membranous sac containing liquid matter(一种内含液体的异常膜囊结构)囊肿:an ovarian cyst 卵巢囊肿 ◇ bone cyst 骨囊肿

cysteine /ˈsɪstɪn/ n.【nutrition】a kind of non-essential amino acid derived from cystine and found in most proteins(由胱氨酸衍生而来的非必需氨基酸)半胱氨酸

cystine /ˈsɪstiːn/ n.【nutrition】a non-essential amino acid found in many proteins(一种非必需氨基酸,存在于多种蛋白中)胱氨酸

cystitis /sɪˈstaɪtɪs/ n. inflammation of the bladder 膀胱炎

cyt(o)- /ˈsaɪtəʊ/ pref. cell 细胞的

cytoplasm /ˈsaɪtəʊplæzəm/ n. the protoplasm outside the nucleus of a cell(细胞核外的原生质)细胞质

cytoplasmic /ˌsaɪtəʊˈplæzmɪk/ adj. of or relating to cytoplasm 细胞质的:cytoplasmic inheritance 胞质遗传

cytosine /ˈsaɪtəʊsiːn/ n.【biochemistry】a pyrimidine base that is an essential constituent of nucleic acid(组成核酸的嘧啶碱)胞嘧啶

D

D /diː/ n. a Dee-ring D 环

dachshund /ˈdæksnd; *AmE* ˈdɑːkshʊnd/ n. (also **badger dog**) a small German dog breed with a long body, a brown or black coat, drooping ears, and very short legs (德国小型犬种,体长,被毛褐色或黑色,两耳下垂,四肢极短)达克斯犬,腊肠犬: The dachshund was formerly used to hunt badgers. 腊肠犬过去常用来猎獾。

daily double n.【racing】(also **double**) a betting option in which the bettor attempts to select the winning horses in two different races on the same program (赌马者须选出赛事日程中两场不同比赛的头马才可赢奖的投注方式)二连赢:To win a daily double, both selected horses must win their respective races. 只有所选的两匹马都在相应的比赛获胜,才能二连赢。 compare running double

daily energy requirement n. the amount of calories required by a horse on a daily basis to maintain weight and health (马匹维持自身体重与健康每天所需的能量)日能量需要

Daily Racing Form n.【racing】(abbr. **DRF**) a tabloid newspaper founded in 1894 in Chicago that published racing news and past performances of racehorses as a statistical service for the bettors (1894 年在芝加哥创办的报纸,报道赛事新闻和马匹以往赛绩,以期为投注者提供信息服务)《每日赛马汇览》

daily sperm output n. the total number of sperm ejaculated by a stallion on a daily basis (公马)日射精子数

daily sperm production n. the total number of sperm produced per day by testes 日产精子数

Daily Triple n. *see* pick three 三连赢

Dales /ˈdeɪlz/ n. *see* Dales pony 戴尔斯矮马

Dales pony n. (also **Dales**) a pony breed originating in the upper dales of North Yorkshire, England; descended from the Celtic pony crossed with Welsh Cob in the 19th century and Clydesdale later; has a small head, wide-set eyes, prominent withers, a strong back, full mane and tail, tough, blue-colored hoofs, and silky feathering; stands up to 14.1 hands and generally has a black coat, although brown, bay, and gray also occur; is tremendously strong, has a remarkable weight-carrying capacity, is calm and sensitive, and a good keeper; formerly used in underground coal mines, farm work, and packing, now used mainly under harness in competitive driving and as an all-purpose riding mount (源于英国北约克夏郡山谷地区的矮马品种,在 19 世纪由凯尔特矮马与威尔士柯柏马杂交繁育形成,后又引入过克莱兹代尔马血统。头小,瞳距大,鬐甲突出,背部发达,鬃尾浓密,蹄质坚硬、呈蓝色,距毛浓密,体高 14.1 掌以下,被毛以黑为主,但褐、骝和青亦有之,体格极其强健,驮载能力强,性情温驯,反应灵敏,易饲喂,过去多用于地下煤矿、农田作业和驮运,现在主要用于驾车赛,同时也是兼用型乘骑马种)戴尔斯矮马

Dales Pony Improvement Society n. an organization founded in England in 1917 to maintain the registry of the Dales pony (1917 年在英国成立的组织机构,旨在对戴尔斯矮马进行品种登记)戴尔斯矮马改良协会

Dales Pony Society n. an organization founded in England in 1957 to promote the breeding and use of the pure-bred Dales pony (1957 年在英国成立的组织机构,旨在促进纯种戴尔斯矮马的育种工作)戴尔斯矮马协会

Dalian Mounted Policewomen n. a mounted police force founded in 1994 in Dalian, China (于

1994年在中国大连成立的骑警队伍）大连女子骑警队

Dalmatian /dæl'meɪʃn/ n. a large, short-haired white dog with black or brown spots（白色被毛上散生黑色或褐色斑点的大型犬种）达尔马提亚犬，大麦町犬，斑点狗：The Dalmatians were historically used to accompany horse-drawn vehicles and protect cargo from theft. 斑点狗过去常用来护卫马车以防货物失窃。 syn coach dog, carriage dog

dam /dæm/ n. the female parent of a horse 母本，母马 compare sire

damage fund n.【hunting】an amount of money paid by some hunting clubs to compensate landowners for damage to their property during hunting（俱乐部为弥补狩猎对庄园造成的损失而偿付的金额）损失补偿

Damalinia equi n.【Latin】*see* biting louse 咬虱

damp /dæmp/ adj. slightly wet; humid 潮湿的
vt. to make damp or moist; moisten 使潮湿；拌湿：damped feeds 潮拌料，湿拌料

damp-proof /'dæmppruːf/ adj. designed to prevent dampness 防潮的：A damp-proof course of floor is recommended to keep the stable dry. 为了保持厩舍干燥，地板建议使用防潮材料铺面。

dam's sire n. the sire of a broodmare; maternal grandsire（繁育母马的父本）外祖父马 syn broodmare sire

dancing competition n. a mounted competition performed on the islands of Sumba and Sumbawa, Indonesia in which ponies are ridden bareback by boys and are lunged by a trainer who directs their movements to the accompaniment of the drum beat（源于印度尼西亚松巴岛和松巴哇岛上的马上竞技项目，其中男孩无鞍策骑矮马在练马师长缰操控下伴随鼓点伴奏行进）舞马比赛

dandy brush n. a stiff bristled brush used for removing the dirt from the horse's body in grooming（用来去除马被毛上尘土的毛刷）硬毛刷

dangler /'dæŋglə(r)/ n. *see* fly terret 额前摆饰

Danish Horse n. *see* Danish Warmblood 丹麦温血马

Danish Sport Horse n. *see* Danish Warmblood 丹麦温血马

Danish Spotted Horse n. *see* Kanbstrup 丹麦斑点马

Danish Warmblood n. a warmblood breed developed in Denmark by crossing local Fredriksborg mares with Anglo-Norman, Thoroughbred, and Polish stallions; stands 16.1 to 16.2 hands and usually has a bay coat, although other coat colors do occur; is naturally well balanced and excels in dressage and show jumping（丹麦采用当地的弗雷德瑞克斯堡马母马与盎格鲁诺曼马、纯血马和波兰马公马杂交培育的温血马品种。体高16.1～16.2掌，毛色以骝为主，兼有其他毛色，体型匀称，在盛装舞步和障碍赛中都有出色表现）丹麦温血马 syn Danish Sport Horse, Danish Horse

Danubian /dæ'njuːbɪən/ n. a warmblood horse breed developed during the early 1900s in Bulgaria by crossing Nonius stallions with Gidran mares; has a compact body, strong neck, powerful quarters, deep chest, and relatively slender legs; stands around 15.2 hands and generally has a black or chestnut coat; used for light draft, riding, and jumping（保加利亚在20世纪初采用诺聂斯马公马和基特兰马母马杂交培育的温血马品种。体质结实，颈肩厚实，后躯发达，胸身宽广，四肢修长，体高约15.2掌，毛色多为黑或栗，主要用于轻挽、骑乘和障碍赛）多瑙河温血马

dapple /'dæpl/ n.【color】a small spot or patch 斑点，花纹
vt. to mark with spots or rounded patches 使带有斑点

dappled /'dæpld/ adj.【color】(of coat) spotted; mottled 有斑点的，有花纹的

dappled gray n.【color】the color of a horse with

black skin and a gray, dappled coat (马皮肤为黑色而青色被毛带花纹的毛色)菊花青,斑点青

Darashouri /ˌdærəˈʃʊəri/ n. one of two types of Persian Arab indigenous to Iran; stands 14.1 to 15.1 hands and has a bay, chestnut, gray, or rarely black coat; is elegant, strong, spirited, energetic, and athletic; used mainly for riding (伊朗本土波斯阿拉伯马的亚型之一,体高14.1~15.1掌,毛色为骝、栗或青,黑色少见,外貌俊美,体格强健,气质高昂,富有活力,运动能力好,主要用于骑乘)达罗舒瑞马

Darby /ˈdɑːbi/ n. *see* Derby 德比赛,打吡赛

dark /dɑːk; *AmE* dɑːrk/ adj.【color】**1** having a deep hue or color (指色调)深的 **2**【racing】(of a race track or a day) having no racing (指赛场)不开赛的

dark bay n.【color】*see* mahogany bay 黑骝毛

dark brown n.【color】a shade of brown that is very close to faded black 黑褐色

dark buckskin n.【color】*see* smutty buckskin 暗沙黄,暗兔褐

dark grullo n.【color】*see* smutty grullo 深蓝青毛

dark horse n.【racing】an underestimated horse or one whose abilities has not been fully recognized (能力尚未被完全认知而被低估了的马)黑马:Mermaid was really the dark horse in this race, with no one ever thinking of her winning beforehand. "美人鱼"可谓这场比赛中的真正黑马,事先根本没有人想到它会获胜。

dark muzzle n.【color】(a horse having) a dark nose 鼻端黑(的马)

Darley Arabian n. (c. 1700 – 1730) one of three foundation sires of the English Thoroughbred; was a purebred Arabian imported into England in 1704 by Thomas Darley for whom it was named (育成纯血马的三大奠基种公马之一,这匹纯种阿拉伯马于1705年由托马斯·达利运至英国,故而得名)达利·阿拉伯:Today, nearly 95% of the Thoroughbreds in the world are descended from Darley Arabian. 当前,世界上约95%的纯血马由达利·阿拉伯繁育而来。— *see also* **Byerley Turk**, **Godolphin Arabian**

dart hole n.【AmE, driving】the hole at the end of the traces through which the trace hook passes (套绳末端的圆口,挂钩由此穿过)套绳孔

Dartmoor /ˈdɑːtmuə(r)/ n. *see* Dartmoor pony 达特姆尔矮马

Dartmoor pony n. (also **Dartmoor**) an ancient pony breed originating in Dartmoor of Devonshire, England; stands below 12.2 hands and generally has a bay, brown, black, and gray coat; has a relatively small head, a well-proportioned neck with a full mane, good back, and slender legs with short cannons and ample bone; is hardy, a good keeper, an ideal mount for children due to its temperament and action (源于英国德文郡达特姆尔山区的古老矮马品种。体高12.2掌以下,毛色多为骝、褐、黑或青,头小,颈姿优美,鬃毛浓密,背部发达,四肢修长,管骨粗短,抗逆性强,易饲喂,性情温驯,步法稳健,是儿童的理想坐骑)达特姆尔矮马

Dartmoor Pony Society n. an organization founded in England in 1899 to promote and encourage the breeding of Dartmoor ponies (1899年在英国成立的组织机构,旨在促进和鼓励纯种达特姆尔矮马的繁育)达特姆尔矮马协会

dash /dæʃ/ n. **1**【racing】an act of running forward suddenly and hastily 冲刺 **2**【racing】a short fast race; sprint 速力赛,短途赛

dashboard /ˈdæʃbɔːd; *AmE* -bɔːrd/ n. a wooden board located in front of a horse-drawn vehicle to protect the driver and passengers from dirt thrown up by the hoofs of the horse (位于马车前侧的木板,可防止马蹄溅起的泥水弄脏车夫和乘客)挡泥板,挡板

day rug n. *see* sheet 日间马衣

day sheet n. *see* sheet 日间马衣

DDFT abbr. deep digital flexor tendon 趾深屈

肌腱

DE abbr. digestible energy 消化能

de Pluvinel, Antoine n. (1555 – 1622) an Italian horse trainer who developed a humane method of teaching dressage (发展了高级马术训练方法的意大利调马师)安托万·德·普鲁维奈尔: Antoine de Pluvinel was once Louis XIII's riding instructor and author of *La Maneige Royale* (1623) and *L'Instruction du Roi* (1626). 安托万·德·普鲁维奈尔曾是法国国王路易十三的骑术教练,著有《皇家驯马》和《国王的教育》。

dead heat n.【racing】(abbr. **dh**) a condition in which two or more horses arriving at the finish line simultaneously (两匹以上赛马同时到达终点的情况)并列头马,并列第一: The 1884 Epsom Derby finished as a dead-heat of two racers. 1884 年英国德比赛最终以两匹赛马并列第一收场。 compare nose, short-neck, neck, short-head, head, length

dead space n.【anatomy】the space of the airways of lung where gas exchange does not happen on their surface (肺内不发生气体交换的区间)无效腔: The dead space is made up of two components, the anatomical dead space and the physiological dead space. 无效腔由解剖无效腔和生理无效腔两部分组成。

dead track n.【racing】a race track with a hard surface that lacks resiliency (缺乏弹性的赛道)硬地赛道

dead weight n.【racing】*see* lead 负重铅块

dead-sided /ded'saɪdɪd/ adj. (of a horse) numb to the action of the leg, spur, or whip aids on its ribs (指马)肋侧麻木的

dealer /'diːlə(r)/ n. one who buys and sells goods 贩子,商贩: feed dealer 饲料商 ◇ horse dealer 马贩子 ◇ The dealer may either own horses or just act as brokers for those who do. 马贩或许自己有马要卖,或许只为马主充当经纪人。

dealer's whip n. a long whip used by horse dealers 马贩用鞭

Debao Pony n. a pony breed indigenous to Guangxi, China; stands below 11 hands and usually has a bay, roan, or gray coat; has strong legs with hard feet and is willing, tough, and quiet; used mainly for packing and pleasure riding (源于中国广西的矮马品种。体高 11 掌以下,毛色多为骝、沙或青,四肢强健,蹄质结实,性情温驯,顽强而安静,主要用于驮运和休闲骑乘)德保矮马

debilitate /dɪ'bɪlɪteɪt/ vt. to weaken the strength or energy of; enervate 使衰弱 | **debilitation** /dɪˌbɪlɪ'teɪʃn/ n.

debris /'debriː; *AmE* də'briː/ n. the fragmented remains of dead or damaged cells or tissue (死亡细胞或组织的残余成分)碎片

Decemjugis /dɪ'semdʒuːgis/ n.【Latin】an ancient Roman chariot driven by ten horses harnessed abreast (古罗马的马车,由 10 匹马并驾拉行)十马并驾[马车]: Nero is said to have driven one such "Decemjugis" at the Olympic games. 据说尼禄曾在古代奥林匹克运动会上驾驭这种十马并驾的马车。

deciduous/dɪ'sɪdʒuəs/ adj. falling off at a certain stage of growth 脱落的

deciduous teeth n. the teeth that would fall off at a certain age 脱落齿,乳齿 syn temporary teeth, milk teeth

declaration /ˌdeklə'reɪʃn/ n.【racing】*see* declaration of runners 参赛声明

declaration of runners n. (also **declaration**) a written statement provided by a horse owner or trainer prior to a race that declares the participation of the horse (赛前由马主或练马师提交的书面申明,宣布马匹将参加比赛)出赛声明

declaration to win n.【racing】a public announcement made by the owner that his horse will win the race (马主对自己马匹将会获胜的公开宣言)获胜宣言

declare /dɪ'kleə(r); *AmE* dɪ'kler/ vt. to announce or state formally and officially 宣布，宣告 PHR V **declare off** to cancel or withdraw a race or competition temporarily 取消（比赛）：It is said that the next week's race meeting was declared off. 听说下周的赛马取消了。**declare a runner** to announce a horse to run in a race 宣布马匹出赛

declared /dɪˈkleəd; *AmE* -ˈklerd/ adj. **1**【racing】（of a horse）confirmed to run a race（指马）宣布出赛的 **2**【AmE，racing】（of a horse）withdrawn from a race（指马）取消参赛资格的

decline /dɪˈklaɪn/ vi. to decrease or fall down 下降，减少

Dee /diː/ n. a dee-ring D 环，D 形环

dee cheek snaffle n. *see* dee-ring snaffle D 环衔铁

deep flexor tendon n.【anatomy】a tendon that connects the muscles of the back of the upper leg to the coffin bone in the foot（连接马上肢肌肉到蹄骨的肌腱）趾深屈肌腱

deep in the girth adj.（also **deep through the girth**）（of a horse）having well-spring ribs and a broad chest capacity（指马）胸身宽广的

deep-chested /ˈdiːpˈtʃestɪd/ adj.（of a horse）having a broad chest capacity（指马）胸深的

deer /dɪ(ə)r/ n. a hoofed grazing animal with antlers borne only by the male（一种单蹄类草食动物，雄性头顶有角）鹿：deer hunting 猎鹿— *see also* **doe，stag，fawn**

deerfly /ˈdɪəflaɪ/ n.（also **deer fly**）a blood-sucking fly of the genus *Chrysops*，whose painful bites often cause localized swelling and inflammation（鹿蝇属血吸虫，被叮咬后常导致局部肿痛和发炎）鹿虻：The deerfly and horsefly are considered the most significant pests in North America. 鹿虻和马蝇被认为是北美最主要的两种害虫。

dee-ring n.（abbr. **D-ring**）a metal，D-shaped ring found on saddles，tack，and harness（马鞍或挽具上的 D 形铁圈）D 环，D 形环

dee-ring bit n. any bit having dee-shaped cheekpieces D 环衔铁

dee-ring snaffle n.（also **dee-cheek snaffle**）a snaffle bit with a single-jointed mouthpiece and D-shaped cheekpieces（一种小铣铁，带单关节口衔和 D 形颊杆）D 环衔铁 syn racing snaffle

defecate /ˈdefəkeɪt/ vi. to discharge feces from the body 排粪

defecation /ˌdefəˈkeɪʃn/ n. the act of defecating 排粪

defect /ˈdiːfekt/ n. a shortcoming or an imperfection that may influence the appearance or performance of a horse（影响马体貌或性能的毛病）缺陷，失格：congenital defect 先天缺陷 ◇ conformation defect 体貌缺陷 ◇ genetic defects 遗传缺陷 | **defective** /-ktɪv/ adj.

defend /dɪˈfend/ vt. to guard against or keep safe from danger，attack，or harm 保卫；防护

defense /dɪˈfens/ n. **1** the act or means of defending against attack，danger，or injury 防御（措施）：defense mechanism of the body 机体的防御机制 **2**【polo】an attempt to defend and frustrate the attack made by an opponent（马球）防守 opp offense

deferent /ˈdefərənt/ adj.（of a duct or vessel）serving or adapted to carry or transport 运送的，输送的：deferent duct 输送管道

deficiency /dɪˈfɪʃnsi/ n. **1** the state of being deficient 缺乏，不足：Zinc deficiency can lead to stunted growth，hair loss，appetite and weight loss，dry skin，and anemia. 锌缺乏可能导致发育迟滞、毛发脱落、食欲和体重下降、皮肤干燥和贫血等症。**2** symptoms caused by deficiency of certain nutritional substances such as vitamins or minerals（由于维生素或矿物质等营养物质缺乏导致的病症）缺乏症

deficient /dɪˈfɪʃnt/ adj. inadequate in amount or degree；insufficient 缺乏的，不足的

deform /dɪˈfɔːm; *AmE* -ˈfɔːrm/ vt. to spoil the natural form of；misshape 损毁；使畸形

deformity /dɪˈfɔːməti; *AmE* -ˈfɔːrm-/ n. the state

D

of being deformed; malformation 畸形;缺陷,失格:hoof deformities 蹄畸形

degenerative joint disease n. (abbr. **DJD**) a joint disease similar to arthritis commonly occuring in horses as a result of repeated stress (与关节炎类似的骨关节病,多由于反复重压所致)退行性关节炎

dehydrate /diː'haɪdreɪt/ vt. to remove water from; dry 使脱水:Horses might become dehydrated when exposed in high temperature without drinking water for a long period of time. 马如果长时间暴露在高温环境且没有饮水,就可能出现脱水症状。

vi. to lose water from the body 失水,脱水

dehydration /ˌdiːhaɪ'dreɪʃn/ n. **1** the process of removing water from a substance 除水,脱干 **2** an excessive loss of water from the body as from illness or fluid deprivation (由疾病或缺水导致体内水分过多丢失的情况)脱水:Dehydration is an excessive loss of water from the body tissue and may follow prolonged sweating in working horses or severe diarrhea in horses suffering from salmonella infection or strongyle infection. 脱水是由于机体组织水分过多丢失造成的,其病因可能是负役马匹出汗过多,或受沙门氏菌、圆线虫感染引发严重的腹泻所致。

deleterious /ˌdelə'tɪəriəs; *AmE* -'tɪr-/ adj. harmful or injurious 有害的,不利的

delivery /dɪ'lɪvəri/ n. **1** the act of conveying or delivering 运送;移交 **2** the act of giving birth; parturition 分娩;生产:difficult delivery 难产 ◇ The forces of delivery comes from the rhythmic muscular contractions of the uterus and pressure exerted by the voluntary straining of abdominal muscles. 分娩力来自子宫平滑肌节律性收缩以及由腹肌自主性伸张产生的压力。

delivery document n. a document given to the buyer to take possession of the horse from the consignor at a sale or auction (马匹拍卖中递交给买方证明其所有权的文件)马匹移交证书

deltoid /'deltɔɪd/ adj.【anatomy】triangular 三角形的

n. (also **deltoid muscle**) a thick, triangular muscle covering the shoulder joint (位于肩胛骨上呈三角形的肌肉)三角肌

demand /dɪ'mɑːnd; *AmE* dɪ'mænd/ n. things required for a specific purpose 需求,需要:The physiological demands of showjumpers are far higher than that of the dressage horses. 障碍赛马的生理需求要远高于盛装舞步马。

demi-passade /'demɪpəˌsɑːd/ n.【dressage】a dressage movement in which the horse traverses back and forth at a fast pace, making a quick reversal and return (一种古典骑术动作,马以快步向前运行后转身返回)半回转

demi-pique /'demɪˌpiːk/ n. *see* demi-piqued saddle 半角柱马鞍

demi-piqued saddle n. (also **demi-pique**) a heavy saddle with half a saddle horn used by 18th-century cavalry and travelers (仅有半个角柱的一种重型马鞍,为 18 世纪骑兵和骑者所用)半角柱马鞍

demi-pirouette /'demɪˌpɪru'et/ n. *see* half-pirouette 旋转半圈,180°[定后肢]旋转

demi-volte /'demɪˌvəʊlt/ n.【dressage】a movement in which the horse makes a half circle on two tracks (马以横步转半圈的动作)半环骑

demodectic mange n. a skin disease caused by the mite *Demodex folliculorum* living in the hair follicles and sebaceous glands of the skin; characterized by blisters, ulcers, and scabs principally around the eyes and forehead (由寄生在毛囊和皮脂腺毛囊的蠕螨所致的皮肤病,症状表现为眼和前额处出现水疱、溃烂和结痂)蠕螨病

den /den/ n. a hole or underground shelter where wild animals, such as foxes or wolves, live and raise their young (狐狸或狼等野兽居住和哺育后代的地洞)洞穴,兽穴:a wolf's den 狼的

洞穴

denerve /diː'nɜːv; *AmE* -'nɜːrv/ vt. (also **nerve, unnerve**) to cut the never of foot to relieve a horse from the pain of chronic lameness (为解除马匹瘸蹄痛苦而割断蹄部神经)割断神经

denerved /diː'nɜːvd/ adj. having the nerve supply to the foot severed 神经割断的

Dennett /'denɪt/ n. *see* Dennett Gig 登尼特马车

Dennett Gig n. (also **Dennett**) a two-wheeled, horse-drawn vehicle of the Gig type introduced in England in the early 1800s (英国在19世纪早期出现的一种双轮轻型马车)登尼特马车

dens /dens/ n.【Latin】(*pl.* **dentes**) tooth 牙齿

dense bone n. *see* compact bone 骨密质

dent- pref. of tooth or teeth 齿的

dental /'dentl/ adj. of or relating to the teeth or dentistry 齿的;牙科的

dental pulp n. the soft interior core of the tooth containing nerves and blood vessels (牙齿内部的软质结构,其中包含神经和血管)牙髓,齿髓

dental star n. (also **tooth star**) a dark mark on the incisor appearing first as a narrow line across the chewing surface, then as dark circle near the center of the tooth in advanced age (切齿上出现的标记,咀嚼面上早期出现黑线,后来随着年龄增长在切齿中心成为黑圈)齿星

dentine /'dentiːn/ n. the hard, dense substance forming the the main part of a tooth beneath the enamel (牙釉质下面的致密成分,构成牙齿的主体)牙质

dentinum /den'taɪnəm/ n.【Latin】dentine 牙质

dentition /den'tɪʃn/ n. **1** the type and arrangement of a set of teeth 齿式 **2** the process of growing new teeth; teething 生牙,长牙

deoxygenate /diː'ɒksɪdʒɪneɪt/ vt. to remove oxygen from sth 去氧,脱氧 | **deoxygenation** /diːɒksɪdʒə'neɪʃn/ n.

deoxyribonucleic acid n.【genetics】(abbr. **DNA**) a protein compound existing in the form of helix in the cells that determines the hereditary characteristics of an individual (以双螺旋形式存在于细胞中的复合物,决定着个体的遗传性状)脱氧核糖核酸 compare ribonucleic acid

Depot Wagon n. (also **station wagon**) a four-wheeled, horse-drawn carriage popular in America during the second half of the 19th century and fitted with two or three rows of forward-facing seats (一种美式四轮马车,在19世纪下半叶比较常见,有2~3排面朝前的座位)车站马车

depressant /dɪ'presnt/ n. an agent that decreases the physiological activities of the circulatory, respiratory, or central nervous systems (可减缓血液循环、呼吸或神经系统等生理活动的药物)抑制剂,镇静剂

depressor /dɪ'presə/ n.【anatomy】any of various muscles that serve to draw down a body part (可将身体某部位下拉的肌肉)降肌 compare levator

depth of girth n. the measurement between the withers and the elbow of a horse (马体鬐甲至肘突的长度)胸深 IDM **have a good depth of girth** (of a horse) deep-chested (指马)胸深的

derby[1] /'dɑːbi; *AmE* 'dɜːrbi/ n. a stiff felt hat with a round crown and a narrow, curved brim; a bowler hat (一种圆顶窄边的硬毡帽)常礼帽: a derby hat 常礼帽

Derby[2] /'dɑːbi; *AmE* 'dɜːrbi/ n. (also **Darby**) **1**【racing】any of the annual horse races held for three-year-old Thoroughbreds (由3岁纯血马参加的比赛)德比赛,打吡赛: Derby racing 德比赛 **2**【BrE】the Epsom Derby 艾普森德比赛 **3**【AmE】the Kentucky Derby 肯塔基德比赛

Derby bank n.【jumping】a bank that drops 10 feet on the downside and is considered one of the most notorious obstacles in show jumping (落地侧直降3米的一道跨堤障碍,被认为是障碍赛中难度最高的跨栏)德比跨堤

Derby, Edward n. (1865 – 1948) a renowned figure of the British racing industry who founded and named the Epsom Derby (英国赛马界大亨,创立并命名了艾普森德比赛)爱德华·德比

Derby Stakes n. *see* Epsom Derby 艾普森德比赛

derma /ˈdɜːmə/ n. another term for dermis 真皮

dermal /ˈdɜːməl/ adj. of or relating to the skin 皮肤的;真皮的:dermal inclusion cyst 皮下包毛性囊肿

dermatitis /ˌdɜːməˈtaɪtɪs; *AmE* ˌdɜːrm-/ n. inflammation of the skin 皮炎

dermatophytosis /ˌdɜːmətəʊfaɪˈtəʊsɪs/ n. *see* ringworm 皮癣

dermic /ˈdɜːmɪk/ adj. of or relating to dermis; dermal 真皮的

dermis /ˈdɜːmɪs; *AmE* ˈdɜːr-/ n. (also **derma**) the sensitive vascular layer of the skin located beneath the epidermis (位于表皮下的结缔组织层,富含血管和神经)真皮 [syn] corium, cutis, true skin [compare] epidermis

descend /dɪˈsend/ vi. **1** to come or inherit from an ancestor or ancestry 遗传,沿袭 **2** to move from a higher to a lower place; drop, fall 下降,降落

descent /dɪˈsent/ n. **1** the origins or ancestry of a horse (马匹的)祖先;血统 **2** the act or an instance of descending 下降:the descent of testes 睾丸下降

desmitis /desˈmaɪtɪs/ n. inflammation of the ligament, as resulting from tearing of the ligament fibrils (韧带拉伤后发炎的症状)韧带炎

destrier /ˈdestrɪə(r)/ n.【French】a war horse used in the Middle Ages (中世纪的)战马,戎马

destroy /dɪˈstrɔɪ/ vt. to humanely kill a horse or dog due to age, illness, or injury (由于年龄、疾病或伤痛而将马或犬人道地杀死)处死,处置:destroy a rabid dog 处死一条疯狗 [syn] put down

deteriorate /dɪˈtɪəriəreɪt; *AmE* -ˈtɪr-/ vt. to become worse; disintegrate 恶化;降解 | **deterioration** /dɪˌtɪərɪəreɪʃn; *AmE* -ˌtɪr-/ n.

detrain /ˌdiːˈtreɪn/ vt. to fail to reach the training intensity or duration to maintain a horse's fitness (训练强度和时间未达到维持马匹健康要求)训练不足 | **detraining** /-nɪŋ/ n.

detrimental /ˌdetrɪˈmentl/ adj. causing damage or harm; harmful, injurious 有害的

develop /dɪˈveləp/ vi. to grow by degrees into a more advanced or mature state 发展,发育 | **development** /-mənt/ n.

developmental /dɪˌveləpˈmentl/ adj. of or relating to development 发育的:developmental abnormality 发育异常 ◇ developmental stage 发育阶段

developmental orthopaedic diseases n. (abbr. **DOD**) any of the nutritional problems that arise in growing horses (马)发育期骨病

devil grass n. *see* Bermuda grass 百慕大草

deworm /dɪˈwɜːm; *AmE* -ˈwɜːrm/ vt. to cure animals of worms; worm 驱虫:Horses should be dewormed every 6-8 weeks by a veterinarian or experienced horse owner. 兽医或有经验的马主应当每隔6~8周就给马匹驱虫一次。

dewormer /dɪˈwɜːmə/ n. an agent used to kill parasites in horses 驱虫剂

dexter /ˈdekstə/ n.【Latin】right 右的

dextra /ˈdekstrə/ adj.【Latin】dexter 右的

dh abbr.【racing】dead heat 并列头马,并列第一

DI abbr. dosage index 剂量索引

diagnose /ˈdaɪəgnəʊz; *AmE* ˌdaɪəgˈnoʊs/ vt. to identify a disease by its symptoms (通过症状来确认疾病)诊断

diagnosis /ˌdaɪəgˈnəʊsɪs; *AmE* -ˈnoʊ-/ n. the act or process of determining the nature and cause of a disease (确定疾病症状和起因的过程)诊断

diagnostic /ˌdaɪəgˈnɒstɪk; *AmE* -ˈnɑːs-/ adj. of, relating to, or used in a diagnosis 诊断的:

Nowadays, the portable X-ray apparatus and fibrescope have become the common diagnostic equipments for nearly every practitioner. 目前，便携式 X 线摄影仪和纤维镜基本已成为每位从业兽医的常用诊断设备。

diagnostic ultrasound n. the use of high-frequency sound waves to image internal structures（采用高频声波成像的技术）超声波诊断术

diagonal /daɪˈægənl/ adj. involving a horse's forefoot and hind foot on the opposite sides（马蹄）对角线的

n. a way of moving when the horse's forefoot moves in pairs with the opposite hind foot at the trot（快步行进中马前肢与对侧后肢步调一致的情况）对角线运步：left diagonal 左对角线运步 ◇ right diagonal 右对角线运步

diagonal aids n. the use of the rein aids applied to one side of the horse's body in combination with the heel on the opposite side（一侧缰绳辅助与对侧脚辅助结合使用）对角辅助：The diagonal aids refer to either the right rein used in combination with the left heel or alternatively, the left rein with the right heel. 对角辅助指右方缰和左脚或左方缰和右脚结合使用的情况。 compare lateral aids

diaphragm /ˈdaɪəfræm/ n.【anatomy】a muscular membrane separating the abdominal and thoracic cavities（胸腔和腹腔之间的肌膜）横膈膜：The lung volume is increased during inspiration by contraction of the diaphragm and expansion of the chest. 吸气时，肺容量的增大通过膈膜收缩和胸腔扩张实现的。| **diaphragmatic** /daɪəfrægˈmætɪk/ adj.

diaphragmatic hernia n. a protrusion of abdominal contents through the diaphragm（腹腔内脏从膈膜脱出的症状）横膈膜疝，膈疝 syn rupture

diaphysis /daɪˈæfəsɪs/ n.【anatomy】(*pl.* **diaphyses** /-siːz/) the shaft of a long bone（长骨的中段）骨干 | **diaphyseal**, **diaphysial** /ˌdaɪəˈfɪzɪəl/ adj.

diarrhea /ˌdaɪəˈrɪə/ n. a condition in which feces are discharged from the bowels too frequently in liquid form（水样粪便排泄过于频繁的症状）腹泻：Diarrhea is often caused by gastrointestinal distress or disorder. 腹泻通常由胃肠疼痛或紊乱导致。

dickey /ˈdɪki/ n.【BrE, slang】(also **dicky**) a folding outside seat for servants at the back of a carriage（马车后面供仆人乘坐的座位）尾座

dicky /ˈdɪki/ n. *see* dickey（马车的）尾座

Dicky Coach n. a horse-drawn coach or carriage in which the box seat was detached from the main bodywork（驾驶座与车身相隔较远的马车）迪克马车

diencephalon /ˌdaɪenˈsefəlɒn/ n.【anatomy】the posterior part of the forebrain that contains the thalamus and hypothalamus（前脑后部包括丘脑和下丘脑的部分）间脑 syn thalamencephalon

diestrum /daɪˈestrəm/ n. *see* diestrus 间情期

diestrus /daɪˈestrəs/ n. (also **diestrum**) a period of sexual inactivity between two estrous cycles（两次发情之间性活动不积极的时段）间情期 | **diestrous** /-trəs/ adj. — *see also* **anestrus, estrus**

diet /ˈdaɪət/ n.【nutrition】the usual food of a horse; ration 日粮；饲料：rich diet 高营养日粮 ◇ horse diet 马匹日粮

dietary /ˈdaɪətəri/ adj. of or relating to diets 日粮的：dietary requirement 日粮需要 ◇ dietary calcium intake 日粮钙摄入量

dietary essential amino acids n.【nutrition】(abbr. **DEAAs**) any of the ten amino acids that are essential for the growth and maintenance of the horse and must be supplied by the ration as they are not internally synthesized（马匹生长和维持机体活动所必需的十种氨基酸，由于自身不能合成，所以从日粮中摄取）日粮必需氨基酸

digest /daɪˈdʒest; *AmE* dɪˈ-/ vt. to convert (food) into simpler chemical compounds that can be absorbed and assimilated by the body (将食物转变成能被身体吸收和同化的简单化合物)消化

digestibility /dɪˌdʒestəˈbɪlɪti/ n. the ratio of the nutrients absorbed and taken in (营养物吸收的比率)消化率

digestible /daɪˈdʒestəbl, dɪˈ-/ adj. readily or easily digested 易消化的,可消化的:digestible energy 消化能

digestion /daɪˈdʒestʃən, dɪˈ-/ n. the act or process of digesting 消化(作用):digestion coefficient 消化系数

digestive /daɪˈdʒestɪv, dɪˈ-/ adj. of or relating to digestion 消化的:digestive system 消化系统 ◇ digestive tract 消化道 ◇ digestive upsets 消化紊乱

digit /ˈdɪdʒɪt/ n. the part of a horse's leg below the fetlock (马四肢球节以下的部分)趾

digital /ˈdɪdʒɪtl/ adj. of or relating to a digit 趾的:deep digital flexor tendon 趾深屈肌腱 ◇ superficial digital flexor tendon 趾浅屈肌腱

digital cushion n. *see* plantar cushion 趾垫

dilate /daɪˈleɪt/ v. to become or make wider or larger; expand (使)膨胀;扩大

diluent /ˈdɪljʊənt/ n. *see* dilutor 稀释液

dilute /daɪˈluːt; *AmE* -ˈljuːt/ vt. **1** to make thinner or less concentrated by adding liquid or water (加溶液或水以降低浓度)稀释,冲淡:diluted semen 稀释精液 **2** to lessen the force or purity by adding other elements 淡化;减弱:color diluting genes 毛色淡化基因 | **dilution** /-luːʃn/ n.

dilutor /daɪˈluːtə(r)/ n. (also **diluent**) a solution used to dilute the raw semen for artificial insemination (人工授精时用来稀释鲜精的溶液)稀释液 [syn] extender

dimple /ˈdɪmpl/ n. a small depression in the flesh of a horse's body (马身体上的凹陷)陷窝,凹痕

dinks /dɪŋks/ n. a pair of fringed chaps that extend only to the knees 须边套裤

dioestrous /daɪˈestrəs/ adj. of or relating to the diestrus 间情期的

dioestrum /daɪˈestrəm/ n.【BrE】diestrum 间情期

dioestrus /daɪˈestrəs/ n.【BrE】diestrus 间情期

Dip WCF abbr. Diploma of the Worshipful Company of Farriers 蹄铁匠公会证书

diphron /ˈdɪfrən/ n. an ancient Greek chariot (古希腊的)马车,战车

Diploma of the Worshipful Company of Farriers n. (abbr. **Dip WCF**) one of three levels of certification awarded to practicing farriers in UK by the Worshipful Company of Farriers on the basis of examinations (由蹄铁匠公会经过考核而颁发给从业蹄铁匠的证书)蹄铁匠公会文凭 [compare] Associate of the Worshipful Company of Farriers, Fellow of the Worshipful Company of Farriers, Registered Shoeing Smith

dipped /ˈdɪpt/ adj. (of horse's back) hollow and arching downwards (马背)凹陷的:dipped back 凹背

dirt-encrusted /ˌdɜːrtnˈkrʌstɪd/ adj. smeared or covered with dirt all over the body 满身尘土的;污浊的

discharge /dɪsˈtʃɑːdʒ; *AmE* -ˈtʃɑːr-/ vt. to release or give off; emit 释放;排出,排泄

n. a substance that is released from inside the body 排放物,排泄物:nasal discharge 鼻涕

discipline /ˈdɪsəplɪn/ n. **1** the training of a horse to obey commands and orders 调教;训练,驯致 **2** a system of rules of conduct 纪律,准则 **3** an activity or sport intended to train the body or mind (训练身心的体育活动)项目:The FEI recognizes eight equestrian disciplines—dressage, show jumping, eventing, endurance riding, combined driving, vaulting, reining, and para-equestrianism. 国际马术联合会认可以下八个马术项目,包括盛装舞步、障碍赛、三日赛、长途耐力赛、驾车赛、马上体操、西部驭马和残疾人马术。

vt. to train a horse to obey orders and commands

训练,驯致

disembowel /ˌdɪsɪmˈbaʊəl/ vt. to remove the internal organs of an animal; eviscerate 取出内脏 | **disembowelment** /-lmənt/ n.

dished face n. (also **dished profile**) a concave profile of a horse's head, usually seen in Arabs (马面貌凹陷的情况,常见于阿拉伯马)面部凹陷 [syn] concave face, stag face

dish-faced /dɪʃˈfeɪst/ adj. (of a horse) having a dished face 凹头的: a dish-faced Welsh pony 一匹凹头的威尔士矮马。

dishing /ˈdɪʃɪŋ/ n. 【BrE】*see* paddling 内向肢势

disinfect /ˌdɪsɪnˈfekt/ vt. to cleanse so as to prevent the growth of disease-carrying microorganisms (为阻止微生物滋生而进行的清洁)消毒,杀菌: disinfected water 消毒液 | **disinfection** /ˌdɪsɪnˈfekʃn/ n.

disinfectant /ˌdɪsɪnˈfektənt/ n. any agent or substances used to disinfect 消毒液;杀菌剂

dismount /dɪsˈmaʊnt/ v. **1** to get off from a horse 下马 **2** (of a stallion) get off from a mating mare after copulation (公马与母马交配后的)下马

disobedient /ˌdɪsəˈbiːdiənt/ adj. (of a horse) refusing to accept the commands of its rider; intractable (指马)拒服命令的,不听话的 | **disobedience** /-diəns/ n.

displaced fracture n. a break in a bone in which the fragments have moved out of alignment 错位性骨折

disqualification /dɪsˌkwɒlɪfɪˈkeɪʃn; *AmE* -ˌkwɑːl-/ n. (abbr. **DSQ**) the act of disqualifying or the state of having been disqualified 取消(参赛)资格: Under the FEI rules, it is forbidden to give any stimulant to a horse and detection of such a substance in the urine, blood, saliva or sweat of a horse will lead to disqualification. 根据国际马术联合会的规则,给马匹注射或服用任何兴奋剂都是违禁的。一旦在马的尿液、血液、唾液和汗液中检测到这些物质,都将取消比赛资格。

disqualify /dɪsˈkwɒlɪfaɪ; *AmE* -ˈkwɑːl-/ vt. to remove a horse or rider from competition due to a serious fault (指马或骑手)取消参赛资格 [compare] withdraw

dissect /dɪˈsekt/ vt. to cut apart or separate specimen for anatomical study (把样本切开进行解剖研究)解剖,剖开

dissection /dɪˈsekʃn/ n. the act or an instance of dissecting 切割;解剖

distaff /ˈdɪstɑːf/ n. the female as a whole 母系 [IDM] **on the distaff side** on the female line of a pedigree 在母系上,母系的

distaff race n. 【racing】 a race in which only female horses are allowed to enter 雌驹赛

distaffer /ˈdɪstɑːfə/ n. a female horse 雌驹,母马

distal /ˈdɪstl/ adj. 【anatomy】 located far from a center or body part 末梢的,远侧的: distal sesamoid bone 远侧籽骨 ◇ distal phalanx 远侧趾骨 [compare] proximal

distal pulse n. the rhythmic beating or throbbing of the arteries in the foot of horse (马四肢上动脉跳动的脉搏)远侧脉搏

distal sesamoid ligament n. 【anatomy】 (also **distal sesamoidean ligament**) the band of fibrous tissue which serves to connect the bottom of the sesamoid bones to the long and short pastern bones (连接籽骨末端和系骨的纤维组织)远侧籽骨韧带

distal sesamoidean ligament n. *see* distal sesamoid ligament 远侧籽骨韧带

distalis /ˈdɪstəlɪs/ adj. 【Latin】distal 末梢的,远侧的 [opp] proximalis

distance /ˈdɪstəns/ n. (abbr. **dst**) the length of a race 赛程: short distance race 短途赛 ◇ long distance riding 长途骑乘

vt. 【racing】to leave far behind; outrun 抛在后面;远远超过

distanced /ˈdɪstənst/ adj. 【racing】 (of a horse) falling behind the rest of the field (赛马)落后的;掉队的

D

distend /dɪsˈtend/ v. (to cause) to swell out or expand from internal pressure 膨胀，肿胀：the abdomen distended rapidly 腹部迅速膨胀起来

distendable /dɪsˈtendəbl/ adj. that can be distended 可扩张的

distension /dɪsˈtenʃn/ n. the act or instance of distending 膨胀：distension colic 胀气性腹痛

disturb /dɪˈstɜːb; *AmE* -ˈstɜːrb/ vt. to interrupt or destroy the settled state; upset 扰乱；搅乱

disturbance /dɪˈstɜːbəns; *AmE* -ˈstɜːrb-/ n. the act of disturbing or state of being disturbed 扰乱；紊乱，失调：disturbance of the electrolyte balance 盐离子平衡失调

disunited /ˌdɪsjuːˈnaɪtɪd/ adj. (of the legs) out of sequence at canter or gallop（指四肢）步调混乱的

disunited canter n. the action of a cantering horse who changes leg sequence 失调跑步，分裂跑步 [syn] cross-canter, broken canter

disunited gallop n. the action of a galloping horse who changes leg sequence 失调袭步 [syn] false gallop

disuse /dɪsˈjuːs/ n. the state of not being used or of being no longer in use 弃用，废弃：use and disuse theory 用进废退学说

ditch /dɪtʃ/ n.【jumping】a jumping obstacle consisting of a narrow trench filled with water 水沟[障碍]

diuretic /ˌdaɪjuˈretɪk/ adj. tending to increase the discharge of urine 利尿的 [opp] antidiuretic

n. (usu. **diuretics**) a drug that tends to increase the discharge of urine 利尿剂，利尿药

diverticular /ˌdaɪvəˈtɪkjʊlə/ adj. of or relating to diverticula 憩室的

diverticulum /ˌdaɪvəˈtɪkjʊləm/ n.【anatomy】(*pl.* -**la** /-lə/) a pouch or sac branching out from a hollow organ or structure（从中空器官或结构分支出来的囊室）憩室：allantoic diverticulum 尿囊憩室 ◇ diverticulum of duodenum 十二指肠憩室 ◇ nasal diverticulum 鼻憩室 ◇ vesical diverticulum 膀胱憩室

divide /dɪˈvaɪd/ vi.【genetics】(of cells) to separate into two identical ones（细胞）分裂

dividend /ˈdɪvɪdend/ n. *see* payoff 红利，奖金

division /dɪˈvɪʒn/ n. the act or process of dividing 分裂（过程）：cell division 细胞分裂

divot /ˈdɪvət/ n. a piece of turf torn up by a horse's hoofs（马蹄掀起的）草皮

DNA abbr. deoxyribonucleic acid 脱氧核糖核酸 [compare] RNA

DNA profiling n.【genetics】(also **DNA testing**) a biotechnical test now adopted by most horse registries to verify the identification or parentage in horses by analyzing the genetic material DNA（通过分析马的DNA进行系谱和亲子鉴定的生物技术，该方法现已被大多登记机构所采用）DNA 检测 [compare] blood-typing

DNA testing n. *see* DNA profiling DNA 检测

dobbin /ˈdɒbɪn/ n.【BrE】a horse used for farm work 农用马，役用马

docile /ˈdəʊsaɪl; *AmE* ˈdɑːsl/ adj. (of a horse) tame or willing; tractable 温驯的，易驾驭的

docility /dəʊˈsɪləti; *AmE* dɑːˈs-/ n. the condition or state of being tame or willing 温顺；温驯

dock[1] /dɒk; *AmE* dɑːk/ n. a hybridized breed of plant, often used as fodder of livestock（一种杂种植物，常作牲畜草料）酸模：curled dock 卷叶酸模 ◇ broad leaved dock 阔叶酸模

dock[2] /dɒk; *AmE* dɑːk/ n. the root or base of a horse's tail 尾根，尾础

vt. to cut off the tail of a horse and burn it by a hot iron to prevent infection（将马尾末端截断后再烙烫以防止感染）断尾 PHR V **dock and set** to cut and set the tail of a horse high 割尾垫高 | **docking** n.：Docking was formerly often practiced on harness horses and cobs to yield a more neat appearance and to prevent the reins from becoming caught under the tail. 过去，挽马和柯伯马常进行断尾，不仅可使外观更加整洁，也可防止马尾与缰绳相缠绕。

dock tail n. a horse having a tail cut short; curtal

断过尾的马 [syn] cocktail

docked /dɒkt/ adj. (of a horse) with part of the tail cut off; curtal (指马)断尾的:docked tail 截短的尾巴

docker /ˈdɒkə(r); *AmE* ˈdɑːk-/ n. one who docks the tail of an animal 断尾工

Docking and Nicking Act n. a law passed in UK in 1948 to prohibit the practice of docking or nicking a horse's tail to improve its appearance for show purposes (英国于 1948 年通过的法案,禁止展览赛中截断马尾以改善体貌的做法)断尾与垫尾法案

docking knife n. a knife used to cut off the tail of a horse 断尾刀

docking iron n. a metal iron when heated used to cauterize the amputated dock 断尾烙铁

dockpiece /ˈdɒkpiːs; *AmE* ˈdɑːk-/ n.【driving】*see* crupper dock 尻兜,尻带

doe /dəʊ; *AmE* doʊ/ n. (*pl.* **doe**) a female deer 雌鹿,母鹿 — *see also* **deer, fawn, stag**

doer /ˈduːə(r)/ n. a horse kept for certain purposes; keeper 饲养的马:a good doer 一匹好养的马 ◇ A poor doer has a picky appetite and often goes off his feed easily. 难养的马挑食且往往食欲不振。

dog cart n. *see* dogcart 轻便马车

dog fox n. a male fox 公狐,雄狐— *see also* **vixen**

dog hound n. a male hound 公猎犬

dogcart /ˈdɒgkɑːt/ n. **1** (also **dog cart**) an open, two-wheeled English carriage drawn by one horse and accommodating two passengers seated back to back (一种英式敞篷双轮马车,由单马套驾,背对背可坐两名乘客)轻便马车 **2** a small cart pulled by dogs 犬拉马车

Dogcart Phaeton n. a four-wheeled, horse-drawn vehicle of the dogcart type 轻便费顿马车 [syn] Four-wheeled Dogcart, Double Dogcart

dogger /ˈdɒgə/ n.【AmE】one who purchases horses, asses, or mules and slaughters them for meat (购买牲口以屠宰获肉者)屠马贩,屠牲口贩

dog-legged whip n. a driving whip with the stock attached at a right angle to the shaft (把柄与鞭杆垂直的马鞭)狗腿马鞭,曲柄马鞭

Døle Gudbrandsdal n. a draft breed originating in Norway's Gudbrandsdal Valley from which the name derives; stands 14.2 to 15.2 hands and generally has a bay, brown or black coat, with palomino and gray occurring rarely; has a heavy head, full forelock, mane, and tail, short back, broad chest, short legs with heavy feathering around the fetlock, and broad feet; used for heavy draft, farm work, and harness racing (源于挪威康伯兰德山谷地区的挽马品种并由此得名。体高 14.2 ~ 15.2 掌,毛色多为骝、褐或黑,银鬃和青毛少见,头重,鬃尾浓密,背短,胸深,四肢粗短,球节处距毛浓密,蹄大,主要用于重挽、农田作业和轻驾车赛)多勒·康伯兰德马

Døle Trotter n. a light version of the Døle Gudbrandsdal bred through heavy introduction of blood from other trotter breeds (在多勒·康伯兰德马的基础上大量引入其他快步马外血育成的轻型马品种)多勒快步马,挪威快步马 [syn] Norwegian Trotter

domador /ˌdɒməˈdɔː(r)/ n.【Spanish】a horse trainer or tamer (南美的)驯马师,调马师

domestic /dəˈmestɪk/ adj. (of animals) kept on farms or domesticated; tame (动物)家养的;驯化的:domestic horse 家马 ◇ domestic donkey 家驴

domesticate /dəˈmestɪkeɪt/ vt. to train and keep wild animals for human use (驯养野生动物为人类所用)驯化,家养:domesticated animals 驯养动物 | **domestication** /dəˈmestɪkeɪʃn/ n.

dominance /ˈdɒmɪnəns/ n. the condition or fact of being dominant 优势;显性:incomplete dominance 不完全显性 [compare] recessiveness

dominant /ˈdɒmɪnənt/ adj. **1** most important or influential in power or control 有优势的;支配的 **2**【genetics】of or relating to the allele that produces the same phenotypic effect in both homozygote and heterozygote（其基因不论纯合体还是杂合体在表型上都相同）显性的：dominant gene 显性基因 ◇ dominant characters 显性性状 opp recessive

D

Don /dɒn；*AmE* dɑːn/ n.（also **Donsky**）a Russian warmblood descended from Turkmene and Karabakh stallions bred to native steppes mares during the 18th and 19th centuries；stands 15.1 to 15.3 hands and generally has light bay, chestnut, or brown coat；has a deep chest, long straight neck and back, and long legs；is hardy, frugal, energetic, and has good stamina；used for riding, long-distance racing, and light harness（俄国温血马品种，于18—19世纪采用土库曼马和卡拉巴赫马公马与当地草原母马杂交繁育而来。体高15.1～15.3掌，毛色多为淡骝、栗或褐，胸深，颈背长而平直，抗逆性强，耐粗饲，体力充沛，耐力好，主要用于骑乘、长途赛和轻挽）顿河马

done /dʌn/ adj.（of a horse）tired and exhausted（指马）精疲力竭的

donkey /ˈdɒŋki；*AmE* ˈdɑːŋ-/ n. the domesticated ass 毛驴，家驴

donkey foot n. *see* club foot 立系，狭蹄

Donsky /ˈdɒnski/ n. *see* Don 顿河马

door-banging /ˌdɔːrˈbæŋgɪŋ/ n. a vice of horse who kicks the stable door with his hoof when hungry or bored（马饥饿或厌烦时用蹄踢门的恶习）踢门癖

doorman /ˈdɔːmən；*AmE* ˈdɔːr-/ n.【BrE】a farrier assistant who helps shoeing horses 蹄铁匠助手

doorway /ˈdɔːweɪ；*AmE* ˈdɔːr-/ n. the entrance to a building（建筑的）入口，门口

dope /dəʊp；*AmE* doʊp/ n. **1** any illegal substance administered to a racehorse to enhance his performance（为了提高赛马成绩而给马服用的非法制剂）违禁药物，兴奋剂 **2**【racing, slang】information about the past performance of a racehorse（赛马以往的）出赛记录，出赛信息 vt. to administer illegal drugs to a racehorse 给（赛马）服违禁药物

dope test n. the test of a horse's blood or urine to examine the existence of illegal drugs（对马的血液或尿液进行的违禁药物检测）兴奋剂检测，药检 syn drug test

doping /ˈdəʊpɪŋ/ n. the practice or process of administering illegal drugs to a horse；nobbling 服违禁药物

dormant /ˈdɔːmənt；*AmE* ˈdɔːrm-/ adj.（of a disease）latent but capable of being activated；potential（疾病）潜在的，潜伏的

dorsal /ˈdɔːsl；*AmE* ˈdɔːrsl/ adj.【anatomy】of the back of a horse（马）背侧的

dorsal stripe n.（also **eel stripe**）a dark stripe found in primitive breeds that runs down from the neck along the back to the tail（原始马种从颈经背延伸至尾的暗色条纹）背线—*see also* **zebra stripes, wither stripe**

dos-à-dos /ˌdəʊzəˈdəʊ/ adv.【French】（of seats）sitting back to back 背对背地 compare vis-à-vis

dose /dəʊs；*AmE* doʊs/ n. a specified quantity of drug or medicine prescribed to be taken at one time or at stated intervals（处方对药物所规定的用量）剂量

double /ˈdʌbl/ n. **1** a jumping obstacle with ditches on both the take-off and landing sides 双侧水沟障碍 **2**【jumping】（usu. **doubles**）*see* double combination 双道组合障碍 **3**【racing】（also **doubles betting**）*see* daily double 双赢，二连赢

double bank n. a bank with ditches on both the take-off and landing sides（起跳侧和落地侧都有水沟的跨堤）双跨堤障碍

double bridle n. an English bridle that has a snaffle bit and a curb bit with separate cheekpieces and reins（一种英式马勒，所带小衔铁和衔链都有各自的颊革和缰绳）双勒，大勒

double combination n.【jumping】（also **doubles**）a jumping obstacle consisting of two consecutive elements with one or two strides between them but numbered and judged as one (由两道跨栏构成的障碍,其间可跨 1 ~ 2 步,但作单个障碍计算)双道组合障碍

double cryptorchid n. *see* rig 双侧隐睾马

Double Dogcart n. *see* Dogcart Pheaton 轻便费顿马车

double fence n. a fence with ditches on both the take-off and landing sides 双水沟跨栏

double gaited adj. (of a horse) capable of either trotting or pacing with good speed (指马能以快步和对侧步快速行进的)二步态的,两步态的:a double gaited horse 一匹二步态的马

double harness n.【driving】the harness for a pair of horses 双马挽具

double horse n. an extra horse used to replace the principal horse when performing stunts or tricks in movie-making (电影中代替主角马出演惊险动作的备用马)代用马,替身马

double muscling n. the characteristics of having pronounced muscling over the croup (臀部肌肉异常发达的特征)双肌性状

double oxer n.【jumping】an obstacle consisting of two oxers positioned with one or two strides between them 双道双横木障碍

double volte n.【dressage】a movement in which the horse makes a six-meter circle twice (盛装舞步动作,马绕行一个 6 米圆圈两次)环骑两圈

double-muscled /ˌdʌblˈmʌsld/ adj. (of a horse or cattle) having pronounced muscling over the croup (马或牛后躯肌肉异常发达)有双肌性状的

doubles /ˈdʌblz/ n.【jumping】*see* double combination 双道组合障碍

doubles betting n.【racing】*see* double 双赢,二连赢

douga /ˈdəugə/ n. (also **duga**) an arched, wooden yoke fixed to the shafts of a Russian carriage high over the horse's neck (俄式马车装在辕杆上的弓形木制项轭)辕弓

doughnut boot n. *see* sausage boot 护腿圈

downgrade /ˌdaʊnˈgreɪd/ vt.【racing】to reduce a race to a lower grade or level 降级:In 2009, the Suburban Handicap was downgraded from a Grade-1 to a Grade-2 event. 纽约市郊让磅赛在 2009 年由一级降为二级赛事。 opp upgrade

Downs /daʊns/ n. *see* Epsom Downs 艾普森·唐斯赛马场

downward transition n. a transition from a faster gait to a slower one (步法由快变慢的换步)下行换步,向下转换 opp upward transition

downwind /ˌdaʊnˈwɪnd/ adv. in the direction in which the wind blows 顺风的,下风向的:The hunter moved downwind toward the fox, lest it will smell him. 猎人从下风向靠近狐狸,以免狐狸嗅出他的气味。 opp upwind

draft[1] /drɑːft; *AmE* dræft/ n. **1** *see* draught 贼风,气流 **2** the act of pulling loads as by horses 拖运,拉送 **3** a team of animals used to pull loads (用来拉运物资的牲畜)畜队,挽队

adj. used or pulling heavy loads 挽用的: a draft horse 一匹挽马

draft[2] /drɑːft; *AmE* dræft/ vt.【hunting】to select or cull a hound from a pack 挑选(猎犬): drafted into the pack 被选入犬队 | **drafting** n.

draft breed n. a heavy horse breed used for transportation and farm work (用于拉运和农田作业的重型马种)挽用马种:The most famous draft horse breeds are Belgian, Clydesdale, Percheron, Shire and Suffolk. 最有名的挽马品种有比利时挽马、克莱兹代尔马、佩尔什马、夏尔马和萨福克马。

draft horse n.【AmE】any heavy horse used to pull heavy loads (用来拉货的)挽马 syn draught horse

drafted /ˈdrɑːftiːd/ adj. **1** (of a hound) being culled from the pack (指猎犬)淘汰的:a drafted hound 一只被淘汰的猎犬 **2** (of a hound) loaned from another hunt (猎犬)借调的,租借的

drafty /ˈdrɑːfti/ adj. having the characteristics of a heavy draft horse 有重型马特征的

drag /dræg/ n. **1**【hunting】the line taken by a fox to its kennel (狐狸回窝的路线)行踪,踪迹 **2**【hunting】a trail of artificial scent left by dragging a strong-smelling substance across the countryside similar to the scent of a fox (人为散播类似狐狸气味的物质在野外留下的踪迹)人工踪迹,气味:The drag is generally laid two or three hours before the hunt. 人工踪迹通常在狩猎前 2 ~ 3 小时铺设。**3** a large horse-drawn vehicle drawn by a team of four horses for hauling heavy loads (驷马联驾用来拖载重物的马车)重型马车

vi.【AmE,racing】(of a horse) to lag behind in a race (指赛马)拖后,落后

drag hound n.【hunting】a hound trained for drag hunt 寻踪猎犬

drag hunt n.【hunting】a hunt conducted on horseback in which the hounds follow an artificial scent previously laid across the countryside similar to that of a fox (马上狩猎项目,其中猎犬追踪事先铺设好的类似狐狸气味的踪迹)寻踪狩猎: Nowadays, drag hunt is gaining popularity in areas where foxes are either protected against hunting or scarce. 现今,寻踪狩猎逐渐在禁猎区和狐狸罕见的地方流行起来。compare live hunt

dragman /ˈdrægmən/ n.【hunting】one responsible for laying the artificial scent used in a drag hunt 铺设人工踪迹者

dragoon /drəˈguːn/ n. a mounted soldier who carried a heavy gun in the past (过去携带机枪的骑兵)重骑兵

vt. PHR V **dragoon sb into sth/doing sth** to force one to do what they do not want to do 逼迫某人做某事

dragsman /ˈdrægzmən/ n. *see* coachman 车夫

drain /ˈdreɪn/ vt. to draw off (water) by a gradual process; empty 排出,排水: well-drained soil 排水良好的土壤

drainage /ˈdreɪnɪdʒ/ n. the action or a method of draining 排水

draught /drɑːft/ n. **1**【BrE】a current of air in an enclosed space (封闭空间内的一股气流)贼风,气流:through-draught 穿堂风 **2**【BrE】*see* draft 拖运;挽力

draught horse n.【BrE】*see* draft horse 挽马

draw /drɔː/ vt. **1**【driving】(of a horse) to pull a carriage with efforts (指马)拉车,挽车 **2** to come near or approach sth PHR V **draw cattle**【cutting】(of a horse) to approach a cow (指马)靠近牛群,接近牛群 **3**【hunting】(of hounds) to search a covert (指猎犬)搜索树丛 PHR V **draw a covert** to hunt every part of a covert in search of the fox 彻底搜索树丛 **draw blank** to search a covert without finding a fox (指猎犬)搜索无果

draw reins n. the long reins that attach to the girth of saddle at one end and pass through the rings of the bit back to the hands of the rider, used to lower and adjust a horse's head position (系在肚带上穿过衔铁环后持在骑手手中的缰绳,主要用来调整马的头姿)低头缰: Draw reins are difficult to use correctly, only very advanced riders should use them. 合理地使用低头缰难度较高,只有高水平的骑手方可采用。

drawing knife n.【BrE】*see* hoof knife 蹄刀

dray /dreɪ/ n.【AmE】(also **transfer dray**) a heavy, low, flat, four-wheeled freight vehicle used for carrying heavy loads, usually drawn by a team of draft horses (一种重型货运平板四轮马车,多由几匹挽马套驾拉行)平板马车 syn flat

vt. to convey or transport by means of a dray 用平板车载送

drench /drentʃ/ vt. to administer liquid medicine orally to an animal (给牲畜)灌药,灌服

n. a large dose of liquid medicine administered orally to livestock (给牲畜灌服的)药液

dress /dres/ v. **1** to arrange or place into alignment according to certain principles（按特定规则）排放，布置 **2** to apply medication, bandages, or other therapeutic materials to a wound 敷药；包扎（绷带）**3** to treat or prepare animal hide in a certain way 处理（生皮）：dressed leather 制好的皮革

dressage /ˈdresɑːʒ/ n. a discipline of Olympic equestrian sports in which a horse is trained to perform a series of classical movements under the gentle manipulation and aids of the rider（奥运会马术项目之一，骑手策马表演多个古典骑术动作）盛装舞步：dressage horse［盛装］舞步马 syn ballet on horseback

dressage arena n.【dressage】a flat, smooth, rectangular area where dressage competitions are held（举行盛装舞步赛的长方形平地）盛装舞步赛场：The dressage arena usually measures 60 m×20 m in international competitions, and the footing is often surfaced with grass, sand, woodchips, plastic, and a variety of other composites. 在国际比赛中，盛装舞步赛场通常为60米×20米，地面材料多为草皮、沙土、木屑、塑胶或其他复合材料铺设。

dresser /ˈdresə(r)/ n. one who dresses wounds 包扎伤口者

dressing /ˈdresɪŋ/ n. **1** the action or process of dressing 布置；处理；包扎 **2** a therapeutic or protective material used to dress a wound, such as bandages（包裹伤口的材料，如绷带）包扎用品，敷料

DRF abbr. *Daily Racing Form*《每日赛马汇览》

D-ring n. *see* Dee-ring D 环

D-ring snaffle n. *see* Dee-ring snaffle D 环衔铁

drive /ˈdraɪv/ vt. **1** (of a coachman) to operate a horse-drawn vehicle 驾车 PHR V **drive with a full hand** to hold two sets of reins in one hand with each rein passing through a different finger（指车夫）单手持缰驾车 **2** (of a herdman) to force a herd of cattle or horses forward in a controlled manner（指牧人）赶牲口，驱赶 **3** (of the rider) to push a ridden horse forward by seat aids（指骑手）赶马前进 **4**【racing】a dash or exertion of the racehorse, such as on the homestretch（终点直道上）冲刺，猛冲 **5**【hunting】to urge the hounds to follow the line of a fox so as not to lose the scent 催促，催赶

driver /ˈdraɪvə/ n. one who drives a horse-drawn vehicle; coachman 驾车者；车夫

driving aids n. any combination of natural and artificial aids used by the driver 驾车辅助

driving blinkers n. *see* blinkers（挽马用的）眼罩

driving hammer n. (also **shoeing hammer**) a farrier's tool used to drive horseshoe nails into the hoof wall（蹄铁匠用来镶蹄钉的工具）钉蹄锤，装蹄锤

driving harness n. harness for horses driving a cart or carriage 驾车挽具

driving whip n. *see* coaching crop 驾车用鞭，车夫用鞭

drooping quarter n. *see* goose rump 尖尻

drop[1] /drɒp; *AmE* drɑːp/ n.【driving】*see* forehead drop 额前垂饰

drop[2] /drɒp; *AmE* drɑːp/ v. **1** to lower or decrease 下降，降落 PHR V **drop down**【racing】(of a horse) to run against lower class horses than it had previously been competed against（指马）降级出赛 **drop on a cow**【cutting】(of a horse) to lower or drop down onto its forelegs prior to working a cow（指截牛马拦截前）俯下前肢 **drop the stirrups** (of a rider) to kick off one's stirrups and continue riding without them（指骑手）脱镫骑乘 **2** to give birth to a foal 产驹，下驹 IDM **drop a foal** 产驹，下驹 **3** (of a stallion) to have its penis fall from the sheath（公马阴茎从包皮）伸出，露出

drop fence n.【jumping】an obstacle in which the landing side of the fence is lower than the take-off side（落地侧比起跳侧低的障碍）下落式跨栏

dropped back n. *see* sway back 背部下凹

dropped sole n. a hoof condition in which the sole drops to the level of the bearing surface of the hoof wall（一种马蹄病症，蹄底下坠至蹄壁负重面水平）蹄底脱出，蹄底下坠 [syn] prolapsed sole, pumiced foot

droppings /ˈdrɒpɪŋz/ n. pl. the excrement of certain animals 粪便，畜粪

Drosky /ˈdrɒski/ n. a four-wheeled, horse-drawn Russian passenger carriage（一种俄式客用四轮马车）特洛斯基马车 [syn] Russian cab

drove /drəʊv; *AmE* droʊv/ n. a large group of animals moving together 畜群

drover /ˈdrəʊvə(r); *AmE* ˈdroʊv-/ n. a rider who drives herds of cattle or sheep to market for sale（将牛羊赶到集市上出售的骑马人）赶牲口者

drug test n.（also **drug testing**）*see* dope test ［违禁］药物检验

dry /draɪ/ adj. **1** free from liquid or moisture; not wet 干的，干燥的 **2** no longer yielding milk or lactating 不泌乳的；干奶的：a dry mare 干奶母马

vt. to make dry; dehydrate 晒干；脱水

dry matter n.【nutrition】（abbr. **DM**）the percentage of feed exclusive of water（饲料除去水分的部分）干物质：absolute dry matter 绝对干物质

dry work n.【cutting】the training of a cutting horse without using cattle（不用真牛进行的截牛训练）干截训练

drylage /ˈdraɪleɪdʒ/ n. semi-wilted silage; haylage 半干青贮饲料

drylot /ˈdraɪlɒt/ n. an enclosure where horses or cattle are confined and fed with fixed amount to control weight or treat a medical condition（将牛马圈起来人工喂养以控制体重或进行治疗的围栏）限饲栏；隔离栏

vt. to confine or isolate horses or cattle in drylot for such purposes（将马或牛）圈入限饲栏，圈入隔离栏

DSO abbr. daily sperm output 日平均射出精子数

DSP abbr. daily sperm production 日产精子数

DSQ abbr. disqualification 取消参赛资格

dub/dʌb/ vt. to fit the hoof to the horseshoe by trimming and rasping the hoof 削锉马蹄装配蹄铁

Dubai World Cup n. a l-1/4 mile horse race held annually since 1996 in Dubai（自1996年起每年在迪拜举行的赛马比赛，赛程2千米）迪拜世界杯：With a purse of $10 million, the Dubai World Cup is undoubtedly the world's richest horse race. 迪拜世界杯奖金1 000万美元，无疑是世界上最富有的赛马比赛。

Dubai World Cup Night n. a series of eight Thoroughbred races and one purebred Arabian race held annually since 2010 at Meydan Racecourse in Dubai（自2010年起每年在迪拜迈丹赛马场举行的系列赛，由8场纯血马比赛与1场纯种阿拉伯马比赛构成）迪拜世界杯赛马之夜

duct /dʌkt/ n.【anatomy】a tubular canal or passage in the body（体内的管状通道）导管：biliary duct 胆管 ◇ deferent duct 输精管

ductus /ˈdʌktəs/ n.【Latin】duct 导管

ductus deferens n. *see* vas deferens 输精管

dude /duːd, djuːd/ n.【AmE】**1** a man, guy 家伙，男人 **2** a city-dweller who spends vacation on a ranch 牧场度假的城里人 **3** one inexperienced in ranch work 没干过农活的人 [syn] greenhorn, tenderfoot

dude ranch n. a ranch in the West which offers accommodations, horseback riding, and other facilities to vacationer（美国西部为度假者提供食宿、骑乘和其他设施的农场）度假牧场，假日农场

duga /ˈdjuːgə; *AmE* ˈduːgə/ n. *see* douga 辕弓

dull /dʌl/ adj.（of coat color）not bright or lustrous（被毛）黯淡的；无光泽的：a dull coat 被毛无光泽

v. make or become dull 使黯淡，变得无光泽

dumb rabies n. another term for paralytic rabies 麻痹性狂犬病

dummy calf n. *see* roping dummy 套牛靶

dummy foal n. *see* barker 患新生适应不良综合征的马驹;呆驹

dummy mare n. *see* phantom 假台畜,假母马

dump /dʌmp/ vt. (of a horse) to unseat his rider or buck him off 撂下马

dun /dʌn/ n.【color】a sandy yellow to reddish brown coat with black or brown points(被毛沙黄或褐红而末梢为黑色或褐色的毛色)兔褐:dusty dun 土兔褐 ◇ golden dun 金兔褐 ◇ red dun 红兔褐 ◇ silver dun 银兔褐 ◇ yellow dun 黄兔褐 ◇ Dun horses always have a dorsal stripe that runs down the middle of their back. 兔褐马的背中央都有背线。

dung /dʌŋ/ n. the feces of livestock; muck, droppings(家畜的粪便)畜粪:Removal of dung in a pasture is an effective method of controlling worm infestation. The dung can be either broken up and spread out by harrowing or picked up by hand or vacuum machine, depending on the size of paddock being cleared. 清理草场畜粪是控制虫害比较有效的方法。根据围场的大小,可采用耙破碎、人工拾拣或真空抽吸等方法。

dung eating n. *see* feces eating 食粪癖

duodenum /ˌdjuːəˈdiːnəm; *AmE* ˌduːə-/ n.【anatomy】the first part of the small intestine(小肠的前段)十二指肠 [compare] jejunum, ileum

dura mater /ˌdjʊərə ˈmeɪtə/ n.【anatomy】the tough fibrous membrane covering the brain and the spinal cord and lining the inner surface of the skull(颅骨内侧覆盖在脑和脊髓上的纤维膜)硬膜:dura mater encephali 硬脑膜 ◇ dura mater spinalis 硬脊膜 [compare] pia mater

duration /djuˈreɪʃn; *AmE* du-/ n. the period of time during which sth continues 持续时期;期间:duration of cycle 发情周期 ◇ duration of daylight 光照时间 ◇ duration of estrus 发情持续期 ◇ duration of lactation 泌乳期 ◇ duration of pregnancy 妊娠期 ◇ duration of work 运动时间

dusty buckskin n.【color】(also **dusty dun**) a type of dun with a brownish to yellow coat and black points(被毛褐黄而末梢为黑的灰毛)土沙黄,土兔褐

dusty dun n.【color】*see* dusty buckskin 土沙黄,土兔褐

Dutch collar n. *see* breast collar 攀胸

Dutch Draught n. a draft breed developed in the Netherlands in the early 20th century by crossing New Zealand-type mares with Brabant stallions and later with Belgian Ardennes; stands up to 16.3 hands and usually has a chestnut, bay, or gray coat; has a massive build, strong legs with heavy feathering, broad joints, and solid hoofs; is active and kind and possesses great stamina; used for draft and farm work(荷兰在20世纪初采用新西兰马母马与布拉邦特马及比利时阿尔登马公马杂交培育的挽马品种。体高16.3掌以下,毛色有栗、骝或青,体型高大,四肢发达且附生距毛,关节粗大,蹄质结实,性情活泼而友好,耐力较好,主要用于重挽和农田作业)荷兰挽马 [syn] Dutch Horse

Dutch Horse n. *see* Dutch Draught 荷兰挽马

Dutch Warmblood n. a warmblood developed in the Netherlands by crossing the Gelderland and Groningen with English Thoroughbred and other French and German warmbloods; stands about 16.2 hands, has a brown, black, chestnut, or gray coat; is quiet and willing and has a supple and flowing action; used for light draft, carriage, riding, and particularly jumping and dressage(荷兰采用格尔德兰马和格罗宁根马与英纯血马及法国和德国温血马品种杂交培育而成的温血马品种。体高约16.2掌,毛色为褐、黑、栗或青,性情安静而温驯,步法流畅,可用于轻挽、驾车和骑乘,在障碍赛和盛装舞步中表现不俗)荷兰温血马:The Dutch Warmblood was originally bred as a carriage horse. The modern Dutch Warmblood is most commonly known for its ability in the show-ring for Jumping, Dressage and Eventing. 荷兰温血马最初被培育为驾车用马，现在的荷兰温血马则以其在跨越障碍赛、盛装舞步以及三日赛中的出色表现而出名。

DVM abbr. Doctor of Veterinary Medicine 兽医学博士

dwell /dwel/ vi.【racing】(of a horse) to be slow to break out of the starting gate (指赛马)出闸缓慢，滞闸：A racehorse who dwells in the starting gate has little chance to win. 一匹滞闸的赛马胜出的概率不大。

dwelling /ˈdwelɪŋ/ n. a pause or suspension of the trick horse's foot in air during movement (运步时马蹄在空中的悬置阶段)空悬

dysfunction /dɪsˈfʌŋkʃn/ n. (also **disfunction**) abnormal or impaired functioning of a bodily system or organ (机体或器官的功能异常或衰退)机能障碍：hormonal dysfunction 激素机能障碍 | **dysfunctional** /-ʃənl/ adj.

dysgenesis /dɪsˈdʒenɪsɪs/ n. defective or abnormal development of an organ, especially of the gonads(尤指性腺)发育不全；不育

dysmaturity /ˌdɪsməˈtʃʊərəti, -ˈtjʊə-; *AmE* -ˈtʃʊr-, -ˈtʊr-/ n. a condition of being immature in fetal development 发育不成熟，发育不全

dyspepsia /dɪsˈpepsɪə/ n. disturbed digestion; indigestion 消化不良

dysplasia /dɪsˈpleɪʒə/ n. abnormal development 发育异常；发育不全 | **dysplastic** /-ˈplæstɪk/ adj.

dysplasia of the growth plate n. *see* epiphysitis 骨骺炎，骨生长板发育不全

dystocia /dɪsˈtəʊʃə/ n. (also **dystokia**) difficult foaling or birth 难产：Dystocia is usually caused by the foal being over-sized, wrongly positioned, or the mare having contractions that are not strong or frequent enough. 难产多半由于胎儿过大、胎位不正或母马子宫阵缩强度不够所致。

Dystrophia ungulae n. *see* seedy toe 蹄匣分离

dystrophy /ˈdɪstrəfi/ n. a degenerative muscular disorder caused by inadequate or defective nutrition, in which the muscles weaken and atrophy (由于营养不良引起的肌肉衰弱和萎缩)肌营养不良；肌萎缩 | **dystrophic** /dɪsˈtrɒfɪk/ adj.

Dzhabe /ˈdʒɑːb/ n. (also **Jabe**) an ancient horse breed originating in Kazakhstan; stands 13.2 to 14 hands and generally has a bay or liver chestnut coat; has a coarse head with thick neck; is hardy and frugal and possesses great stamina (源于哈萨克斯坦的古老矮马品种，体高13.2~14.0掌，毛色多为骝或青栗，面貌粗糙，颈粗，抗逆性强，耐粗饲，耐力持久)扎拜马

Dziggetai /ˈdʒɪgɪtaɪ/ n. a species of wild ass inhabiting the steppes of eastern Asia (生活在东亚高原上的野驴品种)骞驴

EAA abbr.【nutrition】essential amino acids 必需氨基酸

each way n.【racing】a wager in which a bettor wins if the horse he selected finishes in the top three regardless of the positions（所赌赛马跑入前三任何名次即可赢奖的投注方式）入围彩：to bet each way 投入围彩

ear /ɪə(r)；*AmE* ɪr/ n. either of the two pointed organs of hearing on the top of a horse's head（马头顶上的听觉器官）耳：ear tips 耳梢

ear mark n. a notch or mark made in a horse's ear for identification（耳上用于标识的记号）耳号
vt. to cut a notch in a horse's ear for identifying purpose 打耳号

early /ˈɜːli；*AmE* ˈɜːrli/ adj. done before the expected time 提早的，早到的 PHR **early foot**【racing】a racehorse who shows good speed at the beginning 起步发力的赛马 **early to trot/walk**【dressage】premature transition to trot/walk at a marker（舞步马到达某个标识时）提前换为快步/走步

earn /ɜːn；*AmE* ɜːrn/ vt. to get or acquire sth 获得 PHR V **earn a diploma**【racing】(of a horse) to win its first race; break maiden（指赛马）首次获胜 **earn one's tops**【hunting】(of a hunter) to earn a certain status（指猎手）赢得高帮靴地位：The amount of time one must ride to earn one's tops varies between Hunts. 赢得高帮靴地位所需时间各不同猎狐俱乐部有别。

earth /ɜːθ；*AmE* ɜːrθ/ n.【hunting】a den of fox 狐狸洞穴

earth stopper n.【hunting】one employed to block the entrance to the den of a fox to prevent its return（为阻止狐狸返洞而雇佣来堵塞洞穴的人）堵洞者

earth stopping n.【hunting】(also **stopping earth**) the practice or blocking the den of a fox to prevent its return the night preceding the day of hunting（狩猎头天晚上为防止狐狸返洞而堵塞洞穴）堵洞

ease /iːz/ n. a state of relaxation without little effort 轻松，自如：to move with ease 运步轻快
vi. to move or run with little effort 运步轻松，运步自如 PHR V **ease off/up** (of a horse) to reduce the speed by gradually shortening the stride（指马）减速，收步

East Bulgarian n. a warmblood developed in Bulgaria during the late 19th century by crossing local mares with Thoroughbred, English Halfbred, Arab, and Anglo-Arab horses; stands 15 to 16 hands and usually has a chestnut, black, or bay coat; is elegant, well-built and has a small head, straight profile, deep girth and long straight back; is energetic, hardy, fast, and versatile; used for riding, farm work, and equestrian competitions from dressage to cross-country（保加利亚在19世纪末采用当地马匹与纯血马、英国半血马、阿拉伯马、盎格鲁阿拉伯马杂交育成的温血马品种。体高15～16掌，毛色多为栗、黑或骝，面貌俊秀，体格强健，头小，胸深，背长，体力充沛，抗逆性强，速力较好，主要用于骑乘、农田作业、盛装舞步和越野赛等马术项目）东保加利亚马：The East Bulgarian was officially recognized as a breed in 1951. 东保加利亚马于1951年被正式认定为一个品种。

East Friesian n. (also **East Frisian**) a German warmblood descended from Friesian stock crossed with Spanish, Neapolitan, Cleveland Bay, Anglo-Arab, Norman, Thoroughbred, Oldenburg, and recently Hanoverian bloods; stands 15.2 to 16.1 hands and generally has a brown, bay, black, gray, or chestnut coat; has a well-proportioned head, prominent features, a long and straight back and strong legs with broad joints; used for riding and light draft (德国的温血马品种,主要采用弗里斯兰马和西班牙马、那不勒斯马、克利夫兰骝马、盎格鲁阿拉伯马、诺曼马、纯血马、奥尔登堡马繁育而来,最近又引入了汉诺威马血统。体高15.2~16.1掌,毛色多为褐、骝、黑、青或栗,头型适中,面貌特征明显,背长而平直,四肢强健,关节粗大,主要用于骑乘和轻挽)东弗里斯兰马

East Frisian n. *see* East Friesian 东弗里斯兰马

East Prussian horse n. *see* Trakehner 东普鲁士马,特雷克纳马

eastern equine encephalomyelitis n. (abbr. **EEE**) one of three primary types of encephalomyelitis 东方马脑脊髓炎

Eastern horse n. a generic term that refers to the Arab, Barb, Turk, and Syrian breeds (包括阿拉伯马、柏布马、土耳其马和叙利亚马在内的总称)东方马

easy boot n. a light-weight, hoof-shaped boot worn on a horse's hoof to provide protection and traction during transportation (套在马蹄上的轻质靴,在运输中起保护作用并可增加摩擦)简易蹄靴

ECG abbr. electrocardiogram 心电图

echocardiography /ˈekəʊˌkɑːdɪˈɒgrəfi; *AmE* ˈekoʊ-/ n. the use of ultrasound to evaluate the heart muscles to diagnose congestive heart failure (采用超声波诊断阻塞性心力衰竭的方法)超声波心动描记法 | **echocardiographic** /ˌekəʊˌkɑːdɪəˈgræfɪk; *AmE* ˌekoʊ-/adj.

eclampsia /ɪˈklæmpsiə/ n.【vet】a condition in which mares suffer from convulsions and coma after pregnancy due to loss of calcium through heavy milking (母马在妊娠后由于大量泌乳造成缺钙而导致的痉挛和昏迷)惊厥,子痫 | **eclamptic** /-tɪk/ adj.

Eclipse /ɪˈklɪps/ n. (1764 – 1789) a chestnut Thoroughbred foaled in 1764 in England, was the great-great-grandson of Darley Arabian (1764年生于英国的一匹栗色纯血马,为达利·阿拉伯的玄孙)日食:Eclipse kept an undefeated record in his short racing career. 在其短暂的赛马生涯中,"日食"保持了不败战绩。

Eclipse Award n.【racing】an award conferred in the United States since 1971 to the North American divisional racing champions as selected by vote of the Daily Racing Forum, Thoroughbred Racing Associations, and the National Turf Writers Association (自1971年起美国授予北美地区赛马冠军的一个奖项,主要通过每日赛马论坛、纯血马赛马协会以及国家草场作家协会投票选举产生)日蚀奖

Eclipse Stakes n.【racing】a 1.25-mile race for three-year-old horses held annually on July 13th at Sandown Park, England (每年7月13日在英国萨当公园为3岁驹举行的比赛,赛程2千米)日蚀奖金赛

Edelweiss Pony n. *see* Haflinger 哈弗林格矮马,雪绒花矮马

edema /ɪˈdiːmə/ n. an excessive accumulation of serous fluid in the intercellular spaces of the body, resulting in swelling (组织间液异常淤积的情况,常造成肿大)水肿;浮肿

edematous /ɪˈdiːmətəs/ adj. of or relating to edema; swollen 水肿的,浮肿的

EDM abbr. equine degenerative myeloencephalopathy 马萎缩性脑脊髓病

EDTA abbr. ethylenediaminetetraacetic acid 乙二胺四乙酸:The sodium salt of EDTA is used as an antidote for metal poisoning and an anticoagulant. 乙二胺四乙酸钠可用作金属中毒的解毒剂以及抗凝血剂。

Edward Derby n. *see* Derby, Edward 爱德华·德比

EEE abbr. eastern equine encephalomyelitis 东方马脑脊髓炎

eel stripe n. *see* dorsal stripe 背线

efficacious /ˌefɪˈkeɪʃəs/ adj. producing or capable of producing a desired effect; effective (能产生预期效果的)有效的 | **efficaciousness** /-ʃəsnis/ n.

egg bar shoe n.【farriery】(also **egg bar**) a therapeutic horseshoe with the heels joined into an oval bar (蹄铁尾相连的矫正性蹄铁)卵形蹄铁:Egg bar shoe yields more heel support than a conventional shoe and prevents possible interference with the opposing limb. 卵形蹄铁较常规蹄铁不仅能更好地支撑蹄踵,且可有效防止交突发生。

egg butt n. a barrel-shaped joint hinge used to join the bit mouthpieces (衔铁铰链中央的卵形关联)卵柄

egg-but snaffle n. a snaffle bit with an egg-butt joining the mouthpiece 卵柄衔铁

EHRF abbr. Equine Health Research Fund 马健康研究基金会

EHV abbr. equine herpes virus 马疱疹病毒

El Pato /el ˈpɑːtəʊ/ n.【Spanish】(also **Pato**) a mounted sport originating in Argentina around 1610 in which two teams of mounted riders competed to catch a live duck sewn into a piece of leather (于1610年左右源于阿根廷的马上项目,参赛的两队骑手竭力争抢装在皮球中的一只活鸭子)马上夺鸭:El Pato was banned in its original form in 1822. 马上夺鸭的最初形式已于1822年明文禁止。

EIA abbr. equine infectious anemia 马传染性贫血

eighth pole n.【racing】a marking post located on the infield rail exactly one furlong from the finish (立于赛道内栏的标杆,与终点相距200米)1/8英里杆,200米杆 PHR V **kiss the eighth pole**【slang】(of a horse) to finish far behind the field (指马)远远落后,甩尾

eight-horse hitch n.【driving】a team of eight horses harnessed together 八马联驾

eild /eɪld/adj. (of a horse) barren 空怀的;不育的

Einsiedler /ˈaɪnˌsiːdlə(r)/ n. a horse breed developed in Switzerland around the mid-11th century by crossing native stock with Hackney; stands 15 to 16.3 hands and may have any solid coat color with chestnut and bay occurring most often; has a straight back, slightly sloping croup, prominent withers, a slightly convex or straight head, and strong, and well-jointed legs; is docile and suitable for both riding and driving (瑞士在11世纪中叶采用当地马与哈克尼马杂交育成的马品种。体高15~16.3掌,各种单毛色兼有,但以栗和骝最为常见,背部平直,尻稍斜,鬐甲突出,头略凹或平直,四肢发达,性情温驯,骑乘、驾车兼用)艾因斯德勒马 syn Swiss Anglo-Norman

EIPH abbr. exercise-induced pulmonary hemorrhage 运动诱发性肺出血,肺出血

ejaculate /iˈdʒækjuleɪt/ vi. (of a stallion) to eject or discharge semen abruptly in orgasm (公马)射精

n. semen ejected in orgasm from the body (射出的)精液:whole ejaculate 全精液 ◇ An ejaculate of good-quality semen of average volume may be split ten to twenty portions. 公马单次射出的精液若品质好且量适中,可分为10~20份使用。

ejaculation /iˌdʒækjuˈleɪʃn/ n. the act of ejaculating 射精

ejaculatory /iˈdʒækjulətəri/ adj. of or relating to ejaculation 射精的

EKG abbr. electrocardiogram 心电图

ekka /ˈekɑː/ n. *see* hecca 印度马车:an ekka with canopy 带华盖的印度马车

ELA abbr. equine lymphocyte antigen 马淋巴细胞抗原

elastic /iˈlæstɪk/ adj. (of an object) resuming its original shape after being stretched or expanded; flexible (指物体拉伸后易恢复原状)柔韧的,有弹性的:elastic cartilage 弹性软骨 | **elasticity** /ˌiːlæˈstɪsəti/ n.: the elasticity of steps 运步的柔韧性

E

elastin /iˈlæstɪn/ n. a glycoprotein forming the principal structural component of elastic fibers (一种糖蛋白,是构成弹力纤维的主要成分)弹性蛋白

elbow /ˈelbəʊ; *AmE* -boʊ/ n.【anatomy】the upper joint of horse's foreleg 肘关节,肘:point of elbow 肘突

elbow boots n. a padded boots worn on the elbows of a horse's forelegs for protection from injury (戴在马前肢关节上的护具)护肘: Elbow boots are often used on horses having high-stepping foreleg action. 护肘常用在前肢有高蹄动作的马上。

elbow hitting n. a situation in which the horse hits its elbow with the shoe of the same foreleg (前肢蹄铁撞击肘关节的情况)撞肘

electro /ɪˈlektrəʊ; *AmE* -troʊ/ pref. connected with electricity 电的

electrocardiogram /ɪˌlektrəʊˈkɑːdiəʊgræm/ n. (abbr. **ECG**, or **EKG**) a graphic tracing of the electrical impulses produced by the heart muscles (用图像对心脏电脉冲所作的记录)心电图 syn cardiogram

electrocardiograph /ɪˌlektrəʊˈkɑːdiəʊgrɑːf/ n. an instrument used in the detection and diagnosis of heart abnormalities that generates a record of the electrical impulses associated with heart muscle activity (通过记录心脏活动的电脉冲来检测和诊断心脏异常的仪器)心电图仪 syn cardiograph

electrocardiography /elektrəʊˈkɑːdɪəʊgrəfi/ n. the examination of the heart through graphic records of the electrical impulses produced by it (采用心电图对心脏脉冲进行的诊断)心电图描记

electroejaculation /ɪˌlektriːˌdʒækjuˈleɪʃn/ n. (abbr. **EEJ**) the practice of obtaining semen by stimulating the stallion electrically via the rectum (用电极刺激公马直肠采集精液的方法)电刺激射精,电刺激采精: electroejaculation probe 电刺激采精杆 ◇ Electroejaculation has been proven to be a safe and effective means to obtain motile sperm suitable for assisted reproductive techniques. 对于辅助生殖技术而言,电刺激射精被证明是一项安全而有效的采精手段。

electrolyte /ɪˈlektrəlaɪt/ n. any of various inorganic compounds essential for many of the biochemical processes and functioning in regulating the balance of body fluid (对体内生化反应和调节体液平衡有重要作用的多种无机化合物)电解质,离子盐

electronic eye n. *see* automatic timer 电子计时器,电子眼

electronic identification n. a method of horse identification applied by implanting a microchip underneath its skin (通过皮下埋植芯片对马匹进行的鉴定)电子芯片鉴定

electrotherapeutics /ɪˈlektrəʊθerəˈpjuːtɪks/ n. electrotherapy 电疗法

electrotherapy /ɪˌlektrəʊˈθerəpi/ n. the use of electric currents to stimulate healing of an organ or body part (用电流刺激达到治疗目的的方法)电疗法 syn electrotherapeutics

element /ˈelɪmənt/ n.【jumping】one jump that makes up a combination fence (构成组合障碍的一道障碍)单道跨栏,跨栏构件

elephant polo n. another term for pachyderm polo 大象马球,象球

elevated trot n. *see* passage 高蹄快步,也即帕萨基

elf's foot n. *see* flipper foot 蹼状蹄,蹄尖过长

eligible /ˈelɪdʒəbl/ adj. (of a horse or rider) qualified to compete according to established rules (指马或骑手)有参赛资格的 | **eligibility** /ˌelɪdʒəˈbɪləti/ n.

eliminate /ɪˈlɪmɪneɪt/ vt. **1**【physiology】to excrete bodily wastes 排泄 **2** to cancel or disqualify from a competition 取消比赛资格 | **elimination** /ɪˌlɪmɪˈneɪʃn/ n.

ELISA /ɪˈlaɪzə/ abbr. enzyme-linked immunosorbent assay 酶联免疫吸附试验

elk lip n. a conformation defect with a loose and

overhanging upper lip（一种上唇松软下垂的体貌缺陷）麋鹿唇

elve's foot n. *see* flipper foot 蹼状蹄，蹄尖过长

emaciate /ɪˈmeɪʃɪeɪt/ vt. to make extremely thin or weak, esp by starvation 使消瘦，饿瘦

emaciated /ɪˈmeɪʃɪeɪtɪd/ adj.（of a horse）thin or weak（指马）瘦弱的：The pregnant animal gives priority to the fetal needs for nourishment, even if a completely emaciated mare may produce a full-term foal without abortion. 妊娠母畜会优先满足胎儿的营养需要，即使瘦骨嶙峋的母马也能期满产下幼驹而不会流产。

emasculate /iˈmæskjuleɪt/ vt. to castrate or neuter 阉割，去势 | **emasculation** /iˌmæskjuˈleɪʃn/ n.

emasculator /iˈmæskjuleɪtə/ n. **1** a stainless steel tool used to remove the testicles from a horse（阉割马用的）去势器械 **2** one who castrates a male animal 阉割员，去势员

embalm /ɪmˈbɑːm/ vt. to treat a specimen or corpse with preservatives or balm in order to prevent decay（用防腐剂或香脂处理标本或尸体以防止腐烂）防腐处理

embouchure /ˌɒmbʊˈʃʊər/ n.【French】a mouthpiece 口衔

embryo /ˈembriəʊ; *AmE* -briou/ n. a fertilized ovum in its early stages of development in the uterus（受精卵在子宫内的早期发育形式）胚胎

embryo splitting n. the practice of splitting the embryo in two halves 胚胎分割：Since naturally occurring twins in horses are fraternal（non identical）, identical twins in horses can only be obtained by embryo splitting. 由于自然条件下出生的双胞胎马都是异卵双生，因此同卵双生的马只能通过胚胎分割来获得。

embryo transfer n.（abbr. **ET**）the practice of taking an inseminated embryo from a donor mare and implanting it in the uterus of a recipient mare for pregnancy（从供体母马体内取出胚胎植入受体母马子宫以达到妊娠的辅助手段）胚胎移植：equine embryo transfer 马胚胎移植

embryologist /ˌembriˈɒlədʒɪst; *AmE* -ˈɑːl-/ n. one who studies the development of embryos（研究胚胎发育的学者）胚胎学家

embryology /ˌembriˈɒlədʒi; *AmE* -ˈɑːl-/ n. the branch of biology that studies the development of embryos（研究胚胎发育的生物学分支）胚胎学 | **embryological** /ˌembrɪəˈlɒdʒɪkl; *AmE* -ˈlɑːdʒ-/ adj.

embryonic /ˌembriˈɒnɪk; *AmE* -ˈɑːnɪk/ adj. of, relating to, or being an embryo 胚胎的

eminence /ˈemɪnəns/ n.【anatomy】a projection of an organ or body part（机体器官的）隆起，凸起

eminentia /ˈemɪnenʃɪə/ n.【Latin】eminence 隆起

EMND abbr. equine motor neuron disease 马运动神经疾病

emphysema /ˌemfɪˈsiːmə/ n.【vet】a respiratory condition symptomized by labored breathing and an abnormal increase of the air spaces in lung（马所患的呼吸系统疾病，表现为呼吸困难和肺气室异常增大）肺气肿：pulmonary emphysema 肺气肿

empty /ˈempti/ adj.【racing】（of the horse）lacking the strength to complete a fast drive to the finish（马冲刺）乏力的，无力的

emulate /ˈemjuleɪt/ vt. to compete with sb/sth successfully 竞争 | **emulation** /ˌemjuˈleɪʃn/ n.

emulsify /ɪˈmʌlsɪfaɪ/ vt. to make into an emulsion 使乳化：Once fat has been emulsified by bile it is hydrolysed to fatty acids and glycerol by lipase. 脂肪在胆汁的作用下乳化后，被脂肪酶水解为脂肪酸和甘油。

emulsion /ɪˈmʌlʃn/ n. a suspension of small globules of one substance in a liquid（在液体中形成的微颗粒悬液）乳状液

EMV abbr. equine morbillivirus 马麻疹病毒

enamel /ɪˈnæml/ n. the hard covering of a tooth（牙齿的坚硬外层）珐琅质，牙釉质

ENB abbr. equine night blindness 马夜盲症

encephalin /en'sefəlɪn/ n. a morphine-like neurotransmitter secreted by the brain（大脑分泌的一种类吗啡神经递质）脑啡肽

encephalitis /enˌsefə'laɪtɪs/ n. *see* encephalomyelitis 脑脊髓炎

E

encephalomyelitis /enˌsefələʊmaɪə'laɪtɪs/ n.【vet】a viral, epidemic fatal disease resulting in inflammation of the brain and spinal cord; symptoms include high fever, drowsiness, lack of coordination, teeth grinding, partial vision loss, inability to swallow food, and possible paralysis（一种病毒所致的致死性疾病，可导致脑和脊髓炎症，其症状可表现为发热、精神萎靡、运动失调、磨牙、部分失明、吞咽困难和瘫痪）脑脊髓炎：Venezuelan equine encephalomyelitis 委内瑞拉马脑脊髓炎 ◇ equine infectious encephalomyelitis 马传染性脑脊髓炎 [syn] encephalitis, sleeping sickness, brain fever, brain staggers

encephalopathy /enˌsefə'lɒpəθi; *AmE* -'lɑːp-/ n. any of various diseases of the brain 脑病：spongiform encephalopathy 海绵状脑病

enclose /ɪn'kləʊz; *AmE* ɪn'kloʊz/ vt. to surround or close off on all sides 围住；圈起：to enclose the grazing land 将草场围起来

enclosure /ɪn'kləʊʒə(r); *AmE* -'kloʊ-/ n. **1** the act of enclosing or state of being enclosed 圈地；圈围 **2** an enclosed area 围场

endangered /ɪn'deɪndʒəd/ adj.（of a species）faced with the danger of extinction（指物种）濒临灭绝的，濒危的：a highly endangered species 极度濒危物种

endbrain /'endbreɪn/ n.【anatomy】telencephalon 端脑，终脑

endemic /en'demɪk/ adj.（of a certain disease）prevalent in a particular region（指疾病）地方性的：endemic disease 地方性流行病 [compare] epidemic, pandemic

endline /end'laɪn/ n.【polo】(also **end line**) the line that mark the back of a polo field; backline（马球赛场底端的边界）底线：The ball is considered out of bounds if it passes outside of the endline. 球越过底线就被判出界。[compare] sideline

end(o)- pref. inside; within 内；里

endocardium /ˌendəʊ'kɑːdɪəm/ n.【anatomy】(*pl.* **-dia**/-dɪə/) the thin serous membrane that lines the interior of the heart（衬在心脏内面的浆膜）心内膜 [compare] epicardium, pericardium

endochondrial ossification n. a process of bone formation in which bone growth occurs at the epiphyseal plates（在骺板发生的成骨过程）软骨内成骨 [syn] intracartilaginous ossification

endocrine /'endəʊkrɪn; *AmE* -dək-/ adj.【physiology】of or relating to the ductless glands that produce hormones 内分泌的：the endocrine system 内分泌系统

endocrine gland n. any of the ductless glands that secret hormones（分泌激素的无管腺体）内分泌腺

endometrial /'endəu'miːtrɪəl/ adj. of or relating to endometria 子宫内膜的

endometrial cups n.【anatomy】cup-shaped structures formed in the endometrium of a pregnant mare（妊娠母马子宫内膜形成的杯状结构）子宫内膜杯：The formation of the endometrial cups is a unique feature of equine pregnancy. 子宫内膜杯的形成是马匹妊娠的独特表征。

endometritis /ˌendəʊmɪ'traɪtɪs/ n. bacterial infection of the mucous membrane lining of the uterus（子宫内膜由细菌感染所致的炎症）子宫内膜炎：Endometritis is often the leading cause of sterility in mares. 子宫内膜炎常是母马不孕的主要病因。

endometrium /ˌendəʊ'miːtrɪəm/ n.（*pl.* **endometria** /-trɪə/）the glandular mucous membrane that lines the uterus（子宫内壁含分泌腺的黏膜层）子宫内膜

endomysium /ˌendəʊ'mɪzɪəm/ n.【anatomy】a layer of connective tissue interspersed among muscular fibers（肌纤维之间的结缔组织）肌内膜

endorphin /en'dɔːfɪn; *AmE* -'dɔːr-/ n. any of a group of morphine-like proteins produced in nerve tissues to suppress pain and regulate emotional state（大脑神经组织产生的多种吗啡样激素，有镇痛和调节情绪的作用）内啡肽

endoscope /'endəskəʊp; *AmE* -skoʊp/ n. a diagnostic instrument for inspecting the interior part of a hollow organ（用来检查腔体器官内部结构的诊断设备）内窥镜：video endoscope 视频内窥镜

endoscopy /en'dɒskəpi; *AmE* -'dɑːs-/ n. the inspection of a bodily canal or cavity by an endoscope 内窥镜检查，内镜检查

endosteum /en'dɒstɪəm/ n.【anatomy】(*pl.* **endostea** /-tiː/) a thin layer of vascular membrane lining the medullary cavity of a bone（衬在骨髓腔内的脉管膜）骨内膜 compare periosteum

endotoxemia /ˌendəʊ'tɒksɪmiə; *AmE* ˌendoʊ'tɒksɪmɪr/ n. the presence of endotoxins in the blood［内］毒血症

endotoxic shock n. a sudden fall of blood pressure resulting from systemic toxicity as caused by an endotoxin from certain bacteria（由于内毒素中毒导致血压突降的症状）内毒素休克

endotoxin /ˌendəʊ'tɒksɪn/ n. a toxin produced by certain bacteria and released upon destruction of the bacterial cell（由某些细菌产生的毒素，在细菌裂解时释放）内毒素

endurance /ɪn'djʊərəns; *AmE* -dʊr-/ n. the ability or power of a horse to withstand physical pain, stress, or adversity; stamina（马匹经受伤痛、应激或恶劣环境的能力）持久力，耐力：The maximum distance covered in endurance competition is 160 km at average speeds of up to 20 km/h in some desert races. 耐力赛的最长路程为160千米，沙地比赛中其平均速度为每小时20千米以上。

endurance horse n. any horse competing in long-distance race that tests its endurance 耐力赛马

endurance riding n. a speed and endurance race over long distances 耐力赛，长途赛 syn long distance riding

enema /'enəmə/ n. **1** the injection of liquid into the rectum through the anus to cleanse or evacuate the bowels for therapeutic or diagnostic purposes（从肛门将液体注入直肠以清空肠道的治疗方法）灌肠，洗肠 **2** the fluid so injected 灌肠剂：The soap water and liquid paraffin are enemas of common use. 肥皂水和液体石蜡是经常使用的灌肠剂。

energetics /ˌenə'dʒetɪks/ n. the scientific study of the flow and conversion of energy（研究能量流动与转化的学科）力能学，动能学：energetics of exercise 运动力能学

energizer /'enədʒaɪzə/ n. a component of an electric fence 电围栏

energy /'enədʒi; *AmE* 'enərdʒi/ n. the capacity for work or vigorous activity; power 能［量］：energy balance 能量平衡 ◇ energy feeds 能量饲料 ◇ energy metabolism 能量代谢 ◇ energy reserve 能贮

energy cascade n.【nutrition】a scale of feed energy value with each step representing a lower energy value than the one preceding it（食物中能值在消化过程中逐级降低构成的能量梯度）能级：The energy cascade of feed includes gross energy, digestible energy, metabolizable energy, and net energy. 饲料的能级包括总能、消化能、代谢能和净能。

engage /ɪn'geɪdʒ/ vt. **1** (of a horse) to collect and bring his hind legs well under his body（指马将后躯）收拢：to engage its hindquarters 收拢后肢 **2**【racing】(of a horse) to enter and run in race（指马）参赛，出赛

English buggy n. a hooded gig 英式轻型双轮马车

English Connemara Society n. an organization founded in 1947 to promote the breeding of Connemara ponies in England（1947年在英国

成立的组织机构，旨在促进康尼马拉矮马的育种工作）英国康尼马拉矮马协会

English Great Horse n. *see* Great Horse 英格瑞特马

English Jockey Club n. an association founded in 1750 by horse owners and related individuals in England to promote and regulate flat racing (1750年在英国由马主和相关个人成立的组织机构，旨在促进并监管平地赛）英国赛马会，英国骑师俱乐部

English riding n. a popular riding style adopted by most riders in which an English saddle and tack are used（大多骑马者所采用的骑乘方式，使用英式马鞍及相关马具）英式骑乘 compare Western riding

English saddle n. a saddle with long side bars, reinforced cantle and pommel, no horn, and a leather seat（鞍架较长的一种马鞍，前、后鞍桥增固，不带角柱，鞍座为皮制）英式马鞍 compare Western saddle

English stirrup n. any stirrup suspended from the leathers on an English saddle 英式马镫

English Thoroughbred n. *see* Thoroughbred 英纯血马

engorge /ɪnˈgɔːdʒ; *AmE* ɪnˈgɔːrdʒ/ vt. to fill with blood or other fluid; flower 充血，充胀：During estrus the walls of uterus are relaxed and flaccid, the uterine glands secrete moist mucus, and the uterine surface is engorged with blood. 处于发情期的母马，子宫壁变得松弛，其腺体分泌具有润滑作用的黏液，子宫内表面呈充血状态。

enophthalmus /ˌenɒfˈθælməs/ n. a condition in which the eye sinks inward 眼球内陷

ensilage /ˈensɪlɪdʒ/ n. **1** the process of storing and fermenting green fodder in a silo 青贮 **2** fodder preserved in a silo; silage 青贮料

vt. to store and process fodder in a silo; silage 青贮

ensile /ˈensaɪl, enˈsaɪl/ vt. to store fodder in a silo for preservation; ensilage 青贮

enter /ˈentə(r)/ vt. **1**【hunting】to train or introduce a hound to hunt fox（猎犬）参加狩猎 **2** to nominate a horse to participate in a competition or an event（马匹）报名参赛

enteric /enˈterɪk/ adj. of, relating to, or being within the intestine 肠的；肠内的

enteritis /ˌentəˈraɪtəs/ n. inflammation of the intestinal tract marked by colic and diarrhea（肠道发炎的症状，表现为腹痛和下痢）肠炎

enter(o)- pref. intestine 肠的

enterolith /ˈentərəlɪθ/ n. an abnormal concretion of mineral salts in the intestine（肠内矿物质盐异常集结的症状）肠结石 syn intestinal stone

enterotoxemia /ˌentərəʊtɒkˈsiːmiə/ n. a disease caused by large amounts of enterotoxin produced, especially by the *Clostridium perfringens*, often resulting in diarrhea and colic（由产气梭菌分泌大量肠毒素所致的疾病，常造成腹泻和绞痛）肠毒血症

enterotoxin /ˌentərəʊˈtɒksɪn/ n. a toxin secreted by certain bacteria that inhabit the intestines and cause vomiting and diarrhea（寄居在肠道内的某些细菌所分泌的毒素，可引起呕吐和腹泻）肠毒素

entire /ɪnˈtaɪə(r)/ adj.（of a male horse）not castrated or full（指公马）未阉割的

n. an uncastrated male horse; a stallion 公马

entrails /ˈentreɪlz/ n. the internal organs, especially the intestines; viscera 内脏

entry /ˈentri/ n. **1** the action of becoming a member of an organization, etc.（加入机构成为会员）入会，加入：On the 36th Asian Racing Conference 2016 held in Mumbai, China gained entry into the Asian Racing Federation. 2016年在孟买召开的第36届亚洲赛马会议上，中国正式加入亚洲赛马联盟。**2**【hunting】a young unentered hound 未出猎的猎犬 **3** a horse competing in a race 参赛马匹

entry fee n. the amount of money the owner pays to enter a horse in a event, competition, or race（马匹报名参赛所需交纳的金额）参赛费，报

名费

entwine /ɪnˈtwaɪn/ vi. (also **intwine**) to twine or twist together 缠结:The umbilical cord may become entwined around the fetal hind limb, resulting in the hypoxia of fetus. 脐带可能会缠结在胎儿的后肢,从而造成胎儿缺氧。

environmental trait n. any structural or functional defect acquired by a horse after birth and during growth (马匹出生后和生长过程中形成的结构或机能缺陷)环境性状

enzyme assay n. a blood test performed to measure blood enzyme levels (血液)酶检测

enzyme-linked immunosorbent assay n. (abbr. **ELISA**) a diagnostic method of testing used to detect diseases and infection (用来诊断疾病和感染的分析方法)酶联免疫吸附试验

Eohippus /ˌiːəʊˈhɪpəs/ n. (also **Hyracotherium**) the earliest known ancestor of the horse living in the Eocene epoch (已知现今马最早的祖先,生活于始新世)始祖马— *see also* **Hipparion**, **Merychippus**, **Mesohippus**, **Pliohippus**

epicondyle /ˌepɪˈkɒndaɪl/ n.【anatomy】a rounded projection at the end of a bone, located above a condyle and usually serving as a place of attachment for ligaments and tendons (骨末端踝上方的圆形隆起,通常为韧带和肌腱的连接处)上踝: lateral epicondyle of humerus 臂骨外上踝

epidemic /ˌepɪˈdemɪk/ adj. (of a disease) spreading rapidly and extensively and affecting many individuals in an area or a population at the same time; epizootic (在某地区同时出现大量感染群体或迅速传播的)流行性的 compare endemic, pandemic

epidermic /ˌepɪˈdɜːmɪk; *AmE* -ˈdɜːrm-/ adj. of or relating to epidermis 表皮的

epidermis /ˌepɪˈdɜːmɪs; *AmE* -ˈdɜːrm-/ n. the outer, protective, nonvascular layer of the skin of vertebrates that covers the dermis (脊椎动物皮肤最外的保护层,位于真皮之上,无血管分布)表皮 compare dermis

epididymis /ˌepɪˈdɪdɪmɪs/ n.【anatomy】(*pl.* **epididymides** /-mɪdiːz/) a long, narrow convoluted tube connecting testes to the vas deferens; serving as a site for maturation and storage of sperm (连接睾丸与输精管的迂回细管,是精子成熟与储存的场所)附睾

epididymitis /ˌepɪˌdɪdɪˈmaɪtɪs/ n. inflammation of the epididymides 附睾炎

epiglottis /ˌepɪˈglɒtɪs; *AmE* -ˈglɑːtɪs/ n.【anatomy】(*pl.* **epiglottises** /-tɪsiːz/) the thin elastic cartilaginous structure located at the root of the tongue that folds over the glottis to prevent food and liquid from entering the trachea during the act of swallowing (位于舌根后的薄层弹性软骨组织,吞咽食物时可堵住声门以防止食物和液体进入气管)会厌

epihyal /ˌepɪˈhaɪəl/ adj. (also **epihyoid**) of or relating to epihyoideum 上舌骨的
n. the epihyal bone 上舌骨

epihyoid /ˌepɪˈhaɪɔɪd/ adj. & n. epihyal 上舌骨的

epihyoideum /ˌepɪˈhaɪɔɪdəm/ n.【anatomy】a small bone situating between the ceratohyal and the stylohyal bone (一块位于角舌骨与茎状舌骨间的小骨)上舌骨

epimeletic /ˌepɪməˈletɪk/ adj. showing desire for caregiving 护他的,护幼的:epimeletic behavior 护幼行为

epimysium /ˌepɪˈmɪzɪəm/ n.【anatomy】(*pl.* **epimysia** /-zɪə/) the external sheath of connective tissue surrounding a muscle (包围于肌肉外表面的结缔组织膜)肌外膜 compare endomysium

epinephrine /ˌepɪˈnefrɪn/ n. adrenaline 肾上腺素

epiphyseal /ˌepɪˈfɪsɪəl/ adj. (also **epiphysial**) of or relating to epiphysis 骨骺的:epiphyseal cartilage 骺软骨◇epiphyseal line 骺线

epiphyseal cartilage n. a type of cartilage that develops from the epiphyseal plates (由骨骺发育

而来的软骨）骺软骨

epiphyseal plate n. the part where the shaft and epiphysis meet and the immature bone grows in length（骨干与骨骺交接的部分）骺板 [syn] growth plate, metaphyseal plate, physeal plate

epiphysis /ɪˈpɪfɪsɪs/ n. **1** (*pl.* **epiphyses** /-siːz/) the bulged end of a long bone（指长骨膨大的末端）骨骺 **2** pineal gland 松果体

epiphysitis /ˌepɪfɪˈsaɪtɪs/ n. (also **physitis**) inflammation around the growth plate of certain long bones that occurs most often in young horses; characterized by swelling and lameness（长骨生长板发炎的症状，多发于青年马，症状表现为骺端肿大和跛行）骨骺炎，骨生长板发育不全 [syn] physeal dysplasia, dysplasia of the growth plate

episioplasty /ɪˌpiːslə-plæsti/ n. a surgical alteration of the labia performed to correct defective conformation（对阴唇形态缺陷进行的矫正手术）外阴整形术

epistaxis /ˌepɪˈstæksɪs/ n. (*pl.* **epistaxes** /-siːz/) bleeding from the nose 鼻出血

epithalamus /ˌepɪˈθæləməs/ n.【anatomy】a part of the forebrain located above the thalamus（位于丘脑上的部分）上丘脑 [compare] metathalamus

epithelial /ˌepɪˈθiːlɪəl/ adj. of or relating to epithelia 上皮的：epithelial tissue 上皮组织

epitheliogenesis imperfecta n. (abbr. **EI**) a genetic fatal condition in which the foal is born without skin or hoofs（幼驹出生后无皮肤和蹄的致死性遗传病）上皮发育不全

epithelium /ˌepɪˈθiːliəm/ n. (*pl.* **epithelia** /-liə/) membranous tissue composed of one or more layers of cells and forming the covering of most internal and external surfaces of the body and its organs（位于体表或衬在器官内面的膜性组织，由单层或复层致密细胞组成）上皮

epizootic /ˌepɪzəʊˈɒtɪk; *AmE* -zoʊˈɒt-/ adj. (of a disease) affecting a large number of animals at the same time within a particular region; epidemic（同时在特定地区感染大量动物的）流行的

n. an epizootic disease 流行病

EPM abbr. equine protozoal myeloencephalitis 马原虫脑脊髓炎

EPO abbr. erythropoietin 促红细胞生成素

epoch /ˈiːpɒk, *AmE* ˈepək/ n. a unit of geologic time that is a division of a period（地质年代中对纪的划分单位）世：Paleocene epoch 古新世 ◇ Eocene epoch 始新世 ◇ Oligocene epoch 渐新世 ◇ Miocene epoch 中新世 ◇ Pliocene epoch 上新世 ◇ Pleistocene epoch 更新世 [compare] era, period

Epona /iːˈpəʊnɑː/ n.【mythology】the Celtic goddess and protector of horses and cavalry（凯尔特神话传说中马与骑士的守护神）爱波娜

EPSM abbr. equine polysaccharide storage myopathy 马多糖贮积肌病：Rhabdomyolysis is one of the most frequent signs associated with horses affected by EPSM. 横纹肌溶解是患多糖贮积肌病的马匹最常见的症状。

Epsom Derby n. (aslo **Derby**, or **Derby Stakes**) a 1.5-mile horse race held annually on the first Wednesday of June since 1780 at the Epsom Downs course located just 15 miles south of London, England（自1780年起每年6月的首个星期三在英国伦敦南24千米处的艾普森·唐斯赛马场举行的赛马比赛，赛程2 400米）艾普森德比赛，英国德比赛：The Epsom Derby is named for the twelfth Earl of Derby. 艾普森德比赛以德比郡十二世伯爵而命名。

Epsom Downs n. (also **the Downs**) a U-shaped, 1.5-mile racecourse consisting of hills, banks, and turns run counterclockwise; is located 15 miles south of London and considered the most demanding test of jockeys and horses in the world（位于伦敦市南24千米处的U形赛道，长2 400米，主要由山丘、堤坝和弯道构成，赛马以逆时针方向行进，被认为是世界上最具有挑战性的赛道）艾普森·唐斯赛

道,唐斯赛马场:The Downs is the site of the Epsom Derby. 艾普森德比赛在唐斯赛道举行。

Epsom Oaks n. (also **Oaks**) a 1.5-mile stakes race for three-year-old fillies run annually since 1779 at the Epsom Downs course in London, England (自 1779 年起每年在英国伦敦的艾普森·唐斯赛道举行的奖金赛,由 3 岁雌驹参加,赛程 2.4 千米)欧克斯大赛,橡树大赛

Epsom salts n. hydrated magnesium sulfate used in horses as a laxative or purgative (水合硫酸镁,可作马匹泻药使用)泻盐

equestrian /ɪˈkwestriən/ adj. of or relating to horseback riding 骑马的;马术的:equestrian sports 马术运动 ◇ equestrian competition 马术比赛 ◇ Chinese equestrian team 中国马术队

n. one who rides a horses a sport 骑手;马术选手

equestrian vaulting n. *see* vaulting 马上体操

equestrianism /ɪˈkwestriənɪzəm/ n. the sport of riding horses 马术,骑术

equestrienne /ɪˌkwestrɪˈen/ n. a female rider or equestrian 女骑手

equid /ˈekwɪd/ n. the horse family of mammals 马科动物

equine /ˈekwaɪn, ˈiːk-/ adj. **1** of or relating to horses 马的:equine art 马艺术 ◇ equine science 马科学 **2** of or belonging to the horse family 马属的;马科的:equine animals 马属动物

equine arteritis n. *see* equine viral arteritis 马病毒性动脉炎

equine asthma n. *see* chronic obstructive pulmonary disease 马慢性阻塞性肺炎,马哮喘

equine biliary fever n. *see* equine piroplasmosis 马梨形虫病

equine contraception n. a measure adopted by the Bureau of Land Management in the United States as a means of limiting wild horse population (美国土地管理局为了控制野马数量而采取的措施)母马避孕

equine degenerative myeloencephalopathy n. (abbr. **EDM**) a genetically based disease of the spinal cord and brain stem causing weakness and incoordination (马脊髓与脑干所患的遗传性疾病,可导致体弱和运动失调)马萎缩性脑脊髓病,马退行性脑脊髓病

equine dysautonomia n. *see* grass sickness 马自主神经机能障碍,马草瘟

equine encephalitis n. inflammation of the brain 马脑炎

equine exercise physiology n. a branch of equine science dealing with the physiology of horse in exercising 马体运动生理学

equine grass sickness n. *see* grass sickness [马]草瘟

Equine Health Research Fund n. (abbr. **EHRF**) a fund established in the United States in 1985 by the American Horse Shows Association to support research in the fields of equine health and medicine (1985 年由美国马匹展览协会设立的基金,旨在资助马匹健康与医学研究)马健康研究基金会

equine herpesvirus n. (abbr. **EHV**) any of a group of viruses of the family *Herpesviridae* that cause rhinopneumonitis of horses (导致马鼻肺炎的疱疹病毒科病毒)马疱疹病毒

equine infectious anemia n. (abbr. **EIA**) a viral infectious disease that causes anemia of horses 马传染性贫血,马传贫:Equine infectious anemia is usually spread by biting insects, and horses are required by law in some areas to have a Coggins test done a minimum of once per year. 马传染性贫血多通过蚊虫叮咬传播,有些地区法定马匹至少每年进行一次柯金斯检测。 syn swamp fever

equine infectious rhinopneunonitis n. *see* equine viral rhinopenumonitis 马传染性鼻肺炎

equine influenza n. an acute, highly contagious disease characterized by respiratory inflammation and fever (一种急性、高度传染性疾病,临床表现为呼吸道感染和发热)马流感

equine leptospirosis n. a contagious, water-borne disease of horse resulting from infection with various leptospiral organisms and symptomized by fever, depression, anorexia, and abortion of mare（人马共患传染病,多为感染各种钩端螺旋体所致,其症状表现为发热、精神萎靡、食欲不振和母马流产）马钩端螺旋体病

equine lupus n. an immune-mediated disorder of horses（免疫介导的机能紊乱）马狼疮

equine malaria n. *see* equine piroplasmosis 马梨形虫病

equine monocytic ehrlichiosis n. *see* Potomac horse fever 马单核细胞埃利希体病,马波托马可热

equine motor neuron disease n.（abbr. **EMND**）a degenerative disease of horse, characterized by muscle tremors and weight loss（马的一种退行性疾病,症状为肌肉震颤和体重下降）马运动神经疾病

equine night blindness n.（abbr. **ENB**）moon-blindness of horse 马夜盲症

equine piroplasmosis n.（also **piroplasmosis**）a contagious, tick-borne protozoan disease of animals caused by *Babesia caballi* or *Babesia equi*, symptoms may include fever, listlessness, red urine, jaundice, loss of appetite, constipation, and diarrhea（由寄生于蜱上驽巴贝斯虫和马巴贝斯虫所致的传染病,症状表现为发热、萎靡、红尿、黄疸、食欲减退、便秘和下痢）马梨形虫病,马焦虫病 [syn] babesiosis, equine biliary fever, equine malaria, Texas fever, horse tick fever, tristeza

equine podiatrist n. a farrier 蹄铁匠,掌工

equine polysaccharide storage myopathy n.（abbr. **EPSM**）a metabolic disease of horses that results in an abnormal accumulation of glycogen and abnormal polysaccharide in skeletal muscles（马匹的一种代谢性疾病,可导致骨骼肌中糖原和多糖的异常贮积）马多糖贮积肌病

equine protozoal myeloencephalitis n.（abbr. **EPM**）a neurological disease caused by protozoan, symptoms include nerve-cell damage, lameness, loss of coordination and body weakness（由原虫引发的神经性疾病,症状主要表现为神经细胞损伤、跛行、运动失调和体质虚弱）马原虫性脑脊髓炎

equine purpura n. an acute, allergic condition occurring characteristically as a consequence of strangles; symptoms include a sudden onset of subcutaneous edema in the area of the head, eyes, lips, belly, and legs, the appearance of small red spots in the mucous membranes, diarrhea, colic, or in severe cases, anemia（马患腺疫后出现的急性过敏反应,症状包括头、眼、唇、腹部和四肢等处皮下水肿、黏膜出现血斑、腹泻、腹痛甚至贫血）马紫癜

equine research center n.（abbr. **ERC**）an institution for the scientific study of horses 马研究中心

equine respiratory syndrome n. *see* acute equine respiratory syndrome 马急性呼吸综合征

equine rhabdomyolysis n. another term for exercise-related myopathy（运动相关性肌病的别名）马横纹肌溶解症

equine ulcerative lymphangitis n. *see* ulcerative lymphangitis 马溃疡性淋巴管炎

equine variola n. *see* horsepox 马痘

equine vesicular stomatitis n. inflammation of the soft tissues of the mouth characterized by blisters or other lesions（马口腔局部炎症,临床表现为上舌面生水疱和糜烂）马水疱性口腔炎

Equine Veterinary Journal n.（abbr. **EVJ**）an academic journal of veterinary science published by the British Equine Veterinary Association since 1968（英国马兽医协会从 1968 年起出版的马兽医学术期刊）《马兽医学报》

equine viral abortion syndrome n. *see* equine viral rhinopneumonitis 马病毒性流产综合征

equine viral arteritis n.（abbr. **EVA**）（also **equine arteritis**）a highly contagious disease of horses caused by the togavirus transmitted via respiratory tract or by mating; characterized by fever, redness of the eyes, edema of the eyelids and limbs, nasal discharge and congestion, loss of appetite, and abortion in mares (由披膜病毒所致的马传染病,可通过呼吸道或交配传播,症状表现为发热、眼红肿、眼睑和四肢浮肿、鼻塞、食欲下降及母马流产)马病毒性动脉炎

equine viral rhinopneumonitis n. (abbr. **EVR**) a contagious viral infection caused by equine herpes virus Ⅰ, characterized by fever, nasal discharge, cough, mild respiratory infection and abortion of mares (由马Ⅰ型疱疹病毒引起的传染病,症状表现为发热、流鼻、咳嗽、呼吸道感染和母马流产等)马病毒性鼻肺炎 syn equine viral abortion syndrome

equinology /ˌekwiˈnɒlədʒi; *AmE* -ˈnɑː l-/ n. the study of horses; hippology 马学

equipage /ˈekwɪpɪdʒ/ n. a horse-drawn carriage and its attendants 马车及侍从

equirotal /ˌiːkwɪˈrəʊtəl; *AmE* ˌɪkwəˈroʊtəl/ adj. 【Latin】(of the wheels) equal in size or diameter (指车轮)大小相等的

equitape /ˌekwɪˈtæp/ n. a tape which both measures the horse's girth and gives the body weight without the need to convert it (能同时测量马匹胸围并给出体重的皮尺)马用体尺

equitation /ˌekwɪˈteɪʃn/ n. the practice or art of horse riding 骑术: classical equitation 古典骑术

equitation class n. a class of English riding competition in which the rider's position and skills is judged (英式骑术的比赛级别,主要对骑手的骑姿和技术进行评分)骑术级

equivalent /ɪˈkwɪvələnt/ adj. equal in value or meaning; similar, identical 相等的;相当的

Equus /ˈekwəs/ n. **1** 【Latin】 the genus of the horse family 马属动物 **2** a play written by the British dramatist Peter Shaffer in 1973 in which a youth blinds several horses in spite of his love for them (1973年由彼得·谢弗创作的剧本,其中主人公尽管爱马,却将数匹马的眼刺瞎)《恋爱狂》

Equus africanus n. 【Latin】 the scientific name for the African wild ass 非洲野驴

Equus asinus n. 【Latin】 the scientific name for donkey 家驴

Equus burchelli n. 【Latin】 the scientific name for the common zebra 普通斑马

Equus caballus n. 【Latin】 the scientific name for domestic horse 家马

Equus celticus n. 【Latin】 the scientific name for Celtic pony 凯尔特矮马

Equus ferus n. 【Latin】 the scientific name for the wild horse 野马

Equus ferus ferus n. 【Latin】 the scientific name for the Tarpan horse 欧洲野马

Equus ferus przewalskii n. 【Latin】 the scientific name for the Przewalski's horse 普氏野马

Equus grevyi n. 【Latin】 the scientific name for Grevy's zebra 格雷氏斑马

Equus hemionus n. 【Latin】 the scientific name for a subspecies of the Asiatic wild ass 亚洲野驴

Equus hemionus hemionus n. 【Latin】 the scientific name for the Mongolian wild ass 蒙古野驴

Equus hemionus khur n. 【Latin】 the scientific name for the Indian wild ass 印度野驴

Equus hemionus kulan n. 【Latin】 the scientific name for the Turkmenian kulan 土库曼野驴,库兰驴

Equus hemionus onager n. 【Latin】 the scientific name for the Persian onager 波斯野驴

Equus ibericus n. 【Latin】 the scientific name for the Iberian horse 伊比利亚马

Equus kiang n. 【Latin】 the scientific name for the Tibetan wild ass 藏野驴

Equus lambei n. 【Latin】 the scientific name for the Ice Age horse 冰河马

Equus quagga n. 【Latin】 the scientific name for the quagga (非洲)斑驴

Equus zebra n.【Latin】the scientific name for the mountain zebra 山地斑马

Equus zebra hartmannae n.【Latin】the scientific name for Hartmann's mountain zebra 哈特曼山地斑马

E

era /ˈɪərə; *AmE* ˈɪrə/ n. the longest division of geologic time, made up of one or more periods (地质学中最长的时间单位,下划分为纪)代 compare period, epoch

ERC abbr. equine research center 马研究中心

erect /ɪˈrekt/ vi. (of penis) become hard and upright due to filling of the spongy tissue with blood (阴茎海绵组织充血变硬)勃起 | **erection** /ɪˈrekʃn/ n.

erectile /ɪˈrektaɪl/ adj. (of tissue) capable of filling with blood and becoming rigid 勃起的

ergot/ˈɜːgət/ n. a horny growth behind the fetlock joint of horse (马球节上附生的角质)距

Eridge car n. *see* Eridge cart 艾瑞奇马车

Eridge cart n. **1** (also **Eridge car**) a low-slung, four-wheeled vehicle of the dogcart type introduced during the late 19th century, usually drawn by a single horse or large pony (19 世纪末出现的一种低帮四轮马车,常由单马或体壮矮马拉行)艾瑞奇马车 **2** a low-slung, two-wheeled vehicle for two (可坐两人的低帮两轮马车)艾瑞奇双轮马车

erythrocyte /ɪˈrɪθrəsaɪt/ n. red blood cell 红细胞

erythrocyte count n. *see* red blood cell count 红细胞计数

erythropoietin /ɪˌriθrəˈpɔiətɪn/ n. a glycoprotein hormone released from kidneys that stimulates the production of red blood cells in bone marrow (由肾分泌的一种糖蛋白类激素,可刺激骨髓产生红细胞)促红细胞生成素

esophagus /ɪˈsɒfəgəs; *AmE* ɪˈsɑː-/ n.【anatomy】(also **oesophagus**) the muscular tube for passing food from pharynx to the stomach (将食物从咽运送到胃的肌性管道)食管,食道 syn gullet

Esseda /ˈesədə/ n. a light, horse-drawn chariot used by the ancient Britons in war; usually drawn by a pair of horses in pole gear with neck yokes (古布立吞人在战争中使用的轻型战车,多由两匹马套驾杆拉行)埃瑟达战车

essential /ɪˈsenʃl/ adj. basic or indispensable; necessary 基本的;必需的: essential minerals 必需矿物质元素

essential amino acids n.【nutrition】(abbr. **EAA**) the ten amino acids that cannot be made by vertebrates (脊椎动物自身不能合成的 10 种氨基酸)必需氨基酸

Estonian Klepper n. *see* Toric 托利马,托里斯基马

estrogen /ˈestrədʒən/ n.【physiology】(BrE **oestrogen**) any of the several steroid hormones produced chiefly by the ovaries and responsible for promoting estrus and maintenance of female secondary sex characteristics (由卵巢分泌的多种类固醇激素,有刺激发情并维持雌性特征的功能)雌激素

estrous /ˈestrəs/ adj. of or relating to estrus 发情的: estrous cycle 发情周期 ◇ estrous mare 发情母马 ◇ estrous period 发情期

estrus /ˈestrəs/ n.【physiology】the periodic state of sexual excitement before ovulation in the mare during which the mare is most receptive to mating; heat (母马排卵前周期性的性冲动,此间母马交配欲强烈)发情,动情: duration of estrus 发情持续期◇ the strength of estrus 发情强弱程度—*see also* **anestrus, diestrus**

ET abbr. embryo transfer 胚胎移植

etiology /ˌiːtɪˈɒlədʒi; *AmE* -ˈɑːl-/ n. **1** the branch of medicine that deals with the causes or origins of disease (研究发病原因的医学分支学科)病因学 **2** the cause or origin of a disease or disorder as determined by medical diagnosis (经过医学诊断所确定的发病原因)病因

eumelanin /juːˈmelənɪn/ n. any of a group black or brown pigments found in hair and skin or animals (使动物皮肤和毛发呈黑色或褐色的

一类色素)真黑色素 compare pheomelanin

Eurasian wild horse n. *see* Tarpan 欧洲野马

euthanasia /ˌjuːθəˈneɪziə; *AmE* - ˈneɪʒə/ n. the practice of ending the life of an individual suffering from an incurable illness (结束绝症患者生命的措施)安乐死 | **euthanasic** /-zɪk/ adj.

euthanatize /juˈθænətaɪz/ vt. to end the life of an individual suffering from incurable disease by euthanasia 使安乐死,实施安乐死

EVA abbr. equine viral arteritis 马病毒性动脉炎

evade /ɪˈveɪd/ vt. (of a horse) to avoid obeying the rider's commands or aids (马逃避骑手的指令或辅助)逃避,躲避 |**evasion** /ɪˈveɪʃn/ n.

even money n. *see* even money bet 等额赌注

even money bet n.【racing】(also **even money**) a wager in which the bettor will win the same amount of money as wagered if the horse places as anticipated (赌马者所押赛马跑入预定名次后赢得等额奖金的投注方式)同额赌注,等额投注:The betting odds for even money bet are one to one. 等额赌注的赔率为 1:1。compare odds on

evener /ˈiːvənə/ n.【driving】a bar used to equalize the pulling power of a team of horses harnessed abreast (多马并驾中用来平衡挽力的横杆)分力杠,平衡杠

event /ɪˈvent/ n. a contest or discipline in a sports program 项目;赛事

event horse n. *see* eventer 三日赛马;三项赛马

eventer/ɪˈventə(r)/ n. a horse competing in eventing 三日赛马;三项赛马 syn event horse

eventing /ɪˈventɪŋ/ n. **1** an equestrian competition that consists of dressage, show jumping and cross-country tests (包括盛装舞步、跨越障碍赛和越野赛三个项目的马术比赛)综合全能赛;三项赛 **2** *see* three-day event 三日赛

eversion /ɪˈvɜːʃn; *AmE* ɪˈvɜːrʃn/ n. the act or instance of turning outward 外翻

evert /ɪˈvɜːt; *AmE* ɪˈvɜːrt/ v. to turn inside out or turn outward 使外翻;翻转

eviscerate /ɪˈvɪsəreɪt/ vt. to remove the internal organs from the abdominal cavity; disbowel or viscerate 摘除内脏

EVJ abbr. *Equine Veterinary Journal*《马兽医学报》

evolution /ˌiːvəˈluːʃn, ˌev-/ n. a gradual process in which an organism develops into a more complex or advanced form (生物逐渐向复杂而高级的形式发展的过程)进化 | **evolutionary** /ˌiːvəˈluːʃənri; *AmE* -neri/ adj. 进化的: evolutionary chain 进化链

evolve /iˈvɒlv; *AmE* iˈvɑːlv/ vi.【genetics】to develop gradually 进化

EVR abbr. equine viral rhinopneumonitis 马病毒性鼻肺炎

ewe neck n. a conformation fault in which the top line of the neck from the poll to the withers is concave (马匹从项顶至鬐甲弧线下凹的体貌缺陷)羊颈,凹颈 syn upside-down neck

ewe-necked /ˈjuː nekt/ adj. (of horse's neck) concave or upside-down (马颈)凹陷的,羊颈的

exacta /ɪgˈzæktə/ n.【racing】(also **exactor**) a wager in which the bettor will win if he picks the first two finishers in the exact order (博彩者选出前两名赛马的准确次序方可赢奖的投注方式)头二正序彩 compare quinella

exactor /ɪgˈzæktə/ n.【racing】*see* exacta 头二正序彩

excavation /ˌekskəˈveɪʃn/ n.【anatomy】a hole or dent in an organ (机体器官上的)凹陷

excessive angulation of the hock joints n. *see* sickle hocks 跗关节角度过大,镰形飞节

excitability /ɪkˌsaɪtəˈbɪləti/ n. the condition or quality of being excitable 兴奋性;悍性

excitable /ɪkˈsaɪtəbl/ adj. **1** (of horses) easily get excited 易兴奋的,悍性强的 **2** capable of responding to stimuli 可兴奋的: excitable tissue 可兴奋组织

excite /ɪkˈsaɪt/ vt. to stir or arouse to activity; stimulate 使兴奋;激发

excrement /ˈekskrɪmənt/ n. waste matter discharged from the bowels after digestion; feces 排泄物,粪便

excreta /ɪkˈskriːtə/ n. waste matter discharged from the body 排泄物,分泌物

excrete /ɪkˈskriːt/ vt. to discharge waste matter from the body 排泄;分泌

excused /ɪkˈskjuːzd/ adj. 【racing】(of a horse) allowed by racing authorities to withdraw from a race; scratched (指赛马)允许退赛的,取消比赛的

exercise /ˈeksəsaɪz; *AmE* -sərs-/ n. any physical activity performed to develop or maintain health and fitness or increase skill (为保持身体健康或提高技巧而进行的各种活动)运动,训练,锻炼,练习: regular exercise 常规练习 ◇ exercise intensity 运动强度 ◇ exercise physiology 运动生理学

exercise bandage n. a cotton or synthetic bandage wrapped around the leg between the knee or hock and the fetlock to protect the horse's legs while the horse is at work or exercise (缠裹在马前膝或飞节至球节处的绷带,可防止马在运动或训练中受伤)训练绷带 [syn] brace bandage, track bandage

exercise boy n. 【racing】(also **exercise rider**) one responsible for riding racehorses in workout sessions (平时策骑赛马进行训练的人员)练马员,练马骑手

exercise response n. a short-term physiological adaptation made as a result of an increase in the level of muscular activity (机体对肌肉活动强度增加作出的短期生理性适应)运动应答 [compare] training response

exercise rider n. *see* exercise boy 练马员,练马骑手

exercise-induced pulmonary hemorrhage n. (abbr. **EIPH**) the occurrence of hemorrhage in horse's lungs as a consequence of capillary rupturing following strenuous exercise (马超强度训练后,因肺毛细血管破裂而引起的肺部出血)运动诱发性肺出血,肺出血: EIPH occurs most often in racehorses and other high-performance equine athletes. 肺出血在赛马和其他竞技性能较高的马中较为常见。 [syn] lung hemorrhage, pulmonary hemorrhage, exercise related epistaxis, bleeders

exercise-related epistaxis n. *see* exercise-induced pulmonary hemorrhage 运动诱发性肺出血

exercise-related myopathy n. a disease affecting the muscle tissues of horses that results from strenuous physical exertion; often characterized by muscle cramping and straightening, profuse sweat, a tucked-up appearance, stiff gaits, reluctance to move, and red urine in severe cases (马超负荷运动后肌肉组织所患的一种疾病,症状表现为肌肉痉挛和强直、发汗、身体抽搐、步法僵硬和懒于走动,严重时还伴有血尿)运动相关性肌病 [syn] exertion myopathy syndrome, exertional myopathy, equine rhabdomyolysis, over-straining disease

exertion myopathy syndrome n. (also **exertional myopathy**) *see* exercise-related myopathy 运动相关性肌病

exhale /eksˈheɪl/ vt. to breathe air out of lung; expire 呼气 [opp] inhale

exhibitor /ɪgˈzɪbɪtə(r)/ n. one who is involved in the showing of horses (马匹)展览者

Exmoor /ˈeksmʊə(r), -mɔː(r)/ n. (also **Exmoor pony**) an ancient pony breed originating in the Exmoor forest in southern England from which the name derived; stands 11.2 to 12.3 hands and generally has a brown bay or dun coat without marking; has a well-proportioned head with wide-set, toad eye, full mane, long and slightly hollow back, slightly sloping shoulder and croup, deep chest, and sturdy legs; is quiet, docile and possessed of good speed; used as an excellent mount for children (源于英国南部埃克斯穆尔荒野的古老矮马品种,故而得名。体高11.2~12.3掌,毛色多为褐骝或兔褐且无白色别征,头型适中,瞳距大,蛤

蟆眼,鬃毛浓密,背长稍凹,肩尻稍斜,胸深,四肢健壮,性情安静而温驯,速力较好,尤适合儿童骑乘)埃克斯穆尔矮马

Exmoor Pony Society n. an organization founded in England in 1921 to improve and promote the breeding of Exmoor ponies of the moorland type (1921年在英国成立的组织机构,旨在促进沼地型埃克斯穆尔矮马的培育)埃克斯穆尔矮马协会

exostosis /ˌeksɒsˈtəʊsɪs; *AmE* -ˈtoʊsɪs/ n.【vet】(*pl.* **exostoses** /-siːz/) an abnormal bony growth projecting outward on the surface of a bone (骨面上的异常骨质增生)外生骨疣: Exostosis may result from injury, conformation fault, or heredity. 外伤、失格或遗传都可导致外生骨疣的发生。 syn bony growth

exotic wagger n.【racing】any bet other than win, place, or show 外围投注

expel /ɪkˈspel/ vt. to discharge or eject 排出

expire /ɪkˈspaɪə(r)/ vi. to exhale or breathe out 呼气 | **expiration** /ˌekspəˈreɪʃn/ n.

expulsion /ɪkˈspʌlʃn/ n. the act of expelling or the state of being expelled 排出: The third stage of foaling marks the expulsion of the afterbirth. 分娩第三期的特征是排出胎衣。

extend /ɪkˈstend/ vt. **1**【racing】to push a horse to his limits in a race 催马,猛赶 **2** to stretch or lengthen a pace or gait (步幅)伸长: extended canter 伸长跑步 ◇ extended pace 伸长步法 ◇ extended trot 伸长快步 ◇ extended walk 伸长慢步 **3** to increase or enlarge the quantity or volume by adding substance 稀释,扩充: extended semen 稀释精液

extended heel n. a long heel on the outside of a hind-hoof horseshoe (后肢)长蹄铁踵

extended-toe shoe n. a horseshoe in which the shoe is fitted well forward of the natural toe of the foot (前缘超出马蹄前壁的蹄铁)长头蹄铁: Extended-toe shoe is often used on the feet of clubfooted horses or horses with contracted tendons. 长头蹄铁多用于患狭蹄或屈腱炎的马。 syn beaked shoe

extender /ɪkˈstendə/ n. a dilutor 稀释液: seminal extender 精液稀释液

extension /ɪkˈstenʃn/ n. **1** the act of extending or stretching a limb (四肢)伸展 **2** the lengthening of horse's stride in movement (马行进中步幅拉长)伸长,伸展: false extension 步幅虚长 ◇ not enough extension 步幅伸展不足 opp collection

extensor /ɪkˈstensə(r)/ n.【anatomy】(also **extensor muscle**) a muscle that extends a joint or bodily part (可拉伸关节或躯体某部分的肌肉)伸肌 compare flexor

extensor muscle n. *see* extensor 伸肌

extensor process disease n. *see* pyramidal disease 伸肌骨刺

extensor tendon n. a tendon that extends a joint (使关节伸展的肌腱)伸肌腱

exteriorize /eksˈtɪərɪəraɪz/ vt. to put or turn outward by surgery (用手术)外置

extinct /ikˈstɪŋkt/ adj. (of a species) no longer existing or living (物种)绝迹的,灭绝的: At one time widely distributed over Europe and Asia, Przewalski's horse has become extinct in its last wild range in Mongolia during the mid-twentieth century. 曾经在亚欧大陆上分布极其广泛的普氏野马,最终于20世纪中期在蒙古的草场上灭绝了。

extinction /ɪkˈstɪŋkʃn/ n. the state of being extinct 灭绝,绝种

extra- pref. outside, beyond 外

extracapsular /ˌekstrəˈkæpsjulə/ adj.【anatomy】located outside the articular capsule 囊外的: extracapsular ligaments 囊外韧带 opp intracapsular

extracellular /ˌekstrəˈseljulə/ adj. located or occurring outside a cell or cells 细胞外的: extracellular matrix 细胞外间质

extrauterine /ˌekstrəˈjuːtərɪn/ adj. located or occurring outside the uterus 子宫外的: extrauterine pregnancy 宫外孕

extrude /ɪkˈstruːd/ vt. to push or thrust out 挤出,压出

exudate /ˈegzjuˌdeɪt/ n. the fluid discharged from a sore or wound (从伤疤渗出的液体)渗出物,渗出液

E

exude /ɪgˈzjuːd; *AmE* -ˈzuːd/ vi. to discharge or emit gradually; ooze (缓慢地释放或散发)渗出

eye /aɪ/ n. **1** the organ of vision 眼 **2** the uppermost ring on the cheekpiece of a bit to which the bridle cheekpiece is attached (衔铁颊杆上端的孔环,马勒的颊革系于其上)衔铁眼,衔铁孔

eye worm n. an internal parasite that lives in the tear duct and conjunctival sac of horse's eye with face flies serving as intermediate host; heavy infestation often causes mild eye irritation and may result in blindness on rare occasions (生活在马泪管和结膜囊中的内寄生虫,多以面蝇为中间宿主,大量繁殖时可造成眼痛痒,有时甚至可导致失明)眼虫,浑睛虫 syn *Thelazia lacrymalis*

eye-catching /ˈaɪˌkætʃɪŋ/ adj. (of a horse) visually attractive 引人注目的;惹眼的: an eye-catching horse 一匹抢眼的马

F /ef/ abbr.【racing】filly 雌驹

face drop n. *see* forehead drop 额前垂饰

face fly n. a type of daytime fly swarming in large numbers around the eyes and the muzzle of the horse（白天成群聚集在马眼和口鼻周围的一类蝇）面蝇

face marking n.（also **facial marking**）any white mark appearing on the face or head of a horse 面部别征—*see also* **star**, **stripe**, **race**, **blaze**, **bald face**

face piece n. *see* drop 额前垂饰

facial /ˈfeɪʃl/ adj. of or concerning the face 面部的：facial crest 面脊 ◇ facial markings 面部别征

factory-made horseshoe n. *see* machine-made shoe 预制蹄铁

fade /feɪd/ vi. **1**【racing】(of a horse) tire and lose strength in a race（指赛马）乏力，不济：The racer faded near the finish. 这匹赛马接近终点时有点乏力。**2** PHR V **fade back**【cutting】(of a cutting horse) to back off under a clue given by the rider（指截牛马）缓慢后退

faeces /ˈfiːsiːz/ n.【BrE】*see* feces 粪便

failing scent n.【hunting】the scent of a fox that becomes faint（猎物）残留的气味，变淡的气味

fair /feə(r); *AmE* fer/ adj. clean and tidy 整洁的：The bed should be set fair before the horse returns from exercise. 厩床应在马结束训练返回厩舍前清扫干净。

fair catch n.【roping】an instance of roping around the horns, head or neck of a calf（套住牛角、头或颈的情况）有效套牛 syn legal catch

Fairville Cart n. an Australian, horse-drawn cart with detachable shafts（澳大利亚使用的一种双轮马车，带可卸式车辕）费尔维拉马车

Falabella /ˌfɑːləˈbelə/ n. a miniature horse breed developed in Argentina by inbreeding the smallest Shetland ponies; has a full mane and tail, small head, large eyes, prominent withers, short back, and slender legs; stands under 34 inches, and may have any of the coat colors; is quiet, intelligent, and possessed of a graceful action（阿根廷采用小型设特兰矮马近交培育而来的微型马品种。鬃尾浓密，头小，眼大，鬐甲突出，背短，四肢纤细，体高 34 英寸以下，各种毛色兼有，性情安静，反应灵敏，步态优雅）法拉贝拉矮马：According to the Guinness Book of World Records, the smallest horse in the world stood only 12 inches at the withers, and was a Falabella. 根据吉尼斯世界纪录，世界上最小的马正是一匹法拉贝拉矮马，其鬐甲高仅 12 英寸。

fall /fɔːl/ vi. **1** (of a horse) to lose balance and fail to stand on its legs（指马）跌倒，摔倒 **2** (of a rider) to drop down from the back of horse（指骑手）落马，跌落

n. an act or instance of falling down or falling from a horse 跌倒；落马 PHR V **take a fall** 跌倒；跌落，落马

fallen back n. *see* hollow back 凹背

fallen neck n. *see* broken crest 歪颈

falling top n. the folding hood of a horse-drawn carriage（马车的）折叠式车篷

Fallopian tube n. *see* oviduct 输卵管

false canter n. *see* counter canter 反跑步，假跑步

false collar n. a flat leather pad shaped to fit under the harness collar to protect the shoulders of horse（衬在颈圈下的软垫，可保护马肩胛）套包垫，颈圈垫

false favorite n.【racing】a racehorse who is favored by bettors but outrun by other horses of the field（被投注者看好但却失利的赛马）枉被看好的马，爆冷门的马

false gallop n. *see* disunited gallop 仿袭步，假袭步

false ribs n. the ribs that are not attached to the sternum（不与胸骨相连的肋骨）假肋

false rig n. a castrated horse who still demonstrates masculine behaviors（阉割后仍然表现雄性特征的公马）假骟马

false sole n. *see* retained sole 假蹄底，蹄底滞留

false start n.【racing】a condition in which a racehorse runs prematurely before receiving starter's order（赛马在未接收到司闸指令之前起跑）起跑失误，抢跑

falter /ˈfɔːltə(r)/ vi. **1** to move with unsteady or tired steps（走路）摆晃，打战 **2**【racing】(of a horse) tire out badly（指马）力竭

familial /fəˈmɪliəl/ adj. occurring in a family or its members 家族的，家系的

familial trait n.【genetics】a trait that occurs among all members of a family 家族性状：Familial traits may be caused by shared environmental factors but not necessarily inherited. 家族性状可能是由共同的环境因素所致，而并非肯定由遗传决定。

family /ˈfæməli/ n. **1** a taxonomic category ranking below an order and above a genus（地位高于属而低于目的分类单位）科 compare kingdom, phylum, class, order, genus **2** the lineage of a horse（马的）系谱，谱系

fan /fæn/ vt. **1**【rodeo】(of a rider) to slap one's mount with his cowboy hat to encourage a heartier buck（指骑手用牛仔帽拍打坐骑使之跳的更猛）煽动，拍打 **2**【driving】to whip a horse lightly 轻打，轻抽

fan tail n. (also **fantail**) a fan-like tail of horse（马的）扇形尾，扇尾

fancied /ˈfænsɪd/ adj.【racing】*see* favorite（赛马）有望获胜的，被看好的

fancy matched adj.【driving】(of a team of horses) having distinctly different markings and coloring（联驾马匹的别征和毛色截然不同）不搭调的，迥异的

fantail /ˈfænteɪl/ n. *see* fan tail 扇形尾，扇状尾

Fantasia /fænˈteɪzɪə; *AmE* -teɪʒə/ n. a mounted Moroccan sport in which a mock attack is performed by two weapon-bearing teams in a thrilling manner（摩洛哥马上运动项目，两队骑手手持武器模拟打斗，场面异常惊险）马上交锋，马上斗武

FAO abbr. Food and Agricultural Organization 联合国粮农组织

far side n. **1** (also **off side**) the right side of the horse when viewed from behind（后观时马的右侧）远侧，外手 compare near side **2**【racing】the section of a racetrack on the side opposite the homestretch（赛道位于终点直道对面的区段）远侧赛道

far turn n.【racing】the bend of the track off the backstretch（赛道）远侧转弯处

farcy /ˈfɑːsi; *AmE* ˈfɑːrsi/ n. *see* glanders 鼻疽

farm /fɑːm/ n. a place used for growing crops and/or keeping animals 农场，牧场：horse farm 马场 ◇ stud farm 种马场

farm pony n. a sturdy pony used for farm work 农用矮马

farrier /ˈfæriə(r)/ n. a blacksmith who shoes horses 蹄铁匠，蹄师，掌工：head farrier 首席蹄师 syn horseshoer, equine podiatrist

farrier's knife n. *see* hoof knife 削蹄刀

Farriers Registration Act n. an act passed in 1975 in UK that prohibits the shoeing of horses by unqualified practitioners and provides the training and examining of farriers（英国于1975年通过的法案，禁止无蹄铁匠资格者开业装钉蹄铁，并开展蹄铁匠培训和考核）蹄铁匠注册法案

farriery /ˈfærɪəri/ n. **1** the practice of horseshoeing 蹄铁装钉：farriery competition 钉蹄铁比赛 **2** the place or shop where a farrier works 蹄铁匠铺

fascicle /ˈfæsɪkl/ n.【anatomy】(abbr. **fasc.**) a

small of bundle of muscle fibres, nerves, etc 束: atrioventricular fascicle 房室束

fascicular /fəˈsɪkjʊlə/ adj. of, relating to, or composed of fascicles 成束的

fasciculus /fəˈsɪkjʊləs/ n.【Latin】(*pl.* **fasciculi** /-laɪ/) fascicle 束

Fasciola hepatica /ˈfæsɪˌəʊlə hɪˈpætɪkə/ n.【Latin】*see* liver fluke 肝片吸虫

fast /fɑːst/ adj. (of a horse) rapid, swift in movement (指马运步)轻快的,敏捷的

fast martingale n. *see* standing martingale 站立鞅

fast track n.【racing】a dry, hard dirt racing surface on which horses runs faster than on other tracks (地表干硬的泥赛道,赛马在这种赛道上跑得更快)快速赛道

fat /fæt/ n. any of various white, soft tissues stored under the skin of animals; lipid (贮藏在动物皮下的白色软组织)脂肪

Father of Classical Equitation n. *see* Guérinère, François Robichon de la 古典骑术之父,弗朗索瓦·罗比臣·德拉·格伊尼亚

Father of Foxhunting n. *see* John Warde 猎狐之父,约翰·瓦德

fatigable /ˈfætɪgəbəl/ adj. subject to fatigue 疲劳的,劳累的

fatigue /fəˈtiːg/ n. extreme tiredness resulting from physical or mental exertion (身体或精神的)疲惫;疲劳: Depletion of muscular glycogen leads to muscular fatigue. 肌糖原消耗导致肌肉疲劳。

fatigueless /fəˈtiːglɪs/ adj. feeling no fatigue 不知疲倦的

fat-soluble /ˈfætˌsɒljʊbl/ adj. soluble in fats or fat solvents 脂溶性的 compare water-soluble

fat-soluble vitamin n.【nutrition】any vitamin that dissolves in fat or oil 脂溶性维生素

fatten /ˈfætn/ vt. to make livestock fat 育肥

fattening /ˈfætnɪŋ/ adj. (of feeds) used or intending to fatten livestock 育肥的: fattening feeds 育肥饲料

fatty /ˈfæti/ adj.【nutrition】of or relating to fat 脂肪的: fatty acids 脂肪酸 ◇ fatty tissue 脂肪组织

fault /fɔːlt/ n. **1** a weak point or defect in the conformation of a horse (马匹体貌的缺点)缺陷,失格: conformation fault 体格缺陷 **2** a mistake made by a rider or his mount in competition (比赛中骑手或马犯的错误)失误,违例 **3** a scoring unit used to record errors committed by a competitor in an equestrian event (马术项目中用来记录骑手犯规的扣罚分数)罚分: total fault 总罚分

favor /ˈfeɪvə/ vt. to prefer one to another 喜好,偏爱 PHR V **favor a leg 1** (of a horse) to place a large part of its body weight on one leg (指马)偏好某肢负重: A lame horse usually favors a leg to relieve the pain on the injured leg. 瘸马通常会偏好一肢负重以减轻受伤肢的疼痛。**2** (of a horse) to lead with the same front leg at the canter in spite of the direction traveled (指马不论行进方向如何,在跑步时惯用某肢起步)起步偏好某肢

favorite /ˈfeɪvərɪt/ n.【racing】a racehorse considered likely to win a race 被看好的马,有望获胜的马

fawn /fɔːn/ n. a young deer in its first year 幼鹿,小鹿—*see also* **deer, doe, stag**

FBHS abbr. Fellow of the British Horse Society 英国马会会员

feather /ˈfeðə(r)/ n. *see* feathering 距毛

feather edge n. (also **feather-edge**) a sharp side or edge 薄边,锋刃

vt.【driving】to drive a horse-drawn vehicle extremely close to sth 擦边(驶过),削边(越过): The coachman feather edged the huge rock laying in the middle of the road. 车夫擦边驶过挡在马路中央的一块巨石。

feather-edged shoe n. a horseshoe in which the edge of the shoe from the toe to the quarter is squared or rounded off to prevent interference of hoofs (侧边被磨圆的蹄铁,可防止马蹄交突的发生)磨边蹄铁,缺边蹄铁 syn interfering shoe, speedy-cutting shoe, dropped crease shoe, knocked down shoe

feathering /ˈfeðərɪŋ/ n. (also **feather**) the long hairs growing around the fetlock on the lower legs of some draft breeds (某些挽马品种四肢球节上生长的长毛)距毛: a draft horse with light feathering 一匹距毛稀疏的挽马。syn foot-lock

F

featherweight /ˈfeðəweɪt/ n. extremely light weight, as of the jockey 超轻量级

fecal /ˈfiːkl/ adj. (BrE **faecal**) of, relating to, or composed of feces 粪便的: fecal worm egg count 粪便虫卵计数 ◇ The fecal energy is the energy lost from feed in the feces. 粪能是排泄粪便中损失的能量。

fecal exam n. a routine laboratory test of manure sample in which the horse's parasite burden is assessed by counting the worm eggs within (通过粪便虫卵计数来估测马匹寄生虫负荷的实验室检测手段)粪检

feces /ˈfiːsiːz/ n. (BrE **faeces**) the solid dark brown ball-shaped droppings discharged from the anus of horse (马从肛门排出的褐黑色球状物)粪便, 粪球: A full-grown horse will defecate eight to nine times a day, averaging a total of 40 pounds of feces. 成年马每天排泄8~9次, 总共约36千克的粪便。syn manure, dung, droppings, excrement, or horse shit

feces eating n. a vice of horse who eats feces; coprophagy (马匹的恶习)食粪癖

fecund /ˈfiːkənd, ˈfek-/ adj. capable of producing offspring; fertile, productive 能生育的, 可繁育的; 多产的

fecundity /fɪˈkʌndəti/ n. the capacity or power of producing abundantly; fertility; productiveness 生殖力, 繁殖力

Fédération Equestre Internationale n.【French】(abbr. **FEI**) the international governing body founded in Paris in 1921 to organize and regulate all officially recognized equestrian competitions including the Olympic equestrian sports; now is headquartered in Brussels (1921年在法国巴黎成立的国际机构, 旨在组织管理世界上的马术比赛, 其中包括奥运会马术项目, 目前其总部设在布鲁塞尔)国际马术联合会, 国际马联: All nations must be affiliated with the FEI if they wish to participate in official international competitions. 所有国家都只能在成为国际马联的会员后, 方可参加官方组办的国际马术比赛。syn International Equestrian Federation

Federation of International Polo n. (abbr. **FIP**) an international organization founded in 1983 in the United States to facilitate the sport of polo across nations (1983年在美国成立的国际机构, 旨在组办国际马球赛事)国际马球联合会

Federico Caprilli n. *see* Caprilli, Federico 费德里克·卡普瑞利

fee /fiː/ n. **1** the money paid for nominating or entering a horse in a race or other event (马匹参加比赛缴纳的金额)参赛费: entry fee 参赛费, 报名费 **2** the amount paid to a professional rider for competing in a race or other event (支付给职业骑手参赛的费用)劳务费, 骑师费

feed /fiːd/ vt. to give or provide food 饲喂, 饲养 n. food supplied to, or consumed by livestock; fodder 饲料: feed chart 喂料表 ◇ feed room 饲喂间 ◇ feed shed 草料棚 ◇ evening feed 夜饲草料 ◇ high energy feeds 高能饲料 PHR V **go off feed** (of a horse) not eat normally, as due to illness or stress (指马因疾病或应激)食欲减退, 食欲不振

feed bag n. a round canvas bag buckled over the muzzle of a horse to feed it (系在马嘴上用来喂马的帆布袋)喂料袋: horse feed bags 马用喂料袋

feed bin n. a container used for storing feeds 饲料仓, 饲料桶

feed bucket n. a cylindrical container used to feed a horse 喂料桶

feed processing n. any of the methods used to prepare feeds to increase its digestibility（为提高消化率对饲料予以处理的方法）饲料加工：By now, the feed processing methods have become more sophisticated and include steam-ing flaking, micronization and extrusion. 现行的饲料加工工艺已经日趋精细，主要有蒸气压片、饲料微化和膨化技术。

feeder /ˈfiːdə(r)/ n. **1** one who feeds horses 饲养员 **2** a container used to feed the horses 饲喂器具

feeding /ˈfiːdɪŋ/ n. the act or process of providing food to livestock 饲喂，饲养：rules of feeding 饲喂规则 ◇ feeding tube 饲管

feeding time n. the time at which a horse is provided with feed 饲喂时间 [syn] foddering time

feed-shed /ˈfiːdʃed/ n. a building used for storing feeds 饲草棚；饲料间

feedstore /ˈfiːdˌstɔː(r)/ n. a place or room used for storing feeds 饲料贮藏处，饲料贮藏室

feedstuff /ˈfiːdstʌf/ n.【nutrition】(usu. **feedstuffs**) feed provided for livestock; fodder 饲料；草料

FEI abbr. Fédération Equestre Internationale 国际马术联合会，国际马联：The Fédération Equestre Internationale (FEI) is the international governing body for all Olympic equestrian disciplines. 国际马术联合会（FEI）是监管奥林匹克所有马术项目的国际机构。

FEI Show Jumping World Cup n. *see* FEI World Cup 国际马术联合会世界杯障碍赛

FEI World Cup n. (also **FEI Show Jumping World Cup**) an annual international show jumping competition governed by the FEI in which the world's best jumpers and riders compete for championship; established since 1979 and held in a different location every year（由国际马术联合会举办的一项年度国际障碍赛，来自世界各地的顶级赛马和骑手角逐冠军头衔，最初创立于 1979 年且每年在不同地方举办）国际马术联合会世界杯，FEI 世界杯［障碍赛］：From its inception until 1999 the FEI World cup was sponsored by Volvo. 从最初设立至 1999 年，FEI 世界杯障碍赛都由沃尔沃公司赞助。

fell /fel/ n. **1** the hide of an animal; pelt 生皮，兽皮 **2**【BrE】an upland stretch of open country; a moor 高原，荒野 **3** (often **Fell**) *see* Fell pony 费尔矮马

Fell pony n. (also **Fell**) a pony breed originating from the Fells of Cumbria, Britain, from which the name is derived; descended from Celtic ponies, Friesian horses, and Galloway; stands 13 to 14 hands and generally has a black, brown, bay or gray coat without white markings; has a small head, large nostrils, and small ears, luxuriant mane and tail, prominent withers, a long back, short croup, long sloping shoulder, sturdy legs with considerable feathering and bluish, round hoofs; is extremely strong, frugal, a good trotter, and quiet, but lively; used in coal mines for transportation in the past, now used for combined driving, riding, and light draft（源于英国坎布里亚郡费尔高原的矮马品种并由此得名，主要由凯尔特矮马、弗利兹马以及加洛韦马繁育而来。体高 13 ~ 14 掌，毛色多为黑、褐、骝或青，无别征，头小，鼻孔大，耳朵小，鬃尾浓密，鬐甲突出，背长，尻短，肩斜长，四肢粗壮，其上附生距毛，蹄质蓝色且呈圆形，体格强健，耐粗饲，是优秀的快步马，性情安静而活泼；过去曾被用于煤矿运输，现在用于驾车赛、骑乘和轻挽）费尔矮马

Fell Pony Society n. an organization founded in Great Britain in 1927 to promote the breeding and registration of purebred Fell ponies（1927 年在英国成立的组织机构，旨在促进纯种费尔矮马的育种和登记工作）费尔矮马协会

felloe /ˈfeləʊ; *AmE* ˈfeloʊ/ n.【driving】(also **fellowe**, or **felly**) the portion of the carriage wheel rim that holds the spokes in place（轮缘固定辐条的部分）轮辋，轮圈

Fellow of the Worshipful Company of Farriers n. (abbr. **FWCF**) the highest level of certification awarded to a farrier practicing in Britain by the Worshipful Company of Farriers of London (伦敦蹄铁匠公会为在英国从业的蹄铁匠颁发的级别最高的资格证书)蹄铁匠公会会员 compare Associate of the Worshipful Company of Farriers, Registered Shoeing Smith

fellowe /ˈfeləʊ; *AmE* ˈfeloʊ/ n. *see* felloe 轮圈,轮辋

felly /ˈfeli/ n. *see* felloe 轮圈,轮辋

female /ˈfiːmeɪl/ adj. of the sex that gives birth to offspring 雌性的,母的:a female donkey 一头母驴 opp male

femora /ˈfemərə/ n. the plural form of femur [复]股骨

femoral /ˈfemərəl/ adj. of, relating to, or located in the thigh or femur 股骨的:femoral fascia 股筋膜 ◇ femoral region 股部 ◇ femoral trochlea 股骨滑车 ◇ medial ridge of femoral trochlea 股内侧滑车脊

femur /ˈfiːmə(r)/ n.【anatomy】(*pl.* **femurs**, or **femora** /ˈfemərə/) the thigh bone in the hindleg of a horse extending from the hip joint to the stifle (后肢从髋关节至后膝关节的腿骨)股骨:lateral epicondyle of femur 股骨外上踝

fence /fens/ n. **1** a structure made of wood or wire supported with posts or stakes that serves as an enclosure or a boundary (用木板或铁丝以立柱加固搭成围场的)篱笆,围栏,栅栏:The common types of fences used as field boundaries are barbed wire fence, post and rail fence, electric wire fence, and wire-mesh fence, etc. 常见的牧场围栏有刺铁丝围栏、栏杆围栏、电围栏以及铁丝网围栏等。**2**【jumping】a structure or an obstacle that horses jump over in a race or competition (比赛中的)障碍,跨栏:upright fence 直立障碍 ◇ water fence 水渠障碍

v. to jump over an obstacle 跨栏,跨越障碍

fencer /ˈfensə(r)/ n.【jumping】a horse who jumps over fences; jumper 障碍赛马;跨栏马

fencing man n.【hunting】one who repairs hunt jumps and fences damaged during the day's hunting (猎狐结束后修缮障碍和篱笆的人)篱笆维修工,跨栏维护工

fender /ˈfendə(r)/ n. **1** the part of a Western saddle attached with the stirrup and below the seat where the inside of rider's leg rests against (西部牛仔鞍中位于鞍座下方且与马镫相连的皮革,骑手小腿靠于其上)靠腿 **2**【AmE】the splashboard on a horse-drawn vehicle (马车的)挡泥板

feral /ˈferəl/ adj. existing in a wild or untamed state; savage 野生的:feral horses 野生马 ◇ feral species 野生物种

ferment n. /ˈfɜːment; *AmE* ˈfɜːrm-/ any of the substances, such as yeast, that cause fermentation (可引起发酵的酵母等物质)发酵物

v. /fəˈment; *AmE* fərˈm-/ to experience a chemical change because of the action of yeast or bacteria (在酵母或细菌作用下出现的化学反应)发酵

fermentation /ˌfɜːmenˈteɪʃn; *AmE* ˌfɜːrm-/ n. any of a group of biochemical reactions induced by ferments, especially the anaerobic conversion of sugar to acetic acid and alcohol by yeast, mold, etc. (发酵物引起的生物化学反应,尤指酵母或霉菌等将糖转化为乙酸和乙醇的无氧分解过程)发酵

fertile /ˈfɜːtaɪl; *AmE* ˈfɜːrtl/ adj. capable of producing offspring; fecund 可生殖的;有生育能力的

fertility /fəˈtɪləti; *AmE* fərˈt-/ n. the state of being fertile or the ability to produce young 繁殖力,生育力:fertility rate 繁殖率 ◇ fertility test 生育力检测

fertilization /ˌfɜːtəlaɪˈzeɪʃn; *AmE* ˌfɜːrtələˈz-/ n. the union of male and female gametes to form a zygote (雌雄配子结合形成受精卵的过程)受精:fertilization rate 受精率

fertilize /ˈfɜːtəlaɪz; *AmE* ˈfɜːrt-/ vt. (BrE also

fertilise) **1** to cause the fertilization of an ovum; impregnate 使受精；使受孕 **2** to make (soil) fertile by using fertilizer 使肥沃；施肥：highly-fertilized pasture 高肥草场

fescue /ˈfeskjuː/ n. any of various grasses of the genus *Festuca*, often widely cultivated as pasturage (羊茅属草本植物，常作牧草而被广泛种植)羊茅

fetal /ˈfiːtl/ adj. of, relating to, characteristic of a fetus 胎儿的：fetal alignment 胎位 ◇ fetal sac 胚泡 ◇ fetal membranes 胎膜 ◇ There are two fetal fluids, namely, amniotic contained in the amnion, and allantoic contained within the placenta. 胎液包括羊膜中的羊水和胎盘中的尿囊液。

fetlock /ˈfetlɒk; *AmE* ˈfetlɑːk/ n. **1**【anatomy】the fetlock joint 球节 **2** a tuft of hair located behind the fetlock joint of both the fore and hind legs (马前后肢球节上着生的丛毛)距毛 **3** *see* fetlock white 球节白

fetlock boots n. a leg covering used to protect the hind fetlock joint from strikes and blows (一种马用绑腿，可防止后肢球节被撞伤)球节护腿

fetlock joint n. (also **fetlock**) the joint on the horse's legs located between the cannon bone and long pastern (位于管骨与长系骨之间的关节)球节，踝关节 [syn] ankle joint

fetlock ring boot n. (also **fetlock ring**) a hollow rubber ring fitted over the fetlock of the foreleg to protect against brushing (套在马前肢球节处以防被撞伤的橡胶圈)护蹄圈 [syn] brushing ring boot, ring boot

fetlock white n. (also **fetlock**) a leg marking consisting of white that extends from the coronary band up to the bulge of the fetlock joint (马四肢别征之一，其白色区域从蹄冠延伸至球节处)球节白

fetus /ˈfiːtəs/ n. (BrE also **foetus**) the unborn foal from about one and half months after fertilization until birth (从受精后一个半月至出生前的形态)胎儿

fever in the feet n. *see* laminitis 蹄叶炎

fever rings n. *see* hoof rings 蹄环纹

few spot leopard n. a coat color pattern of white horse with a few colored spots and areas of colored skin (白马的毛色类型，被毛和皮肤仅有几个深色斑点)零星猎豹斑

FFAs abbr. free fatty acids 自由脂肪酸

fiacre /fiːˈɑːkrə; *AmE* -kə/ n.【French】**1** a four-wheeled, horse-drawn carriage formerly used for public hire in Paris (过去巴黎使用的四轮公用马车)出租马车 **2** *see* Fiakr 菲阿克马车

Fiakr /fiːˈɑːkr/ n. (also **Fiacre**) a horse-drawn carriage used mainly for public hire in Central Europe during the 19th century (19 世纪在中欧国家使用的公用马车)菲阿克马车

fiber /ˈfaɪbə(r)/ n. (BrE also **fibre**) **1** the thick, coarse plant tissue used for making ropes or cords (制作绳索用的植物组织)纤维 **2** the rough part of feed that helps digestion; roughage (饲料中有助于消化的粗质成分)纤维素，纤维饲料：dietary fiber 日粮纤维

fiber shoe n. *see* rope shoe 绳制蹄铁

fiber-optic endoscope n. (also **fiber-optic scope**) a medical instrument consisting of a flexible tube containing light-conducting fiber bundles which reflect an image into an eyepiece to view the interior of a body cavity (由光导纤维管制成的医疗器械，可根据反射到眼孔的图像来观察体腔内部结构)光纤内窥镜

fiber-optic scope n. *see* fiber-optic endoscope 光纤内窥镜

fibre /ˈfaɪbə(r)/ n. *see* fiber 纤维

fibrescope /ˈfaɪbrəskəʊp; *AmE* -koʊp/ n. an endoscope made of flexible optical fiber 纤维内窥镜

fibrillation /ˌfɪbrɪˈleɪʃn/ n. **1** the forming of fibers 纤维形成 **2** a rapid quivering of heart muscle fibers due to uncoordinated contraction (由于收缩不齐造成的心肌纤维震颤)纤颤：atrial fibrillation 心房纤颤 ◇ ventricular fibrillation 心室纤颤

fibrin /ˈfaɪbrɪn/ n. an elastic, insoluble protein derived from fibrinogen during the process of coagulation of blood (一种不溶性弹力蛋白，血液凝固时由纤维蛋白原转变而来)纤维蛋白

fibrinogen /faɪˈbrɪnədʒən/ n. a protein present in the blood plasma that is essential for blood coagulation (血浆中存在的凝血所必需的蛋白质)纤维蛋白原

fibro- pref. fiber or fibrous tissue 纤维;纤维组织

fibrocartilage /ˌfɪbrəʊˈkɑːtɪlɪdʒ/ n. a type of cartilage found in the nose and some supporting joints of the horse (存在于鼻腔和其他支持关节中的一类软骨)纤维软骨

fibro-fatty /fɑɪbrəuˌfæti/ adj. consisting of muscular fibers and fat 脂肪纤维的: fibrofatty tissue 脂肪纤维组织

fibrosis /faɪˈbrəʊsɪs/ n. an abnormal accumulation of excessive fibrous tissue in certain organs (机体器官中纤维组织异常增生的情况)纤维化 | **fibrosic** /-ˈbrɒsɪk/ adj.

fibrous /ˈfaɪbrəs/ adj. containing, consisting of, or resembling fibers 纤维的，含纤维的；纤维状的

fibula /ˈfɪbjələ/ n.【anatomy】(*pl.* **fibulae**, or **fibulas**) a long, slender bone of the lower hindleg extending from the stifle joint to the hock (后肢位于膝关节和跗关节之间的长骨)腓骨

fibular /ˈfɪbjʊlə(r)/ adj. of or relating to the fibula 腓骨的: head of fibula 腓骨头

fibularis /fɪbˈjʊlærɪs/ adj.【Latin】fibular 腓骨的；腓侧的

fiddle[1] /ˈfɪdl/ n.【slang】whip 马鞭

fiddle[2] /ˈfɪdl/ vi. to alter or meddle aimlessly or in a dishonest way 随意更改，背地更换: Generally speaking, it is better not to fiddle with the horse's feed too much and often, you may just cause a number of nutritional problems; although necessary adjustment of diet is also very important to the well-being of horses depending on its condition and the work he is subjected to. 一般来说，根据马的体况和负役种类对马匹日粮做适度调整对于马的健康是非常重要的；但是，如若随意过频更换日粮将会引发许多代谢性疾病。

fiddle head n. a large, coarse head of a horse 提琴头

fidget /ˈfɪdʒɪt/ vi. to behave or move nervously or restlessly 烦躁不安

fidgety /ˈfɪdʒɪti/ adj. restless and nervous 站立不安的；烦躁不安的: a fidgety horse 一匹站立不安的马

field /fiːld/ n. **1**【racing】the entire group of horses entered in a race (全体)参赛马匹；参赛马数: In his record-breaking win in the 1973 Belmont Stakes, Secretariat left the field 76m behind him. 在破纪录夺得 1973 年贝尔蒙特大赛冠军时，"秘书处"将其他参赛马匹甩开了 76 米的距离。**2** *see* pari-mutuel field 赌马场 **3**【hunting】the hunt field 猎场 PHR V **reverse field** (of hounds) to change the direction traveled in pursuit of a fox (猎犬)改变追踪方向 **4**【hunting】(also **hunt field**) the mounted followers led by a Field Master 猎场随从，猎狐队员 **5** a piece of land suitable for farming or pasture (适合耕作或放牧的土地)田野，土地

field boarder n. a horse kept and grazed in a pasture for a fee 牧场寄养马

field boots n. a tall riding boot made of either black or brown leather (由黑色或褐色皮革制成的高帮靴)牧场马靴，牧靴

Field Master n.【hunting】one appointed by the hunt committee to lead and control the field (由狩猎委员会任命带领和管理猎场的人)猎场主 syn master of the pack

fig /fɪg/ n.【colloq】the physical condition or figuration of a horse (马的)体况，体貌
v. (of a horse) to trot with lively tail carriage (指马)昂扬运步

figging /ˈfɪgɪŋ/ n. *see* gingering 肛门塞姜

figure[1] /ˈfɪgə(r)/ n.【racing】the handicapper's rating number that identifies the winning chance of a horse（评磅员对赛马获胜可能进行的估计）评分，评估

figure[2] /ˈfɪgə(r)/ vi.【racing】(of a horse) to have the chance of winning（指马）有望胜出

figure eight n.【dressage】a presentation in which the horse moves in the shape of 8（盛装舞步中的）8-字运步

figure eight bandage n.（also **figure eight wrap**）a way to wrap the leg bandage（缠护腿绷带的方式）8-字形缠裹

figure eight noseband n. *see* grackle noseband 8-字形鼻革，X 形鼻革

figure eight wrap n. *see* figure eight bandage 8-字形缠裹

file[1] /faɪl/ n. a line of animals positioned one behind the other 纵列，纵队

vi. to march or walk in a line 纵列行进

file[2] /faɪl/ n. a hardened steel tool with cutting ridges for smoothing or reducing a surface（用来磨边或打光表面的金属工具）锉刀

vt. to smooth or rasp with a file 锉平，锉光：to file teeth 锉牙 | **filing** /ˈfaɪlɪŋ/ n.

filiform /ˈfɪlɪfɔːm/ adj.【physiology】having the form of or resembling a thread or filament 丝状的：filiform papillae 丝状乳头

fillet strap n. *see* fillet string 马衣固定带

fillet string n.（also **fillet strap**）a string or strap passed under the horse's tail which connects the two rear flaps of a rug or horse blanket to hold it in place and prevent it from slipping（系在马衣后端两侧的细绳，用来固定马衣以防滑落）马衣固定带，马衣系绳 [syn] rug fillet

Fillis, James n.（1850－1900）a 19th century influential British riding master who wrote *Breaking and Riding*（19 世纪英国颇具影响力的骑术大师，著有《调教与骑乘》等书）詹姆士·菲力斯

filly /ˈfɪli/ n. a female horse under three years old（3 岁以下的母马）雌驹 [compare] colt

film patrol n.【racing】the crew that records a horse race on film or tape（赛马）摄制组

filum /ˈfaɪləm/ n.【anatomy】(*pl.* **fila**/-lə/) a thread-like structure; filament 丝，细丝：terminal filum 终丝

fimbria /ˈfɪmbrɪə/ n.【anatomy】(*pl.* **fimbriae** /-brɪiː/) a funnel-shaped structure with fringes, as at the opening of the fallopian tubes（带穗的漏斗状结构，如输卵管开口处）输卵管伞：ovarian fimbria 卵巢伞 ◇ fimbriae of the oviducts 输卵管伞 ◇ After shed from the follicle, the egg enters the fimbriae of the oviduct and travels to the isthmus where it is met by a swarm of spermatozoa and fertilized by one. 从卵泡中排出后，卵子落入输卵管伞并移行到峡部，在那里与成群的精子相遇并与其中一个受精。| **fimbrial** /-rɪəl/ adj.

finish /ˈfɪnɪʃ/ n. **1**【racing】the terminal or end of a race 终点 **2** a smooth surface texture; polish 光泽；涂饰，抛光

vt. to reach the end of a race; complete 跑到终点，获得：The colt finished third in his debut race. 这匹雄驹首次出赛跑了第三名。

PHR V **finish fast**【racing】(of a horse) to gain on the leader on the last stretch of a race（指马在最后直道上逼近头马）奋力直追，终点夺冠

finish line n.【racing】a red ribbon tied between two posts across the racetrack that marks the end of a race（系在赛道两侧立杆上的红带，用来标示比赛的终点）终线，终点线

finish post n.【racing】the post on which a red ribbon is tied to mark the end of a race（用来系红带以标示比赛终点的立杆）终点杆 [syn] winning post

Finnish /ˈfɪnɪʃ/ n. *see* Finnish Horse 芬兰马：Finnish horse is the only native breed of horse in Finland. 芬兰马是芬兰唯一的本土马种。

Finnish Draft n.（BrE **Finnish Draught**）the heavy type of Finnish horse 挽用芬兰马

F

Finnish horse n. (also **Finnish**) a Finnish cold-blood historically bred in the heavy and the light-heavy types; descended from native ponies crossed with many imported warm and cold-blood types; stands 14.3 to 15.2 hands and usually has a chestnut coat with white markings, although brown, bay, or black coats also occur; is strong with good bone, docile, good-natured, and hard working; is a good all-round horse used for driving, riding, and trotting (源于芬兰的冷血马品种,历来育有重挽和轻驾两个类型,主要由当地矮马引入诸多温血马及冷血马杂交培育而来。体高 14.3～15.2 掌,被毛以栗为主且带白色别征,但褐、骝、黑也较为常见,体格强健,管围较粗,性情温驯,工作能力强,为优秀的兼用型马种,主要用于驾车、骑乘和快步赛)芬兰马—*see also* **Finnish Draft**, **Finnish Universal**

Finnish Universal n. the light-heavy type of Finnish horse 兼用芬兰马

FIP abbr. Federation of International Polo 国际马球联合会

fire /ˈfaɪə(r)/ n.【racing】a burst of speed by a horse running in a race (赛马的)猛冲,冲刺

fire brand n. *see* hot brand 火烙印记

fireman /ˈfaɪəmən; *AmE* ˈfaɪərmən/ n. a blacksmith or farrier who hot shoes horses 热装蹄铁匠

firing /ˈfaɪərɪŋ/ n. a method of producing scar tissue and hastening reparative process by burning the skin over an injury or wound (烫烧伤口使之结痂以加速修复的治疗方法)火疗法,火烫法

firm track n.【racing】a firm turf track on which horses run faster than other surfaces (表面坚实的草皮赛道,马匹在这种赛道上速度较快)坚实赛道

first incisors n. *see* central incisors 第一切齿,中切齿

first jockey n.【racing】the principal jockey hired by an owner or trainer to ride his racehorses in competition (马主或练马师优先雇佣参加比赛的骑手)首要骑师,首席骑师

first leg n.【racing】the first half of a double event (一场比赛中的前半部分)上半场,前场

first milk n. *see* colostrum 初乳

first phalanx n. *see* long pastern bone 第一趾骨,长系骨

first premolars n. *see* wolf teeth 第一前臼齿,犬齿

first-aid /ˌfɜːstˈeɪd; *AmE* ˌfɜːrst-/ adj. serving for emergent circumstances 急救的:first-aid kit 急救箱 ◇ first-aid procedure 急救措施

fish eye n. *see* wall eye 玻璃眼,玉石眼

fishmeal /ˈfɪʃmiːl/ n.【nutrition】a protein supplement produced from crushed fish (干鱼粉碎后制成的蛋白添加料)鱼粉: Though a good source of amino acids, calcium, and phosphorus, fishmeal is seldom fed to horses because of cost and lack of palatability. 尽管鱼粉是氨基酸、钙和磷的良好来源,但由于成本原因且适口性差,故很少用来喂马。

fistulous withers n. inflammation of the bursae of the withers characterized by swelling and lesion of skin (鬐甲处黏液囊发炎的症状,表现为肿胀和溃烂)鬐甲瘘

fit[1] /fɪt/ adj. (of a horse) in good physical condition (指马)体况好的,膘情好的

fit[2] /fɪt/ vt.【farriery】to shoe a horse (给马)装蹄,钉掌

fitness /ˈfɪtnəs/ n. the state or condition of being fit 体况良好,健康;适应性:fitness training 适应性训练

five-eighths pole n.【racing】a marking pole located on the inside rail exactly five furlongs from the finish line (立于赛道内栏的标杆,与终点相距 1 000 米)5/8 英里杆,一千米杆

five-gaited adj. (of a horse) capable of performing the stepping pace and rack in addition to the gaits of walk, trot, and canter (指马除了走慢步、快步和跑步以外还善走高蹄对侧步和单蹄快步)五步态的:a five-gaited horse 五步

态马

fixed head n. *see* top pommel 上鞍桥

fixed race n. a horse race in which the winner has been determined in advance（赛前头马既已确定的比赛）黑哨赛：Though judged illegal by the Jockey Club, fixed race does occur even today. 尽管英国赛马会规定黑哨赛是违法的，但即便今天也仍有出现。 syn boat race

Fjord /fjɔːd; *AmE* fjɔːrd/ n. *see* Norwegian Fjord Horse 挪威峡湾马

flaccid /ˈflæ(k)sɪd/ adj. (of muscles) not firm; slack（肌肉）松弛的

flag /flæg/ n. **1** a piece of cloth attached at one edge to a staff and used as a sign, symbol, or signal（一块单边和木杆相连的织物，用来起标识或警示作用）旗，旗子：red flag 红旗 ◇ white flag 白旗 ◇ to jump between the flags 从两旗间跳过障碍 **2**【vaulting】a compulsory exercise in individual vaulting in which the astride performer hops to his or her knees, extends the right leg straight out behind parallel to the horse's spine and stretches the left arm straight forward（单人马上体操规定动作，其中运动员双膝移至马背后，右腿后展与马背平行并向前伸开左臂）迎风展臂—*see also* **basic seat, flank, mill, scissors, stand**

vi. to place a flag or use it to signal 挥旗

PHR V **flag down**【racing】to signal an exercise rider with a flag to slow down 挥旗减速

flagman /ˈflægmən/ n.【rodeo】one who signals the end of a timed event with a flag（挥旗示意计时项目时间已到的人）执旗仲裁，执旗裁判

flak jacket n. (also **flak vest**) a vest made of heavy fabric worn by riders for protection against injury（骑手穿的厚料背心，可保护身体免受伤害）防护背心，防护马甲

flak vest n. *see* flak jacket 防护背心，防护马甲

flake /fleɪk/ n. a flat, thin piece or layer; a chip 薄片；屑

vi. to come off in flat, thin pieces or layers; chip off 剥落；掉屑

flaky /ˈfleɪki/ adj. **1** made of or resembling flakes 层状的 **2** (of hoof) forming or tending to form flakes（指马蹄）分层的，剥落的：Dry, flaky and brittle hoofs can be an indication of dietary imbalance, although the effects of shoeing and wearing can not be ignored. 马蹄干而脆且易分层是日粮不平衡的一种指征，但钉蹄和磨损的影响也不容忽视。

Flanders horse n. *see* Brabant 弗兰德斯马，布拉邦特马

flank[1] /flæŋk/ n. **1** the part of a horse's body between the ribs and the hip; the side（体侧从最末肋骨至腰角的部分）腹侧，肷 **2**【vaulting】a compulsory exercise in individual vaulting in which the performer swings both legs forward then backward（个人马上体操规定动作，选手前后摆动两腿）侧摆—*see also* **basic seat, flag, mill, scissors, stand**

flank[2] /flæŋk/ vt.【roping】to throw a calf to the ground by hand（将牛）放倒，撂倒：After the calf is flanked, the roper ties any three legs together with a pigging string. 在牛被放倒后，套牛者用一根捆蹄绳将牛的三个蹄子捆在一起。| **flanking** n.

flap /flæp/ n. *see* saddle flap 鞍翼

flapper /ˈflæpə(r)/ n.【racing】a horse who runs in an unofficial race 跑非正式比赛的马

flapping /ˈflæpɪŋ/ n.【racing】(also **flapping meeting**) a race meeting not authorized by the Jockey Club（未得到赛马会授权举行的比赛）非官方赛马，黑市赛

flapping meeting n. *see* flapping 非官方赛马，黑市赛

flare /fleə(r); *AmE* fler/ n. **1** an outward curving or distortion of the hoof wall 蹄壁外张 **2**【vaulting】a compulsory movement performed in vaulting competition（马上体操动作）跪立展臂

flask /flɑːsk/ n. a small, flat metal container for liquor 小酒壶

flat /flæt/ adj.【racing】(of a track) having an even, level surface 平地的:flat race 平地赛 n.【BrE】*see* dray 平板马车

flat bone n. the cannon bone that appears wide and flat when viewed from the side (侧看时马管骨宽而扁平)管骨扁平 compare round bone

flat canter n. a canter lacking sufficient suspension between strides 悬空期过短的跑步

flat foot n. a foot lacking normal concavity in the sole (底部无正常弧度的马蹄)平蹄:The flat feet are often present in some draft breed and much common in the fore than hind feet. 平蹄常见于一些重型马匹品种,且前蹄较后蹄更为普遍。

flat racehorse n. a horse competing in flat racing 平地赛马

flat racing n.【racing】a horse race conducted on flat ground without obstacles, hurdles, or fences (在没有篱笆、跨栏和障碍的平地上举行的比赛)平地赛 syn racing on the flat

flat-sided /ˌflætˈsaɪdɪd/ adj. (of a horse) having a narrow chest or ribcage 狭胸的,肋部浅平的 syn slat-sided, slab-sided

flatulence /ˈflætjʊləns; *AmE* -tʃə-/ n. a build-up of excessive gas in the digestive tract (消化道内积气过多的症状)肠胃胀气,肠胃臌气

flatulent /ˈflætjʊlənt; *AmE* -tʃə-/ adj. of, afflicted with, or caused by flatulence 肠胃胀气的;胀气引起的:The flatulent colic usually occurs because the intestines are overloaded with rich food, leading to gas build-up. 胀气性腹痛常由于肠内食糜淤积过多而产生大量的气体所致。

flatwork /ˈflætwɜːk/ n. exercises performed on even ground 平地训练:flatwork exercises 平地训练

flax /flæks/ n. a widely cultivated plant of the genus *Linum*, having blue flowers, seeds that yield linseed oil, and slender stems from which a textile fiber is obtained (亚麻属植物,种植广泛,花蓝色,种子可榨油,茎皮可做纤维)亚麻

flaxen /ˈflæksən/ adj. having the pale grayish-yellow color of flax fiber 浅黄的,亚麻色的:a horse with flaxen mane and tail 一匹鬃尾淡黄的马

flaxen chestnut n.【color】a chestnut coat with a significantly lighter mane and tail (体色为栗色,但鬃尾颜色较淡的毛色)金栗毛 compare light chestnut

flaxseed /ˈflæksiːd/ n. the seed of flax; linseed 亚麻籽

flea-bitten /ˈfliːˌbɪtn/ adj. **1** covered with fleas or fleabites 满身跳蚤的;被跳蚤咬的 **2**【color】(of a horse) having reddish-brown speckles over the entire coat (指马全身被毛生有褐色斑点)带蚤斑的,带斑点的

fleabitten gray n.【color】a light gray coat with little speckles of black and/or brown (青色被毛中生有黑色或棕色斑点)斑点青

fleck /flek/ n.【color】a small spot or flake; speckle 斑点

flecked /flekt/ adj. randomly distributed with irregular spots of white hairs throughout the coat (被毛中散生不规则白色斑点的)有斑点的:flecked blanket 斑点披毯型

flecked roan n. *see* sabino 散沙毛

Flehmen posture n. (also **Flehmen**) the posture taken by a sexually aroused stallion when courting a mare in which the stallion stretches its head and neck with the upper lip curling upward (发情公马追求母马时头颈伸直、上唇外翻的姿势)性嗅姿势,反唇姿势

Flehmen response n. the physiological reaction of the horse to certain odors or sexual arousal characterized by the Flehmen posture (马对某种气味或性冲动时做出的仰天姿势)性嗅反射

flesh /fleʃ/ n. **1** the muscles of a horse 肌肉;膘情 PHR V **lose flesh** (of a horse) to lose weight due to diseases or poor nutrition (指马由于疾病或营养不良)消瘦,掉膘 **2**【hunting】the meat fed to hounds (喂猎犬的)肉

flesh hovel n.【hunting】a room where meat is hung until fed to the hounds（悬挂喂猎犬所用肉的房间）挂肉棚，挂肉间

flesh mark n.【color】a patch of skin lacking pigment（皮肤上缺少色素的斑块）肉色斑

flex /fleks/ vt. to bend (a joint) 屈曲（关节），弯曲

flexible /ˈfleksəbl/ adj.（of a horse） capable of bending freely 柔韧的，有弹性的，可屈伸的

flexibility /ˌfleksəˈbɪləti/ n. the condition or state of being flexible 柔韧性，屈伸

flexion /ˈflekʃn/ n.【anatomy】（also **flection**） the bending of a joint or limb in the body by the action of flexors（身体关节或四肢在屈肌作用下的弯曲）屈体，屈曲 | **flexional** /-nl/ adj.

flexion test n.【vet】a test the veterinarian performs to diagnose lameness（关节）屈曲检测

flexor /ˈfleksə(r)/ n.【anatomy】（also **flexor muscle**） a muscle in the body that acts to bend a joint or limb when contracted（收缩时可弯曲身体关节或四肢的肌肉）屈肌 compare extensor

flexor tendon n. a tendon at the back of the horse's leg; back tendon 屈肌腱：deep digital flexor tendon 指深屈肌腱 ◇ superficial digital flexor tendon 指浅屈肌腱 ◇ flexor tendon deformities 屈腱炎

flexura /ˈfleksjʊrə/ n.【Latin】flexure 弯曲

flexure /ˈflekʃə/ n.【anatomy】a curve, turn, or fold, such as a bend in a tubular organ（管状器官的弯曲处）弯曲，皱褶：pelvic flexure 骨盆曲 ◇ sternal flexure 胸骨曲

flier /ˈflaɪə/ n.【vaulting】（also **flyer**） a performer in vaulting who is supported by others in the air（马上体操中由队友托于空中的表演者）空中飞人，马上飞人

flight /flaɪt/ n.【jumping】the act or process of moving through the air in jumping（跳跃时人马在空中的滑行）腾空

flipper foot n. a horse's hoof with its toe overgrown and flared（蹄尖生长过长且向上翘起）蹼状蹄 syn elve's foot, elf's foot

float /fləʊt; *AmE* floʊt/ n. a two-wheeled, horse-drawn low cart pulled by a single horse（单马套拉的双轮低帮马车）平板马车，低帮马车
vt. to file off the sharp or uneven edges a horse's teeth by using a rasp（锉平马牙齿边缘锋棱）锉掉，锉平：A horse's teeth should be floated at least once every year. 马的牙齿应该每年锉整一次。| **floating** n.

flog /flɒg; *AmE* flɑːg/ vt. to beat or strike severely with a whip or rod 鞭打，鞭笞 IDM **flog a dead horse** to waste one's efforts in a fruitless undertaking 鞭打死马；枉费力气

flooring /ˈflɔːrɪŋ/ n. the material used in making floors; floor 地板；地板材料：The most commonly found flooring materials under stalls are concrete, wood, clay, sand, asphalt and brick. 厩舍最常见的地面材料有混凝土、木板、黏土、沙子、沥青和砖块。

flower /ˈflaʊə(r)/ n. the engorgement of penis with blood to increase in size（阴茎）充血；充胀

fluke /fluːk/ n. any of several trematode flatworms that affect horses and other livestock（寄生于马和其他家畜身上的多种扁形吸虫）片吸虫

flute bit n. a bit in which the straight-bar mouthpiece is perforated with holes（杆上钻有孔洞的衔铁）带孔衔铁，笛杆衔铁：Flute bit is often used on wind-suckers. 笛杆衔铁多用在有咽气恶习的马上。

flutter /ˈflʌtə(r)/ v.（of one's heart） to beat very quickly and irregularly（指心脏）乱跳，悸动：a fluttering pulse 悸动的脉搏
n. **1**【BrE, colloq】a small bet 小额赌注：to have a flutter on the horses 在赛马上小额投注 **2** a rapid and irregular heartbeat（心脏快速而无规律的跳动）颤动，悸动

fly[1] /flaɪ/ n. any of a large group of two-winged insects including the horsefly, stable fly, housefly, and deerfly (双翅目多种昆虫,包括马蝇、厩螫蝇、家蝇和鹿虻)蝇

fly[2] /flaɪ/ vi.【jumping】(of a horse) to jump over a fence without touching the obstacle (指马)腾空跃起: to fly over the obstacle 腾空越过障碍

fly bonnet n. a crocheted hood covering the ears and the forehead of horse to protect it from flies and gnats (戴在马头上的针织护罩,可驱除苍蝇、蚊虫的骚扰)驱蝇罩 syn fly cap

fly cap n. *see* fly bonnet 驱蝇罩

fly fence n. *see* flying fence 腾空跨过的障碍

fly mask n. a covering over a horse's face worn for protection from flies and gnats (用来防止蝇虫叮咬的面罩)驱蝇面罩

fly sheet n. a thin covering for the body of the horse, used in the summer months to protect the horse from flies (马夏天罩在身上用来防止蝇虫叮咬的轻薄外套)防蝇马衣

fly spray n. an insecticide used to kill or repel flies and other summer-time pests (用来杀死或夏季蝇虫的药物)灭蝇灵,驱蝇喷雾剂

fly terret n. a small ring attached to the browband of the bridle, used for decoration on heavy harness horses (马勒额革上系的环形饰物,主要用作挽具装饰)驱蝇摆饰

fly whisk n. (aso **fly wisk**) a tool made of horsehair attached to a wooden handle, often used to keep flies off horses by riders (用马尾制成的带柄长刷,常用来驱赶马体上的苍蝇)蝇刷;拂尘: In Chinese culture, the fly-whisk is frequently seen as an attribute of both Daoist and Buddhist deities. 在中国文化中,拂尘常被视为道家和佛家的一件法器。

fly wisk n. *see* fly whisk 驱蝇刷;拂尘

flying angel n.【vaulting】a freestyle exercise in which the flier is supported by another vaulter at the shoulders in a horizontal position with arms and legs outstretched (马上杂技中的自选动作,空中表演者被队友从肩托举至空中,两臂开展)展翅天使,天使展翼

flying change of leg n. *see* flying lead change 空中换腿,空中换肢

flying coach n. a horse-drawn passenger coach which traveled between London and Exeter, York and Chester, Great Britain in four days time (过去往返于英国伦敦、埃克塞特、约克和切斯特之间的客用马车,历时4天)快速马车,快客马车

flying fence n.【jumping】(also **fly fence**) an obstacle or fence that can be cleared at a gallop 可腾空跨过的障碍

flying lead change n. an advanced movement in which the horse changes leading leg at the canter during the moment of suspension (跑步悬空期中马匹更换领先肢的情况)空中换领先腿,空中换领先肢 syn flying change of leg, change in the air, change of leg in the air, change of leg at the canter

flying mount n. *see* running-quick mount 飞身上马

foal /fəʊl; *AmE* foʊl/ n. **1** a young horse or donkey of either sex from the time of birth until weaning (从出生至断奶期间的马驹或驴驹)[幼]驹: an orphan foal 无母幼驹 IDM **in foal**(of a mare) pregnant (母马)怀驹的: a mare in foal 怀孕母马 **with foal at foot** (of a mare) giving birth to a live foal (指母马)产驹的,带驹的 **2** the offspring to either a male or female parent 子女,后代: He was the first foal of doctor Cage. 他是凯奇医生的第一个儿子。vt. to give birth to a foal 产驹,下驹: Eclipse was foaled in 1764. 名马"日蚀"生于1764年。syn drop

foal heat n. the estrus period occurring approximately nine to eleven days after foaling, during which the mare shows signs of heat (母马产驹后9~11天出现的发情迹象)产后发情 syn nine-day heat

foal heat scours n. diarrhea that occurs in the mare seven to nine days after foaling (母马产驹后7~9天出现的腹泻)产后腹泻

foaling /ˈfəʊlɪŋ; *AmE* ˈfoʊ-/ n. instance or process of giving birth to a foal 产驹，分娩：difficult foaling 难产 ◇ foaling box 产驹舍 ◇ foaling rate 产驹率 ◇ foaling stall 产驹栏

focus /ˈfəʊkəs; *AmE* ˈfoʊ-/ n.【vet】the frequently affected region of a bodily infection or disease （机体感染或发病的主要部位）病灶

fodder /ˈfɒdə(r); *AmE* ˈfɑːd-/ n. the feed for horses and farm animals, especially chopped hay or straw （饲喂家畜的草料，尤指铡草或干草）草料，饲草

vt. to feed with fodder 饲喂 | **foddering** n.

foddering time n. *see* feeding time 喂料时间，饲喂时间

foetal /ˈfiːtl/ adj.【BrE】fetal 胎儿的

foetus /ˈfiːtəs/ n.【BrE】fetus 胎儿

foil /fɔɪl/ vt.【hunting】(of a hunted animal) to obscure or confuse a trail or scent so as to evade pursuers （猎物为逃避追捕将踪迹搅乱）破坏，搅乱：The fox may foil its own scent by crossing over its tracks. 狐狸会来回穿行以此搅乱自己留下的气味踪迹。

foiled /fɔɪld/ adj. (of the scent) obliterated or confused by livestock which cross its line （指猎物踪迹）被搅乱的，被破坏的

foiled line n.【hunting】the line of a fox when confused by the scent of livestock 被搅乱的踪迹，被破坏的踪迹 syn stained line

folacin /ˈfɒləsɪn/ n.【nutrition】an intestinally synthesized B-complex vitamin necessary for cell metabolism and normal blood formation, deficiency is characterized by anemia and poor growth.（由肠道微生物合成的维生素B复合物，是细胞代谢和血液生成所必需的物质，缺乏表现为贫血和发育不良）叶酸 syn folate, folic acid

folate /ˈfəʊleɪt; *AmE* ˈfoʊl-/ n. *see* folacin 叶酸

folding head n. *see* folding hood 折叠式顶篷

folding hood n.【driving】(also **folding head**) a convertible or folding leather top found on some horse-drawn vehicles 折叠式顶篷 syn falling top

folic acid n.【nutrition】*see* folacin 叶酸

follicle /ˈfɒlɪkl; *AmE* ˈfɑːl-/ n. **1** (also **ovarian follicle**) a small sac or cavity in the ovary that contains a developing ovum surrounded by its encasing cells（卵巢上的小囊腔，内含被滋养层细胞包围且处于发育期的卵细胞）卵泡：primordial follicle 原始卵泡 ◇ primary follicle 初级卵泡 ◇ secondary follicle 次级卵泡 ◇ mature follicle 成熟卵泡 **2** a small anatomical cavity or depression; sac 囊；滤泡：hair follicle 毛囊 ◇ thyroid's follicle 甲状腺滤泡 ◇ follicle cell 滤泡细胞

follicle-stimulating hormone n.【physiology】(abbr. **FSH**) a protein hormone formed in the anterior lobe of the pituitary gland that stimulates the growth and development of follicles in the ovary and activates sperm-forming cells（由垂体前叶所分泌的蛋白质激素，可促进卵泡的生长发育并刺激精子细胞生成）促卵泡素，卵泡刺激素

follicular /fəˈlɪkjʊlə(r)/ adj. of or relating to a follicle or follicles 卵泡的；滤泡的：follicular atresia 卵泡闭锁 ◇ follicular antrum 卵泡腔 ◇ follicular cyst 卵泡囊肿 ◇ follicular fluid 卵泡液 ◇ follicular hormone 卵泡激素 ◇ follicular stigma 卵泡斑，排卵点

follicular stage n. the period of the estrous cycle, including proestrus and estrus, during which follicles are formed and the mare becomes receptive to the stallion （发情周期中包括前情期和动情期的时段，这时卵泡逐渐发育成熟，母马交配欲望强烈）卵泡期

followers /ˈfɒləʊərz/ n.【hunting】*see* field 猎场随从

foment /fəʊˈment; *AmE* foʊˈm-/ vt. to apply hot and moist cloth to the body to lessen pain or discomfort（用湿热棉布敷在体表以减轻疼痛）热敷，热罨

fomentation /ˌfəʊmenˈteɪʃn; *AmE* ˌfoʊm-/ n. **1** the application of hot, moist materials to the body to ease pain 热敷，热罨 **2** anything used for fomenting 热敷物，热罨剂

font /fɒnt; *AmE* fɑːnt/ n. (also **fount**) **1** an abundant source of water 水源，泉源 **2** a basin for holding water 饮水盆：stock font 牲畜饮水处

food wagon n. *see* chuck wagon 野餐四轮马车

foot /fʊt/ n. **1** the lower extremity of a horse's leg; hoof（马四肢的最末端）蹄：The old adage "no foot, no horse" should be always kept in mind when selecting a horse. 选马时，"无蹄则无马"的谚语要牢记在心里。**2** the way or manner of moving; step 步子，步法：to walk with a light foot 步法较轻 **3** (abbr. **ft**) a unit for measuring length equal to 12 inches or 30.48 centimeters（长度测量单位，相当于12英寸或30.48厘米）英尺

foot mange n.【vet】*see* chorioptic mange 足螨病

foot rest n. *see* footboard 踏脚板

footboard /ˈfʊtbɔːd/ n.【driving】(also **foot rest**) a board or platform of a horse-drawn vehicle on which the driver and passengers rest their feet（马车上供车夫和乘客放脚的木板）踏脚，脚踏板

footboard lamp n.【driving】a small lamp attached to the footboard of a horse-drawn vehicle for illumination（马车脚踏板上安装的照明车灯）马车灯；马灯

footfall /ˈfʊtfɔːl/ n. the sound of a footstep or footsteps 蹄音

footing /ˈfʊtɪŋ/ n. the surface condition of a track or arena（赛场的）地况：The footing was wet and dangerous. 赛场地况潮湿而危险。

foot-lock /ˈfʊtlɔk/ n. the long hairs around the fetlock of certain horses; feathering 距毛

foot-stool /ˈfʊtstuːl/ n. **1**【farriery】a wooden block upon which a farrier rests the foot of the horse in fitting the horseshoe（蹄铁匠装蹄时用来放置马蹄的木墩）装蹄台 [syn] shoeing block **2**【driving】a low wooden box of a horse-drawn carriage upon which passengers rested their feet（马车上安置的矮凳，乘客可借此歇脚）脚凳

footwear /ˈfʊtweə(r); *AmE* -wer/ n. outer coverings for the feet, such as shoes or boots 鞋，靴

forage /ˈfɒrɪdʒ/ n. the grass and grains fed to horses; fodder 饲料，饲草

vi. to wander in search of food; graze 觅食，采食

forage poisoning n. *see* botulism 肉毒杆菌中毒，饲草中毒

forbidden substance n. any illegal drug or substance that might affect the performance of the horse, such as stimulant, depressant, tranquilizer, or local anesthetics（可影响马匹成绩的任何制剂）违禁药物，违禁物品

fore /fɔː(r)/ n. the front part of a horse（马体）前部，前躯

fore- pref. before or in front of 前面的

forearm /ˈfɔːrɑːm; *AmE* ˈfɔːrɑːrm/ n. the upper part of the foreleg above the knee（马前肢以上的部分）前臂

forecarriage /fɔːˈkærɪdʒ/ n. the front section of a horse-drawn carriage（马车的前部）前架

forecast /ˈfɔːkɑːst; *AmE* ˈfɔːrkæst/ n.【racing】the bookmaker's odds on each horse running in a race as based on past performance（赌马经纪人根据马匹以往赛绩估测的赔率）估算赔率，预测赔率

forefoot /ˈfɔːfʊt; *AmE* ˈfɔːrfʊt/ n. either of the front feet of a horse（马的）前蹄 [syn] front foot

forefooted /fɔːˈfʊtɪd/ adj.【roping】(of a steer) roped by the front foot（指牛）套住前蹄的

forehand /ˈfɔːhænd; *AmE* ˈfɔːrh-/ n. **1**【polo】a way of striking the ball in the direction of the travel（朝运行的方向击球）正向[击球]，正手[击球]：forehand shot 正手击球 [compare] backhand **2** the front section of a horse's body including the head, neck, shoulders and forelegs（马体包括头、颈、肩和前肢的部分）前躯 [syn] front, forequarters [IDM] **on its forehand**

(of a horse) placing most of its weight on its forehand (指马)前躯负重过多

forehead /ˈfɔ:hed; *AmE* ˈfɔ:rhed/ n. the part of horse's face above the eyes (马面部位于两眼之上的部分)前额

forehead drop n. (also **forehead piece**) a decorative piece of the show harness which lies on the horse's forehead (展览赛挽具上位于马前额上的饰件)额前垂饰 [syn] drop, face drop, face piece

forehead piece n. *see* forehead drop 额前垂饰

foreleg /ˈfɔ:leg; *AmE* ˈfɔ:rleg/ n. (also **front leg**) either one of the two front limbs of a horse below the knee (马前膝以下的部分)前肢

forelimb /ˈfɔ:lɪm; *AmE* ˈfɔ:rlɪm/ n. either one of the front legs of a horse (马的)前腿;前肢

forelock /ˈfɔ:lɒk; *AmE* ˈfɔ:rlɑ:k/ n. the part of a horse's mane extending between its ears and falling onto the forehead; foretop (生于两耳间垂在前额上的鬃毛)额发;门鬃 [compare] bang

foremilk /ˈfɔ:mɪlk; *AmE* ˈfɔ:rm-/ n. *see* colostrum 初乳

forequarters /ˈfɔ:rkwɔ:təz/ n. the forehand of a horse (马的)前躯体 [compare] hindquarters

Forest /ˈfɒrɪst; *AmE* ˈfɔ:r-/ n. *see* New Forest Pony 新福里斯特矮马

foretop /ˈfɔ:tɒp/ n. the forelock of a horse (马的)额发,门鬃

forewale /ˈfɔ:weɪl/ n.【driving】the foremost rim of a collar 颈圈前缘

forfeit /ˈfɔ:fɪt; *AmE* ˈfɔ:rfət/ n.【racing】the sum of entry fee that is not refundable to the owner if the horse is withdrawn from competition (赛马取消比赛资格后不予退还的参赛费)损失,丧失

vt. **1** to withdraw a horse from competition; scratch 取消比赛资格 **2** to sacrifice an entry fee by withdrawing a horse from a race (因取消赛马参赛造成的报名费损失)损失,丧失

forfeit list n.【racing】a list of horses who are announced ineligible to compete in a race (所宣布的无参赛资格的马匹名单)禁赛名单

forge[1] /fɔ:dʒ; *AmE* fɔ:rdʒ/ n. **1** a farrier's workshop where iron is heated and shaped into horseshoes and other metal pieces; smithy 铁匠铺 **2** a furnace where horse shoes and other metal pieces are heated or wrought (用来加热锻造蹄铁等金属部件的火炉)锻炉,熔炉

vt. to shape horseshoe or other metal object by heating and hammering 锻造(蹄铁)

forge[2] /fɔ:dʒ; *AmE* fɔ:rdʒ/ vi. (of a horse) to move with increased speed and power, as in the homestretch of racetrack (指赛马)稳步加速: Man O' War forged ahead, leaving other horses behind. "战神"加速跑到了最前面,将其他马匹远远抛在后面。

forging /ˈfɔ:dʒɪŋ; *AmE* ˈfɔ:rdʒɪŋ/ n. a gait fault in which the toe of the hind shoe strikes against the sole of the forefoot on the same side (马匹运步缺陷,后肢蹄铁头撞击同侧前肢蹄底)追突: Forging may be recognized clearly by the clicking noise with each stride as the hind foot strikes the sole of the fore foot in shod horses. 钉掌的马如发生追突,则在每步中都可清晰听到后蹄撞击前蹄底的咔嗒声。[syn] click

forhoss /ˈfɔ:hɔ:s; *AmE* ˈfɔ:rhɔ:s/ n.【dated】the leader in a team when driven at length (马纵队中的)首马,头马

fork /fɔ:k; *AmE* fɔ:rk/ n. **1** (also **pitchfork**) a long-handled tool used for pitching hay, etc. 铁叉: two-tined fork 双股叉 ◇ three-pronged fork 三股叉 **2** the saddle fork (马鞍的)鞍叉

vt. to sit on a horse with legs astride; bestride, straddle 叉骑,跨骑: The sidesaddle was still being used as late as in the 1920s, when proper women did not fork a horse. 侧鞍的使用一直持续到20世纪20年代,因为过去端庄的妇女不跨骑马匹。

form /fɔ:m; *AmE* ˈfɔ:rm/ n. **1**【racing】the past performance of a racehorse (赛马以往的)出赛记录 **2** *see* daily racing form 每日赛事预告

3【hunting】a hollow in the ground where a hare rests and hides (野兔的)洞穴

form sheet n.【racing】a list of horses competing in a race and their past racing performances (记录参赛马匹及其以往赛绩的表格)出赛表,出赛单

F

formula /ˈfɔːmjələ/ n. (*pl.* **formulae** /-jəliː/) a method or fixed procedure of doing sth 程序;配方

formulate /ˈfɔːmjuleɪt; *AmE* ˈfɔːrm-/ vt.【nutrition】to prepare or mix feedstuff according to a specified formula 配合(饲料)

formulation /ˌfɔːmjuˈleɪʃn; *AmE* ˌfɔːrm-/ n.【nutrition】the preparation and mixing of feeds according to formula (饲料)配合: ration formulation 饲料配合

forrard /ˈfɔːrəd; *AmE* ˈfɔːrərd/ interj.【hunting】(also **forrard-on**) a cheer given by the huntsman to encourage the chase of the hounds (鼓励猎犬追捕猎物的喊声)追啊! 前进!

forty horse hitch n. a hitch of forty horses harnessed together 四十匹马联驾,卌马联驾

forward deviation of the carpal joints n. *see* bucked knees 凸膝

forward seat n.【jumping】a position in which the rider stands in the stirrups, bending slightly forward at the waist to put his weight over the shoulders of his mount (骑手站立马镫,腰以上前倾至马肩胛处的姿势)前倾坐姿,前倾骑姿: Forward seat enables the rider to stay over the center of gravity more easily when jumping. 在跨越障碍时,前倾骑姿可使骑手将身体调整至马体重心上。

foster /ˈfɒstə(r); *AmE* ˈfɔːs-/ vt. to bring up or nurture the offspring of others 收养,抚养

adj. of or related with fostering 收养的,寄养的: If a foster mare cannot be found for an orphan foal, it must be bottle-fed. 如果找不到合适的寄养母马,无母幼驹就必须用奶瓶人工喂养。

foul[1] /faʊl/ adj. having an offensive smell or taste; disgusting 恶臭的,难闻的: The stable should be mucked out daily, otherwise the accumulated ammonia gases emitted from the manure will develope a very foul smell, which is harmful both to equines and humans. 马房每天都应当进行清扫;否则,粪尿中散发出的氨气会导致空气污浊并危害马匹和人员健康。

foul[2] /faʊl/ n. **1** un unfair stroke or play in a game 违例 **2** an instance of breaking the rules of a game or sport 犯规

adj. contrary to the rules of a game or sport played 犯规的,违例的: In the game of polo, charging, intimidation, foul hooking, or crossing the line of the ball are all considered as foul play. 马球比赛中,冲撞、有威胁性动作、钩杆违例或越过球线都将视为犯规。

vi. to commit a foul act against an opponent 犯规,违例

foul hooking n.【polo】a foul act in which a player hooks an opponent's mallet at a level higher than the withers of the horse (马球比赛中在马鬐甲以上使用钩杆)钩杆犯规

found /faʊnd/ vt. to establish or set up the basis of sth 奠基;建立

foundation /faʊnˈdeɪʃn/ n. **1** the basis on which sth is established 基础 **2** a group of animals from which a certain breed is developed 基础群

foundation mare n. one of the original, primary mares used to establish the characteristics of a breed (育成一个马种最初使用的母马)基础母马,奠基母马 syn tap root mare

foundation sire n. one of the original, primary stallions used to establish the characteristics of a particular horse breed (育成一个马种最初使用的公马)奠基公马

founder /ˈfaʊndə(r)/ n. inflammation of the laminae of the hoof 蹄叶炎

v. **1** (of horses) to stumble and go lame (指马)跌绊,跛行 **2** (of horses) to be afflicted with laminitis (指马)患蹄叶炎

founder rings n. *see* hoof rings 蹄环纹

founder stance n. the standing posture assumed by a horse afflicted with acute laminitis in which the hind and fore feet are placed well forward of their usual positions（患蹄叶炎的马所采用的站姿，其前后蹄所处位置较正常姿势更靠前）蹄叶炎站姿

Four Horsemen n.【mythology】(also **Four Horsemen of the Apocalypse**) the four riders of horses described in *Revelation* which represent the four plagues of humankind—war, famine, pestilence, and death（基督教《圣经》启示录中记述的四名骑士，分别代表战争、瘟疫、鼠疫和死亡四种灾难）天启四骑士

Four Horsemen of the Apocalypse n. *see* Four Horsemen 天启四骑士

four time n. the foot-falls occurring independently of each other marked by four hoof beats at each stride（马完成一步四蹄分别触地的声响）四蹄音

four-beat stepping pace n. *see* stepping pace [四蹄音]破对侧步

fourgon /fʊər'gəʊŋ/ n. a long covered wagon, used in the late 19th centuries for carrying baggage, supplies, etc.（19世纪末使用的一种带篷四轮马车，用来运送行李和物资）货运马车，辎重马车

Four-H Club n. (abbr. **4-H, 4-H Club**) a youth organization sponsored by the US Department of Agriculture to instruct young people in rural areas in modern farming and other useful skills（由美国农业部所创立的青年组织，旨在向乡村地区青少年提供现代农业种植、养殖以及其他技能的教育）4-H 俱乐部

four-in-hand /ˌfɔːɪn'hænd; *AmE* ˌfɔːrɪn'hænd/ n. **1** (also **four-in-hand team**) a four-horse team consisting two wheelers and two leaders（由两匹辕马和两匹头马组成的四马）四马联驾，驷马联驾 **2** a carriage drawn by a team of four horses with one pair harnessed in front of the other（由前后两对马套拉的马车）四马联驾马车

Four-in-Hand Driving Club n. a driving club founded in Great Britain in 1856 to promote the recreational practice of carriage driving（1856年在英国成立的驾车俱乐部，旨在促进休闲驾车的普及）四马联驾俱乐部

four-in-hand team n. *see* four-in-hand 四马联驾

four-point trim n.【farriery】a hoof trimming technique 四点削蹄法

Four-wheeled Dogcart n. *see* Dogcart Pheaton 四轮费顿马车

four-wheeler /ˌfɔː'wiːlə/ n. **1** any of the four-wheeled, horse-drawn vehicles 四轮马车 **2** (usu. **Four-Wheeler**) a four-wheeled, horse-drawn cab formerly used in London, England（过去英国伦敦使用的）四轮出租马车

fox /fɔːks/ n. any of a group of wild carnivorous mammals of the genus *Vulpes* having a bushy tail and a reddish-brown or gray fur（犬科狐属食肉类哺乳动物，尾巴大，被毛多为褐红色或灰色）狐狸：red fox 赤狐◇ gray fox 灰狐 ◇ Fox is often noted for its cunning and speed. 狐狸常以狡猾和速度著称。PHR V account for the fox【hunting】(of a hound) to hunt and kill a fox（猎犬）捕杀狐狸 syn Charley, Charlie

fox den n. (also **fox earth**) a cave used by a fox as its hiding place 狐穴，狐洞

fox dog n. a foxhound 猎狐犬

fox earth n. *see* fox den 狐洞，狐穴

fox kennel n.【hunting】the hiding place of a fox located above the ground level（狐狸位于地面之上的藏身处）地上狐窝 compare fox den

fox sense n.【hunting】the natural ability of a hound to trace the line of a fox（猎犬追踪狐狸踪迹的本能）猎狐天性，猎狐本能

fox terrier n. a short-haired terrier formerly used to unearth foxes or other burrowing animals（一种耳朵较短的㹴犬，常用来将狐狸赶出洞窝）猎狐㹴犬

fox trot n. a slow, short, unevenly spaced, four-beat artificial gait of horse（马的一种四蹄音

人工步法,运步慢,步幅短而不匀)狐步

foxhound /ˈfɒkshaʊnd; *AmE* ˈfɑːks-/ n.(also **fox dog**) any of various hounds with a keen scent and sight trained to hunt foxes(嗅觉和视觉敏锐、用于追捕狐狸的猎犬)猎狐犬: English foxhound 英国猎狐犬 ◇ American foxhound 美国猎狐犬

Foxhound Kennel Stud Book n. a breeding record of foxhounds first compiled in 1841 in Britain(英国1814年首次对猎狐犬繁育进行记录的登记簿)猎狐犬登记册

fox-hunt /ˈfɒkshʌnt; *AmE* ˈfɑːks-/ n. the sport or activity of hunting a fox across country with a pack of hounds by mounted riders(骑马带领猎犬在野外追捕狐狸)猎狐

vi. to hunt fox as a sport or for pleasure 猎狐 | **fox-hunting** n.:Fox-hunting was once a traditional sport of the English landed gentry. 猎狐曾是英国乡绅阶层的传统运动项目。

fox-hunter /ˈfɒkshʌntə(r); *AmE* ˈfɑːks-/ n. any horse ridden in pursuit of fox and other hunted game, usually possessed of good stamina and jumping ability(追捕狐狸或其他猎物所骑的马,通常耐力和跨越能力极佳)猎狐用马

fox-hunting club n.(also **hunt**, or **hunt club**) an organization founded by a group of hunters to promote the sports of fox-hunting(由狩猎者创建的组织机构,旨在促进猎狐运动的发展)猎狐俱乐部

F_r abbr. respiratory frequency 呼吸频率

fractional time n.【racing】the time taken at various points between the start and finish of a race(赛马从起点到终点不同距离段所用的时间)分段用时

fractious /ˈfrækʃəs/ adj. difficult to control; unruly 难驾驭的;暴烈的:A muzzle twitch may be a helpful and necessary means of restraint for fractious horses. 对于性情暴烈的马来说,鼻捻子是一种非常有用也十分必要的管制手段。

fracture /ˈfræktʃə(r)/ n. an instance of breaking a bone or the state of being broken 折断,骨折: comminuted fracture 粉碎性骨折 ◇ compound fracture 复合性骨折 ◇ sesamoid fracture 籽骨骨折 ◇ slab fracture 断层骨折 ◇ simple fracture 单纯性骨折 ◇ stress fracture 应力性骨折 [syn] break

vi.(of bone) break or crack 折断,骨折

Franches-Montagnes n. a heavy warmblood originating in the Jura region of Switzerland by putting imported Anglo-Norman stallions to native mares; stands about 15 hands and may have a bay or chestnut coat; has a strong build, a heavy head, a full forelock falling over a broad forehead, small ears, a muscular, arched neck, short back, and slightly sloping croup; is steady, sure-footed, active, and calm; is quite versatile and suitable for both driving and riding(源于瑞士汝拉山区的重型温血马品种,主要采用进口的盎格鲁诺曼马和当地母马杂交繁育而来。体高约15掌,毛色为骝或栗,体格强健,头重,额发浓密,耳小,颈肌发达且呈弓形,背短,尻稍斜,运步稳健,性情活泼而安静,用途广泛,可用于驾车和骑乘)弗朗什—蒙泰涅马

Francis Barlow n. *see* Barlow, Francis 弗朗希斯·巴罗

François Robichon de la Guérinère n. *see* **Guérinère, François Robichon de la** 弗朗索瓦·罗比臣·德拉·格伊尼亚

Frederico Grisone n. *see* Grisone, Frederico 弗雷德里克·格瑞森

Frederiksborg /ˈfredərɪksˌbɔːrg/ n. a horse breed developed in 1562 at the Royal Stud in Frederiksborg, Denmark by crossing Neapolitan and Andalusians; stands 15.1 to 16.1 hands and has a chestnut coat; has a well-proportioned, but slightly convex head, strong shoulder, deep chest, straight back, and small feet; is quiet and willing and used as a military charger in the past and mainly for light harness and riding nowadays(1562年在丹麦腓特烈堡皇家种马

场培育的马品种,主要采用那不勒斯马和安达卢西亚马杂交繁育而来。体高 15.1 ~ 16.1 掌,被毛栗色,头型适中,面颊稍凹,肩部发达,胸身宽广,背部平直,蹄小,性情温驯,过去主要用作战马,现在主要用于轻挽和骑乘)腓特烈堡马:Frederiksborg has contributed greatly to the development of the Lipizzaner and the Orlov Trotter. 腓特烈堡马对利比扎马和奥尔洛夫快步马的育成贡献较大。 [syn] Danish Horse

free handicap n.【racing】a race in which no nominating fees are required 无参赛费的比赛

free jump vi.【jumping】(of a horse) to jump over obstacles free of rider (马在无骑手驾驭下跨过障碍)自由跨越,空跳 [syn] loose jump | **free jumping** n.

free martin n.【genetics】a filly twin of a colt (与雄驹双生的雌驹)自由马丁

free walk n. a walk on a loose rein with the head free and the back relaxed (马不受缰绳控制而放松地走步)自由慢步

freelance /ˈfriːlɑːns; *AmE* -læns/ adj.【racing】(of a jockey) riding for different stables or horse owners rather than being permanently employed by one (骑师为不同马场或马主效力而非固定受雇于一处)自由职业的:a freelance jockey 自由[职业]骑师

n. a freelance jockey or rider 自由骑师

vi. to ride in a race as a freelance 以自由骑师身份参赛

free-ranging adj. (of horses) wandering or roaming freely (指马)自由驰骋的

free-roaming adj. (of horses) roaming freely; free-ranging (指马)自由驰骋的

Free-Roaming Wild Horse and Burro Act n. a law passed in the United States in 1971 to protect wild horses and burros from poaching, death, and capture (美国 1971 年通过的法案,旨在保护野马和野驴不受盗猎和捕捉的威胁)野外驴马保护法

freestyle exercises n.【vaulting】(also **kür exercises**) any of the optional exercises performed by the vaulters (马上杂技中可以选择表演的动作)自由动作,自选动作 [compare] compulsory exercises

freeze brand n. (also **freeze-marking**) a brand marked by applying a pre-cooled iron on horse's skin for several seconds (将液氮预冷的烙铁短时间压在马皮肤上形成的印记)冷冻烙印,冻号:freeze brand iron 冷冻烙铁,冻号烙铁 ◇ The freeze-brand is only applied long enough to kill the chromocytes in the hair follicle so as to not damage the skin or kill the hair follicle. 冻号烙印持续时间刚好能杀死毛囊中的色素细胞,但却不能损伤皮肤或破坏毛囊。

vt. to mark with a branding iron pre-cooled in liquid nitrogen (用液氮预冷的烙铁)打冻号 | **freeze-branding** n.

freeze-marking n. *see* freeze-branding 冷冻烙印,冻号

Freidberger /ˈfriːdˌbɜːgə(r); *AmE* -ˌbɜːrgə(r)/ n. a new warmblood breed developed recently at the Avenches Stud in Switzerland based on the Franches-Montagnes improved with heavy infusions of Shagya Arabian and Norman blood; stands 15.2 to 16 hands and has an Arabian-type head, good shoulders and quarters, short back, deep girth, and strong legs with plenty of bone; is possessed of a good disposition and used mainly for riding (瑞士阿旺什种马场最近培育的温血马品种,主要通过在弗朗奇斯—蒙泰葛尼斯马的基础上大量引入沙迦阿拉伯马和诺曼马外血育成。体高 15.2 ~ 16 掌,面貌与阿拉伯马相似,肩膀和后躯发达,背短而胸深,四肢粗壮,性情温驯,主要用于骑乘)弗里德堡马

French Anglo-Arab n. a horse breed developed in the 1840s at the Pompadour stud farm; descended from a nucleus of broodmares of Oriental origin interbred with local breeds and the addition of Arab and Thoroughbred blood; has a solid build with good conformation and powerful hindquarters; stands 15.2 to 16.3 hands and generally has a bay, brown, black, or chestnut coat; used for riding, racing, and other competitions（法国蓬巴杜种马场在19世纪40年代育成的温血马品种，主要采用东方母马核心群与当地品种杂交，再引入阿拉伯马和纯血马外血培育形成。体格强健，体貌俊秀，后躯发达，体高15.2～16.3掌，毛色多为骝、褐、黑或栗，主要用于骑乘、赛马和其他竞技项目）法国盎格鲁阿拉伯马

French brushing boot n. a light duty brushing boot used to protect the fetlock joint（用来保护球节的轻便护腿）法式防蹭护腿

French Cavalry School at Saumur n. a riding school established in France in the early 18th century to teach cavalry the classical equitation and other forms of competitive sport（法国在18世纪初建立的骑术学校，旨在向骑兵传授古典骑术和其他马术竞技技巧）法国索米尔骑兵学校 syn School of Saumur, School of Mounted Troop Instruction

French Phaeton n. a French horse-drawn carriage of the Phaeton type, usually drawn by a single horse in shafts or a pair in pole gear（一种法式马车，多由单马驾辕或双马套驾杆拉行）法式费顿马车

French Saddle horse n. *see* Selle Français 法兰西乘用马，塞拉法兰西马

French Saddle pony n. a pony breed developed recently in France through the selective crossing of native mares with Arab, Connemara, New Forest and Welsh stallions; stands 12.1 to 14.2 hands and may have any of the coat colors; has a small head, long neck, straight back, a wide, deep chest, and strong legs with well-formed joints; is quiet, yet energetic; used for jumping, dressage, and harness（法国最近采用本土母马与阿拉伯马、康尼马拉矮马、新福里斯特矮马和威尔士矮马公马杂交育成的矮马品种。体高12.1～14.2掌，各种毛色兼有，头小，颈长，腰背平直，胸身宽广，四肢健壮，性情温驯，体力充沛，主要用于障碍赛、盛装舞步和驾车赛）法国乘用矮马

French tie n. *see* mud tie 尾巴打结

French Tilbury Tug n. a French version of the Tilbury Tug 法式提尔堡马车

French trotter n. a trotting breed developed in France in the late 1830s by putting Thoroughbred, halfbred English Hunters, and Norfolk Trotters to Norman mares; is light in build and has a fine head, prominent withers, strong back, and sloping hindquarters; stands 15.1 to16.2 hands and may have a bay, black, chestnut, or occasionally gray coat; an athletic and fast horse used for harness racing, riding, and cross-breeding（法国在19世纪30年代末采用雄性纯血马、英半血狩猎马和诺福克快步马与诺曼马母马交配繁育而来的快步马品种。体格轻盈，面貌俊秀，鬐甲突出，背部有力，斜尻，体高15.1～16.2掌，毛色为骝、黑或栗，青毛少见，是优秀的竞技马种，步速快，主要用于驾车赛、骑乘和其他马种的改良）法国快步马 syn Norman Trotter

fresh /freʃ/ adj. **1** recently made, produced, or harvested 新鲜的；刚采获的：fresh grass 青草 ◇ fresh semen 鲜精 **2**（of a horse）high-spirited and energetic（指马）有活力的，活泼的：a fresh colt 一匹活泼的雄驹

fresh cattle n.【cutting】cattle that have not previously been used in cutting events or competitions（尚未参加过截牛或其他项目的牛）新牛

fresh fox n.【hunting】a fox that has not been pursued by the hounds in a hunt 没被捕猎过的狐狸

fresh line n.【hunting】any new scent left by a fox or other prey（猎物新近留下的气味）新鲜踪迹

fresh-catched coachman n.【driving】a newly trained driver 驾车新手，新任车夫

freshen /ˈfreʃn/ vt.【racing】to let a tired horse to

rest and restore its energy 使恢复体力,使焕发活力

fret /fret/ vt. to gnaw or wear away 啃咬;磨损

Frideriksborg /ˈfrɪdəriksbɔːrg/ n. *see* Frederiksborg 腓特烈堡马

Friendship Stakes n. an American stakes race held at Louisiana Downs for two-year-olds (美国路易斯安那唐斯赛道为 2 岁驹举行的奖金赛)友谊大赛

Friesian /ˈfriːʒn/ n. (also **Frisian**) an old horse breed indigenous to the Netherlands; has a compact and strong build, a short, but well-arched neck, straight and short back, broad chest and loins; stands about 15 hands and commonly has a black coat; has a long, wavy mane and tail, feathered legs, large, strong feet, and high action in trot; used for driving, farming, and dressage (源于荷兰本土的古老马品种。体格强健,颈短且曲线优美,背短平直,胸身宽阔,体高约 15 掌,被毛多为黑色,鬃尾长而卷曲,带距毛,蹄大而结实,运步举蹄较高,主要用于驾车、农田作业和盛装舞步)弗里斯兰马 [syn] Harddraver

Friesian Chaise n. a two-wheeled, horse-drawn vehicle of the Gig type pulled by two Friesian horses harnessed on either side of a pole (一种轻型双轮马车,由两匹弗里斯兰马套驾杆拉行)弗里斯兰马车

Frisian /ˈfriːʒn/ n. *see* Friesian 弗里斯兰马

frock coat n.【hunting】a knee-length coat worn by members of the hunt (猎狐者所穿的齐膝长大衣)狩猎大衣: Staff members will wear square-skirted frock coats while members of the field will wear round-skirted ones. 工作人员穿方襟大衣,而场内猎狐会员则穿圆襟的。

frog /frɒg; *AmE* frɔːg/ n. the wedge-shaped area on the underside of the horse's hoof that acts as a shock absorber (马蹄底的 V 形区域,可起缓冲作用)蹄叉: frog cleft 蹄叉沟 ◇ sensitive frog 蹄叉敏感区

front /frʌnt/ n. the fore part of a horse 前躯: good front 前躯匀称 [opp] hind [PHR] **down in front** (of a horse) with forehand lower than the hindquarters when viewed horizontally (指马)前躯过低: Obviously, it is a conformation fault for any horse to be down in front. 显然,任何马匹前躯过低都是一种体格缺陷。

front boot n. a wooden storage box located between the driving seat and the front axle of a horse-drawn coach (位于马车驾驶座和前车轴之间的存货木箱)车前行李箱

front foot n. the forefoot of a horse 前蹄

front leg n. *see* foreleg 前肢

front runner n.【racing】a horse who prefers to run in the front of the field 抢着领跑的马

frontal /ˈfrʌntl/ adj.【anatomy】of or relating to the forehead 额的: frontal bone 额骨

frost /frɒst; *AmE* frɔːst/ n. a coat color pattern of the Appaloosa with white speckles on a dark base coat (阿帕卢萨马毛色类型之一,深色被毛上散生有白色斑点)霜斑 [compare] leopard, snowflake, blanket, marble

frost nails n. a special horseshoe nail designed to provide temporary hard surface traction (一种特制蹄钉,可增加蹄铁与地面的摩擦)防滑蹄钉,防霜蹄钉

frosty /ˈfrɒsti; *AmE* ˈfrɔːsti/ adj.【color】(of horse's coat) having white hairs around certain body parts (马身体某些部位上散生白毛)霜白的: a frosty coat 霜白毛 [syn] skunk tail

frozen track n.【racing】a race track with a frozen surface 冰冻赛道,冻土赛道

frugal /ˈfruːgl/ adj. (of a horse) able to maintain its body condition on very little ration; thrifty (指马采食很少日粮便可维持体况)耐粗饲的

frusemide /ˈfruːsəmaɪd/ n. *see* furosemide 呋塞米,利尿磺胺,速尿灵

frush /frʌʃ/ n. **1**【farriery】the frog of a horse's foot 蹄叉 **2**【dated】*see* thrush 蹄叉腐烂

FSH abbr. follicle-stimulating hormone 促卵泡素,卵泡刺激素

ft abbr. foot 英尺

full /fʊl/ adj.（of a horse）male, entire（指马）公的，未阉的：a full horse 一匹公马

full blood n.（also **fullblood**）*see* hotblood 热血马，全血马

full book n. the maximum number of mares allowed to be bred by a stallion in any given year（每年允许种公马予配母马的最大数目）配种额度

full bridle n. *see* double bridle 双勒

full brother n. a relationship of male horses having the same sire and dam（公马之间同父同母的亲缘关系）全同胞兄弟 [compare] half brother

full clip n. a type of clip to remove off the entire coat of the horse（对马周身被毛进行的修剪）全修剪：In full clip, the mane may be clipped or left full. 全修剪时，鬃可剪也可留着。

full cry n.【hunting】a loud cry of the whole pack when pursuing the line of a fox（追踪狐狸时所有猎犬发出的嚎叫）狂吠

full halt n. a complete stop of the horse achieved by pulling reins（骑马者拽缰后）停下，停止：The horse was reined to a full halt. 马被拽着停了下来。[compare] half halt

full hand n.【driving】a method of holding the four reins of a team by the left hand（四马联驾中用左手握四根缰绳的方法）单手持缰

full mouth n. **1** the mouth of a horse at five or six years of age when it has grown all its permanent teeth（马在5～6岁时永久齿换齐的状况）齐口 PHR V **to have a full mouth**（of a horse）to grow all its permanent teeth（指马）牙齐口：A horse usually has a full mouth at the age of six. 马通常在六岁时齐口。**2** a horse six years of age or older 齐口马

full pass n.【dressage】a dressage movement performed on two tracks in which the horse moves laterally without forward movement（一种盛装舞步动作，马以双蹄迹横向行进而不前行）横向行进，横步 [syn] full travers

full pastern n. *see* pastern 系全白，全系白

full port n. the part of the mouthpiece that curves upward in the center about one inch（衔杆中段约1英寸的凸起）高衔凸 [syn] high port

full sister n. a relationship of female horses by the same sire and out of the same dam（雌马之间同父同母的亲缘关系）全同胞姐妹 [compare] half sister

full stocking n. a leg marking consisting of white that extends from the coronet along the cannon bone, and includes the knee（马四肢别征之一，其中白色区域从蹄冠沿管骨延伸至膝盖以上）全管白

full travers n. *see* full pass 横向行进，侧向运行

fuller /fʊlə(r)/ n. **1**【farriery】a groove made on horseshoes; crease（蹄铁上的）槽沟 **2** a hammer used by a blacksmith for grooving barstock（用来给蹄铁条开槽的铁锤）开槽锤

vt. to cut grooves in the ground-side of barstock before it is forged into horseshoes（在蹄铁条着地侧）开槽 | **fullering** n. 开槽

fullered shoe n. a horseshoe having a narrow groove cut into the ground-side surface（着地侧）开槽蹄铁，带槽蹄铁：Fullered shoe could provide better traction and prevent slipping when the groove fills with dirt. 开槽蹄铁的槽上滋上泥土后，不但能增加摩擦力，还可防滑。

fullering iron n. *see* crease iron 开槽钢模

full-mouthed /ˌfʊlˈmaʊðd/ adj.（of a horse）having a full set of permanent teeth（指马）齐口的：A horse usually becomes full-mouthed at five or six years of age. 马通常在5～6岁长为齐口。

fully headed adj.（of a carriage）fully enclosed（指马车）封闭的，带顶篷的

function /ˈfʌŋkʃn/ n. the physiological action a body organ or tissues（机体器官或组织的生理作用）功能，机能

vi. to have or perform a function; serve 起作

用,发挥功能 | **functioning** n.

functional residual capacity n.【physiology】(abbr. **FRC**) the volume of air in the lung after breathing out at rest(静息呼气后肺内的残气量)机能残气量

funeral horse n. *see* Black brigade 殡仪马,丧葬马

funk /fʌŋk/ vi. (of a horse) to get frightened onto its toes or show other signs of nervousness (指马)受惊,畏缩

furacilin /fjʊrəˈsɪlɪn/ n. *see* furacin 呋喃西林

furacin /ˈfjʊrəsɪn/ n. (also **furacilin**) a brand of topical antibacterial medication with nitrofurazone as its active ingredient (一种外用抗菌药物,活性成分为硝呋醛)呋喃西林

Furioso /ˌfjʊrɪˈəʊsəʊ; *AmE* -ˈoʊsoʊ/ n. *see* Furioso-North Star 弗雷索马

Furioso-North Star n. a warmblood descended from two English stallions, Furioso and North Star, imported to the Mezöhegyes Stud, Hungary in about 1840; has a muscular body, straight back, sloping hindquarters, and low-set tail; stands 16 to 16.2 hands and has a brown, bay, or black coat with white markings the exception; is versatile and used in both riding and harness driving (由1840年左右从英国进口到匈牙利梅兹海吉斯种马场的两匹公马——弗雷索和北极星繁育而来的温血马品种。体格强健,背部平直,斜尻,尾础较低,体高16~16.2掌,毛色多为褐、骝或黑,无白色别征,为兼用型马种,可用于骑乘和挽车)弗雷索—北极星马

furious driving n. a reckless racing of coach drivers and wagoners (车夫们举行的驾车赛,行径鲁莽冒险)肆意驾车,飚驾:Furious driving was alleged a chargeable offense in 19th-century England. 恣意飚驾在19世纪的英国将会遭到指控。

furlong /ˈfɜːlɒŋ; *AmE* ˈfɜːrlɔːŋ/ n.【racing】(abbr. **fur**) a unit for measuring distance equal to 201 meters or 1/8 mile (长度测量单位,相当于201米或1/8英里)弗隆 [syn] furrow long

furosemide /fjʊəˈrəʊsəmaɪd/ n. (also **frusemide**) a medication used in the treatment of bleeders and hypertension of horses, commonly known by the trade name Lasix (治疗马匹肺出血和高血压的药物,商品名为速尿灵)呋喃苯氨酸,利尿磺胺,呋塞米

furrier /ˈfʌrɪə(r)/ n. **1** one that deals in furs 皮货商 **2** one who prepares furs 毛皮加工者

furrow /ˈfʌrəʊ; *AmE* ˈfɜːroʊ/ n. a long, narrow, shallow trench made in the earth by plowing (犁地时在土地上划出的一道长而窄的沟槽)犁沟:to open a furrow 起沟

furrow long n.【racing】(abbr. **furlong**) the length of a plowed field, used as a unit of measure equal to 1/8 mile in horse race (耕地犁沟的长度,赛马中相当于200米)犁沟长;弗隆

fuse /fjuːz/ vi. to become mixed or united; merge 融合;合并

fusion /ˈfjuːʒn/ n. the act or process of uniting together or merging 融合;合并

futchell /ˈfʊtʃel/ n.【driving】the longitudinal pieces of wood that support the splinter bar, pole, or shaft at one end and attach to the axletree bed and to the sway bar on the other (马车底盘上的长木,一端用来支撑衡杆与辀或辕,而另一端则与轴座和防摆杆固定)通柱

futchell stay n. an iron plate used to strengthen a wooden futchell on a horse-drawn carriage (用来加固的铁板)通柱加固铁板

fuzztail /fʌzˈteɪl/ n. *see* mustang 生马,野马

fuzztail running n. the herding and catching of wild horses or mustangs 生马捕捉,野马捕捉

FWCF abbr. Fellow of Worshipful Company of Farriers 蹄铁匠公会会员

G /dʒiː/ abbr.【racing】grade（赛事的）级别：G-1 stakes 一级奖金赛

gad /gæd/ n. **1** a pointed stick used for driving cattle（赶牲口用的）尖头棒 **2** a spur 马刺 **3**【racing】a jockey's whip 马鞭

vt. to drive cattle or livestock with a pointed stick（用尖头棒）驱赶

gag /gæg/ n. **1** *see* gag bit 转环衔铁 **2** the cheekpieces of a gag snaffle（衔铁上的）转环，颊片

GAG abbr. glycosaminoglycan 黏多糖

gag bit n.（also **gag snaffle**）a snaffle bit with holes cut into its cheekpieces, through which an extra rein passes to exert pressure against the corners of the mouth（颊杆上带孔洞的衔铁，额外一条缰绳从中穿过，借此可以施力于口角）转环衔铁

gag bridle n. a bridle with a gag snaffle 转环勒

gag reins n. pl. a pair of reins that pass through the holes on the gags 转环缰

gag snaffle n. *see* gag bit 转环衔铁

gait /geɪt/ n. the way how a horse moves with ordered footfalls in each stride（行进中马蹄落地的方式）步法，步伐：artificial gait 人工步法 ◇ natural gaits 自然步法 ◇trained gait 训练步法 ◇ All horses possess these four natural gaits—walk, trot, canter and gallop, while some horses have a fifth or sixth natural or trained gait. 所有马都会以慢步、快步、跑步和袭步四种自然步法行进，而有些马还会第五或第六种人工步法。

gaited /ˈgeɪtɪd/ adj.（of a horse）having artificial gaits 步态的：a five-gaited horse 一匹五步态马

gaited horse n. a horse having trained gaits besides the four natural gaits（除四种自然步法外还善走其他人工步法的马）步态马：Examples of gaited horse breeds are the Tennessee walkers, Saddlebreds, Paso Finos and Icelandic Horse. 步态马品种有田纳西走马、美国乘用马、帕索·菲诺马和冰岛马。

Galiceño /ˌgælɪˈsiːnəʊ/ n. a pony breed descended from the Portuguese Garrano brought to Mexico in the 16th century by the conquistadors; has an average head, short and muscular neck, pronounced withers, short back, narrow chest, relatively straight shoulder, strong long legs, and small feet; stands 12 to 13.2 hands, usually has a bay, black, or chestnut coat; is docile, intelligent, versatile, and possessed of great endurance and speed; often used as a riding and jumping pony for children（由16世纪西班牙殖民者运入墨西哥的葡萄牙加里诺矮马繁育而来的矮马品种。头型适中，颈粗短，鬐甲突出，背短，胸腔狭窄，直肩，四肢长而健壮，蹄小，体高12～13.2掌，毛色以骝、黑或栗常见，性情温驯，反应敏快，用途广泛，速力好，常用于儿童骑乘和跨栏）加利西诺矮马

Galician pony n. an ancient pony breed originating in Galicia of Spain from which the name derived; stands about 13 hands and has usually a brown or black coat which grows notably longer and thicker in winter; has a heavy head, small ears, long neck, a thin, flowing mane, low-set tail, and tough, short legs; is very hardy, docile, and frugal; used for riding and light draft（源于西班牙加利西亚地区的古老矮马品种，故而得名。体高约13掌，毛色多为褐或黑，冬季被毛厚而浓密，头重，耳小，颈长，鬃毛稀疏而光滑，尾础低，四肢粗短，抗逆性强，性情温驯，耐粗饲，多用于骑乘和轻挽）加利西亚矮马

gall /gɔːl/ n. **1** *see* bile 胆汁 **2** a sore or swelling on the horse's skin caused by rubbing or abrasion as by ill-fitting tack（皮肤由于磨蹭造成的疼痛）肿痛，擦伤：a saddle gall 鞍伤

vt. to cause sore or swelling by friction or abrasion; chafe 使肿痛，擦伤

vi. to become sore or swollen by friction or abrasion 肿痛

gall bladder n.【anatomy】a small, pear-shaped muscular sac located near the liver where the bile secreted by the liver is stored until needed by the body for lipid digestion（位于肝附近的梨状囊，可储存肝脏分泌的胆汁以供消化脂肪使用）胆囊：It is worthy of noting that the horse doesn't have a gallbladder in which to store bile; instead bile trickles continuously into the duodenum. 值得提及的是，马没有储存胆汁的胆囊，而是连续不断地分泌到十二指肠中。

gallop /ˈgæləp/ n. the fastest, four-beat gait of the horse in which the feet strike the ground separately（马步速最快的四蹄音步法，其四蹄分别着地）袭步：right lead gallop 右领袭步 ◇ left lead gallop 左领袭步 ◇ The stride of gallop is more extended and the moment of suspension longer than that of the canter. 袭步的步幅和滞空时间都较跑步长。IDM **at full gallop** running or riding at full speed 飞奔；疾驰—*see also* **walk**, **trot**, **canter**

vi. to go or ride at a gallop 疾驰；策马飞奔 IDM **gallop up** 疾驰，飞奔

Galvayne, Sydney n.（1846－1913）a renowned British horse breaker and trainer active in the late 1800s（19世纪末英国著名的调马师）希德尼·格尔瓦尼

Galvayne's groove n.（also **Galvayne's mark**）a dark line which appears on the upper corner incisor of horses between 8 and 10 years of age, named for the British horse trainer Sydney Galvayne who wrote a book on horse dentition（马8～10岁时上颌隅齿出现的黑线，以撰写马齿式著作的英国练马师希德尼·格尔瓦尼命名）牙沟：Galvaynes' groove may extend downwards gradually, and is often used to estimate the age of a horse. 牙沟会逐渐从齿根向齿冠延伸，常可作为判定马匹年龄的依据。

Galvayne's mark n. *see* Galvayne's groove 牙沟

game /geɪm/ n. **1** a form of play or sport 游戏，运动 **2**（also **wild game**）any of the wild animals that are hunted for sport or profit; quarry 野味，猎物

gamete /ˈgæmiːt/ n. a reproductive cell, especially a mature sperm or egg（成熟生殖细胞，尤指精子或卵子）配子：female gamete 雌配子 ◇ male gamete 雄配子

gammy legged adj.（also **gummy legged**）（of horse's legs）swollen due to strain or hard work; stocked up（指马腿）肿胀的，水肿的

gap /gæp/ n.【racing】an opening in the outside rail where horses enter and leave the racetrack（赛场位于外栏的开口，马匹由此进出）开口，进出口

Garrano /ˈgærənəʊ/ n. an ancient pony breed originating in the Portuguese regions bordering with Spain; stands 10 to 12 hands and almost always has a chestnut coat with luxuriant mane and tail; is lightly built, strong, hardy, and sure-footed; used for packing, light farm work, riding, and trotting races（源于葡萄牙与西班牙交界地区的矮马品种。体高10～12掌，毛色为栗，鬃尾浓密，体格轻而强健，抗逆性强，步法稳健，主要用于驮运、轻役、骑乘和快步赛）加里诺矮马

Garrison finish n.【racing】a victory from a come-from-behind horse 加里森式取胜；后发制胜：Garrison finish is named after Snapper Garrison who commonly finished in such a fashion. 加里森式取胜得名于骑手斯耐伯·加里森，此人常以此方式冲过终点。

Garron /ˈgærən/ n. one type of Highland pony in Scotland（苏格兰的高地矮马）加隆马

Garry /ˈgæri/ n. *see* Gharry 甘瑞马车

garter /ˈgɑːtə(r); *AmE* ˈgɑːr-/ n. a narrow

leather strap formerly used to hold the breeches constant in the boot (过去用来固定马裤的皮绳)马裤带 syn jodhpur strap

gas /gæs/ n. *see* flatulence (肠胃)胀气

gaseous /ˈgæsiəs, ˈgeɪsiəs/ adj. of, relating to, or existing as a gas 气体的: The gaseous energy is the energy lost from the feed in gaseous products, such as methane, carbon dioxide, etc. 产气能是饲料代谢中所产气体, 如 CH_4、CO_2 等损失的能量。

gaskin /ˈgæskɪn/ n.【anatomy】the part of a horse's hind leg between the stifle and the hock (马后膝至飞节的部分)小腿; 胫 syn second thigh

gasp /gɑːsp; *AmE* gæsp/ vi. to breathe with difficulty and pain 喘息

n. the act of breathing in this way 喘息

gastric /ˈgæstrɪk/ adj. of, relating to, or associated with the stomach 胃的: gastric juices 胃液

gastrocnemius /ˌgæstrɒkˈniːmɪəs/ n.【anatomy】(*pl.* **gastrocnemii** /-ˈniːmiːi/) the largest and most prominent muscle of horse's gaskin (小腿上的)腓肠肌

gastrointestinal /ˌgæstrəʊinˈtestənl/ adj. of or relating to the stomach and intestines 胃肠的: gastrointestinal parasites 胃肠寄生虫 ◇ gastrointestinal tract 胃肠消化道

Gastrophilus /ˌgæstrəˈfaɪləs/ n.【Latin】the genus of botfly 马胃蝇属: *Gastrophilus haemorrhoidalis* 痔马胃蝇 ◇*Gastrophilus intestinalis* 肠马胃蝇 ◇ *Gastrophilus nasalis* 鼻马胃蝇

gate /geɪt/ n. **1**【jumping】a single vertical jumping obstacle consisting of wooden planks supported by standards (由木板经立杆加固构成的单道竖立障碍)门板障碍 **2**【racing】the starting gate 起跑闸, 马闸: mechanical gate 机械起跑闸 ◇ electric gate 电动起跑闸

gate card n.【racing】a card issued by racing authorities to verify that a horse is correctly schooled in starting gate procedures (由赛事机构发给参赛者的卡片, 借此证明赛马已接受过出闸训练)入闸卡, 出闸卡: Gate cards are required of all competing horses. 所有参赛马都须有出闸卡。

gaucho /ˈgaʊtʃeʊ; *AmE* -tʃoʊ/ n. (*pl.* **gauchos**) a cowboy of the South American plains 南美牛仔: Gauchos are thought by many to be the world's finest rough riders. 南美牛仔一直被认为是世界上最优秀的野骑选手。

gauge /geɪdʒ/ n. (also **gage**) the distance between two wheels on an axle (两车轮之间的距离)轮距, 轨距

gay /geɪ/ adj.【dated】(of a horse) carrying its head and tail well with airs (指马)趾高气扬的

n. a horse who carries its head and tail well with airs 趾高气扬的马

Gayoe /ˈgeɪjəʊ/ n. (also **Gayoe pony**) a pony breed native to the Gayoe Hills, Northern Sumatra, Indonesia from which the name derived; stands around 12.2 hands and may have a coat of any solid color with bay occurring most common; is sturdy and has short legs; used primarily for packing (源于印度尼西亚苏门答腊北部盖欧耶山区的矮马品种, 并由此得名。体高约 12.2 掌, 被毛多为单色, 以骝最为常见, 体格健壮, 四肢粗短, 主要用于驮运)盖欧耶矮马

Gayoe pony n. *see* Gayoe 盖欧耶矮马

gear /gɪə(r); *AmE* gɪr/ n. the equipment and accessories used in harness driving 驾车挽具

vt. to equipe a horse with driving harness 套驾车挽具, 套驾 PHR V **gear up** 套马驾车

gee /dʒiː/ interj. used to command a horse to turn to the right or hurry up (右转或加速的驱马声)右转! 驾!

vi. to turn to the right or hurry up 右转; 驾 IDM **gee up**! 驾

gel /dʒel/ n. the jelly-like portion of the semen produced by the accessory glands of the male (由雄性附属腺体产生的胶状液体)前液

geld /geld/ vt. to castrate a male horse; neuter 阉

割,去势

Gelderland /ˈgeldəˌlænd/ n. (also **Gelderlander**) a breed of warmblood horse developed in the Gelderland province of the Netherlands by crossing native mares with Andalusian, Neapolitan, Norman, and Norfolk stallions; stands 15.2 to 16 hands and has a chestnut, bay, black, or gray coat commonly with white markings; is elegant and has a long head, crested neck, high-set tail, and short-coupled body; used for light draft, driving, leisure riding, and jumping (荷兰格尔德兰省培育的温血马品种,主要采用当地母马与安达卢西亚马、诺曼马和诺福克马公马杂交培育而成。体高15.2～16掌,毛色多为栗、骝、黑或青,白色别征常见,体貌俊秀,头长,颈弧形,尾础高,背腰短,主要用于轻挽、娱乐骑乘和障碍比赛)格尔德兰马

Gelderlander /ˈgeldəˌlændə(r)/ n. *see* Gelderland 格尔德兰马

gelding /ˈgeldɪŋ/ n. a castrated male horse 去势公马;骟马

gelding donkey n. a castrated male donkey 骟驴 [syn] john

general anesthesia n. an artificially induced insensibility to the sense of pain all over the body (人工使全身失去痛觉)全身麻醉,周身麻醉

general anesthetics n. an agent capable of depriving the whole body of sensation of pain (可使全身失去痛觉的药剂)全身性麻醉剂

general purpose saddle n. *see* all-purpose saddle 兼用鞍,综合鞍

general service wagon n. (abbr. **GS wagon**) an open, four-wheeled, horse-drawn vehicle with a single seat for the driver and a back boot used to convey goods (一种敞篷四轮马车,车后带的贮物箱可用来运送货物)兼用四轮马车

General Stud Book n. (abbr. **GSB**) the registration book for English Thoroughbreds 英国纯血马登记册

generate /ˈdʒenəreɪt/ vt. to reproduce or give birth to offspring 生育,繁育

generation /ˌdʒenəˈreɪʃn/ n.【genetics】**1** a group of individuals born and living about the same time (生活于同时代的群体)代,同代 **2** the average interval of time between the birth of parents and the birth of their offspring (亲代与子代出生时的间隔)世代

generation interval n.【genetics】the average age of the parents when their offspring are born (父母代繁育子代时的平均年龄)世代间隔: Horses have a relatively long generation interval of 9-11 years compared with other domestic animals. 马的世代间隔为9～11年,与其他家畜相比更长些。

generous /ˈdʒenərəs/ adj. (of a horse) giving its best when competing in a race (指马比赛中)不惜力的,尽全力的

genet /ˈdʒenɪt/ n. *see* jennet 母驴,雌驴

genetic /dʒəˈnetɪk/ adj. of or relating to genetics or genes 基因的; 遗传的: genetic disease 遗传病 ◇ genetic diversity 遗传多样性 ◇ genetic marker 遗传标记 ◇ Genetic testing validates horse pedigrees for breed registry authorities, sales companies and race track. 遗传检测可为马匹登记机构、买卖公司以及赛马场证实马匹的系谱。

genetics /dʒəˈnetɪks/ n. **1** a branch of biology that deals with heredity of living organisms (研究遗传机制的生物学分支学科)遗传学: horse genetics 马匹遗传学 **2** the genetics properties or features of an organism 遗传特征,遗传

genital /ˈdʒenɪtl/ adj. of or relating to the reproductive organs 生殖的: genital tract 生殖道 n. (usu. **genitals**) the reproductive organs, especially the external sex organs 外生殖器

genitalia /ˌdʒenɪˈteɪliə/ n. the genitals 外生殖器

genome /ˈdʒiːnəʊm; *AmE* -noʊm/ n.【genetics】the complete set of genes in a cell of a living organism (细胞或有机体的基因构成)基因组: Some horsemen may worry that once the horse genome is sequenced, all the mystery and magic will be gone from horse breeding. 有些马业人士担心，一旦马的基因组被完全测序，马匹育种的所有秘密和魔力都将不复存在。

genotype /ˈdʒenətaɪp, dʒiːn-/ n.【genetics】the genetic constitution of an organism（生物体遗传物质的组成）基因型 compare phenotype

vt. to determine the genotype of an animal by molecular biological methods（通过分子生物学方法确定个体的基因型）基因分型：Now, genotyping has been widely used in parentage testing. 目前，基因分型在亲子鉴定中使用较广。

genotype selection n. the selection of breeding stock based on the genetic constitution of the animals rather than appearance（根据基因型而不是外貌性状进行的选育）基因型选育

gentleman jockey n.【racing】an amateur rider, generally in a steeplechase（越野障碍赛的）业余骑手

gentleman rider n. *see* amateur rider 业余骑手

genu /ˈdʒiːnjuː/ n.【Latin】(*pl.* **genua** /-njuə/) the knee or knee-like structure of body part; stifle 膝；膝关节 | **genual** /ˈdʒenjʊəl/ adj.

genus /ˈdʒiːnəs/ n. (*pl.* **genera** /ˈdʒenərə/) a taxonomic category ranking below a family and above a species（分类学上介于科和种之间的划分）属 compare kingdom, phylum, class, order, family, species

George Edward Arcaro n. *see* Eddie Arcaro 乔治·爱德华·阿克洛

Géricault, Théodore n. (1791 - 1824) a noted French painter of equestrian scenes（法国著名画家，擅长马术场景写生）西奥多·热里科

germ /dʒɜːm; *AmE* dʒɜːrm/ n. **1** a microorganism or bacterium 微生物，细菌 **2** the earliest form of an organism 生殖细胞；胚芽

German Coldblood n. *see* Rhineland Heavy Draft 德国冷血马，莱茵兰德重挽马

German silver n. a white alloy of copper, zinc, and nickel used for making some ornamental tack and bits（铜、锌和镍的合金，用于制作马具和衔铁饰物）德国银，白铜 syn albata, white brass

German wagon n. a horse-drawn vehicle with a half hood covering the rear seat only and a raised box seat for the driver located well above the body（一种四轮马车，半敞式车篷位于后座之上，车夫驾驶座离车身较高）德式马车

germinal /ˈdʒɜːmɪnl/ adj. **1** of, relating to, or having the nature of a germ cell 生殖细胞的；胚的 **2** of, relating to, or occurring in the earliest stage of development 早期的，初期的

gestation /dʒeˈsteɪʃn/ n. the development of embryo in the uterus from conception until birth; pregnancy（胚胎在子宫内从受孕至分娩的发育过程）怀孕；妊娠：The gestation period for colts is generally a few days longer than for fillies. 雄驹的妊娠期通常要比雌驹多几天。

get[1] /get/ n. the offspring or progeny of a stallion（公马的）后代：A foal is the get of a stallion and the produce of a mare. 马驹是公马和母马的后代。compare produce

get[2] /get/ vt. to do in a certain way 行事 PHR V

get a bite | **get hanged**【driving】(of the whip) become caught in the harness or on the bars（指皮鞭）被缠住 **get (down) into the ground**【cutting】(of a horse) to execute a good sliding stop, setting deep onto his haunches（指马在滑停时后躯）蹲入地面 **get under**【jumping】(of a horse) to take off too near to the jump（指马）起跳过近

getaway /ˈgetəweɪ/ n.【racing】the start of a race 起跑，开始

getaway day n.【racing】the last day of a race meeting 赛马节最后一天

get-up /getʌp/ interj. (also **giddy-up**) an exclamation used to command horses to go ahead or at a faster pace（赶马前进或加速时用的叫喊）驾！

GH abbr. growth hormone 生长激素

Gharry /ˈgæri/ n. (also **Garry**) an open, four-wheeled carriage used in parts of India and the Middle East, usually drawn by a single horse and driven from an elevated seat (在印度和中东等地使用的敞开式四轮马车,多由单马拉行,车夫驾驶座较高)甘瑞马车

giddap /gɪˈdæp/ interj. *see* get-up 驾!

giddy-up /gɪdɪˈʌp/ interj. (also **giddap**) *see* get-up 驾!

Gidran /ˈgɪdrən/ n. *see* Gidran Arabian 基特兰[阿拉伯]马:Middle European Gidran 中欧基特兰马 ◇ Southern and Eastern European Gidran 东南欧基特兰马

Gidran Arabian n. (also **Gidran**) a horse breed developed in Hungary in 1816 from an Arab Stallion of the Siglavy strain named Gidran crossed with Thoroughbreds and Arabs; has Arab characteristics and is now classified into two types - Middle European Gidran and Southern and Eastern European Gidran according to their build and substance; stands 16.1 to 17 hands and usually has a chestnut coat, although bay and black do occur; used mainly for riding, light draft and jumping (匈牙利 1816 年采用一匹名为基特兰的阿拉伯公马与纯血马和阿拉伯马杂交育成的马品种,目前根据其体型分中欧吉卓马和东南欧吉卓马两个类型。阿拉伯马特征明显,体高 16.1 ~ 17 掌,毛色以栗为主,但骝和黑亦有之,主要用于骑乘、轻挽和障碍赛)基特兰阿拉伯马 [syn] Hungarian Anglo-Arab

gig /gɪg/ n. **1** a light, two-wheeled passenger vehicle drawn by a single horse (一种单马拉行的轻型双轮客用马车)双轮马车 **2**【harness racing】*see* sulky 轻驾车赛用马车

gimp /gɪmp/ n.【slang】a lame horse 瘸马;跛马

gimpy /ˈgɪmpi/ adj.【slang】(of a horse) lame or limping (马)跛的,瘸的:a gimpy horse 一匹瘸马

gin horse n. a horse used for draft work and milling 挽马;拉磨马

ginger /ˈdʒɪndʒə(r)/ vt. to insert ginger or other irritants into the anus of a horse to achieve a high tail carriage (将生姜或其他刺激物塞入马的肛门,使其尾巴高翘)塞姜

gingering /ˈdʒɪndʒərɪŋ/ n. the illegal practice of inserting ginger into horse's anus to achieve a high tail carriage (在马肛门内塞生姜等刺激物使尾高翘的违禁做法)塞姜:Gingering is considered both illegal and cruel to horses. 给马肛门塞姜的做法既违例又残忍。[syn] figging

ginney /ˈgɪni, ˈdʒɪni/ n. *see* guinea 马夫

girth /gɜːθ; *AmE* gɜːrθ/ n. **1** (also **heart girth**) the circumference of a horse's body measured from behind the withers around the barrel (马体躯位于鬐甲后方的周长)胸围:girth measurement 胸围测量 **2** a body part located just behind the horse's front elbow (马位于前肘后的部位)带径部:A deep girth always goes in the horse's favor as it indicates plenty of heart room. 胸深对马来说总是有利的,因为这说明它心室容量大。**3** a strap encircling a horse's barrel in order to secure a load or saddle on its back; cinch (系在马胸腔处的皮带,用来固定背上的马鞍或驮载之物)肚带:girth mark 肚带压迹 ◇ girth groove 肚带沟

vt. to secure or hold in place with a girth 系肚带:tightly girthed saddle 用肚带系牢的马鞍

IDM **girth up** to tighten the girth on a horse 系紧肚带

girth gall n. a gall that develops in the belly where the girth generally passes (马腹部系肚带处形成的蹭伤)带径处伤迹:Horses with thin, sensitive skin are most prone to girth galls. 皮肤薄且敏感的马匹易被肚带蹭伤。

girth place n. the place on the belly of the horse just behind its front elbow where the girth is fitted (马前肘后侧部分,肚带系于此处)系肚带处

girth safes n. flat, single layered pieces of leather with horizontal slots cut into them and through which the girth straps pass, used to protect the saddle flap from being rubbed by the girth buckles（马鞍上的单片皮革，其上有狭缝肚带扣带穿过，可防止肚带扣蹭破鞍翼）肚带护扣革

girth shy adj. *see* cinch bound（指马）怕肚带的，抗拒肚带的

girth straps n. narrow pieces of leather attached vertically to the English saddle tree on both sides to which the girth are buckled when saddling the horse（系在鞍架两侧的细皮带，备鞍时用来扣系肚带）肚带扣带：There are usually three girth straps on each side of the saddle，although lightweight saddles may have only one. 马鞍两侧通常各有3根肚带扣带，而轻型鞍则可能仅有1条。

girth weight tape n. a measuring tape calibrated in pounds rather than inches which is placed around the girth of the horse to measure his weight（一种测量皮尺，刻度为磅，绕在马的带径处用来测量其体重）胸围—体重测量皮尺

girthy /ˈɡɜːθi；*AmE* ˈɡɜːrθi/ adj. *see* cinch bound（指马）紧肚带时不安的，抗拒紧肚带的

Giuseppe Castiglione n. *see* Castiglione，Giuseppe 朱塞佩·伽斯底里奥内；郎世宁

gland /ˈɡlænd/ n.【anatomy】an organ that produces particular chemical substances for use elsewhere in the body（机体内的器官，可分泌特殊化学物质作用于其他部位）腺体：adrenal gland 肾上腺 ◇ mixed gland 混合腺 ◇ serous gland 浆液腺

glanders /ˈɡlændəz；*AmE* -dərz/ n. a contagious，fatal disease of horses and other equine species caused by the bacterium *Pseudomonas mallei*，characterized by nasal discharge，swollen lymph nodes，and ulceration of the upper respiratory tract，lung and skin（马属动物所患的致死性传染病，由鼻疽杆菌引起，其症状为流鼻，淋巴结肿大以及上呼吸道、肺和皮肤溃疡）马鼻疽：Glanders has been largely eliminated from horse populations today. 目前，鼻疽在马群体中已基本消灭。 syn farcy，malleus

glandule /ˈɡlændjuːl/ n.【physiology】a small gland 小腺

glans /ɡlænz/ n.【anatomy】（*pl.* **glandes** /ˈɡlændiːz/）the glans clitoris or glans penis 阴蒂头；阴茎头

glans clitoris n.【anatomy】the anterior part of clitoris 阴蒂头

glans penis n.【anatomy】the bulbous or terminal end of the penis；balanus 阴茎头，龟头

Glass Coach n. a four-wheeled，horse-drawn vehicle introduced in the 17th century，named for its glass panels above the elbow line（出现在17世纪的一种四轮马车，由于装有玻璃窗而得名）玻窗马车

glass eye n. *see* wall eye 玻璃眼，玉石眼

gleet /ɡliːt/ n. *see* nasal gleet 鼻腔黏液，鼻腔脓液

globulin /ˈɡlɒbjʊlɪn/ n.【physiology】a group of soluble proteins present in blood plasma，milk，and muscles（存在于血浆、乳汁和肌肉中的可溶性蛋白质）球蛋白：immune globulin 免疫球蛋白

glomerulus /ɡləʊˈmeərjʊləs；*AmE* ɡloʊˈmer-/ n.【anatomy】（*pl.* **glomeruli** /-laɪ/）a small cluster or mass of capillaries or nerve fibers（由毛细血管或神经纤维束构成的结构）小球：glomeruli of kidney 肾小球 ◇ renal glomeruli 肾小球

gloss /ɡlɒs；*AmE* ɡlɔːs/ n. shine or luster on the surface of the coat（被毛的）光泽，光亮

glossy /ˈɡlɒsi；*AmE* ˈɡlɔːsi/ adj. having a smooth，shiny，lustrous surface；sleek 光滑的，有光泽的：a glossy coat 光亮的被毛

glottal /ˈɡlɒtl；*AmE* ˈɡlɑːtl/ adj. relating to or articulated in the glottis 声门的：glottal cartilage 声门软骨

glottis /ˈɡlɒtɪs/ n.【anatomy】（*pl.* **glottises**；or **glottides**）the opening between the vocal cords at the upper part of the larynx（喉腔两声带间

的开口)声门

glucose /ˈgluːkəʊs; *AmE* -koʊs/ n. a monosaccharide found widely in most plant and animal tissue; is the principal circulating sugar in the blood and the major energy source of the body (广泛存在于动植物的单糖,随体液循环至周身各处,是机体主要的能量来源)葡萄糖: blood glucose concentration 血糖浓度

glucostat /ˈgluːkəˌstæt/ n. any chemical substance, such as hormones, that functions to regulate the blood glucose level (能调节血糖水平的药物)血糖调节剂

glue-on shoe n. a plastic horseshoe glued to the hoof with epoxy or other adhesives (用环氧树脂粘在马蹄上的橡胶蹄铁)粘合蹄铁,胶粘蹄铁: Glue-on shoes generally last no longer than three to four weeks. 胶粘蹄铁通常只能维持3~4周。

gluteal /ˈgluːtiːl/ adj. of, near or situated at the buttocks of a horse 臀的,近臀的: gluteal region 臀部 ◇ gluteal muscles 臀肌

glycosaminoglycan /ˌglaɪkəʊsəˌmiːnəʊˈglaɪkæn/ n. (abbr. **GAG**) any of the compounds found in cartilage and the synovial fluid of the joints (见于软骨和关节黏液中的一种复合物)黏多糖

GnRH abbr. gonadotropin-releasing hormone 促性腺激素释放激素

go /gəʊ/ vi. to move in a certain way 运动,行进 PHR V **go amiss**【racing】(of a mare) to come into season at the time when she is due to race (指母马在出赛时发情)发情时机不当,发情不逢时 **go big**【harness racing】(of a driver) to drive as fast as possible at the beginning of the race (轻驾车赛中)先发制人,先声夺人 **go into the cow**【cutting】(of a horse) to step forward up to the cow when prompted by the rider (指马)突入牛群,朝牛突进 **go on**【racing】(of a horse) to win another race of longer distance (指马)再度获胜,再次折冠 **go short**(of a horse) to move with uncertain, restricted stride due to lameness or other limb problems (马患肢体疾病后)运步不稳,步幅较短 **go to ground**【hunting】(of a prey) to take refuge into the ground or other shelter (指猎物)钻入地洞 **go under the wire**【racing, dated】(of a horse) to win a race (指马)获胜,夺冠 **go well into the bridle** (of a horse) to accept the bit without pulling sideways (指马)受勒控制

goal /gəʊl; *AmE* goʊl/ n. **1**【polo】a specified area of polo field into which players attempt to hit the ball to score (马球赛场的特定区域,选手将球击入此处方可得分)球门 **2**【polo】a point scored when the ball is hit into the goal (球射入球门后的)得分: To equalize wind and turf conditions, the teams change sides after each goal is scored. 为了平衡刮风和场地的影响,射门得分后两队要交换场地。**3**【polo】the value of a player to the team (球员在队中的)价值,分数: high-goal player 高分值球员 intermediate-goal player ◇ 中分值球员 low-goal player 低分值球员

goal handicap n.【polo】*see* polo handicap (马球)分级体系

goal post n.【polo】either of the two vertical poles located on the short ends of a polo field that define the goal (球场上用来标识球门的两根立柱)门柱: In polo, the goal posts should be a minimum of 3 m and set 7 m apart. 马球比赛中,球门门柱至少应高于3米,且两根门柱间距要在7米以上。

goal rating n. *see* polo handicap 等级评分体系

goat grabbing n. *see* Buzkashi 叼羊大赛,马背夺羊

goat knees n. *see* bucked knees 凸膝

goat snatching n. *see* Buzkashi 叼羊大赛,马背夺羊

Godolphin Arabian n. (c. 1723 – 1753) one of three foundation sires of the English Thoroughbred; was a purebred Arabian imported into England in 1728 by Edward Coke and later bought by the Earl of Godolphin for whom it was named（1728 年由爱德华·考克运人英国的一匹纯种阿拉伯马，后被哥德芬伯爵购得故而得名，是英纯血马三大奠基公马之一）哥德芬·阿拉伯 — *see also* **Byerley Turk, Darley Arabian**

Godolphin Barb n. another name for Godolphin Arabian（哥德芬·阿拉伯的别名）哥德芬·柏布

going /ˈɡəʊɪŋ/ n.【BrE】the condition of a racetrack or other surface（赛道或地面的状况）地况，路面：In raining season, the going was very wet. 在雨季，赛场地表非常的潮湿。 syn footing

goiter /ˈɡɔɪtə(r)/ n.【vet】an abnormal enlargement of the thyroid gland, often caused by dietary iodine excess or deficiency（甲状腺的异常肿大，多由日粮中碘过多或缺乏引起）甲状腺肿大

Gold State Coach n.（also **Gold Coach**）a custom-made, horse-drawn vehicle of the coach type built in 1953 for the coronation procession of Queen Elizabeth II of England（1953 年英国伊丽莎白女王二世加冕仪式上专用的四轮马车）［仪仗］鎏金马车，［仪仗］黄金马车：On certain occasions, the gold coach will be used to convey the foreign leaders to Buckingham Palace. 有些场合下，会用黄金马车将外国领导人运至白金汉宫。

golden age of coaching n. the first half of 19th century during which horse-drawn carriages reached the height of their popularity（19 世纪上半叶马车流行的盛期）车马黄金时代

golden age of fox hunting n.【hunting】the period from 1820 to 1890 during which fox hunting reached its greatest popularity and there were no fumes of motors to confuse the scent（1820—1890 年猎狐运动盛行的时期，那时尚无马达油烟混淆狐狸的气味）猎狐黄金时代

golden age of racing n.【racing】the second half of the 19th century during which horse racing reached its peak（19 世纪后半叶赛马业兴旺发达的时代）赛马黄金时代

golden dun n.【color】a type of dun having yellow coat hairs mixed with white（黄色被毛中散布有白毛的兔褐毛）金兔褐 syn creamy dun

golden hoof n. the soft, golden, hoof undersurface of newborn foal, consisting of overgrown horn（新生马驹蹄底多余的柔软呈金黄色的角质组织）黄金蹄

golden horse n. the Palomino horse 帕洛米诺马，银鬃马

golden horse of the West n. the Palomino horse 帕洛米诺马，银鬃马

Golden Horse Society n. an organization founded in Britain in 1947 to maintain the Palomino breed registry（1947 年在英国成立的组织机构，旨在对帕洛米诺马进行品种登记）银鬃马协会

gonad /ˈɡəʊnæd; *AmE* ˈɡoʊ-/ n.【anatomy】the male or female reproductive gland that produces gametes, especially a testis or an ovary（动物体内产生配子的生殖腺，尤指睾丸或卵巢）性腺，生殖腺

gonadotropin /ˌɡəʊnəˈdɒtrɒpɪn; *AmE* ˌɡoʊnædəˈtrɒpɪn/ n.【physiology】any hormone that stimulates the growth and activity of the reproductive glands, especially the FSH and LH（促进性腺生长和活动的多种激素，尤指促卵泡素和促黄体素）促性腺激素

gonadotropin-releasing hormone n.【physiology】（abbr. **GnRH**）a hormone secreted by the hypothalamus responsible for the release of FSH and LH from the pituitary to the blood stream（由下丘脑分泌的一种激素，可调节促卵泡素和促黄体素的释放）促性腺激素释放激素

Gondola of London n. *see* Hansom cab 伦敦冈朵拉，汉瑟姆出租马车

gone away interj.【hunting】a call announced by the huntsman indicating the fox has left his covert（狐狸）出窝了！

gonitis /gəˈnaɪtɪs/ n. inflammation of the stifle joint caused by injuries to the ligaments or bacterial infection of the joint due to puncture wounds（由于韧带拉伤或刺伤后细菌感染导致马膝关节发炎的症状）膝关节炎

good bottom n.【racing】a racetrack that is firm under the surface（指赛马场）底层坚实

good cow n.【cutting】a cow who does not panic when cutted from the herd（从群中拦出时不惊慌失措的牛）好截的牛

good curl n.【roping】a well thrown lariat or rope 投准的套索；投的好！

good mouth n. a mouth of horse that is responsive to the action of the bit（指马）口角灵敏

good roof n.【dated】（also **good top**）a good top line 背部线条优美

good-boned /ˌgudˈbəʊnd/ adj.（of a horse）having a large bone measurement（指马）管围较粗的

goose rump n. a conformation fault of a horse in which the hindquarters slopes too sharply from the point of the croup to the point of buttocks（马后躯从腰角到臀端过于倾斜的体貌缺陷）斜尻；尖尻 [syn] drooping quarters, jumping rump

goose-rumped /ˌguːsˈrʌmpt/ adj.（of a horse）having the conformation fault of goose rump 斜尻的；尖尻的

Gorst Gig n. a hooded, horse-drawn vehicle of the Gig type having a boot under the seat; hung on sideways-elliptical springs（一种带斗篷的轻便双轮马车，车座下带行李箱，车身悬于半椭圆形减震弹簧上）格斯特马车

Gotland /ˈgɔtlənd/ n.（also **Swedish Gotland**）an ancient pony breed originating from the Swedish island Gotland from which the name derived; has a broad forehead, wide-set eyes, muscular and crested neck, strong back, rounded croup, and a full mane and tail with light fetlock feathering; stands 12 to 14 hands and may have any of the solid coat colors with black, bay, and chestnut occurring most common; is long-lived, intelligent, lively, and docile; used for riding, driving, and jumping（源于瑞典哥特兰岛上的古代矮马品种并由此得名。额宽，瞳距大，颈粗且呈弧形，背部有力，后躯浑圆，鬃毛浓密，球节处距毛稀疏，体高12～14掌，各种单毛色兼有，但以黑、骝和栗最为常见，寿命长，反应敏快，性情活泼、温驯，主要用于骑乘、驾车和跨越障碍）哥特兰马 [syn] Forest Pony

gouge /gaʊdʒ/ vt.（of a rider）to strike one's mount hard with spurs（用靴刺）猛踢

Governess car n. *see* Governess cart 家教马车

Governess cart n.（also **Governess car**）a small, two-wheeled vehicle with a rounded shape introduced around 1880, formerly designed to carry a governess and the children in her charge（1880年前后出现的小型双轮马车，过去多用来运送家庭教师及其所教的儿童）家教马车

GRA abbr. Girls Rodeo Association 女子牛仔竞技协会

Grackle /ˈgrækl/ n. *see* Grackle noseband X形鼻革

Grackle noseband n.（also **Grakle noseband**）a noseband consisting of two straps which cross over the bridge of the nose diagonally（由两条皮带交叉系在马鼻梁上的鼻革）X形鼻革，叉形鼻革 [syn] figure eight noseband, cross noseband, Mexican noseband

grade /greɪd/ n. a level of competition or race（赛事）级别，等级：There are several grades of eventing starting with the easiest intro classes, and progressing through to pre-novice, novice, intermediate and advanced classes. 综合全能比赛分好几个等级，从最简单的入门级，到预备级、初级、中级以及高级比赛。

vt. to classify or arrange in degrees 分级

Grade Ⅰ n.【racing】（abbr. **G-1**）the highest of three levels of stakes races held in a country（赛事级别最高的奖金赛）一级赛

Grade Ⅱ n.【racing】(abbr. **G-2**) the middle of three levels of stakes races held in a country (赛事级别居中的奖金赛)二级赛

Grade Ⅲ n.【racing】(abbr. **G-3**) the lowest of three levels of stakes races held in a country (赛事级别最低的奖金赛)三级赛

graded race n.【racing】a stakes race held in a country as designated by racing authorities, with requirement on the minimum purse and restriction on eligibility by sex or age only (由赛事监管机构对国内举行的奖金赛进行的划分,对奖金设置有特定要求,且仅对赛马的性别和年龄有所限制)分级赛: In the United States, the graded race constitutes 15% of all stakes races and are divided into Grade Ⅰ, Ⅱ, and Ⅲ as based on the quality of the competing horses. 在美国,分级赛占所有奖金赛比例的15%,根据赛马的性能通常将赛事划分为Ⅰ、Ⅱ、Ⅲ三个等级。 syn group race

graduate /ˈɡrædʒuət/ n. **1**【racing】a horse or jockey who wins his first race 初次获胜的马;初次获胜的骑师 **2**【racing】a racehorse moving up into allowance, stakes, or handicap racing 晋级赛马

grain /ɡreɪn/ n. the fruit or seed of a cereal grass such as wheat, oats, barley, and corn (禾本科作物如小麦、燕麦、大麦、玉米等的种子)谷粒

grain founder n. (also **grain overloading**) founder caused by ingestion of excessive amounts of grain which results in inflammation of the sensitive laminae of the hoof (因过量采食谷物导致蹄叶发炎的症状)谷食性蹄叶炎

grain overloading n. *see* grain founder 谷食性蹄叶炎

Grakle noseband n. *see* Grackle noseband X形鼻革

Gram stain n. a staining method for the preliminary identification of bacteria (对细菌进行初步鉴定的一种染色方法)革兰氏染色

gram-negative bacteria n. the families of bacteria that do not stain blue by Gram's method in laboratory examination (不能被革兰氏染色法染成蓝色的细菌)革兰氏阴性菌

gram-positive bacteria n. the families of bacteria that stain blue by Gram's method in laboratory examination (被革兰氏染色法染成蓝色的细菌)革兰氏阳性菌

Grand Breton n. *see* Heavy Draught Breton 布雷顿重挽马

Grand Circuit n. a program of harness racing on the tracks 轻驾车赛路线

Grand Liverpool Steeplechase n. *see* Grand National Steeplechase 利物浦越野障碍大奖赛,全英越野障碍大奖赛

Grand National n. *see* Grand National Steeplechase 全英越野障碍大奖赛

Grand National Steeplechase n. a handicap steeplechase race held at the 7.22-km Aintree racecourse located in Liverpool, England since 1839, the course has 30 fences and a water jump 15 feet wide (自1839年起在英国利物浦艾翠赛道上举行的越野障碍赛,赛程7.22千米,内设30道跨栏和宽度为15英尺的水沟障碍)全英越野障碍大奖赛 syn Grand National, Liverpool Steeplechase, Grand Liverpool Steeplechase

Grand Pardubice n. a steeplechase race held annually in Czech on the second Sunday in October since 1874 and run on a 4.5-mile course over plowed fields with 31 fences (自1874年起每年10月的第2个星期日在捷克斯洛伐克举行的越野障碍赛,难度极高,赛程7.2千米,内设31道障碍)帕尔杜比策大奖赛

Grand Prix n. **1** a horse race of the first class (赛马)大奖赛: Grand Prix de Paris 巴黎大奖赛 **2**【dressage】the highest level of international dressage competition (国际盛装舞步比赛的最高级别)大奖赛—*see also* **Prix St. Georges, Intermediare Ⅰ, Intermediare Ⅱ**

granddam /ˈɡrændæm/ n. (also **grand dam**) the mother of a horse's dam 祖母马: paternal granddam 父系祖母马 [syn] second dam

grandsire /ˈgrændˌsaɪə/ n. (also **grand sire**) the father of a horse's sire (父系)祖公马,祖父马:maternal grandsire 母系祖父马—*see also* **granddam**

grandstand /ˈgrændstænd/ n. 【racing】 the roofed seating area for spectators at a racetrack (赛马场上为观众搭建的看台)观众席,看台

Grant's zebra n. one of the sub-species of common zebra (普通斑马的亚种)格兰特斑马

granula /ˈgrænjuːlə/ n. 【Latin】granule 颗粒

granular /ˈgrænjələ(r)/ adj. composed of, containing, or resembling granules 颗粒的;颗粒状的

granulate /ˈgrænjuleɪt/ v. to become or form granules 颗粒化,形成颗粒

granule /ˈgrænjuːl/ n. a small particle, especially within a cell (细胞中的)颗粒;微粒

granulocyte /ˈgrænjuləˌsaɪt/ n. any of a group of white blood cells having granules in the cytoplasm (胞质中含有颗粒的白细胞)粒细胞: neutrophilic granulocyte 中性粒细胞 ◇ acidophilic granulocyte 嗜酸性粒细胞 ◇ basophilic granulocyte 嗜碱性粒细胞

grapes /greɪps/ n. 【vet】 characteristic fungal growth developing as a result of scratches (皮肤擦伤后遭真菌感染引起的增生)葡萄疮

grass /grɑːs/ n. any wild plant grazed by herbivorous animals (草食动物采食的野生植物)饲草,牧草: fresh grass 青草 ◇ grass hay 干草 [compare] hay [IDM] **at grass**(of horses) turned out on a pasture (指马)在草场上

grass belly n. *see* hay belly 草包肚,草腹

grass clippings n. the vegetative materials resulting from grass mowing operations (刈草机修剪草场后产生的植物碎屑)草屑,草渣:Grass clippings are sometimes used as a food source of horses. 草屑有时也用作马匹饲草。

grass cutter n. 【polo】a polo ball hit with such force by the player that it grazes the surface of the grass field (击打时可以切断牧草的马球)割草击球

grass founder n. inflammation of the sensitive laminae of horse's hoof caused by the ingestion of excessive amounts of protein found in lush pastures (马匹采食大量高蛋白牧草后导致蹄叶发炎的症状)草食性蹄叶炎

grass hay n. the cut and dried grass 干草:Grass hay may be fed freely with little likelihood of overfeeding, colic, or founder resulting. 干草可随意饲喂,且很少会导致马匹暴食、腹痛或蹄叶炎等症。

grass pen n.【hunting】an area where bitches in season are confined (母犬发情时圈养的围场)草场围栏

grass sickness n. a fatal disease of unknown origin that results in a drastic slowdown of the horse's digestive system (马患的一种起因不明的致死性疾病,可导致消化系统机能急剧衰退)马草瘟 [syn] grass tetany, equine grass sickness, equine dysautonomia

grass slip n.【racing】a written permission granted by the racetrack authorities to a jockey, trainer, or owner to exercise a horse on the turf course (赛场主管获准在草场上训练马匹而授予的)草地训练证

grass tetany n. *see* grass sickness 草瘟

grassland /ˈgrɑːslænd; *AmE* ˈgræs-/ n. an area of grass or grasslike vegetation 绿地,草场: grassland management 草场管理

gray /greɪ/ n.【color】(also **grey**) a coat color of horse having dark skin and white hairs mixed with black ones (马匹毛色类型之一,皮肤为深色,被毛中混有黑、白两色毛发)青毛: light gray 淡青 ◇ dapple gray 斑点青 ◇ steel gray 铁青 ◇ rose gray 玫瑰青 ◇ Gray horses are born black and become progressively whiter with each shedding. 青毛马出生时被毛为黑,其色调会随着每次换毛逐渐变浅。

vi. (of the coat) to change from any color to gray(被毛)变浅;褪色 | **graying** n. 褪色

gray fox n. a carnivorous mammal of the dog family native to North America; has a pointed muzzle, erect ears, long bushy tail, gray coat and is noted for its cunning and alertness（分布于北美的犬科食肉哺乳动物，鼻尖，耳朵直立，尾大，被毛灰色，以其狡猾、机敏著称）灰狐：Gray fox is the primary quarry of foxhounds. 灰狐是猎狐犬的主要猎物。[syn] *Urocyon cinereoargenteus*

G

gray overo n.【color】a gray horse with large ragged patches of white（带大片锯齿状白斑的青毛马）花青毛马

gray ticked adj.【color】(of coat color) sparsely distributed with white hairs throughout the coat on any part of the body（马体部位上散生白毛）斑点青的

graze[1] /greɪz/ v. **1** to feed on growing grasses in a field or on pastureland 吃草，采食 **2** to put livestock on pasture to feed on grass 放牧：rotational grazing 划区轮牧 ◇ zero grazing 零牧

graze[2] /greɪz/ v. to scrape or scratch slightly; abrade 擦伤，蹭破

n. a minor scratch or abrasion 擦伤

grazing bit n. a curb bit designed to enable the horse to graze with the bit in its mouth（一种铰链式衔铁，马带口衔时也可采食牧草）采食衔铁

grease /griːs/ n. the oily substance secreted by the sebaceous gland（皮脂腺分泌的）油脂，油渍

greasy /ˈgriːsi,-zi/ adj. coated with grease or containing too much grease 油污的；油腻的

greasy heel n. *see* scratches 蹄踵炎

great coat n. a heavy overcoat worn by the postillion to protect him from rain and cold（左马驭者用来挡雨御寒的外套）大衣 [syn] postboy's coat

Great Horse n.（also **English Great Horse**）a strong medieval war horse developed in England in the early 10th century; was able to carry a fully-armored knight bearing heavy weapons and agile enough to charge ahead in combat（于10世纪初在英格兰育成的中世纪战马，体格强健，可载身穿铠甲手持兵刃的骑士，且动作敏捷适于战场冲锋）英格瑞特马

green /griːn/ adj. **1**（of a horse）broken but not fully trained（马）尚未调教好的：a green horse 一匹未调教好的马 ◇ Some people will want to bring a green horse along, but others will want to buy a "made" animal. 有些人想自己训练调教马匹，但也有人乐意买调教好的马。**2**（of a rider）inexperienced（指骑手）无经验的，新的：a green rider 无经验的骑手；新手

n. a green horse 一匹未调教好的马

green meat n. the basic feed of horses consisting of hay, grass, and straw 饲草，草料

greenbroke /ˈgriːnˌbrəʊk; *AmE* -ˌbroʊk/ adj.（of a horse）recently broken but still inexperienced（指马）刚调教的 [compare] broke

n. a greenbroke horse 刚调教的马：A greenbroke horse should only be ridden by advanced horse-person until it's training is completed. 在调教尚未完成前，刚调教的马匹只能由高级骑手来策骑。

greenhorn /ˈgriːnhɔːn; *AmE* -hɔːrn/ n. an inexperienced or immature rider 生手，新手

gregarious /grɪˈgeərɪəs; *AmE* -ˈger-/ adj.（of animals）tending to form a group with others 群居的，集群的：By nature, the horse is a gregarious animal used to living in herds of six to twenty animals, each group with its own lead stallion. 马在本性上是群居动物，经常以6～20匹组成一群，每群都有各自的领头公马。

Grevy's zebra n. one kind of zebra that lives on the plains and in the bushy and semi-desert areas of northern Kenya, Ethiopia and Somalia and in the southern Sudan, characterized by its narrow stripes all over the body（栖息在肯尼亚北部、埃塞俄比亚、索马里和苏丹南部的一种斑马，身上的斑纹较细）格雷氏斑马，细纹斑马 [syn] *Equus grevyi*

grey /greɪ/ n.【color】*see* gray 青毛(马)

greyhound /ˈgreɪhaʊnd/ n. a breed of tall, slender dog with smooth coat and long thin legs, used in racing and hunting (一种身体瘦长的犬,被毛光滑,腿细长,多用于赛跑和狩猎)灰狗,灵猩

greyhound-like /ˌgreɪhaʊndˈlaɪk/ adj. like a greyhound 似灰狗的,狭瘦的:greyhound-like sides 胸肋狭瘦

grid /grɪd/ n.【jumping】a series of obstacles 格栅障碍,系列障碍

grind /graɪnd/ vt. (of a horse) to grate its teeth together, as due to boredom, anger, resistance to the bit, or tension (马由于烦躁、愤怒、抗衔或紧张)磨牙:to grind teeth 磨牙

grinder /ˈgraɪndə(r)/ n. a molar tooth 臼齿

gripe /graɪp/ vi. to have or cause sharp pains in the bowels (肠道)绞痛

n. (usu. **gripes**) spasmodic pains in the bowels; colic 腹痛;绞痛

Grisone, Frederico n. a 16th-century Italian equestrian who established his riding school in Naples and was one of the first to promote the use of combined aids and the leg rather than the spurs (16 世纪意大利的骑师,在那不勒斯建立了自己的骑术学校,首先倡导联合采用多种辅助而非马刺)弗雷德里克・格瑞森

grissel /ˈgrɪsl/ n. *see* rount 沙灰毛马

grit /grɪt/ n. small particles of sand or stone 砂砾,砂石:The stud holes must be clean and free from grit in order to be able to fit the studs easily. 为了安装防滑钉之便,蹄钉孔内必须没有砂石。

grizzle /grɪzl/ n. *see* rount 沙灰毛马

groin /grɔɪn/ n.【anatomy】the hollow on either side of the horse's body where the thigh meets the abdomen (马大腿与腹相连处的凹陷处)腹股沟,鼠蹊 [syn] lisk

Groningen /ˈgrəʊnɪŋən/ n. (also **Groningen horse**) a Dutch warmblood breed developed from Friesians and Oldenburgs; has a deep chest, powerful quarters and shoulders, and short legs; stands 15.2 to 16 hands and generally has a black, dark brown, or bay coat; is an easy keeper and used for light draft and riding (荷兰温血马品种,由弗里斯兰马和奥尔登堡马培育而来。胸深,后躯与肩胛发达,四肢短,体高 15.2 ~ 16 掌,毛色多为黑、褐黑或骝,易饲养,主要用于轻挽和骑乘)格罗宁根马

Groningen Horse n. *see* Groningen 格罗宁根马

groom /grum/ n. one employed to brush, clean and take care of horses (刷拭料理马匹的雇员)马夫,马倌 [syn] guinea, ginney

vt. to clean and brush a horse 刷拭,清洁(马体):The optimum time to groom a horse is when he returns to the stable after exercise as soon as he is dry. 马匹结束训练回到厩舍汗液刚干的时候,便是刷拭被毛的最佳时间。

groomer /ˈgrumə(r)/ n. the brush and tools used in grooming horses 刷拭工具:electrical groomer 电动刷 ◇ As well as keeping the coat clean an electric groomer massages the skin better than one could by hand. 除了能清洁被毛外,电动刷还可取得比手工刷拭更好的按摩效果。

grooming /ˈgrumɪŋ/ n. the act or process of cleaning and brushing a horse (马体)刷拭,梳理:mechanical grooming 电动修理 ◇ grooming behavior 梳理行为 ◇ grooming tools 刷拭工具

grooming kit n. the brushes, combs, and other equipment collectively used to groom horses 刷拭工具[箱]

grooming stable rubber n. *see* stable rubber 厩用擦布

groom's coat n. a single-breasted coat customarily worn by the groom (马夫穿的单襟大衣)马夫大衣,马夫外套

groove /gruːv/ n. (also **Galvayne's groove**) the dark line appearing on the upper corner incisors of horse over the age of ten years (10 岁以上的老马上颌隅齿出现的黑线)纵沟:At the age of 15, the Galvayne's groove extends halfway down the upper corner incisor. 马 15 岁时,上颌隅齿表面纵沟会延伸至半个牙齿的长度。

vt.【farriery】to cut groove or grooves 开槽

gross feeder n. a hound or horse who overeats 暴食的猎犬;暴食的马

ground[1] /graʊnd/ n. **1** the surface of land 地面 PHR **near/well to the ground** (of a horse) having short cannon bone (指马)四肢较短的 **gain ground** (**on sb/sth**)【racing】to run close to or overtake sb/sth 紧追;赶超 **save ground**【racing】to cover the shortest distance in a race by running straight on the stretches and hugging the inside rail on the turns (赛马在直道和转弯处)省路,抄近道 **stand on the ground**【racing】(of a jockey) be suspended from competition (指骑师)被停赛的:This jockey has stood on the ground for two years because of doping. 由于服用兴奋剂,这名骑师已被停赛两年。**2**【driving】the area traveled by a coachman (车夫的行程) 行程: The coachman's ground was forty-five miles between Madrid and Toledo. 车夫从马德里到托莱多的行程为 72 千米。

ground[2] /graʊnd/ v. (of the hoof) touch the ground after raising (指马蹄)落地,着地 | **grounding** n. 落地,着地

ground line n.【jumping】a line drawn on the ground in front of a fence to help the horse and rider judge the take-off point (划在障碍前地面上的直线,用来帮助马和骑手确定最佳起跳点)地面起跳线

ground man n.【jumping】one who assists to set fences, raise and lower rails, and provide instruction on riding style or technique on the ground (为骑手设定障碍、升降横栏或给予技术指导的人员)地面指导;地面助理

ground manners n. the behavior of a horse when handled, groomed, saddled, or lunged on the ground (马在牵行、刷拭、备鞍或打圈时的行为)地面姿态

ground pole n.【jumping】a round wooden pole laid on the ground to assist the horse in regulating its stride during jumping (放置在地面上的木杆,用来调整马跨栏时的步伐)地杆 syn cavaletti, ground rail, guard rail, trot pole

ground rail n. *see* ground pole 地杆

grounded /ˈgraʊndɪd/ adj.【racing】(of a jockey) suspended from competition by racing authorities for rule infraction (骑手由于违例而被赛事主管机构禁赛)停赛的,禁赛的

group /gruːp/ n.【BrE, racing】a level of stakes races, often capitalized when used in a race title (奖金赛的赛事级别,用于赛事名前时要大写)级别,等级:Group-1 Epsom Derby 一级艾普森德比赛

group race n.【BrE, racing】a method established in 1971 by racing organizations to classify stakes races run in Europe into grades 1, 2, or 3 (1971 年赛事机构确立的方法,将欧洲奖金赛划分为 1、2、3 三个等级)分级赛 syn graded race

grow /grəʊ/ vi. to increase or develop in size or degree 生长;增长 PHR V **grow out** (of a horse) to exceed the age limit for certain races (指马)超过年龄限制

growth /grəʊθ/ n. the act or process of growing 生长:growth curve 生长曲线

growth cartilage n. *see* epiphyseal plates 生长软骨,骺板

growth plate n. *see* epiphyseal plate 生长板,骺板

grub /grʌb/ n. the worm-like larva of certain insects 幼虫

gruel /ˈgruːəl/ n. a thin, watery mash of oatmeal 燕麦粥

grullo /ˈgruːləʊ; *AmE* -loʊ/ n.【color】(also **slate grullo**) a slated-colored coat with slight variations of shade and a dark head, commonly has wither, dorsal stripes and zebra stripes on the legs (被毛如石板的毛色,全身色调有细微差异,头部色调深,多带鹰膀和背线,且

四肢上有斑马纹)石板青,青兔褐 syn blue dun

grunt /grʌnt/ n. a deep, guttural sound made by a horse when threatened or frightened (指马受胁迫时发出的低声喉音)哼叫

vi. (of a horse) to utter a deep, guttural sound when threatened or frightened (指马在受惊时)哼叫:Grunting is usually considered as a a potential sign of wind unsoundness. 一般认为哼叫是马匹呼吸性能不良的指征。

grunter /ˈgrʌntə(r)/ n. a horse who grunts when threatened or frightened (受到惊吓后)哼叫的马

GS Wagon abbr. general service wagon 兼用四轮马车

GSB abbr. General Stud Book 英国纯血马登记册

Guaga /ˈgwɑːgə/ n. a small, four-wheeled, horse-drawn public carriage popular in Cuba throughout the 19th century; usually drawn by four ponies abreast and entered from the rear (19 世纪流行于古巴的一种小型公共四轮马车,多由四匹矮马并驾,乘客从后门上车)古伽马车

Guajira /gwɑːˈhɪərə/ n. a Colombian-bred Criollo descended from Spanish stock brought to South America by the Conquistadors in the 16th century (哥伦比亚育成的克里奥罗马,由 16 世纪西班牙殖民者从运到南美的西班牙马繁育而来)瓜希拉马

Guanzhong n.【Chinese】(also **Guanzhong horse**) a breed of horse developed in the 1960s in Shaanxi, China by crossing native horses with Budyonny, Don, and Ardennais; stands about 15 hands and generally has a chestnut or bay coat; has a medium-sized head, pricking ears, sloping shoulders, straight back, well-formed legs with little or no feathering, and hard feet; is frugal and used mainly for driving and farming (20 世纪 60 年代在中国陕西省育成的马品种,主要由当地马经布琼尼马、顿河马以及阿尔登马改良培育而成。体高约 15 掌,毛色以栗、骝为主,头型中等,耳竖立,背腰平直,斜肩,四肢端正,无距毛或距毛很少,蹄质坚韧,耐粗饲,主要用于驾车和农田作业)关中马

guard rail n. *see* ground pole 地杆

Guérinère, François Robichon de la n. (1688 – 1751) a French riding master who developed classical equitation and established the basis for modern equitation; often renowned as the Father of Classical Equitation (法国骑术大师,变革了古典骑术并为现代骑术奠定了基础,常被誉为古典骑术之父)弗朗索瓦・罗比臣・德拉・格伊尼亚

Guerney /ˈgɜːrni/ n. a horse-drawn vehicle of the cab type popular in the United States around the turn of the 19th century (19 世纪之交流行于美国的一种出租马车)格恩尼马车

guinea /ˈgɪni/ n. **1** a former British coin worth 21 shillings (过去的英国硬币,相当于 21 先令)几尼 **2**【racing, slang】(also **ginney**) a groom 马夫:The term guinea was originated from the fact that the winning owners of racehorse usually tip the groom a guinea. "Guinea"这个术语源于赛马获胜的马主通常付给马夫一几尼的小费。

guinea hunter n.【BrE, slang】one who acts as an intermediary between the seller and buyer of a horse for payment (充当马匹买卖中介赚钱的人)贩马商,牙子

Guizhou Pony n.【Chinese】a pony breed originating in Guizhou province, China; stands about 11 hands and generally has a chest or bay coat; has a relatively small and compact build, an elegant profile, and small pricking ears; is frugal and docile and used mainly for packing in mountain areas (产于中国贵州省的矮马品种。体高约 11 掌,毛色以栗、骝为主,体格较小,体质结实,面貌清秀,耳小直立,耐粗饲,性情温驯,多用于山区驮运)贵州矮马

gullet /ˈgʌlɪt/ n. **1** the esophagus 食管 **2** the arch-shaped iron fork underneath the inner side of the saddle（马鞍内侧的弓形铁叉）鞍沟

gulp /gʌlp/ vi. to swallow air audibly, as in nervousness; gasp 吞咽；喘气

gumbo /ˈgʌmbəʊ/ n.【racing】heavy, sticky mud, as on the racetrack 烂泥；泥泞

gummy legged adj. *see* gammy legged（指马下肢）水肿的，肿胀的

G

Guoxia pony n.【Chinese】an ancient pony breed originating in Southwest China; has an unrefined profile, short neck and back, straight shoulders, and strong legs with hard feet; stands below 11 hands and usually has a bay, gray, or roan coat; is willing, tough, and quiet; used for packing, riding and pleasure（源于中国西南的古老矮马品种。面貌粗糙，颈背短，直肩，四肢强健，蹄质结实，体高 11 掌以下，毛色多为骝、青或沙，性情温驯、顽强而安静，主要用于山地驮运、骑乘及娱乐）果下矮马，果下马

Gustav Rau n. *see* Rau, Gustav 格斯塔夫·劳

gut /gʌt/ n.【anatomy】the internal alimentary organs, especially the intestines or stomach; entrails（体内的消化器官）肠胃；内脏

guttural /ˈgʌtərəl/ adj. of or relating to the throat 喉的，咽喉的：guttural pouch 喉囊

gymkhana /dʒɪmˈkɑːnə/ n. **1** a horse show where various mounted contests or games are held 马上竞技，马术大会：Gymkhana was evolved from mounted exercises performed by Indian soldiers during the British colonial period. 马术大会源于殖民地时期英国驻印士兵的马上训练。**2** the place where such an event is held 表演场；竞技场

gynecology /ˌgaɪnəˈkɒlədʒi; *AmE* -ˈkɑːl-/ n.（BrE also **gynaecology**）the scientific study of the treatment and diseases concerning child birth or female reproductive system（研究分娩和雌性生殖系统相关疾病和治疗的科学）产科学 | **gynecological** /ˌgaɪnəkəˈlɒdʒɪkl/ adj.：gynecological examination 产科检查

gyp /dʒɪp/ n. a female foxhound 雌猎犬

Gypsy Wagon n. an enclosed, horse-drawn wagon used by gypsies 吉普赛马车

habit /ˈhæbɪt/n. (also **riding habit**) **1** a woman's formal clothing for riding side-saddle in the past (过去女性侧鞍骑乘的装束)骑马套装,骑装 **2** the clothing worn by a rider 骑装:The fashion of riding habit had kept evolving during the past centuries. 在过去几个世纪,骑装样式一直在变。

habitat /ˈhæbɪtæt/n. the natural environment of plants or animals (动植物生长的自然环境)生境,栖息地

Habronema /ˌhæbrəˈnemə/n. 【vet】(*pl.* **habronemae** /-nemiː/) a species of internal parasite found in the mucosa of the stomach (吸附在马胃黏膜上的内部寄生虫)马柔线虫,马胃线虫 syn stomach worm

habronemiasis /ˌhæbrəˈnemɪəsɪs/n. 【vet】a condition resulting from heavy infestation of *Habronemae*, often characterized by ulceration of itchy nodules and gastritis (马遭胃线虫感染后所患的病症,表现为瘙痒、糜烂和胃炎等症)马胃虫线病:cutaneous habronemiasis 皮肤线虫病 ◇ gastric habronemiasis 胃线虫病

hack[1] /hæk/n. **1** a horse used for riding 乘用马:a good hack 好骑的马 **2** a ride on a horse 骑马,骑乘 **3** a worn-out horse for hire; a jade 出租的老马;驽马

vi. to ride a horse for pleasure or exercise 骑马

hackamore /ˈhækəˌmɔː(r)/n. a bitless bridle widely used by Western riders for teaching young horses proper flexion of neck and manners through control achieved by applying pressure to the horse's nose, poll, and its chin (美国西部骑手在调教青年马时常用的一种无口衔马勒,主要通过施力于马的鼻梁、项顶和下颌进行控制)调马勒,驯马勒:Hackamores provide a gentler way to communicate with the horse without sticking a metal object in their mouth. 调马勒控马方式轻柔,不带戳刺口腔的衔铁。

hackamore bit n. a bit without mouthpiece that consists of a curved metal noseband and shank that apply pressure to horse's nose and chin respectively (一种无嚼铁的口衔,主要由施力于鼻梁和下颌的弧形鼻杆和颌杆构成)调马衔铁:Hackamore bits are used particularly for the rough riding often required in rodeo events, and since there is no bit in the horse's mouth, it is impossible to injure that area with quick jerking motions. 调马衔铁在牛仔竞技等野骑项目中尤其常用,由于它不带口衔,所以骑手猛力拽扯也不会伤及马的口角。 syn mechanical bit

hacking /ˈhækɪŋ/adj. of or related with riding horses 骑马的,乘用的:hacking clothes 骑装 ◇ a hacking jacket 乘用夹克 ◇ a hacking outfit 一套骑装

n. instance of riding horses 骑马,策马

hackles /ˈhæklz/n. pl. the erectile hairs along the back and neck of a dog (犬颈部竖立的毛发)竖毛,立毛 IDM **have/get hackles up** (of a hound) have hair along the neck and back stand upright due to excitement (指猎犬由于兴奋)颈毛直立:The whole pack of hounds are having their hackles up and barking when they catch sight of a red fox. 看到一只赤狐后,猎犬个个都颈毛直立、一片狂吠。

Hackney /ˈhækni/n. **1** (also **Hackney horse**) a light draft horse developed in England by crossing Norfolk Trotters with native and Arab horses; has a flat shoulder and distinctive high-stepping action with flexion of the knee in trot; stands 14 to 15. 2 hands and may have any of the solid colors; used mainly for driving (英国采用诺福克快步马与当地马和阿拉伯马杂交育成的轻型挽用马品种。肩膀平直,快步行进时举蹄高且屈膝动作明显,体高 14 ~ 15.2 掌,各种单毛色兼有,主要用于驾车)哈克尼马 **2** *see* Hackeny Coach 哈克尼马车

Hackney Coach n. (also **Hackney**) a four-wheeled, springless, six-passenger public service vehicle used throughout European during the 17th and l8th centuries, usually drawn by a pair of horses in pole gear (17—18 世纪在欧洲使用的一种公共客用四轮马车,无减震弹簧,可乘 6 人,多由双马套驾杆拉行)哈克尼马车

Hackney horse n. *see* Hackney 哈克尼马

Hackney Horse Society n. an organization founded in Great Britain in 1884 to promote and encourage the breeding of Hackneys, harness horses, cobs, and ponies (1884 年在英国建立的组织机构,旨在促进哈克尼马、挽马、柯柏马和矮马的育种工作)哈克尼马协会

Hackney Pony n. a pony breed developed in England in the late 19th century by crossing the Hackney with Fell and Welsh ponies; has a light head, arched neck, prominent withers, a long and rounded croup, and slender but strong legs; stands 12.1 to 14 hands and may have a brown, black, gray or roan coat; is fiery, energetic, and fast; formerly used to pull delivery vehicles, now used chiefly for show purpose (英国在 19 世纪末采用哈克尼马与费尔矮马和威尔士矮马杂交培育的矮马品种。头轻,颈弓形,鬐甲突出,尻部浑圆,四肢修长、强健,体高 12.1 ~ 14 掌,毛色多为褐、黑、青或沙毛,性情激昂,体力充沛,步速快,过去多用来运货,现在主要用于展览赛)哈克尼矮马: The Hackney pony is often referred to as the prince of ponies because of his elegance and friskiness. 哈克尼矮马由于姿态优雅、活泼好动,经常被冠以"矮马王子"的称号。

h(a)em- pref. of blood 血的

haemolysis /hɪˈmɒlɪsɪs/n. 【BrE】 *see* hemolysis 溶血症

haemorrhage /ˈhemərɪdʒ/n. 【BrE】 *see* hemorrhage 大出血,过量失血

Haflig /ˈhæflɪg/n. *see* Haflinger 哈弗林格马

Haflinger /ˈhæflɪŋə(r)/n. (also **Haflig**) a cold-blood breed originating in the southern Austrian around the mountain village of Hafling from which the name derived; has a broad head, large eyes, open nostrils, small ears, long back, muscular loins and quarters, tough hoofs, short legs, and free action with long strides; stands up to 14 hands and always has a chestnut coat with flaxen mane; is long-lived, frugal, sure-footed, tough, hardy, sound, and kind; used for packing and draft (源于奥地利南部哈弗林格山区的冷血马品种并由此得名。额宽,眼大,鼻孔开张,耳小,背长,后躯发达,蹄质结实,四肢粗短,运步流畅,步幅大,体高约 14 掌,被毛栗色,鬃毛淡黄,寿命长,耐粗饲,抗逆性强,性情温驯,主要用于驮运和重挽)哈弗林格马: Haflinger is often branded with Austria's native flower, the Edelweiss with the letter H in the center. 哈弗林格马身上常烙有奥地利国花雪绒花的图案,图案中间有字母"H"。 syn Edelweiss Pony

HAHS abbr. Hooved Animal Humane Society 有蹄动物人道主义协会

hair /heə(r); *AmE* her/n. the fine thread-like strands growing over the body of a horse (马体身上的)被毛

vi. to grow hair over its body 长毛 PHR V

hair up (of a horse) to grow a thick winter coat 长厚毛,长粗毛: Thoroughbreds do not hair up much even in winter. 即使在冬天,纯血马也不太长厚毛。

hair color n. the coat color of a horse 毛色

hairworm /ˈheəˌwɜːm; *AmE* ˈherˌwɜːrm/n. (also **horsehair worm**) any of various slender parastic nematode worms that live in the stomach and small intestine of animals (寄生在家畜胃

和小肠内的多种线虫)毛细线虫,铁线虫

hairy /ˈheəri; *AmE* ˈheri/adj. 【slang】(of a horse) having heavy feathering around its fetlock (指马)距毛浓密的:a hairy Shire 一匹距毛浓密的夏尔马。

hairy bullfinch n.【jumping】an obstacle consisting of a live and untidy hedge jumped over not through (由成簇的活篱笆构筑的障碍)毛树篱障碍

half brother n.【genetics】a relationship of male horses out of the same dam but by different sires (公马之间母本相同而父本不同的亲缘关系)半同胞兄弟 compare full brother

half cannon n. *see* sock 半管白

half chaps n. *see* leggings 护腿

half halt n. a maneuver performed by the rider to demand the horse's attention for a change of direction or gait (骑手在变向或换步时为提高马匹注意而采取的举动)半停,暂停

half moon bit n. *see* Mullen mouth bit 半月形口衔

half moon mouth n. *see* Mullen mouth bit 半月形口衔

half pass n.【dressage】a lateral dressage movement performed on two tracks in which the horse moves sideways and forwards at the same time (马以双蹄迹向斜前方行进的盛装舞步动作)斜横步 compare side pass

half pastern n. a leg marking of the horse which extends from the coronary band half way up the pastern (马四肢别征之一,其白色区域从蹄冠延伸至系部中段)半系白

half sister n.【genetics】a relationship of female horses out of the same dam but by different sires (雌驹之间母本相同而父本不同的亲缘关系)半同胞姐妹 compare full sister

half stocking n. a leg marking of the horse which extends from the coronary band to the lower half of the cannon bone just above the fetlock joint (马四肢别征之一,其白色区域从蹄冠延伸至球节以上管骨中段处)1/2 管白,半管白 syn half cannon,sock

half volte n.【dressage】one half of a 6-meter circle 半圈

half-blood /ˈhɑːfˌblʌd; *AmE* ˈhæf-/n. (also **halfblood**) another term for the warmblood 半血马,温血马 |**half-blooded** /ˌhɑːfˈblʌdid/adj.

half-bred /ˈhɑːfˌbred; *AmE* ˈhæf-/adj. half-blooded; hybrid 半血的;混血的

half-breed /ˈhɑːfˌbriːd; *AmE* ˈhæf-/n. **1** a horse of mixed blood strains, having parents of different races or breeds (双亲来源于不同品种或系谱的马)混血马 **2** a horse having only one parent that is purebred (仅单亲是纯种的马)半血马

adj. (of a horse) half-blooded; hybrid 半血的,温血的;混血的

half-mile pole n.【racing】a marking pole located on the infield rail half mile from the finish (立于赛道内栏的标杆,与终点相距 800 米)1/2 英里杆,半英里杆

half-miler n. **1**【racing】a track of half mile 半英里赛道 **2**【racing】a horse who excels at running a half mile race 擅跑半英里的马

half-pass counter change n. *see* counter half-pass 变向斜横步

half-pirouette n.【dressage】a dressage movement in which the horse makes a 180 degree turn, with the forehand scribing a half circle around the inside hind leg (一种盛装舞步动作,马以内侧后肢为轴旋转半圈)旋转半圈,180°[定后肢]旋转

half-struck /ˈhɑːfstrʌk; *AmE* ˈhæf-/adj. (of a carriage) with hood folded halfway (车篷)半敞的

halfway /ˌhɑːfˈweɪ; *AmE* ˌhæf-/adj. midway between a length of distance 中途的

adv. in the middle of a way or distance 中途:to halt halfway 中途停下

halter /ˈhɔːltə(r)/n. a bitless leather or nylon headstall fitted over a horse's head to lead or control him (套在马头上的皮革或尼龙制笼

头,不带衔铁,用来牵控马匹)笼头:Yorkshire halter 约克夏笼头 syn head collar

vt. to put a halter on the head of a horse(给马)戴笼头

halter puller n. a horse who pulls the halter rope when tied 扯缰的马

halter pulling n. a stable vice of horse who pulls backward hard when tied up (马被拴时使劲拽缰绳的恶习)扯缰,拽缰

halter rope n. *see* lead line 牵马绳,牵马缰

halter-break /ˈhɔːltəbreɪk/vt. to train a horse to wear a halter and to be led by a rope attached to it 戴笼头调教

halter-broke /ˈhɔːltəbrəʊk; *AmE* -broʊk/adj. (of a horse) trained and accustomed to wear a halter (指马)戴惯笼头的

hame /heɪm/n.【driving】(usu. **hames**) one of the two curved wooden or metal pieces of a harness that fit over the collar around the neck of a draft horse and to which the traces are attached (挽马颈圈上的木质或铁制构件,套车挽绳系于其上)枷板,颈轭

hame chain n.【driving】a small-linked chain used to connect the bottom ends of the hames (用来拴系轭底端两头的细链)轭链

hame strap n.【driving】a leather strap used to hold the tops and/or the bottom ends of the hames together (用来拴系颈轭上下两端的皮革)轭带

hames collar n. *see* harness collar 颈圈

ham-fisted /ˌhæmˈfɪstɪd/adj. (also **mutton-fisted**) *see* ham-handed 手法重的,手法拙的

ham-handed /ˌhæmˈhædɪd/adj. (also **ham-fisted**) (of a rider) clumsy or heavy-handed (指骑手)手法重的,手法拙的

hammer cloth n. *see* hammercloth 镶边坐垫

hammer head n. the head of horse that is coarse and unrefined 面貌粗糙;钝头:a horse with a hammer head 一匹钝头马

hammercloth /ˈhæmərˌklɒθ/n. (also **hammer cloth**) an ornamental covering for the coachman's seat (车夫座位上的饰用罩布)镶边坐垫

Hammock Wagon n. a horse-drawn vehicle used in 10th century Anglo-Saxon England, with its body resembling a hammock hooked between two wooden posts (英国在10世纪使用的一种马车,车体形似吊床,用铁钩悬于两根木柱之间)吊床马车

hamshackle /ˈhæmʃækl/vt. to restrain a cow or horse by tying a rope around the head and one of the forelegs; hobble (将家畜的头与单个前肢捆起来)保定,绊蹄

hamstring /ˈhæmstrɪŋ/n.【anatomy】the tendon that joins the muscles of the gaskin to the point of the hock (连接小腿肌与跗关节上的筋腱)跟腱 syn Achilles tendon

vt. to cripple a horse by cutting the hamstrings 割断跟腱;致残

hamstrung /ˈhæmstrʌŋ/adj. disabled or crippled by an injury to hamstring above the hock 跟腱受伤的,致残的

hand /hænd/n. **1** the end part of a rider's arm used to hold the reins (持缰的)手 PHR **between hand and leg** (of a horse) under control of the rider's leg and hand aids (指马)被驾驭的,听从辅助 **good hands** (of a rider) light and sensitive on manipulating the reins with hands (指骑手)控缰手法轻巧 **in hand** ①(of a horse) being led or handled by the hand (指马)被牵着的 ②【racing】(of a racehorse) running under restraint (指赛马)受牵制的 PHR V **give with the hands** (of the rider) to relax the grip on the reins to release bit pressure on the bars of the horse's mouth (指骑手为降低衔铁在马口角的施力而松手)放松缰绳 **2** a unit of length, equal to 4 inches, used to measure the height of a horse (马体高的测量单位,相当于4英寸)掌,掌高

hand breeding n. (also **hand mating**) the mating of a stallion and mare as controlled by a handler (由配种员控制公马与母马配种)辅助繁育,人工繁育

hand canter n. a semi-extended canter between a gallop and a canter (介于袭步和跑步之间的跑步)伸长跑步

hand gallop n. a collected gallop or extended canter 收缩袭步

hand gate n.【hunting】*see* hunting gate 猎场围栏的门

hand mating n. *see* hand breeding 辅助繁育,人工繁育

hand piece n. the part of the whip held in the hand of the user (马鞭的)把手

hand rub vt. to rub or massage the legs of the horse to improve blood circulation and prevent swelling (用手按摩马的四肢以促进血液循环并消除肿胀)用手按摩

Handbook on Horse Care and Management n. a handbook written on horse management and veterinary medicine by Li Shi (? – 845), a court official of the Tang dynasty in ancient China(一本有关马政和兽医的著作,由中国古代唐朝官员李石撰写)《司牧安骥集》

handful /ˈhændfʊl/n.【driving】the reins that the driver's hand holds 手持缰绳 PHR V **have a handful** (of the driver) to pick up the reins of a team 抓起缰绳,两手持缰

handicap /ˈhændɪkæp/n. **1** the advantages or disadvantages of weight, distance, or time placed upon competitors to equalize their chances of winning (为了让参赛者有相同的获胜机会,而给他们规定的负重、距离、时间等有利或不利的条件)让磅 **2** (abbr. **HCP**) a handicap race 让磅赛

vt. to assign handicaps to competitors by evaluating their past performance (根据参赛者以往成绩给予负重磅数)设置障碍,评定磅数

handicap race n.【racing】(also **handicap**) a race in which horses carry different weights assigned by the racing secretary to equalize their winning chances (通过让赛马肩负不同磅数的重量来平衡获胜机会的比赛)让磅赛 compare conditions race

handicap system n.【racing】a racing system in which horses carry different weights based on their previous performance, thus giving equal chance of winning (根据赛马以往赛绩设定不同负重以均衡获胜概率的比赛制度)让磅制

Handicap Triple Crown n.【AmE racing】*see* New York Handicap Triple Crown 纽约让磅赛三冠王

handicapper /ˈhændɪˌkæpə(r)/n. **1**【racing】a racing official responsible for assigning the weights to be carried by each horse running in a handicap race (在让磅赛中负责给每匹赛马设定负重的人员)评磅员 **2**【racing】one who bets on horses based on a thorough study of their past performances 赌马行家,投注行家

handle[1] /ˈhændl/vt. to train, lead or ride a horse 训练;牵马

handle[2] /ˈhændl/n.【racing】the total amount bet on a race, in a day, a meeting, or a season 赌金总额,投注总额

handled /ˈhændld/adj. (of a young horse) that has been touched, brushed, haltered, and led prior to breaking (指马在调教之前进行过基本训练的)初步训练过的

handler /ˈhændlə(r)/n. one who trains or leads a horse (训练、骑乘或牵马的人)练马师;牵马者

handling /ˈhændlɪŋ/n. the initial phase of training of a young horse (马匹的)基本训练

handpiece /ˈhændpiːs/n.【driving】part of the rein held in the hands of the driver (缰绳的手持部分)手中的缰绳

hand-reared /ˈhændˌrɪəd/adj. (of a foal) bottle-fed by a person (指幼驹)人工喂养的

handy /ˈhændi/adj. **1** readily accessible; convenient 顺手的;便捷的:Always store stable tools and equipment in a safe place which is both handy and out of the reach of horse. 马厩用具和装备必须放在安全的地方,既要顺手又让马够不到。**2** (of a horse) nimble, light-footed, and agile in movement（指马）轻快的，敏捷的 | **handily** /-dəli/adv.: a horse pacing handily 一匹运步轻快的马

Hanover /ˈhænəʊvə(r)/n. *see* Hanoverian 汉诺威马

Hanoverian /ˌhænəuˈvɪərɪən; *AmE* ˌhænəˈverɪən/ n.（also **Hanoverian horse**）a warmblood horse breed developed from the crossing of German horses with Thoroughbreds; stands 16 to 17.2 hands and generally a chestnut, brown, black or gray coat with white markings; is quiet, athletic and mainly used for equestrian competitions（由德国马和纯血马杂交获得的温血马品种。体高 16～17.2 掌，毛色多为栗、褐、黑或青且带白色别征，性情温和，体形矫健，主要用于马术竞技项目）汉诺威马

Hanoverian Cream n. *see* Isabella 淡银鬃

Hanoverian horse n. *see* Hanoverian 汉诺威马

Hansom /ˈhænsəm/n. *see* Hansom Cab 汉瑟姆出租马车

Hansom cab n.（also **Hansom**）a two-wheeled, covered, horse-drawn vehicle with the driver's seat positioned behind, patented in 1834 by Joseph Hansom for whom it was named（一种双轮公用马车，车夫驾驶座位于后面，于 1834 年由约瑟夫·汉瑟姆申请专利并因此得名）汉瑟姆出租马车 [syn] gondola of London

Happy Valley Racecourse n. one of the two racecourses used by the Hong Kong Jockey club for horse racing meets（香港赛马会组织赛马比赛的两个赛场之一）欢乐谷马场: First built in 1845 to provide horse racing for British people living in Hong Kong, the Happy Valley Racecourse was rebuilt in 1995 and became a world-class horse racing facility. 欢乐谷马场始建于 1845 年，以满足在香港英国人的赛马需要。1995 年，经重建成为世界级的赛马场地。—*see also* **Sha Tin Racecourse**

hard /hɑːd; *AmE* hɑːrd/adj.（of a horse）resistant to illness and poor nutrition; hardy（指马）抗性强的；耐粗饲的: a hard horse 一匹抗性强的马

hard hat n. a safety helmet 安全帽，头盔

hard mouth n. a horse who is not sensitive to the pressure on the bit（对衔铁施力反应不灵敏的马）口硬的马，口不灵的马 | **hard-mouthed** adj.

hard palate n. the anterior bony arch of the roof of the mouth（口腔的）硬腭 [compare] soft palate

hard pressed adj.【hunting】（of hunted animals）under the chasing of hounds（指猎物）被紧追的

hard track n.【racing】a turf track lacking resiliency（弹性差的赛道）硬赛道

Harddraver /hɑːˈdreɪvə(r)/n. *see* Friesian 弗里斯兰马

hardel /hɑːdl/vt.【hunting, dated】to couple hounds 给猎犬配种

hardiness /ˈhɑːdɪnəs; *AmE* ˈhɑːrd-/n. the state or quality of being hardy 抗逆性

hard-to-control adj.（of a horse）difficult to handle or control（指马）难驾驭的，不羁的

hardy /ˈhɑːdi; *AmE* ˈhɑːrdi/adj. capable of surviving unfavorable and adverse conditions 抗逆性强的

hare /heə(r); *AmE* her/n. a fast-running wild mammal that resembles rabbit but has longer ears and hind legs adapted for leaping（一种貌似家兔的野生哺乳动物，耳和后肢较长，善于奔跳）野兔

vt.【hunting】to hunt rabbits or hares 猎野兔: go haring 猎野兔

hare coursing n. a sport in which hares are chased by hounds in an enclosed area（猎犬在围场追捕野兔的狩猎项目）逐野兔，猎野兔

hare hunting n. the sport of hunting hares or rabbits 猎野兔

harelip /ˈheəlɪp; *AmE* ˈherlɪp/n. *see* cleft lip 唇裂，兔唇

harem /ˈhɑːrəm; *AmE* ˈhærəm/n. the group of mares occupied by a stallion 母马群: In the natural state, the stallion roams with his harem wherever they can find food. 在自然条件下，公马和母马群会四处游走以便觅食。

Harma /ˈhɑːmə; *AmE* ˈhɑːrmə/n. a two-wheeled hunting or war chariot used by the ancient Persians, usually drawn by two or four horses harnessed abreast (古代波斯人使用的一种双轮狩猎马车或战车，多由两匹或四匹马并驾拉行)波斯马车

harness /ˈhɑːnɪs; *AmE* ˈhɑːrnɪs/n. the combination of leather straps and metal pieces used to put a horse to a carriage or plow, generally including a bridle, collar, hames, tugs or traces, back band, belly band, and breeching (马套车或犁地用具的总称，通常包括马勒、颈圈、颈轭、鞘绳、背带、肚带和后鞧)挽具：American harness 美式挽具 ◇ double harness 双马挽具 ◇ driving harness 驾车挽具 ◇ English harness 英式挽具 ◇ Hungarian harness 匈牙利挽具 ◇ plowing harness 犁地挽具 ◇ Russian harness 俄式挽具 ◇ trotting harness 轻驾车赛挽具 PHR V **break to harness** to train a horse to drive a vehicle 调马驾车

vt. to put a horse to a vehicle 套车，套驾

harness collar n.【driving】an oval-shaped leather ring fitted around the neck of a horse in driving or plowing (驾车或犁地时套在马颈上的卵圆形皮圈)套包，颈圈 syn neck collar, hames collar, collar

harness crupper n.【driving】*see* crupper 尻带，尻兜

harness hopples n.【racing】(also **hopples**) a set of leather straps used in harness racing to restrain the movement of the legs on trotting and pacing horses and to maintain their desired gaits (轻驾车赛中拴系在马四肢上的皮带，用来控制马四肢运步并维持比赛中规定的步法)驾车赛绊腿

harness horse n. a horse harnessed to pull a cart or wagon 挽马，舆马

harness loops n. *see* keepers 革环，鞘环

harness martingale n.【driving】a leather strap attached on one end to the girth and on the other to the lowest end of the collar to keep the collar in place (一根挽用皮带，其一端系于肚带，另一端系在颈圈下部，起到固定颈圈的作用)驾车鞅，挽用鞅

Harness Race Cart n. an American sulky of the Gig type 赛用马车

Harness Race Sulky n. a modern, two-wheeled racing cart popular in the United States and Australia (流行于美国和澳洲的赛用双轮马车)赛用马车

harness racing n. a sport in which competing horses pull two-wheeled carts under the command of their drivers and race over a distance of one mile with either the gait of trot or pace (马拉车在车夫驾驭下进行比赛的运动项目，赛程为1.6千米，马以快步或对侧步行进)轻驾车赛：Harness racing is generally considered a native American sport, although it actually traces back more than 3,000 years. 驾车赛普遍被认为是一项美国本土运动，但其历史可以追溯到3 000多年前。syn Standardbred racing—*see also* **pacing, trotting**

harness rack n. a wooden rack on which harness tack is hung when not in use (用来挂挽具的木架)挽具架

harness trotting n. *see* trotting 快步驾车赛

harras /ˈherəs/n. an enclosed area or establishment where the mares are bred by the stallion; stud farm (种公马为母马配种的围栏)配种圈；种马场

harrier /ˈhærɪə(r)/n. any of a breed of small hounds used for hunting hares (用来追捕野兔的小型猎犬)猎兔犬：harrier pack 猎兔犬队

harrow /ˈhærəʊ; *AmE* -roʊ/n. a farm implement consisting of a heavy frame with sharp spikes or upright disks, used to break up and even off plowed ground (一种带尖齿或垂直圆盘的重型农具，用来破碎土块和平整犁地)耙：disc harrow 圆盘耙

vt. to break up and level soil or land with a harrow 耙地:Harrowing helps to aerate the soil, it also breaks up and spreads out the droppings of horses, thus reducing the larvae on the pasture. 耙地不仅可增加土壤透气性,同时也可破碎马粪,从而杀死草场中的部分幼虫。

hart /hɑːt; *AmE* hɑːrt/n. (*pl.* **hart**, or **harts**) a male red deer 雄赤鹿,公赤鹿 —*see also* **hind**

Hartmann's mountain zebra n. the smallest and southernmost species of zebra living in Africa (生活于非洲南端的斑马品种,体型很小)哈特曼山地斑马 [syn] *Equus zebra hartmannae*

harvest mite n. *see* chigger 恙螨,秋螨

haul /hɔːl/v. **1** to pull or drag forcibly; tug, tow 拽,拖 **2** to transport with a cart or truck (用马车或卡车)拖运

haulage /ˈhɔːlɪdʒ/n. **1** the act or process of hauling 拖运 **2** a charge made for hauling 拖运费

hauler /ˈhɔːlə(r)/n.【AmE】*see* haulier 货运车夫;货运工

haulier /ˈhɔːlɪə(r)/n. (also **hauler**) one who drives wagon for carrying goods 货运车夫

hauling test n. a test of the pulling power of a horse 拉力测验

haunch /hɔːntʃ/n. (usu. **haunches**) the buttock and upper thigh in the body 腰部,臀部

haunches-in n.【dressage】*see* travers 腰[向]内斜横步

haunches-out n.【dressage】*see* renvers 腰[向]外斜横步

Haute Ecole n.【French】*see* high school 高等骑术学校

haw /hɔː/interj. a voice command given to a horse to turn left (吆喝马向左转)嘶! [compare] gee

hay /heɪ/n. grass or other legumes cut and dried 干草:hay meal 干草粉 ◇ baled hay 干草捆 ◇ Excessive dampness of hay is a major source of the hay fever. 水分过高是造成干草热的主要因素。 [PHR V] **hit the hay**【colloq】to go to bed 休息,就寝 [IDM] **make hay while the sun shines** to make good use of an opportunity while it lasts 趁着太阳晒干草;抓住有利时机

hay bale n. any bundle of hay bound with twine or wire into rectangular shape (用绳或铁丝捆扎的干草)干草捆

hay belly n. a horse having a distended belly 草包肚,草腹 [syn] grass belly, potbelly

hay burner n. **1**【racing】a racehorse who does not win enough prize money to pay for its feed (所挣奖金不足饲喂成本的马)亏本喂的马,白喂的赛马 **2** an old and worn-out horse 老马;乏马 [syn] oatburner

hay waferer n. *see* waferer 干草制块机

haycock /ˈheɪˌkɒk/n. a small cone-shaped pile of hay in the field (田野里的小型圆锥形草堆)草垛

haylage /ˈheɪleɪdʒ; *AmE* ˈheɪlɪdʒ/n. roughage made of grass cut and dried to 50 - 60 percent dry matter, then baled to blocks and packed into tough airtight plastic sacks (牧草收割晾至干物质含量为50%~60%时,压扎成捆,然后装在密封塑料带中制成的粗料)半干青贮[料]:Haylage lies between hay and silage in its feeding value and digestibility. 半干青贮料的饲养价值和消化率介于干草和青贮饲料之间。 [syn] drylage, semi-wilted forage

hayloft /ˈheɪlɒft; *AmE* -lɔːft/n. a loft over a stable or barn used for storing hay or straw (用来储藏干草的顶棚)干草棚

haymaker /ˈheɪmeɪkə(r)/n. a machine used to make hay automatically 干草机

haymaking /ˈheɪmeɪkɪŋ/n. the procedure of making hay, such as cutting and drying (牧草的收割和风干等过程)干草制作

haymow /ˈheɪmaʊ/n. **1** a hayloft 干草棚 **2** the hay stored in a loft; haystack 干草堆

haynet /ˈheɪnet/n. (also **hay net**) a corded net holding hay and hung in a stall or trailer from which horses feed (挂在马房或拖车内的网兜,内装干草供马采食)干草兜,草料兜

hayrack /ˈheɪræk/n. (also **hay rack**) a wooden or metal rack for holding hay and feeding livestock（用来盛草以喂牲畜的木质或金属架）干草架

hayrake /ˈheɪreɪk/n. a rake used to ted or gather dried hay（用来摊晒干草的耙）干草耙

hayrick /ˈheɪrɪk/n. *see* haystack 草堆，草垛

haystack /ˈheɪstæk/n. (also **hayrick**) a large pile of hay stacked outdoors 草堆，草垛

haytea /ˈheɪˌtiː/n. a drink for sick horses made by infusing hay in hot water（将干草浸在热水中为患病马制作的饮品）草茶

haywire /ˈheɪwaɪə/n. wire used to bind hay into bales 捆草铁丝

haze /heɪz/vt.【rodeo】(of a cowboy) to drive or keep a steer running in a straight line parallel to a wrestler（摔牛项目中牛仔赶牛与摔牛者平行沿直线行进）驱赶

hazer /ˈheɪzə(r)/n.【rodeo】a cowboy who rides along the steer opposite to the wrestler to keep it running in a straight line（策马赶牛与摔牛者成直线行进的牛仔）摔牛副手，摔牛搭档

Hb abbr. hemoglobin 血红蛋白

HB abbr. hot-blood 热血马

HBPA abbr. Horsemen's Benevolent and Protective Association 骑师慈善保护协会

HCG abbr. human chorionic gonadotropin 人绒毛膜促性腺激素

Hcp abbr.【racing】handicap race 让磅赛

Hct abbr. hematocrit 红细胞比容

hd abbr.【racing】head 头长

head /hed/n. **1** the forwardmost body part of a horse（马体最前端的部分）头：The head of a horse should be in proportion with the rest of the body. 马头的大小要与身体其他部位成比例。**2** the front or forward part of sth 前端 PHR **head of the saddle** the pommel of a saddle 前鞍桥 **3**【racing】(abbr. **hd**) a winning margin approximately equal to the length of the horse's head（赛马中胜出的距离，长度与马头相当）头长 compare dead-heat, nose, short-head, short-neck, neck, length

vt. to turn in a certain direction 转向 PHR V **head up** (of a horse) to raise its head suddenly（指马）猛然抬头 **head a cow**【cutting】to ride his mount in front of a cow to force it to change direction or stop 策马与牛对峙 **head a fox**【hunting】to force a fox to turn back 逼狐狸返身

head bumper n. a lined covering put over the top of a horse's head to provide protection during transportation（戴于马头顶在运输中起保护作用的头套）护头，头套

head collar n. (also **headcollar**) an alternative to a halter 笼头

head groom n. (also **head lad**) one in charge of the welfare and general condition of horses（负责料理马匹总体事务的人员）领班马夫

head lad n. *see* head groom 领班马夫

headcollar /ˈhedkɒlə(r); *AmE* -kɑːlə(r)/n. *see* head collar 笼头

header /ˈhedə(r)/n.【roping】one who ropes the head or horns of a calf in team roping（团体套牛中负责套牛头或牛角的人）套牛头者 compare heeler

headland /ˈhedlənd/n. **1** a strip of land left unplowed at the end of a furrow（在犁沟两端未耕的土地）枕地，畦头地 **2**【hunting】the edge or margin of a field 猎场边缘

headpiece /ˈhedpiːs/n. a strap of the bridle fitted across the poll behind the horse's ears, combined with the cheekpieces to support the bit in the horse's mouth（马勒中绕在马耳后项顶处的皮带，与颊革相连以承接衔铁）顶革 syn crown piece

head-shy /ˈhedʃaɪ/adj. (of a horse) afraid of having his head touched（马头）闪躲的，怕触头的

headstall /ˈhedstɔːl/n. the section of a bridle that fits over a horse's head 笼头 syn bridle head

health certificate n. a document issued by a licensed veterinarian to prove that the horse has been examined and is sound and free from any contagious condition（由注册兽医检查并证明

马匹健康、无传染性疾病的文书)健康证明

hearse /hɜːs; *AmE* hɜːrs/n. a four-wheeled, horse-drawn vehicle formerly used for transporting coffins to a church or cemetery(过去用来将棺材运送至教堂或墓地的四轮马车)柩车,灵车

heart /hɑːt; *AmE* hɑːrt/n.【anatomy】the hollow, muscular organ that pumps blood around the whole body by means of rhythmic contractions and dilations(体内的一个中空肌质器官,可通过节律性收缩与舒张维持体内的血液循环)心脏

heart brass n. a heart-shaped brass ornament attached on the harness to ward off the evil eye(马具上的铜制心形缀饰,认为可以驱邪)心形铜饰

heart girth n.(also **girth**) the circumference of the chest just behind the withers(马鬐甲后的胸腔周长)胸围

heart muscle n. *see* cardiac muscle 心肌

heart rate n.(abbr. **HR**) the number of times the heart beats in a minute(每分钟的心跳次数)心率:Horses' heart rates vary from between about 25 beats/min at rest to 250 beats/min during maximal exercise. 马的心率可以从静息时的25次/分到最大运动强度时的250次/分之间变动。

heart room n. the space through the horse's girth and chest for the heart and lungs(胸腔内容纳心肺的空间)胸室

heart sound n. the sound made by the closing of the valves during cardiac cycle(心动周期中由心脏瓣膜闭合产生的声音)心音:first heart sound 第一心音 ◇ second heart sound 第二心音

heartbeat /ˈhɑːtbiːt; *AmE* ˈhɑːrt-/n. the beat of the heart 心跳,心律:heartbeat irregularity 心律不齐

heat /hiːt/n. **1** the estrus 发情:be in heat 发情 ◇ go out of heat 停止发情 ◇ silent heat 安静发情 **2** the quality of being warm or hot 热量,热:The bigger the horse, the more difficult it becomes for him to lose heat, as big horses have a relatively lower surface area to body mass ratio. 马匹体型越大,散热也就越困难,因为体格高大的马其身体表面积与体重的比率相对较小。

vi. to become warm or hot 发热,发热 |**heating** n.

heat bumps n.(also **heat rash**) small reddish swellings occurring on the skin that result from high temperature during hot summer(由于夏天高温导致皮肤出现的红肿斑点)热疹

heat cramps n. a spasm or cramping of the muscles caused by electrolyte and fluid imbalance due to heat exhaustion(由于热虚脱致使体液电解质失衡引起的肌肉痉挛)热痉挛

heat exhaustion n.(also **heat prostration**) a condition in which the body fails to maintain its heat regulatory mechanism due to overheating and depletion of body fluids; symptoms include weakness, dizziness, rapid breathing, elevated body temperature, heavy sweating, and rapid heart rate(由于体温过高和体液流失导致机体丧失温度调节能力的症状,临床表现为体虚、眩晕、呼吸急促、体温升高、发汗并伴有心率加快等)热虚脱,中暑衰竭:Heat exhaustion may result from high environmental temperature, poor ventilation, high humidity, or overexertion. 气温过高、通气差、湿度过高或训练过度都可诱发热虚脱。

heat increment n.(abbr. **HI**) the heat produced during the process of ingestion and digestion(机体在进食和消化过程中产生的热量)热增耗;增生热

heat prostration n. *see* heat exhaustion 热虚脱

heat-loss /ˈhiːtlɔːs/n.(also **heat loss**) the loss of heat by sweating 散热

heatrash n. *see* heat bumps 热疹

heatstroke /ˈhiːtstrəʊk; *AmE* -stroʊk/n.(also **sunstroke**) a failure of the heat regulatory mechanism at the central nervous system level caused by overexposure to the sun and symptomized by elevated temperature, dehydration, convulsions, rapid breathing, shock, coma, and even death (由于曝晒导致机体中枢神经系统热调节机制紊乱的情况,症状表现为体温过高、虚脱、痉挛、呼吸急促、休克、昏迷甚至死亡)中暑,日射病

heaves /hiːvz/n. the chronic pulmonary emphysema of horses [慢性]肺气肿

heavy /ˈhevi/adj. (of a horse) having relatively greater weight or build (指马)重的,高大的: a horse with a heavy head 头重的马 ◇ heavy top 头颈过重

heavy boned adj. (of a horse) having a large and heavy bone structure (指马)体格粗壮的

Heavy Draught Breton n. one of three distinct morphological types of Breton that existed more than 4,000 years ago in the present Brittany, France (布雷顿马的三个类型之一,在现今法国的布列塔尼已有 4000 多年历史)布雷顿重挽马 [syn] Grand Breton

heavy horse n. a large draft horse, such as the Shire and Percheron (体型高大的挽马,如夏尔马和佩尔什马)重型马 [compare] light horse

heavy-headed adj.【racing】(of a horse) responding slowly to guidance (指马)头重的

heavyweight /ˈheviweɪt/n. a horse capable of carrying heavy weights (能驮载重物的马)重载型马 [compare] lightweight, middleweight

Hecca /ˈhekə/n. (also **ekka**) a two-wheeled, horse-drawn passenger cart popular in 19th century India (19 世纪在印度使用的一种客用双轮马车)海克马车

heel /hiːl/n. the posterior part of a horse's hoof 蹄踵

heel avulsion n. *see* avulsion of the hoof wall at the heel 蹄踵裂

heel crack n. *see* avulsion of the hoof wall at the heel 蹄踵裂

heel horse n.【roping】(also **heeler**) a horse one rides to rope the hind leg of a calf or steer in team roping (团体套牛中套后腿者所骑的马)套腿马

heel-bug /ˈhiːlbʌg/n. a disease caused by wet muddy condition and symptomized by the swelling of fetlock joint and lameness (马在泥泞环境中出现的肢蹄病,症状表现为系关节肿胀和跛行)蹄踵炎

heeler /ˈhiːlə(r)/n. **1**【roping】(also **tailer**) one who ropes the hind leg of a calf in team roping (团体套牛中负责套牛后腿的人)套后腿者 [compare] header **2** *see* heel horse 套腿马

heifer /ˈhefə(r)/n. a young female cow 小母牛

height /haɪt/n. (also **body height**, or **withers height**) the tallness of a horse as measured perpendicularly from the highest point of its withers to the bottom of the hoof, generally defined in hands (马从鬐甲最高处至蹄底的垂直高度,通常以掌为单位)体高

Heihe n.【Chinese】(also **Heihe horse**) a horse breed developed in Heilongjiang province, China; descended from Mongolian horses improved with other foreign bloods; stands 14 to 15 hands and usually has a bay, chestnut, gray or black coat; has a medium head, bold eyes, prominent withers, sloping shoulders, and strong legs; is docile and active and used for driving and farm work (分布于中国黑龙江省的马品种,在蒙古马的基础上引入外血培育而成。体高 14 ~ 15 掌,毛色以骝、栗、青或黑为主,头型适中,眼大,鬐甲明显,斜肩,四肢健壮,性情温驯,气质活泼,用于驾车和田间作业)黑河马

Heilongjiang horse n.【Chinese】(also **Heilongjiang**) a horse breed developed in Heilongjiang province of China by crossing native breeds with Ardennes, Soviet Heavy Draught, and Orlov Trotter in the 1950s; stands 14.2 to 15 hands and generally has a bay or chestnut coat; has a straight head, well-arched neck, prominent withers, broad chest, straight back, strong legs sometimes with sickle hocks; used mainly for draft and farm work (中国黑龙江省于 20 世纪 50 年代采用当地马与阿尔登马、

苏维埃重挽马、奥尔洛夫快步马杂交改良培育的马品种。体高 14.2 ~ 15 掌,毛色以栗、骝为主,直头,颈姿良好,鬐甲明显,胸深宽广,背部平直,四肢强健,时有曲飞,多用于挽车和农田作业)黑龙江马

hell cart n.【slang, dated】a curse of pedestrians when splashed with mud by a passing horse-drawn vehicle (行人在被驶过的马车溅上泥水后发出的咒骂)该死的马车! 见鬼的马车!

helmet /ˈhelmɪt/n. a hard head covering worn by riders to protect the head from injury (骑手为保护头而戴的硬壳帽)头盔: polo helmet 马球头盔 ◇ racing helmet 赛用头盔

hematinic /ˌheməˈtɪnɪk; *AmE* ˌhiːm-/adj. acting to increase the amount of hemoglobin in the blood (可增加血液中血红蛋白含量的)补血的

n. any agent that increases the amount of hemoglobin and erythrocyte levels in the blood (可增加血液中血红蛋白含量和红细胞数量的药物)补血药,补血剂

hematocrit /ˈhiːmətəʊkrɪt; *AmE* hiːˈmætəkrɪt/n. (abbr. **Hct**) the percentage by volume of packed red blood cells in a given sample of blood after centrifugation (血样离心后红细胞占全血体积的百分比)红细胞比容

heme /hiːm/n.【physiology】the deep red ferrous component of hemoglobin (血红蛋白含铁的组成成分,呈深红色)血红素

hemicellulose /ˌhemɪˈseljʊləʊs/n.【nutrition】a kind of polysaccharide found in plant cell walls (存在于植物细胞壁中的多糖)半纤维素

hemisphere /ˈhemɪsfɪə(r); *AmE* -sfɪr/n.【anatomy】either of the lateral halves of the cerebrum 大脑半球: cerebellar hemisphere 小脑半球 ◇ cerebral hemisphere 大脑半球 | **hemispherical** /ˌhemɪˈsferɪkl; *AmE* -ˈsfɪr-/adj.

hemispherium /ˈhemɪsfəriəm/n.【Latin】(*pl.* **hemispheria** /-riə/) hemisphere 大脑半球

hemlock /ˈhemlɒk; *AmE* -lɑːk/n. any of several poisonous plants of the genera *Conium* and *Cicuta* (毒芹属和铁杉属的多种有毒植物)毒芹

hemo- /ˈhiːməʊ/pref. blood 血

hemocyte /ˈhiːməsaɪt/n. any of the cells found in blood, including the red blood cells and the white blood cells (存在于血液中的细胞,包括红细胞和白细胞)血细胞

hemoglobin /ˌhiːməʊˈgləʊbɪn; *AmE* ˈhiməˌglobɪn/ n.【physiology】(abbr. **Hb**) an iron-containing protein found in red blood cells of vertebrates that functions in conveying oxygen (脊椎动物红细胞中的含铁蛋白,负责运送氧气)血红蛋白: Around 99.9 percent of oxygen is carried bound to hemoglobin in the red blood cells. 大约99.9%的氧气是与红细胞中血红蛋白结合而被运送。

hemoglobinemia /ˌhiːməˌgləʊbɪˈniːmɪə/n. a condition in which the level of hemoglobin is excessively high in the blood plasma (血浆中血红蛋白水平过高的病症)血红蛋白血

hemoglobinuria /ˌhiːməˌgləʊbɪˈnʊərɪə/n. a condition in which hemoglobin is present in the urine of a horse (马尿液中出现血红蛋白的病症)血红蛋白尿,血尿[症] syn red urine

hemolysis /hɪˈmɒlɪsɪs/n. (also **haemolysis**) the destruction or dissolution of red blood cells, with subsequent release of hemoglobin and usually causing red urine (因血细胞破裂或溶解引起血红蛋白的释放,常造成血尿)溶血症

hemolytic /ˌhiːməˈlɪtɪk/adj. of or related with hemolysis 溶血的: hemolytic anemia 溶血性贫血 ◇ hemolytic disease 溶血病 ◇ hemolytic jaundice 溶血性黄疸

hemopoiese /ˌhiːməpɔɪˈiːsɪs/n. (also **hematopoiesis**) the process of forming blood cells 造血作用,血细胞发生 | **hemopoietic** /-pɔiˈetɪk/adj.: hemopoietic stem cell 造血干细胞

hemorrhage /ˈhemərɪdʒ/n. (also **haemorrhage**) an excessive loss of blood from the

blood vessels; profuse bleeding 大出血;失血过多:internal hemorrhage 内脏出血;肠出血 | **hemorrhagic** /ˌheməˈrædʒɪk/adj.
vi. to bleed profusely 大出血

hemostasis /ˌhiːməˈsteɪsɪs/n. (*pl.* **hemostases** /-siːz/) the stoppage of bleeding or hemorrhage 止血

hemp /hemp/n. **1** any of various annual plants of the genus *Cannabis sativa* 大麻 **2** the tough, coarse fiber of this plant used to make cordage 大麻纤维:hemp rope 麻绳

hepar /ˈhiːpɑː/n.【Latin】liver 肝

heparin /ˈhepərɪn/n. an organic compound found especially in lung and liver tissue and having the ability to prevent the clotting of blood (存在于肺和肝组织中的一种有机化合物,有抗凝血作用)肝素

heparinize /ˈhepərɪnaɪz/vt. to treat blood with heparin to prevent it from clotting 用肝素处理,抗凝处理:heparinized blood sample 肝素处理过的血样 | **heparinization** /hiːpɑːrɪnaɪˈzeɪʃn/n.

hepat(o)- prefix. of liver 肝

hepatic /hɪˈpætɪk/adj. of, relating to, or resembling the liver 肝的; 肝状的

Hephaestus /hɪˈfiːstəs/n.【mythology】the son of Zeus and Hera in Greek mythology; was the god of fire and metalwork who built chariots and shod with brass the horses of celestial breed in Olympus (希腊神话中宙斯和赫拉之子,身为火神和铁匠,他为奥林匹亚山的众神打造座驾并给天马装钉铜制蹄铁)赫菲斯托斯;火神

Hequ n.【Chinese】(also **Hequ horse**) an all-purpose horse breed originating in the Southwest of China; stands about 13.6 hands and generally has a black, bay, or gray coat; has a heavy head and sturdy build; is docile, frugal and used mainly for farm work and light draft (主产于中国西南地区的挽乘兼用型马品种。体高约 13.6 掌,毛色主要为黑、骝或青,头较重,体质结实,性情温驯,耐粗饲,主要用于农田作业和轻挽)河曲马

herb /hɜːb; *AmE* hɜːrb/n. any of various plants used in medicine 草药,药草

herbal /ˈhɜːbl; *AmE* ˈhɜːrbl/adj. of or relating to herbs 草药的:Chinese herbal medicine 中草药学 ◇ herbal treament 草药治疗

herbal practitioner n. *see* herbalist 草药医生

herbalism /ˈhɜːblɪzəm; *AmE* ˈhɜːrbl-/n. the use and application of herbs for the purpose of healing (采用中草药治疗疾病的方法)草药用法;草药学 [syn] herbal medicine

herbalist /ˈhɜːbəlɪst; *AmE* ˈhɜːrb-/n. a doctor who studies and practises in using herbs to treat illness (研究并采用草药治病的医生)草药大夫,草药医生 [syn] herbal practitioner

herbivore /ˈhɜːbɪvɔː(r); *AmE* ˈhɜːrb-/n. an animal that eats only plants 草食动物 [compare] carnivore, omnivore

herbology /həˈbɒlədʒi/n. the study and practice of treating physical ailments and illnesses by using herbs and natural agents (使用草药或天然物质治疗疾病的方法)草药学

herd /hɜːd; *AmE* hɜːrd/n. a group of livestock kept or living together 牧群,畜群: herd life 群居生活 ◇ herd leader 畜群首领 ◇ Horses naturally live in herds. 马天生营群居生活。
vt. to gather or drive livestock as a herd 赶拢,驱赶

herd obedience n. the tendency of an individual to do what the group does (个体的)随群性,合群性

herdsman /ˈhɜːdzmən; *AmE* ˈhɜːrd-/n. (*pl.* **herdsmen**) one who keeps or takes care of a group of livestock 牧人:Lush grasslands beckoned the herdsman. 丰茂的草地召唤牧人前来。

hereditary mutiple exostosis n. (abbr. **HME**) a rare skeletal dysplasia consisting of mutiple benign bone neoplasms occurred most commonly at sites of active bone growth such as the metaphyseal regions of long bones, scapula, pelvis and vertebrae (一种罕见的骨骼发育异常疾病,表现为长骨、肩胛骨、髋骨和椎骨的干骺端多处出现良性骨瘤)遗传性多发外生骨疣

heritable /ˈherɪtəbl/ adj.【genetics】(of a trait) capable of being inherited 可遗传的: heritable traits 遗传性状

heritability /ˌherɪtəˈbɪləti/ n.【genetics】the capability of being inherited against genetic differences 遗传力: The heritability of racehorse performance has been shown to be approximately 33 percent in a related study. 一项有关的研究表明,赛马体能的遗传力约为33%。

hermaphrodite /hɜːˈmæfrədaɪt; *AmE* hɜːrˈm-/ n.【genetics】one having the reproductive organs and the secondary sex characteristics of both sexes (同时具有两性生殖器官和第二性征)两性体;雌雄同体

hermaphroditic /hɜːˌmæfrəˈdɪtɪk; *AmE* hɜːrˌm-/ adj. of or relating to hermaphrodite 雌雄同体的,两性的

hermaphroditism /hɜːˈmæfrədaɪtɪzəm; *AmE* hɜːrˈm-/ n. (also **hermaphrodism** /-dizəm/) the condition of being a hermaphrodite 雌雄同体,两性畸形: Hermaphroditism in horses is a very rare occurrence. 雌雄同体在马中出现的概率很小。

hernia /ˈhɜːniə; *AmE* ˈhɜːrniə/ n. the protrusion of an internal organ or tissue through the wall that contains it (体内器官或组织突出体壁的情况)疝气: umbilical hernia 脐疝 ◇ inguinal hernia 腹股沟疝 ◇ scrotal hernia 阴囊疝 ◇ ventral hernia 腹疝

herpes /ˈhɜːpiːz; *AmE* ˈhɜːrpiːz/ n. any of several viral diseases causing the eruption of small blisterlike vesicles on the skin or mucous membranes (多种病毒性疾病,表现为皮肤黏膜出现水疱)疱疹: herpes simplex virus 单纯疱疹病毒

herring-gutted adj. (of a horse) with a chest that narrows sharply behind the girth (指马)狭胸的

hessian /ˈhesiən; *AmE* ˈheʃn/ n. a strong fabric made from hemp or jute (由亚麻或黄麻纤维织成的粗布)麻布: hessian sack 麻袋

heterozygote /ˌhetərəˈzaɪgəʊt; *AmE* -goʊt/ n.【genetics】an organism that has different alleles at a particular gene locus on homologous chromosomes (在同源染色体的某个特定基因位点上有不同等位基因的生物体)杂合体 compare homozygote

heterozygous /ˌhetərəˈzaɪgəs/ adj.【genetics】**1** having different alleles at one or more corresponding chromosomal loci (在一个或多个染色体位点上有多个等位基因的)杂合的 **2** of or relating to a heterozygote 杂合体的

Hh abbr. hands high 掌高

hiatus /haɪˈeɪtəs/ n.【anatomy】(*pl.* **hiatus**, or **hiatuses** /-siːz/) an aperture or a fissure in an organ or a body part (身体器官或部位中的开口或裂缝)裂孔;孔: hiatus of esophagus 食管裂孔 ◇ aortic hiatus 主动脉裂孔 ◇ venacaval hiatus 腔静脉孔

Hickstead /ˈhɪksted/ n.【jumping】one of the greatest show jumping centers in the world, located in England and built on the grounds of Hickstead Place in 1960 (世界上著名的障碍赛场,于1960年建于英国的希克斯特德庄园)希克斯特德

Hickstead Place n. the mansion of Douglas Bunn who in 1960 started the All England Jumping Course there to provide a continental-style jumping facility with permanent obstacles for British horses and riders (道格拉斯·波恩的私人庄园,此人于1960年在那里建造了全英障碍赛场,以供英国骑手使用)希克斯特德庄园

hidden ride n. a riding technique in which the rider slips from the back of the horse and ride along its side in a prone position (一种策马技巧,其中骑者侧身伏卧在马匹体侧)隐式骑乘,隐身骑乘: The hidden ride was originally adopted by the Roman cavalry to reduce the size of the target to the enemy. 隐身骑乘最早由罗马骑兵采用,主要用来减少被敌方射中的概率。

hide /haɪd/n. **1** a shelter used to view wild animals closely (狩猎观望的)隐身处 **2** the skin of an animal, usually with fur on it (带毛的兽皮)毛皮;生皮: riding boots made from buffalo boots 用水牛皮做的马靴

hidebound /ˈhaɪdbaʊnd/adj. (of a horse) having abnormally dry, stiff skin that adheres tightly to the underlying flesh resulting from malnutrition or internal parasite infestation (马由于营养不良或寄生虫感染导致皮肤异常干硬且无弹性)皮包骨的,瘦骨嶙峋的

hierarchy /ˈhaɪərɑːki/n. the order of dominance within a herd of horses living together; pecking order (马群内部的)等级;次序

high airs n. 【dressage】 the classical dressage movements performed with either the forelegs or all four feet off the ground (马两前肢或四肢同时离地表演的高等骑术动作)空中动作 [syn] airs above the ground, schools above the ground [compare] low airs

high ringbone n. a bony growth occurring within the joint linking the long and short pastern bones (在长系骨和短系骨关节间附生的赘骨)上位环骨瘤 [compare] low ringbone

high roller n.【rodeo】 a horse who leaps high into the air when bucking 腾空跳起的马

high school n. **1** a riding school of advanced equitation 高等骑术学校 [syn] haute ecole **2** classical equitation trained of horses in high schools 高等骑术

high school horse n. a horse trained in classical riding and able to perform classical or high airs 高等骑术用马

high weight n.【racing】 the top weight assigned to a jockey or carried by a horse in a race (赛马中骑手体重或马匹负重的最高限额)重量上限

high-blowing /haɪˈbləʊɪŋ/adj. (of a horse) making an audible noise by flapping of the nostrils when galloping in fast paces (指马全速奔跑时鼻孔发出声响)呼吸急促的,喘息的: a high-blowing horse 一匹呼吸急促的马

Highland /ˈhaɪlənd/n. *see* Highland Pony 高地矮马

Highland Pony n. (also **Highland**) an ancient pony breed originating in the highlands of Scotland, which is one of the nine pony breeds native to Great Britain; has primitive markings including a dorsal stripe and zebra marks on the legs; stands 12.2 to 14.2 hands and may have a bay, gray, mouse dun, palomino, brown, black, or chestnut coat; is long lived, docile, affectionate, and hardy; used for riding, packing, light draft, and farm work (源于苏格兰高地的矮马品种,是英国九个本土矮马品种之一,带原始别征,其中包括背线和四肢上的斑马纹。体高 12.2 ~ 14.2 掌,毛色为骝、青、鼠褐、银鬃、褐、黑或栗,寿命长,性情温驯、活泼,抗逆性强,主要用于骑乘、驮运、轻挽和农田作业)高地矮马

Highland Pony Society n. an organization founded in 1923 in Britain to maintain the registry and promote the general interests of the breeders and owners of Highland Ponies (1923 年在英国成立的组织机构,旨在对高地矮马进行登记并增进育种者和马主的利益)高地矮马协会

high-strung /ˌhaɪˈstrʌŋ/adj. (of a horse) tending to be very nervous and easily excited (指马)易激动的,易兴奋的;悍性高的

hilus /ˈhaɪləs/n. 【Latin】(*pl.* **hili** /-laɪ/) the entrance of blood vessel or lymph duct into an organ or body part (血管或淋巴管至机体器官的入口)门: hilus of kidney 肾门 ◇ hilus of lung 肺门 ◇ hilus of spleen 脾门

hind /haɪnd/n. **1** the back part of a horse (马体)后部,后躯 **2**【hunting】(*pl.* **hind**, or **hinds**) a female red deer 母赤鹿 [compare] hart

hind- pref. after or at the back 后的;后面的

hind cannon n. the large metatarsal 大跖骨,后管骨

hind hunting n.【hunting】the sport of hunting female red deer on horseback(骑马)猎赤鹿

hindleg /ˈhaɪndleg/n.(also **hind leg**) either one of the back legs of a horse(马的)后肢 syn hindlimb

hindlimb /ˈhaɪndlɪmb/n.(also **hind limb**) *see* hindleg 后肢

hindquarters /ˌhaɪndˈkwɔːtəz; *AmE* -ˈkwɔːrtərz/ n.(also **quarters**) the rear part of a horse(马的)后躯 compare forequarters

hinny /ˈhɪni/n. the sterile, hybrid offspring of a female donkey and a male horse(公马与母驴交配所生的不育杂交后代)驴骡:Hinny usually has a horse-like head with shorter ears than a mule and a horse-like tail. 驴骡面貌和尾巴与马相似,耳朵则比骡短些。compare mule

hip /hɪp/n.【anatomy】a projection of the pelvis(骨盆的突起)髋:the hip joint 髋关节 ◇ point of hip 腰角

hip number n. the identification number attached to the hindquarters of a horse at auction(拍卖时贴在马后躯的号码)后躯号码

hippalectryon /ˌhɪpəˈlektrɪɒn/n.【mythology】a mythical four-legged beast with the foreparts of a horse and the hind parts of a rooster(希腊神话中的四足怪兽,前半身为马、后半身为公鸡)半鸡马,鸡马兽

hipparch /ˈhɪpɑːk/n. a cavalry commander of ancient Greece(古希腊的)骑兵司令

Hipparion /ˈhɪpərɪən/n. one of the early ancestors of horse who lived in the Pliocene epoch and had three toes on both the fore and hind legs(马的远祖之一,生活于上新世,前后肢均为三趾)三趾马 —*see also* **Merychippus**, **Mesohippus**, **Pliohippus**

hippiater /ˈhɪpɪeɪtə/n. *see* hippiatrist 马兽医

hippiatrics /hɪpɪˈætrɪks/n. equine veterinary medicine 马兽医学 | **hippiatrical** /-trɪkl/adj.

hippiatrist /hɪˈpaɪətrɪst/n.(also **hippiater**) an equine veterinarian 马兽医

hipp(o)- pref. horse 马

hippo /ˈhɪpəʊ; *AmE* ˈhɪpou/n.【colloq】a hippopotamus 河马

hippocampus /ˌhɪpəˈkæmpəs/n. **1**【mythology】a sea beast with the foreparts of a horse and the hind parts of a fish, ridden by Poseidon and other sea gods(希腊神话中的海怪,前半身像马、后半身似鱼,是海神波赛顿的坐骑)半鱼马,鱼马兽 **2**【anatomy】(*pl.* **hippocampi** /-paɪ/) a ridge in the floor of each lateral ventricle of the brain that has a central role in memory processes(大脑两侧脑室上的皱褶,在记忆过程中起主要作用)海马

Hippocrene /ˈhɪpəkriːn/n. **1**【mythology】the mythical fountain which sprung from Mount Helicon following a strike of Pegasus's hoof(希腊神话中由佩加索斯用蹄踩后从赫利孔山涌出的泉水)灵泉 syn horsewell **2** a source of inspiration 灵感的源泉

hippodrome /ˈhɪpəˌdrəʊm/n. an arena for chariot or horse races in ancient Greece or Rome(古希腊或古罗马时期驾车赛或赛马的场地)赛马场

hippogriff /ˈhɪpəgrɪf/n.【mythology】(also **hippogryph**) a mythical animal with the hind parts of a horse and the foreparts of a griffin(神话中的怪兽,后体似马,前躯为鹫)半鹰马,鹰马兽

hippologist /hɪˈpɒlədʒɪst; *AmE* -ˈpɑːl-/n. one who studies the science related with horses 马学专家,马学家

hippology /hɪˈpɒlədʒi; *AmE* -ˈpɑːl-/n. the scientific study of horses; equinology 马科学,马学

hippophagist /hɪˈpɒfədʒɪst; *AmE* -ˈpɑːf/n. one who eats horseflesh 食马肉者

hippophagy /hɪˈpɒfədʒi; *AmE* -ˈpɑːf-/n. the habit or practice of eating horseflesh 食马肉

hippophile /ˈhɪpəʊfaɪl/n. one who loves horses

马匹爱好者;爱马人

hippopotamus /ˌhɪpəˈpɒtəməs; *AmE* -ˈpɑːt-/n. (abbr. **hippo**) a large, aquatic African herbivorous mammal having thick dark skin, short legs with four toes, and a broad, wide muzzle (非洲水栖草食哺乳动物,体型大,皮厚,四肢粗短,蹄带四趾,口鼻宽大)河马 [syn] river horse

hipposandal /ˌhɪpəˈsædl/n. a type of primitive hoof protector with a metal sole tied to the horse's hoof with leather straps, used by Romans before the invention of horseshoes (古罗马人在蹄铁发明前使用的铁制护蹄用具,通过皮绳将其系在马蹄上)马鞋

hireling /ˈhaɪəlɪŋ; *AmE* ˈhaɪər-/n. a horse that is hired out for riding 出租用马

Hispano /hɪˈspænəʊ/n. (also **Hispano Arabian**) a horse breed developed in Spain by putting local Arab mares to English Thoroughbred stallions; stands 14.3 to 16 hands and usually has a bay, chestnut, or gray coat; is quiet, but energetic, agile, intelligent, and has great courage and spirit; a versatile horse well suited to jumping, dressage, and bullfighting (西班牙采用本土阿拉伯马母马与英纯血马公马繁育而来的马品种。体高 14.3 ~ 16 掌,毛色多为骝、栗或青,性情安静而富有活力,反应敏快,性格刚毅,是优秀的兼用型马种,可用于障碍赛、盛装舞步及斗牛比赛)西班牙阿拉伯马 [syn] Spanish Anglo-Arab, Spanish Arab

Hispano Arabian n. *see* Hispano 西班牙阿拉伯马

histamine /ˈhɪstəmiːn/n. 【biochemistry】a physiologically active substance produced in the body in response to an allergy or infection, often causes pruritis, urticaria, and bronchoconstriction (在机体产生过敏反应或遭感染后释放的一种生理活性物质,常引起皮肤瘙痒、麻疹及气管收缩)组胺

histidine /ˈhɪstɪdiːn/n. 【nutrition】 an essential amino acid important for the growth and repair of tissues (一种必需氨基酸,有助于组织生长和修复)组氨酸

hitch /hɪtʃ/n. **1** the fastening devices used to harness the horse(s) to a vehicle (套马车用的器件)拴系装置 **2** the loop or hook used to connect the horse trailer to the vehicle pulling it (将拖车挂到马车所用的环或钩)挂钩 **3** a team of horses used to pull a carriage or plow (驾车或拉犁的马队)联驾马匹,马队: an eight-horse hitch 八马联驾

vt. **1** to fasten with a loop, hook or other devices 挂住;拴住: a conestoga with a trailer hitched on to it at the back 一辆后面挂着拖车的草原篷车 **2** to tie a horse to a rail, post, etc 拴马,系: to hitch a horse to a fence 将马拴在栅栏上 **3** to harness one or more horses to a carriage 套车 PHR V **hitch up** 套车

hitching /ˈhɪtʃɪŋ/n. the act or instance of fastening, harnessing a horse 拴系;套车

hitching post n. a post to which horses are tied 拴马桩,拴马杆 [syn] horse post

Hittie Handbook for the Treatment of the Horse n. the earliest known treatise on the care and use of the horse for battle and chariot racing, which was inscribed on six clay tablets around 1360 BC (现今已知最早的一部关于战马治疗和管理的著作,于公元前 1360 年左右镌刻于 6 块泥碑之上)《希提疗马手册》

hives /haɪvz/n. an allergic reaction caused by irritants or an infection characterized by bumps or welts on the skin (由刺激物引起的免疫过敏反应,其症状为皮肤瘙痒和红肿)荨麻疹 [syn] nettle rash, urticaria

HKJC abbr. Hong Kong Jockey Club 香港赛马会

HME abbr. hereditary mutiple exostosis 遗传性多发外生骨疣

hobble /ˈhɒbl; *AmE* ˈhɑːbl/n. (also **hopple**) a piece of rope or strap tied to the legs of a horse in order to restrain and immobilize him (系在马腿上用来保定的绳索)绊腿,蹄绊: breeding hobbles 配种绊腿

vt. to restrain a horse by using a hobble 绊(腿)

hobby horse n. *see* rocking horse 摇摆木马

hock /hɒk; *AmE* hɑ:k/n. 【anatomy】 the large joint halfway up the hindleg of a horse; tarsus (马后肢中段处的关节)飞节,跗关节:point of hock 跗结节

hock boot n. a leather covering on the hock for protection 飞节护腿

H

hock spavin n. *see* spavin 飞节肿胀

hog /hɒg; *AmE* hɔ:g/vt. to clip a horse's mane close to the neck 剪鬃:Some establishments still prefer to see all manes hogged, but it is both unnatural and inelegant for the horse to be without his mane and forelock. 目前,有些地方仍然喜欢将马鬃剪掉,但马没有鬃和额发既不自然也不雅观。

hog back n. *see* roach back 凸背

hog backed adj. *see* roach backed 凸背的

hog mouth n. *see* undershot jaw 下颌突出,地包天

hogtie /ˈhɒgtaɪ; *AmE* ˈhɔ:g-/vt. to tie the four feet of a horse or cow as a restraint 捆绑四肢

hold[1] /həʊld/vi. to continue for a certain time 保持,维持:The scent of a fox does not hold long after rain. 雨后狐狸的气味不会保持太久。

hold[2] /həʊld/n. 【hunting】 a covert in which a fox rests (狐狸的)藏身处

holding /ˈhəʊldɪŋ; *AmE* ˈhoʊ-/adj. 【racing】 (of a racetrack) soft or heavy (赛场)松软的

holding scent n. 【hunting】 the strong scent of a quarry 猎物的浓烈气味

holla /ˈhɒlə/interj. 【hunting】 a cry of the hunter to indicate that a fox has been seen (打猎时看到狐狸的叫声)吆喝!

hollow /ˈhɒləʊ; *AmE* ˈhɑ:loʊ/adj. deeply indented or concave 凹的,凹陷的

hollow back n. a conformation defect in which the horse's back curves downwards in an unnatural way (马背部下凹过多的体格缺陷)凹背

hollow knee n. a conformation defect of horse when its knee is placed behind the vertical line which passes down the elbow joint to the fetlock of horse (马前膝位于由肘端和球节构成的垂线后方的体格缺陷)凹膝 [syn] calf-kneed

hollow wall n. *see* seedy toe 蹄匣分离

holly /ˈhɒli; *AmE* ˈhɑ:li/n. any of various shrubs from which the whip shafts are made (用来制作鞭杆的灌木)枸骨;冬青

Holstein /ˈhɒlstaɪn/n. (also **Holsteiner**) a German-bred warmblood developed in the early 14th century as a war horse and later improved with heavy infusion of Thoroughbred blood; stands 16 to 17 hands and generally has a bay, brown, black, chestnut, or gray coat; historically used under harness, but now used for dressage, show jumping, and eventing (德国在14世纪早期育成的温血马品种,后经大量引入纯血马外血予以改良。体高16~17掌,毛色可为骝、褐、黑、栗或青,过去以挽用为主,目前主要用于盛装舞步、障碍赛和三日赛)荷斯坦马,霍士丹马

Holsteiner /ˌhɒlˈstaɪnə(r)/n. *see* Holstein 荷斯坦马,霍士丹马:The Holsteiner has proven to be desired as a sport horse, excelling in disciplines like dressage, show jumping and eventing. 霍士丹马被证明是优秀的体育竞技用马,在盛装舞步、跨越障碍赛以及三日赛中都表现十分出众。

homebred /ˈhəʊmˌbred; *AmE* ˈhoʊm-/adj. **1** 【racing】 (of a horse) foaled in the country where it races; domestic (赛马)本土繁育的,本土出生的 **2** (of a horse) bred or raised by its owner (指马)自繁自养的

homestretch /ˈhəʊmˌstretʃ; *AmE* ˈhoʊm-/n. 【racing】 (also **home stretch**) the straight part of the racetrack from the last turn to the finish line (赛道从最后转弯处至终点的部分)终点直道

homologous /həˈmɒləgəs; *AmE* ˈhoʊˈmɑ:l-/adj. **1** corresponding or similar in position, value, structure, or function (结构或功能)相似的;

类似的 **2**【genetics】having or belonging to the same evolutionary origin 同源的: homologous chromosomes 同源染色体 ◇ homologous pairing 同源配对

homologue /ˈhɒməˌlɒg/n. (also **homolog**) **1** a homologous thing or part 类似物; 同源物 **2**【genetics】homologous chromosome 同源染色体

homozygote /ˌhɒməˈzaɪgəʊt; *AmE* ˌhɑːməˈzaɪgoʊt /n. 【genetics】an organism that has the same alleles at a particular gene locus on homologous chromosomes (同源染色体特定位点上等位基因相同的个体)纯合子 compare heterozygote

homozygous /ˌhɒməˈzaɪgəs/adj. 【genetics】**1** having the same alleles at a particular gene locus on homologous chromosomes (同源染色体特定位点上等位基因相同的)纯合的 **2** of or relating to homozygote 纯合子的

honda /ˈhɒndə/n.【roping】a small eye or loop at the end of the lariat (索套末梢的)索套眼，索套孔

Hong Kong Champions & Chater Cup n. a 2,400-meter turf race held annually at Sha Tin Racecourse for Thoroughbreds over three years old (每年在香港沙田马场为 3 岁以上纯血马举行的草地赛,赛程 2 400 米)香港冠军暨遮打杯: The Hong Kong Champions & Chater Cup was first run in 1870 as "The Champion Stakes" but became known by its current name since 1955 in honor of renowned Hong Kong businessman and racehorse owner, Paul Chater. 香港冠军暨遮打杯最初于 1870 以"冠军奖金赛"为名开跑,后来为纪念香港商人和马主保罗·遮打,自 1955 年起更为现名。

Hong Kong Cup n. a 2,000-meter race run annually since 1988 in Hong Kong (自 1988 年起每年在香港举行的比赛,赛程 2 000 米)香港杯

Hong Kong Gold Cup n. a 2,000-meter turf race held annually since 1979 at Sha Tin Racecourse for Thoroughbreds over three years old (自 1979 年起每年在香港沙田马场为 3 岁以上纯血马举行的草地赛,赛程 2 000 米)香港金杯赛

Hong Kong Jockey Club n. (abbr. **HKJC**) a non-profit organization founded in 1846 in Hong Kong to regulate the horse racing, Mark Six, and other betting entertainment in Hong Kong (1846 年在香港成立的组织机构,旨在组织管理香港的赛马、六合彩和其他博彩产业)香港赛马会

Hong Kong Mile n. a one-mile race run annually in Hong Kong since 1991 (自 1991 年起每年在香港举行的比赛,赛程 1.6 千米)香港一哩赛

Hong Kong Racing Museum n. a museum established at the Happy Valley Racecourse in 1995 to trace the history of horse racing in Hong Kong since 1840 (1995 年在欢乐谷赛马场建立的博物馆,旨在展示自 1840 年以来香港的赛马历史)香港赛马博物馆

Hong Kong Sprint n. a 1,000-meter race run annually in Hong Kong since 1999 (自 1999 年起每年在香港举行的比赛,赛程 1 000 米)香港短途赛

Hong Kong Stewards' Cup n. a 1,600-meter turf race held annually at Sha Tin Racecourse for Thoroughbreds over three years old (每年在香港沙田马场为 3 岁以上纯血马举行的比赛,赛程 1 600 米)香港董事杯

Hong Kong Triple Crown n. the racing series consisting of the Hong Kong Stewards' Cup, the Gold Cup, and the Champions & Chater Cup (由香港董事杯、金杯赛和冠军暨遮打杯构成的系列赛事)香港三冠王

Hong Kong Vase n. a 1.5-mile annual race run annually in Hong Kong since 1994 (自 1994 年起每年在香港举行的比赛,赛程 2.4 千米)香港瓶

hood /huːd/n. *see* carriage hood (马车)顶篷

hooded stirrup n. *see* tapadera 带套马镫

hoof /huːf/n. (*pl.* **hoofs**, or **hooves** /huːvz/) the horny part of a horse's foot 马蹄:hoof oil 蹄油 ◇ hoof pick 蹄钩

hoof angle n. the degree of slope at which the dorsal line of the hoof intersects with the plane of its ground-side surface (蹄背线与蹄底面相交形成的斜线角度)蹄角度 [syn] toe angle

hoof bar n. (also **bar**) the part of the hoof wall that turns inward at the heel (蹄壁在蹄踵处内敛的部分)蹄支 [syn] bar of the hoof

hoof beat n. the sound of a hoof striking against the ground (马蹄撞击地面时发出的声音)蹄音

hoof black n. an oil applied to make the horse's hoofs black and shiny [亮]蹄漆 [compare] hoof oil

hoof care n. (also **hoofcare**) care given to the foot of horse 护蹄

hoof cutters n.【farriery】a farrier tool used to trim off the hoofs 剪蹄钳

hoof gauge n.【farriery】a farrier tool used to measure hoof angle (用来测量蹄角度的工具)蹄角度仪

hoof head n. the area where the hoof joins the leg at the coronary region 蹄冠

hoof horn n. the horny, tough, insensitive parts of the hoof, such as the hoof wall 蹄角质层

hoof knife n.【farriery】a wood-handled knife with a slightly curved blade and upturned end, used to pare away the dead sole or ragged parts from the hoof (蹄铁匠使用的木柄弯刀,用来削除马蹄上的死亡组织或破损之处)蹄刀 [syn] farrier's knife, drawing knife

hoof leveler n.【farriery】a tool used to determine the hoof angle and if the hoof is level to the ground (用来测量蹄角度和水平程度的工具)蹄水平仪

hoof nippers n.【farriery】a pritchel-type farrier tool with long handles and a wide cutting tip, used to remove the surplus growth of the hoof wall (蹄铁匠使用的长柄钳嘴工具,用来剪除蹄壁的多余部分)蹄钳

hoof oil n.【farriery】a polish applied to smooth or shine horse's hoofs 蹄油 [compare] hoof black

hoof pick n.【farriery】a hooked tool used to pick out the horse's hoofs (抠蹄用的弯钩)蹄钩:a hoof pick with a brush 带毛刷蹄钩

hoof rasp n. a rasp used by a farrier to level the bearing surface of hoof in horseshoeing (装钉蹄铁时用来锉平蹄底的工具)蹄锉

hoof rings n. the characteristic horizontally distorted rings that appear on the hoofs of a horse affected with laminitis (马患蹄叶炎后在蹄壁上出现的环形曲纹)蹄环纹 [syn] founder rings, laminitic rings, fever rings

hoof sealant n. (also **sealant**) any of the artificial varnishes applied to the hoof wall to limit moisture loss (涂在马蹄上以防止水分散失的清漆)封蹄漆

hoof tester n. a pincer-type tool used to detect bruises or injuries of horse's hoof (用来检测马蹄伤情的工具)检蹄钳

hoof wall n. the horny part of the hoof (蹄的角质部分)蹄壁:The hoof wall could be divided into the toe, the quarters, and the heel. 蹄壁又可细分为蹄尖壁、蹄侧壁和蹄踵壁。

hoofbound /ˈhʊfˌbaʊnd/adj. (of a horse) having a dry and contracted hoof that results in pain and lameness (由于马蹄质干缩导致疼痛和跛行)干缩蹄的

hoofed /huːft/adj. having hoofs; ungulate 有蹄的;有蹄类的

hoofless /ˈhuːfləs/adj. without or having no hoofs 无蹄的

hooflet /ˈhuːflɪt/n. the small hoof of deer, pig, etc. (鹿、猪等的)小蹄

hooflike /ˈhuːfˌlaɪk/adj. resembling a hoof 蹄状的

hoofprint /ˈhuːfˌprɪnt/n. the impression left by a hoof on a surface (蹄在地表踩出的凹痕)蹄迹,蹄印

hook[1] /hʊk/n. **1** a piece of metal curved or bent

back at an angle (一端回折弯曲的金属件) 弯勾;钩 **2** a small curve appearing on the exterior surface of the upper corner incisors of an old horse (老年马上隅齿外侧出现的曲缘) 燕尾:Hooks on teeth are generally removed by floating. 牙齿上的燕尾通常都要锉掉。

hook[2] /hʊk/ vt. **1** to fasten two pieces together with a hook 勾住,拴系 PHR V **hook up** to harness a horse to a carriage 套马车 **2**【polo】to attempt to interfere or spoil an opponent's shot by placing one's mallet in the way of his striking mallet (用球杆破坏对方射门) 钩杆 n. the action or practice of interfering an opponent's shot by conduting this way 钩杆: cross hook 交错钩杆 ◇ high hook 高钩杆

hooked /hʊkt/ adj.【cutting】(of a cutting horse) focused on the cow being worked (指截牛马) 盯紧牛的,死盯着的: The cutting horse was hooked on the cow before him. 截牛马紧盯着它面前的牛。

hooved /huːvd/ adj. (of animals) having hoofs; ungulate 有蹄的:hooved animals 有蹄动物

Hooved Animal Humane Society n. (abbr. **HAHS**) an organization founded in 1971 in the United States to promote the humane treatment of hooved animals through education, legislation, investigation and legal intervention (1971年在美国成立的组织机构,旨在通过教育、立法、调查和司法来倡导人类善待有蹄类动物) 有蹄动物人道主义协会

hooves /huːvz/ n. . the plural form of hoof [复]蹄

hop /hɒp; *AmE* hɑːp/ vt.【racing, slang】to administer illegal drugs to a horse in race; dope (给马) 服违禁药物 PHR V **hop up** to administer illegal drugs to a horse 给马服兴奋剂

hopped /hɒpt; *AmE* hɑːpt/ adj.【racing】(of a horse) illegally drugged (指马) 服用违禁药物的

hopple /ˈhɒpl; *AmE* ˈhɑːpl/ n. & vt. *see* hobble 绊腿:harness hopples 轻驾车赛绊腿

hormone /ˈhɔːməʊn; *AmE* ˈhɔːrmoʊn/ n.【physiology】any of the biochemical substances produced in the endocrine glands and conveyed by the bloodstream or lymph stream to organs or body part to exert its physiological activity (由机体内分泌腺产生的多种化学物质,经血液或淋巴循环运送至身体其他器官和部位以发挥其生理功能) 激素,荷尔蒙 | **hormonal** /-nl/ adj.

hormone assay n. a blood test performed to assess hormone levels in the blood (对血液中激素水平进行的检测) 激素检测,荷尔蒙分析

horn /hɔːn; *AmE* hɔːrn/ n. **1**【hunting】*see* hunt horn (猎狐) 号角 PHR V **touch the horn** to blow the hunting horn 吹响号角 **2** *see* saddle horn 鞍角,鞍头 **3** the uterine horn 子宫角 **4** the hard outer covering of the horse's hoof (马蹄表的) 角质层

horse /hɔːs; *AmE* hɔːrs/ n. a large, single-toed, herbivorous mammal of the genus *Equus* with a long mane and tail; used for packing, riding, and driving, etc. (一种大型草食马属哺乳动物,单蹄,鬃、尾长,常用于驮运、骑乘和驾车) 马: cold-blooded horse 冷血马 ◇ draft horse 挽马 ◇ gift horse 礼品马 ◇ halfbred horse 半血马 ◇ heavy horse 重型马 ◇ harness horse 车马,舆马 ◇ horse flesh 马肉 ◇ hot-blooded horse 热血马 ◇ horse breeding 马匹育种 ◇ light horse 轻型马 ◇ pack horse 驮马 ◇ police horse 警马 ◇ riding horse 乘用马 ◇ saddle horse 乘用马 ◇ war horse 战马,戎马 ◇ warm-blooded horse 温血马—*see also* **colt**, **filly**, **foal**, **gelding**, **mare**, **stallion** PHR **lot of horse in a little room**【BrE】a short-coupled, short, compact, horse having good bone 短小精悍的马 IDM **eat like a horse** to eat lots of food 吃得多,食量大 (**straight**) **from the horse's mouth** (of information) given by one who is directly involved; reliable (指消息) 可靠的,直接的 **hold your horses** to be patient and reconsider before taking action 三思而行

horses for courses【BrE】the act of matching people with suitable tasks or jobs 知人任善 **not a boy's horse** a horse that can only be controlled by a strong and experienced rider 非孩童可骑的马;烈马 **pay for a dead horse** to pay for sth not worth the money 花冤枉钱,枉费钱财 **tell that to the Horse Marines**【dated】a phrase of derision said to sb guilty of gross exaggeration(嘲笑夸大其词者的用语)鬼信你说的话!

vi. to ride horseback 骑马 PHR V **horse about/around** to act in a noisy rough playful way 起哄,瞎闹

horse apples n.【slang】the droppings of horses 马粪

horse bean n. *see* broad bean 蚕豆

horse blanket n.(also **blanket**) a cotton or fabric covering worn over a horse to provide warmth and protection from biting insects, bad weather, dirt, etc.(穿在马身上的罩衣,用来保暖、防蚊虫叮咬或抵御恶劣天气)马衣 syn clothes, rug

horse block n. *see* mounting block 上马台

horse boat n. a boat used to transport horses and cattle 运马船

horse bot n. *see* bot 马蝇蛆

horse botfly n. *see* botfly 马蝇

horse box n.(also **horsebox**) an enclosed vehicle for transporting horses 运马车

horse brass n.(also **horsebrass**) a piece of brass ornament fastened on a horse for decoration(佩在马身上的黄铜饰物)马用铜饰:Collecting horse brasses itself has become a hobby of many people, and they are just strapped on for certain decorative reasons. 马用铜饰收藏已经成为许多人的爱好,尽管佩在马身上也仅出于装饰目的。syn brass

horse bus n. an early passenger carriage pulled by horses 公共马车

horse cavalry n. soldiers who fought on horseback 骑兵部队

horse coper n.【BrE】a horse dealer who conducts underhand methods to influence the sale or purchase(采用欺诈或不良手段干涉马匹交易价格的人)牙子,马贩子

horse cutting n.【rodeo】(also **cutting**) a rodeo event in which one rides to separate a calf from a herd(牛仔竞技项目之一,骑手策马将牛从牛群中隔离出来)截牛,卡停

horse dealer n.(also **horse trader**) one who trades horses and ponies for profit 马贩子,牙子

horse devil n. a wild indigo common to the southern United States that rolls around by the wind when dried thus sometimes frightens horses(一种野生木蓝属植物,多见于美国南部,枯萎后随风滚动,故而有时会吓到马匹)风滚草 syn tumbleweed

horse doctor n. an equine veterinarian 马兽医

horse drawing n. a competition in which the pulling power of draft horses is tested(比试马挽力大小的竞赛)马拉力赛,挽力比赛

horse feathers n.【slang】nonsense 废话

horse float n.【AusE】*see* horse box 运马车

Horse Guards n. a group of English cavalry riders selected for the sovereign 英国皇家骑兵卫队

horse hinny n. a male hinny 公骡

horse hoof n. *see* hoof 马蹄

horse identification n. the action or process of marking a horse by hot brands, acid brands, freeze brands, tattoos, and microchips to identify its ownership(采用火烙印、酸蚀标记、冷烫标记、刺纹和芯片植入等方法对马匹进行的身份标识)马匹鉴定

horse keeper n. one employed to take care of horses 马匹料理员

horse knacker n. *see* knacker 马贩子,屠马贩

horse louse n.(also **louse**) an external blood-sucking parasite hosting the skin of horse(见于马皮肤上的吸血性寄生虫)马虱

horse lover n.(also **horselover**) one who likes horses very much 爱马者,爱马人

horse marine n. **1** a marine assigned to the cavalry 骑兵团的水手 **2** a cavalryman serving in a ship 船上的骑兵 **3** one who is out of place; misfit 不合时宜者

horse mule n. a male mule 公骡 syn john mule

horse nail n.【farriery】*see* horseshoe nail 蹄铁钉

horse parlor n.【racing】a hall where betting on horse races is conducted 投注厅,赌马厅

horse pistol n. a long pistol formerly carried by horsemen 旧时骑兵用的步枪

horse post n. **1** a hitching post to which horses are tied 拴马桩 **2** mail service performed on horseback 骑马邮递,马背邮递

horse race n. *see* race 赛马

horse racer n.【racing】a jockey 骑师

horse racing n.【racing】the sport in which horses and their riders compete in races (骑手策马进行速度比赛的运动)赛马 syn sport of kings

horse riding n.【BrE】the sport of riding horses 骑马,乘骑

horse room n.【racing】a bookmaker's office where information on horse races is provided and the bets are placed 赌马厅,赌马室

Horses and Ponies Protection Association n. (abbr. **HAPPA**) a charity association founded in 1937 in UK to protect horses from slaughter and to promote all aspects of equine welfare (1937 年在英国成立的慈善机构,旨在保护马匹免于屠宰并促进马匹保健事业)马与矮马保护协会

horse's ass n.【slang】a silly or stupid person 笨蛋,蠢货

horse sense n.【colloq】common sense 常理,常识

horse shit n. **1** the feces of horses 马粪 **2** nonsense 蠢话,废话

horse show n. a competition held to display the qualities and capabilities of horses and their riders 马匹展览会,马匹选美

horse sickness n. *see* African horse sickness 非洲马瘟

horse standard n. *see* measuring stick 测杖

Horse Stepping on Flying Swallow n. a bronze statuette of a galloping horse on a flying swallow, unearthed in 1969 from a tomb of the Eastern Han dynasty (AD 25 – 220) at Wuwei county of Gansu province, China (1969 年从中国甘肃省武威县一座古代东汉墓中出土的一尊青铜马雕像,瑞兽昂首腾空,踩于飞燕之上)马踏飞燕

horse tick n. any tick that feeds on horses 马蜱,马虱

horse tick fever n. *see* equine piroplasmosis 马焦虫病,马梨形虫病

horse trade n. the practice of buying and selling horses 马匹交易

v. to engage in horse trade 进行马匹交易

horse trader n. a horse dealer 马贩子,牙子

horse trailer n. a trailer used to transport horses and pulled by another vehicle 运马拖车 compare horsebox

horse trials n. *see* combined training 马匹综合赛

horse whisperer n. **1** (also **whisperer**) a person who uses the technique of whispering to train and calm horses (通过低语来调教和安抚马的人)马语者 **2** (esp. ***The Horse Whisperer***) a best-selling novel written by British author Nicolas Evans (1950-) about horses in 1995 (英国作家尼古拉斯 · 伊文思于 1995 年创作的有关马的一部小说)《马语者》

horse-and-buggy /ˈhɔːs-ænd-ˈbʌgɪ/adj. **1** of or relating to the era before the advent of the industrial revolution and automobiles (工业革命尚未来临之前的)车马时代的:the horse-and-buggy age 车马时代 **2** sticking to dated ideas or conventions; old-fashioned and conservative 守旧的,老式的:horse-and-buggy educational methods 老式教育方法

horseback /ˈhɔːsbæk; *AmE* ˈhɔːrs-/n. the back of a horse 马背 IDM **on horseback** sitting on a horse 骑着马:a soldier on horseback 骑着马

的士兵

adj. sitting on a horse 骑着马的：a horseback tour 骑马旅行 ◇ horseback riding 骑马，骑乘

adv. on the horseback 马背上：to ride horseback 骑马

horsebacker /ˌhɔːsˈbækə(r)；*AmE* ˌhɔːrs-/n. a person who rides on horseback 骑马者

horseboy /ˈhɔːsbɔɪ；*AmE* ˈhɔːrs-/n. a groom who works in a stable 马夫

horsebreaker /hɔːsˈbreɪkə；*AmE* ˈhɔːrs-/n. one who breaks and trains horses；trainer 调马师，练马师

horsecar /ˈhɔːscɑː(r)；*AmE* ˈhɔːrs-/n. **1** an American horse-drawn vehicle formerly operated over a light railway 有轨马车，轨道马车：The horsecar was invented in the United States in 1831. 轨道马车于1831年在美国发明。**2** a vehicle used to transport horses；horsebox 运马车

horsecorser /ˈhɔːsˌkɔːsə(r)；*AmE* ˈhɔːrsˌkɔːrsə(r)/n.【dated】a cunning horse dealer 马牙子，马贩子

horse-drawn adj. (of a vehicle) pulled by a horse or a team of horses (车)马拉的：a horse-drawn hearse 马拉柩车

horseface /ˈhɔːsfeɪs；*AmE* ˈhɔːrsf-/n. a person having a long and homely face (脸长而面相普通的人)马脸

horsefaced /ˈhɔːsfeɪst；*AmE* ˈhɔːrsf-/adj. (of a person) having a long and homely face (指人)马脸的

horse-fall /ˈhɔːsfɔːl；*AmE* ˈhɔːrs-/n. a fall performed by a stunt horse in the movie (电影中)马的摔倒，马的倒地

horseflesh /ˈhɔːsfleʃ；*AmE* ˈhɔːrs-/n. the flesh of a horse 马肉 [syn] horsemeat

horsefly /ˈhɔːsflaɪ；*AmE* ˈhɔːrs-/n. any of various large insects that bite livestock (一种叮咬家畜的昆虫)马蝇

horsehair /ˈhɔːsheə(r)；*AmE* ˈhɔːrsher/n. **1** the hair of a horse, especially from the mane or tail 马鬃；马尾；马毛 **2** a piece of cloth made of horsehair 马毛织物

horsehide /ˈhɔːshaɪd；*AmE* ˈhɔːrs-/n. **1** the skin or hide of a horse 马皮 **2** the leather made from the hide of a horse 马革 **3** a baseball 棒球：The baseball was named "horsehide" because, until replaced by cowhide in the late 20th century, the traditional covering of a baseball was horsehide. 棒球过去之所以称为"马革"，是因为在20世纪末被牛皮取代之前，其外层都由马革制成。

horselaugh /ˈhɔːslɑːf；*AmE* ˈhɔːrslæf/n. a loud, coarse laugh (放声)狂笑：When I asked for more time I just got the horselaugh. 我提出再多给些时间，得到的回答却是一阵狂笑。

horseleech /ˈhɔːsliːtʃ；*AmE* ˈhɔːrs-/n. **1** an aquatic blood-sucking worm of the genus *Haemopis*, historically used by veterinarians to treat common diseases of horse (水生蚂蟥属吸血虫，过去兽医常用来治疗马的某些疾病)蚂蟥，马蝗 **2**【BrE, dated】one who is extremely greedy or insatiable 贪婪者 [IDM] **the daughter of the horseleech** a very greedy person 贪心之人，贪婪之徒

horselike /ˈhɔːslaɪk；*AmE* ˈhɔːrs-/adj. resembling a horse 似马的，像马的

horseman /ˈhɔːsmən；*AmE* ˈhɔːrs-/n. (*pl.* **horsemen** /-mən/) **1** a man riding on horseback；horse rider 骑者 **2** one skilled in horse riding 骑师，骑手 **3** one who raises, manages, or trains horses 马业人员

horsemanship /ˈhɔːsmənʃɪp；*AmE* ˈhɔːrs-/n. the art of training and riding horses；equitation 马术；骑术

Horseman's Sunday n. (also **Horseman's Sunday Service**) an annual religious celebration held since 1968 in London to bless horses and promote horse riding around Hyde Park (自1968年起每年在伦敦举行的庆祝节日，旨在为马匹祈福并促进海德公园周围的骑马活动)骑手礼拜日：The Horseman's Sunday event takes place each year on the penultimate Sunday in September. 骑手礼拜日庆祝活动，在每年9月的倒数第二个星期天举行。

horsemeat /ˈhɔːsmiːt; *AmE* ˈhɔːrs-/n. *see* horseflesh 马肉

Horsemen's Benevolent and Protective Association n. (abbr. **HBPA**) an association founded in the United States in 1940 to provide services and information for owners and trainers of racehorses (于1940年在美国成立的协会，旨在为马主和练马师提供服务和信息)骑师慈善保护协会

horseowner n. one who owns a horse 马主

horseplay /ˈhɔːspleɪ; *AmE* ˈhɔːrs-/n. a rough, noisy play 瞎闹，嬉戏

vi. to engage in horseplay 打闹，嬉戏

horseplayer /ˈhɔːsˌpleɪə(r); *AmE* ˈhɔːrs-/n. 【racing】(also **horse player**) one who regularly bets on horse races 赌马客，赛马迷

horseplaying /ˈhɔːsˌpleɪŋ; *AmE* ˈhɔːrs-/n. 【racing】the act of betting on horse races 赛马投注；赌马

horsepower /ˈhɔːspaʊə(r); *AmE* ˈhɔːrs-/n. **1** (abbr. **hp**) a unit of power equal to 750 watts (功率单位，相当于750瓦特)马力：The term horse-power was first conceived by Scottish engineer James Watt to explain the work rate of his newly invented steaming engines in the 18th century. 马力由18世纪苏格兰机械师詹姆士·瓦特首次使用，用来解释他刚发明的蒸汽机的工作效率。**2** the power exerted by a horse in pulling 马的挽力

horsepower-hour n. (abbr. **hph**) a unit of work or energy equal to 0.7457 kWh (功或能的单位，等于0.7457千瓦时)马力时

horsepox /ˈhɔːspɒks; *AmE* ˈhɔːrs-/n. a viral disease of horses characterized by vesiculopustular eruption of the skin, especially on the pasterns (马患的一种病毒性疾病，症状表现为蹄系上出现大量的疱疹)马痘 [syn] equine variola

horseproof /ˈhɔːspruːf; *AmE* ˈhɔːrs-/adj. (of a device) made or designed to prevent horses from breaking through or running away (设备)防马的：a horseproof latch 防马门闩

horseshoe /ˈhɔːsʃuː; *AmE* ˈhɔːrs-/n. **1** (also **shoe**) a U-shaped piece of iron plate nailed to the horse's foot to provide protection from rough surfaces (钉在马蹄底的U形钢条，可起到保护马蹄的作用)蹄铁，马掌 IDM **lost a horseshoe**【slang】(of a girl) been seduced (指姑娘)被人诱奸，不幸失身 **2** anything shaped like a horseshoe 马蹄形物件

vt. to put shoes on a horse 钉蹄，挂掌 | **horseshoeing** n. 钉蹄

horseshoe brass n. a horseshoe-shaped brass ornament attached to the harness and other tack to ward off the evil (系在马具上的马蹄形铜饰，认为可以辟邪)半月形铜饰

horseshoe nail n.【farriery】a thin, pointed nail used to fix a horseshoe to the hoof 蹄铁钉 [syn] horse nail

horseshoe pad n. a horseshoe-shaped leather or rubber lining that is placed between the horseshoe and hoof to provide protection to the sole as a buffer (衬在马蹄和蹄铁间的皮革或橡胶垫，有保护蹄底的作用)蹄铁垫 [syn] pad, shoe pad

horseshoer /ˈhɔːsʃuːə(r); *AmE* ˈhɔːrs-/n. one who shoes horses 蹄铁匠，钉蹄师，掌工

horse-sick adj. (of pasture) spoiled by heavy grazing and infested with parasites (指草场)采食过度的；虫害泛滥的

horsetail /ˈhɔːsteɪl; *AmE* ˈhɔːrs-/n. **1** the tail of a horse 马尾巴 **2** any of various nonflowering plants of the genus *Equisetum*, having a jointed, hollow stem and narrow, sometimes much reduced leaves (木贼属多种隐花植物，茎中空分节，叶狭长)问荆 [syn] equisetum

horse-tamer n. *see* horse-trainer 驯马师，调马师

horse-trading n. **1** the practice of making a horse trade 马匹交易，马匹买卖 **2**【AmE】shrewd bargaining or clever business dealing 精明的交易

horse-trainer n. one who breaks and trains horses 调马师，练马师 syn horse-tamer

horseway /ˈhɔːsweɪ; *AmE* ˈhɔːrs-/n. a trail or passage reserved only for riders and horse-drawn vehicles（供骑马者或马车通行的道路）马道

horsewell /ˈhɔːswel; *AmE* ˈhɔːrs-/n. *see* Hippocrene 灵泉

horsewhip /ˈhɔːswɪp; *AmE* ˈhɔːrs-/n. a whip used for riding or driving horses 马鞭

vt. to beat or flog with a whip 用马鞭抽打

horsewoman /ˈhɔːswʊmən; *AmE* ˈhɔːrs-/n. (*pl.* **-women** /-wɪmɪn/) a woman who rides on horseback 女骑手

horseword /ˈhɔːswɜːd; *AmE* ˈhɔːrswɜːrd/n. (also **horse word**) any word or phrase about horses 马用术语

horsey /ˈhɔːsi; *AmE* ˈhɔːrsi/adj. *see* horsy 似马的；爱马的

horsiness /ˈhɔːsɪnɪs; *AmE* ˈhɔːrs-/n. the quality or state of being horsey 马的性情；爱马情结

horsy /ˈhɔːsi; *AmE* ˈhɔːrsi/adj. (also **horsey**) **1** of, relating to, or resembling horses or a horse 马的，似马的：Jack had a long, rather horsy face. 杰克的脸长得有点像马脸。**2** interested in or devoted to horses and equitation 喜爱马的，爱好马术的 **3** characteristic of horsemen and horsewomen 骑手特征的

hoss /hɔːs/n.【slang】a horse 马

host /həʊst; *AmE* hoʊst/n. an animal or plant on which another organism (such as parasites) lives（供他种生物寄居的动植物个体）宿主，寄主：intermediate host 中间宿主 ◇ ultimate host 终末宿主

hostile /ˈhɒstaɪl; *AmE* ˈhɑːs-/adj. unfavorable or adverse 不利的；恶劣的：In fact, the mare's reproductive tract is a particularly hostile environment for the survival of sperm. 其实，母马生殖道内环境对精子存活是非常不利的。

hostler /ˈhɒslə(r); *AmE* ˈhɑːs-/n. (also **ostler**) one employed to tend the horses of people staying at an inn（旅馆照料马匹的人）马夫 syn horseboy

hot /hɒt; *AmE* hɑːt/adj. **1** having a degree of temperature that is higher than normal 发烫的，发热的 **2**【racing】close to success or expected to win 有望获胜的；热门的 **3** (of a horse) easily excited 易兴奋的；悍性强的

vt. to cause (a horse) to increase in intensity or excitement 使兴奋，使激动 IDM **hot up** (of a horse) to get or become excited 兴奋起来

hot brand n. an identifying mark of symbol or numbers found on the hip or the shoulder of a horse, usually applied to the hide by using a hot iron（马后臀和肩上的特征性标记，多由烙铁烫留于皮肤上）火烫烙印：The hot brand does not hold up in a court of law as proof of ownership as it is easily altered. 由于火烫烙印容易被更改，因而不能在法庭上作为所有权证据使用。

vt. to brand a horse by applying a red hot iron with mark, symbol, or numbers to its hide（用带标记、符号或数字的烙铁在马皮肤上）打烙印 syn fire brand, hot iron brand

hot fit vt.【farriery】(also **hot set**) to press a hot shoe against the prepared bottom of the hoof until it scorches sufficiently to make the surface level（将烧红的蹄铁压在蹄底将蹄面烫平装蹄）热装蹄 compare cold fit

hot horse n. an excitable horse 易兴奋的马，悍性高的马

hot iron brand n. *see* hot brand 火烫烙印

hot quit vt.【cutting】(of the cutter) to quit working a cow when it tries to return to the herd（指截牛者）中途停止：The cutter will be given a three-point penalty if he hot quits, and he may only quit a cow when that cow is obviously turned away from his horse or when he comes to a dead stop in the arena. 截牛者中途停止将

罚 3 分，一般只有在牛明显跑离拦截马匹或停住不动时方可停止拦截。

hot set vt. 【farriery】*see* hot fit 热装蹄

hot shaping n. the process of modifying the shape of a horseshoe to fit the hoof after heating it in a forge（将蹄铁烧热后塑形装蹄的过程）热成型

hot shoe n. 【farriery】a horseshoe made or shaped as in a forge 热蹄铁 [compare] cold shoe

vt. to shoe a horse with horseshoes shaped and fitted to the hoof using the heat of a forge; hot fit（将蹄铁烧热后塑型装蹄）热装蹄

hot shoeing n. the practice of fitting hot shoes 热装蹄

hot shoer n. a farrier who hot fits horseshoes to a horse 热装蹄匠 [compare] cold shoer

hot shot n. 【rodeo】 an electric charge illegally given by a rider to his mount in bucking or jumping events to shock the horse into bucking more strongly thus to improve scores（在骑野马项目中骑手通过电击使马奔跳更激烈以提高得分的非法手段）电击，电刺激

hot walk vt. to walk a horse to cool down after exercise or competition（训练或赛后）遛马降温

hot walker n. **1** one employed to walk horses to cool them down after exercise or competition（训练或赛后牵马慢行使其降温的雇员）遛马员 **2** a mechanical device that leads horses in circles at a slow walk to cool them down after training（一种电动走步机，可引导受训马匹转圈行走以降低体温）走步机

hotblood /ˈhɒtblʌd; *AmE* ˈhɑːt-/n.（also **hot blood**）a classification of horse breeds, generally referring to the high-spirited Arabians or Thoroughbred（马匹种类划分之一，多指阿拉伯马或纯血马等悍性较强的马）热血马 [syn] fullblood [compare] coldblood, warmblood

hot-blooded /ˌhɒtˈblʌdɪd; *AmE* ˌhɑːt-/adj. **1** of hotblood horses 热血的 **2**（of a horse）easily excited; high-strung（指马）易兴奋的，悍性强的

hound /haʊnd/n. a dog trained and used for hunting 猎犬：drag hound 寻踪狩猎犬 ◇ first season hound 首次出猎的猎犬 ◇ fox hound 猎狐犬 ◇ harrier hound 猎兔犬 ◇ stag hound 猎鹿犬

hound couples n. 【hunting】a harness consisting of two collars used to connect two hounds together（连接两只猎犬的套具，带有两个颈圈）双犬套索，双犬套具

hound glove n. 【hunting】a glove with rubber studs on the inner side, worn by a handler to massage hounds（内面带有橡胶钉的手套，借此可按摩猎犬）猎犬按摩手套

hound jog n. *see* hound pace 犬步

hound pace n.（also **hound jog**）the pace at which a hound normally travels on the road 猎犬对侧步

hound trot n. *see* jog 慢快步

Hounds Please interj. 【hunting】a verbal warning to the field to take care not to override the hounds（警示打猎者策马躲开猎犬的呼喊）小心猎犬！

house /haʊs/vt. to keep or store in a house 储藏：For large establishments, it is economical to make enough space for housing large quantities of bedding materials and fodder. 对大型养马场来说，留出能储藏大量垫料和饲草的空间是比较经济的。

housefly /ˈhaʊsflaɪ/n. a swarming, daytime fly that prefers the interior of buildings and feeds on moist or decaying organic matter（一种白天在室内集群活动的苍蝇，以水分高的腐殖质为食）家蝇

hovel /ˈhɒvl; *AmE* ˈhʌvl/n. an open shed in a field or grassland to shelter livestock turned out on pasture（野外或草场上的半敞式牲口棚，吃草家畜可借此乘凉）凉棚，茅草棚

HPA abbr. Hurlingham Polo Association 英国马球总会

HR abbr. heart rate 心率

hub /hʌb/n. 【driving】（also **wheel hub**）the

center piece of a wheel to which the spokes attach (车轮中心部件,辐条与之相连)轮毂 syn nave

Hucul /ˈhʌkl/n. (also **Huzul**) a pony breed originating in the Carpathian region of Poland; has a short head and a low-set tail; stands 12 to 13 hands and may have any of the coat colors with dun and bay occurring most often; is very hardy, frugal, calm, sure-footed, and has good endurance; used chiefly for packing, light draft, and farm work (源于波兰喀尔巴阡山区的矮马品种。头短,尾础低,体高 12 ~ 13 掌,各种毛色兼有,以兔褐和骝毛常见,逆性强,耐粗饲,性情安静,步法稳健,耐力强,主要用于驮运、轻挽和农田作业)胡克尔矮马 syn Carpathian pony

hue /hjuː/n. 【color】a particular gradation or shade of color 色彩;色调

hull /hʌl/n. **1** a saddle 马鞍 **2** the dry outer covering of a grain or seed; chaff (谷物或种子的)外壳,外稃

vt. to remove the outer covering of grain or cereal 去壳

hulled /hʌld/ adj. having hull removed 去壳的

hulled peanut meal n.【nutrition】a high protein, high energy concentrate produced from peanuts hulled before oil extraction (花生去壳榨油后的副产品,可作高蛋白、高能量的精料使用)去壳花生粕

human chorionic gonadotropin n.【physiology】(abbr. **HCG**) a hormone found in the urine of women during the first 50 days of pregnancy, sometimes used to stimulate ovulation in mares (孕妇妊娠期头 50 天的尿液中含有的一种激素,有时用来促进母马排卵)人绒毛膜促性腺激素

humeral /ˈhjuːmərəl/adj. of or near the humerus 肱骨的,肱部的

humerus /ˈhjuːmərəs/n.【anatomy】(*pl.* **humeri** /-raɪ/) the long bone of the forelimb extending from the shoulder to the elbow (从前肢肩胛至肘的长骨)肱骨

humid /ˈhjuːmɪd/adj. containing a high amount of water or vapor; damp 潮湿的

humidity /hjuːˈmɪdəti/n. the state or quality of being humid; dampness (空气)湿度:absolute humidity 绝对湿度 ◇ relative humidity 相对湿度

hummel /ˈhʌməl/adj. (of a stag) having no antlers (指雄鹿)无角的

n.【Scots】a stag without antlers; not stag 无角雄鹿

humor /ˈhjuːmə(r)/n. **1**【dated】the swelling of a horse's leg; stock-up 下肢肿胀 **2**【anatomy】a body fluid 体液:aqueous humor 水样液,眼房水

hunch seat n. an incorrect riding position taken by inexperienced riders with arched spine (新手骑马时脊柱呈弓形的不正姿势)弓背骑姿

Hungarian Anglo-Arab n. *see* Gidran Arabian 匈牙利盎格鲁阿拉伯马,基特兰阿拉伯马

Hungarian Arabian n. *see* Shagya Arabian 匈牙利阿拉伯马,沙迦阿拉伯马

Hungarian Shagya n. *see* Shagya Arabian 沙迦阿拉伯马

hunt[1] /hʌnt/ vt. to chase and kill game for food or as a sport 打猎,狩猎 PHR V **hunt a cow**【cutting】(of a horse) to follow the movement of a cow with its eyes (指马)紧盯牛的动向

n. **1** the act or sport of hunting, usually with horses and hounds 打猎,狩猎:hunt season 狩猎季节 **2** the sport of fox hunting 猎狐

adj. (of dogs) trained for hunting (指犬)打猎的:a hunt dog 一只打猎用犬,一只猎犬

Hunt[2] /hʌnt/n. **1**【hunting】(also **Hunt Club**) a fox-hunting club 猎狐俱乐部 **2**【hunting】(also **hunt country**) a country area used for hunting 猎狐用地;猎场

hunt ball n.【hunting】a social gathering held during the hunt season 狩猎舞会

hunt boots n. (also **hunting boots**) tall black riding boots worn by participants in a Hunt 猎狐

靴;狩猎靴

hunt cap n. 1 (also **hunting cap**) a hard-shell black helmet worn by participants when hunting (打猎人员头戴的黑色硬壳帽)猎狐帽,狩猎头盔 syn hunting helmet 2 a black helmet worn by riders in equestrian competitions 安全帽,头盔

Hunt Club n. a fox-hunting club 猎狐俱乐部

hunt coat n.【hunting】a coat worn by participants of a hunting activity 猎狐大衣,狩猎大衣: Men generally wear scarlet or black hunt coats, while gray, black, and dark blue are acceptable for women. 男士通常会穿紫色或黑色猎狐大衣,而灰色、黑色和深蓝色则更为女士们所钟爱。

hunt colors n. the coat color worn by participants in a particular fox-hunting club (一个猎狐俱乐部的服装颜色)制服颜色

Hunt Committee n.【hunting】the annually elected Board of Directors of a hunt club (猎狐俱乐部每年换届选举产生的董事会)猎狐委员会

hunt country n. *see* Hunt 猎狐用地;猎场

hunt field n. 1 an open area used for hunting 狩猎场地 2 the mounted followers led by a Hunt Master 狩猎随员,猎狐队员

hunt horn n. (also **hunting horn**) a cylindrical copper instrument used to signal the hounds and the field (用来警示猎犬和狩猎成员用的铜质号角)狩猎号角,猎狐号角

hunt kennel n. an enclosed place for fox-hunting 猎狐围场

hunt livery n.【hunting】the distinctive uniform worn by the staff and/or members of a particular Hunt (狩猎俱乐部成员穿的制服)狩猎制服

Hunt Master n.【hunting】(also **Master**) one appointed by the hunt committee to organize and manage all aspects of the hunt (由猎狐委员会任命管理狩猎各种事务的人)猎主,猎狐主 syn Master of Foxhounds

hunt races n. *see* point-to-point 分站越野赛

hunt seat n. (also **hunting seat**) the position of a rider who sits back into the saddle with feet and legs pushed forward (骑者坐于马鞍后端且两脚前蹬的姿势)狩猎乘姿

Hunt Secretary n.【hunting】one responsible for keeping the notes, files and records of the Hunt and coordinating relations (负责猎狐俱乐部资料工作并协调各方关系的人)猎狐秘书长

hunt servant n.【hunting】one employed to work at a Hunt 猎狐随员

hunt staff n.【hunting】the officers responsible for the management of the hounds and operation of the hunt (负责猎犬管理和猎场运营的)猎场职员

Hunt subscription n.【hunting】the annual fee paid by members of a Hunt for participation 猎狐会员费

hunt terrier n. a small, short-legged dog used to drive foxes from its hiding places inaccessible to the hounds (一种体型短小的犬种,常用来将狐狸赶出藏身处)猎用㹴犬

Hunt Treasurer n.【hunting】one responsible for maintaining the financial records of the Hunt 猎场财务专员,猎场会计

hunter[1] /ˈhʌntə(r)/ n. 1 a man who hunts 猎人;狩猎者 2 any horse bred or trained for hunting 狩猎马: Hunters should possess good stamina and jumping ability and are generally Thoroughbred or Thoroughbred crosses. 狩猎马须具备较好的耐力和跨越能力,通常为纯血马或其杂交种。

Hunter[2] /ˈhʌntə(r)/ n. a specific type of horse originating in Ireland (源于爱尔兰的马匹类型)亨特马: Irish Hunter 爱尔兰亨特马

hunter clip n. a type of clip in which the entire coat of the horse's body is removed with the exception of the saddle patch to offer some protection against saddle sores (马匹被毛修剪类型之一,除备鞍处不予修剪外,其他部位都进行修剪)狩猎型修剪

Hunters Improvement and National Light Horse Breeding Society n. an organization founded in Great Britain in 1885 to improve and promote the breeding of hunters and other horse breeds used for riding and driving（1885年在英国成立的组织机构，旨在促进狩猎马和其他乘用或挽用马种的改良和育种工作）国家狩猎马及轻型马改良育种协会

hunter trials n.【hunting】a competitive event held by most Hunts during the hunting season in which horses are ridden over a course of natural obstacles in hunt field within a specified amount of time（猎狐俱乐部在狩猎季节期间举行的项目，骑手们策马跨越狩猎田野中的天然障碍）狩猎马越野赛

Hunter-Type Pinto n. a type of Pinto horse suitable for hunting and fencing（适合狩猎和跨栏的花马）狩猎型花马

hunting /ˈhʌntɪŋ/n. **1** the activity or sport of hunting wild animals or game 打猎，狩猎：cub hunting 猎幼狐 ◇ deer hunting 猎鹿 ◇ fox hunting 猎狐 ◇ mink hunting 猎貂 ◇ hare hunting 猎野兔 ◇ stag hunting 猎雄鹿 **2** the sport of fox hunting 猎狐

hunting boots n. *see* hunt boots 猎狐靴

hunting box n.【hunting】a small house occupied primarily during the hunting season 猎狐小屋

hunting crop n. a crop used by rider 短马鞭

hunting flask n.【hunting】(also **flask**) a small flattened metal container used to carry liquor while hunting（狩猎时用来盛酒的扁平金属容器）猎用酒壶

hunting gate n.【hunting】a small gate of a hunting enclosure 猎场的门 [syn] bridle gate, hand gate

hunting head n. *see* top pommel 上鞍桥

hunting helmet n. *see* hunt cap 猎狐帽，狩猎头盔

hunting horn n. *see* hunt horn 狩猎号角

hunting iron n. (also **plain hunting iron**) a common English-style stirrup made of stainless steel（一种不锈钢制的英式马镫）狩猎马镫

Hunting Phaeton n. *see* Beaufort Phaeton 狩猎费顿马车，博福特费顿马车

hunting saddle n. *see* close contact saddle 紧贴式马鞍，狩猎鞍

hunting season n.【hunting】the period of time during which organized hunts are conducted 猎狐季节

hunting seat n. *see* hunt seat 狩猎乘姿，狩猎骑姿

huntress /ˈhʌntrɪs/n. a woman who hunts 女猎人，女猎手

huntsman /ˈhʌntsmən/n. **1** a man who hunts 猎人；狩猎者 **2**【hunting】a man who is in charge of the hounds during the hunt 猎犬看管人

hurdle /ˈhɜːdl; *AmE* ˈhɜːrdl/n. any one of a series of fences over which a horse must jump in hurdle racing（跨栏越野赛中的障碍）跨栏，障碍

hurdle race n. a mounted race in which participants race over a cross-country course with a series of hurdles 跨栏越野赛

vi. to participate in a hurdle race 参加跨栏赛 | **hurdle racing** n.

hurdler /ˈhɜːdlə(r); *AmE* ˈhɜːrd-/n. a horse that runs in races over hurdles 跨栏越野赛马

Hurlingham Polo Association n. (abbr. **HPA**) an association founded in 1925 in UK to regulate and promote the sport of polo in both UK and Ireland（1925在英国成立的协会，旨在监管并促进英国和爱尔兰马球运动的开展）英国马球总会

hurried /ˈhʌrid; *AmE* ˈhɜːrid/adj. (of a horse) moving or pacing in a rush, unrhythmic way（指马）步法匆乱的，混乱的

husk /hʌsk/n. the dry outer covering of grain seeds; hull 种皮；谷壳

hussar /hʊˈzɑː/n. **1** a Hungarian light horseman（匈牙利的）轻骑兵 **2** a cavalryman dressed in uniform（穿制服的）轻骑兵

Huzul /huːzl/n. *see* Hucul 胡克尔矮马

hyaline /ˈhaɪəlɪn/adj. 【anatomy】 resembling glass; translucent or transparent 玻璃似的;透明的:hyaline cartilage 透明软骨

hyaluronic acid n.【biochemistry】a viscous fluid found within a joint and acting as a lubricating agent (关节中起润滑作用的黏液)透明质酸

hybrid /ˈhaɪbrɪd/n.【genetics】an offspring produced by breeding animals of different varieties, species, or races; cross (不同种类或品系的动物繁衍而来的后代)杂种:a hybrid species 杂交品种 ◇ The species hybrid between domestic horse and zebra is called zorse. 马与斑马的杂交后代称为杂交斑马。

hybridization /ˌhaɪbrɪdaɪˈzeɪʃn; *AmE* -dəˈzeɪ-/ n. the process or practice of hybridizing 杂交:interspecific hybridization 种间杂交

hybridize /ˈhaɪbrɪdaɪz/v. to produce or cause to produce hybrids; crossbreed 杂交

hydrochloric acid n. the acid solution of hydrogen chloride used in medicine and to brand (医用或酸蚀烙印中所用的氯化氢溶液)盐酸

hydrochloride /ˌhaɪdrəˈklɔːraɪd/n. a compound resulting from the reaction of hydrochloric acid with an organic base (盐酸与有机碱反应生成的化合物)盐酸盐:thiamine hydrochloride 盐酸硫胺素

hydroponic /ˌhaɪdrəˈpɒnɪk; *AmE* -ˈpɑːn-/adj. of or relating to hydroponics 无土栽培的:hydroponic grass 无土栽培饲草

hydroponics /ˌhaɪdrəˈpɒnɪks; *AmE* -ˈpɑːn-/n. the cultivation of plants in nutrient solution rather than in soil (利用营养液而非土壤进行植物栽培)无土栽培

hydrotherapy /ˌhaɪdrəʊˈθerəpi; *AmE* ˌhaɪdroʊ-/n. the use of the massaging action of water stream to treat certain diseases, such as arthritis (利用水流按摩治疗关节炎等疾病的方法)水疗

hydroxyapatite /ˌhaɪdrɒksɪˈæpətaɪt/n. the principal bone salt that provides hardness to skeleton (骨中主要的矿物质)羟磷灰石:The hydroxyapatite crystals contain calcium and phosphorus in a ratio of approximately 2∶1, which is the rationale for providing calcium and phosphorus in the horse's diet. 羟磷灰石结晶的钙、磷比约为2∶1,这个比率是马匹日粮配合中钙、磷供给量的参考值。

hygiene /ˈhaɪdʒiːn/n. the practice of keep living areas clean in order to maintain good health and prevent diseases (为保持健康并防止疾病发生而保持生活区域清洁)卫生:A poor level of hygiene endangers the horse's health and may lead to contagious diseases. 卫生状况差有损于马的健康,而且会诱发多种传染病。

hygienic /ˈhaɪˈdʒiːnɪk; *AmE* -dʒenɪk/adj. of or related with hygiene 卫生的,清洁的:a stable with hygienic conditions 一栋卫生清洁的马厩

hyperimmune plasma n.【vet】the blood plasma containing high level of antibodies against certain diseases (含高浓度抗体的血浆,可抵抗特定疾病)高免血浆

hyperkalemia /ˌhaɪpəkəˈliːmiə/n. an abnormal high level of potassium ions in the blood plasma (血液中钾离子水平过高)高钾血症

hyperkalemic periodic paralysis n. (abbr. **HYPP**) an inherited muscle disease occurring in Quarter Horses, characterized by muscle weakness and tremors, sweating and difficulty in breathing (一种见于夸特马的遗传性肌病,症状表现为肌肉乏力、打战、发汗以及呼吸困难)高血钾周期性瘫痪

hyperlipemia /ˌhaɪpəlɪˈpiːmiə; *AmE* -pəlaɪˈpiː-/ n. (also **hyperlipidemia**) an abnormal high level of fat or lipids in the blood 血脂过高,高血脂

hyperlipidemia /ˌhaɪpəlɪpɪˈdiːmiə/n. *see* hyperlipemia 血脂过高,高血脂

hyperparathyroidism /ˌhaɪpəˌpærəˈθaɪrɔɪdɪzəm/ n. *see* big head disease 甲状旁腺功能亢进,大头病

hypersensitivity /ˌhaɪpəˌsensəˈtɪvəti/n. an overreaction to any irritant substances that contact the skin, characterized by reddening, heat, weeping

H

sores（皮肤接触刺激物后反应过度敏感的病症，临床表现为患处发热和红肿）超敏性，超敏反应

hyperthermia /ˌhaɪpəˈθɜːmiə; *AmE* -ˈθɜːrmiə/ n. the condition of having unusually high body temperature as caused by disease, overexertion, or high environmental temperature; symptoms include heat exhaustion and heat stroke（由于疾病、过度劳累、气温过高所致的体温过高，症状表现为热性虚脱和中暑）体温过高 syn elevated body temperature, raised body temperature

hyperthyroidism /ˌhaɪpəˈθaɪrɔɪdɪzəm/ n. overactivity of the thyroid gland, usually characterized by an increased rate of metabolism, weight loss, and enlargement of the gland（甲状腺机能过于活跃的病症，临床表现为代谢率提高、体重下降和甲状腺肿大等症状）甲状腺功能亢进，甲亢 compare hypothyroidism

hypodermic /ˌhaɪpəˈdɜːmɪk; *AmE* -ˈdɜːrm-/ adj. of or relating to the layer just beneath the epidermis; subcutaneous 皮下的：hypodermic injection 皮下注射

hypogastric /ˌhaɪpəˈgæstrɪk/ adj. near or located at the lower part of the abdomen 腹下的：The hypogastric region includes the inguinal and the pubic regions. 腹下区域包括腹股沟部和耻骨部。

hypomagnesemia /ˌhaɪpəˌmægnəˈsiːmiə/ n. the condition of having deficiency of magnesium in the blood, often symptomized by muscle incoordination and nervousness（血液中镁含量不足的病症，症状表现为运动失调和神经过敏）低镁血症

hypophyseal /haɪˈpɒfəsiːl/ adj.【anatomy】of or relating to the pituitary gland 垂体的

hypophysis /haɪˈpɒfəsɪs/ n.【Latin】(*pl.* **hypophyses** /-siːz/) the pituitary gland 垂体：The hypophysis is usually divided into two parts, the adenohypophysis and the neurohypophysis. 垂体可分为腺垂体和神经垂体两部分。| **hypophysial** /ˌhaɪpəʊˈfɪzɪəl/ adj.

hypothalamus /ˌhaɪpəˈθæləməs/ n.【anatomy】a region of the brain below the thalamus that coordinates the automatic nervous system and regulates body temperature and secretion of hormones（大脑中位于丘脑之下的区域，负责调节自主神经系统并调控体温和激素分泌）下丘脑 compare epithalamus, metathalamus

hypothermia /ˌhaɪpəˈθɜːmiə; *AmE* -ˈθɜːrmiə/ n. the condition of having abnormally low body temperature as caused by viral infection, profuse bleeding, or low environmental temperatures（由于感染病毒、大量失血或气温低下致使体温过低的症状）低体温，体温过低

hypothyroidism /ˌhaɪpəʊˈθaɪrɔɪdɪzəm/ n. abnormal low activity of the thyroid gland; symptoms include decreased metabolic rate, weight gain, and development of a crested neck（甲状腺机能活动过低的病症，临床表现为代谢率降低、体重增加和颈峰突起）甲状腺机能减退 compare hyperthyroidism

hypovolemic shock n. a bodily collapse caused by severe blood or fluid loss, symptoms include low pulse pressure, irregular respiratory rate, low body temperature, pale mucous membranes, and ataxia（由于严重失血或体液流失导致的休克，症状表现为脉动降低、呼吸紊乱、体温降低、黏膜泛白和共济失调）低血容量性休克

hypoxemia /haɪˈpɒksiːmiə; *AmE* -ˈpɑːk-/ n. (BrE also **hypoxaemia**) the condition of having abnormally low concentration of oxygen in the arterial blood（动脉血氧浓度过低的症状）血氧过低 | **hypoxemic** /-mɪk/ adj.

hypoxia /haɪˈpɒksiə; *AmE* -ˈpɑːk-/ n. a shortage or deficiency of oxygen in body tissues（机体组织氧含量不足或缺乏）氧不足，缺氧症 | **hypoxic** /-sɪk/ adj.

HYPP abbr. hyperkalemic periodic paralysis 高血钾周期性瘫痪：The genetic defect in HYPP has been identified to be a single nucleotide change that results in the replacement of the amino acid phenylalanine with leucine in the sodium channel gene of muscle protein. 遗传性疾病高血钾周期性瘫痪已经被确认是由于肌细胞钠离子通道蛋白基因发生单核苷酸变异,而造成亮氨酸取代苯丙氨酸所致。

Hyracotherium /ˌhaɪərəkəuˈθɪəriəm/n. *see* E-ohippus 始祖马

hysterectomy /ˌhɪstəˈrektəmi/n. *see* uterectomy 子宫切除术

hysteritis /ˌhɪstəˈraɪtɪs/n. inflammation or uterus; uteritis or metritis 子宫炎

I

IAD abbr. inflammatory airway disease 呼吸道炎性疾病,呼吸道炎症

IAHA abbr. International Arabian Horse Association 国际阿拉伯马协会

Iberian horse n. an ancient horse breed native to the Iberian peninsula (源于伊比利亚半岛的古老马种)伊比利亚马:The Iberian horses are thought to be among the oldest types of domesticated horses. 伊比利亚马据信是最早被人类驯化的马种之一。[syn] *Equus ibericus*

ice /aɪs/n. a piece of frozen water used to reduce swelling and inflammation 冰块:ice packs 冰袋 ◇ canvas ice boots 帆布冰套腿

vt. to reduce swelling and inflammation by applying ice (用冰块消除肿胀和炎症)冰敷 | **icing** /-sɪŋ/n.

Ice Age horse n. an extinct species of horse native to the tundra steppes of Eurasia until about 8,000 BC; stood about 14 hands and had a black brown coat with long, flaxen mane and dorsal stripe (约1万年前驰骋在欧亚大陆冻原上的原始马种,现已灭绝,体高约14掌,被毛黑褐,鬃浅黄且带背线)冰河马 [syn] *Equus lambei*

ice skid n. a braking device used to prevent a horse-drawn carriage from sliding on an icy slope (用来防止马车在冰面上打滑的制动设备)冰面防滑楔;冰上制轮器

Icelandic horse n. (also **Icelandic pony**) a pony breed descended from the ancient English Pacers brought to Iceland by the Norse who settled there between 860 and 935 AD; is small, stocky, has a large head, a short, thick neck, deep girth, short back and legs, thick mane and tail, and feathering on the heels; stands up to 13.2 hands and usually has a gray or dun coat; is extremely intelligent, docile, hardy, and sure-footed, has great endurance and is noted for its homing instinct; often kept in semi-feral conditions; used for mining, farming, and transportation, as (公元860—935年由移居冰岛的古挪威人带到那里的古英吉利侧步马繁育而来的矮马品种。体型小,体质结实,头重,颈粗短,胸深,背和四肢较短,鬃尾浓密,蹄踵附生距毛,体高13.2掌以下,毛色多为青或兔褐,反应敏快,性情温驯,抗逆性强,步法稳健,耐力好,归巢本能强,常营半野生放养,用于煤矿开采、农田作业和运输)冰岛[矮]马:The Icelandic horse is a versatile riding horse commonly known for its strength comparative to its small size and for its gaits which include the running walk and tølt. 冰岛马是骑乘兼用型马种,不仅因个小力大而出名,而且以其独特的快慢步和托特步而为人熟知。

Icelandic pony n. *see* Icelandic horse 冰岛矮马

ICF abbr. Intern Classification Farrier 见习蹄铁匠

icterus /ˈɪktərəs/n. *see* jaundice 黄疸

identical /aɪˈdentɪkl/adj. of or relating to twins developed from the same fertilized ovum and having the same genetic makeup and closely similar appearance (由单个受精卵发育而来、具有相同遗传组成且外貌相似)同卵双生的,孪生的:identical twins 同卵双生

Ig abbr. immunoglobulin 免疫球蛋白

ileocecal /ˌɪlɪəʊˈsiːkəl/adj. of or between the ileum and cecum (位于回肠和盲肠间的)回盲的:ileocecal junction 回盲结

ileocolonic /ˌaɪliːəʊkəˈlɒnɪk/adj. of the ileum and colon 回结肠的:ileocolonic aganglionosis 回结肠无神经节

ileum /ˈɪliəm/n.【anatomy】(*pl.* **ilea** /ˈɪliə/) the terminal portion of the small intestine extending from the jejunum to the cecum (小肠从空肠延至盲肠间的部分)回肠 [compare] duodenum, jejunum

ileus /ˈɪliəs/n. a painful obstruction of the ileum or other part of the intestine, causing colic, vomiting, and constipation (回肠或肠道其他部位阻塞的症状,可引起腹痛、呕吐和便秘)肠梗阻

iliac /ˈɪlɪæk/adj. of, relating to, or situated near the ilium 髂骨的

ili(o)- pref. 髂骨的

iliocostal /ˌɪlɪəˈkɒstl/adj. of, relating to, or situated near the ilium and ribs 髂肋的:iliocostal muscle 髂肋肌

iliocostalis /ˌɪlɪəʊˈkɔstəlɪs/n.【Latin】the iliocostal muscle 髂肋肌

ilium /ˈɪliəm/n.【anatomy】(*pl.* **ilia** /-liə/) the largest of the three bones constituting the pelvis and forming part of the croup and the point of hip (组成骨盆的骨块,构成马体的尻和腰角)髂骨

IM abbr. intra-muscular 肌内的:an IM injection 肌内注射

imbibe /ɪmˈbaɪb/vt. to absorb or take in; drink or suckle 吸收;吮吸

immature /ˌɪməˈtjʊə(r); *AmE* -ˈtʃʊr/adj. (of a foal) not fully grown or developed; not sexually mature (指马驹)不成熟的;未发育完全的

immune /ɪˈmjuːn/adj. of or relating to immunity 免疫的:immune depression 免疫抑制 ◇ immune response 免疫应答

immune stimulant n. an agent that stimulates the body's immune responses against infection or illness (通过激活机体免疫应答来抵御感染的制剂)免疫激活剂

immune system n. the system of the body that provides protection against infection and diseases (机体保护自身免受感染的系统)免疫系统:The immune system is a self-defensive barrier to fight against the invasion of pathogens. It includes the lymphoid organs such as the thymus, spleen, bone marrow and lymph nodes, and the lymphoid tissue and the lymphocytes existing all over the body, etc. 机体免疫系统是抵抗病原入侵、保护自身的屏障,主要由胸腺、脾脏、骨髓、淋巴结等淋巴器官、淋巴组织和周身各处的淋巴细胞等组成。

immunity /ɪˈmjuːnəti/n. the ability to resist infection caused by a specific pathogen (对特定病原的抗感染能力)免疫力,免疫性:active immunity 主动免疫 ◇ passive immunity 被动免疫

immuno- /ɪˈmjuːnəʊ/pref. immune, immunity 免疫的

immunoglobulin /ɪˌmjuːnəʊˈglɒbjʊlɪn/n. (abbr. **Ig**) any of a group of glycoproteins secreted by plasma cells that functions as antibody in immune response (由浆细胞产生的多种糖蛋白,作为机体免疫应答中的抗体发挥其功能)免疫球蛋白

immunological /ˌɪmjʊnəˈlɒdʒɪkl/adj. of or relating to immune system, reaction, etc. 免疫学的

immunology /ˌɪmjuˈnɒlədʒi; *AmE*- ˈnɑːl-/n. the branch of biomedicine that is concerned with the structure and function of the immune system (生物医学分支学科,研究免疫系统的结构和功能)免疫学

impact /ˈɪmpækt/n. the contact or striking of one thing against another 撞击力,碰撞:The hoofs of a horse can absorb most of the impact when striking on the ground. 马落地时,马蹄可以吸收大部分撞击力。

impaction /ɪmˈpækʃn/n. the obstruction of food in the intestine 阻塞:impaction colic 阻塞性腹痛

impair /ɪmˈpeə(r); *AmE* -ɪmˈper/vt. **1** to injure 伤害,损伤 **2** to make less in strength, value, quality, etc. 削弱

Imperial Riding School of Vienna n. *see* Spanish Riding School of Vienna 维也纳皇家骑术学

校，西班牙骑术学校

impermeable /ɪmˈpɜːmiəbl; *AmE* -ˈpɜːrm-/adj. impossible to permeate 不通透的，密闭的

implant /ɪmˈplɑːnt; *AmE* -ˈplænt/vt. to insert or embed surgically（用手术方法）植入

vi.【anatomy】(of a fertilized egg) to become attached to and embedded in the uterine lining（受精卵）附植

implement /ˈɪmplɪment/n. a tool or piece of equipment used for a particular purpose 工具，用具：agricultural implements 农具

implicate /ˈɪmplɪkeɪt/vt. to involve or associate closely 使牵连，导致：Whilst a deficiency of vitamin E has been implicated in fertility problems, feeding extra does not guarantee extra fertility. 缺乏维生素 E 可导致马匹繁殖力下降，但添加量过多并不能提高繁殖力。

import /ɪmˈpɔːt; *AmE* ɪmˈpɔːrt/vt. to bring in or transport a horse from a foreign country for trade or sale（从国外通过买卖引入马匹）进口，引入：All registered imported horses are identified by either an asterisk or by the abbreviation Imp. in front of the name and the country of export. 所有进口的注册马匹名字前都有"*"或"Imp"及出口国家名称。

impost /ˈɪmpəʊst/n.【racing】the weight carried by a horse in a handicap race（评磅赛中赛马所载的重量）负重

impotence /ˈɪmpətəns/n. inability of a stallion to perform sexual intercourse（公马）阳痿；性功能低下

impotent /ˈɪmpətənt/adj. (of a stallion) incapable of performing sexual intercourse; sterile（指公马）阳痿的；无性功能的

impregnate /ˈɪmpregneɪt; *AmE* ɪmˈpreg-/vt. to make a mare pregnant by means of artificial insemination or natural mating（通过人工授精或自然交配使母马怀孕）使怀孕，使受孕

impression /ɪmˈpreʃn/n.【anatomy】a mark produced on the surface of an organ or body part by pressure（机体器官表面受压产生的痕迹）压痕，压迹：cardiac impression 心压迹 ◇ esophageal impression 食管压迹 ◇ renal impression 肾压迹

imprinting /ɪmˈprɪntɪŋ/n. the imposition of a behavior pattern in the early life of a social animal by stimuli or through association with others（群居动物幼年在接受外界刺激或与其他个体交往的过程中形成的行为模式）印记

improve /ɪmˈpruːv/vt. to increase or refine the qualities of a breed by introducing other foreign bloods（通过引入外血来提高马种品质）改良：improved breeds 改良品种

improved Maremmana n. a modern Maremmana breed crossed with Thoroughbred to increase the stature and refine the appearance（为了增强体格和改良体貌，采用摩雷曼纳马与纯血马杂交培育的现代品种）改良型摩雷曼纳马：Unfortunately, the improved Maremmana lost the hardiness and the exceptional stamina which characterized the breed. 可惜改良型摩雷曼纳马丢失了原品种特有的强健和耐力。

impulsion /ɪmˈpʌlʃn/n. the force or energy generated by the hindquarters of a horse to move forward（后躯推动马匹前进的力量）驱动力

PHR V **lack impulsion** (of a horse) lack sufficient energy to move forward at a regular gait or pace（指马）动力不足，推动力不足：A horse can have a great amount of impulsion while performing the piaffe, and at the same time a horse at great speed may lack impulsion. 有的马在表演原地快步时可能动力强劲，而有的马即使在高速驰骋中也仍然动力不足。

in vitro /ɪn ˈviːtrəʊ; *AmE* ˈviːtroʊ/adj.【Latin】(of processes) taking place outside a living body（指过程）体外的：in vivo experiments 体外试验

adv. outside a living body 体外：an egg fertilized in vitro 体外受精卵 compare in vivo

in vitro fertilization n. (abbr. **IVF**) an artificial method of fertilization in which the ovum, surgically acquired from the mare, is fertilized in a test tube with sperm from a donor male, then returned to the uterus of the original egg donor or a surrogate mother to complete the gestation period (卵细胞的人工授精技术,从母马获取的卵子与公马提供的精子在试管中完成受精过程,然后植入到供体母马或代孕母马子宫内完成妊娠期过程)体外受精

in vivo /ɪn ˈviːvəʊ; *AmE* -ˈviːvoʊ/adj.【Latin】(of processes) taking place in a living body (指过程)体内的

adv. within a living body 体内 compare in vitro

in-and-out n.【jumping】an obstacle consisting of two or three jump elements in which the individual elements are separated by one or two strides (由两三道跨栏构成的障碍,其中每个跨栏间距一步或两步)进出跨栏

in-and-outer adj. & adv.【racing】(of a horse) running inconsistently in races (指马)赛绩不稳定的:an in-and-outer racehorse 一匹赛绩不稳定的马 ◇ to run in-and-outer 发挥不稳定

inattentive /ˌɪnəˈtentɪv/adj. (of a horse) not listening to the rider's aids due to distraction (指马)分神的,注意力不集中的

in-blood /ɪnˈblʌd/adj.【hunting】(of hounds) having made a recent kill (猎犬)杀生的

inbred /ˌɪnˈbred/adj.【genetics】(of a horse) bred by closely related dam and sire (指马由亲缘关系较近的母本和父本繁育)近亲繁育的;近交的:Not only have Thoroughbreds become more inbred over the past 40 years, the research shows, but the rate of inbreeding has accelerated over the years. 研究表明,纯血马在过去40年间,不仅近交程度有所增加,而且近交率也呈增加趋势。

inbreed /ˈɪnbriːd/vt. to breed by mating of closely related dams and sires, especially to preserve desirable traits in a stock (为固定期望性状而在近亲个体之间实施配种)近交

inbreeding /ˈɪnbriːdɪŋ/n. the mating of closely related individuals to preserve a particular trait 近交:In natural conditions, the departure of the young maturing fillies from the group greatly limited the amount of inbreeding that might otherwise have occurred. 在自然情况下,青年雌驹的离群大大降低了近交发生的可能性。

inch /ɪntʃ/n. (abbr. **in.**) a unit for measuring length equal to 1/12 foot or 2.54 centimeters (长度测量单位,相当于1/12英尺或2.54厘米)英寸

incisor /ɪnˈsaɪzə(r)/n.【anatomy】(also **incisor tooth**) any of the 12 front teeth located in pairs at the front of both the upper and lower jaws of a horse, used for cutting and biting (成对位于马上、下颌前方的12颗牙齿,主要用于切割和撕咬)切齿:central incisors 中切齿,门齿 ◇ middle incisors 中切齿 ◇ corner incisors 隅齿 syn pincers

incisor tooth n. *see* incisor 切齿

Incitatus /ˌɪnsɪˈteɪtəs/n. an alias for emperor Julius Caesar's famous racing stallion Porcellus, meaning swift and speeding (古罗马凯撒大帝的有名公马"小猪"的别名)迅驰:Incitatus was ultimately appointed a citizen of Rome and a Senator. "迅驰"后来被任命为罗马市民兼议会议员。

incomplete fracture n.【vet】a fracture which does not extend completely through the bone 不完全骨折

incus /ˈɪŋkəs/n.【anatomy】(*pl.* **incudes** /ɪnˈkjuːdiːz/) an anvil-shaped bone between the malleus and the stapes in the middle ear of mammals (位于哺乳动物中耳锤骨与镫骨间的砧形骨)砧骨 syn anvil

index /ˈɪndeks/n.【racing】a list of the recent races printed in racing newspapers or magazines (赛马报刊上印刷的比赛目录)赛事目录,赛事索引

India-bred n. a breed of horse originated in India, originally used as war horse (源于印度的马种,过去常作战马使用)印度马

Indian broke adj. (of a horse) trained to allow mounting from the off side (指马)可从右侧跨

骑的

Indian martingale n. a type of martingale used to control the position of the horse's head and neck (一种用来控制马头颈姿势的鞅)印第安鞅

Indian pony n. the Pinto horse 印第安马,花马

Indian style n. **1** a method of mounting adopted by Indians in which the rider stands at the shoulder of the horse facing the haunches, grabs hold of the horse's mane with both hands and swings up on the back of the horse with the right leg(印第安人采用的上马方法,骑手面朝马体后躯站于马肩胛处,双手抓鬃,右腿用力跨骑到马背上)印第安式[上马] [syn] swing-up **2** *see* bareback riding 光背骑马,无鞍骑乘

indigenous /ɪnˈdɪdʒənəs/adj. (of breeds) originating and living in an area or environment 当地的;本土的

indigestible /ˌɪndɪˈdʒestəbl/adj. difficult or impossible to digest 难消化的,不消化的

indirect rein vt. *see* neck rein 间接控缰,颈部压缰

indoor /ˈɪndɔː(r)/adj. of or situated in the interior of a building 户内的,室内的:The indoor polo has also gained great popularity in recent years. 近几年来,室内马球也倍受欢迎。[opp] outdoor

indoor arena n. a large area indoors where horses can be ridden or exercised 室内赛场,室内练马场:The footing in indoor arenas is usually made of sand, shredded rubber or a special mix. 室内赛场的地面材料通常选用沙土、碎橡胶或特制混合材料。

infect /ɪnˈfekt/vt. **1** to affect with a disease-causing microorganism or agent 感染(病菌) **2** to communicate a disease to another 传染

infection /ɪnˈfekʃn/n. the process or act of being infected 传染;感染

infectious /ɪnˈfekʃəs/adj. of or relating to infection; contagious 传染的:infectious diseases 传染病

infectious arthritis n. inflammation of the joint capsule caused by the introduction of infectious organisms(由于病菌侵入导致关节囊发炎的症状)传染性关节炎:The infectious arthritis could destroy the joint cartilages and underlying bone and cause new bone growth which ultimately results in ankylosing arthritis or osteoarthritis. 传染性关节炎能破坏关节软骨及关节下的骨组织,造成骨质增生,并最终恶化为僵直性关节炎或骨关节炎。

infective /ɪnˈfektɪv/adj. capable of causing infection; infectious 感染的;传染性的:infective larvae 感染性幼虫 | **infectivity** /ˌɪnˈfektɪvəti/n.

inferior /ɪnˈfɪəriə(r); *AmE* -ˈfɪr-/adj.【anatomy】located beneath or directed downward 下部的,下方的:inferior check ligament 膝下支撑韧带

infertile /ɪnˈfɜːtaɪl; *AmE* -ˈfɜːrtl/adj. incapable of producing offspring; sterile 不育的,不孕的

infertility /ˌɪnfɜːˈtɪləti; *AmE* -fɜːrˈt-/n. the state or condition of being infertile 不育,不孕

infest /ɪnˈfest/vt. to inhabit or live in large numbers, often causing disease or damage to the health of host(害虫大量寄生给宿主造成疾病或损伤)寄生,滋生:livestock that were infested with tapeworms. 寄生着大量绦虫的家畜。

infestation /ˌɪnfeˈsteɪʃn/n. the condition or state of being infested with large numbers of parasites and vermins 大量寄生;虫害:heavy infestation of large roundworms. 寄生着大量的马蛔虫。

infield /ˈɪnfiːld/n.【racing】the area within the inside rail of a racetrack(赛道内栏围成的区域)内场区

infield rail n.【racing】*see* inside rail 内栏

inflammation /ˌɪnfləˈmeɪʃn/n. a localized condition in which a body part becomes red, swollen and sore due to infection or injury(机体部位由于遭受感染或受伤后出现的红肿和疼痛反应)炎症,发炎

inflammatory /ɪnˈflæmətri; *AmE* -tɔːri/adj. characterized or caused by inflammation 发炎的；发炎引起的：inflammatory response 炎症反应

inflammatory airway disease n.（abbr. **IAD**）an irritative airway disease caused by viral infection or the chronic pulmonary stress and inhalation of foreign substance, symptoms include a slight nasal discharge and occasional light coughing early in exercise（由于病毒感染、肺部应激或吸入异物诱发的呼吸道疾病，其症状表现为流鼻和训练中轻嗽）呼吸道炎症，炎性呼吸道疾病

inflate /ɪnˈfleɪt/ vt. to fill with air or gas so as to make it swell 充气，使膨胀
vi. to become inflated 膨胀：Immediately after birth, the foal's lung is inflated and oxygen concentration in the blood stream climbs dramatically. 出生后不久，幼驹的肺便开始充气，其血氧含量也急剧上升。| **inflation** /ɪnˈfleɪʃn/ n.

influenza /ˌɪnfluˈenzə/ n. an acute contagious viral infection generally characterized by fever and respiratory involvement（一种急性病毒性传染病，主要症状是发热和呼吸道感染）流感：equine influenza 马流感

in-foal /ˈɪnˌfəʊl/ adj.（of a mare）carrying fetus; pregnant（母马）怀驹的：an in-foal mare 怀驹母马，妊娠母马

infraorbital /ˌɪnfrəˈɔːbɪtl; *AmE* -ˈɔːrb-/ adj.【anatomy】located below the orbit of the eye 眶下的：infraorbital foramen 眶下孔

infraspinatus /ˌɪnfrəspɪˈneɪtəs/ n.【anatomy】the shoulder muscle on the lower side of the scapular spine（位于肩胛冈下侧的肩肌）冈下肌

infraspinous /ˌɪnfrəˈspɪnəs/ adj.【anatomy】located below the scapular spine 冈下的：infraspinous fossa 冈下窝

infringe /ɪnˈfrɪndʒ/ vt. to transgress or exceed the limits of; break or violate 超限；违反，违例
vi.（~ **on sth**）to encroach on sth 侵害，侵犯：Herbal treatments are gaining increasing popularity among horse owners, as they could support the horse's health and his ability to cope with demanding activities without infringing any of the rules and regulations governing competitions. 饲草中添加草药正越来越受到马主们的青睐，因为它在不违反任何竞赛规则的同时，可增进马体健康并使之有体力完成高强度运动。

infundibulum /ˌɪnfʌnˈdɪbjʊləm/ n.（*pl.* **infundibula** /-lə/）**1**【anatomy】any of various funnel-shaped bodily openings or structures 漏斗状开口；漏斗状结构：infundibulum of uterine tube 输卵管漏斗部 **2** the funnel-shaped depression in the biting surface of the incisor teeth（切齿咀嚼面上的漏斗状凹痕）齿坎

infuse /ɪnˈfjuːz/ vt. to introduce fluid into an organ or body 灌注，输入：Mare's uterus is sometimes infused with a sterile saline solution post-foaling to clean it out. 母马产驹后，有时可用无菌生理盐水灌注子宫进行彻底清洁。| **infusion** /ɪnˈfjuːʒn/ n.

ingest /ɪnˈdʒest/ vt. to take feed into the body for digestion or absorption 摄食，采食

ingesta /ɪnˈdʒestə/ n.【nutrition】food taken into the body as nourishment（摄入的）食物

ingestion /ɪnˈdʒestʃən/ n. the act or process of ingesting 摄食；摄取

ingestive /ɪnˈdʒestɪv/ adj. of or relating to ingestion 摄食的：ingestive behavior 摄食行为

inguinal /ˈɪŋɡwɪnəl/ adj. of, relating to, or located in the groin 鼠蹊的，腹股沟的：During the later part of fetal development, the testis migrates through the inguinal canal and descends into the scrotum. 在胚胎发育后期，睾丸沿腹股沟管迁移并下降到阴囊中。

inguinal hernia n. a condition in which the tissue of the groin area descends into the inguinal canal with the descension of testicles into the scrotum（疝囊组织伴随睾丸下降落入腹股沟所致的疝气）腹股沟疝

inhalant /ɪnˈheɪlənt/ n. a medication taken by inhalation in vapor or aerosol form 吸入药，嗅剂

inhale /ɪnˈheɪl/ vt. to bring air into lung by

breathing; inspire 吸气,吸入 opp exhale

in-hand line n. a cotton or synthetic rope attached to the bridle by a hook and snap to handle or work a horse on ground (扣在勒上的绳索,用来在地面操控马匹)牵马绳,调教索

inherent /ɪnˈhɪərənt; *AmE* -ˈhɪr-/adj. existing as an essential constituent; inherited or intrinsic 固有的;内在的:inherent sexual rhythm 固有性节律

inherit /ɪnˈherɪt/vt. to receive from an ancestor 继承;遗传:inherited defects 遗传缺陷 ◇ inherited disease 遗传性疾病

inheritance /ɪnˈherɪtəns/n. the process of genetic transmission of characteristics from parents to offspring (特征从亲代传递给子代的过程)遗传:Mendelian inheritance 孟德尔遗传

inject /ɪnˈdʒekt/vt. to introduce a drug solution into a body part by using a syringe 注射 | **injection** /-ʃn/n.

inositol /ɪˈnəʊsətəʊl; *AmE* ɪˈnoʊsɪtoʊl/n.【nutrition】a B-complex vitamin essential for proper metabolism (一种B族维生素,有助于维持机体正常代谢)肌醇

insecticide /ɪnˈsektɪsaɪd/n. any substance used to kill insects 杀虫剂

inseminate /ɪnˈsemɪneɪt/vt. to introduce semen into the reproductive tract of a mare in order to make her pregnant (将精液输入母马生殖道内使之受孕)授精

insemination /ɪnˌsemɪˈneɪʃn/n. the process or action of inseminating 授精:Artificial insemination is not at present allowed to be used on Thoroughbreds, it is largely because of the fear that one stallion would receive an inordinate number of mares in any covering season. 目前繁育纯血马禁止使用人工授精,主要是担心公马在配种期会为过多的母马配种。

inside /ˌɪnˈsaɪd/n. **1** the side of the horse on the inner side of any circle or track being travelled (马运行弧线的里侧)内侧 opp outside **2** the left side of a horse (马匹的)左侧,里手 **3**【racing】the inside position (赛道的)内侧

inside leg n. the leg of the rider or horse on the inside of the movement (马的)内侧肢;(骑手的)内方腿:In a circle to the right, the rider's inside leg would be the right leg. 向右转圈时,骑手的内方腿即他的右腿。

inside position n.【racing】the position closest to the inside rail (赛道)内侧位

inside rail n.【racing】(also **infield rail**, or **running rail**) the rail or fence separating the racing strip from the infield (将赛道和内场地分开的栏杆)内场栏杆,内栏 compare outside rail

insoluble /ɪnˈsɒljəbl; *AmE* -ˈsɑːl-/adj. (of a substance) incapable of being dissolved in a liquid (指物质)不可溶的:Vitamin D is insoluble in water. 维生素D不溶于水。opp soluble

inspect /ɪnˈspekt/vt. to examine or check carefully 检查

inspection /ɪnˈspekʃn/n. a visual examination of the horse to evaluate its quality and condition (对马匹体况的视检)检查:health inspection 健康检查 ◇ horse inspection 马匹检验 ◇ rectal inspection 直肠检查 ◇ veterinary inspection 兽医检查

inspector /ɪnˈspektə(r)/n. one authorized to examine a horse visually 检查员

inspire /ɪnˈspaɪə/v. to draw in air by inhaling; inhale 吸气 | **inspiration** /ˌɪnspəˈreɪʃn/n. 吸气:When the volume of the lung is increased during inspiration, the pressure within the lung decreases. 在吸气过程中,当肺容量增大时,肺内压开始降低。

instinct /ˈɪnstɪŋkt/n. an inborn behavior or capability that is characteristic of a species (个体与生俱来的行为或能力)本能:It is by behavioral instinct that the newborn foal reaches the teat for imbibing milk. 新生幼驹贴近乳头吮奶是它的本能行为。| **instinctive** /ɪnˈstɪŋktɪv/adj. 本能的

instruct /ɪnˈstrʌkt/vt. **1** to teach one the skills of riding, etc. 指导，传授 **2** to give orders to; direct 命令；指示

instruction /ɪnˈstrʌkʃn/ n. **1** the act or practice of instructing 指导，传授 **2** detailed information 说明：feeding instruction 饲喂说明

instructor /ɪnˈstrʌktə(r)/ n. *see* instructor 骑术教练，骑术指导

insula /ˈɪnsjʊlə/ n.【Latin】island 岛

insular /ˈɪnsjələ(r); *AmE* ˈɪnsələr/ adj.【anatomy】of or relating to isolated tissue or an island of tissue 胰岛的；脑岛的

insulate /ˈɪnsjʊleɪt/ vt. **1** to put in a detached or isolated position; isolate 隔离 **2** to prevent the passage of heat, electricity, or sound by surrounding with a nonconducting material（用绝缘材料阻碍热、电或声的传输）绝缘，隔离

insulation /ˌɪnsjʊˈleɪʃn/ n. the act of insulating or the state of being insulated 隔离；绝缘

insulin /ˈɪnsjəlɪn; *AmE* -səl-/ n.【physiology】a hormone secreted by the pancreas that functions to regulate the blood-sugar levels（一种由胰腺分泌的激素，负责调节血糖水平）胰岛素

intake /ˈɪnteɪk/ n. the amount of food or drink taken in 采食量：daily intake 日采食量 ◇ voluntary intake 自由采食量

integument /ɪnˈtegjʊmənt; *AmE* -ˈtegjəm-/ n. the skin and hair that covers the horse's body; coat（覆盖马体的毛皮）体被；被毛

intercostal /ˌɪntəˈkɒstl; *AmE* ˌɪntərˈkɑːstl/ adj. located between the ribs 肋间的：external intercostal muscle 肋间外肌 ◇ intercostal artery 肋间动脉

n. a space, muscle, or part situated between the ribs 肋间隙；肋间肌：The major inspiratory muscles are the diaphragm and the external intercostals. 呼吸肌主要指横膈膜和肋间外肌。

interdigitate /ˌɪntəˈdɪdʒɪteɪt/ vi. **1**（of laminae）to interlock（蹄叶）交错，并错 **2**（of a leg）to collide with another leg（马蹄）冲撞，撞击 | **interdigitation** /ˌɪntɜːrdɪdʒɪˈteɪʃn/ n.

interfere /ˌɪntəˈfiə(r); *AmE* ˌɪntərˈfir/ vi. **1**【racing】（of a horse）to run into another horse in a race（指赛马）阻碍，妨碍 **2**（of horse's feet）to collide or strike with each other（马蹄互相碰撞）交突 [compare] overreach

interfering /ˌɪntəˈfɪərɪŋ; *AmE* ˌɪntərˈfɪrɪŋ/ n. a gait fault in which one foot strikes the inside of the opposite leg（马蹄撞击的一种运步缺陷）交突 [compare] overreaching

interfering shoe n. *see* feather-edged shoe 防交突蹄铁

intermediare /ˌɪntəˈmiːdiər; *AmE* -tərˈmiːdir/ n.【French】*see* intermediate 中级［赛］

Intermediare Ⅰ n.【dressage】（also **Intermediate Ⅰ**）a level of international dressage competition between Prix St. Georges and Intermediare Ⅱ（国际盛装舞步赛介于圣・乔治奖与Ⅱ号中级赛之间的级别）Ⅰ号中级赛—*see also* **Prix St. Georges, Intermediare Ⅱ, Grand Prix**

Intermediare Ⅱ n.【dressage】（also **Intermediate Ⅱ**）a level of international dressage competition ranking between Intermediare I and Grand Prix（国际盛装舞步赛介于Ⅰ号中级赛与大奖赛之间的级别）Ⅱ号中级赛 —*see also* **Prix St. Georges, Intermediare I, Grand Prix**

intermediate /ˌɪntəˈmiːdiət; *AmE* -tərˈm-/ n.（also **intermediare**）a level of equestrian competition between the novice class and the advanced class（介于初级和高级之间的赛事级别）中级［赛］—*see also* **intro**, **pre-novice**, **novice**, **advanced**

intermittent /ˌɪntəˈmɪtənt; *AmE* -tərˈm-/ adj. happening or starting at intervals 间断的，断续的 | **intermittence** /-əns/ n.

intern /ˈɪntɜːn; *AmE* ˈɪntɜːrn/ n.（also **interne**）a rider gaining practical knowledge under supervision of one more experienced（在经验更丰富的骑师指导下学习的骑手）见习骑师

Intern Classification Farrier n.（abbr. **ICF**）a level of ability of student and novice farriers conferred by the American Farriers Association on the basis of written examination（美国蹄铁匠协会根据笔试成绩对见习蹄铁匠进行的水平认证）见习蹄师水平资格

internal /ɪnˈtɜːnl; *AmE* ɪnˈtɜːrnl/ adj. of, relating to, or located inside the body or an organ（位于机体或器官内部）内侧的，内部的

International Alpha System n. a system of symbols applied by the US Bureau of Land Management for branding the mustangs（美国国家土地管理局标识北美野马的标号体系）国际 α 体系

International American Albino Association n.（abbr. **IAAA**）an organization founded in 1985 to replace the American Albino Association to maintain breed registries for the American Cream and American White（1985 年取代美国白马协会成立的组织机构，旨在对美国乳色马和美国白马进行品种登记）国际美国白马协会

International Arabian Horse Association n.（abbr. **IAHA**）an organization founded in 1950 in Westminster, Colorado, USA to register Anglo-Arab and Half-Arab horses and to regulate competition and shows of Arab Horses（1950 年在美国科罗拉多州的威斯敏斯特市成立的组织机构，旨在对盎格鲁阿拉伯马和半血阿拉伯马进行登记并监管相关的赛事）[美国]国际阿拉伯马协会

International Equestrian Federation n. *see* Fédération Equestre Internationale 国际马术联合会

International Equine Reproduction Symposia Committee n.（abbr. **IERSC**）an international organization set up in Cambridge in 1974 to arrange meeting every four years on problems of equine breeding（1974 年在剑桥大学成立的国际机构，每四年组织一次马育种工作会议）国际马匹繁育研讨委员会

International Federation of Pony Breeders n. an organization founded in England in 1951 to develop international markets for different pony breeds and to serve as a liaison between breeders and buyers（1951 在英国成立的组织机构，旨在开展矮马国际交易并充当买卖中介）国际矮马育种者联合会

International Horse Show n. *see* Royal International Horse Show [皇家]国际马匹展览赛

International Society for the Protection of Mustangs and Burros n.（abbr. **ISPMB**）an organization founded in the United States in 1965 to register and protect the wild horses and burros in North America（1965 年在美国成立的组织机构，旨在对北美野马和野驴进行登记并予以保护）国际野马野驴保护协会

International Symposium on Equine Reproduction n.（abbr. **ISER**）an international meeting held over a period of six days every four years since 1974 to discuss issues concerning equine reproduction（自 1974 年起每四年一届举行的为期 6 天的国际会议，旨在讨论马匹繁育的议题）国际马匹繁育研讨会

International Union of Journeymen Horseshoers of the United States and Canada n.（abbr. **IUJH**）an organization founded in 1893 to replace the Journeyman Horseshoers National Union to organize tests for horseshoers（1893 年取代美国国家职业蹄铁匠联盟而成立的组织机构，旨在举办蹄铁匠资格考试）北美国际蹄铁匠联盟，北美国际钉蹄师联盟 [syn] Platers Union

International Veterinary Acupuncture Society n.（abbr. **IVAS**）an international association founded in the United States in 1974 to train and certify veterinarians worldwide in the use of acupuncture（1974 年在美国成立的组织机构，旨在对兽医针灸治疗进行培训和资格认定）国际兽医针灸协会

interne /ˈɪntɜːn; *AmE* ˈɪntɜːrn/ n. *see* intern 见习骑师，见习骑手

interosseous /ˌɪntərˈɒsɪəs/ adj.【anatomy】located between the bones 骨间的

interosseous splint n. a splint occurring between

the cannon and splint bones characterized by a bump which is usually not accompanied by lameness (管骨和小掌骨间附生的赘骨,有肿块出现,多无跛行并发)骨间内生赘骨

interphalangeal arthritis n. *see* ringbone 趾间关节炎,环骨瘤

interrupted stripe n. a white marking on the horse's face in which a long, narrow stripe runs from the forehead, ending at eye level, then continues to the top of the nostrils (马面部别征之一,白色条纹始于额前止于眼间,后又延伸至鼻端)断流星

interscapular /ˌɪntəˈskæpjʊlə; *AmE* ˌɪntərˈ skæpjələ/ adj. located between the two scapulae 肩胛间的

interspecies /ˌɪntəˈspiːʃiːz; *AmE* ˌɪntərˈs-/adj. arising or occurring between species; interspecific 种间的:an interspecies hybrid 种间杂种 ◇ interspecies competition 种间竞争

interstitial /ˌɪntəˈstɪʃl/ adj. 【anatomy】 relating to or situated within the spaces between tissues or parts of an organ 组织间的,间质的:interstitial cells 间质细胞 ◇ interstitial fluids 组织间液

intertrack wager n.【racing】a bet placed at one racetrack on horses running in a horse race at another track 场间投注,场际投注
vt. to place a wager at one racetrack on horses running in a horse race at another track 场间投注 | **intertrack wagering** n.

intervertebral /ˌɪntəˈvɜːtɪbrəl/adj. located between vertebrae 椎间的:intervertebral foramen 椎间孔

intestinal /ɪnˈtestɪnl/adj. of or relating to intestines 肠的:intestinal impaction 肠阻塞 ◇ intestinal torsion 肠扭转

intestinal catastrophe n. a dramatic and serious colic in which the intestines become twisted or rotated about its mesentery, thus obstructing the blood supply (由于肠和腹膜缠结导致血液供应阻塞而导致的急性腹痛)肠绞痛 [syn] twisted gut

intestinal stone n. *see* enterolith 肠结石

intestine /ɪnˈtestɪn/n. 【anatomy】the portion of the alimentary canal extending from the stomach to the anus (从胃部延伸到肛门的消化道部分)肠,肠道:large intestine 大肠 ◇ small intestine 小肠

intra- pref. within 内的,内部的

intra-alveolar /ˌɪntrəˈælvɪələr/adj. within the alveoli or lung 肺泡内的;肺内的:intra-alveolar pressure 肺内压

intra-articular /ˌɪntrəˈɑːtɪkjʊlə; *AmE* -ˈɑːrtɪkjələ/ adj. within the joint space (关节腔内的)关节内的

intracapsular /ˌɪntrəˈkæpsjʊlə/adj. within the articular capsule (关节)囊内的:intracapsular ligaments 囊内韧带

intracartilaginous ossification n. *see* endochondrial ossification 软骨内成骨 [opp] extracapsular

intracellular /ˌɪntrəˈseljʊlə; *AmE* -ljələ/adj. within or inside of the cell 细胞内的

intractable /ɪnˈtræktəbl/adj. (of a horse) difficult to manage or control; stubborn (指马)难驾驭的;倔强的

intracytoplasmic sperm injection n. (abbr. **ICSI**) the practice of injecting a sperm into the egg cell which is then implanted into the uterus of a mare (将单个精子注入卵细胞内然后再植入母马子宫内的辅助生育手段)胞浆单精子注射

intradermal /ˌɪntrəˈdʒːməl; *AmE* -ˈdʒːrm-/adj. (of a medicine) administered directly into the skin 皮内的:intradermal injection 皮内注射

intramuscular /ˌɪntrəˈmʌskjələ(r)/adj. (abbr. **IM**) within or administered directly into muscle tissues 肌内的,肌内注射的:intramuscular fat 肌内脂肪 [compare] intravenous

intramuscular injection n. the administering of a medication by injecting directly into the muscle tissue 肌内注射

intrapleural /ˌɪntrəˈplʊərəl/adj. occurring or situated within the pleural cavity 胸膜腔内的:The intrapleural pressure averages −0.5 to −0.3 kPa in horses at rest. 马静息时胸膜腔内压平均在 −0.5～−0.3 千帕之间。

intrathoracic /ˌɪntrəθəˈræsɪk/adj. occurring or situated within the thoracic cavity 胸腔内的

intrauterine /ˌɪntrəˈjuːtəraɪn/adj. occurring or situated within the uterus 子宫内的

intravenous /ˌɪntrəˈviːnəs/adj. (abbr. **IV**) administered directly into the bloodstream through a vein 静脉注射的:intravenous injection 静脉注射

I

intro /ˈɪntrəʊ; *AmE* ˈɪntroʊ/n. (also **intro class**) the lowest class of equestrian competition in which only some basic equitation are assessed (马术比赛的最低级别,仅对基本骑术有所考核)入门级 —*see also* **pre-novice**, **novice**, **intermediate**, **advanced**

intro class n. *see* intro 入门级

intromission /ˌɪntrəˈmɪʃn/n. the insertion of a stallion's penis into the vagina of a mare (公马阴茎进入母马阴道)插入

intumesce /ˌɪntjuːˈmes; *AmE* -tuːˈm-/vi. to swell 肿大,膨胀

intumescence /ˌɪntjuːˈmesəns; *AmE* -tuˈm-/ n. **1** the process of swelling or the state of being swollen 肿大,肿胀 **2**【anatomy】a swollen organ or body part 膨大:cervical intumescence 颈膨大 ◇ lumbar intumescence 腰膨大

intumescentia /ˌɪntjuːməˈsenʃɪə; *AmE* -tuˈm-/n.【Latin】(*pl.* **intumescentiae** /-ʃɪiː/) an intumescence of an organ or body part 膨大

invaginate /ɪnˈvædʒɪneɪt/vi. to infold or become infolded as in the formation of the prepuce (如包皮)内褶,内陷

invagination /ɪnˌvædʒɪˈneɪʃn/n. the process of invaginating or the condition of being invaginated 内褶,内陷:The prepuce is a double invagination of the skin which covers the free portion of the penis when not erect. 包皮为皮肤内褶形成的双层结构,包裹在未勃起的阴茎头部。

inventory /ˈɪnvəntri; *AmE* ˈɪnvəntɔːri/n. a detailed list of things in one's possession 清单:It is helpful to maintain an inventory of all saddlery and equipments. 对所有的马具和装备列个清单是非常有用的。

investigate /ɪnˈvestɪgeɪt/vt. to explore or inquire into certain circumstance 探究:Horses like to explore and investigate a new environment by using their senses of sight, hearing, smell, taste, and touch. 马喜欢用它们的视觉、听觉、嗅觉、味觉和触觉等感官探究新的环境。| **investigation** /ɪnˌvestɪˈgeɪʃn/n.

investigative /ɪnˈvestɪgətɪv; *AmE* -geɪtɪv/adj. of or characterized by investigation or exploration out of curiosity 探究的:investigative behavior 探究行为

invitational /ˌɪnvɪˈteɪʃənl/n. a competitive event restricted to invited riders and horses (仅限受邀骑手和马匹参加的比赛)邀请赛

involuntary /ɪnˈvɒləntri; *AmE* ɪnˈvɑːlənteri/adj. not subject to control of the volition 不随意的

involuntary muscle n. any muscle fibers that are not controlled by will, such as the muscles involved in breathing and digestion (收缩不受意念控制的肌纤维,比如与呼吸和消化相关的肌肉)不随意肌

involute /ˈɪnvəluːt/vi. to curl to inward or return to a normal or former condition (内卷或恢复原状)内卷;复原

involution /ˌɪnvəˈluːʃn/n.【anatomy】the process by which an organ returns to its former size and condition (器官恢复至原形的过程)回缩,恢复:uterine involution 子宫回缩

iodine /ˈaɪədiːn; *AmE* -daɪn/n. a mineral component of the hormone thyroxin (构成甲状腺素的矿物质元素)碘:iodine tincture 碘酒 ◇ iodine deficiency 碘缺乏症

Iomud /ˈaɪəməd/n. (also **Yomud**) a Russian warmblood developed by the Iomud Turkoman tribe from the ancient Turkmene; has a compact, sinewy body, and long legs; stands about 15 hands and generally has a gray coat, although bay and chestnut also occur; is exceptionally resistant to heat and has tremendous stamina and speed; historically used as cavalry mounts and now for riding and long distance races (俄国的温血马品种,由约穆德土库曼部落采用古老的土库曼马培育而来。体格结实,肌腱发达,四肢修长,体高约15掌,毛色多为青,但骝和栗亦有之,耐热性强,耐力好,速度快,过去主要作军马使用,目前用于骑乘和长途赛)约穆德马

iris /ˈaɪrɪs/ n. the round, contractile membrane that surrounds the pupil of eye (围在瞳孔周围的伸缩性膜结构)虹膜: Iris regulates the amount of light entering the eye. 虹膜可调节入眼光线的亮度。

Irish car n. *see* Jaunting car 出游马车

Irish Cob n. a breed type developed in Ireland since the 18th century by crossing Connemara, Irish Draft, and English Thoroughbred; stands 15 to 15.3 hands and may have a bay, brown, black, gray, or chestnut coat; has a short neck and back, and short strong legs; used for riding and light draft (爱尔兰于18世纪采用康尼马拉马、爱尔兰挽马和英国纯血马杂交培育而来的马匹类型。体高15~15.3掌,毛色多为骝、褐、黑、青或栗,颈背短,四肢粗短,主要用于骑乘和轻挽)爱尔兰柯柏马: The breed characteristics of Irish Cob are not stable and it often exceeds the height limits for cobs. 爱尔兰柯柏马的品种特征不甚稳定,且体高常超过柯柏马的限制。

Irish Draft n. (also **Irish Draught**) a draft breed developed in Ireland by crossing Connemara with Clydesdale during the 19th century; stands 15 to 17 hands and may have a brown, bay, chestnut, or gray coat; has a strong yet heavy shoulder, sound, strong legs with little feathering, free and straight natural action; used for jumping and breeding of Irish Hunters (爱尔兰于19世纪采用康尼马拉矮马与克莱德谷挽马杂交培育而成的挽马品种。体高15~17掌,毛色多为褐、骝、栗或青,肩部发达,四肢有力,距毛稀少,步法流畅,主要用于障碍赛和爱尔兰亨特马的培育)爱尔兰挽马

Irish Draught n. *see* Irish draft 爱尔兰挽马

Irish Half bred n. *see* Irish Hunter 爱尔兰亨特马;爱尔兰半血马

Irish hobby n.【Irish, dated】any equestrian sport including racing (包括赛马在内的)马术运动

Irish Hunter n. a horse breed developed in Ireland by crossing Thoroughbred with Irish Draft; stands 16 to 17.1 hands and may have a bay, brown, black, gray, or chestnut coat; used for hunting, show jumping, and eventing (爱尔兰采用纯血马与爱尔兰挽马杂交培育的马品种,体高16~17.1掌,毛色多为骝、褐黑、青或栗,主要用于狩猎、障碍赛和三日赛)爱尔兰亨特马 [syn] Irish Halfbred

Irish Jaunting Car n. *see* Jaunting Car 出游马车

Irish martingale n. (also **Irish rings**) an auxiliary rein consisting of two metal rings connected by a short leather strap through which the bridle reins run, used to prevent the reins from entangling (由皮带连接两金属环构成的辅缰用具,左、右方缰由此穿入以防止缠结)爱尔兰鞅

Irish rings n. *see* Irish martingale 爱尔兰鞅

Irish Sport Horse n. a relatively new horse breed developed in Ireland by crossing Irish Hunter with Thoroughbred (爱尔兰近年采用爱尔兰亨特马与纯血马杂交培育的新的马品种)爱尔兰竞技马,爱尔兰运动马

iron /ˈaɪən; *AmE* ˈaɪərn/ n. **1**【nutrition】a metal mineral element essential in the nutrition of the horse (马营养中的必需金属矿物质元素)铁: Iron deficiency often causes anemia of horses. 铁缺乏常导致马匹贫血。**2** a stirrup iron 马镫 **3** a branding iron 烙铁

iron cart n. a two-wheeled, box-shaped, horse-drawn cart constructed of cast iron plates; often used to carry water or liquid manure for agricultural purposes（由铁板制成的两轮厢式马车,主要用来运水或农用粪肥）铁箱马车

iron gray n.【color】*see* steel gray 铁青毛

iron horse n.【slang】a locomotive 火车头;铁马

iron shot v.【cutting】to encourage the horse into action by spurring 用马刺踢

irreducible hernia n. a kind of hernia in which the protruding organ has become attached to other body parts and cannot be pushed back into the cavity（疝气类型之一,脱出器官由于和其他组织粘连而不能推回到原腔体内）不可复疝,难复性疝 [compare] reducible hernia

irregular /ɪˈregjələ(r)/adj. not even or uniform in strides 步幅不匀的

irregular pacing walk n. a walk in which both legs on the same side move forward almost together as in pacing（马同侧前后两肢几近同时移动的慢步,看似对侧步）不规则对侧慢步

Isabella /ˌɪzəˈbɛlə/n.【color】(also **Y'sabella**) a very light, cream-colored shade of palomino, with non-blue eyes and white or flaxen mane and tail（色调较淡的银鬃毛色,眼睛不带蓝色,鬃尾为白色至淡黄不等）淡银鬃 [syn] Hanoverian Cream

Isabella quagga n. an extinct wild ass of a pale yellow color which existed in large herds throughout Africa until the late 1800s（19世纪之前驰骋于非洲的一种野驴,现已灭绝,被毛为灰黄色）伊莎贝拉野驴 [syn] *Asinus isabellinus*

ISAG abbr. International Society for Animal Genetics 国际动物遗传学会

ISBC abbr. International Stud Book Committee 国际纯血马登记委员会

ischemia /ɪsˈkiːmiə/n. a shortage or blockage in the blood supply to an organ or tissue caused by constriction or obstruction of the blood vessels（身体器官或组织因血管收缩或栓塞而出现的供血不足或阻滞）局部缺血

ischium /ˈɪskɪəm/n.【anatomy】(*pl.* **ischia** /-kɪə/) one of the major bones that constitute the pelvis（构成骨盆的主要骨头之一）坐骨 | **ischial** /-kɪəl/adj.

ISER abbr. International Symposium on Equine Reproduction 国际马匹繁育研讨会

islet /ˈaɪlət/n.【anatomy】a functional unit of pancreas 胰岛

isolate /ˈaɪsəleɪt/vt. to separate from others 隔离

isolation /ˌaɪsəˈleɪʃn/n. the act of isolating or state of being isolated 隔离;隔离状况:geological isolation 地域隔离

isolation barn n. a facility where sick horses are quarantined from healthy ones（用于分离病马与健康马的）隔离栏

isoleucine /ˌaɪsəˈluːsiːn/n.【nutrition】an essential amino acid that is isomeric with leucine（一种必需氨基酸,是亮氨酸的同分异构体）异亮氨酸

isometric /ˌaɪsəˈmetrɪk/adj.【physiology】of or involving muscular contraction in which the length of the muscle remains the same（肌肉收缩时长度保持不变的）等长收缩的:isometric contraction 等长收缩

isosmotic /ˌaɪsɒzˈmɒtɪk/adj. of or exhibiting equal osmotic pressure; isotonic 等渗的

isotonic /ˌaɪsəʊˈtɒnɪk; *AmE* ˌaɪsoʊˈtɑːn-/adj. **1**【physiology】of or involving muscular contraction in which the muscle remains under relatively constant tension while its length changes（指肌肉收缩时长度变化而张力不发生变化的）等张的:isotonic contraction 等张收缩 **2** having the same concentration or isosmotic 等浓度的;等渗的

isovolumetric /ˌaɪsəʊˌvɒljʊˈmetrɪk; *AmE* -soʊˌv-/ adj.(of chambers of the heart) with the volume remaining the same（心室）等容的: isovolumetric ventricular contraction 等容心室收缩 ◇ isovolumetric ventricular relaxation 等容心室舒张

ISPMB abbr. International Society for the Protection of Mustangs and Burros 国际野马野驴保护协会

isthmus /ˈɪsməs/n. 【anatomy】(*pl.* **isthmuses**, or **isthmi** /-maɪ/) a narrow strip of tissue or passage joining two organs or parts together (连接机体两器官或部分的狭长组织或细管)峡部:isthmus of the oviduct 输卵管峡部

Italian Heavy Draft n. a medium-sized, cold-blood, draft breed developed in Ferrara, Italy during the 1860s by crossing Breton with Ardennes and Percheron; stands 14.2 to 15.3 hands and usually has a chestnut coat, although bay and red roan also occur; has a light head set on a short neck, broad and deep chest, sloping shoulders, straight back, short legs with broad joints and small hoofs; formerly used for farm work and driving (意大利费拉拉地区于19世纪60年代采用布雷顿马与阿尔登马和佩尔士马杂交培育而成的挽用冷血马品种。体高14.2~15.3掌,毛色以栗为主,骝和红沙毛亦有之,头轻而颈短,胸身宽广,斜肩,直背,四肢粗短而蹄小,过去多用于农田作业和驾车)意大利重挽马

Italian Saddle n. *see* close contact saddle 意大利鞍

Italian seat n. *see* forward seat 意大利骑姿;前向骑姿

itch /ɪtʃ/n. **1** an irritating skin sensation 痒,瘙痒 **2** any of various skin disorders, such as scabies, marked by intense irritation and itching 疥癣 vi. to feel or have an itch 发痒,瘙痒

itchy heel n. *see* chorioptic mange 蹄疥癣,足螨

itchy leg n. *see* chorioptic mange 蹄疥癣

ITW abbr. intertrack wagering 场间押注

IU abbr. international unit 国际单位

IUJH abbr. International Union of Journeyman Horseshoers 职业蹄铁匠国际联盟

IV abbr. intravenous 静脉的;静脉注射

IVAS abbr. International Veterinary Acupuncture Society 国际兽医针灸学会

ivermectin /ˌaɪvəˈmektɪn/n. a generic name for an antiparasitic agent (一种抗寄生虫药物)伊维霉素

IVF abbr. in vitro fertilization 体外授精

J

jab /dʒæb/vt. **1** to poke or strike as with a spur (用马刺等)踢,捅 **2** (of the rider) to jerk the bit through the reins (指骑手)猛拽 PHR V **jab the mouth** (骑手)猛拽马嘴

Jabe /ˈdʒeɪb/n. *see* Dzhabe 扎拜马

jack /dʒæk/n. **1** (also **jackass**) a male donkey or ass 公驴:jack stock 种用公驴 syn stallion donkey **2** *see* bone spavin 飞节内生赘骨 **3** a horse mule 马骡

Jack Cart n. a flat, horse-drawn cart used in the mountain areas of southwest of England (英国西南山区过去使用的平板双轮马车)杰克马车

jack spavin n. any hard swelling on the lower inside of the hock 飞节内下肿

jackass /ˈdʒækæs/n. a male donkey; jack 公驴

jack-knife /ˈdʒæknaɪf/n. a knife with a folding blade 折刀

vi. 【rodeo】(of a horse) to form a V-shape by clicking his fore and hind legs together when bucking in the air (指马腾空跳跃时前后肢相撞成 V 形)折刀跳起,折刀姿势 | **jack-kniving** n.

jackpot /ˈdʒækpɒt; *AmE* -pɑːt/n. the most valuable prize in a particular event (比赛中最重要的奖项)头奖:to win the jackpot 赢得头奖

jade /dʒeɪd/n. (also **jadey**) a worthless, worn-out horse 老马,驽马

jadey /ˈdʒeɪdi/n. *see* jade 老马,驽马

Jaf /dʒæf/n. a light breed indigenous to Kurdistan derived from the Persian Arab; is well accustomed to harsh desert conditions and has Arab characteristics; stands about 15 hands and generally has a bay, chestnut, or gray coat; is spirited but gentle, hardy, tough, wiry, possessed of great stamina, and particularly noted for its tough, hard hoofs, and used chiefly for riding and light draft (源于库尔德斯坦的本土轻型马品种,由波斯阿拉伯马繁育而来,尤适应恶劣的沙漠环境。带有阿拉伯马特征,体高约 15 掌,毛色多为骝、栗或青,有悍威,性情温驯,抗逆性强,体质结实,耐力极好,蹄质结实,主要用于骑乘和轻挽)杰弗马

jag /dʒæg/n. a small load or portion; pack 驮载物,负载物

jagger /ˈdʒægə(r)/n. a peddler's pack horse (商贩用的)驮马

jail /dʒeɪl/n.【racing】a condition in which a horse is required to remain idle for a month when moving to a new barn (赛马换厩后 1 个月之内不能出赛的情况)禁赛 :be in of jail 处于禁赛期

jaivey /ˈdʒeɪvi/n. *see* jarvey 车夫

jam /dʒæm/n.【racing】(also **traffic jam**) a crush or congestion of horses on the racetrack when competing in a race (指赛马)扎堆,拥挤

vi. (of horses) to get crowded together 扎堆,挤

jammed heel n. a situation in which a horse's heel was pushed up into the hoof by incorrectly fitted horseshoe (由于装蹄不当而使蹄踵镶入蹄内的情况)蹄踵内陷

japan /dʒəˈpæn/n. a black coating put on leather to give it a glossy surface (涂在皮革上的黑色涂层)亮漆,黑漆面

vt. to coat leather with a glossy finish (给皮革)上漆面 | **japanning** n. 涂漆面[工艺]

japanned leather n. *see* patent leather 漆皮

japanner /dʒəˈpænə(r)/n. one employed to prepare and glaze leather (给皮革上漆面的雇工)漆面工

jargon /ˈdʒɑːgən; *AmE* ˈdʒɑːrgən/n. words or phrases used by a particular group or profession

(特定行业团体所使用的术语)行话

jarvey /ˈdʒɑːvi; *AmE* ˈdʒɑːrvi/n.【Irish】the driver of a jaunting car(驾驭游览马车的人)车夫 syn jaivey

jaundice /ˈdʒɔːndɪs/n. a condition in which the skin, sclera of the eye, and mucous membranes become yellow due to the accumulation of bile pigment(由于胆汁色素沉积造成皮肤、巩膜和黏膜组织发黄的症状)黄疸: hemolytic jaundice 溶血性黄疸 ◇ obstructive jaundice 阻塞性黄疸 syn icterus

jaunt /dʒɔːnt/n. a short trip or excursion made for pleasure 短途游览,出游

vi. make a short journey 短途游览,出游

Jaunting Car n.(also **Jaunty Car**) a horse-drawn, two-wheeled passenger vehicle first appeared in Dublin around 1813(于 1813 年前后出现在都柏林的双轮客用马车)游览马车,出游马车 syn Irish Car, Outside Car

Jaunty Car n. *see* Jaunting Car 游览马车,出游马车

Java /ˈdʒævə/n. a pony breed indigenous to the island of Java, Indonesia; stands about 12 hands and may have a coat of any color; has a heavy head, short, thick and muscular neck, a straight, long back, and long, well-muscled legs; is docile, willing, strong and hardy(源于印度尼西亚爪哇岛的本土矮马品种。体高约 12 掌,各种毛色兼有,头重,颈粗短,背长且平直,四肢修长,性情温驯,抗逆性强)爪哇矮马: The Java pony is similar in conformation to the Timor, but is slightly taller and stronger. 爪哇矮马的体貌与帝汶矮马较为相似,但稍高且更为健壮。

jaw /dʒɔː/n.【anatomy】either of the two bony structures in vertebrates that form the framework of the mouth and contain the teeth(构成脊椎动物口腔的两块骨质结构,其上附生牙齿)颌,颌骨: lower jaw 下颌骨 ◇ upper jaw 上颌骨

JBM abbr. just beat maiden 处女赛获胜

jejunum /dʒiˈdʒuːnəm/n.【anatomy】the section of the small intestine between the duodenum and the ileum(小肠位于十二指肠和回肠之间的区段)空肠

jennet /ˈdʒenɪt/n. **1** (also **jenny**, or **genet**) a female ass; she-ass 母驴 syn mare donkey **2** (also **Spanish Jennet**) a small Spanish riding horse of the Middle Ages; has a compact and well-muscled build and was noted for its smooth ambling gait and good disposition(西班牙中世纪的乘用马品种。体格紧凑,善走溜花蹄,性情温驯)詹妮特马: The Spanish Jennet horse is generally thought to have descended from the Andalusian. 普遍认为,詹妮特马由安达卢西亚马繁育而来。

jennet jack n. a male ass used to breed females to produce stock(用于繁育种群的公驴)种公驴,配种公驴

jenny /ˈdʒeni/n. **1** a female donkey; jennet 母驴 **2** (usu. **Jenny**) a small open carriage drawn by a small pony or donkey(一种敞开式小型马车,由矮马或驴拉行)珍妮马车

jerk /dʒɜːk; *AmE* dʒɜːrk/v. to pull or stop with a sudden sharp movement 猛拉,急停: to jerk the reins 猛拽缰绳 ◇ to jerk slack 拉紧绳索

n.【roping】the abrupt stop of the roped calf by the horse 猛拽,猛扯: The calf was thrown to the ground with a jerk. 猛扯之后,小牛便被放倒在地

jerk line n. **1**【AmE】a large team of horses strung out two abreast(两两并驾构成的一队马)马队 **2**【roping】the long rope attached to the saddle horn that is pulled taut by the horse when the calf is roped(系在牛仔鞍角柱上的长绳,小牛被套住后由马扯紧)扯牛套绳

Jerky /ˈdʒɜːki; *AmE* ˈdʒɜːrki/n. a light, four-wheeled topless buggy pulled by one horse(由单马套拉的四轮无盖轻型马车)杰凯马车

jet black n.【color】a pure black coat with points of the same color(体色和末梢都呈深黑的毛色类型)乌黑,漆黑

J

JHNU abbr. Journeyman Horseshoers National Union 全国职业蹄铁匠联盟

Jianchang n.【Chinese】(also **Jianchang horse**) a pony breed originating in Sichuan province, China; stands about 12 hands and usually has a bay coat; is lightly built, hardy, and frugal; used for packing in mountain areas (产于中国四川省的马品种。体高约12掌,毛色以骝为主,体型轻巧,抗逆性强,耐粗饲,主要用于山地驮运)建昌马

jib /dʒɪb/vi. (of a horse) to stop short and turn restively from side to side; balk (指马)徘徊,逡巡不前

jibbah /ˈdʒɪbə/n. the bulge in the forehead of an Arabian horse (阿拉伯马前额的突起)凸起,隆突

jibber /ˈdʒɪbə(r)/n. a horse who stops short and turns from side to side or moves back 逡巡不前的马

jig /dʒɪɡ/n. an irregular, four-beat jog trot performed by some horses in place of walking (一种不规则的四蹄音快步,用以代替慢步)小快步 | **jigging** n.

jiggle /ˈdʒɪɡl/vt. (of the rider) to shake the reins in short, quick movements to encourage the horse to pick up the bit (骑手为鼓励马受衔而轻快地抖动缰绳)轻抖,轻扯

Jilin n.【Chinese】(also **Jilin horse**) a draft horse breed developed in Jilin province, China by crossing Mongolian horse with Ardennes and Don; stands about 15 hands and may have a bay or chestnut coat; is strongly built and has pronounced withers; used for riding and driving (分布于中国吉林省的挽马品种,由蒙古马经阿尔登马及顿河马杂交改良培育而成。体高约15掌,毛色以骝、栗为主,体格健壮,鬐甲明显,用于骑乘和驾车)吉林马

jingle /ˈdʒɪŋɡl/n. **1** a horse-drawn cart having long shafts 辕车 **2** an Irish, public-hire vehicle pulled by a single horse (爱尔兰由单马套驾的公用马车)出租马车

jink /dʒɪŋk/n. a quick, sharp twist or turn 扭曲;急转

vi. to make quick, sharp turns; twist 急扭,急转

jinked back n. a condition in which the backbone becomes malformed due to injury or trauma, often results in pinched nerves and atypical movement (由于创伤导致马脊柱变形的情况,常造成神经疼痛和运动失调)脊背扭伤,脊背扭曲 [syn] chinked back

jinked neck n. a condition in which the vertebrae in the neck become malformed due to injury or infection, often results in pinched nerves and atypical movement (由于创伤或感染导致马颈椎变形的情况,常造成神经疼痛和运动失调)颈椎扭伤,颈椎扭曲

Jinker /ˈdʒɪŋkə(r)/n. a horse-drawn cart popular in Victoria, Australia during the mid-19th century (19世纪中叶流行于澳大利亚维多利亚州的双轮马车)贞克马车

jinking /ˈdʒɪŋkɪŋ/n. a condition in which the vertebral column become malformed or twisted due to injury or, in some cases, normal movement (由于创伤或自然移位导致的椎骨变形或移位)椎骨扭曲

Jinzhou n.【Chinese】(also **Jinzhou horse**) a breed of horse developed in certain areas of Liaoning province, China by crossing Mongolian horses with Hackney, Anglo-Norman and Orlov Trotter; stands about 14 hands and generally has a bay coat, although chestnut and black also occur; has a refined conformation and good speed, and is one of the most common breeds used in China's national equestrian sports and events (分布于中国辽宁省的马品种,主要采用蒙古马与哈克尼马、盎格鲁诺尔曼马以及奥尔洛夫快步马杂交培育而成。体高约14掌,毛色以骝为主,栗和黑少见,体貌俊秀,速力快,是中国开展马术运动的主要品种之一)金州马

job horse n. *see* rental horse 租用马,出租马

job master n.【dated】one who hires out horses, vehicles, and harness 租马贩;租车贩

jockey /ˈdʒɒki; *AmE* ˈdʒɑːki/n. **1**【racing】a rider who rides horse in races as a profession [职业]骑师: freelance jockey 自由职业骑师 ◇ Jockeys are small and lightweight to allow the horse to run faster. 骑师都身材小、体重轻,以便赛马可以跑的更快。**2**【dated】an unreliable person 不可靠的人 **3**【dated】(also **jockie**) a dishonest horse dealer 牙子,马贩子
vi. **1** to ride a horse in a race(在赛马中)骑马 **2** PHR V **jockey for position** (of a jockey) to maneuver his mount to gain a favorable position in a race (骑师)抢占有利位置

jockey agent n.【racing】one who arranges races for a professional rider 骑师经纪人: Jockey agent generally charges a fee of 20 percent or more on the rider's earnings. 骑师经纪人通常会从骑手所赢奖金中抽取20%以上作为佣金。

jockey apprentice n. *see* apprentice jockey 见习骑师

jockey club n. **1** an association founded by individuals and horse owners to maintain the stud book and registry of Thoroughbreds, and regulate horse racing and formulate rules in an area or country(某个地区或国家由个人和马主组成的团体机构,旨在保管纯血马登记册并对其进行注册工作,组织监管赛马比赛并制定相应的规则)赛马会: Hong Kong Jockey Club 香港赛马会 **2** (usu. **the Jockey Club**) the English Jockey Club 英国赛马会 **3** an area of a racetrack reserved for club members and consisting of box seats, a restaurant, and lounges(跑马场为会员保留的场所,内设包间、餐厅和休息室)骑师俱乐部

jockey fee n.【racing】a sum of money paid to a jockey for riding in a race(骑师参加赛马比赛应得的报酬)骑师酬金,骑师出赛费

Jockeys' Guild n. an association founded in the United States in 1940 with dedication to the health and safety of professional jockeys(于1940年在美国成立的行会,旨在确保职业骑手的健康和安全)美国骑师公会

jockey's room n.【racing】a room located besides a racetrack where jockeys store their tack and prepare for a race(赛场边供骑师存放马具并做赛前准备的房间)骑师间

jockeyship /ˈdʒɒkɪʃɪp/n. **1** the practice of participating in horse races 赛马[活动] **2** the state or attitude of being a jockey 骑师身份;骑师精神

jodhpur boots n. *see* paddock boots 牧场靴,马靴

jodhpur strap n. *see* garter 马裤带

jodhpurs /ˈdʒɒdpəz; *AmE* ˈdʒɑːdpərz/ n. pl. (also **jods**) full-length riding pants with reinforced patches on the inside of the leg; breeches(骑马穿的长裤,腿内侧带衬垫)马裤

jods /dʒɒdz; *AmE* dʒɑːdz/n. pl.【colloq】jodhpurs or breeches 马裤

jog /dʒɒg; *AmE* dʒɑːg/n. **1** (also **hound jog**, or **jog trot**) a relaxed, slow trot with a shortened stride performed in western riding disciplines(西部骑乘中所用的慢快步,步幅较短)缓慢快步 **2**【racing】a slow warm-up for horses before the race(赛前的)慢跑热身

Jog Cart n. (also **Jogging cart**) a light, horse-drawn American cart used for training trotting horses(一种美式轻型双轮马车,主要用来训练快步马)慢跑马车

jog trot n. *see* jog 缓慢快步

Jogging Cart n. *see* Jog Cart 慢跑马车

john /dʒɒn; *AmE* dʒɑːn/n.【colloq】*see* gelding donkey 骟驴,叫驴

john mule n. a male mule 公骡

John Solomon Rarey n. *see* Rarey, John Solomon 约翰·所罗门·瑞

John Warde n. *see* Warde, John 约翰·瓦德

Johnson grass n. a tall, coarse perennial grass, *Sorghum halepense*, used for horse and cattle feed(一种多年生草本,学名石茅高粱,主要

J

用作牛和马的饲草）约翰逊草，强生草：Johnson grass hay 约翰逊干草 ◇ Johnson grass is generally cut at the early-bloom stage to yield two to three cuttings. 约翰逊草多在早花期收割 2～3 茬。

joint /dʒɔɪnt/n.【anatomy】the point where two or more bones join together（两块以上骨之间的连接处）关节

joint capsule n.【anatomy】the sac-like membrane enclosing the joint space and which secretes synovial fluid 关节囊

J

joint cartilage n. *see* articular cartilage 关节软骨

joint evil n. *see* joint ill 关节病

joint fluid n. *see* synovial fluid 关节液，滑液

joint ill n.（also **joint evil**）a serious and often fatal disease of foals caused by infection of *Actinobacillus equuli* that enters the umbilical opening shortly after birth; symptoms may include swelling of the leg joints, fever, anemia, and an elevated white blood cell count（新生幼驹脐带开口遭马放线杆菌入侵感染导致的致死性疾病，症状包括四肢关节肿痛、发热、贫血和白细胞增多）关节病 [syn] navel ill, navel infection

joint oil n. *see* synovial fluid 关节液，滑液

jointed bit n. any bit having a jointed mouthpiece 关节衔铁，链式衔铁

jointed snaffle n. any snaffle bit having a joint in the middle to allow the bit to bend or flex 单关节普通衔铁

Joint-Master n.【hunting】one appointed by the hunt committee to share the responsibilities of the Hunt Master（被委员会任命与猎场主共同负责相关事务的人）副猎主，联席猎主

Jomud /ˈdʒɒmuːd/n. *see* Iomud 约穆德马

jostle /ˈdʒɒsl; *AmE* ˈdʒɑːsl/vi.【racing】**1**（of a jockey or horse）to make physical contact with or bump another horse during a race（指骑手或赛马间）撞击 **2** to compete with others to gain an advantageous position 抢位，挤位

jostling stone n.【dated】a mounting block 上马台

joule /dʒuːl/n.（abbr. **J**）the international unit of energy（功和能的国际单位）焦耳

journeyman /ˈdʒɜːnimən; *AmE* ˈdʒɜːrn-/n. **1**【racing】a professional jockey 职业骑师 **2** *see* journeyman farrier 职业蹄铁匠

journeyman farrier n. a farrier who has fully served an apprenticeship in a farrier shop and is qualified to conduct his own business（在铁匠铺见习期满可独立谋业的蹄铁匠）职业蹄铁匠，职业蹄师

Journeyman Horseshoers National Union n.（abbr. **JHNU**）an association of draft horse farriers founded in Philadelphia, USA in 1874, renamed the International Union of Journeyman Horseshoers of the United States and Canada in 1893（1874 年在美国费城由蹄铁匠成立的组织机构，1893 年更名为美国和加拿大国际职业蹄铁匠联盟）美国职业蹄师联盟

joust /dʒaʊst/vi.（of knights）to fight on horseback with long lances in medieval times（指中世纪身披盔甲的武士）策马刺枪

n. **1** a combat between two mounted knights using lances 马上刺枪 **2** a mounted competition in which two armoured knights try to stab each other with long, wooden lances on galloping horses（一种马上竞技比赛，两名身穿铠甲的骑士手持长矛试图将对手挑落下马）马上刺枪比赛

jowl /dʒaʊl/n.【anatomy】the lower portion of jaw bone 下颌，下颌骨

jughead /ˈdʒʌɡhed/n. **1** a stupid horse 驽马，劣马 **2** a horse with a large, ugly head 钝头马

jugular /ˈdʒʌɡjələ(r)/adj. of or located in the region of the neck 颈的，颈部的：jugular groove 颈沟

jugular vein n. one of two large veins located on either side of the horse's windpipe（位于马气管两侧的大静脉）颈静脉

Jules Charles Pellier n. *see* Pellier, Jules Charles 朱利叶斯·查尔斯·佩里埃

jump /dʒʌmp/v. **1** (of a horse) to spring off the ground or over an obstacle; leap (指马跳离地面或越过障碍)跳跃;跨 PHR V **jump clean** 顺利跨过障碍 **jump flat** (of a horse) to jump over an obstacle with a flat back and head held high 直背跨栏 **jump free** (of a horse) jump over fences without a rider (指马)空跨 **2** to pass over sth by jumping 越过,跨过:His horse fell as it jumped the last hurdle. 他的坐骑在跨越最后一道障碍时跌倒了。
n.【jumping】an obstacle to be jumped 跨栏,障碍:jump standard 跨栏标杆 ◇ vertical jump 垂直障碍 ◇water jump 水碍

jump designer n.【jumping】*see* course designer (场地障碍赛)路线设计师

jump jockey n. a jockey excelling in races over hurdles or steeplechases 越野障碍赛骑师

Jump Seat Carriage n. *see* Jump Seat Wagon 弹座马车

Jump seat saddle n. *see* jumping saddle 障碍赛用鞍

Jump Seat Wagon n. (also **Jump Seat Carriage**) a four-wheeled, horse-drawn American wagon that could be converted into a single seat vehicle by folding the front seat down and jumping the rear seat forward (一种美式四轮马车,可将前座折叠而将后座支起来改成单座马车)弹座马车 syn Shift Seat Wagon

jump stand n.【jumping】(also **stand**) a metal or wooden post used to support horizontal rails over which horses jump (障碍赛中用来支撑横栏的金属或木制立杆)支架,标杆:Two jump stands are required to support each rail. 每个跨栏需要两根立杆支撑。syn jump standard

jump standard n.【jumping】*see* jump stand 跨栏标杆

jumper /ˈdʒʌmpə(r)/n. **1**【jumping】any horse trained for jumping over fences 障碍赛马 **2** a horse who runs in steeplechases or hurdle races 越野障碍赛马 **3** one who rides a horse jumping over obstacles 障碍赛骑手

jumper's bump n. a conformation defect of horse who has a bump at the point of the croup (马尻部隆起的体貌缺陷)腰角隆凸:Horse with jumper's bumps is often erroneously thought to have greater jumping ability. 人们常错误地认为,腰角有隆凸的马跨越能力更强。

jumping /ˈdʒʌmpɪŋ/n. **1** the action of leaping off the ground 跳跃;跨越 **2** the sport of show jumping 障碍赛

jumping chute n. (also **jumping lane**) a narrow chute with high walls on either side, through which a horse may be ridden or free jumped over a series of obstacles for training purpose (两侧设有高墙的巷道,马匹通过跨越内设障碍以达到训练目的)跨栏巷

jumping Derby n. a jumping event in which the horse jumps natural fences over a longer course than in show jumping (赛程长且障碍天然的跨栏比赛)跨栏德比赛

jumping lane n. *see* jumping chute 跨栏巷

jumping order n.【jumping】the order in which a horse is allowed to participate in show jumping (障碍赛中马出赛的次序)起跳次序:Jumping order is usually determined in advance of an event by drawing the lots, and riders near the end of the order have the advantage of seeing how the first riders complete the course. 障碍赛的起跳次序通常经抽签决定,次序靠后的骑手可通过观察前面的骑手如何跨越而占一定优势。syn starting order

jumping rump n. *see* goose rump 尖尻

jumping saddle n. a variation of English saddle used in showing jumping 障碍赛鞍 syn jump-seat saddle

jump-off /ˈdʒʌmpɒf/n.【jumping】(also **ride-off**) an extra round held for the riders who have completed the course without faults previously (为顺利跨越而无罚分的骑手额外举行的比赛)附加赛,争时赛:In jump-off, the competitor with the least number of jumping faults in the fastest time is the winner. 在附加赛中,罚

分最少且用时最短的选手即为冠军。syn barrage, jump against the clock

juncture /ˈdʒʌŋktʃə(r)/n.【anatomy】the connection between two bones or body parts（骨间或机体内的）连接；联结 | **junctural** /-rəl/adj.

junior /ˈdʒuːniə(r)/n.【jumping】**1** a rider under 18 years of age（18 岁以下的）青少年骑手 **2** a distinction of competitors determined by age（根据年龄对参赛者划分的组别）青少年组

Junky /ˈdʒʌŋki/n. a two-wheeled, horse-drawn vehicle formerly used in some parts of Australia（澳大利亚某些地方过去使用的双轮马车）扎凯马车

Just Beat Maiden n.【racing】(abbr. **JBM**) a horse who has only won a maiden race 处女赛获胜马匹

Justin Morgan n. the foundation sire of the Morgans; was a small bay-colored horse foaled in the United States in 1790（摩根马的奠基公马，于 1790 年出生在美国，体型小，骝毛）贾斯汀·摩根：Justin Morgan was sired by a Thoroughbred and out of a mare of mixed breeds. 贾斯汀·摩根是由一匹纯血公马和一匹混血母马交配所生。

jute /dʒuːt/n. a coarse fiber used for making sacks, blankets and cords（用来制作麻袋、毯子和绳索的粗质纤维）黄麻

Jutland /ˈdʒʌtlənd/n. an ancient heavy coldblood indigenous to Jutland Island, Denmark; has a heavy head, great depth of chest and girth, short and heavily feathered legs; stands 15.2 to 16 hands and generally has a chestnut coat with flaxen mane and tail, although roan, bay, and black coats also occur; is strong, docile, and gentle possessed of great endurance; used for heavy draft and farm work（源于丹麦日德兰岛的重型冷血马品种。头重，胸身宽广，四肢粗短且距毛浓密，体高 15.2～16 掌，被毛多为栗色且鬃尾淡黄，但沙、骝和黑亦有之，性格顽强，性情温驯，耐力出色，主要用于重挽和农田作业）日德兰马：The Jutland horse was often used in the Middle Ages for jousting and carrying heavily armored knights into battle. 在中世纪，日德兰马常被用于马上刺枪和战场上重甲骑士的坐骑。

juvenile /ˈdʒuːvənaɪl; *AmE* -vənl/n. a two-year-old horse 两岁驹

Kabachi /kə'bɑ:tʃɪ/n. a Russian mounted sport in which the rider throws his spear through a small hoop fitted at the top of a post 3 meters tall on a galloping horse（俄国马上运动项目，骑手策马疾驰时掷矛穿过3米高木杆顶端上的铁环）马上掷矛

Kabardin /kə'bɑ:dɪn/n. a Soviet-bred warmblood descended from indigenous Caucasian mountain stock crossed with Arab, Turkmene, and Karabakh blood; is sure-footed, agile, strong, and possessed of great endurance; stands 14.1 to 15.1 hands and usually has a bay, brown, or black coat; has a long neck, strong legs, and excellent feet, but weak quarters and sickle hocks; used for riding, equestrian competitions, and packing（苏联采用高加索山区马与阿拉伯马、土库曼马和卡拉巴赫马杂交繁育而来的温血马品种。步法稳健，动作敏快，体格健壮，耐力强，体高14.1～15.1掌，毛色多为骝、褐或黑，颈长，四肢粗壮，蹄质结实，后躯乏力，镰形飞节普遍，主要用于骑乘、马术项目和驮运）卡巴金马

Kadir Cup n. an annual mounted pig-sticking competition first held in 1874 in India（自1874年起每年在印度举行的骑马猎野猪比赛）卡迪尔杯猎野猪大赛

Karabair /ˌkærə'beə(r); *AmE* -'ber/n. an ancient warmblood indigenous to the Central Asian mountains; stands 14.2 to 15 hands and generally has a gray, bay, or chestnut coat; is hardy, agile, sure-footed, and fast; used for light draft, driving, packing, riding, and mounted sports（源于中亚山区的古老温血马品种。体高14.2～15掌，毛色多为青、骝或栗，抗逆性强，反应敏快，步法稳健，步速较快，主要用于轻挽、驾车、驮运、骑乘和马上竞技项目）卡拉巴伊尔马

Karabakh /ˌkærə'bɑ:k/n. a Soviet-bred warmblood originating in the mountains of Karabakh; stands 14 to 14.1 hands and generally has a golden dun coat, although chestnut, bay, and gray also occur; has a small, fine head with the dished profile typical of the Arab, low-set tail, and good feet; is calm, energetic, tough, and sure-footed; used for riding, equestrian games, racing, and limited harness work（苏联育成的温血马品种，源于伊卡拉巴赫山区。体高14～14.1掌，毛色多为金兔褐，但栗、骝和青亦有之，头小，面目俊秀，带有阿拉伯马凹形面貌特征，尾础低，蹄质结实，性情安静而富于活力，性格顽强，步法稳健，主要用于骑乘、马术娱乐项目、赛马和驾挽）卡拉巴赫马

Karacabey /'kærəˌkəbi/n. a warmblood developed at the beginning of the 20th century at the Karacabey Stud of Turkey by crossing local mares with Nonius stallions; is tough, versatile and has good conformation; stands 15.1 to 16.1 hands and may have a coat of any solid color; previously used by the Turkish cavalry, and now mainly for riding and light draft（20世纪初土耳其卡拉贾贝伊种马场采用当地母马与诺聂斯马公马交配培育而来的温血马品种。体格粗壮，用途广，体型优美，体高15.1～16.1掌，各种单毛色兼有，过去常被土耳其骑兵使用，现在用于骑乘和轻挽）卡拉贾贝伊马

karozzin /kə'rɒzɪn/n. a four-wheeled, horse-drawn passenger carriage used in Malta, usually pulled by a single horse and seated four passengers face to face under a high canopy（马耳他岛上使用的四轮客用马车，由单匹马或矮马拉行，面对面可坐四名乘客，所带顶篷较高）卡罗金马车

Kathiawari /ˌkæθiə'wɑ:ri/n. a pony breed indigenous to the state of Kathiawari, India; descended from small, frugal native ponies crossed with Arabs; stands 14 to 15 hands and may have any of the coat colors; has a weak neck and quarters, low-set tail, sickle hocks, and ears that curve distinctively inwards; is light in build, hardy, tenacious, and extremely frugal; used for riding, packing, farming, light draft, and racing (源于印度卡提阿瓦省的矮马品种,主要由当地矮马与阿拉伯马杂交繁育而来。体高 14 ~ 15 掌,各种毛色兼有,颈细,后躯弱,尾础低,镰形飞节,耳朵内敛,体格纤小,抗逆性强,性情顽烈,耐粗饲,主要用于骑乘、驮运、农田作业、轻挽和赛马)卡提阿瓦矮马

kave /keɪv/vi. (of a horse) to paw or scrape with its front legs; port (马用前肢)刨趴

Kazakh /kɑːˈzɑːk/n. an ancient horse breed originating in Kazakhstan, thought to have descended from the Asiatic wild horse refined by significant infusions of Don blood; stands 12.1 to 13.1 hands and generally has a bay, chestnut, or black coat; has a full mane and tail, strong limbs and hard feet; is willing, quiet, frugal, strong, and possessed of great stamina and hardiness; used mainly for riding, herding, and farm work and have contributed substantially to the breeding of Yili horse (源于哈萨克斯坦的古老矮马品种,普遍认为由亚洲野马引入大量顿河马外血改良而来。体高 12.1 ~ 13.1 掌,毛色以骝、栗和黑为主,鬃尾浓密,四肢健壮,蹄质结实,性情温驯而安静,耐粗饲,速力优秀,抗逆性强,主要用于骑乘、放牧和农田作业,对伊犁马的育成贡献较大)哈萨克马

keep /kiːp/vt. to maintain or confine a horse in a fenced area or pasture 圈养;放牧: to keep a horse at grass 在草场上牧马 PHR V **keep the horse to the track** to follow a previously made path 沿老路行进;循规蹈矩

keeper /ˈkiːpə(r)/n. **1** a horse kept on a farm or pasture 饲养的马: a poor keeper 一匹难养的马 ◇ An easy keeper could maintain good body condition even on small rations, while a hard keeper might be susceptible to digestive disturbances as the quality of the feed changes. 一匹好养的马即便日粮很少也可以维持良好体况,而一匹难养的马在饲料质量有所变化时消化系统可能会出现紊乱。 syn doer **2** (usu. **keepers**) any of the leather loops used on saddlery to keep the ends of a strap in place (马具上的皮革圈,用来固定皮带末梢)革环,鞘环 syn runners, harness loops

Keiger Horse n. *see* Keiger Mustang 凯格野马

Keiger Mesteño Association n. an organization founded in Oregon, USA in 1988 to protect and preserve the remaining Keiger Mustangs in captivity from extinction (1988 年在美国俄勒冈州成立的组织机构,旨在保护现存的凯格野马免于灭绝)美国凯格野马协会

Keiger Mustang n. (also **Keiger horse**) a group of wild horses first discovered in 1978 in the Steens Mountains of Southern Oregon, USA; stands 14 to 15.2 hands and usually has a dun coat with zebra and dorsal stripes; has a short back, strong legs and hoofs, a willing disposition, and is extremely sure-footed (1978 年在美国俄勒冈州南部的斯汀山脉首次发现的野马品种,体高 14 ~ 15.2 掌,毛色兔褐,带斑纹和背线,背短,四肢健壮,蹄质结实,性情温驯,运步稳健)凯格野马: The Keiger Mustang is commonly believed to have descended from the Spanish horses brought to North America by the conquistadors in the 16th century. 人们普遍认为,凯格野马由 16 世纪西班牙殖民者带到北美的西班牙马繁衍而来。

kennel /ˈkenl/n. **1**【hunting】the dwelling of a fox or other wild animal, especially above the ground (野兽在地面上的居所)窝,巢穴 **2** a shelter for a dog or hound 犬窝,狗圈 **3** a building where hounds are boarded, bred or trained 犬场,狗场: kennel coat 犬场制服 **4** a pack of hounds 一群猎犬

vt. to keep in a kennel (将犬)圈养

kennel huntsman n.【hunting】one employed by the Hunt to take care of the hounds (猎狐俱乐部照料猎犬的雇员)猎犬主管

kennel man n.【hunting】one who works in the hunt kennels 犬场工人,犬场饲养员

Kentucky Derby n.【racing】a 1.25-mile race for three-year olds run annually since 1875 at Louisville racetrack in Kentucky, USA (自1875年起在美国肯塔基州路易斯维尔赛马场为3岁驹举行的比赛,赛程2千米)肯塔基德比大赛

Kentucky Saddlebred n. *see* American Saddlebred 美国骑乘马

Kentucky Saddler n. *see* American Saddlebred 美国骑乘马

kept-up /kept'ʌp/ adj. (of horses) kept in stable during the summer months (指马夏季)圈养的,厩饲的

keratin /'kerətɪn/ n. a tough, insoluble protein forming the main structural constituent of hair and hoofs (一种不溶性硬质蛋白,是毛发和蹄的主要组成成分)角蛋白

keratinize /'kerətɪnaɪz/ v. to change into a form containing keratin 角质化 | **keratinization** /ˌkerətɪnaɪ'zeɪʃn/ n.

keratoma /ˌkerə'təʊmə/ n. (also **keratosis**) a rare tumor developing in the deep layer of the hoof wall (在马蹄壁深层内形成的罕见肿瘤)角质瘤

keratosis /ˌkerə'təʊsɪs; *AmE* -'toʊ-/ n. *see* keratoma 角质瘤

Kerry Bog Pony n. *see* Kerry Pony 克雷[沼地]矮马

Kerry Pony n. (also **Kerry Bog Pony**) a mountain pony breed native to the county of Kerry, Ireland, from which the name derived; stands 10 to 12 hands and may have a coat of any solid color; is robust and hardy and used mainly for riding and driving (源于爱尔兰克雷郡的本土矮马品种并由此得名。体高10~12掌,各种单毛色兼有,富于活力,抗逆性强,主要用于骑乘和驾车)克雷矮马

kersey /'kɜːzi; *AmE* 'kɜːrzi/ n. a coarse, ribbed fabric formerly used for making ankle boots and trousers (一种有棱的粗织物,过去常用来制作绑腿和马裤)克尔赛呢

Kersey protection boots n. *see* ankle boots 克尔赛护腿

ketamine /'ketəmiːn/ n.【vet】a highly effective anesthetic (一种高效麻醉剂)氯胺酮,开他敏

key horse n.【racing】a racehorse used in multiple betting combinations (在多重组合彩投注中的赛马)主马

keys /kiːz/ n. pl. (also **bit keys**) pieces of shaped metal attached to the mouthpiece of some bits to encourage the horse to accept the bit (某些口衔上附带的缀铁,有促进马受衔的作用)衔缀:Young horse will play with the keys using its tongue. 青年马会用舌头玩弄衔缀。 syn players

kiang /kiː'æŋ/ n. (also **kyang**) a subspecies of the Asiatic wild ass mainly living on the Tibetan plateau (亚洲野驴的亚种,主要栖息于青藏高原)藏野驴:The kiang is the largest of the wild asses. 藏野驴在野驴品种中体型最大。 syn *Equus kiang*, Tibetan wild ass

kibitka /kɪ'bɪtkə/ n. a covered, horse-drawn wagon used formerly in Russia (俄国过去使用的)带篷马车

kick /kɪk/ v. **1** (of a horse) to strike out with either of his hind legs (指马)踢,蹶:In hunting, the tail of a horse who kicks is tagged with a red ribbon. 猎狐时,有踢蹶恶习马的尾巴上会系条红带。**2** to win in a wager, competition, or bet (比赛)获胜;(投注)押中

kicker /'kɪkə(r)/ n. a horse that kicks (爱)踢人的马

kicking /'kɪkɪŋ/ n. a vice of horse who kicks 踢蹶:kicking strap 防踢带 ◇ kicking boots 防踢

护腿

kidney /ˈkɪdni/n.【anatomy】either one of the two bean-shaped organs in the body functioning to maintain water and electrolyte balance and filter the blood of metabolic wastes to produce urine (位于身体背部两侧的豆状器官,有维持水和电解质平衡、过滤血液代谢物以及泌尿的作用)肾脏

kill /kɪl/vt. to deprive of life; slaughter, destroy 杀生,杀死

n. **1** animals killed in hunting (狩猎中)杀死的猎物 **2** the raw meat fed to the hounds (喂猎犬的)生肉

kilocalorie /ˈkɪləʊˌkæləri; *AmE* ˈkɪləˌk-/n. (abbr. **Kcal**) a unit of energy equal to 1,000 calories (能量单位)千卡

kilojoule /ˈkɪlədʒuːl/n. (abbr. **kJ**) an energy unit equal to 1,000 joules (能量单位)千焦

Kimberwick /ˈkɪmbəwɪk/n. *see* Kimblewick 双缰衔铁

Kimblewick /ˈkɪmblwɪk/n. (also **Kimberwick**) a bit with a single pair of reins to control the mouthpiece 双缰衔铁 [syn] Spanish snaffle bit

kinesiology /kɪˌniːsɪˈɒlədʒi; *AmE* kəˌniːsɪˈɑːl-/ n. the scientific study of muscles, joints and the mechanics of body motion (研究肌肉、关节及身体运动机理的科学)运动机能学: equine kinesiology 马运动机能学

kinesipathy /kɪˌniːsɪˈɒpəθi; *AmE* kəˌniːsɪˈɑːp-/n. the functional pathology of movement characterized by hypomobility or hypermobility of the motor units (运动机能出现病变的情况,表现为肌肉运动单位活力过低或过高)运动病理学,运动机能障碍

Kineton noseband n. (also **lever noseband**) a severe noseband used for horses who pull or jerk reins (一种较为苛刻的鼻革,用于拽缰的马匹)杠杆鼻革 [syn] lever noseband

King George V (Gold) Cup n. an annual individual show jumping competition held under the FEI rules at the Royal International Horse Show in London, England (在英国伦敦皇家国际马匹展览赛上举行的年度个人障碍赛)乔治五世杯[障碍赛]: In 1911 the International Horse Show received royal patronage and the King George V Gold Cup was awarded for the first time. 1911年,国际马术比赛获得了皇家赞助,而乔治五世杯也首次颁奖。

kingdom /ˈkɪŋdəm/n. the highest taxonomic classification into which organisms are grouped (生物学分类体系中最高的划分类别)界 [compare] phylum, class, order, family, genus, species

kink /kɪŋk/n. a tight curl or twist of hair, reins, etc. (毛发或缰绳的缠绕或打结)缠结,纽结

v. to form kinks 缠结

kinked tail n. the twisted tail of a horse 马尾扭曲

Kiplingcotes Derby n. (also **Kiplingcotes Race**) the oldest annual horse race in England run on a 4-mile course in the Yorkshire Wolds since 1519 on the third Thursday of every March (英国最早的赛马比赛,从1519年起每年3月第3个星期四在英国约克夏的丘陵赛道上举行,赛程6.4千米)吉卜林科兹德比赛 [syn] Yorkshire Derby

Kirghiz /kɪəˈgiːz; *AmE* kɪrˈgiːz/n. an ancient horse breed indigenous to Kirghizia, Russia; stands 12.3 to 14 hands, and usually has a chestnut, grey or bay coat; has a straight profile with short legs and hard hoofs; is extremely hardy and well-suited to harsh climates; used mainly for packing and riding (源于俄国吉尔吉斯地区的古老的马品种。体高12.3～14掌,毛色以栗、青、骝常见,面貌平直,四肢粗短,蹄质结实,极耐粗饲,可适应恶劣环境,主要用于骑乘和驮运)吉尔吉斯马

Kladrub /kləˈdruːb/n. *see* Kladruber 克拉德鲁比马

Kladruber /kləˈdruːbə(r)/n. (also **Kladrub**) a warmblood developed in the 16th century at the Royal Stud in Kladruby, Czechoslovakia by putting Andalusian stallions to native mares; stands 16.2 to 18 hands and may have a black or gray coat; used for harness work, riding, and dressage (捷克克拉德鲁比皇家种马场于16世纪采用安达卢西亚马公马与当地母马繁育而来的温血马品种。体高16.2~18掌，毛色多为黑或青，主要用于挽车、骑乘和盛装舞步)克拉德鲁比马

Klepper /ˈklepə(r)/n. a native pony breed descended from native Latvian and Estonian mares crossed with Norfolk Roadster; stands 13 to 15 hands and generally has a black, bay, chestnut, or gray coat; is extremely hardy and possessed of great endurance and power; noted for its smooth trot and used mainly for riding, light draft and farm work (由拉脱维亚马和埃斯托尼亚马母马与诺福克快步马公马杂交培育而来的矮马品种。体高13~15掌，毛色多为黑、骝、栗或青，抗逆性极强，耐力出色，快步流畅，用于骑乘、轻挽和农田作业)克莱波矮马

klibber /ˈklɪbə(r)/n. a wooden saddle specifically designed for ponies to carry peats in the Shetland Islands (过去设特兰岛上矮马驮载泥炭用的木质马鞍)设特兰驮鞍，木驮鞍

Kluge Hans n. *see* Clever Hans 聪明马汉斯

kmph abbr. kilometers per hour 千米/小时

Knabstrup /ˈnæbstrʌp/n. (also **Knabstruper**) a Danish horse breed with spotted Appaloosa patterns on a roan coat; stands about 15 hands and is used predominantly for circus (丹麦的花斑马品种，多为沙毛，体高约15掌，主要用于马戏表演)纳普斯特鲁马 [syn] Danish Spotted Horse

Knabstruper /ˈnæbstrʌpə(r)/n. *see* Knabstrup 纳普斯特鲁马

knacker /ˈnækə(r)/n. one who slaughters old or unwanted horses 屠马贩 [IDM] **go to the knackers** (of animals) to be butchered (动物)被人屠宰

knacker's yard n. *see* knackery 屠马场

knackery /ˈnækəri/n. 【BrE】a horse slaughter yard 屠马场 [syn] knacker's yard

knee /niː/n. the part of a horse's foreleg between the forearm and the cannon (马前肢介于前臂与管骨的部分)前膝 [syn] carpus: [PHR] **back at the knee** (of the knees) bending backward 凹膝: The fault of being back at the knee places additional strain on the tendons running down the back of the lower forelegs. 凹膝这种缺陷造成马前肢下部后侧的筋腱必须承担额外的负荷。**over at the knees** (of the knees) bending forward 凸膝 **knees and hocks to the ground** (of a horse) having short cannon bones (马)管骨短的;腿短的: a horse having knees and hocks to the ground 一匹四肢很短的马 [syn] near/well to the ground

knee boots n. *see* knee caps 护膝

knee caps n. **1** a protective covering for the knees of horses [马用]护膝 [syn] knee boots, knee protectors, knee pads **2** *see* knee guards (骑手用)护膝，护腿

knee guards n. a full-front padded leather covering used to protect the rider's knee from bruises and collision (系在膝盖上的护垫，可保护骑手膝盖被擦破和撞伤)护腿，护膝 [syn] knee caps, knee pads

knee pad n. *see* knee guards 护腿，护膝

knee protectors n. *see* knee caps [马用]护膝

knee roll n. a pad attached to the underside of the saddle flap at about knee height to provide the rider with improved grip and security in the saddle (鞍翼下方齐膝高的衬垫，有助于骑手的骑乘安全)膝垫，挡膝

knee spavin n. a bony growth occurring at the back inside of the knee caused by a blow or strain (由于遭受撞击或扭伤而在前膝内侧附生赘骨的症状)膝内肿

knee-sprung /ˈniːsprʌŋ/adj. (of knees) projecting forwards 凸膝的: Should the knee project forwards, the horse is said to be "knee-sprung" or "over at the knee". 如果马的前膝向前突出，我们便说这匹马"凸膝"。

knight /naɪt/n. (abbr. **Knt**, or **Kt**) a gentleman serving as a mounted soldier in medieval times

(中世纪的)骑士

knobbler /ˈnɒblə(r); *AmE* ˈnɑːb-/n. *see* brocket 两岁雄鹿

knock /nɒk; *AmE* nɑːk/vi. **1**【jumping】(of a horse) to strike the rail of an obstacle in jumping (指马)撞击(栏杆): to knock a rail down 撞落跨栏 **2** (of a horse) to strike its one hoof or ankle or with another while running (指马)撞踝;撞蹄 **3**【polo】to strike the polo ball with mallet (用球杆)击球 PHR V **knock in** to initiate by hitting the polo ball into play (马球)击球开赛 compare throw-in

knock knees n. a conformation defect in which the horse's knees deviate towards each other (马前膝偏向内侧的体格缺陷)膝内向,X 形膝 syn carpus vulgus, medial deviation of the carpal joints

knockdown /ˈnɒkdaʊn/n.【jumping】the act of knocking a fence down in jumping an obstacle 撞落(栏杆): Penalties for a knockdown in the showjumping are 4 for the first knockdown and 8 for the second. 在障碍赛中,首次撞落跨栏给予 4 分的处罚,第二次撞落则罚 8 分。

knocker /ˈnɒkə(r); *AmE* ˈnɑːk-/n.【slang】a horse with cow and/or sickle hocks X 形飞节的马;镰形飞节的马

knock-kneed /ˌnɒkˈniːd/adj. (of knees) turning towards each other 膝内向的 syn ox-kneed

knot /nɒt; *AmE* nɑːt/n. a fastening made by tying together two string, ropes, etc. (用两根绳系在一起)结: release knot 活结 ◇ bowline knot 单套结

knot head n. an unruly horse who is a potential threat to himself or his rider 顽劣的马,危险的马

Knt abbr. knight 骑士

Konik /ˈkɒnɪk/n. a pony breed descended from the Tarpan horse and refined by infusions of Arab blood in Poland; stands 12 to 13.3 hands and may have a bay, dun, or palomino coat; is hardy, willing, frugal, and extremely resistant to hunger and cold; has a slightly heavy head, long mane and tail, and a short straight back; used for riding and packing (由塔畔马繁衍而来的波兰矮马品种,后引入阿拉伯马外血进行过改良。体高 12 ~ 13.3 掌,毛色可为骝、兔褐或银鬃,抗逆性好,性情温驯,耐粗饲,抗饿耐寒能力强,头偏重,鬃尾浓密,背短而平直,主要用于骑乘和驮运)考尼克矮马

Kt abbr. knight 骑士

Kuhailan /kuːˈhaɪlən/n. one of the three primary sub-breeds of the Assil (阿西利阿拉伯马的三个亚种之一)库海兰阿拉伯马 —*see also* **Muniqi**, **Siglavy**

kulan /ˈkuːlən/n. (also **Turkmenian kulan**) a subspecies of the Asiatic wild ass native to central Asia (亚洲野驴的一个亚种,主要生活在中亚地区)土库曼野驴,库兰野驴: The Turkmenian kulan was declared endangered in 2008. 土库曼野驴于 2008 年被定为濒危物种。syn Transcaspian wild ass, *Equus hemionus kulan*

kumel /ˈkʌml/n.【hunting】a fox covering located above the ground; kennel (地面以上狐狸的隐身处)藏身处;狐窝

kumiss /ˈkuːmɪs/n. (also **koumiss**) a drink made from fermented mare's milk by some nomads of central Asia (中亚游牧民族用母马乳汁发酵制成的饮品)马奶酒,酸马奶

kür exercises n.【vaulting】*see* freestyle exercises 自选动作

kür program n.【vaulting】an event consisting of kür exercises 自选项目

Kustanair /ˈkustəneə(r); *AmE* -ner/n. an ancient Oriental horse breed originating in Kazakhstan; stands 14 to 15.1 hands and may have a coat of any solid color; has a light head and long, well-muscled legs; used for riding and light draft (源于哈萨克斯坦的古老的东方马品种。体高 14 ~ 15.1 掌,各种单毛色兼有,头轻,四肢修长,肌腱发达,用于骑乘和轻挽)库斯塔奈马 syn Russian Steppe Horse

kyang /kiːˈæŋ/n. *see* kiang 藏野驴

L

L / el / abbr.【racing】length 身长:1/2 L 半个身长 ◇ to win by 3/4 L 以3/4 身长优势获胜 ◇ 1 L 一个身长

labium /ˈleɪbɪəm/n. **1**【Latin】a lip-like structure 唇状结构 **2** (*pl.* **labia** /-biə/) any of the folds of a mare's external genitals (母马外生殖器的皱褶)阴唇

labor /ˈleɪbə(r)/n. (BrE **labour**) **1** the physical work a horse undertakes 使役,作业 **2** the physical efforts of delivery; parturition 努责,阵痛;分娩:first-stage labor 一期阵痛

labor-saving /ˈleɪbəˌseɪvɪŋ/adj. (BrE **labour-saving**) designed to save the labor or effort needed for certain work 省力的;节省劳力的:It is labor-saving to locate the feed shed near the main stable complex. 饲料棚位于厩舍主体建筑附近比较省力。

labyrinth /ˈlæbərɪnθ/n.【anatomy】a group of complex interconnecting anatomical cavities, such as of inner ear (错综相连的解剖腔,如内耳)迷路:ethmoid labyrinth 筛骨迷路 ◇ osseous labyrinth 骨迷路 ◇ membranous labyrinth 膜迷路

laced reins n. reins with outer edges laced with leather strands to provide improved grip (外缘用细皮条编织而成的缰绳)花边缰绳

lacerate /ˈlæsəreɪt/vt. to injure by cutting or tearing 割伤,撕裂:The jumper's left foreleg had been badly lacerated. 这匹障碍赛马的左前肢已被严重拉伤。

laceration /ˌlæsəˈreɪʃn/n. any injury that penetrates the skin 撕裂,拉伤:superficial laceration 浅层拉伤 ◇ deep laceration 深度拉伤

lacing /ˈleɪsɪŋ/n. *see* rope walking 拧绳姿势

lactate[1] /ˈlækəteɪt/n. any salt of lactic acid 乳酸盐:Blood lactate is indicative of the significant anaerobic contribution. 血液中的乳酸含量是无氧酵解的重要指标。

lactate[2] /lækˈteɪt/vt. to secrete or produce milk 泌乳

lactate dehydrogenase n.【biochemistry】(abbr. **LDH**) an enzyme that catalyzes the conversion of pyruvate to lactic acid (催化丙酮酸转变为乳酸的酶)乳酸脱氢酶

lactating /lækˈteɪtɪŋ/adj. (of a mare) producing milk (母马)泌乳的

lactating mare n. a mare who produces milk and has a foal at foot 泌乳母马

lactation /lækˈteɪʃn/n. **1** the secretion of milk by the mammary glands 泌乳 **2** the period during which the mare secretes milk 哺乳期,泌乳期:lactation period 哺乳期

lactational /lækˈteɪʃnl/adj. of or relating to lactation 泌乳的;泌乳期的

lactational anestrus n. the period of sexual inactivity of the nursing mare during lactation (母马在泌乳过程中性活动消极的阶段)泌乳乏情期

lactic /ˈlæktɪk/adj. of, relating to, or derived from milk 乳的,乳汁的

lactic acid n.【biochemistry】a metabolite of the anaerobic breakdown of glucose (葡萄糖无氧酵解的代谢产物)乳酸:lactic acid bacteria 乳酸菌 ◇ Excessive accumulation of lactic acid often results in muscle fatigue, inflammation, and pain. 乳酸积累过多常导致肌肉疲劳、酸痛和发炎。

lactic acidosis n. an abnormal increase of lactic acid in the bloodstream resulting from overexertion, symptomized by stiffness and pain of muscles (马匹训练过度后血液中乳酸浓度异常升高的情况,症状表现为身体肌肉僵直和疼痛)乳酸酸中毒

lactobacillus /ˌlæktəʊbəˈsɪləs; *AmE* -toʊ-/n. (*pl.*

lactobacilli /-laɪ/) a rod-shaped, aerobic bacterium that produces lactic acid from the fermentation of carbohydrates (一种杆状需氧细菌,可将碳水化合物酵解产生乳酸)乳酸[杆]菌

lactose /ˈlæktəʊs; *AmE* -toʊs/n.【biochemistry】a disaccharide found in milk (一种存在于乳汁中的双糖)乳糖

lad /læd/n. *see* groom 马夫

lade /leɪd/vt. to load with sth; burden 负载;负重

laden /ˈleɪdn/adj. heavily loaded with weight 负重的,满载的: a fully laden cart 一辆满载的马车

lady's saddle n. *see* side-saddle 淑女鞍,侧鞍

lady's Phaeton n. *see* park Phaeton 女式费顿马车,公园费顿马车

laity /ˈleɪəti/n. a general term for those who are not members of a profession or specialized field; lay people (不属于某职业或专业领域人的总称)外行

lame /leɪm/adj. unable to walk well because of an injury or illness to the leg or foot (因受伤或疾病造成行走困难)跛的,瘸的: His horse went lame. 他的马走路跛行。

lame hand n.【driving, dated】a poor coachman 拙手车夫

lameness /leɪmnəs/n. the condition or state of being lame 跛行,瘸

lamina /ˈlæmɪnə/n. (*pl.* **laminae** /-niː/) one of the thin layers of vascular tissue in the hoof of a horse (马蹄中的脉管组织层)蹄叶: insensitive laminae 非敏感蹄叶 ◇ sensitive laminae 敏感性蹄叶

laminitic rings n. *see* hoof rings 蹄环纹

laminitis /ˌlæmɪˈnaɪtɪs/n. inflammation of the sensitive layers of tissue inside the hoof (马蹄敏感组织层发炎)蹄叶炎: Very minor laminitis occurs almost daily in horses, by stumbling, stepping on rocks or by hitting the hoof against another hoof or object, and often does not create lameness in the horse. 由于跌绊、踩到岩石或两蹄相撞,马几乎每天都会患轻度蹄叶炎,但这通常不会造成马匹跛行。[syn] founder | **laminitic** /-tɪk/adj.

lampas /ˈlæmpəs/n. (also **lampers**) the swelling of the hard palate due to inflammation of the mucous membrane (由口腔黏膜发炎导致硬腭红肿的病症)硬腭炎,硬腭红肿: Lampas often occurs in young horses while they transit from mare's milk to hard feed and grain or during eruption of the permanent incisors. 硬腭炎在幼驹从吮乳过渡到采食硬质饲料或青年马在长出切齿时较为常见。[syn] palatitis

lampers /ˈlæmpərz/n. *see* lampas 硬腭红肿,硬腭炎

lance /lɑːns; *AmE* læns/n. a long weapon with a wooden shaft and a sharp metal head that was formerly used by mounted knights and cavalrymen (古代骑士与骑兵使用的一种木制长杆武器,尖端带金属利刃)长矛,刺枪

lancer /ˈlɑːnsə(r); *AmE* ˈlæn-/n. a cavalryman armed with lances 持矛骑兵

land /lænd/vi. **1**【vaulting】(of a vaulter) to dismount from horseback to settle onto the ground (指马上杂技表演者)下马,落地 **2** (of thehorse's hoof) to touch ground following suspension (马蹄)落地

Landais /ˈlændeɪ/n. (also **Landais pony**) a pony breed indigenous to the Landes region in southeast France; stands 11.1 to 13 hands and may have a brown, black, bay, or chestnut coat; has a small head, short ears, straight profile, a long neck thick at the base, short back, hard hoofs, and a thick mane and tail; is hardy, easy to keep, and resistant to harsh weather; used for riding and light draft (源于法国东南部朗德地区的本土矮马品种。体高11.1~13掌,毛色多为褐、黑、骝或栗,头小,耳短,面貌平直,颈长而颈础粗,背短,蹄质坚实,鬃尾浓密,抗逆性好,易饲养,对恶劣气候适应性强,主要用于骑乘和轻挽)朗德矮马

Landau /ˈlædɔː/n. a four-wheeled, horse-drawn carriage introduced in the late 16th century in Landau, Bavaria from which the name derived; was hung on semi-elliptical springs and had opposite seats for four passengers; usually drawn by a pair of horses in pole gear (16 世纪末出现于巴伐利亚兰道市的四轮马车并由此得名。车身悬于半椭圆形减震弹簧之上,可供四名乘客对坐,多由两匹马套驾杆拉行)兰道马车

Landau barouche n. *see* Barouche Landau 巴洛齐兰道马车

Landaulet /ˌlædɔːˈleɪt/n. a smaller version of the Landau 小兰道马车

landing side n.【jumping】the far side of an obstacle onto which the horse lands after jumping over the obstacle (一道障碍的远侧,马跨越后落于此处)落地侧 compare take-off side

Landseer, Edwin Henry n. (1802 – 1873) a Victorian painter well known for his paintings of animals (英国维多利亚时期的画家,擅长动物写生)埃德温·亨利·兰西尔

landship /ˈlædʃɪp/n. a roughly-made, horse-drawn wagon with wickerwork sides and high wheels, formerly used by the Spanish settlers on the South American pampas and was usually drawn by teams of horses or oxen (南美大草原上早期西班牙殖民者使用的粗制四轮马车,车帮由藤条编制,车轮较高,常由多匹马或牛联驾拉行)陆运马车,陆地之舟

lap[1] /læp/n.【racing】one round of a racetrack (赛道的)一圈

vt. to get ahead of a competitor in a race by one or more laps (在比赛中领先对手一或多圈)套圈

lap[2] /læp/vt. to place one thing over another; overlap 交叠,交错 PHR V **lap traces**【driving】to cross the traces of a four-in-hand team to keep the horses close together (为靠拢联驾马匹交叠缰绳)交错缰绳

lap[3] /læp/vt. (of an animal) to take up a liquid or food with the tongue (动物)用舌舔食:the puppy was lapping up a saucer of milk 幼犬在舔盛牛奶的碟子

large colon n.【anatomy】the section of the large intestine located between the cecum and small colon (大肠介于盲肠和小结肠之间的区段)大结肠:The large colon in the horse usually measures 3 to 3.6 meters in length. 马的大结肠多为 3 ~ 3.6 米长。

large intestine n.【anatomy】the section of a horse's alimentary canal including the cecum, colon, and rectum (马匹消化道包括盲肠、结肠和直肠)大肠 compare small intestine

large metacarpal n. *see* cannon bone 管骨,大掌骨

large metacarpal bone n. *see* cannon bone 管骨,大掌骨

large roundworm n. a stout, whitish worm up to 30 cm in length that lives in the intestines of horses and may result in unthriftiness, loss of energy, and even colic with heavy infestation (寄生在马肠道内的灰白色蠕虫,成虫长约 30 厘米,大量繁殖可导致马体况下降和腹痛等症)马蛔虫:Foals are extremely susceptible to large roundworms. 幼驹对马蛔虫极其易感。 syn equine ascarid, parascaris equorum

large standard donkey n. a donkey standing 12 to 14 hands tall (体高在 12 ~ 14 掌的)大型标准驴

large strongyles n.【vet】a type of parasite of the genus *Strongylus* that infests the cecum and colon of horses (寄生于马盲肠和结肠的圆线虫属寄生虫)大圆线虫 syn bloodworm, redworm

lariat /ˈlærɪət/n. a rope with a noose at one end, used for catching cattle or horses in a ranch; lasso (末端带活结的长绳,多用来套牛或马)套索,套绳

vt. to catch with a lariat; rope 套捉

lariat neck ring n. (also **neck ring**) a rigid rope circled around the base of the horse's neck to

L

guide it without the aid of reins and bit（套在马颈础来操控马匹的绳索）颈套索

lark /lɑːk; *AmE* lɑːrk/vi.【hunting】to ride or jump carelessly in a frolic manner（骑马）嬉戏；蹦跳 | **larking** n.

larva /ˈlɑːvə; *AmE* ˈlɑːrvə/n.（*pl.* **larvae** /-viː/）a worm-like form of many insects before metamorphosis（昆虫变态前的蠕虫形态）幼虫

larval /ˈlɑːvl; *AmE* ˈlɑːrvl/adj. of or relating to larvae 幼虫的；幼虫期的：larval development 幼虫发育

larvicide /ˈlɑːvɪsaɪd/n. any of the chemicals used to kill larvae of insects 杀幼虫药

laryngal /ləˈrɪŋgl/adj. *see* laryngeal 喉的

laryngeal /ləˈrɪnʒiəl/adj.（also **laryngal**）of, relating to, affecting, or near the larynx 喉的

laryngeal hemiplegia n. *see* laryngeal paraplegia 喉偏瘫

laryngeal paraplegia n. a paralysis of the muscles of vocal cords in larynx that often results in a roaring sound in breathing（喉部声带肌肉瘫痪，呼吸时常导致喘鸣）喉偏瘫 [syn] laryngeal hemiplegia

laryngeal ventriculotomy n. a surgical operation performed on the larynx to alleviate roaring（为了减缓喘鸣症状而对喉部实施的手术）喉室切开术 [syn] ventricle stripping

laryngic /ləˈrɪndʒɪk/adj. laryngeal 喉的

laryngoplasty /ləˈrɪŋgəˈplæsti/n. a surgical operation performed on the larynx to correct laryngeal hemiplegia（为了矫正喉偏瘫而实施的手术）喉成形术 [syn] tie-back surgery

larynx /ˈlærɪŋks/n.【anatomy】part of the respiratory tract between the pharynx and the trachea that contains the vocal cords（呼吸道位于咽与气管之间且包含声带的区段）喉

laserpuncture /ˌleɪzəˈpʌŋktʃə/n. the use of low-energy laser beams to stimulate the acupuncture points for healing purposes（采用低能激光刺激针灸穴位达到治疗目的方法）激光针灸

lash /læʃ/n. the flexible part of a whip, such as a plaited leather or thong 鞭梢

vt. to strike with a whip 鞭打

Lasix /ˈlæsɪks/n. the registered trademark for furosemide, often used for the treatment of hypertension and bleeders of horses（呋喃苯氨酸的商标名，常用于治疗马高血压和肺出血）呋塞米，利尿灵

lasso /ˈlæsəʊ; *AmE* ˈlæsoʊ/n.（*pl.* **lassos**）a lariat used for catching horses or cattle 套索

vt. to catch horses or cattle with a lasso 套捉

late double n.【racing】a second daily double offered during the later part of a race program（赛事日程中）晚场二连赢

lateral /ˈlætərəl/adj. situated away from the center line of the body 体侧的，外侧的：lateral flexion 侧屈

n.（usu. **laterals**）the lateral incisor 侧切齿

lateral aids n. aids given by the rider's hand and leg on the same side of the horse for instruction（骑手用来向马传达指令的同侧手脚）同侧扶助 [compare] diagonal aids

lateral cartilage n. cartilage attached to the coffin bone（与蹄骨相连的软骨）外侧软骨

lateral deviation of the carpal joints n. *see* bowed knees 膝弓形，膝O形

lateral deviation of the metacarpal bones n. *see* offset knees 管骨外偏，膝外偏

lateral gait n. a gait in which two feet on the same side leave and touch the ground simultaneously; pace（同侧两肢同时离地与落地的运步方式）对侧步

lateral incisor n.（also **lateral**）any of the four incisor teeth located on either side of the central incisors top and bottom（上、下颌位于门齿两侧的四颗切齿）侧门牙，侧切齿 [syn] second incisor

lateral movement n. *see* lateral work 侧向运步，侧向行进

lateral wedge shoe n. a horseshoe thicker on one side, used to alter the weight balance of the foot

(单侧加厚的蹄铁,用来调整马蹄负重平衡)单侧楔形蹄铁

lateral work n. (also **lateral movement**) any movement in which the horse moves sideways and the hind feet do not follow in the path of the forefeet (马侧向运行时前后肢蹄迹不重叠的动作)侧向运步,侧向行进 [syn] work on two tracks, side steps

lateralis /ˈlætərəlɪs/adj.【Latin】lateral 侧的,外侧的

lather /ˈlɑːðə(r); *AmE* ˈlæð-/n. a white foam formed on the coat of horse after profuse sweating (马大量出汗后体被上形成的汗沫)汗渍

vi. to sweat profusely or become coated with lather 发汗,大量出汗 PHR V **lather up** to sweat profusely following strenuous exercises 大量发汗

latigo /ˈlætɪgəʊ/n. a leather strap attached to the metal ring in a Western saddle (系在西部牛仔鞍铁环的皮革)皮绳

latissimus /laːˈtɪsɪməs/adj.【Latin】widest or broadest 最宽阔的:latissimus dorsi muscle 背阔肌

Latvian /ˈlætvɪən/n. (also **Latvian Harness Horse**) a warmblood developed by crossing the native Latvian stock with the Oldenburg, Finnish Draught and Ardennes during the 17th century; stands 15.1 to 16 hands and may have a bay, brown, or chestnut coat; has rather short legs, with good bone and some feathering; is strong, calm, docile, and possessed of good endurance; is an all-purpose horse used mainly for driving and riding (17 世纪拉脱维亚先后采用当地马种与奥尔登堡马、芬兰重挽马及阿尔登马杂交培育而成的温血马品种。体高 15.1 ~ 16 掌,毛色多为骝、褐或栗,四肢较短,管围粗壮且附生距毛,体格强健,性情安静温驯,耐力突出,是优秀的兼用马种,多用于驾车或骑乘)拉脱维亚马

Latvian Harness Horse n. *see* Latvian 拉脱维亚轻挽马

Latvian Heavy Draft n. another term for the Lithuanian Heavy Draft horse when bred in Latvia (拉脱维亚培育的立陶宛重挽马)拉脱维亚重挽马

lavage /ˈlævɪdʒ, læˈvɑːʒ/n. the washing out of a body cavity, such as the stomach or lower bowel, by flushing with fluid (用液体冲洗体腔的方法,如胃和下部肠道)灌洗

lavender roan n. *see* lilac roan 紫沙毛

lawn clippings n. (also **turf grass clippings**) the cut remnants of fresh grass from a lawn (草地刈割产生的碎草)草屑:Lawn clippings should not be used as horse feed, since they may cause colic and flatulence of horses. 草坪刈割的草屑不应用作马的饲草,因为这会引起马腹痛和肠胃胀气。

Lawton gig n. a two-wheeled, horse-drawn vehicle de luxe manufactured by the British firm of Lawton in the second half of the 19th century (19 世纪下半叶由英国劳顿公司制造的一种双轮豪华轻型马车)劳顿马车

lax /læks/adj. (of bowels) loose and not easily retained or controlled 腹泻的

n. excessive and frequent evacuation of watery feces resulting from gastrointestinal distress or disorder; diarrhea (由于肠胃疼痛或紊乱导致机体排出大量水样粪便)下痢,腹泻

laxative /ˈlæksətɪv/adj. stimulating evacuation of the bowels; purgative 轻泻的

n. a food or drug that stimulates evacuation of the bowels 轻泻剂,泻药

lay[1] /leɪ/vt. **1** to keep or put in a particular position 放置,安放 PHR V **lay up** to confine a horse to recover from an injury or illness (圈养受伤或患病马匹使之康复)调养,康复:That horse was laid up in the stable for a month after delivery. 那匹马产驹后在厩舍调养了 1 个月。**2** to produce and deposit 生产,沉积:lay down muscle and fat 沉积肌肉和脂肪 **3**【racing】to place an amount of money on a horse race; wager 下注,押注 **4**【racing】(of a jockey) to occupy a certain running position to make a strategic move (指骑手策略性地)卡位,占位

lay[2] /leɪ/adj. non-professional 外行的: the lay man 外行

layout /ˈleɪaʊt/n. a particular way of arrangement or design 规划,设计: There are many things need to be thoroughly considered in the layout of a stable. 在规划设计马厩时,要综合考虑多种因素。

lay-up /ˈleɪʌp/n. the act or instance of laying up 休养,调养

lazy back n. the backrest of a seat in a horse-drawn vehicle (马车座位的)靠背

lb abbr. libra or pound 磅

lea /liː/n. (also **ley**) an open area of land covered in grass; grassland 草场;牧地: The lowing herd winds slowly over the lea. 低吟的牛群徐徐绕过草场。

lead[1] /liːd/vt. **1** to guide or direct a horse to move in certain course 牵马,牵行 [saying] You can lead a horse to water, but you can't make him drink. you can show people the way to do things, but you cannot force them to act. <谚> 马不喝水焉能强按头;师父领进门,修行在个人。**2** (of one front leg) to strike the ground first and extend farther forward than the other in canter or gallop (跑步或袭步中前肢)领步,起步

lead[2] /liːd/n. **1**【racing】a piece of metal placed in the pockets of saddle pad (置于鞍垫中的金属块) [负重] 铅块 [syn] dead weight **2** *see* leading leg 起步肢: Horses often change lead in galloping. 马袭步行进时常更换起步肢。

lead bag n. *see* lead pad 铅块袋

lead harness n. a set of leather pieces and metals used to connect the lead horses to a carriage (套驾头马所需的皮革和配件)头马挽具: Lead harness generally consists of a bridle, collar, hames, traces, back band, belly band, and breeching. 头马挽具多包括勒、颈圈、枷板、挽绳、背带、肚带和坐鞧。

lead horse n. **1** *see* leader 首马,领头马 **2** the highest ranking horse in a herd; alpha horse (马群中优势序列最高的马)领头马

lead hound n.【hunting】the hound who runs in front of the pack in hunting (狩猎时跑在前面的猎犬)领头犬

lead line n. *see* lead rope (牵马的)缰绳

lead pad n.【racing】(also **lead bag**) a saddle bag loaded with slabs of lead to achieve the minimum weight required by a race (用来增加负重内装铅块的鞍袋)铅块袋 [syn] weight cloth

lead rope n. (also **lead line**) a rope attached to the halter to lead or tie a horse (系于笼头上用来牵马的绳子)缰绳

leader /ˈliːdə(r)/n. **1** a horse who leads in a tandem (双马纵驾中位于前面的马)首马,领头马 **2** either one of the two head horses who lead a team to pull a carriage (多马联驾中的)领头马,首马: near leader 左侧头马 ◇ off leader 右侧头马 [syn] lead horse [compare] center, wheeler

leaders /ˈliːdərz/n. the pair of horses harnessed in front of a multi-horse hitch pulling a vehicle (多马联驾中套在前面的马)首马,领头马 [compare] center pairs, wheelers

leading leg n. the front leg that leads the gait at canter or gallop (在跑步或袭步中领先的前肢)领先肢: In the canter and gallop, the leading leg has to take more weight and do more work than its opposite number. So the tired horses often change legs in mid-gallop for this reason. 在跑步和袭步中,领先肢比其他三肢的负重和活动量都大。因而,马用袭步跑累时,经常在运步中换蹄。

lean /liːn/adj. thin with veins and bony projections clearly visible 清瘦的: a horse with a lean head 一匹面部清瘦的马

leaping head n. (also **leaping pommel**) *see* lower pommel 下鞍桥

leash /liːʃ/n. **1** n. a rope or strap attached to the collar of a dog（系在犬项圈上的）皮带，皮条 **2**【hunting】a set of three animals 三只动物：leash of foxes 三只狐狸

leather /ˈleðə(r)/n. **1** the prepared or tanned hide of animals（鞣制处理后的兽皮）皮革：The best leather for tack comes from slow-maturing cattle. 制作马具最好的皮革来源于那些生长缓慢的牛。**2** any of various articles or parts made of tanned hide, such as a saddle or whip; leatherwork 皮革制品，皮具 **3** a whip 马鞭 **4**（usu. **leathers**）a type of riding breeches made of leather 皮马裤

leatherware /ˈleðəweə(r); *AmE* -wer/n. *see* leatherwork 皮具

leatherwork /ˈleðəwɜːk; *AmE* -wɜːrk/n.（also **leatherware**）tack or harness made of leather 皮具；马具

leery /ˈlɪəri; *AmE* ˈlɪri/adj.【dated】(of a horse) lacking diligence in working; cunning（指马）不出力的；奸猾的

left-hand course n.【racing】a racetrack in which the horses run counter clockwise 逆向跑道，左手赛道

left-lead /ˈleftliːd/adj.（of canter or gallop）with the left foreleg extending farther forward（指跑步或袭步）左起的，左领跑的：left-lead canter 左领跑步 ◇ left-lead gallop 左起袭步 opp right-lead

leftover /ˈleftəʊvə(r); *AmE* -toʊv-/adj. remaining or left as unused 剩余的，吃剩的
n. remains of feed 吃剩的饲料

leg /leg/n. **1** the part of the fore or hind limb of the horse below the knee or hock（马前肢膝下或后肢飞节以下的部分）下肢，腿 PHR **both legs out of the same hole**（of a horse）with a very narrow chest and forelegs set too close together（指马胸腔狭窄且前肢靠得过近的）狭踏的；狭胸的 **on the leg**（of a horse）having disproportionately long legs（指马）四肢过长的 **2** the portion of a tall riding boot between the ankle and the knee（高统马靴的）靴帮 **3** a race constituting a series of competition（系列赛中的）一项，一场：The Macau Guineas is the first leg of the newly formed Triple Crown for four-year-old Thoroughbreds. 澳门几尼赛是最近为4岁纯血马设立的澳门三冠王中的第一场比赛。
vi. to go on foot; run 步行；奔跑 PHR V **leg up 1** to help one mount a horse by lifting up his bent left leg with your hands to boost him up and into the saddle（双手支撑上马者的屈膝左腿，以协助骑手跃起上马）协助上马 **2**【racing】to increase the speed of a horse with work 让马加劲，快马加鞭 **3** to bring a horse back to fitness following a period of inactivity（马）恢复体能训练

leg aids n. the use of the rider's legs to communicate direction to the horse 腿辅助，下肢辅助 syn lower aids

leg barring n.（also **leg bars**）*see* zebra stripes 斑马纹；四肢斑纹

leg brace n. *see* surgical leg brace 四肢夹板

leg lock n.【racing】an illegal maneuver performed by a jockey to impair others by hooking legs（骑师通过钩腿阻碍他人的违例动作）钩腿阻挡

leg mange n. *see* chorioptic mange 蹄疥癣，足螨

leg marking n. any white mark on the horse's leg defined by the pattern and location（马四肢上的白色标记，可根据样式和位置进行划分）四肢别征 —*see also* **face marking**

legal hook n.【polo】any hook of an opponent's shot conducted below the level of the withers of the opponent's mount（对方球员射门时，在低于对手坐骑鬐甲的位置以钩杆干扰）合法钩杆，正当钩杆

Legend of the Eight Horses n.【mythology】an ancient Chinese legend originating in the 9th century BC that tells the story of the eight horses of the Emperor Mu of the Chow Dynasty who traveled around 10,000 miles in one day, thus enabling him to hurry back to suppress a coup d'état and save his throne (公元前 9 世纪中国古代的传说,讲述了周穆王驾八匹马日行三万里,及时赶回宫廷镇压叛乱以保王位的故事)穆王八骏

leggings /ˈlegɪŋz/n. pl. a leather or canvas wrap used to protect the lower leg of the rider from rubbing and to provide a better grip in riding (骑手裹在小腿上的护具,可防止蹭伤并使得坐骑更稳固)皮护腿 syn half chaps

leggy /ˈlegi/adj. having long legs 四肢细长的:a leggy filly 一匹四肢细长的雌驹

legume /ˈlegjuːm/n. any plant of the pea family that has seeds in pods (带豆荚的)豆科植物:legume hay 豆科干草

leg-up /ˈlegʌp/n. a method of assisting others in mounting horse 抬腿上马 PHR V **give sb a leg-up** **1** to help one mount a horse by lifting up his bent left leg with your hands to boost him up and into the saddle ((用双手抬住他人屈膝的左腿,以协助对方上马入鞍)抬腿协助上马 **2** to help one improve their situation 助人提高,助人改善

leg-wrap /ˈlegræp/n. a bandage wrapped around the lower legs of a horse for protection against injury (缠在马下肢起保护作用的绷带)四肢绷带,绑腿

length /leŋθ/n. **1** (also **body length**) the extent from the point of a horse's shoulder to the point of its buttocks (马从肩端到臀端的斜线长度)体长 **2**【racing】(abbr. **L**) a winning margin roughly equal to the body length of a horse (赛马胜出的距离,大致相当于马的体长)身长 : to win by half a length 以 1/2 身长优势获胜 ◇ In 1973, Secretariat won the Belmont Stakes by 31 lengths (76 m). 在 1973 年,"秘书处"以 31 个身长优势夺得贝尔蒙特大赛的冠军。 compare dead-heat, nose, short-head, head, short-neck, neck

lens /lenz/n. 【anatomy】 a transparent, biconvex body of the eye that focuses light rays entering through the pupil to form an image on the retina (眼球中的透明双凸面体,光线进入瞳孔后在此聚集并在视网膜上形成图像)晶状体

lenticular /lenˈtɪkjʊlə/adj. of or shaped like lens 晶状体的;豆状的:lenticular bone 豆状骨

lentiform /ˈlentɪfɔːm; *AmE* -fɔːrm/adj. lenticular 晶状体的;豆状的

leopard /ˈlepəd; *AmE* -ərd/n. **1** one of the color patterns of the Appaloosa with various size of dark spots on a light colored coat all over the body (阿帕卢萨马毛色类型之一,马匹浅色被毛中散布有大小不等的深色斑点)猎豹斑 compare blanket, frost, marble, snowflake **2** a horse with leopard spotting coat pattern 豹纹花斑马

leopard spotting n. (also **leopard marking**) the coat color pattern of leopard 豹纹花斑

leptospirosis /ˌleptəʊspaɪˈrəʊsɪs/n. an infectious disease of domestic animals caused by spirochetes of the genus *Leptospira*, often characterized by jaundice and fever (由钩端螺旋体引发的家畜传染病,症状多表现为黄疸和发热)钩端螺旋体病 syn swamp fever

lesion /ˈliːʒn/n. a wound or an injury of body tissue 损伤,溃烂

lespedeza /ˌlespəˈdiːzə/n. any of various plants of the pea family, having compound leaves with three leaflets and various colored flowers and often grown for forage or soil improvement (豆科的多种植物,三小叶复生,开各种颜色的花,常种植用作饲草或用来改良土壤)胡枝子

lespedeza hay n. a legume hay made from cut and dried lespedeza (胡枝子收割晒干后制成的牧草)胡枝子干草

let /let/vt. to allow or give permission to sb/sth 允许,让 PHR V **let down** to take a horse out of work for rest 使歇息,使休息

lethal /ˈliːθl/adj. capable of causing death; fatal 致死的:partial lethal dose 半致死量

lethal dominant white n.【genetics】the dominant gene that controls the white color of foals

(控制幼驹白色被毛的显性基因)致死显性白化基因:All white horses with the lethal dominant white gene are heterozygotes as the homozygous foals have already died in uterus before birth. 所有携带致死显性白化基因的白马都是杂合子个体,因为纯合子个体在出生前就在子宫内夭折。

lethal gene n.【genetics】a gene that leads to death of a foal at or shortly after birth, such as the albino gene (可导致幼驹出生后死亡的基因,如白化基因)致死基因

lethal white foal syndrome n. (abbr. **LWFS**) a genetic disorder prevalent in the American Paint Horses that leads to death of a foal shortly after birth (在美国花马中常见的遗传疾病,幼驹在出生后不久便死亡)致死白化驹综合征

lethargic /ləˈθɑːdʒɪk; *AmE* -ˈθɑːrdʒ-/adj. lacking energy or vitality; dull 无精打采的,无活力的

lethargy /ˈleθədʒi; *AmE* -θərdʒi/n. a state of inactivity or apathy 无精打采,懒散

leucine /ˈluːsiːn/n.【nutrition】an essential amino acid (一种必需氨基酸)亮氨酸

leucocyte /ˈluːkəsaɪt/n. (also **leukocyte**) the white blood cell 白细胞,白血球

leucoderma /ˌluːkəˈdɜːmə; *AmE* -ˈdɜːrmə/n. *see* leukoderma 白斑病

leucopenia /ˌluːkəˈpiːnɪə/n. (also **leukopenia**) an abnormal reduction in the number of leukocytes in the circulating blood (血液中白细胞数量异常减少的情况)白细胞减少 | **leucopenic** /-nɪk/adj.

leu-in /lju-ɪn/vt.【hunting】to send the hounds into a covert of fox 让犬入狐穴

leukocyte /ˈluːkəsaɪt/n. *see* leucocyte 白细胞,白血球

leukoderma /ˌluːkəˈdɜːmə; *AmE* -ˈdɜːrmə/n. (also **leucoderma**) an acquired, permanent whitening of skin or hair as a result of trauma from badly fitting harness, saddlery, bits, cryogenic surgery, etc. (由于马具、马鞍和衔铁装备不当或进行低温手术后造成皮肤或毛发发白的症状)白斑病

leukopenia /ˌluːkəˈpiːnɪə/n. *see* leucopenia 白细胞减少 | **leukopenic** /-nɪk/adj.

levade /ləˈveɪd, ləˈvɑːd/n.【dressage】a classical dressage movement in which the horse stands up on his deeply bent hind legs and hold for that position (古典骑术动作之一,马以后肢直立并保持这种静立姿势)静立:The levade is considered as the base for the courbette, and the longer it is held, the more difficult for the horse to perform. 静立被认为是直立跳跃的基础动作,马匹站立的越久,其完成的难度也越大。 [syn] controlled rear

levator /ləˈveɪtə(r)/n.【anatomy】a muscle that raises a bodily part (提举身体某部位的肌肉)提肌:levator muscle of upper lip 上唇提肌 [compare] depressor

level /ˈlevl/n. a relative degree of proficiency or intensity 水平,级别:novice level 初级 ◇ top level 顶级 ◇ blood glucose level 血糖水平

level mover n. a horse who moves evenly in a balanced manner 运步平稳的马

level pack n.【hunting】a pack of hounds matched in color, size, conformation, and ability 毛色、体格、外貌和能力等相匹配的猎犬群;齐整的猎犬群

lever noseband n. *see* Kineton noseband 杠杆鼻革

ley /leɪ/n. *see* lea 草地;牧地

LH abbr. luteinizing hormone 促黄体生成素

LHLT abbr. low heel, long toe 低蹄踵,长蹄尖

liberty /ˈlɪbəti; *AmE* ˈlɪbərti/n. the smooth movement of a horse (马运动)自如,流畅:to run at liberty 运步流畅

liberty horse n. a circus horse who runs freely on command of the tamer (在驯兽师口令下自由行进的马戏马)自由马

Libyan /ˈlɪbɪən/n. *see* Barb 利比亚马

Libyan Barb n. *see* Barb 利比亚柏布马

lice /laɪs/n. the plural form of louse [复]马虱

Lichuan n.【Chinese】(also **Lichuan horse**) a breed of horse originating in Hubei province, China; stands about 12 hands and usually has a gray, bay, chest, or black coat; has a heavy head, large belly, prominent withers, sloping shoulders, and strong hoofs; used primarily for riding, packing, and farm work (产于中国湖北省的马品种。体高约12掌,毛色以青、骝、栗、黑为主,头重,腹大,鬐甲突出,斜肩,四肢强健,主要用于骑乘、驮运和农田作业)利川马

Lichuan horse n. *see* Lichuan 利川马

lick /lɪk/ vt. to pass the tongue over sth 舔
n. the act of licking 舔(食): salt lick 舔盐 ◇ mineral lick 矿物质舔砖

life cycle n. the course of development through which a living organism changes from a fertilized zygote to its mature state (生命有机体从受精卵发育为成体的过程)生活周期;生活史: the life cycle of ascarid 马蛔虫生活史

lifeline /ˈlaɪflaɪn/ n. the umbilical cord 生命索,脐带

lifespan /ˈlaɪfspæn/ n. (also **life span**) the average or maximum period of time one can be expected to live (个体存活最长的时间周期)寿命: A horse's lifespan, usually ranging from 20 to 35 years, varies according to its breed, health care, nutrition, and environment. 马的寿命基本为20~35岁,因品种、保健、营养和环境而有所差异。

lig. abbr. ligament 韧带

ligament /ˈlɪgəment/ n. (abbr. **lig.**) a band of tough, fibrous, flexible tissue that connects one bone to another (链接骨与骨的纤维弹性组织)韧带

light[1] /laɪt/ vi. to get down from a horse; dismount 下马

light[2] /laɪt/ adj. **1** not dark in color (毛色)淡的,浅色的: a light coat color 浅毛色 **2** (of a horse) weighing little (指马)轻型的: light horse breed 轻型马种

light brown n.【color】a light shade of brown 淡褐色,浅褐

light chestnut n.【color】a light reddish-brown coat 淡栗毛

light gray n.【color】**1** the color of a horse with black skin and white coat (马肤色为黑而被毛为白的毛色)淡青毛 **2** a horse with such a coat color 淡青毛马: The light gray is the sort of horse that people often mistake for "white." 人们常常将淡青毛马误认为是白马。

light harness n. a horse show class for the American Saddlebred in which demonstration of the walk and animated trot are required (美国骑乘马表演赛级别,主要展示马匹的慢步和快步)轻挽级

light horse n. a horse light in weight and suitable for riding or hunting (体重轻且适合骑乘或狩猎的马)轻型马

light seat n. the position of a mounted rider who carries his weight on his thighs rather than on his seat when posting the trot (骑者起浪时的姿势,体重主要由两股而非鞍座承载)轻骑姿 [syn] crotch seat

light-boned /ˈlaɪtbəʊnd/ adj. (of a horse) having a small bone measurement (指马)管骨细的,管围细的: A light-boned horse usually has a limited weight carrying capacity. 一匹管围细的马通常载重能力也有限。

lightweight /ˈlaɪtweɪt/ adj. weighing relatively little; not heavy 重量轻的: Racing lads being lightweight are easy to spring up and sit into the saddle. 骑手体重轻的话比较容易跨骑到鞍中。
n. a horse capable of carrying light weights as determined by it's bone and substance (载重能力较差的马)轻载型马,轻型马 [compare] middleweight, heavyweight

lilac dun n.【color】a variation of dun with lilac-colored body and chocolate points (马被毛为淡紫色而末梢为暗褐色的兔褐类型)深兔褐,紫兔褐

lilac roan n.【color】a roan pattern with white hairs uniformly mixed into the coat of a dark or liver chestnut（暗栗色被毛中散生白毛的沙毛类型）紫沙毛 syn lavender roan

limb /lɪm/n. either of the fours leg of a horse（马的）四肢:front limb 前肢 ◇ rear limb 后肢

limited-age event n.【cutting】a cutting event restricted to horses between three and six years of age（参赛马年龄限于3～6岁的截牛比赛）限龄级项目

Limousin /ˈlɪməziː n/n.（also **Limousin Halfbred**）a heavy halfbred horse developed in the Limousin area of France by crossing native mares with Thoroughbred, Arab, and Anglo-Norman stallions; stands 16 to 17 hands and generally has a chestnut or bay coat; is docile and has large bone measurement; used mainly for driving and light draft（法国利木赞地区采用当地母马与纯血马、阿拉伯马和盎格鲁诺尔曼马公马杂交培育而成的重型半血马品种。体高16～17掌，毛色则多为栗或骝，性情温驯，管围较粗，主要用于驾车和轻挽）利木赞马

Limousin Halfbred n. *see* Limousin 利穆赞半血马，利穆赞马

linchpin /ˈlɪntʃpɪn/n.【driving】（also **lynchpin**）a locking pin inserted at the end of an axle of a vehicle to keep the wheel in place（马车轴末端起固定车轮作用的辖钉）轮辖，轮楔

line[1] /laɪn/n. **1** the ancestry of a horse on the male side（马父辈祖先的系谱）品系，血系 **2** the profile of a horse（马的）面貌，外形 **3** the direction of travel 行进方向，路线 **4**【hunting】the scent trail left by an animal（动物留下的）踪迹 PHR V **hit the line**【hunting】（of the hounds）to pick up the scent of the quarry（指猎犬）嗅到踪迹 **hit off the line**【hunting】（of the hounds）to lose the the scent of the fox（指猎犬）嗅丢踪迹

line[2] /laɪn/vt. to cover the inner surface with a layer of different material 装衬里

line gaited adj.（of a horse）moving with hind feet following directly in line with its corresponding forefeet on the same side（马在行进时后蹄印与前蹄印重叠成直线）直线行进的，步法线性的

line of the ball n.【polo】an imaginary line created by the travelling ball after being hit（马球被击打后假想的行进路线）球运行路线: Crossing the line of the ball is perhaps the most common foul in polo. 横穿球运行路线或许是马球比赛中最常见的违例。

line of the fox n.【hunting】the scent trail left by a fox（狐狸留下的）踪迹

line-back /ˈlaɪnbæk/n.【color】a horse having a dark dorsal stripe 有背线的马

line-backed /ˈlaɪnbækt/adj.（of a horse）having a dark dorsal stripe on its back（指马）有背线的:a line-backed horse 一匹有背线的马

line-breeding /ˌlaɪnˈbriːdɪŋ/n. the mating of horses of common ancestry to preserve certain traits or characteristics（为了稳固特定性状而让祖先相同的马匹进行交配）线性育种:To lessen the dangers inherent to the highly intensified inbreeding programs, several generations are often removed in line-breeding. 为了降低近交选育强度过高的风险，线性育种时常将中间几个世代的个体剔除掉。

lingua /ˈlɪŋgwə/n.【Latin】tongue 舌

lingual /ˈlɪŋgwəl/adj. of or relating to the tongue 舌的

lingula /ˈlɪŋgjʊlə/n.【Latin】a small tongue-like structure 小舌

lining /ˈlaɪnɪŋ/n. a covering or coating for an inside surface 衬里；内壁:lining of riding boots 马靴衬里 ◇ lining of gut 肠内壁

linseed /ˈlɪnsiːd/n.（also **flaxseed**）the small, flat seed of flax with a high protein and fat content（亚麻的籽实，蛋白和脂肪含量较高）亚麻籽:linseed cake 亚麻籽饼 ◇ Linseed is usually fed to horses as a hot or cold mash after being soaked and boiled. 通常亚麻籽浸泡熬煮

后,饲喂马匹。

linseed mash n. a wet mash made by boiling linseed meal or linseed together with bran and water (用亚麻籽粕或亚麻籽加麸皮、水熬成的稀粥)亚麻糊,亚麻粥:Linseed mash must be fed within 24 hours to avoid spoiling. 为防止变质,熬好的亚麻粥应在24小时内喂完。

linseed meal n. 【nutrition】(also **linseed oil meal**) a high protein, high energy concentrate produced from the residue of the flaxseed following extraction of the oil (亚麻籽榨油后的残渣,是一种高能量、高蛋白的精饲料)亚麻籽粕

linseed oil n. the oil extracted from the flaxseed 亚麻油

linseed oil meal n. *see* linseed meal 亚麻籽粕

lip /lɪp/ n. either of two soft folds at the opening of horse's mouth (马口的两片皱褶)唇

vt. to touch with its lips 用唇玩弄 PHR V **lip the bars** (of a horse) to play with the bit shanks with his lips (指马)用唇玩弄衔铁杆

lip marking n. any colored mark or spot on the lip of the horse (马口唇上的标记或斑点)口唇别征

lip tattoo n. a tattoo made on the inside lip of a racehorse for registration (赛马注册时在唇内侧蚀刻的花纹)唇刺标,唇刺青

Lipizzan /ˈlɪpɪtsɑːn/ n. *see* Lipizzaner 利比扎马

Lipizzaner /ˌlɪpɪˈtsɑːnə(r)/ n. (also **Lipizzan**, or **Lippizaner**) an Austrian-bred warmblood descended from Andalusian stock put to native Karst horses in the late 16th century; stands 14.2 to 16 hands and may have a dark black, brown, or mousy-gray coat; has a heavy head, small ears, crested neck, compact body, and full mane and tail; bred and used extensively at the Spanish Riding School of Vienna to perform classical dressage movements (16世纪末奥地利采用安达卢西亚马公马与当地的喀斯特马母马繁育而来的温血马品种。体高14.2~16掌,毛色可为黑、褐黑、褐或鼠灰,头重,耳小,颈峰突起,体质结实,鬃尾浓密,在西班牙骑术学校大量培育并用于古典盛装舞步表演)利比扎马

Lippizaner /ˌlɪpɪˈzɑːnə(r)/ n. *see* Lipizzaner 利比扎马

lisk /lɪsk/ n. *see* groin 腹股沟,鼠蹊

lists /lɪsts/ n. pl. **1** the barriers enclosing an area for a jousting tournament in ancient times (古代举行马上刺枪的)围栏,围场 **2** an arena or scene of combat or conflict 战场,杀场:to enter the lists 参加争斗,投身战场

Lithuanian Heavy Draft n. a coldblood breed descended from the local Zhmud put to the Finnish Horse and Swedish Ardennes in Lithuania during the late 1890s; stands 15 to 16 hands and usually has a chestnut coat, although black, roan, bay, or gray also occur; has a massive build, a short, thick, muscular neck, full mane, short legs with little feathering, and a smooth, fast action; is quiet, strong, and possessed of great powers of traction; used mainly for heavy draft (立陶宛在19世纪90年代末采用当地的泽慕德马公马与芬兰马和瑞典阿尔登马母马杂交繁育而来的冷血马品种。体高15~16掌,毛色以栗为主,但黑、沙、骝和青亦有之,体格强健,颈粗短,鬃毛浓密,四肢粗短,距毛较少,步伐流畅轻快,性情安静而顽强,挽力惊人,多用于重挽)立陶宛重挽马

litter /ˈlɪtə(r)/ n. **1** unwanted objects or trash lying about in a public place (公共场所四处丢弃的废物)垃圾 **2** straw or other materials used as bedding for animals (给动物垫窝的稻草或其他材料)垫草,垫料 **3** the group of young offspring born to an animal at one birth (动物一次所生的幼崽)一窝:a litter of puppies 一窝幼犬 ◇ litter size 一窝幼崽数 **4** a couch or seat on shafts carried on the backs of bearers; stretcher (固定在架杆上用人抬行的座椅)抬轿;担架:carry a wounded soldier on a litter 用担架抬受伤士兵

v. **1** to make untidy with scattered rubbish 使凌乱,使脏乱: Apart form the unaesthetic appearance, a yard littered with waste fodder, manure and bedding will quickly harbour vermin and become a health risk. 除了不大雅观之外,一个到处撒有饲草、粪便和垫料的围场很快就会成为虫害的栖息地而危害人畜健康。**2** to set straw, etc. as bedding for an animal 铺垫料,垫窝 **3** (of fox, hound, etc) to bring forth young 下崽,产仔

live foal guarantee n. a provision in a breeding contract which guarantees the owner of the bred mare to produce a live foal as a result of a paid breeding fee (配种合同中的条款,在支付配种费后须保证与配母马产下活驹)产驹担保,活驹担保:The live foal guarantee generally gives the owner of the mare the right to re-breed to the same stallion in the following season in the event the mare fails to produce a live foal as a result of the initial breeding. 如果母马初次配种没有产下马驹,活驹担保可使马主有权在下一个配种季节免费让同匹公马为母马再次配种。

live hunt n.【hunting】a hunt in which the hounds trail the scent left by a live animal or quarry (猎犬追踪活猎物所留踪迹的捕猎活动)活猎追捕,活物狩猎 compare drag hunt

live weight n.【racing】the body weight of a jockey 骑师体重 compare dead weight

liver /ˈlɪvə(r)/n. a large glandular organ in the horse's body that cleans the blood and produces bile (马体内清洁血液并产生胆汁的腺体器官)肝脏

liver chestnut n.【color】the darkest shade of chestnut colors consisting of red and black hairs with no black points (色调最深的栗毛,被毛由红、黑毛发构成但无黑色末梢)褐栗色

liver fluke n. a parasitic trematode flatworm that infests the liver of certain animals and may cause abdominal pain, anemia, and poor performance (一种寄居在动物肝脏的扁形吸虫,可导致腹痛、贫血和体能下降等症)肝片吸虫 syn *Fasciola hepatica*

Liverpool /ˈlɪvəpuːl; *AmE* ˈlɪvərp-/n.【jumping】an obstacle consisting a body of shallow water beneath an oxer or in front of a vertical jump (一道位于双横栏下面或竖直障碍前面的水沟障碍)[利物浦]水渠障碍

Liverpool bit n. a kind of driving bit 利物浦衔铁

Liverpool gig n. a fully enclosed, two-wheeled, horse-drawn vehicle first introduced in the mid-19th century in Liverpool, England from which the name derived (一种封闭式双轮马车,于19世纪中叶出现在英国利物浦故而得名)利物浦轻便马车

Liverpool Steeplechase n. *see* Grand National steeplechase 利物浦越野障碍赛,英国越野障碍大奖赛

livery /ˈlɪvəri/n. **1** a special uniform worn by professional members of the Hunt 猎狐制服;骑装 **2** the boarding and feeding of horses [马匹]寄养 **3** the hiring out of horses and carriages 车马出租 **4** a livery stable 寄养马厩:a full livery 全寄养马厩 ◇ a half livery 半寄养马厩 ◇ a DIY livery 自助式寄养马厩 PHR **at livery** (of a horse) kept a livery stable (指马)寄养的

livery stable n. (also **livery**) a stable where horses are boarded, cared, or hired for fee (料理寄宿马匹或出租马匹的收费马厩)寄养马厩

liveryman /ˈlɪvərɪmən/n.【BrE】(also **livery man**) one employed to keep a livery stable (寄养马厩的员工)马夫

livestock /ˈlaɪvstɒk; *AmE* -stɑːk/n. domestic animals kept on a farm; stock (农场养殖的家畜)牲畜,农畜

Llanero /ljɑːˈneɪrəʊ/n. a Venezuelan-bred Criollo descended from Spanish horses brought to South America by the conquistadors in the 16th century; has a head similar to a Barb with a rather convex profile, stands approximately 14 hands and may have a dun, white, yellow cream, or pinto coat; used for ranch work and riding（由16世纪殖民者带到南美的西班牙马繁育而来的委内瑞拉克里奥罗马，面貌与柏布马较为相似，面颊呈凹形，体高约14掌，毛色多为兔褐、白、乳黄或花毛，主要用于牧场作业和骑乘）拉牙纳诺马

load /ləʊd; *AmE* loʊd/n. a weight or mass carried by a pack animal; burden（驮畜所载的重量）驮量，负重

vt. **1** to put a load on the back of an animal 驮运，驮载 **2** to put a load in a vehicle for transportation 运载，拖运：Young horse should be taught to load in a box or trailer at an early age. 青年马匹应当较早学会在装运箱和拖车中被运载。

loaded shoulder n. an excessively thick shoulder of a horse（指马）肩部臃粗

loaf /ləʊf; *AmE* loʊf/vi. to idle one's time away 闲逛，懒散

loafer /ˈləʊfə(r); *AmE* ˈloʊf-/n.【racing】a horse reluctant to run hard unless being urged by its rider（骑手不驱赶便不愿猛跑的马）懒散的马，不卖力的马

loafing hound n.【hunting】a hound who is not dedicated to hunting 懒散的猎犬

lob /lɒb; *AmE* lɑːb/n.【racing】a racehorse restrained by its rider from running well（被骑手扼制无法猛跑的马）受约束的马

vi. to run slowly and heavily（指马）慢跑，费力地跑

lobo dun n.【color】a darker shade of dun in which the slate-colored body coat is mixed with black hairs（石板青被毛中混有黑毛的兔褐毛，色调由此显得更深）灰兔褐

local anesthesia n. an artificially induced, localized loss of the sense of pain（局部丧失痛觉的麻醉方法）局部麻醉

local anesthetic n. an agent that deprives the sensation of pain to a specific area in the body（可使身体局部丧失痛觉的药物）局部麻醉剂

lock /lɒk; *AmE* lɑːk/n.【racing】a guaranteed win 必然胜出：This horse has a lock on the race. 这匹马必定胜出。[syn] sure thing

lockjaw /ˈlɒkdʒɔː; *AmE* ˈlɑːk-/n. *see* tetanus 牙关紧闭，破伤风

loco /ˈləʊkəʊ; *AmE* ˈloʊkoʊ/n. (*pl.* **locos**, or **locoes**) **1** the locoweed 疯草 **2** the loco disease 疯草病

vt. to poison with locoweed 疯草中毒

loco disease n. *see* locoism 疯草病

locoism /ˈləʊkəʊɪzəm; *AmE* ˈloʊkoʊ-/n.（also **loco disease**）a disease of livestock caused by locoweed poisoning and characterized by weakness, lack of coordination, trembling, partial paralysis, and ultimately, death（家畜采食疯草所患的中毒症状，临床表现为乏力、动作失调、战栗、偏瘫，并最终死亡）疯草症，疯草病

locoweed /ˈləʊkəʊwiːd; *AmE* ˈloʊkoʊ-/n.（also **loco**）any of the several plants of the legume family, genera *Astragalus and Oxytropis*, that cause severe poisoning when ingested by livestock（黄芪属和棘豆属的多种豆科植物，家畜采食后可产生严重的中毒反应）疯草：locoweed poisoning 疯草中毒 [syn] crazyweed

lodge /lɒdʒ; *AmE* lɑːdʒ/vi. to become fixed or retained in a place（被固定或滞留在某处）卡住：In delivery, the forelegs appear relatively easily but the elbows may become lodged on the pelvic brim. 分娩时，前肢产出相对比较容易，但肘关节很可能会卡在母马的骨盆边缘。

loft /lɒft; *AmE* lɔːft/n. a floor or space directly under the roof of a building used for storage（建筑物的顶层，多用于储存物品）顶楼，阁楼

loin /lɔɪn/n. (*pl.* **loins**) the body part on either side of the backbone between the false ribs and hip bone（脊柱两侧位于假肋和髋骨之间的部分）腰：hollow loins 凹腰 ◇ The loins are the weakest part of the horse's back. 腰是马背部最弱的部位。

loin cloth n. a waterproof cloth covered over the

loins of carriage horses in the rain (下雨时遮盖在挽马腰部的防水罩布)背腰雨披,背腰罩布

Lokai /ˈlɒkaɪ/n. a warmblood of mixed ancestry originating in southern Tadzhikistan; stands over 14 hands and usually has a golden chestnut coat; has a long neck, short back, sloping croup, and notably tough hoofs; used for riding, packing, and mounted competitive sports (源于塔吉克斯坦南部地区的温血马品种,血统混杂。体高 14~14.2 掌,毛色以金栗为主,颈长,背短,斜尻,蹄质坚硬,主要用于骑乘、驮运和马上竞技项目)洛卡伊马

London Cart Horse Parade Society n. an organization founded in London, England in 1890 to improve the general condition and treatment of cart horses (1890 年在英国伦敦成立的组织机构,旨在改善挽马的境况和待遇)伦敦车马仪仗协会

London International n. *see* Royal International Horse Show 皇家国际马匹展览赛

London tan n. a yellow-brown tanned leather used to make saddlery (制作马鞍用的黄褐色鞣制皮革)黄褐革

London Van Horse Parade Society n. an organization founded in UK in 1904 to improve the general condition and treatment of van and light draft horses (1904 年在英国成立的组织机构,旨在改善篷车挽马和轻挽马的总体境况)伦敦篷车马仪仗协会

Long Wagon n. (also **medieval Long Wagon**) a luxurious, horse-drawn, dead-axle traveling carriage used throughout Europe in the 13th and 14th centuries; usually drawn by teams of six or more horses controlled by postilions (欧洲 13—14 世纪使用的死轴豪华马车,多由左马驭者策六匹以上的马队拉行)长形马车

long-coupled /ˌlɒŋˈkʌpld/adj. (of a horse) having a long back and loins (指马)背腰长的 opp short-coupled

longe /lɒŋdʒ/n. a long rein used to lunge a horse 打圈长缰,调教索

vt. to lunge a horse 打圈 | **longeing** n.

longe rein n. *see* lunge rein 打圈长缰

longe whip n. *see* lunge whip 打圈长鞭

longeing cavesson n. *see* lungeing cavesson 打圈调教索

longeing whip n. *see* lunge whip 打圈用鞭

longer /ˈlɒŋdʒə(r)/n. *see* lunger 打圈者

long-rein /ˌlɒŋˈreɪn/vt. to lunge a horse by using a long rein 用长缰训练马;打圈

Lonsdale Wagonette n. a luxurious wagonette designed by the 5th Earl of Lonsdale after whom it was named; has a low-hung, rounded body, and falling hood (一种小型豪华四轮马车,由朗斯代尔伯爵五世设计并因此得名,车身低且呈圆形,带折叠式顶篷)朗斯代尔马车

look /lʊk/n. the expression of a horse (马的)神情,表情 PHR **look of eagles** 【racing】 a proud look in a racehorse as if he were the winner (获胜赛马)骄傲的神情

looker /ˈlʊkə(r)/n. 【BrE, dated】 one hired to watch horses or cattle grazing on grassland (照看牛马在草场采食的人)放牧工,看马人

loop /luːp/n. **1** a circle-shaped piece 圆环,圆圈 **2** 【dressage】 a circle a horse makes in movement 转圈,圆圈 PHR **loops not equal** circles scribed by a dressage horse that are not equal in size or shape (指马)转圈不等

loose /luːs/adj. (of a horse) not tied up or unbridled (马)未拴的;无羁的:a loose horse 无羁之马

loose box n. a stall used for stabling horse, foaling, etc. (用来圈养马匹或产驹的房舍)马房

loose jump n. *see* free jump 空跃

vi. (of a horse) to jump over obstacles without a rider on its back (指马)空跃,自由跳跃:In loosing jumping, the horse is controlled from the ground by voice commands or long lines. 空骑跨跃中,马通过地面人员给予的指令或长缰进行操控。

lop /lɒp; *AmE* lɑːp/vi. to hang loosely; droop 下

垂,耷拉

lop neck n. a conformation defect in which a horse's neck falls to one side(马颈歪向一侧的体格缺陷)颈峰歪斜,歪颈 [syn] broken crest

lope /ləʊp; *AmE* loʊp/n. a slow canter, usually performed in Western riding disciplines(西部骑乘中的慢跑)小跑步,慢跑步

vi. (of a horse) to move with the three-beat gait of lope(指马)小跑

lop-eared /ˈlɒpɪərd/adj. having bent or drooping ears(耳朵)下垂的,耷拉的:a lop-eared hound 耷拉着耳朵的猎犬

loppy /ˈlɒpi; *AmE* ˈlɑːpi/adj. hanging down or limp; pendulous 下垂的,松软的;耷拉的:loppy ears 耷拉的耳朵

lorimer /ˈlɒrɪmə(r)/n. *see* loriner [马具]铁匠

loriner /ˈlɒrɪnə(r)/n. (also **lorimer**) a blacksmith who makes the metal parts of saddle and harness tack such as bits, spurs, and stirrups(制作马鞍和挽具上金属部件的铁匠)[马具]铁匠

Loriners Company n. a guild organized in England around 1245 to represent loriners(1245年左右在英国成立的行会,代表马具铁匠的利益)马具铁匠公会

lorry /ˈlɒri; *AmE* ˈlɔːri/n. (also **lurry**) a large, heavy horse-drawn dray or truck used for general delivery(用来运送物资的重型平板马车)马拉货车;平板马车

louse /laʊs/n. (*pl.* **lice** /laɪs/) *see* horse louse 马虱

low /ləʊ; *AmE* loʊ/n. a sound made by the cattle; moo(牛)低吟,哞

vi. (of a cow) to make a deep sound(牛)低吟,哞叫:the lowing of cattle 牛群的哞叫声

low airs n. the classical movements performed by a horse on the ground, such as the piaffe and pirouette(马在地面上表演的古典骑术动作,如原地快步和定后肢旋转)地面动作 [compare] high airs

low heel, long toe n. (abbr. **LHLT**) a condition of the hoof in which the heels are excessively low due to trimming or wear and the toe often too long(马蹄踵由于削剪或磨损显得较低,而蹄尖过长的状况)低蹄踵,长蹄尖

low ringbone n. a type of ringbone that develops within the joint connecting the short pastern and pedal bone(在短系骨和蹄骨关节间附生的赘骨)下位环骨瘤 [compare] high ringbone

low withers n. a conformation defect of horse in which the croup is higher than the withers(马尻部高于鬐甲的体格缺陷)低鬐甲,鬐甲低平

lower aids n. *see* leg aids 下身辅助;腿脚辅助 [compare] upper aids, voice aids

lower pommel n. (also **leaping pommel**) the lower of the two padded pommels on the side-saddle(侧鞍上位置较低的鞍桥)下鞍桥 [compare] top pommel

low-slung /ˌləʊˈslʌŋ/adj. (of a vehicle) having low sides(指马车)侧帮低的

lozenge /ˈlɒzɪndʒ; *AmE* ˈlɑːz-/n. *see* bit guard 护颊套

LSD abbr. long, slow distance work 长途慢速作业

lubricant /ˈluːbrɪkənt/n. any substance, such as Vaseline or oil, that reduces friction when applied on a surface of moving parts(涂在活动器件表面以减少摩擦用的物质,如凡士林或油脂)润滑剂

lubricate /ˈluːbrɪkeɪt/vt. to apply a lubricant to make slippery or smooth 润滑

lucerne /luːˈsɜːn; *AmE* -ˈsɜːrn/n.【BrE】alfalfa [紫花]苜蓿

lug /lʌg/vi. **1** to move or drag with efforts 拖动 PHR V **lug in**【racing】(of a horse) to bear in(赛马)侧向内栏 **lug out**【racing】(of a horse) to bear out(赛马)侧离内栏 **2** (of a horse) to lean on the bit when ridden or driven(指马)拖衔铁,拽衔铁

lugger /ˈlʌgə(r)/n. a horse who leans on the bit

when ridden or driven 拖衔铁的马,拽衔铁的马

lumber /ˈlʌmbə(r)/n.【AmE】wood used in building; timber 木料,木材
vi. to move in a slow, heavy and awkward way 艰难移动;缓慢移动:A truck lumbered past. 一辆卡车缓缓驶过。

lung hemorrhage n. *see* exercise-induced pulmonary hemorrhage [运动诱发性]肺出血

lunge /lʌndʒ/n. (also **longe**) the training of a horse in a circle ring by using a long rein attached to the cavesson (马)打圈,长缰训练 PHR V **jump on the lunge** to train a horse to jump over obstacles guided by a lunge line 长缰驯马跨栏
vt. to exercise or train a horse in a circle ring by using a long rein attached to the cavesson(马)打圈,长缰训练 |**lungeing** n.

lunge circle n. the track scribed by the horse as he is worked around the lunger (马打圈时划出的轨迹)圆圈:In vaulting the lunge circle must be at least 42 feet (13 m) in diameter. 在马上体操中,打圈直径至少应在 13 米以上。

lunge line n. (also **lunge rein**) a long cotton or synthetic rope attached to the halter, bridle, or lungeing cavesson by a snap to train or exercise the horse in a circle ring (用扣环系在笼头或调教索上的长缰,用来在圈内调教或训练马匹)打圈长缰

lunge rein n. *see* lunge line 打圈长缰

lunge whip n. (also **longe whip**) a long whip used for lungeing horses 打圈长鞭:The lunge whip enables the lunger to reach the horse as much as 6 m away while working him in a circle. 打圈长鞭使练马员在圈中离马 6 米以上仍可控制马匹。

lungeing /ˈlʌndʒɪŋ/n. (also **lunging**) a method of exercising or training a horse in a circle ring by using a long rein that is attached to the cavesson (用带长缰的调教索训练马绕圈跑步)打圈,长缰训练 :lungeing ring 调教圈,练马圈 ◇ lungeing cavesson 调教索

lunger /ˈlʌndʒə(r)/n. (also **longer**) one who lunges a horse 打圈练马员

lungworm /ˈlʌŋwɜːm; *AmE* -wɜːrm/n. any of several nematode worms that are parasitic in the lungs of mammals and may result in bronchitis and pneumonia of horses (寄生在哺乳动物肺内的线虫,可引发马气管炎和肺炎)肺线虫,肺丝虫

lungworm infection n. an infection of the lower respiratory track by the lungworms; symptoms include coughing, increased respiratory rate, bronchitis, or pneumonia (呼吸道感染肺线虫的病症,临床可表现为干咳、呼吸急促、气管炎和肺炎)肺线虫感染

lurry /ˈlʌri/n. *see* lorry 平板马车,马拉货车

Lusitano /ˌluːsiˈtɑːnəʊ/n. (also **Pure Blood Lusitano**) a horse breed indigenous to Portugal and similar to the Andalusian in profile; stands 15 to 16 hands and usually has a gray coat, although bay, black, chestnut and dun also occur; has a fine, small head, thick, well-set-on neck, substantial girth depth, short back, rounded croup, powerful shoulders; is frugal, intelligent, agile, and docile; used for light farming, light draft, riding, and in the bullfighting ring by the rejoneadores (源于葡萄牙的本土马品种,体貌与安达卢西亚马较为相似。体高15~16掌,毛色以青为主,但骝、黑、栗和兔褐亦有之,头小,面貌俊秀,颈姿优雅,胸身宽广,背短,后躯浑圆,肩部发达,耐粗饲,反应敏捷,性情温驯,多用于农田作业、轻挽、骑乘及斗牛者的坐骑)卢西塔尼亚马,卢西塔诺马

luteal /ˈluːtɪəl/adj. of or relating to the corpus luteum 黄体的

luteinize /ˈluːtənaɪz/v. (cause to) to develop into or become part of corpus luteum 黄体化

luteinizing hormone n.【physiology】(abbr. **LH**) a protein hormone released by the pituitary gland that primarily causes ovulation (垂体所

分泌的一种蛋白质激素,其作用主要是引发排卵)黄体生成素:luteinizing hormone releasing factor 黄体生成素释放因子

luteolysin /ˌluːtɪəˈlaɪsɪn/ n. any substance, such as prostaglandin, that causes luteolysis 黄体溶解素

luteolysis /ˌluːtɪəˈlaɪsɪs/ n. the degeneration of the corpus luteum caused by prostaglandin(由前列腺素引起的黄体溶解退化)黄体溶解

luteolytic /ˌluːtɪəˈlaɪtɪk/ adj. of or relating to luteolysis 黄体溶解的

LWFS abbr. lethal white foal syndrome 致死白化驹综合征

L

lymph /lɪmf/ n. a clear, yellowish fluid derived from body tissues that contains white blood cells and circulates throughout the body(机体组织产生的一种水样淡黄色液体,内含白细胞并在体内循环)淋巴:lymph nodes 淋巴结 ◇ lymph vessel 淋巴管 ◇ Lymph acts to remove bacteria and certain proteins from the tissues, transport fat from the small intestine, and supply mature lymphocytes to the blood. 淋巴的作用是清除组织中细菌和某些蛋白质,从小肠中输送脂肪,并为血液提供成熟的淋巴细胞。

lymphangitis /ˌlɪmfænˈdʒaɪtɪs/ n. inflammation of the lymphatic vessels that usually occurs in horse's hind legs and causes swelling and pain(淋巴管发炎的症状,多发生于马的后肢,常引起肿胀和疼痛)淋巴管炎:epidemic lymphangitis 流行性淋巴管炎

lymphatic /ˌlɪmˈfætɪk/ adj. of or relating to lymph, a lymph vessel, or a lymph node 淋巴的:lymphatic system 淋巴系统 ◇ lymphatic vessels 淋巴管

lymphocyte /ˈlɪmfəsaɪt/ n. any of the nearly colorless cells formed in lymphoid tissue that function in the development of immunity against infection(一种见于淋巴组织中近似无色的细胞,有调节免疫抵抗感染的功能)淋巴细胞

lymphoid /ˈlɪmfɔɪd/ adj. of or relating to lymph or the lymphatic tissue where lymphocytes are formed 淋巴的,淋巴样的

lymphonodulus /ˌlɪmfəˈnɒdjʊləs/ n.【Latin】(*pl.* **-duli** /-laɪ/) a small lymph node 淋巴小结

lysine /ˈlaɪsiːn/ n.【nutrition】an essential amino acid required by the body for optimum growth(一种必需氨基酸,为体格生长所必需)赖氨酸

Macau Derby n. a 1,800-meter G-1 race held at the Taipa Racecourse of Macau for four-year-old Thoroughbreds in May each year (每年5月在澳门氹仔马场为4岁马匹举行的一级平地赛,赛程1 800米)澳门德比赛

Macau Gold Cup n. a 1,800-meter G-1 race held at the Taipa Racecourse of Macau in late May each year for Thoroughbreds above three years old (每年5月底在澳门氹仔马场为3岁以上马匹举行的一级平地赛,赛程1 800米)澳门金杯赛

Macau Guineas n. a 1,500-meter G-2 race held at the Taipa Racecourse of Macau for four-year-old Thoroughbreds in March each year (每年3月在澳门氹仔马场为4岁马匹举行的二级平地赛,赛程1 500米)澳门几尼赛

Macau Jockey Club n. (abbr. **MJC**) an organization founded in 1989 to replace the Macau Trotting Club and regulate horse racing and other betting events in Macau (于1989年成立以取代澳门快步赛俱乐部的组织机构,旨在组织管理当地的赛马和投注产业)澳门赛马会

Macau Triple Crown n. a racing series consisting of the Macau Guineas, the Macau Derby, and the Macau Gold Cup (由澳门几尼赛、德比赛和金杯赛构成的系列赛事)澳门三冠王:To date, no horse has won the Macau Gold Cup as well as the first two Triple Crown races. 迄今为止,还没有马在赢得澳门三冠王前两场比赛的同时,也在澳门金杯赛中获胜。

Macau Trotting Club n. an organization founded in 1980 in Macau to introduce harness racing in Asia, was eventually renamed Macau Jockey Club to organize Thoroughbred flat races in 1989 (于1980年在澳门成立的机构,旨在将驾车赛引入亚洲,最终于1989年更名为澳门赛马会并组办纯血马平地赛)澳门赛马车会

macerate /ˈmæsəreɪt/ vt. to make soft by soaking or steeping in water (在水中浸泡使之变软)浸软:Sugar beet pulp should be macerated before feeding to prevent it swelling in the gullet, stomach or small intestine and causing problems. 甜菜渣必须浸软后方可饲喂,这可防止其在食管和胃肠中膨胀引起消化道疾病。

machine-made shoe n. any ready-made horseshoe that is available in different sizes (一种机器制造的蹄铁,有各种尺寸规格可选)预制蹄铁 [syn] factory-made shoe

mackintosh /ˈmækɪntɒʃ; *AmE* -tɑːʃ/ n. (abbr. **mac**) **1** a lightweight, waterproof fabric that was originally made of rubberized cotton (由胶化棉布制成的轻质防水织物)防水胶布 **2** a rain coat made of such material 胶布雨衣

macrophage /ˈmækrəfeɪdʒ/ n. any of the large phagocytic cells that engulf microorganisms entering into the body (可吞噬侵入体内微生物的细胞)巨噬细胞

macs /mæks/ n.【BrE】riding breeches made of mackintosh 胶布马裤,防水马裤

mad-dog syndrome n. the abnormal and aggressive behavior demonstrated by a horse affected with rabies (患狂犬病的马所表现出的典型攻击性行为)疯狗综合征

made /meɪd/ adj. (of an animal) well-trained and schooled (动物)调教好的:a made field hunter 一匹调好的狩猎马 ◇ made pack 训好的猎犬群

madrina /ˈmædrɪnə/ n. *see* bell mare 系铃母马

magnesium /mægˈniːziəm/ n.【nutrition】an essential mineral required by horse in small amounts for proper bone and tooth development (马匹所需微量矿物质元素,为马匹骨骼和

牙齿正常发育所必需)镁:magnesium deficiency 镁缺乏

magnesium sulfate n. a colorless, crystalline compound used in horses as a laxative and as a general cathartic(一种无色晶体化合物,可作马匹泻药使用)硫酸镁 syn Epsom salts

magnetic field therapy n. *see* magnetic therapy 电磁疗法

magnetic therapy n.(also **magnetic field therapy**) the use of magnetic fields to treat soft tissue and bony injuries(采用磁场治疗软组织和骨损伤的疗法)电磁疗法;磁疗

mahogany /məˈhɒgəni; *AmE* -ˈhɑːg-/n.【color】a reddish-brown color 红褐毛

adj. of a reddish-brown color 红褐色的

mahogany bay n.【color】a dark shade of bay with its red coat mixed either with black hairs or red hairs with black tips(红色被毛中混有黑毛或被毛红色末梢为黑的骝毛)黑骝毛 syn dark bay

mahout /məˈhaʊt/n. one who rides and controls an elephant 骑象者,驯象者

maiden /ˈmeɪdn/adj.(of mare) not foaling yet(指母马)未产驹的

n. **1** a maiden mare 处女马 **2**【racing】*see* maiden horse 尚未获胜的马

maiden allowance n.【racing】a weight concession given to a racehorse who has not won in any races(对没有获胜的赛马给予的负重让磅)处女马让磅

maiden class n. a horse show class for maiden horses(由处女马参加的比赛级别)处女级

maiden horse n.【racing】(also **maiden**) a racehorse who has not won a race 未赢过比赛的马

maiden mare n.(also **maiden**) a mare who has never been bred to produce a foal(尚未配种产驹的母马)处女母马:The maiden mares and those who suckle a foal may refuse to accept the stallion even though they appear to be in estrus from their signs at the teasing board and from a visual inspection of the cervix. 通过试情栏观测哺乳母马和处女母马的表现以及子宫颈状态来看,即使它们正处于发情期,也可能会拒绝接收公马交配。

maiden race n.【racing】a race held for horses who have not previously won any races(为那些尚未在比赛中获胜的赛马举行的比赛)处女赛

mail cart n. a light, two-wheeled, horse-drawn cart used to deliver mail(一种轻型双轮马车,主要用来投递邮件)邮递马车,邮差马车

Mail Coach n. a horse-drawn public coach used in England during the 18th and early 19th centuries to deliver the Royal Mail(英国18—19世纪初用来运送皇室信件的四轮马车)邮递马车

Mail Phaeton n. a horse-drawn carriage of the Phaeton type popular in the 1820s used mainly for mail delivery(19世纪20年代流行的一种费顿马车,多用于邮递业务)邮递费顿马车

main bar n. the largest of three bars hanging from the singletree of a horse-drawn vehicle to which the traces are connected(与挂杠相连的三根杆中最粗的横杠,驾车挽绳系于其上)主杆,主杠

maintain /meɪnˈteɪn/vt. to keep sth in good condition; continue; preserve 维持,保持

maintenance /ˈmeɪntənəns/n. the act of maintaining or the state of being maintained 保持,维持:maintenance requirements 维持需要 ◇ net energy for maintenance (NE_m) 维持净能 ◇ The NE_m is the basic energy requirement of feedstuff, a non-working, stabled horse that is maintaining a stable bodyweight is satisfying its requirement for maintenance. 维持净能是饲料的基本能量需要,马匹既不服役也不活动,并维持衡定体重是满足维持净能的条件。

maize /meɪz/n.【BrE】*see* corn 玉米

major penalty n.【cutting】a three-or five-point penalty infraction(截牛中罚3~5分的违例)

严重违例 [compare] minor penalty

make /meɪk/vt. **1** to cause to exist or come about 造成;导致 PHR V **make a noise** (of a horse) to have audible respiratory problem in breathing (指马)呼吸急促,喘鸣 **make and break** to break a young horse 调教马匹 **2** to gain or earn money or profit 获得,获利 PHR V **make a check**【racing】(of a horse) to run into the first five places and thus earn prize money for the owner (指马跑入前五名而获奖)赢得支票,比赛获奖

malabsorption /ˌmæləbˈsɔːpʃn/n. inadequate absorption of nutrients from the gastrointestinal tract (肠胃消化道对营养物吸收不充分)吸收不良

malalignment /ˌmæləˈlaɪnmənt/n. an abnormal position or presentation of fetal foal in delivery (分娩时体位异常的情况)胎位异常,胎位不正

Malapolski /ˌmæləˈpɒski/n. a Polish warmblood breed developed recently by crossing Oriental stock with Thoroughbred, Furioso, and Gidran Arab; stands 15.3 to 16.2 hands and may have any of the solid colors; is calm, good-tempered, and mainly used for riding, light draft, and jumping (波兰最近采用东方马和纯血马、弗雷索马及基特兰阿拉伯马杂交培育的温血马品种。体高15.3~16.2掌,各种单毛色兼有,性情安静、温驯,主要用于骑乘、轻挽和障碍赛)曼拉波斯基马

male /meɪl/adj. of or relating to the sex that has organs to produce sperm 雄性的,公的:a male horse 一匹公马 [opp] female

malignant /məˈlɪgnənt/adj. (of a disease) threatening to life; virulent (疾病)致命的;恶性的:malignant tumor 恶性肿瘤 [opp] benign

mallein /ˈmælɪɪn/n. an effective medicine used in curing malleus [马]鼻疽菌素

malleolus /məˈliːələs/n.【anatomy】ankle 踝:lateral malleolus 外踝 ◇ medial malleolus 内踝

mallet /ˈmælɪt/n.【polo】a polo mallet (马球)球杆

malleus[1] /ˈmælɪəs/n. the glanders 马鼻疽

malleus[2] /ˈmælɪəs/n.【anatomy】(*pl.* **mallei** /-lɪaɪ/) the hammer-shaped bone that is the outermost of the three small bones in the mammalian middle ear (哺乳动物中耳内三块小骨中最外侧的锤形骨)锤骨 —*see also* **incus, stapes**

malocclusion /ˌmæləˈkluːʒn/n. a conformation defect of horse in which the upper and lower teeth do not meet well when the jaws are closed (马颌骨闭合时上、下牙齿咬合不齐的体貌缺陷)口齿不正,错位咬合

malpostured /ˌmælˈpɒstʃəd/adj. (of fetus) having an abnormal posture in uterus 胎势不正的

malt /mɔːlt/n. germinated barley grain used chiefly in brewing and distilling (酿造用的)麦芽:malt sugar 麦芽糖

maltose /ˈmɔːltəʊz; *AmE* -toʊz/n. a white crystalline disaccharide formed in the sprout of grain (种子发芽过程中产生的双糖,呈白色结晶状)麦芽糖 [syn] malt sugar

mamilla /mæˈmɪlə/n.【anatomy】(also **mammilla**) the nipple or teat of mammals 乳头

mamillate /ˈmæmɪleɪt/adj. of or resembling mamilla 乳头状的;乳头的

mamma /ˈmæmə/n.【Latin】(*pl.* **mammae** /-miː/) the mammary gland 乳腺

mammal /ˈmæml/n. any of various warm-blooded vertebrate animals characterized by a covering of hair on the skin and milk-producing mammary glands for nourishing the young in the female (多种温血脊椎动物,其共同特点是皮肤上附有被毛,雌性乳腺分泌乳汁哺育幼崽)哺乳动物

mammalian /mæˈmeɪliən/adj. of or relating to mammals 哺乳动物的

n. any of the mammals 哺乳动物

mammary /ˈmæməri/adj.【anatomy】of or relating to the milk-secreting organs of mammals 乳腺的

mammary gland n. the gland of mare that secrets

milk after foaling (母马产驹后分泌乳汁的腺体)乳腺

mamm(o)- pref. of mammary gland 乳腺

mammoth jack stock n. a breed of donkey known for it's large size and height 大型驴

Man O'War n. (1917 – 1947) a chestnut Thoroughbred foaled in 1917 in Kentucky, USA, who was beaten only once in 21 starts and believed by many to be the greatest American-bred racehorse of all times (1917 年在美国肯塔基州出生的一匹栗色纯血马,出赛 21 场仅负 1 场,被普遍认为是美国有史以来最杰出的赛马)战神 —*see also* **Big Red**

manada /məˈnɑːdə/n.【AmE】a herd of wild mares led by a mustang stallion (雄性北美野马率领的一群母马)母马群

Manchester team n. *see* tandem 双马纵驾

mandible /ˈmændɪbl/n.【anatomy】the lower jaw of a vertebrate animal 下颌骨

mandibula /mænˈdɪbjʊlə/n.【Latin】(*pl.* **mandibulae**/-liː/) mandible 下颌骨 compare maxilla

mandibular /mænˈdɪbjʊlə(r)/adj. of or relating to the mandible 颌下的 compare maxillary

mandibular gland n. one of the salivary glands located at the underside of the jaw (位于颌骨下方的唾液腺)颌下腺 syn submaxillary gland

mane /meɪn/n. the long hair growing along the neck of the horse between the poll and withers (马项顶至鬐甲之间沿颈峰生长的长毛)鬣,鬃:mane comb 鬃梳

mane and tail comb n. a metal or plastic comb used to untangle or remove debris from the mane and tail (一种铁质或塑料梳,用来梳理和清洁鬃尾)鬃尾梳

manège /mæˈneʒ, mɑːˈneʒ/n.【French】**1** an enclosed area used for training and schooling horses (训练或调教马匹的场地)练马场 **2**【dated】the exercises or training of a horse in the manège (练马场进行的训练)马术训练,驯马

manganese /ˈmæŋgəniːz/n.【nutrition】a trace mineral element essential in the formation of cartilage (一种微量矿物质元素,为软骨形成所必需)锰:Deficiency of manganese might cause abnormal skeletal development and reproductive failure. 锰缺乏可导致骨骼发育不良和繁殖力低下。

mange /meɪndʒ/n. a skin disease caused by parasitic mites and characterized by skin lesion, incessant itching and loss of hair (由螨虫所致的皮肤病,症状为皮肤溃烂、奇痒和脱毛)疥癣 syn acariasis

manger /ˈmeɪndʒə(r)/n. a trough or an open box from which livestock take feed; crib (饲喂牲畜的木槽)饲槽;料槽

manipulate /məˈnɪpjuleɪt/vt. to operate or control by skill; handle 操纵,驾驭

manipulation /məˌnɪpjuˈleɪʃn/n. the act or practice of manipulating 操纵,驾驭

Manipuri /ˌmʌnɪˈpʊərɪ/n. (also **Manipuri pony**) an ancient pony breed indigenous to India; stands 11 to 13 hands and may have a bay, brown, gray, or chestnut coat; has a light, well-proportioned head, full mane and tail, and sturdy legs; is strong, fast, energetic, and hardy; used mainly for riding and polo (印度本土的古老矮马品种,体高11 ~ 13 掌,毛色可为骝、褐、青或栗,头轻而匀称,鬃尾毛浓密,四肢粗壮,体格强健,速力较好,抗逆性强,主要用于骑乘和马球运动)曼尼普尔矮马

manners /ˈmænərz/n. the way that a horse obeys and accepts the commands of its rider (马接受骑手指令的方式)举止,姿态

manure /məˈnjʊə(r); *AmE* məˈnʊr/n. the droppings of farm animals; feces 粪便,粪肥:manure disposal 粪便处理 ◇ manure pile 粪堆

manure spray n. an insecticide sprayed over stored manure to kill larvae of insects and parasites (喷在粪堆上的杀虫剂,可杀死昆虫或寄生虫的卵)粪便喷雾[杀虫]剂

marathon /ˈmærəθən/n.【racing】any horse race

longer than 1-1/4 miles（赛程在2千米以上的比赛）长途赛

marble /ˈmɑːbl; *AmE* ˈmɑːrbl/n.【color】a color pattern in the Appaloosa who has dark sprinkles on a light base coat（阿帕卢萨马毛色类型之一，其浅色被毛上散生有暗色斑纹）大理石斑纹 compare blanket, frost, leopard, snowflake

mare /meə(r); *AmE* mer/n. **1** an adult female horse about four years of age or older（4岁以上的成年雌性马匹）母马：barren mare 空怀母马 ◇ foaling mare 产驹母马 ◇ in-foal mare 妊娠母马 ◇ lactating mare 哺乳母马 **2** any female horse who has borne a foal 产驹母马

mare donkey n.【BrE】a jennet 母驴

mare hinny n. a female hinny 母驴骡

mare immunological pregnancy test n.（abbr. **MIP-TEST**）an immunological test that detects the PMSG in a mare's blood stream to determine its pregnancy（通过免疫检测母马血液中孕马血清促性腺激素的水平以断定其受孕状况）母马妊娠免疫诊断

mare mule n.（also **molly mule**）a female mule 母骡

Maremmana /ˈmærəˌmænə/n. a breed of horse indigenous to Italy; stands 15 to 15.3 hands and may have any solid coat color with bay, chestnut, and black occurring most often; is solidly built, hardy, adaptable, and easy to keep; used as a light draft or farm horse and as a mount for the Italian police（意大利本土马品种。体高15～15.3掌，各种单毛色兼有，但以骝、栗和黑最为常见，体格健壮，抗逆性强，适应性好，易饲养，主要用于轻挽、农田作业，并用作意大利骑警的坐骑）摩雷曼纳马

Marengo /məˈreŋgəu/n.（c. 1793 – 1831）the gray Arab stallion ridden by Bonaparte Napoleon during his Austrian and Italian campaigns and at the Battle of Waterloo in 1815（拿破仑的一匹青毛阿拉伯公马，他曾策骑此马出征奥地利和意大利，并参加了1815年的滑铁卢战役）马伦戈

mare's month n.【racing】the month of September during which the mares do not run well（母马赛绩不佳的9月）母马失利月

mare's nest n. **1** a discovery that seems interesting but is found to be worthless 无价值的发现 **2** a complicated and difficult situation 复杂的局势：This area of the law is a veritable mare's nest. 这一法律领域真是复杂难料。

mareyeur /marjɜːr/n.【French, dated】a horse formerly used in France to transport fish from Boulogne to Paris（法国过去用来从布洛涅往巴黎运送水产品的马匹）水产运输马

margin /ˈmɑːdʒɪn; *AmE* ˈmɑːrdʒən/n.【anatomy】the border of an organ or certain tissue（特定器官或组织的边缘）缘：ciliary margin 睫状缘 ◇ coronary margin 冠状缘

margo /ˈmɑːgəʊ; *AmE* ˈmɑːgoʊ/n.【Latin】margin 缘

mark /mɑːk; *AmE* mɑːrk/n. **1** a spot on a horse's body（马体上的）斑点 **2** remains of the infundibulum in teeth wearing（牙齿磨损中黑窝残留的痕迹）齿坎痕

marker /ˈmɑːkə(r); *AmE* ˈmɑːrk-/n.【genetics】a segment of DNA sequence that can be used as an indicator for locating genes or testing other parameters（用于基因定位或检测的特征性DNA片段）标记：genetic marker 遗传标记

marking /ˈmɑːkɪŋ; *AmE* ˈmɑːrk-/n. *see* white marking 别征，白彰：body marking 身体别征 ◇ facial marking 面部别征 ◇ leg marking 四肢别征

marksman /ˈmɑːksmən/n.（*pl.* **marksmen**）one skilled in shooting 神射手，神枪手

Marocco /məˈrɒkəʊ; *AmE* məˈrɑːkoʊ/n. a performing horse active in Europe during the late 16th century who could count numbers, rear, and dance on command from his owner Thomas Bankes（在16世纪末活跃于欧洲的一匹表

演用马,在马主托马斯·班克斯的口令下,可以计数、直立并跳舞)玛洛克

martingale /ˈmɑːtɪŋgeɪl/n. a piece of equipment used commonly in English riding to prevent a horse from raising his head too high, consisting of a strap which buckles around the horses neck, and another strap which attaches to the girth at one end, passes between the forelegs and attaches to either the noseband (standing martingale) or the reins (running martingale) at the other end (英式骑乘中使用的马具,主要用来防止马抬头过高,由系于马颈的皮带和穿过前肢与鼻革或缰相连的另一条皮带构成)低头革,马丁革:standing martingale 站立低头革 ◇ running martingale 跑步低头革 ◇ bib martingale 围嘴低头革 ◇ Irish martingale 爱尔兰低头革

Marwari /məːˈwɑːri/ n. (also **Marwari horse**) a horse breed indigenous to the state of Jodhpur, India; stands 14 to 15 hands and may have a coat of any color including skewbald and piebald; has a weak neck and quarters, low-set tail, sickle hocks, and ears that curve distinctively inwards; is tough, hardy, frugal, and tenacious, and used for riding, packing, light draft, and farm work (源于印度焦特布尔州的本土马品种。体高 14 ~ 15 掌,包括花毛在内的各种毛色兼有,颈和后躯不甚发达,尾础低,镰形飞节,耳稍内敛,体格粗壮,抗逆性强,耐粗饲,性情顽强,主要用于骑乘、驮运、农田作业和轻役)马尔瓦尔马

Maryland Hunt Cup n. a celebrated steeplechase race run annually in Glyardson, Maryland, USA since 1896 over a permanent course with fences constructed of solid timber up to 1.7 meters high (美国久负盛名的越野障碍赛,自 1896 年起每年在马里兰州格里亚德森的赛道上进行,障碍高度可达 1.7 米)马里兰狩猎杯障碍赛

masculine /ˈmæskjəlɪn/adj. of, relating to, or characteristic of the male 雄性的 | **masculinity** /ˌmæskjuˈlɪnəti/n.

masculinize /ˈmæskjəlɪnaɪz/vt. (BrE also **-ise**) to give a masculine appearance or character (使具有雄性外貌或特征)雄性化

mash /mæʃ/n. a mixture of ground grain and nutrients fed to livestock and poultry (由谷类与营养物质调成的粥状混合物,用来饲养牲畜和家禽)粥,糊:bran mash 麸皮粥

mask /mɑːsk/n.【hunting】the head of a fox 狐狸的头

massa /ˈmæsə/n.【Latin】(*pl.* **massae**/-siː/) a mass or block 块:massae laterals 侧块

massage /ˈmæsɑːʒ; *AmE* məˈsɑːʒ/n. the rubbing or kneading of body muscles and tissues for therapeutic purposes (为达到治疗目的而对身体肌肉和组织进行的揉捏或推拿)按摩

vt. to rub and knead to relieve tension or pain 按摩

masseter /mæˈseɪtə/n.【anatomy】a thick muscle in the cheek that functions to open and close the jaws during chewing (位于面颊的一块肌肉,负责咀嚼时颌骨的开合)咬肌:masseter muscle 咬肌 | **masseteric** /ˌmæsəˈterɪk/adj.

master /ˈmɑːstə(r); *AmE* ˈmæs-/n. **1** one experienced enough to train others 师傅,大师,能手, **2**【hunting】(usu. **Master**) the Hunt Master 猎主,狩猎主

vt. to become adept at or expert in sth 掌握,精通

Master of Foxhounds n. (abbr. **MFH**) *see* Hunt Master 猎主

Master of Hounds n. (abbr. **MH**) one responsible for management of hounds used in the hunt 猎犬照料者

Master of the Foxhounds Association of America n. an organization founded in 1907 to promote the sport of fox and drag hunting in North America (1907 成立的组织机构,旨在促进北美地区猎狐和寻踪狩猎的发展)美洲猎犬主协会

Master of the pack n. *see* field master 猎场主，狩猎主

mastership /ˈmɑːstəʃɪp/n.【hunting】the period of time during which a Master reigns over a specific Hunt 猎主任期

masticate /ˈmæstɪkeɪt/vt. to chew food 咀嚼（食物）：The two main functions of the mouth is to masticate feed and wet it with saliva. 口腔的主要功能是咀嚼并用唾液湿润食物。| **mastication** /ˌmæstɪˈkeɪʃn/n.

mastitis /mæˈstaɪtɪs/n. inflammation of the mammary gland resulting from bacteria-related infection（由细菌感染导致的乳腺炎症）乳房炎，乳腺炎：acute mastitis 急性乳腺炎

matador /ˈmætəˌdɔː(r)/n.【Spanish】one who fights and kills the bull in bullfighting performance（西班牙斗牛表演中挑斗并屠杀公牛的人）屠牛士；斗牛士 —*see also* **picador**

match /mætʃ/n. **1** a game or contest in which two players or teams compete with each other（两人或两队相互角逐的游戏）比赛，竞赛 **2** a polo match 马球比赛

match race n.【racing】a race between two horses with no prize（两马之间没有奖金的比赛）对抗赛，对决赛

mate /meɪt/n. either of a pair of farm animals brought together for breeding（用于配种目的牲畜）配偶，种畜

vt. ~ **sth**（**to/with sth**）to put animals together for breeding purposes 配种

vi. ~（**with sth**）to copulate and produce young 交配：In natural state, wild horses seldom mate with burros. 在自然情况下，野马很少会与野驴交配。

mater /ˈmeɪtə(r)/n.【Latin】a layer of membrane enclosing the brain and spinal cord（包裹大脑和脊髓的膜层）脑膜，脊膜：dura mater 硬脑膜 ◇ pia mater 软脑膜

maternal /məˈtɜːnl; *AmE* məˈtɜːrnl/adj. of or relating to the dam 母系的 compare paternal

maternal grandsire n. the sire of a horse's dam 母系祖公马，外祖父公马

mating hobbles n. *see* breeding hobbles 配种绊腿

mating posture n. the position assumed by the mare in preparation to receive the stallion during breeding（母马在接受公马爬跨时的姿势）配种姿势，交配姿势：When the mare takes the mating posture, both her forelegs and hindlegs are extended outward and her back is slightly arched. 在母马采取交配姿势时，其前后肢呈外张姿势，背稍微下凹。

matron /ˈmeɪtrən/n. a mare who has produced a foal 产驹母马

mature /məˈtʃʊə(r), -ˈtjʊə(r); *AmE* -ˈtʃʊr, -ˈtʊr/adj.（of a horse）having reached full natural growth with his mouth full（指马生长发育完全且牙口齐全）成年的，成熟的：Horse usually becomes mature at the age of five years old. 马通常在5岁步入成年。opp immature

mature horse n. a horse of either sex who has a full mouth and reached maturity（齐口且性成熟的马）成年马

maturity /məˈtʃʊərəti, -ˈtjʊə-; *AmE* -ˈtʃʊr-, -ˈtʊr-/n. **1** the state or quality of being mature 成熟，成年 **2**【racing】a horse race for four-year-old horses, usually entered before their birth（由4岁马参加的比赛，通常在出生之前报名）成年赛

maverick /ˈmævərɪk/n. an unbranded young calf 未打烙印的犊牛

maxilla /mækˈsɪlə/n.【anatomy】(*pl.* **maxillae** /-liː/) the upper jaw bone of horse 上颌骨 syn maxillary bone

maxillary /mækˈsɪləri/adj. of the upper jaw 上颌骨的：maxillary bone 上颌骨

maximal oxygen uptake n.【physiology】the oxygen taken up by muscle（肌肉的）最大吸氧量，最大摄氧量

Mb abbr. myoglobin 肌红蛋白

MCHC abbr. mean corpuscular hemoglobin concentration 平均红细胞血红蛋白浓度

M

MCV abbr. mean corpuscular volume 平均红细胞容积: When reading blood test results, the MCV indicates the size of the blood cells. 在血液化验结果中,平均红细胞容积是血细胞多少的指标。

ME abbr. metabolizable energy 代谢能

meadow /ˈmedəʊ; *AmE* -doʊ/n. a stretch of grassland; lea 草地;草场

meadow grass n. one kind of blue grass of the genus *Poa* 草地早熟禾,草坪草

meal /miːl/n. **1** the grain or plant seeds ground coarsely to powder (碾碎后的籽实)粉;片;粕: oatmeal 燕麦片 ◇ soybean meal 大豆粕 **2**【hunting】a cooked porridge fed to the hounds 犬食

mealy /ˈmiːli/adj. (of the muzzle) having oatmeal-colored dapples or spots (指口鼻)带粉斑的: The Exmoor pony is an example of a horse which has a mealy muzzle. 埃克斯穆尔矮马是带有粉斑鼻的典型马种。

mean corpuscular hemoglobin concentration n. (abbr. **MCHC**) a measure of the concentration of hemoglobin in a given volume of paced red blood cells (定量压缩红细胞体积中的血红蛋白所测的含量)平均红细胞血红蛋白浓度

mean corpuscular volume n. (abbr. **MCV**) a measure of the average volume of a red blood corpuscle (血液红细胞所测体积大小的平均值)平均红细胞容积

meander /miˈændə(r)/vi. to move aimlessly and idly without fixed direction; amble, wander (无固定方向和目标的移动)漫步,散步

measure /ˈmeʒə(r)/vt. to determine the dimensions, quantity, or capacity of an object (对物体大小、体量和容积的估测)测量

measurement /ˈmeʒəmənt; *AmE* ˈmeʒərm-/n. the act or procedure of measuring 测量: body measurement 体尺测量 ◇ bone measurement 管围测量 ◇ girth measurement 胸围测量

measuring stick n. a straight rigid stick calibrated in inches or centimeters and used to measure the body height of horses (用来测量马匹体高的木杆,其上标有英寸或厘米刻度)测杖 [syn] horse standard

measuring tape n. a plastic tape marked in inches or centimeters used to measure the heart girth or circumstance of cannon bone of the horse (带有英寸或厘米刻度的皮尺,用来测量马的胸围和管围等指标)卷尺,皮尺

meat and bone meal n.【nutrition】a protein supplement made from flesh and bones of animals (用动物的肉和骨制成的蛋白质补充料)肉骨粉: Although a good source of protein, calcium, and phosphorus, the meat and bone meal is not commonly added to horse rations. 尽管肉骨粉是补充蛋白质、钙和磷的良好来源,但马匹日粮中通常不予添加。

meatus /mɪˈeɪtəs/n.【Latin】(*pl.* **-tuses** /-siːz/) a body opening or passage (机体内的开口)道;开口: common nasal meatus 总鼻道 ◇ dorsal nasal meatus 背侧鼻道 ◇ external auditory meatus 外耳道 ◇ internal acoustic meatus 内耳道 ◇ meatus ethmoidales 筛道

mechanical bit n. *see* hackamore bit 调马衔铁

Mecklenburg /ˈmeklənbɜːg/n. a German warmblood breed similar to the Hanoverian in appearance but smaller in build; stands 15.3 to 16 hands and may have a brown, bay, chestnut, or black coat; is strong, courageous, and used mainly as a riding and cavalry mount (德国的温血马品种,体貌与汉诺威十分相似但体型稍小。体高15.3～16掌,毛色多为褐、骝、栗或黑,体格健壮,悍性强,主要用于骑乘和骑兵坐骑使用)梅克伦堡马

meconium /məˈkəʊnɪəm; *AmE* -ˈkoʊ-/n. the blackish waste material accumulated in the intestines of a newborn foal and discharged shortly after birth (蓄积在胎儿肠道中的黑色代谢废物,出生不久后排出体外)胎粪: meconium colic 胎粪性腹痛 ◇ meconium retention 胎粪滞留

medial /ˈmiːdiəl/adj.【anatomy】relating to, situated in, or extending toward the middle 中间的;内侧的,向内的

medial deviation of the carpal joints n. *see* knock knees 膝内向,X 形膝

medial deviation of the hock joints n. *see* cow hocks 飞节内向;X 形飞节

medialis /ˈmiːdɪəlɪs/adj.【Latin】medial 内侧的,向内的

median /ˈmiːdiən/n.【anatomy】of or situated in the middle part of body (身体)中部的,正中的

mediastinum /ˌmiːdɪəˈstaɪnəm/n.【anatomy】(*pl.* **-tina** /-nə/) the connective tissue between the two pleural sacs in mammals (哺乳动物左、右胸膜腔之间的结缔组织)纵隔 | **mediastinal** /-ˈtaɪnl/adj.

medicament /mɪˈdɪkəmənt; *AmE* ˈmedɪ-/n. an agent that promotes recovery from injury or ailment; a medicine (可促进个体从伤病中恢复的制剂)药物;药剂

medicine /ˈmedɪsɪn/n. any agent used to treat disease or relieve pain (用来治病或镇痛的制剂)药物,药品

medicine ball n. (also **medicine pill**) a medicine capsule administered through mouth (口服的)药片,药丸

medicine hat n.【color】a color pattern in pinto horses who have a white face with dark hair covering the poll and sometimes ears (花马毛色类型,面部为白色,项顶和耳朵为其他深色)医生帽,大夫帽: It is believed by the Indians that a medicine hat horse can bring good luck. 印第安人认为"大夫帽"能带来好运。

medicine pill n. *see* medicine ball 药片,药丸

medieval Long Wagon n. *see* Long Wagon (中世纪)长形马车

medulla /mɪˈdʌlə/n.【anatomy】(*pl.* **medullae** /-liː/) the core of certain organs or body structure, such as the marrow of bone (身体器官或机体结构的内核,如骨髓)髓质 : renal medulla 肾髓质 ◇ medulla ossium 骨髓 ◇ medulla spinalis 脊髓 compare cortex

medulla oblongata n.【anatomy】(*pl.* **-dullae -gatae**) the lowermost portion of the vertebrate brain responsible for the control of respiration, circulation, and certain other bodily functions (脊椎动物脑最末端的部分,有调控呼吸、血液循环及其他身体机能的作用)延髓

medullary /meˈdʌləri, ˈmedəl-/adj. of or relating to medulla 髓质的;骨髓的: medullary cavity 骨髓腔

meek /miːk/adj. (of horses) submissive, tractable (指马)听话的;温驯的

meet /miːt/n. **1**【racing】a race meeting 赛马节 **2**【hunting】the location where the hunt servants, hounds, and field assemble before the hunt (出猎前随员、猎犬和打猎者集合的地方)打猎集会点 **3**【hunting】the hunt 狩猎,猎狐

vi. (of hunters) to come and bring their hounds together at a certain location to form one pack (指打猎者带着猎犬在某处汇集)集合,会合

megajoule /ˈmegdʒuːl/n. (abbr. **MJ**) an energy unit equal to one million joules (能量单位)兆焦

megrim /ˈmiːgrɪm/n. *see* staggers 眩晕症

meiosis /maɪˈəʊsɪs; *AmE* -ˈoʊs-/n.【genetics】the process of cell division in sexually reproducing organisms that reduces the number of chromosomes in gamete cells (有性生殖生物体的细胞分裂过程,所产生的配子细胞染色体数目减半)减数分裂 | **meiotic** /-ˈɒtɪk/adj. compare mitosis

melanin /ˈmelənɪn/n. any of a group of natural dark pigments existing in skin, hair, fur, and feathers (存在于皮肤、头发、毛皮及羽毛中的天然色素)黑色素: Melanin occurs as pigment granules in the hair, skin, iris and some internal tissues in two related forms: eumelanin and pheomelanin. 黑色素主要以真黑色素和褐黑色素两种形式存在于毛发、皮肤、巩膜及体内某些组织的色素颗粒中。

M

melanocyte /ˈmelənəsaɪt/n. an epidermal cell capable of synthesizing melanin（一种可生成黑色素的上皮细胞）黑素细胞

melanoma /ˌmeləˈnəʊmə; *AmE* -ˈnoʊ-/n. a tumor of melanocyte in the skin（皮肤黑素细胞所患的肿瘤）黑素细胞瘤：Malignant melanomas are usually larger and softer than benign growths. 恶性黑素细胞瘤通常比良性瘤大且软些。 syn black cell tumor

Melbourne Cup n. a 2-mile handicap race held annually since 1861 in Australia（澳大利亚自1861年起每年举行的让磅赛，赛程3.2千米）墨尔本杯

melton /ˈmeltn/n. a heavy, tightly woven cloth used chiefly for making overcoats and hunting jackets（质地紧密的厚毛料织物，多用于制作外套和狩猎大衣）麦尔登呢

membrana /memˈbreɪnə/n.【Latin】membrane 膜

membrana nictitans n.【Latin】*see* nictitating membrane 瞬膜

membrane /ˈmembreɪn/n.【anatomy】a thin layer of tissue covering surfaces or connecting structures or organs of the body（包裹或连接机体器官的薄层组织）膜：basement membrane 基膜 ◇ thyrohyoid membrane ◇ 甲状舌骨膜 tympanic membrane 鼓膜 ◇ vitreous membrane 玻璃体膜

Mendel, Gregor Johann n.（1822–1884）Austrian botanist who discovered the principle of inheritance through years of experiments with garden peas（奥地利植物学家，通过多年的豌豆试验发现了遗传定律）孟德尔，格雷戈·约翰：Mendel is generally regarded as the founder of the science of genetics. 人们普遍将孟德尔视为遗传学的奠基人。

Mendelian /menˈdiːlɪən/adj. of or relating to Gregor Mendel or his theories of genetics 孟德尔的，孟德尔遗传学的：Mendelian genetics 孟德尔遗传学

meningitis /ˌmenɪnˈdʒaɪtɪs/n. inflammation of the membranes enclosing the brain and the spinal cord; often caused by bacterial or viral infection and characterized by fever, vomiting, intense headache, and stiff neck（由细菌或病毒感染引起脑膜发炎的症状，临床表现为发热、呕吐、头痛及颈部僵硬）脑膜炎

meninx /ˈmiːnɪŋks/n.【anatomy】(*pl.* **meninges** /mɪˈnɪndʒiːz/) one of the three membranes enclosing the brain and spinal cord in vertebrates（包裹脊椎动物大脑和脊髓的膜层）脑膜，髓膜

mental /ˈmentl/adj. of or relating to the chin 下颌的

Merck Veterinary Manual n.（abbr. **MVM**）a concise and reliable animal health reference book used by veterinarians and other animal health professionals all over the world（全球兽医和动物保健专家使用的一本有关动物疾病诊断的简明参考书）《默克兽医手册》：First published in 1955, the *Merck Veterinary Manual* has now been adopted by the veterinary profession as the most practical and comprehensive resource worldwide.《默克兽医手册》首版于1955年发行，是目前兽医行业最适用也最全面的参考工具书。

Mèrens /mɛrã/n.【French】(also **cheval de Mèrens**) an ancient horse breed originating at the foothills of the Pyrenean mountains of France; is similar in appearance to the Fell and Dale; stands 13 to 14.1 hands and always has a solid black coat; has a relatively coarse head, heavy mane and forelock, small ears, a short neck which is thick at the base, a long back, well-muscled croup, long, full tail, and short legs; is sure-footed, docile, hardy, and frugal; used for farm work, riding, and packing（源于法国比利牛斯山区的古老矮马品种，体貌与费尔矮马和山谷矮马相像。体高13～14.1掌，毛色始终为黑，面貌粗糙，鬃毛和额发浓密，耳小，颈粗短，背长，臀部肌肉发达，尾长而浓密，四肢较短，步法稳健，性情温驯，抗逆性强，耐粗饲，主要用于农田作业、骑乘和

驮运)梅隆马 [syn] Ariégeois pony

meridian /məˈrɪdiən/ n. any of the longitudinal lines or pathways on the body along which the acupuncture points are distributed (分布于身体上的脉络,针灸穴位位于其上)经脉,经络

Merychippus /məˈraɪkɪpəs/ n. one of the early ancestors of the modern horse who lived in the Miocene Epoch and had three toes in both fore and hind feet (马的远祖之一,生活于中新世,前后肢皆为三趾)原马 —*see also* **Eohippus**, **Hipparion**, **Mesohippus**, **Pliohippus**

mesair /ˈmeseə(r); *AmE* -ser/ n.【dressage】*see* mezair 多次直立

mesencephalon /ˌmesenˈsefəlɒn/ n.【anatomy】the middle portion of the vertebrate brain (脊椎动物脑的中段)中脑 [syn] midbrain

mesentary /ˈmesəntəri/ n. *see* mesentery 肠系膜

mesenteric /ˌmesənˈterɪk/ adj. of or relating to mesenteries 肠系膜的

mesenteric artery n. a blood vessel that supplies blood to the intestines (为肠道供应血液的血管)肠系膜动脉

mesentery /ˈmesəntəri/ n.【anatomy】(also **mesentary**) any of several folds of the peritoneum that connect the intestines to the dorsal abdominal wall, especially such a fold that envelops the jejunum and ileum (连接肠道与背侧壁的腹膜,尤指包裹空肠和回肠的皱褶)肠系膜

mesogastric /ˌmesəuˈgæstrik/ adj. near or located at the middle part of the abdomen 腹中部的: mesogastric region 腹中部

Mesohippus /ˌmesəˈhɪpəs/ n. one of the early ancestors of the modern horse who lived in the Oligocene Epoch (马的远祖之一,生活于渐新世)中马 [syn] Middle Horse —*see also* **Eohippus**, **Hipparion**, **Merychippus**, **Pliohippus**

metabolic /ˌmetəˈbɒlɪk; *AmE* -ˈbɑːl-/ adj. of or relating to metabolism 代谢的: metabolic disorder 代谢性紊乱

metabolism /məˈtæbəlɪzəm/ n. the complex physical and chemical reactions occurring within a living cell or organism that are necessary for the maintenance of life (活细胞或生物体内维持生命所进行的复杂的物理化学反应)[新陈]代谢

metabolite /məˈtæbəlaɪt/ n. any substance produced in metabolism (在代谢过程中产生的物质)代谢物

metabolize /məˈtæbəlaɪz/ vt. (BrE also **-ise**) to turn feed or food in the body into particles, energy, and waste by biochemical processes (将食物通过生化过程转变为分子、能量和废物)代谢

metacarpal /ˌmetəˈkɑːpl; *AmE* -ˈkɑːrpl/ adj. of or relating to the metacarpus 掌骨的
n. any one of the bones of the metacarpus 掌骨: large metacarpal 大掌骨 ◇ metacarpal region 掌部 ◇ small metacarpal 小掌骨 ◇ lateral small metacarpal 外侧小掌骨 ◇ medial small metacarpal 内侧小掌骨

metacarpus /ˌmetəˈkɑːpəs; *AmE* -ˈkɑːrp-/ n.【anatomy】a group of five bones that forms the part of the foreleg between the knee and fetlock (由5块骨组成的整体,构成马前肢介于前膝与球节的部分)掌骨

metal curry n. (also **metal curry comb**) a metal, handled grooming tool with roughly jagged teeth, used to scrape dirt and debris from a brush (铁制带齿刷拭工具,用来清理体刷上的污垢和杂物)铁梳刨,铁篦子 [compare] rubber curry

metal curry comb n. *see* metal curry 铁梳刨,铁篦子

metal horse n.【vaulting】*see* vaulting barrel 铁马;马上体操练习桶

metaphyseal plate n. *see* epiphyseal growth plate 骺生长板

metaphysis /mɪˈtæfɪsɪs/ n.【anatomy】(*pl.* **metaphyses** /-siːz/) part of a long bone connecting the epiphysis and the diaphysis (长骨中连接骨骺与骨干的区段)干骺端 [compare] diaphysis, epiphysis

M

metatarsal /ˌmetəˈtɑːsl; *AmE* -ˈtɑːrsl/adj. of or relating to the metatarsus 跖骨的

n. any of the bones of the metatarsus 跖骨：large metatarsal 大跖骨 ◇ lateral small metatarsal 外小跖骨 ◇ medial small metatarsal 内小跖骨

metatarsus /ˌmetəˈtɑːpəs; *AmE* -ˈtɑːrp-/n.【anatomy】the group of five bones that forms part of the hindleg of horse between the tarsus and fetlock（马后肢位于跗骨和球节间的部分，由五块骨构成）跖骨

metestrus /metˈiːstrəs; *AmE* -ˈestrəs/n.【physiology】the period of time between estrus and diestrus（介于发情期和乏情期之间的时段）后情期

methionine /mɪˈθaɪəniːn/n.【nutrition】a sulfur-containing essential amino acid（一种含硫的必需氨基酸）蛋氨酸，甲硫氨酸

methylsulfonylmethane /ˌmeθɪlˈsʌlfənɪlˈmeθeɪn/ n.（abbr. **MSM**）an oral medicine derived from DMSO with the same anti-inflammatory properties, but without the unpleasant odor（由二甲基亚砜衍生而来的口服消炎药物，无类似的难闻气味）二甲基砜

Métis Trotter n. a Russian warmblood developed in the early 20th century by crossing Orlov with imported American Standardbreds; stands 15.1 to 15.3 hands and usually has a gray, black, or chestnut coat; has a flowing, far-reaching action and is much faster than the Orlov; used for riding and trotting races（俄国 20 世纪初采用奥尔洛夫快步马和进口的美国标准马杂交培育的温血马品种。体高 15.1～15.3 掌，毛色通常为青、黑或栗，运步流畅，步幅长，步速比奥尔洛夫快，用于骑乘和快步赛）梅第快步马：The Métis Trotter was officially recognized as a breed in 1949. 梅第快步马于 1949 年被正式认定为一个品种。

metritis /mɪˈtraɪtɪs/n. inflammation of the muscular and endometrial layers of the uterus caused by the introduction of microorganisms during parturition（分娩时由于微生物侵入导致子宫肌层和内膜层发炎的症状）子宫炎：acute metritis 急性子宫炎 ◇ contagious equine metritis 马传染性子宫炎

mew[1] /mjuː/n. the high-pitched cry of a cat or bird（猫或鸟的）尖叫

vi. to utter a high-pitching cry or noise 尖叫

mew[2] /mjuː/v.【hunting】（of a stag）to shed its antlers; molt（鹿角）脱落 | **mewing** n.

Mexican noseband n. see Grackle noseband X 形鼻革

mezair /ˈmezeə(r); *AmE* -zer/n.【dressage】（also **mesair**）a High School movement in which the horse performs a series of levades in succession（一种高级花式骑术，马连续多次直立的古典骑术动作）多次直立

MF abbr. Master of Hounds 猎犬主人

MFHs abbr. Master of Foxhounds 猎狐犬主：In fox-hunting, the MFHs is in charge of the entire operation. 在猎狐中，猎狐犬主全权负责一切事务。

MH abbr. Miniature Horse 微型马

microbe /ˈmaɪkrəʊb; *AmE* -kroʊb/n. any of the microorganisms, such as fungi, bacteria or viruses（包括真菌、细菌或病毒等在内的生物体）微生物

microbial /maɪˈkrəʊbiəl; *AmE* -ˈkroʊb-/adj. of or caused by microbe 微生物的：microbial infection 微生物感染

microchip /ˈmaɪkrəʊtʃɪp; *AmE* -kroʊ-/n. a minute piece of integrated circuit embedded beneath the skin of a horse that contains his registration information（埋植在马皮肤下的小块集成电路板，其中储存着马匹的注册信息）芯片

vt. to implant a microchip under the skin of an animal as a way of identifying it（为识别身份而在动物皮下植入芯片）埋植芯片，芯片植入

microchip implantation n. the practice of embedding microchip beneath the skin of a horse for identifying purposes（为鉴定身份而在马皮肤下面埋植芯片）芯片植入，芯片埋植

micronization /ˌmaɪkrənaɪˈzeɪʃn/n. 【nutrition】a feed processing method in which the grain is microwaved by infrared radiation to make the starch swell and gelatinize through rapid internal heating（饲料加工工艺，谷物经红外线辐射后，由于内温急剧上升而使谷物淀粉膨胀并发生糊化）[饲料]微化技术

micronize /ˈmaɪkrənaɪz/vt. to reduce to small particles 微粉化

middle distance n. 【racing】a horse race longer than 7 furlongs but less than 1.25 miles（赛程在1 400～2 000米的比赛）中程赛

middle distance horse n. a horse competing in flat racing around 1,600－2,400 meters（参加1 600～2 400米平地赛的马匹）中程赛马

Middle European Gidran n. one of two types of Gidran Arabian that is heavier than the other and thus used predominantly in harness（两种基特兰阿拉伯马类型之一，体型较另一个类型稍重，因而主要用于挽车）中欧基特兰阿拉伯马 —*see also* **Southern and Eastern Gidran**

Middle Horse n. *see* Mesohippus 中马

middleweight /ˈmɪdlweɪt/adj.（of a horse）medium in body weight（指马）体重中等的，体重适中的：a middleweight horse 体重适中的马匹

n. a horse capable of carrying medium weights as determined by its bone and substance（由于管围和体格所限负载能力适中的马）中型马 compare heavyweight, lightweight

middling /ˈmɪdlɪŋ/n.（usu. **middlings**）coarsely ground wheat mixed with bran（混有麸皮的粗制麦粉）粗面粉：coarse middlings 粗面粉 ◇ fine middlings 精粉

midsection /ˈmɪdˌsekʃn/n. the portion of the horse's body between the forehand and the hindquarters（马体介于前躯与后躯之间的部分）中段，躯体中部

Mierzyn /ˈmɪəzɪn; *AmE* ˈmɪrz-/n. a medium-sized pony breed native to Poland; stands about 14 hands and usually has a bay coat; is easy to keep, resistant to cold and hunger, and possessed of amazing power and endurance for its size（波兰的本土中型矮马品种。体高约14掌，毛色以骝为主，易饲养，抗饥寒能力强，挽力和耐力极佳）梅尔兹马

migrate /ˈmaɪɡreɪt/vi. to move from one place and settle in another（离开某地而迁居别处）迁移，迁徙

migration /maɪˈɡreɪʃn/n. the act or instance of migrating 迁移，移行

mildew /ˈmɪldjuː; *AmE* -duː/n. any of various minute parasitic fungi that produce a whitish coating or discoloration on plants and grains（一种寄生性真菌，常在植物和谷物表面形成白层或使之变色）霉菌

mile /maɪl/n. a unit for measuring distance equal to 1,609 meters（长度测量单位，相当于1 609米）英里

mile pole n. 【racing】a post located on the infield rail exactly one mile from the finish line（立于赛道内栏的标杆，与终点相距1.6千米）一英里杆

militaire /ˈmɪlɪteər; *AmE* -ter/n. 【French】*see* combined training event 综合全能赛

military trails n. *see* combined training 综合赛

milk /mɪlk/n. a whitish liquid produced by the mammary glands of mare after foaling, containing proteins, fats, lactose, vitamins, and minerals（母马分娩后乳腺分泌的乳白色液汁，内含蛋白质、脂肪、乳糖、维生素和矿物质）乳汁，奶：mare milk 母马乳汁 ◇ milk protein 乳蛋白 compare colostrum

milk teeth n. *see* deciduous teeth 脱落齿，乳齿

mill /mɪl/n. 【vaulting】a compulsory exercise in which the vaulter performs a complete circle above the horizontal line of the withers by swinging his leg over the horse's back or neck（马上杂技表演规定动作之一，表演者在马背鬐甲处摆腿完成旋转动作）马背回旋 —*see also* **basic seat, flag, scissors, stand, flank**

Miller's disease n. *see* Big head disease（大头症的别名）米氏症

milo /ˈmaɪləʊ/n. a kind of sorghum（高粱的一种）蜀黍

mimic /ˈmɪmɪk/vt. to imitate or simulate 模仿，模拟：Our sound knowledge of hormones has enabled us to synthesize drugs which mimic the action of natural hormones often in a more powerful fashion. 对激素的深入了解，已让我们能够合成一些模仿天然激素作用机理的药物，而且功效也更强。

mimicry /ˈmɪmɪkri/n. the act or instance of mimicking 模仿，模拟：mimicry behavior 模仿行为

mineral /ˈmɪnərəl/n. 【nutrition】an inorganic element, such as calcium, iron, potassium, sodium, or zinc, that is essential to the nutrition of animals and plants（动植物营养必需的钙、铁、钾、钠和锌等无机元素）无机元素；矿物质：mineral deficiency 矿物质缺乏症 ◇ mineral feed 矿物质饲料 ◇ mineral imbalance 矿物质失衡 ◇ mineral lick 矿物质舔砖 ◇ mineral mixture 矿物质混合料 ◇ mineral requirement 矿物质需要量

mineralocorticoid /ˌmɪnrələˈkɔːtɪkɔɪd/n. any of a group of steroid hormones, such as aldosterone, that are secreted by the adrenal cortex and regulate the balance of water and electrolytes in the body fluid（由肾上腺皮质分泌的多种类固醇激素，如醛固酮，起调节体液中水、电解质平衡的作用）盐皮质激素

miniature /ˈmɪnətʃə(r); *AmE* -tʃʊr/n. an animal of very small size 微型动物，迷你动物：a rare breed of miniature horses 一个稀有微型马种

miniature donkey n. a donkey stands below 9 hands or less at withers when fully grown; is thrifty and adaptable to any climate or altitude（鬐甲处体高 9 掌以下的驴，耐粗饲，对各种气候和海拔适应性都很强）微型驴：Today, more than 10,000 miniature donkeys exists in the United States. 现今，美国约有 1 万多头微型驴。

Miniature Landau n. a small horse-drawn vehicle of the Landau type, generally drawn by a pair of ponies（一种小型兰道马车，多由两匹矮马拉行）微型兰道马车

Miniature Mediterranean Donkey n. a donkey breed native to the Mediterranean islands of Sicily and Sardinia; is identified as either Sicilian or Sardinian according to their ancestry（源于地中海岛屿上的本土驴品种，根据其产地可细分为西西里微型驴和撒丁尼亚微型驴两种类型）地中海微型驴：The Miniature Mediterranean Donkey is now becoming nearly extinct in the land of their origin, but are extensively bred in the United States as miniature donkeys. 目前，地中海微型驴在当地的岛屿已基本绝迹，但在美国却被大量用于微型驴的培育。

Miniature Mediterranean Donkey Association n.（abbr. **MMDA**）a non-profit organization founded in UK to ensure the welfare of the miniature donkeys and to assist breeders and pet owners（成立于英国的非营利组织，旨在促进微型驴的福利并为育种者和宠物爱好者提供支持）地中海微型驴协会

Miniature Sardinian Donkey n. *see* Sardinian donkey 撒丁尼亚微型驴

Miniature Sicilian Donkey n. *see* Sicilian donkey 西西里微型驴

mini-clipper /ˈmɪnɪˌklɪpə(r)/n. a battery-powered clipper used for cutting the coat hairs of horses（用来修剪马匹被毛的）电动剃刀

minimum wager n.【racing】the minimum amount of money bet on a wager in horse race（赛马投注允许的最少金额）最低投注额

minor penalty n.【cutting】a one-point penalty infraction（截牛中罚 1 分的违例）轻度违例 compare major penalty

minus pool n.【racing】the amount of money provided by the racetrack from its own funds to cover insufficient funds in pari-mutuel betting when the pool is not enough to pay winning ticket holders（当赌金池不足以为获奖者返款时，赛马场为了弥补不足而拿出的资金）资金负额，投注赤字

minute ventilation n. *see* minute ventilation volume 每分通气量：At rest an average sized Thoroughbred horse might have a minute ventilation of somewhere around 50 - 60 litres/min, produced by taking breaths of 5 - 6 litres, ten times a minute. 一匹体型适中的纯血马静息时的通气量为 50 ~ 60 升/分，期间呼吸 10 次左右，每次呼吸量为 5 ~ 6 升。

minute ventilation volume n.（also **minute ventilation**）the total volume of atmospheric air that moves either into or out of the lung each minute（肺每分钟吸入或呼出空气的总量）每分（钟）通气量

MIP-TEST abbr. mare immunological pregnancy test 母马免疫妊娠诊断

miscarry /ˌmɪsˈkæri/ v. to abort or slip 引产；早产

misfit /ˈmɪsfɪt/ adj. **1**（of a horse）not representative of its breed type（指马）无品种特征的 **2**（of a horse）unsuitable for its work（指马）不相称的，不适合的

n. a horse that is not representative of its breed or unsuitable for its work 无品种特征的马；不相称的马

mishandle /ˌmɪsˈhændl/ vt. to mismanage or treat roughly; maltreat 使用不当；虐待

miss /mɪs/ vt. **1**【polo】（of a polo player）to take stroke on the ball but fail to make contact（指马球中击球）失误，未击中 **2**【cutting】（of a horse）to overrun a cow（指截牛马）跑过，错过（牛）**3**【roping】to fail to rope the calf, steer, or other animal 未套中

Missouri Fox Trotter n. a warmblood descended from Spanish Barb stock put to Morgans and Thoroughbreds followed by infusions of Saddlebred and Tennessee Walking Horse during the 1820s in Missouri, USA; stands 14 to 16 hands and all coat colors may occur; has a strong, compact body, intelligent head, tapered muzzle, long neck, low-set tail; is sure-footed and moves in a smooth, broken, four-beat gait known as the fox trot; usually ridden in Western tack（美国密苏里州在 19 世纪 20 年代采用西班牙柏布马与摩根马和纯血马杂交培育而成的温血马品种，后引入美国乘用马和田纳西快步马外血进行改良。体高 14 ~ 16 掌，各种毛色兼有，体格结实，面貌聪颖，口鼻突兀，颈长，尾础低，步法稳健，以善走四蹄音狐式快步著称，主要用于西部骑乘）密苏里狐步马：The stud book for the Missouri Fox Trotter was established in 1948. 密苏里狐步马的品种登记始于 1948 年。 [syn] Missouri Fox Trotting Horse

Missouri Fox Trotting Horse n. *see* Missouri Fox Trotter 密苏里狐步马

mistaught /mɪsˈstɔːt/ adj.（of a horse）improperly or badly broken and trained（指马）调教不当的

misteach /ˌmɪsˈtiːtʃ/ vt. to break or train a horse improperly or badly（指马）调教不当

mite /maɪt/ n. any of several species of small external parasites that cause mange（引发疥癣的多种外寄生虫）螨虫 [syn] acarus

mitochondrion /ˌmaɪtəʊˈkɒndriən; *AmE* ˌmaɪtoʊˈkɑːn-/ n.【biochemistry】（*pl.* **-dria** /-driə/）any of the organelles within the cell on whose inner membrane a series of chemical reactions convert the energy of oxidation into the chemical energy of ATP（细胞器结构，通过其内膜上发生的生化反应，可将氧化能转化为 ATP 的化学能）线粒体

mitosis /maɪˈtəʊsɪs/ n.【genetics】（*pl.* **mitoses** /-siːz/）the process in which one body cell divides into two cells of the same（体细胞一分为二的过程）有丝分裂 [compare] meiosis

mixed gait n. a condition in which a horse does not move with any one true gait at a time（指马行进时步法不固定的情况）混合步法

M

mixed meeting n.【racing】a race meeting in which both flat and steeplechase races are held on the same day（同时举行平地赛和越野障碍赛的节日）混合赛马节

mixed pack n.【hunting】a group of hounds consisting of both bitches and dog hounds（由公、母猎犬混合组成的队伍）混合犬队

mixed sale n. a horse sale in which mares, yearlings, stallions of one breed are offered for sale（由单个马种的母马、1岁驹和公马构成的交易会）混合拍卖会，混合交易会

MJC abbr. Macau Jockey Club 澳门赛马会

MMDA abbr. Miniature Mediterranean Donkey Association 地中海微型驴协会

mob /mɒb; *AmE* mɑːb/n.【AusE】a flock or herd of animals 畜群；兽群

vt.【hunting】(of hounds) to attack a quarry in groups 围攻，群攻：to mob a fox 围攻狐狸

model horse n. an authentic, small-scale sculpture of a horse made in wood, plastic, etc.（小型木质或塑料制仿真马雕塑）模型马；马雕塑

Mohammed's Ten Horses n. the ten horses who, according to legend, formed the foundation stock of the Prophet Strain（传说中穆罕默德选来培育"先知家系"的十匹奠基马）穆罕默德十骏

moist /mɔist/adj. slightly wet; damp or humid 潮湿的

moisten /mɔisn/v. to make or become moist 使潮湿；受潮

moisture /ˈmɔistʃə(r)/n. the vapor in the atmosphere or condensed liquid on the surfaces of objects; dampness（空气中的水汽以及凝结在物体表面的液滴）潮湿；水分：moisture content 水分含量

molar /ˈməʊlə(r); *AmE* ˈmoʊ-/adj. of or relating to molar teeth 臼齿的

n. (also **molar tooth**) any of the 12 permanent grinding teeth located behind the premolars on each side of the upper and lower jaws（上、下颌两侧位于前臼齿后面的12颗永久齿）臼齿，槽牙 [syn] grinders, cheek teeth

molar tooth n. *see* molar 臼齿，槽牙

molasses /məˈlæsɪz/n. a thick, dark brown by-product produced in refining raw sugar extracted from sugar cane or sugar beets（甘蔗或甜菜榨取的粗糖在精炼过程中产生的深褐色副产品）糖蜜

molly mule n. a mare mule 母骡

moment of suspension n. the moment when a horse has all four feet off the ground in movement（马行进中四蹄离地的时期）悬空期

Monday morning complaint n. *see* stock-up 下肢水肿

Monday morning disease n. another term for azoturia（氮尿症的别名）周一清晨病

Monday morning evil n. *see* stock-up 下肢水肿

Monday morning leg n. *see* stock-up 下肢水肿

monensin /mʌˈnensɪn/n. a broad-spectrum antibiotic used chiefly as an additive to beef cattle feed（一种广谱抗生素，主要用作肉牛饲料添加剂）莫能菌素：Take care to check that the feed designed for horse do not contain monensin or lincomycin, as they are poisonous to horses. 要仔细检查，马的饲料配方中不能含有莫能菌素和林可霉素，因为这两种抗生素对马是有毒的。

money /ˈmʌni/n.【racing】any sum a racehorse wins（赛马赢得的金额）奖金 PHR V **run in the money** (of a horse) to finish in the top four positions（指马）进入前四：It does not make sense that a player will win a bet placed on a horse who runs in the money. In fact, he will be only paid if the horse places in the first three positions. 赌马者所押赛马跑入前四就想赢奖有点荒唐，其实只有这匹马跑入前三才会赢奖。**run out of money** (of a horse) to finish in any position other than first three placings; off the board（指马）未入三甲：to finish out of the money 未跑入前三

money rider n.【racing】a jockey who excels in

high-stakes races（在高额奖金赛中表现不俗的骑师）奖金骑师，赢奖骑师

Mongolian /mɒŋˈɡəʊliən；*AmE* mɑːŋˈɡoʊ-/n. *see* Mongolian horse 蒙古马

Mongolian horse n.（also **Mongolian**）an ancient breed originating in Mongolia that had widespread influence on all Asian horse breeds; stands 12 to 14 hands and may have a brown, black, mouse dun or palomino coat; has a heavy head, short ears, thick forelock, mane, and tail, low withers, strong back, and sturdy, well-boned legs; is extremly hardy, frugal, and active; used for riding, packing, light draft, and farm work（源于蒙古的古老的马品种，对亚洲其他马种影响较大。体高12～14掌，毛色可为褐、黑、鼠灰或银鬃，头重，耳短，额发及鬃尾毛发浓密，低鬐甲，背短，四肢粗壮，抗逆性强，耐粗饲，性情活泼，主要用于骑乘、驮运、轻役和农田作业）蒙古马

Mongolian wild horse n. a subspecies of the Asiatic wild horse 蒙古野马 [syn] *Equus heminous*

monkey crouch n.【racing】a riding position popularized by Tod Sloan in which the jockey rides with short stirrups and his body bent forward over the withers of the horse（由托德·斯隆所推广的骑马姿势，骑手采用短镫策马，同时身体前倾至马鬐甲处）猴蹲式［策骑］ [syn] monkey-on-a-stick

monkey mouth n. *see* undershot jaw 下颌突出，地包天

monkey-on-a-stick n. *see* monkey crouch 猴蹲式

monocular /məˈnɒkjʊlə；*AmE* -kjələ/adj. using or relating to only one eye 单眼的，单目的：monocular vision 单眼视觉

monocyte /ˈmɒnəsaɪt；*AmE* ˈmɑːn-/n. a large, phagocytic white blood cell, having a single well-defined nucleus and very fine granulation in the cytoplasm（存在于血液中的吞噬性白细胞，细胞核明晰可辨，胞浆内有微小颗粒）单核细胞

monodactyl /ˌmɒnəˈdæktɪl/n. an animal having a single toe on each foot（每个蹄仅有单趾的动物）单趾动物

monorchid /məˈnɔːkid；*AmE* -ˈnɔːrk-/n. a stallion with only one testicle descended into the scrotum; unilateral cryptorchid（仅单侧睾丸下降到阴囊的公马）单睾马，隐睾马 [compare] cryptorchid

monovular /məˈnɒvjʊlə/adj. of or derived from one egg 单性的；同卵的

monovular twin n. one of two foals brought forth at birth resulting from the fertilization of one egg splitting in two（由单个受精卵一分为二发育出生的两匹马驹）同卵双生，同卵双胞胎：Monovular twins are an extremely rare occurrence in horses. 同卵双生在马中极为罕见。 [syn] identical twin

moon blindness n.（also **night blindness**）a recurrent inflammation of horse's eye that leads to blindness（马眼所患的一种再发性炎症，可导致马匹失明）夜盲症 [syn] periodic ophthalmia, recurrent uveitis

mope /məʊp；*AmE* moʊp/vt.【driving】to place shields over a horse's eyes to deprive him of vision（为遮蔽视线而给马戴眼罩）蒙眼，遮眼

Morab /mɔːˈrɑːb/n. a breed of horse developed in the early 20th century in the United States through the cross-breeding of Morgan and Arabian horses; stands 14.2 to 15.2 hands and usually has a bay, chestnut or gray coat; has a compact body, sloping shoulders, and a wide deep chest; is intelligent, curious and quick to learn; used for riding, farm work, and show purpose（美国20世纪初采用摩根马和阿拉伯马杂交育成的马品种。体高14.2～15.2掌，毛色多为骝、栗或青，体格强健，斜肩，胸身宽广，敏快而好奇心强，用于骑乘、牧场作业和展览比赛）摩拉伯马：By the 1920s the term Morab had been coined to describe this unique cross of breeds and their popularity grew. 时至20世纪20年代，术语摩拉伯马就已被创造出来特指这种杂交马品种，而且受欢迎

M

程度也与日俱增。

Morgan /ˈmɔːgən; *AmE* ˈmɔːrg-/n. an American warmblood descended from a cross of Thoroughbreds, Arabs, Welsh Cobs, and Fjord horses; stands 14.1 to 15 hands and may have a bay, chestnut, or black coat with white markings; has a short, broad back, strong shoulders, short, strong legs, well-crested neck, and a full mane and tail; is versatile, tough, and even-tempered with comportable and elastic gaits; used for hunting, dressage, riding, and driving (美国培育的温血马品种,主要采用纯血马、阿拉伯马、威尔士柯柏马和挪威峡湾马杂交繁育而来。体高14.1~15掌,毛色多为骝、栗或黑且带白色别征,背宽短,肩胛发达,四肢粗短,颈姿优美,鬃尾浓密,用途广泛,体质粗实,性情温和,步法舒适且富有弹性,主要用于狩猎、盛装舞步、骑乘和挽驾)摩根马: Morgan has contributed to the development of several American breeds, such as the Standardbred, Saddlebred, and Tennessee Walking Horse. 摩根马对美国标准马、美国乘用马和田纳西走马等多个马品种的育成都起了重要作用。

morning glory n.【racing, slang】a horse who runs well in the morning exercises but fails to do the same when racing (晨练时成绩不俗但在比赛中却难以发挥的马)晨跑马;牵牛花 [syn] morning horse

morning horse n. *see* morning glory 晨跑马

morning line n.【racing】the line or column in racing publications that predicts the probable the odds for horses beforehand (赛前预测发布马匹赔率的新闻栏目)早报预测栏: morning line makers 早报预测人

morning line odds n.【racing】the probable odds for each racehorse as calculated by an experienced track handicapper prior to betting (经验丰富的评磅员在赛前为马匹计算的可能赔率)预测赔率

morning stable n.【BrE】the morning feeding time 晨喂时间

Morocco saddle n. another name for the Burford saddle (伯弗德鞍的别名)摩洛哥鞍

morphology /mɔːˈfɒlədʒi; *AmE* mɔːrˈfɑːl-/n. **1** the form and structure of organisms 形态 **2** the branch of biology that deals with the form and structure of organisms (研究机体形态与结构的生物学分支)形态学

mort /mɔːt; *AmE* mɔːrt/n.【hunting】a horn signal denoting the death of the fox or other quarry (狩猎中通告猎物已死信息的号角)杀生号角

mortality /mɔːˈtæləti; *AmE* mɔːrˈt-/n. the condition of being dead; death 死亡: mortality rate 死亡率

mortality insurance n. insurance covering financial losses due to the death of a horse [马匹]死亡保险

morula /ˈmɒrələ; *AmE* ˈmɔːrʊlə/n.【biology】(*pl.* **morulae** /-liː/) a solid mass of embryonic cells resulting from division of a fertilized ovum (由受精卵分裂形成的实心胚)桑葚胚 [compare] blastula

motile /ˈməʊtɪl; *AmE* ˈmoʊtaɪl/adj. moving or having the power to move spontaneously; mobile 能动的;游动的

motility /məʊˈtɪləti; *AmE* moʊ-/n. the capability of sperm to move spontaneously (精子的)游动,活力: progressive motility 前向游动 ◇ backward motility 后向游动 ◇ circular motility 环形游动

mottled /ˈmɒtld; *AmE* ˈmɑːtld/adj. (of coat) spotted with different colors (被毛)斑驳的,带斑点的: mottled skin 斑点皮肤

mould /məʊld; *AmE* moʊld/n. (AmE also **mold**) any of various fungi that often cause disintegration of organic matter (可分解有机物的多种真菌)霉菌

vi. to become moldy 发霉,霉变

mouldy /ˈməʊldi; *AmE* ˈmoʊ-/adj. (also **moldy**) covered with or containing mold; musty or stale

发霉的:mouldy feeds 发霉饲料

mount /maʊnt/n. **1** a horse being ridden 乘用马;坐骑 **2** an instance or act of riding a horse 骑乘,骑马

v. **1** to get on a horse to ride it 上马,骑马 **2** (of male animals) to climb onto (a female) for copulation (公畜)爬跨 | **mounting** n.

mount money n.【rodeo】a sum of money paid to a rodeo rider in an exhibition (付给牛仔竞技骑手的劳务费)上马费,策骑费

Mountain Bashkir n. (also **Mountain Bashkir Curly**) one type of the Bashkir Curly originating in Bashkiria, Russia; stands 13.1 to 14 hands and may have a bay, chestnut or palomino coat; has a distinctive thick, curly winter coat and thick mane, tail, and forelock, a short neck, low withers, elongated hollow back, a wide and deep chest, short and strong legs, and small feet for its size; is docile, strong, quiet, and hardy; used for packing, light draft, and riding (源于俄国巴什基尔地区的矮马品种,体高13.1~14掌,毛色多为骝、栗或银鬃,被毛厚而卷曲,鬃、尾和额发浓密,颈短,鬐甲低,背长而凹陷,胸身宽广,四肢粗短,蹄小,性情温驯而安静,抗逆性强,主要用于驮运、轻挽和骑乘)山地巴什基尔卷毛马:Due to its exceptionally hard hoofs, Mountain Bashkir is generally left unshod. 山地巴什基尔卷毛马蹄质异常坚硬,通常并不需要挂掌。compare Steppe Bashkir

Mountain Bashkir Curly n. *see* Mountain Bashkir 山地巴什基尔卷毛马

mounted /ˈmaʊntɪd/adj. riding on horseback 马背上的:mounted games 马背比赛,马上项目

mounted archery n. a sport or practice in which a rider shoots arrows on a running horse (骑手策马射箭的训练)骑射

Mountie /ˈmaʊnti/n. (also **Mounty**) a member of the Royal Canadian Mounted Police 加拿大皇家骑警

mounting block n. a stone or wooden block upon which a rider stands to get on a horse (骑者站在上面上马的石块或木墩)上马台 syn horsing stone, jostling stone, horse block, pillion post

Mounty /ˈmaʊnti/n. *see* Mountie 加拿大皇家骑警

moustache /məˈstɑːʃ/n. *see* mustache 触须

mouth /maʊθ/n. **1** the opening of head through which a horse intakes fodder and water (马头用于采食和饮水的开口)口;口腔:bad mouth 口齿不正 **2** the mouthpiece of a bit 口衔

vt. **1** to fit bit in the mouth of a horse 上口衔 **2** to determine the age of a horse by examining its teeth (通过查看牙齿判断马匹年龄)看牙口

mouth speculum n. an instrument used to hold a horse's mouth open (用来打开马口腔的工具)开口板,开口器

mouthing /ˈmaʊθɪŋ/n. **1** the practice of determining the age of a horse by examining its teeth wear (通过观察马匹牙齿磨损程度估测其年龄)看牙口 **2** the fitting of bits to control a horse 上口衔

mouthing bit n. *see* breaking bit 调教用衔铁

mouthpiece /ˈmaʊθpiːs/n. part of the bit that rests on the tongue of the horse's mouth (衔铁靠在马舌头上的部分)嚼子,口衔

mouthy /ˈmaʊθi, -ði/adj. **1**【hunting】(of a hound) likely to bark (指犬)爱叫的 **2** (of a horse) likely to bite or nibble (指马)爱啃的,爱咬的

move /muːv/v. to go in a certain direction 运步,行进 PHR V **move off** to move forward from a standing position 起步前进 **move off the leg** (a horse) to respond and move away from the rider's leg pressure (指马)被踢后动身 **move up 1** to increase in speed 加速 **2**【racing】(of a racehorse) to run in a higher class than previously entered (指赛马进入高级别比赛)晋级

movement /ˈmuːvmənt/n. the act of moving or change in step or position 运步;移动 PHR

behind the movement (of a rider) with his body riding behind the imaginary vertical line drawn perpendicular to the back of the horse (骑手身体落在马背垂线后面)骑乘靠后,坐骨乘:To ride behind the movement may result in too much weight placed on the loins of the horse thus impeding the action of the hind legs and movement of the horse's back. 坐骨乘常造成马后躯负重过多,从而妨碍其后肢运步和背部活动。

movements off the ground n.【dressage】*see* airs above the ground 空中[高难]动作

movements on the ground n.【dressage】any of the classical movements performed by the horse on the ground (马在地面上完成的各种高等骑术动作)地上[高难]动作:Movements on the ground include simple and flying changes, turns on the forehand and haunches, side steps, shoulder in, shoulder out, haunches in, haunches out, passage, pirouette, piaffe, levade and pesade. 地面高难动作包括简单和空中换腿、定前/后肢旋转、斜横步、肩向内横步、肩向外横步、臀靠里横步、臀靠外横步、高蹄快步、原地旋转、原地快步、后肢静立和直立。

mow[1] /maʊ/vt. to cut down (grass) with a scythe or mechanical device (用镰刀或器械割草)刈,割

mow[2] /maʊ/n. a stack of hay or other feed stored in a barn 干草堆;料堆:hay-mow 干草堆

mowburnt /ˈməʊˌbɜːnt; *AmE* ˈmoʊˌbɜːr-/adj. (of hay) damaged or heating due to high moisture content (指草堆因水分过多而产热)发热的,积热的

mower /ˈməʊə(r); *AmE* ˈmoʊ-/n. **1** (also **mowing machine**) a machine used to cut grass 刈草机,割草机 **2** one who cuts grass with a scythe or mechanical device 割草工,刈草工

mowing machine n. *see* mower 刈草机

moxa /ˈmɒksə/n. a downy material prepared from the dried leaves of certain plants, such as mugwort, burned on the skin to alleviate pain and muscular problems (用艾属植物干叶制成的絮状物,在皮肤表面焚烧可以缓解疼痛和治疗肌病)艾草,艾叶

moxibustion /ˌmɒksɪˈbʌstʃən/n. a type of acupuncture practiced by burning a herb known as moxa over acupuncture points to alleviate pain and muscular problems (在穴位处针灸后烧艾,以缓解疼痛和治疗肌病)艾灼针灸

mph abbr. mile per hour 英里/小时

muck /mʌk/n. waste matter from farm animals; manure (牲畜的)粪便:muck heap 粪堆 ◇ muck sack 清粪袋

vt. to remove droppings and soiled bedding from a stall or stable 起粪;清粪 PHR V **muck out** to clean out a stable 打扫(畜栏),清扫(马厩)

mucking /ˈmʌkɪŋ/n. the act or process of removing droppings and soiled bedding from a stable 起粪;清粪 syn muck-out

muck-out /ˈmʌkaʊt/n. *see* mucking 清粪;清扫

mucosa /mjuːˈkəʊsə; *AmE* -ˈkoʊsə/n.【Latin】the mucous membrane 黏膜 | **mucosal** /-sl/adj. 黏膜的

mucous /ˈmjuːkəs/adj. of, containing, or secreting mucus 黏液的;黏膜的:mucous secretions 黏膜分泌物

mucous membrane n.【anatomy】a membrane that lines the body passages and cavities and is rich in mucous glands and produces mucus which moisturizes and protects the lining (肠道或腔体内衬的膜组织,其中富含黏液腺分泌的黏液,有湿润并保护底层组织的作用)黏膜 syn mucosa

mucus /ˈmjuːkəs/n. the viscous, slippery liquid secreted by cells and glands of the mucous membranes as a protective lubricant coating (由黏膜细胞或腺体分泌的液体,具有润滑保护内壁的作用)黏液

mud brush n. a grooming brush with very stiff bristles, used to remove the dried mud on the coat of a horse (刷拭用的硬毛刷,用来去除

马体上的泥巴)泥刷

mud fever n. a skin condition caused by the dermatophilus organism and characterized by small crusty patches of raised hair on the lower body of a horse (由嗜皮菌所致的皮肤病,症状表现为马体下肢表面着生结痂小斑块)泥瘟,泥土热

mud runner n. *see* mudder 善跑泥赛道的马

mud sticker n. *see* calk (蹄铁上的)尖刺,铁刺

mud tie n. (also **French tie**) the practice of folding up and tying the tail of the horse to prevent it from entangling in the harness or collecting mud (为了防止马尾与挽具缠结或沾带泥水而将其捆扎的做法)尾结,扎尾

mudder /ˈmʌdə(r)/n.【racing】a racehorse who runs well on a muddy or sloppy track 善跑泥赛道的马 [syn] mud runner,mudlark

muddy dun n.【color】a light brownish red or brownish yellow coat with chocolate brown points (体色红褐或褐黄而末梢为褐的毛色)褐兔褐:Muddy dun is the darkest of the red dun shades. 褐兔褐是红兔褐色调最深的毛色。

muddy track n.【racing】a wet racetrack with mud 泥赛道

mudlark /ˈmʌdlɑːk; *AmE* -lɑːrk/n. *see* mudder 善跑泥赛道的马

Mug's horse n. a horse who seems difficult to handle but in fact is easy to ride 看似暴躁而好骑的马

mule /mjuːl/n. the sterile offspring of a male donkey and a female horse,characterized by its short mane,long ears and fine-boned legs with small hoofs (母马与公驴杂交后所生的不孕后代,鬃毛短、耳朵长,四肢修长而蹄较小)骡 :mare mule 母骡 ◇ male mule 公骡 ◇ Although a sterile hybrid,a small percentage of female mules may produce foals. 尽管骡是不育的杂种,但有一小部分母骡则可能产驹。 [compare] hinny

mule ear n. a leather loop stitched to the heel of a riding boot to help the rider pull it on (缝在马靴后跟上方便提靴的皮圈) [syn] rat tail

mule foot n. (*pl.* **mule feet**) a long,narrow hoof with straight quarters and high heels (蹄长而窄且蹄踵较高)直狭蹄;骡蹄

mule footed adj. (of a horse) having a long,narrow hoof with high heels (指马蹄)狭小的;骡蹄的

mule jack n. a male donkey used to breed mares to produce mules (给母马配种的公驴)产骡公驴

mule shoeing contest n. a timed contest held in the United States in which farriers trim and shoe the feet of mules,judged principally on speed and skills (在美国举行的由蹄铁匠为骡钉掌的计时比赛,主要根据钉蹄所用时间和技术进行评分)钉骡掌大赛:On the mule shoeing contest,it often takes a proficient farrier less than five minutes to trim and shoe all four feet. 在钉骡掌大赛上,熟练蹄工给骡装钉四蹄所用时间通常不超过5分钟。

mullen mouth n. *see* mullen mouth bit 半月形口衔

mullen mouth bit n. (also **mullen mouth**) a snaffle bit with a straight bar mouthpiece that is slightly bent or curved to accommodate the shape of the tongue (衔杆稍弯以容纳马舌的斯浪弗衔铁)半月形口衔 [syn] half moon bit,half moon mouth,mulling mouth

mulling mouth n. *see* mullen mouth bit 半月形口衔

multi-jointed snaffle n. a snaffle bit in which the mouthpiece has more than one joint,which is more severe than a regular snaffle (口衔由几个链接构成的衔铁,比普通勒要厉害)多节勒

Muniqi /ˈmjunɪki/n. one of the three primary sub-breeds of the Assil (阿西利阿拉伯马的三个亚种之一)慕尼奇阿拉伯马 —*see also* **Kuhailan**,**Siglavy**

Murakoz /məˈrɑːkəʊz/n. (also **Murakozi**,**Mu-**

rakozer) a coldblood developed in the early 20th century by crossing native mares with Percheron, Belgian Ardennes, and Noriker stallions in Murakoz, Hungary from which the name derived; stands about 16 hands and usually has a chestnut coat with a flaxen mane and tail, although bay, brown, and black do occur; has a strong frame, small withers, hollowed back, round hindquarters, and some feathering; is swift-moving and possessed of a good temperament; used for draft and farm work (20 世纪初匈牙利穆拉克兹镇育成的冷血马品种并由此得名,主要采用本土母马与佩尔什马、比利时阿尔登马以及诺里克马公马交配繁育而成。体高约 16 掌,毛色多为栗且鬃尾金黄,但骝、褐和黑兼有,体格结实,鬐甲小,凹背,后躯发达,距毛稀疏,运步较快,性情温驯,主要用于挽车或农田作业)穆拉克兹马: By the 1920s, nearly one fifth of all horses in Hungary were Murakoz. 时至 20 世纪 20 年代,匈牙利近 1/5 的马匹都是穆拉克兹马。

Murgese /ˈmɜːɡiːz; *AmE* ˈmɜːrɡ-/n. (also **Murghese**) an Italian warmblood descended from Arab and Barb crosses during the 19th century; stands 14 to 16 hands and usually has a chestnut coat, although brown, black, and gray also occur; has a prominently jawed head, broad neck, full mane, pronounced withers, straight back, and large leg joints; is docile but lively and used for riding, farm work, and light draft (意大利温血马品种,由阿拉伯马和柏布马在 19 世纪杂交繁育而来。体高 14 ~ 16 掌,毛色以栗为主,但褐、黑和青亦有之,头颌骨突出,颈粗,鬃毛浓密,鬐甲突出,背部平直,四肢关节粗大,性情温驯而活泼,主要用于骑乘、农田作业以及轻挽)穆尔格斯马

Murghese /ˈmɜːɡiːz; *AmE* ˈmɜːrɡ-/n. *see* Murgese 穆尔格斯马

muscle /ˈmʌsl/n. a bundle of elongated fibrous tissue in the body that contracts on stimuli to produce bodily movement (体内的一种束状纤维组织,受到刺激后可通过收缩产生运动)肌肉, 肌肉组织: muscle bundle 肌束 ◇ muscle contraction 肌肉收缩 ◇ muscle fiber 肌纤维 ◇ red muscle fiber 红肌纤维 ◇ white muscle fiber 白肌纤维 ◇ superficial muscle 浅层肌 ◇ deep muscle 深层肌 ◇ skeletal muscle 骨骼肌 ◇ cardiac muscle 心肌 ◇ smooth muscle 平滑肌 ◇ involuntary muscle 不随意肌 ◇ skin muscle 皮肌 ◇ striated muscle 横纹肌 ◇ visceral muscle 内脏肌 PHR **on the muscle** (of a horse) fit in condition (指马)体况好的,上膘的

muscle atrophy n. a reversible diminution of muscle due to a decrease in the size of individual muscle fibers; occurs when a muscle is not subjected to sustained periods of tension (由于肌纤维变细而使肌肉体积减小的情况,在肌肉不受持续张力时出现)肌萎缩

muscle fibre n. a multinucleated single muscle cell formed by fusion of myoblasts (成肌细胞融合形成的多核肌细胞)肌纤维: The blood supply for muscle fibres is supplied by an extensive capillary network that runs between the fibres. 肌纤维的血液供应由分布其中的丰富的毛细血管网完成。

muscular /ˈmʌskjələ(r)/adj. **1** of, relating to, or consisting of muscle 肌肉的: muscular contraction 肌肉收缩 ◇ muscular system of horse 马体肌肉系统 **2** (of a horse) having well-developed muscles (指马)肌肉发达的,强壮的: a muscular build 体格健壮

muscularity /ˌmʌskjʊˈlærəti/n. the quality or state of being muscular 肌肉发达;健壮程度

musculature /ˈmʌskjələtʃə(r)/n. the system or arrangement of muscles in a body (机体的肌肉分布情况)肌肉组织;肌肉系统

musculoskeletal /ˌmʌskjʊləʊˈskelətl/adj. of, relating to, or involving both the muscles and the skeleton 肌肉骨骼的

musculoskeletal system n. the body structure of the horse including the bones, muscles, ligaments, tendons, and joints（包括骨骼、肌肉、韧带、筋腱和关节在内的马体结构）肌肉骨骼系统

mustache /mə'stɑːʃ/n. (also **moustache**) the long bristles growing near the mouth of a horse（马口角周围生有的长硬毛）髭，触须

Mustang /'mʌstæŋ/n. a wild horse roaming in the West of America and Mexico, descended from European-bred horses introduced by the Spanish conquistadors in the 16th century; served as the foundation stock for a large number of the American breeds; stands 13.2 to 15 hands and may have a coat of any color; has a heavy head, a light frame, good bone, and notably tough feet; is sturdy, tough, and highly adaptable（驰骋在美国西部和墨西哥的野马，由16世纪西班牙殖民者带到那里的欧洲马种繁衍而来，是美国多个马匹品种的奠基种。体高13.2～15掌，各种毛色兼有，头重，体型轻，管围粗，蹄质坚硬，体格强健，抗逆性好，适应性强）北美野马，穆斯唐马：At the beginning of the 20th century, the numbers of Mustang exceeded two million, herds are now reduced to less than 300,000. 20世纪初，北美野马的数目在200万匹以上，而现在其种群数量不到30万匹。 [syn] fuzztail, wild horse

muster /'mʌstə(r)/n.【AusE】a cattle round-up（牛群）赶拢，聚拢

vt. to round up a group of cattle, sheep, ect. 赶拢

mutate /mjuː'teɪt/v. to undergo or cause mutation 变异；突变

mutation /mjuː'teɪʃn/n.【genetics】a structural change of DNA sequence within a gene or chromosome of an organism that often results in the creation of a new character or trait（生物体基因或染色体内DNA结构发生的变化，可导致新性状的出现）变异，突变

mute /mjuːt/adj.【hunting】(of the hounds) without giving tongue when hunting the line of a fox（猎犬追踪狐狸时不出声）不出声的，不吠叫的

n.【hunting, dated】a pack of hounds used to hunt a fox, coyote, or stag 猎犬队；猎犬群

mutton withers n. a wide, flat withers of horse, often seen in Quarter Horses and pony breeds（宽而平的鬐甲，常见于夸特马或矮马）鬐甲低平，鬐甲不明显 [compare] prominent withers

mutton-fisted /ˌmʌtn'fɪstɪd/adj. *see* ham-fisted 手法重的，手法拙的

mutton-withered /ˌmʌtn'wɪðərd/adj. (of a horse) having a low, flat withers like that of a sheep（指马像绵羊那样鬐甲又低又平）鬐甲低平的

mutual /'mjuːtʃuəl/adj. held in common or benefited by each other 共有的；相互的

mutualism /'mjuːtʃʊəlɪzəm/n. an association between organisms of two different species that is beneficial to each other（两种不同种类的有机体之间互利协作的情况）互生 [compare] parasitism, symbiosis

mutuel /'mjuːtjʊəl/n. *see* pari-mutuel 同注分彩；赌金计算器

mutuel pool n. the pari-mutuel pool 赌金池；投注总额

mutuel ticket n. a ticket that shows the wager a bettor has made（标示投注金额的票据）赌马彩票

mutuel window n.【racing】a booth located in racecourse where the mutuel tickets are sold（赛马场上出售赌票的窗口）售票窗口

muzzle /'mʌzl/n. **1** the lower part of a horse's head, including the nostrils, lips and chin（马头包括鼻孔、唇和下颌的部分）口鼻部 IDM **muzzle in the pint pot** (of a horse) having a small and refined muzzle（指马）口鼻小巧 **2** a leather or mesh covering fitted over a horse's muzzle to prevent him from biting, eating blan-

M

ket, etc.(戴在马口鼻上的皮套或网罩,以防马咬人或啃食披毯等)口套 : leather muzzle 皮制口套

MVM abbr. *Merck Veterinary Manual*《默克兽医手册》

my(o)- pref. muscle 肌[肉]

myoclonus /maɪˈɒklənəs; *AmE* -ɑːˈklə-/ n. a quick, spasmodic involuntary muscle contraction occurring in various brain disorders(由多种脑神经紊乱引起的肌肉迅速且不自主的间歇性收缩)肌痉挛,肌阵挛 | **myoclonic** /-nɪk/ adj.

myofibril /ˌmaɪəˈfaɪbrəl/ n.【physiology】the contractile filament of a muscle cell(肌细胞中的收缩性纤丝)肌原纤维

myoglobin /ˌmaɪə(u)ˈgləʊbɪnləʊ-/ n.【physiology】(abbr. **Mb**) a heme-containing protein that carries and stores oxygen in muscle fibers(肌纤维中含血红素的蛋白,负责运送和贮存氧气)肌红蛋白: Myoglobin is structurally similar to a subunit of hemoglobin, but with a higher affinity for oxygen. 肌红蛋白与血红蛋白的亚基结构类似,但与氧的亲和力更高。

myoglobinuria /ˌmaɪəˈgləʊbənjʊəriə; *AmE* -ˈgloʊbə-/ n. the presence of myoglobin in the urine 肌红蛋白尿: equine myoglobinuria 马肌红蛋白尿 ◇ paralytic myoglobinuria 麻痹性肌红蛋白尿

myometrium /ˌmaɪəˈmetrɪəm/ n.【anatomy】the middle muscular coat of the uterine wall(子宫壁中层的肌组织)子宫肌层

myopathy /maɪˈɒpəθi; *AmE* maɪˈɑːp-/ n. a disease of muscle tissue caused by damage or infection(由创伤或感染所致的肌肉组织病变)肌病: exertional myopathy 运动性肌病 | **myopathetic** /ˌmaɪəʊpəˈθetɪk/ adj.

myosin /ˈmaɪəsɪn/ n.【physiology】a fibrous protein that forms together with actin the contractile filaments of muscle cells(与肌动蛋白构成肌细胞收缩带的纤维蛋白)肌球蛋白

myositis /ˌmaɪəʊˈsaɪtɪs; *AmE* ˌmaɪəˈs-/ n. inflammation of muscle tissue characterized by pain and sometimes spasm in the affected area(肌肉组织发炎的症状,表现为患处疼痛甚至痉挛)肌炎

myotonia /ˌmaɪəˈtəʊnɪə/ n. a spasm or temporary rigidity of voluntary muscles caused by various muscular disorders(由多种肌紊乱引发的肌肉强直性痉挛或暂时性僵硬)肌强直 | **myotonic** /-ˈtɒnɪk/ adj.

N

Naadam /ˈnɑːdəm/n. an equestrian festival held annually in July or August in Mongolia during which events like horse racing, archery, and wrestling are conducted (蒙古国每年 7 ~ 8 月份举行的全国性马术活动,主要进行赛马、套马以及骑射等项目的比赛)那达慕

Naadam of the Country n. the national equestrian festival held annually in the Mongolian capital Ulan Bator during the National Holiday from July 11th to 13th in the National Sports Stadium (蒙古国首都乌兰巴托每年于 7 月 11 ~ 13 日在国家体育场举行的全国性马术活动)全国那达慕

nag /næg/n. **1**【dated】a horse suitable for riding 乘用马 **2** an old worthless horse 老马,驽马 [syn] yaboo **3**【slang】a racehorse 赛马

vt. to train a horse for riding or hunting 调教,训练

nagging /ˈnægɪŋ/n. the act or practice of training and schooling a horse for riding (马匹的)调教,训练

nagsman /ˈnægzmən/n.【BrE, dated】a man who made his living by breaking and schooling horses (过去以调教马匹为生的人)驯马师

nail /neɪl/n.【farriery】(also **horseshoe nail**) a small thin pointed piece of metal with a flat head, used for shoeing a horse (装钉蹄铁用的尖头细钉)蹄钉

vt. to fit horseshoe to the hoof with nails 装钉蹄铁

nail bind vt.【farriery】to nail too close to the sensitive part of the foot when shoeing a horse (装蹄铁时蹄钉镶入马蹄敏感部位)镶钉过深

nail cutter n.【farriery】a long-handled, pritchel-type tool used to cut off clinched nail tips (一种长柄鸭嘴钳,用来夹断敲弯的蹄钉尖)夹钉钳 [syn] nail nipper

nail nipper n.【farriery】*see* nail cutter 夹钉钳

nail prick vt.【farriery】*see* quick 钉伤(马蹄)

nail quick vt.【farriery】*see* quick 钉伤(马蹄)

nap /næp/n. **1** a horse who disobeys the aids of its rider by bucking, rearing, etc. (通过弓背跳或起仰等方式拒不服从骑手辅助的马)倔强的马 **2**【racing】a reliable betting tip 可靠的赌马消息

nape /neɪp/n. the poll of a horse (马的)项顶

nappy /ˈnæpi/adj. (of a horse) refusing to obey the aids of the rider; intractable (指马)拒听命令的;倔强的

Naqu n.【Chinese】(also **Naqu horse**) a horse breed indigenous to western China; stands about 14 hands and usually has a dun coat with black mane; has a small head, thin neck, and strong bones; used for riding, packing, and farm work (中国西部地区的本土马品种,体高约 14 掌,毛色多为兔褐,鬃毛黑色,头小,颈细,骨骼粗壮,用于骑乘、驮运和农田作业)那曲马

NARHA abbr. North American Riding for the Handicapped Association 北美残疾人骑乘协会

narrow /ˈnærəʊ; *AmE* -roʊ/adj. **1** being close in position 狭窄的: base narrow 狭踏 **2** lacking muscle 狭瘦的 [PHR] **narrow behind** (of a horse) lacking muscle in the hindquarters (指马)后躯狭窄,后躯狭瘦

narrow-chested /ˌnærəʊˈtʃestid/adj. (of a horse) having a narrow chest (指马)狭胸的 [opp] wide-chested

nasal /ˈneɪzl/adj. of, in, or relating to the nose 鼻的: nasal peak 鼻端 ◇ nasal cavity 鼻腔 ◇ nasal passage 鼻腔

nasal gleet n. (also **gleet**) a colored discharge from the nostrils resulting from inflammation of the nasal cavity (由于鼻腔发炎而从鼻孔流出的有色黏液)鼻腔黏液,鼻腔脓液

nasal hemorrhage n. bleeding from the horse's nose; epistaxis 马鼻出血

NASFHA abbr. North American Selle Français Horse Association 北美塞拉 · 法兰西马协会

nasogastric /ˌneɪzəʊˈgæstrɪk; *AmE* ˌneɪzoʊ-/ adj. of or relating to the canal connecting the nasal cavity and stomach 鼻胃的

nasogastric feeding n. the practice of feeding an animal by delivering liquid food through a flexible tube inserted from the nose into the stomach (将流质食物通过软管由鼻孔输至胃的饲喂方法)鼻饲

nasogastric tube n. a long flexible tube inserted from the nose to the stomach of the horse for medical use (用于医疗目的从鼻孔插到胃里的软管)鼻胃管,鼻饲管

natal /ˈneɪtl/ adj. of or pertaining to birth 出生的,新生的

National Cutting Horse Association n. (abbr. **NCHA**) an association founded in 1946 in Fort Worth, Texas, USA to organize and regulate cutting horse contests (1946 年在美国得克萨斯州沃思堡成立的组织机构,旨在组织监管美国的截牛比赛项目)美国截牛协会

National Equestrian Federation n. the national governing body of equestrian events and activities in any country affiliated with the FEI (监管国内马术比赛项目和相关活动的机构,隶属于国际马术联合会)国家马术联合会

National Finals Rodeo n. (abbr. **NFR**) an annual rodeo competition held in Texas, USA (美国得克萨斯州每年举办的牛仔竞技比赛)美国牛仔竞技总决赛

National Horse Association of Great Britain n. an organization founded in Great Britain in 1922 to promote the welfare of horses and the interests of horse owners and breeders, was later replaced by the British Horse Society in 1947 (1922 年在英国成立的组织机构,旨在提高马匹和矮马福利以及维护马主和育马者的权益,于 1947 年由英国马会取代)英国国家马匹协会

National Horse Show n. a premier horse show held annually in Madison Square Garden, New York, USA (美国每年在纽约麦迪逊方形花园举行的马匹展览)美国马匹展览会,美国马匹选美赛:The National Horse Show was first organized in the United States in 1883. 美国马匹展览会于 1883 年首次举行。

National Master Farriers and Blacksmiths Association n. an organization founded in UK in 1905 to protect the welfare of farriers and blacksmiths (1905 年在英国成立的组织机构,旨在保护蹄师和铁匠的权益)英国大师级蹄铁匠协会

National Museum of Horse Culture n. a museum founded in Beijing in 2003 to dislay the rich horse culture of China throughout the history (于 2003 年在北京成立的博物馆,旨在展现中国从古至今的丰富马文化)中国马文化博物馆

National Museum of Racing and Hall of Fame n. a museum founded in 1950 in Saratoga Springs, New York, to honor the achievements of American Thoroughbred racehorses, jockeys, and trainers (1950 年在纽约州萨拉托加温泉市成立的博物馆,旨在表扬美国历史上优秀的赛马、骑师和练马师)美国赛马博物馆暨名人堂

National Pony Society n. (abbr. **NPS**) an organization founded in 1893 in UK to promote the breeding and registration of British pony breeds (1893 年在英国成立的组织机构,旨在促进矮马的育种及其登记工作)英国矮马协会

National Reining Horse Association n. (abbr. **NRHA**) an organization founded in Ohio, USA in 1966 to organize, promote, and establish rules for the sport of reining (1966年在美国俄亥俄州成立的组织机构，旨在组织、促进并建立驭马比赛规则)美国驭马协会，美国雷宁协会

National Research Council n. (abbr. **NRC**) an American institution established to examine current practices in the nutrition and feeding of horses and publishes recommendations on horse nutrition (美国的科研机构，考察当前马匹营养情况并发布营养推荐参考值)美国科学研究委员会

National Show Horse n. (abbr. **NSH**) a new horse breed developed in the United States by crossing Arabian and American Saddlebred in the 1970s; stands 14.3 to 16.2 hands and may have any of the coat colors; has a small refined head, small ears, high-set, swan-like neck without pronounced crest, and high tail carriage; often used for riding but versatile in many disciplines (美国在20世纪70年代采用阿拉伯马和美国骑乘马杂交培育而来的新的马品种。体高14.3~16.2掌，各种毛色兼有，头小，面貌清秀，耳小，颈础高但颈峰低，尾巴高翘，主要用于骑乘，且在其他马术项目也表现不俗)美国马术马，美国表演马

National Trotting Association of Great Britain n. an organization founded in Great Britain in 1952 to promote trotting races (1952年在英国成立的组织机构，旨在促进快步赛)英国快步赛马协会

National Veterinary Medical Association of Great Britain and Ireland n. an organization founded in UK in 1881 to promote and supervise the practice of veterinary medicine in Great Britain and Ireland (1881年在英国成立的组织机构，旨在促进和监管英国和爱尔兰的兽医行业)英国及爱尔兰兽医协会

Nations Cup n. *see* Prix des Nations 民族杯国际障碍赛

native pony n. any one of the nine pony breeds indigenous to Great Britain including the Welsh Pony, Shetland Pony, Dartmoor, New Forest, Highland Pony, Exmoor, Fell, Dales, and Connemara (英国本土的九个矮马品种的统称，包括威尔士矮马、设特兰矮马、达特姆尔矮马、新福里斯特矮马、高地矮马、埃克斯穆尔矮马、费尔矮马、戴尔斯矮马和康尼马拉矮马)英国本土矮马

NATRC abbr. North American Trail Ride Conference 北美越野骑乘协会

natural aids n. any means by which the rider conveys his instruction to the horse by use of the legs, hands, body-weight and voice (骑手在传达命令时用腿、手、体重以及声音与马沟通的方式)自然辅助 [compare] artificial aids

natural allures n. the natural gaits of a horse 自然步法

natural cover n. the breeding of the mare by a stallion under natural conditions (公马在自然条件下为母马配种)自然交配，自然配种

natural fence n. any natural obstacle found in a hurdling course (越野障碍赛中的)天然障碍: Piled logs, stone walls, gates, and hedges could be all used as natural fences. 圆木堆、石墙、门栏和篱笆都可以作为天然障碍使用。

natural gait n. any of the gaits that a horse has when he was born, such as the walk, trot, canter and gallop (马匹出生后便具有的步法，如慢步、快步、跑步和袭步)自然步法 [compare] artificial gait

naturopathy /ˌneɪtʃəˈrɒpəθi; *AmE* -ˈrɑːp-/ n. a system of therapy that uses natural remedies and physical means to treat certain illness (采用自然手段或物理方法来治疗特定疾病)自然疗法 | **naturopathic** /-θɪk/ adj.

nave /neɪv/ n.【driving】the hub of a wheel 轮毂

navel /ˈneɪvl/ n. the hollow mark in the middle of a horse's abdomen where the umbilical cord was cut at birth (马腹部中央的凹痕，为分娩后脐带割断处遗留)脐 [syn] umbilicus, belly button

navel ill n. (also **navel infection**) *see* joint ill

N

(幼驹所患的)关节病,脐病

navicular /nəˈvɪkjələ/n.【anatomy】(also **navicular bone**) a small, flat, boat-shaped bone located behind the coffin joint in the hoof (位于蹄骨后的一块舟状扁骨)舟骨

navicular bone n. *see* navicular 舟骨

navicular bursitis n. *see* navicular disease 舟骨[滑膜囊]炎

navicular disease n. (also **navicular syndrome /bursitis**) a chronic degenerative condition of the horse's navicular bone symptomized by stumbling, shortened stride, and a characteristic pointing of the toe of the forefoot when at rest (马舟骨所患的慢性退行性疾病,临床表现为跛行、步幅缩小,站立时仅前肢蹄尖点地)舟骨炎,舟骨病: Navicular disease may be managed in some horses through corrective shoeing and in some cases nerving of the affected limb. 舟骨炎在有些马匹中可通过矫正性装蹄或割断患病蹄的神经而得以控制。 syn podotrochilitis

navicular syndrome n. *see* navicular disease 舟骨综合征

NCHA abbr. National Cutting Horse Association 美国截牛协会

Neapolitan /ˌnɪəˈpɒlɪtən/n. (also **Neapolitan horse**) an ancient warmblood breed originating in the plains around Naples, Italy; was similar to the Andalusian in appearance and deemed extinct by the 1950s (源于意大利那不勒斯平原地区的温血马品种,体貌与安达卢西亚马相似,至20世纪中叶便已绝种)那不勒斯马

Neapolitan School n. a classical riding school founded in Naples, Italy in 1532 by Federico Grisone who, after Xenophon, is considered the first of the classical masters (1532年意大利人弗雷德里克·格里森在那不勒斯创建的骑术学校,他被认为是继色诺芬之后的又一个古典骑术大师)那不勒斯骑术学校

near /nɪə; *AmE* nɪr/adj. **1** close in degree or space 近的 **2** being on the left side of a horse or a vehicle (马/车的)近侧的;左侧的

near horse n.【driving】any of the horses on the left of a team (联驾中处于左侧的马)左马,近侧马: near wheeler horse 左侧辕马

near center n.【driving】the horse on the left of a pair of centers 内侧中马,左侧中马 —*see also* **off center**

near leader n.【driving】the leader horse on the left on a team (联驾中位于左侧的头马)左侧头马

near pommel n. (also **near head**) *see* top pommel 上鞍桥

near side n. the left side of the horse when viewed form the rear of a horse (后观时马匹的左边)内侧,左侧,里手 compare far side

nearside /ˈnɪəsaɪd; *AmE* ˈnɪrs-/adj. of or situated at the near side of a horse 内侧的,左侧的,里手的: nearside horse 左马

nearside backhand shot n.【polo】a way of striking the ball across the neck of the horse in the reverse direction to the travel (挥杆至马颈左侧朝与运行相反的方向击球)左侧反手击球

nearside forehand shot n.【polo】a way of striking the ball across the neck of the horse in the same direction as the travel (挥杆至马颈左侧朝运行相同的方向击球)左侧正手击球

nearside horse n.【driving】the left-hand horse in a pair used to pull a vehicle (一对拉车挽马中左侧的马匹)内侧马匹,里手马匹

neck /nek/n. **1** the body part joining the head to the shoulders (身体连接头与肩的部位)颈: base of the neck 颈础 ◇ Conformation types of neck include ewe neck, bull neck, and swan neck. 马的颈型有羊颈、牛颈和鹅颈。**2**【racing】(abbr. **nk**) a winning margin approximately equal to the length of the horse's neck including the head (赛马中胜出的距离,长度与马颈相当)颈长: to win a race by a neck 以颈长优势胜出 compare dead heat, short-neck, short-head, head, length

neck clip n. *see* belly clip 颈腹修剪

neck collar n. *see* harness collar (挽用)颈圈,颈圈

neck rein vt. (also **indirect rein**, or **opposite rein**) to guide or direct the horse by applying rein contact against the neck on the side opposite to which a turn is required, e. g. to turn the horse to the right by applying pressure to the neck with the left rein (通过缰绳向马的颈部对侧施压来操控马匹,如右转弯时以左方缰靠向马颈施力)颈部压缰 : to neck rein a horse 颈部压缰控马 | **neck reining** n.

neck ring n. *see* lariat neck ring 颈索,颈环革

neck strap n. a narrow leather strap fitting around the base of the horse's neck to keep the martingale in place (系在马颈上的窄皮带,用来固定低头革)项革,颈带

neck stretcher n. an elastic cord attached to the girth which passes the left bit ring, over the horse's poll, through the right bit ring, and again connects back to the girth to encourage the horse to lower his head and stretch his neck (系于肚带上穿过左衔铁环,然后绕过项顶后穿过右衔环再系于肚带的绳索,用来使马低头展颈)展颈革,展颈索

neckshot /ˈnekʃɒt/n. **1**【hunting】a shot in the neck of an animal 朝颈开枪 **2**【polo】a shot in which the player strikes the ball by reaching under the neck of the horse from either side (球员俯身在马颈下任意两侧击球)颈侧击球,颈下击球

necrosis /neˈkrəʊsɪs; *AmE* -ˈkroʊ-/n. (*pl.* **necroses** /-siːz/) the death of tissues caused by injury or disease (因创伤或疾病引起的组织死亡)坏死,坏疽 | **necrotic**/neˈkrɒtɪk/adj.

negative reinforcement n. a negative reward used to modify behavior, such as punishment (为矫正马匹行为而给予的惩罚)负面强化

negotiate /nɪˈgəʊʃieɪt; *AmE* -ˈgoʊ-/vt.【jumping】to succeed in clearing obstacles; complete 越过,跨越(障碍): In the showjumping, the riding horse is required to negotiate a succession of obstacles, style and grace is irrelevant. 在障碍赛中,乘骑马匹要按规定跨越多道障碍,对其跨栏方式和姿态则不加要求。

neigh /neɪ/n. the long, high-pitched sound made by a horse (马)嘶鸣声
vi. to utter the characteristic sound of a horse; whinny 嘶鸣

nematode /ˈnemətəʊd; *AmE* -toʊd/n. any of several worms having unsegmented, cylindrical bodies (线虫纲的多种寄生虫,虫体呈柱状而不分节)线虫

neonatal /ˌniːəʊˈneɪtl; *AmE* ˌniːoʊ-/adj. of or relating to newborn infants 新生的,初生的: neonatal care 初生护理

neonatal isoerythrolysis n. (abbr. **NI**) an acute hemolytic disease of newborn foals caused by immunologically mediated red blood cell destruction due to the incompatibility of maternal-fetal blood group (一种见于新生幼驹的急性溶血性疾病,由于母仔血型不匹配而在免疫介导下引发红细胞裂解所致)新生幼驹同种红细胞溶解症

neonatal maladjustment syndrome n. (abbr. **NMS**) a condition of newborn foals characterized by gross behavioral disturbances including loss of affinity for the mare and sucking reflex, apparent blindness, and aimless wandering (新生幼驹所表现的综合性行为失调,其症状包括缺乏母马依赖和吮乳反射、失明和四处摇晃)新生幼驹适应不良综合征: The mortality of neonatal maladjustment syndrome is approximately 50 percent. 新生幼驹适应不良综合征的死亡率约为50%。

neonate /ˈniːəʊneɪt; *AmE* ˈniːoʊ-/n. a newborn foal 新生幼驹

nerve /nɜːv; *AmE* nɜːrv/n. **1** any of the long fibers in the body that transmits impulses between the brain and body parts (在大脑和身体部位之间传输冲动的细长纤维)神经: the optic nerve 视神经 ◇ nerve endings 神经末梢 **2** (usu. **nerves**) an agitated state or a feeling of worry or anxiety 紧张;恐慌 **3** the courage to do sth dangerous or difficult 勇气,气魄

vt. **1** to cut a nerve or nerve group surgically to eliminate pain (通过手术割断神经以消除疼痛)神经切除 [syn] denerve, unnerve **2**【hunting】(of a hunter) to display courage in the hunting field (指狩猎者)表现神勇

nerve block n. the practice of blocking the passage of impulses along a nerve, especially by administration of local anesthetics (通过使用局部麻醉剂阻断神经脉冲的传递)神经阻断

vt. to block the sensation of feeling by cutting or anesthetizing a nerve 阻断(神经)

nerving /ˈnɜːvɪŋ; *AmE* ˈnɜːrv-/n. the practice of cutting a nerve or nerve group to eliminate pain; neurectomy 切除神经:Nerving is not always an effective way to treat pain and may result in the development of neuromas. 切除神经并不总能消除疼痛,相反有时还会导致神经瘤的发生。

nervous /ˈnɜːvəs; *AmE* ˈnɜːrvəs/adj. (of a horse) easily agitated or distressed; high-strung (指马)神经紧张的;易受惊的

net /net/adj. (of an amount) remaining after a deduction has been made 纯的;净剩的:net weight 净重

net energy value n.【nutrition】the energy fraction of the feed remaining after subtracting waste energy from the total energy (饲料总能中扣除代谢能损失之外余留的能量份额)净能[值]

nettle /ˈnetl/n. a plant having toothed leaves that cause skin irritation on contact (一种叶片呈锯齿状的植物,皮肤与其接触后会引起刺痒)荨麻

vt. to irritate or simulate 刺激

nettle rash n. skin irritation caused by contact with nettle; hives (皮肤接触荨麻后引起的刺痒症状)荨麻疹

neurectomy /njʊˈrektəmi; *AmE* nʊˈ-/n. surgical removal or cutting of a nerve 神经切除,神经切割

neur(o)- pref. of nerves or nervous system 神经的

neuropathy /ˌnjʊəˈrɒpəθi; *AmE* ˌnʊˈrɑːp-/n. any pathology or abnormality of the nervous system (神经系统的病变或异常症状)神经疾病

neurotransmitter /ˌnjʊrəʊtrænzˈmɪtə(r); *AmE* ˌnʊroʊ-/n.【physiology】a chemical substance, such as acetylcholine or dopamine, that transmits electrical impulses across nerve cells (负责传递神经细胞间电脉冲的化学物质,如乙酰胆碱或多巴胺)神经递质

neuter /ˈnjuːtə(r); *AmE* ˈnuːtə(r)/n. a castrated animal 阉畜;去势动物

vt. to castrate or spay 阉割;摘除卵巢

New Forest n. *see* New Forest Pony 新福里斯特矮马

New Forest Hound n. a British hound breed formerly used in hunting (英国过去用于狩猎的犬种)新福里斯特猎犬

New Forest Hunt Club n. a club founded in 1789 in England to organize and oversee hunting in the region of New Forest (1789 年在英国成立的俱乐部,旨在监管新福里斯特地区的狩猎运动)新福里斯特狩猎俱乐部

New Forest Pony n. (also **New Forest**) an ancient pony breed originating in the New Forest region of southern England for which it was named; stands 12 to 14 hands and may have any coat color except pinto; has a long and sloping shoulder, long and low action; is intelligent, cunning, and easily trained; used for riding, harness driving, and show competitions (源于英国南部新福里斯特地区的古老矮马品种,故而得名。体高 12 ~ 14 掌,除花毛外各种毛色兼有,肩斜,步幅长而举蹄低,反应敏快,易训练,用于骑乘、驾车和展览比赛)新福里斯特矮马,新森林矮马

New Forest Pony Breeding and Cattle Society n. an organization founded in Great Britain in

1938 to promote the breeding and registry of the New Forest Pony（1938 年在英国成立的组织机构，旨在促进新福里斯特矮马的培育和品种登记工作）新福里斯特矮马育种协会

New Kirgiz n. *see* Novokirghiz 新吉尔吉斯马

New York Handicap Triple n.【AmE，racing】a series of three handicap races-the Metropolitan Handicap，the Brooklyn Invitational Stakes，and the Suburban Handicap-held annually at Belmont Park，New York（美国纽约贝尔蒙特公园每年举行的三项让磅赛，包括大都会让磅赛、布鲁克林邀请赛和市郊让磅赛）纽约让磅三项赛

New York Handicap Triple Crown n.【AmE，racing】（also **Handicap Triple Crown**）an award given to a racehorse who wins all three handicap races held in New York in one season（对在单个赛季赢得纽约三项让磅赛的马匹授予的称号）纽约让磅赛三冠王

New Zealand Cobb & Co coach n.（also **Cobb & Co coach**）a horse-drawn carriage used for long-distance passenger transport and mail delivery in New Zealand during the second half of the 19th century（于 19 世纪下半叶在新西兰使用的四轮马车，用于长途客运和邮递业务）新西兰考伯四轮马车：The New Zealand Cobb & Co coach could seat up to 18 passengers，together with mail and luggage. 除邮件和行李之外，新西兰考伯四轮马车最多可坐 18 名乘客。

New Zealand Horse Society n.（abbr. **NZHS**）an organization founded in 1950 in New Zealand to organize and promote equestrian competitions under international rules（于 1950 年在新西兰成立的组织机构，旨在组织监管国内的马术运动项目）新西兰马会

New Zealand rug n. an extremely durable，waterproof canvas rug covered over a horse turned outdoors to protect it from bad weather（一种帆布制耐用防水马衣，马匹在户外时可以抵御恶劣天气）新西兰马衣

newborn /ˈnjuːbɔːn；*AmE* ˈnuːbɔːrn/ adj. very recently born 新生的：a newborn foal 新生幼驹

Newhouse，Charles n.（1805 – 1877）a British painter known for his coaching and road scenes（英国著名画家，以描绘马车和路景见长）查尔斯·纽豪斯

Newmarket /ˈnjuːmɑːkɪt；*AmE* -mɑːrkɪt/ n. the center of horse racing in England renowned for the famous studs and racehorses bred there since the early 17th century（英国赛马业的中心，早在 17 世纪初就以当地多个著名种马场和所培育的赛马而闻名于世）纽马基特：As the headquarters of the Jockey Club，Newmarket is also the location of the National Stud and and two primary racecourses，the Round Course and the Summer Course where two of the country's five Classic Races-the 2，000 and 1，000 Guineas，and numerous other Group races are run. 纽马基特不仅是英国赛马会总部驻地，同时也是国家种马场及圆形赛道和夏日赛道两个主要赛马场的所在地，而五项经典大赛中的两千几尼赛和一千几尼赛以及诸多等级赛也在此举行。

Newmarket boots n. a tall riding leather boot with the leg made of waterproof canvas（一种高帮马靴，靴筒由防水帆布制成）纽马基特马靴

Newmarket breastplate n. *see* Aintree breast girth 纽马基特胸带，安翠胸带

Newmarket girth n. *see* Aintree breast girth 纽马基特胸带，安翠胸带

Newmarket Town Plate n.（also **Town Plate at Newmarket**）a horse race run annually on the Round Course at Newmarket since 1665（自 1665 年起每年在纽马基特圆形赛道上举行的赛马比赛）纽马基特奖杯赛：Instituted in 1664 by King Charles II，the Newmarket Town Plate was the first race of its kind to be staged under written rules. 纽马基特奖杯赛于 1664 年由英国国王查理二世创立，是首场根据文

N

书规定举行的赛事。

NH abbr.【racing】Northern Hemisphere 北半球

NI abbr. neonatal isoerythrolysis 新生幼驹同种红细胞溶解症

nibble /ˈnɪbl/ vt. to bite gently or eat in small quantity 轻咬，啃：When horses are turned out together in paddocks they can often be seen nibbling each other's withers and backs. 当把马成群放在牧场时，可经常见到它们相互啃咬鬐甲和背部。

nick[1] /nɪk/ n. **1** a small cut 刻痕，切口 **2**【BrE, slang】the condition of a horse 体况 IDM **in poor/good nick** (of a horse) in good physical condition or health（指马）体况差/好的

vt. to cut the tendons and muscles under the tail of a horse surgically to achieve an elevated tail carriage（手术割断马尾下的肌腱以提高尾座）割断，切割 | **nicking** n.：Nicking is practiced almost exclusively on American Saddlebreds. 切断尾下肌腱垫高尾巴的做法几乎仅见于美国骑乘马。

nick[2] /nɪk/ n. an exceptional horse bred by mating two different blood lines 适配所产的好马

v.【genetics】to breed horses by mating two different blood lines to achieve certain desired traits or qualities, such as greater speed, good bone, etc.（采用两个不同血统的马品种交配以获得预期性状）适配 | **nicking** n.

nicker /ˈnɪkə(r)/ n. a low-pitched sound made by a horse to show pleasurable anticipation（马柔和的嘶鸣声，表示喜悦地期待）轻轻嘶鸣

vi. to neigh softly; whinny 轻轻嘶鸣

nictitating membrane n. (also **membrana nictitans**) a thin sheath of tissue underlying the upper and lower eyelids of certain animals（某些动物位于上、下眼睑下面的薄层鞘膜）瞬膜：protrusion of the nictitating membrane 瞬膜脱出 syn third eyelid

night rug n. a lined, cotton blanket covered on the horse to protect it from cold during the night（披在马身上的棉毯，可防止马匹夜间着凉）夜用马衣 syn stable rug, stable blanket

nightcap /ˈnaɪtkæp/ n.【racing】the last race listed in a racing program（比赛日程中的最后一场比赛）末场比赛，终场比赛

night-eye n. the chestnut on the legs of a horse（马四肢上附蝉的俗称）夜眼

Nightingale, Basil n. (1880 – 1910) a renowned British painter who excelled in hunting scenes and racehorses（英国著名画家，擅长狩猎和赛马写生）巴兹尔·南丁格尔

nip /nɪp/ n. a sharp pinch or bite 夹；啃

vt. to pinch or bite sharply 猛夹；咬紧：Frank winced as the dog nipped him on the leg. 弗兰克被狗咬住了腿，疼得龇牙咧嘴。

nippers /ˈnɪpəz; *AmE* ˈnɪpərz/ n. **1** first incisors; centrals 中切齿；门齿 **2**【farriery】a tool used to remove extra hoof wall（用来剪除蹄壁的工具）剪蹄钳

nipping /ˈnɪpɪŋ/ n. a vice of horse who bites gently 啃咬

nit /nɪt/ n. the egg or young of a louse or other parasitic insect 虱卵，幼虫

nitrofurazone /ˌnaɪtrəfjʊrəˈzəʊn; *AmE* -ˈzoʊn/ n. an antibacterial medication（一种抗菌药物）呋喃西林

nitrogen /ˈnaɪtrədʒən/ n.【nutrition】a nonmetallic element that is the major constituent of protein（一种非金属元素，是组成蛋白质的主要成分）氮

nitrogen-free /ˈnaɪtrədʒənˈfriː/ adj. having or containing no nitrogen 无氮的，不含氮的：nitrogen-free extract (NFE) 无氮浸出物

nitrogenous /naɪˈtrɒdʒənəs/ adj. of or containing nitrogen 氮的，含氮的

nk abbr.【racing】neck 颈长

NMS abbr. neonatal maladjustment syndrome 新生幼驹适应不良综合征

no time interj.【rodeo】an exclamation called when a rodeo rider exceeds the time allowed; signaled by the flagman waving his flag（牛仔竞技项目中骑手超过规定时间后给予的提

示，执旗者通过挥旗示意）时间到！

nobble /ˈnɒbl; *AmE* ˈnɑːbl/vt. to prevent a racehorse from winning a race by drugging it; dope（给赛马服药以阻止其获胜）下药 | **nobbling** n.

noble science n.【hunting】the art and science of hunting with foxhounds（用猎犬狩猎的艺术和知识）贵族学问；猎狐

nod /nɒd; *AmE* nɑːd/v.（of a horse）to drop and raise its head repeatedly（指马头不停地低垂和上抬）点头，晃头

nodal /ˈnəʊdl; *AmE* ˈnoʊ-/adj. of or relating to node or nodes 结的

nodding /ˈnɒdɪŋ; *AmE* ˈnɑːd-/n. a vice of a horse who drops and raises its head repeatedly（马不停地上下晃动头部的恶习）晃头；点头

node /nəʊd; *AmE* noʊd/n.【anatomy】a knot-like structure of a body organ（机体器官的结状突起）结，结节：atrioventricular node 房室结 ◇ hemolymph nodes 血淋巴结 ◇ lymph nodes 淋巴结 ◇ sinuatrial node 窦房结

nodular /ˈnɒd jʊlə; *AmE* ˈnɑːdʒələ/adj. of or resembling a node 结的，小结的

nodule /ˈnɒd juːl; *AmE* ˈnɑːdʒuːl/n.【anatomy】a small knot-like protuberance of tissue（机体组织的结节状突起）小结：lymphatic nodule 淋巴小结 ◇ nodule of vermis 蚓部小结

noisy /ˈnɔɪzi/adj.【hunting】（of hounds）barking noisily; mouthy（猎犬）吵闹的，吠叫的

nomad /ˈnəʊmæd, *AmE* ˈnoʊ-/n. a member of people who have no fixed home and move from place to place in search of food, water, and grazing land（没有固定住所并到处迁移以寻找食物、水和草场的人群）游牧部落，游牧民族

nomadic /nəʊˈmædɪk; *AmE* noʊ-/adj. of or relating to nomads 游牧民族的；游牧的：nomadic tribes 游牧部落

nominate /ˈnɒmɪneɪt; *AmE* ˈnɑːm-/vt. to propose and pay a fee to make a horse eligible to compete in an event（提议并缴纳相关费用让马匹有资格参加比赛）报名

nomination /ˌnɒmɪˈneɪʃn; *AmE* ˌnɑːm-/n. the act of nominating a horse to compete in a race 报名参赛

nomination fee n. an amount of money paid to make a horse eligible to compete in an event 报名参赛费

nominator /ˈnɒmɪneɪtə(r); *AmE* ˈnɑːm-/n.【racing】one who owns a horse at the time it is nominated to compete in a race（马匹报名参赛时的所有人）参赛马主

non-articular fracture n. a break in a bone which does not involve a joint 非关节性骨折

non-articulating ring bone n. a bony growth not attached to the joints（不与关节相连的增生赘骨）非结合性赘骨

noncontender /ˌnɒnkənˈtendə(r); *AmE* ˌnɑːn-/n.【racing】a horse who is not expected to win in a race 无望获胜的马

nondescript /ˈnɒndɪskrɪpt; *AmE* ˈnɑːn-/adj.（of a horse）having no distinctive qualities or character（指马品质特征不突出）低下的，普通的：No amount of skill training will turn a nondescript dressage horse into a medal winner. 再多的技术训练也不能把一匹水平低下的舞步马变成奖牌得主。

nondisplaced fracture n. a fracture of bone in which the bone fragments remain in alignment（骨头断端未发生错位的骨折）非错位性骨折

non-essential /ˌnɒnɪˈsenʃl; *AmE* ˌnɑːn-/adj. not completely necessary 非必需的：non-essential amino acids 非必需氨基酸

non-handfed /nɒnˈhædfed/adj.（of horses）living or kept in a natural or untamed state and not fed artificially（指马在自然状态下生活而非人工喂养）野生的；非人工喂养的：non-handfed horses 非人工喂养的马匹

nonhorsey /nɒnˈhɔːsi; *AmE* -ˈhɔːrsi/adj. not owning or relating to horses 不养马的；与马无关的

non-infectious /nɒnɪnˈfekʃəs/adj.（of a disease）

N

not infectious or contagious (疾病)非传染性的:noninfectious diseases 非传染性疾病

Nonius /ˈnəʊnɪəs; *AmE* ˈnoʊ-/n. a warmblood breed developed in Hungary during the early 1800s by putting the French stallion Nonius to Andalusian, Arab, Norman, and English half-bred mares; stands 14.2 to 16.2 hands and may have a bay, black, or brown coat; has an elegant head, long neck, and strong back; is strong, hardy, docile, and lively, and used mainly for harness and riding (匈牙利在 19 世纪初育成的温血马品种,主要采用一匹名为诺聂斯的法国种公马与安达卢西亚马、阿拉伯马、诺曼马和英国半血马母马交配繁育而来。体高 14.2 ~ 16.2 掌,毛色多为骝、黑或褐,面貌俊秀,颈长,背腰有力,体格强健,抗逆性强,性情温驯而活泼,主要用于驾车和骑乘)诺聂斯马

N

nonlactating /nɒnˈlækteɪtɪŋ/adj. not producing milk; yeld 不泌乳的;干乳的:nonlactating mare 干乳母马

nonleading leg n. the foreleg that doest not lead in canter 非起步前肢,滞后前肢 syn trailing foreleg

nonpathogenic /ˈnɒnˌpæθəˈdʒenɪk/adj. causing no diseases; not pathogenic 不致病的;非病原的

non-pro /ˈnɒnprəʊ; *AmE* ˈnɑːnproʊ/n. & adj. *see* nonprofessional 非职业的;非职业骑手

non-professional /ˌnɒnprəˈfeʃənl; *AmE* ˌnɑːn-/ n. (abbr. **non-pro**) a rider who does not accept payment or prize money for competing in any equestrian event (参加马术项目不收报酬或不赢得奖金的骑手)非职业骑手 compare professional

non-progressive color pattern n. a coat color remaining the same from birth through old age (从出生到老年保持不变的毛色)非渐进式毛色类型 compare progressive color pattern

non-ruminant /nɒnˈruːmɪnənt/adj. (of an animal) without a functional rumen; monogastric (指动物)非反刍的:non-ruminant herbivore 非反刍草食动物

n. an animal having no functional rumen 非反刍动物

non-sweater /nɒnˈswetə(r)/n. a horse who cannot sweat normally 不能发汗的马 compare anhydrosis

Norfolk cart n. a two-wheeled, horse-drawn cart used in Norfolk, England in the mid-19th century, usually hung on semi-elliptical springs with either curved or straight shafts (19 世纪中叶出现在英国诺福克郡的一种双轮马车,车身悬于半椭圆形减震弹簧上,车辕弧形或平直)诺福克马车

Norfolk Roadster n. *see* Norfolk Trotter 诺福克快步马

Norfolk Trotter n. an extinct horse breed native to Norfolk, England from which the name derived; was a strong, short-legged harness trotter with exceptional speed, fortitude, and endurance (源于英国诺福克郡并因此得名的马品种,现已灭绝。体格强健,四肢较短,速力极佳)诺福克快步马:The Norfolk Trotter had influenced the breeding of many trotters in the world. 诺福克快步马对世界上其他快步马的育成影响较大。 syn Norfolk Roadster, Roadster

Norichorse n. *see* Noriker 诺里克马

Noriker /ˈnɒrɪkə(r)/n. (also **Noric horse**) an Austrian draft breed indigenous to the central Alpine region of Europe-once known as the Roman province of Noricum from which the name derived; stands 15.2 to 16.1 hands and generally has a bay, chestnut, or spotted coat; has a slightly heavy head, a short neck, wavy mane, broad withers, a long, slightly-hollow back, and sturdy legs with broad joints and feathering; used for heavy draft and farm work and is well-suited for work in the mountains (源于阿尔卑斯山区中部的奥地利挽马品种,此地在罗马时期称诺里克姆省,故而得名。体高 15.2 ~ 16.1 掌,毛色多为骝、栗或花斑色,

头稍重,颈短,鬃毛浓密,鬐甲宽广,背长而稍凹,四肢粗壮且附生距毛,主要用于重挽和农田作业,尤适合山地作业)诺里克马 syn Pinzgauer Noriker, Oberlander, South German Coldblood

Norman /ˈnɔːmən; *AmE* ˈnɔːrm-/n. (also **Norman Horse**) an ancient French heavy draft breed originating in Normandy of France; descended from a mix of German, Arab, and Barb blood and later crossed with English Thoroughbred and Norfolk Trotter in the 18th and 19th centuries; stands about 16 hands and usually has a bay or chestnut coat; used mainly for riding and jumping (源于法国诺曼底地区的重挽马品种,主要由德国马、阿拉伯马和柏布马杂交培育而成,并于18和19世纪通过与英纯血马和诺福克马杂交进行过改良。体高约16掌,毛色多为骝或栗,主要用于骑乘和障碍赛)诺曼马 syn Anglo-Norman, French Saddle Horse

Norman Cob n. a breed of draft horse originating in Normandy, France (源于法国诺曼底的挽马品种)诺曼柯柏马

Norman Horse n. *see* Norman 诺曼马

Norman Trotter n. *see* French Trotter 诺曼快步马,法国快步马

Norman-Percheron Association n. an organization founded in the United States in 1876 to maintain the stud book and promote the breeding and use of Percheron horses, the name was changed to the Percheron Association in the next year (1876年在美国成立的组织机构,旨在进行佩尔什马的登记并推动其育种工作,次年更名为佩尔什马协会)诺曼佩尔什马协会

North American Riding for the Handicapped Association n. (abbr. **NARHA**) an organization founded in the United States in 1981 to organize riding events, facilities, and programs for the physically disabled individuals (1981年在美国成立的组织机构,旨在为残疾人组办骑乘活动、提供设施和项目)北美残疾人骑乘协会

North American Selle Français Horse Association n. (abbr. **NASFHA**) an organization formed in Winchester, Virginia, USA in 1990 to promote the breeding of Selle Français and keep its registry in the United States (1990年在美国弗吉尼亚州温彻斯特市成立的组织机构,旨在推进美国塞尔法兰西马的登记和育种工作)北美塞尔法兰西马协会

North Swedish n. *see* North Swedish Horse 北方瑞典马

North Swedish Horse n. a heavy, medium-sized coldblood originating in Sweden; stands 15.1 to 15.3 hands and may have a dun, brown, chestnut, or black coat; has a heavy head, short neck, powerful build, short, strong legs with plenty of bone; is hardy, easy to keep, good tempered and possessed of tremendous pulling power; used mainly for heavy draft and farm work (源于瑞典的重型冷血马品种。体高15.1~15.3掌,毛色多为兔褐、褐、栗或黑,头重,颈短,体格高大,四肢粗短,抗逆性强,耐粗饲,性情温驯,挽力惊人,主要用于重挽和农田作业)北方瑞典马 syn North Swedish

North Swedish Trotter n. a lighter version of the North Swedish Horse; has a long and active stride and used mainly in harness racing (源于北方瑞典马的轻型马种,步幅长而轻快,常用于轻驾车赛)北方瑞典快步马 syn North-Hestur

Northern Dales Pony Society n. an organization founded in Great Britain in 1957 to promote the breeding of the Dales Pony (1957年在英国成立的组织机构,旨在推动纯种戴尔斯矮马的育种工作)北方戴尔斯矮马协会

Northern Dancer n. (1961–1990) a Canadian-bred Thoroughbred racehorse that won the 1964 Kentucky Derby and Preakness Stakes and then became one of the most successful sires of the

20th century（一匹加拿大繁育的纯血马，于1965年获得美国肯塔基德比赛和普瑞克尼斯大赛，并成为20世纪最成功的种公马之一）北方舞蹈家

Northern Hackney Horse Club n. an organization founded in Great Britain in 1945 to promote interest in Hackney horses, cobs, and ponies（1945年在英国成立的组织机构，旨在促进公众对哈克尼挽马、柯柏马和矮马的兴趣）北方哈克尼马俱乐部

North-Hestur n. *see* North Swedish Trotter 北方瑞典快步马

Northlands /ˈnɔːθləndz; *AmE* ˈnɔːrθ-/n. a relatively rare pony breed indigenous to Norway; descended from the Asiatic wild horse and the Tarpan; stands about 13 hands and may have a chestnut, bay, brown, or gray coat; has a well-proportioned head, short neck, long back, rounded croup, sturdy legs, and full forelock, mane and tail; is frugal, quiet and energetic, and used mainly for riding and light draft（挪威本土较为稀有的矮马品种，主要由亚洲野马和塔盼马繁衍而来。体高约13掌，毛色多为栗、骝、褐或青，头型匀称，颈短，背长，后躯浑圆，四肢粗短，额发、鬃和尾毛发浓密，耐粗饲，性情安静而富有活力，主要用于骑乘和轻挽）诺斯兰德兹矮马

Norwegian Fjord Pony n. an ancient pony breed indigenous to Norway and similar in appearance to Przewalski's horse; stands 13 to 14 hands and usually has a dun coat with erect mane and black dorsal stripe; is gentle, hardy, and sure-footed; used for riding and fram work（源于挪威的古老矮马品种。体貌与普氏野马较为相似。体高13～14掌，被毛兔褐，鬃毛直立，带黑色背线，性情温驯，抗逆性强，步法稳健，用于骑乘和农田作业）挪威峡湾马：The Norwegian Fjord horses still retain some characteristics of the primitive wild horses, and are often known to have zebra stripes on their legs. 挪威峡湾马仍然保留着原始马种的某些特征，四肢上常有斑马纹。

Norwegian Trotter n. *see* DØle Trotter 挪威快步马

nose /nəʊz; *AmE* noʊz/n. **1** the forward portion of a horse's head including the muzzle and mouth（马头前端的部分）鼻 **2**【racing】(abbr. **ns**) the narrowest winning margin in a race（赛马胜出的最短距离）鼻长：The horse won by a nose. 这匹马以一鼻之长险胜。 compare dead-heat, short-head, head, short-neck, neck, length

noseband /ˈnəʊzbænd; *AmE* ˈnoʊ-/n. (also **nosepiece**) the strap of a bridle or halter that passes over the bridge of a horse's nose（马勒或笼头绕过马鼻梁的部分）鼻革

nosebleed /ˈnəʊzbliːd; *AmE* ˈnoʊ-/n. bleeding from the nose; epistaxis 鼻出血

nostril /ˈnɒstrəl; *AmE* ˈnɑːs-/n. either of two external openings of the nasal cavity（鼻腔的两个外开口）鼻孔：nostril dilator muscle 鼻孔开张肌 ◇ Wide nostrils permit unimpeded breathing. 鼻孔大则呼吸通畅。

nott stag n.【BrE】a stag without antlers 无角雄鹿 syn hummel

Nottingham cart n. a light, two-wheeled, horse-drawn vehicle popular in the mid-19th century in Nottingham, England, from which the name derived（一种轻型双轮马车，于19世纪中叶风行于英国诺丁汉郡故而得名）诺丁汉马车 syn Nottingham cottage cart

Nottingham cottage cart n. *see* Nottingham cart 诺丁汉双轮马车

nourish /ˈnʌrɪʃ; *AmE* ˈnɜːrɪʃ/vt. to provide nutrients necessary for life and growth; nurture（为生命成长提供必需的养分）滋养；养育

nourishment /ˈnʌrɪʃmənt; *AmE* ˈnɜːr-/n. **1** the act of nourishing or state of being nourished 滋养 **2** things that nourish; foods 营养品；食物

novice /ˈnɒvɪs; *AmE* ˈnɑːv-/n. **1** an inexperienced rider 新手：a novice fox hunter 猎狐新手 **2** (also **novice class**) a class of competition higher than the pre-novice class and lower than the intermediate class（高于预备级但低于中级的比赛级别）初级：novice class 初级

Novokirghiz /nəʊˌvɒkɪrˈgiːz/n.（also **New Kirghiz**）a warmblood developed in Kirghizia in 1930s by breeding the old Kirghiz stock with Thoroughbred and Don；stands 14.1 to 15.1 hands and may have a coat of any solid color with bay occurring most often；has a longish back，straight shoulders，short strong legs with good bone，and strong feet；is good tempered，active，sure-footed and tough；used for riding，harness，and packing（俄国吉尔吉斯地区育成的温血马品种，20 世纪 30 年代通过引入纯血马和顿河马改良吉尔吉斯马培育而成。体高 14.1～15.1 掌，各种单毛色兼有，以骝最为常见，背长，直肩，四肢粗短，蹄质结实，性情温驯，富有活力，步法稳健，主要用于骑乘、挽车和驮运）新吉尔吉斯马

ns abbr.【racing】nose 鼻长

NSH abbr. *see* National Show Horse 美国马术马，美国表演马

NT abbr. Nuclear Transfer 核移植

nullify /ˈnʌlɪfaɪ；*AmE* ˈnʌləˌfaɪ/vt. to make sth invalid；invalidate 使无效

numb /nʌm/adj.【cutting】（of a cow）not threatened by a horse and showing little desire to return to the herd（指所截之牛即不惧怕马也不无心返群）麻木的，无动于衷的

number ball n.【racing】a small，numbered ball selected in a blind draw from a box to determine post positions of the racehorses（带有数目的小球，用来从盒中抓阄以决定赛马的闸位）数字球，闸位球 [syn] pill

number board n.【racing】an electronic board upon which post positions are identified after the number balls are drawn（赛马场上的电子公告牌，用以显示抽签后的赛马闸位）闸位公告牌

number box n.【racing】a box containing numbered balls used to determine post positions（装有决定赛马闸位抓阄小球的盒子）位次盒，闸位盒

number cloth n.【racing】a piece of square，white linen cloth placed under the saddle，which bears the number corresponding to the horse's program number（衬在马鞍下的方形白布，其上印有赛马出场编号）号码布 [syn] saddle cloth

number four n.【polo】the polo player in the number four position 四号球员

number four position n.【polo】one of the four positions on a polo team responsible for defending the goal and guarding the number one position on the opponent's team（马球比赛中的占位，主要负责看守球门并阻止对方一号球员的进攻）四号位 [syn] the number four，the back

number one n.【polo】the polo player assuming the number one position 一号球员

number one position n.【polo】one of the four positions on a polo team responsible for playing offense and scoring goals（马球比赛中的占位，主要负责进攻和射门得分）一号位

number three n.【polo】the polo player assuming the number three position 三号球员

number three position n.【polo】one of four positions on a polo team responsible for both offense and defense（马球比赛中的占位，担当攻守兼备的角色）三号位：Number three position is often played by the highest rated and the most experienced player on the team. 马球三号位多由积分最高且经验最丰富的球员担当。

number two n.【polo】the polo player assuming the number two position 二号球员：Number two player often alternates positions with number one and three players where they are not occupied. 在一号和三号球员不到位时，二号球员常常与他们交换位置。

number two position n.【polo】one of four positions on a polo team responsible primarily for scoring goals or passing the ball to number one player(马球比赛中的占位,主要负责得分或传球给一号球员射门)二号位

numdah /ˈnʌmdɑː/n. *see* numnah 鞍垫

numnah /ˈnʌmnɑː/n.(also **numdah**) a saddle-shaped cotton pad placed under a saddle to absorb sweat and provide protection(衬在马鞍下的棉垫,有吸收汗液和保护马背的作用)鞍垫:The numnah are used for several reasons, of which the most common reasons are: to wick away moisture from the horse's skin due to sweating, to absorb some of the shock exerted in the saddle while being ridden, and to reduce pressure points on the horse's back caused by an ill-fitting saddle. 使用鞍垫有诸多目的,其中最主要的有以下几个:吸收马皮肤上的汗液,消减骑乘运行中的冲力,此外还可减小由于马鞍不合适而造成的压力。syn saddle pad

nurse /nɜːs; *AmE* nɜːrs/vt.(of a mare) to breast-feed the foal(母马)哺乳

vi.(ot a foal) to suckle milk from the mare's udder(指驹)吮奶:If a newborn foal is not nursing within two hours, the vet should be alerted to give it and the mare an examination. 如果新生幼驹在两小时内还没有吮奶,就应赶快叫来兽医给幼驹和母马做一个检查。

nursery handicap n.【racing】a handicap race for two-year old horses(由2岁驹参加的让磅赛)幼驹让磅赛

nurture /ˈnɜːtʃə(r); *AmE* ˈnɜːrtʃ-/vt. to nourish or feed; cultivate 滋养,喂养;抚育

nut /nʌt/n. **1** a cube of fodder 饲料块,颗粒料 **2**【slang】(*pl.* **nuts**) testicles 睾丸

nutcracker /ˈnʌtkrækə(r)/n. a horse who grinds his teeth 磨牙的马

nutrient /ˈnjuːtriənt; *AmE* ˈnuː-/n. a source of nourishment, especially an ingredient in a food(营养物质的来源,尤指食物中的营养成分)营养品;养分:nutrient analysis 营养成分分析 ◇ nutrient composition 营养组成 ◇ nutrient value 营养价值 ◇ nutrient digestibility 饲料养分消化率 ◇ nutrient requirements 营养需要

nutrition /njuˈtrɪʃn; *AmE* nu-/n. **1** food or nourishment; nutrients 营养物,营养品:nutrition trials 营养测定 **2** the process of providing or obtaining food necessary for health and growth(获取生长所需食物的过程)营养,滋养 **3** the branch of science that deals with food and nutrients 营养学

nutritional /njuˈtrɪʃənl; *AmE* nu-/adj. of or relating to nutrition 营养的:nutritional value 营养价值

nutritional anemia n. a type of anemia caused by a deficiency in protein, mineral, and/or vitamins; often characterized by decreased count of red blood cells(由于日粮中缺乏蛋白质、矿物质或维生素而导致的贫血,临床症状常表现为红细胞减少)营养性贫血

nutritional deficiencies n. deficiency of any necessary substance that provides nourishment for the body(为机体提供养分的必需物质的缺乏)营养缺乏症

nutritional hyperparathyroidism n. another term for osteodystrophia fibrosa(纤维性骨营养不良的别名)营养性甲状旁腺功能亢进

nutritional muscular dystrophy n. another term for white muscle disease(白肌病的别名)营养性肌萎缩

nutritionist /njuˈtrɪʃənɪst; *AmE* nu-/n. an expert in the field of nutrition 营养学家:equine nutritionist 马匹营养学家

nutritious /njuˈtrɪʃəs; *AmE* nu-/adj. of or having nutrition 有营养的

NZHS abbr. New Zealand Horse Society 新西兰马会

Oaks /əʊks; *AmE* oʊks/ n.【racing】*see* Epsom Oaks 欧克斯大赛,橡树大赛

oakum /ˈəʊkəm; *AmE* ˈoʊk-/ n. loose hemp or jute fiber used for packing cracked hoofs (捆绑马蹄开裂的麻絮)麻丝,麻线

oat /əʊt; *AmE* oʊt/ n. **1** a cereal grass widely cultivated as a feed for animals (一种广泛种植的禾本科作物,可用作动物饲料)燕麦: oat grass 燕麦草 ◇ oat hay 燕麦干草 **2** (usu. **oats**) the edible grain of this cereal plant used as fodder for animals (燕麦的可食谷粒,常用作动物饲料)燕麦粒: naked oats 裸燕麦

oatburner /əʊtˈbɜːnə(r); *AmE* oʊtˈbɜːrnə(r)/ n. *see* hay burner 白喂的马;乏马

obedience /əˈbiːdiəns/ n. the condition of being obedient 温驯

obedient /əˈbiːdiənt/ adj. (of a horse) willing to accept the commands and aids of the rider (指马)温驯的

Obel Lameness Grades n.【vet】(also **Obel Scale**) a system developed by Dr. Niles Obel in 1948 to assess degrees of lameness resulting from laminitis (1948 年由兽医博士奈尔斯·奥贝尔制定的体系,用来对蹄叶炎诱发的跛行进行等级划分)奥贝尔跛行等级 compare AAEP Lameness Scale

Obel, Niles n. an American veterinary who introduced the so-called Obel Lameness Grades to assess degrees of lameness of horses (建立了奥贝尔跛行等级的美国兽医)奈尔斯·奥贝尔

Obel scale n. *see* Obel Lameness Grades 奥贝尔跛行等级: The Obel scale classifies lameness from I to IV, with I indicating mild soreness and IV designating extreme pain with reluctance to move. 奥贝尔等级将跛行分为 I ~ IV 四个等级, I 级为轻微酸痛, IV 级疼痛剧烈且马不愿运步。

Oberlander /ˌəʊbəˈlædə(r); *AmE* ˌoʊbərˈləndə(r)/ n. *see* Noriker 奥伯兰德马

obese /ˈəʊbiːz; *AmE* ˈoʊbiːz/ adj. extremely fat or overweight 肥胖的

obesity /əʊˈbiːsəti; *AmE* oʊˈ-/ n. the condition or state of being fat; overweight 肥胖;过重: Obesity is generally thought to reduce fertility of mares and certainly can make foaling more difficult. 人们普遍认为,肥胖会使母马繁殖力下降,同时也易造成难产。

objection /əbˈdʒekʃn/ n.【racing】a complaint submitted to the governing body against unfair ruling in racing (针对判决不公而向监管机构提出的异议)申诉;异议: objection overruled 申诉否决 ◇ objection sustained 申诉成立 syn protest

objection flag n.【racing】a red flag raised to indicate that an objection has been filed against a race (示意对比赛有异议而出示的红旗)反对旗,异议旗

oblique /əˈbliːk/ adj. situated in a slanting position 倾斜的: external abdominal oblique muscles 腹外斜肌

oblique fracture n. a bone fracture at an angle 斜骨折

oblique shoulder n. *see* sloping shoulder 斜肩

obstetric /əbˈstetrɪk/ adj. (also **obsterical**) of or relating to obsterics 产科的: obstetric chains 产科链 ◇ obstetric crutch 产科梃 ◇ obstetric hook 产科钩

obstetrical /əbˈstetrɪkl/ adj. *see* obstetric 产科的: obstetrical diseases 产科疾病

obstetrics /əbˈstetrɪks/ n. pl. the branch of medicine dealing with delivery or birth of the young

(有关幼体分娩的医学分支)产科

obstreperous /əb'strepərəs/ adj. (of a horse) difficult to control or train; fractious (指马)难驾驭的

obstructive jaundice n. a jaundice condition caused by obstruction of the flow of bile from the liver to the duodenum (从肝到十二指肠的胆汁分泌堵塞后引发的黄疸)阻塞性黄疸

obstructive pulmonary disease n. *see* chronic obstructive pulmonary disease 阻塞性肺病

occult spavin n. an arthritis of the inner side of hock joint without visible signs (跗内侧关节炎症,无明显症状)隐性飞节内肿

odd board n.【racing】*see* totalizator board 赔率板,赌金显示板

odd colored adj.【color】(of a horse's coat) consisting of large irregular patches of more than two colors (马被毛有两种以上颜色的大斑块构成)多毛色的

O

odds /ɒdz; *AmE* ɑːdz/ n.【racing】**1** the chance of wining a bet (投注获利的)概率 **2** the ratio between the amount won by a bettor and the amount he bets on a race (投注者赢钱数目与投注数目的比率)赔率:minimum odds 最低赔率 ◇ odds of ten to one 10:1 的赔率

odds man n.【racing, dated】one formerly employed to calculate the betting odds and amounts of bets staked in any given race before totalizators were adopted (在赌金计算器出现之前,过去负责计算赛马赔率及投注总额的雇员)赔率计算师

odds-on /ˌɒdz'ɒn; *AmE* ˌɑːdz'ɑːn/ adj.【racing】(of a bet) having a good chance to win money (指投注)有胜算的,有望赢奖的:an odds-on bet 胜算较大的投注

odor /'əʊdə(r); *AmE* 'oʊ-/ n. (also **odour**) a distinctive smell given off by an individual (个体散发的特殊味道)气味,体味:The odor of a mare in heat is a more certain sexual signal for mating and one which carries some distance. 发情母马的气味是个更为可靠的交配信号,并可传播较长距离。

odour /'əʊdə(r); *AmE* 'oʊ-/ n.【BrE】*see* odor 气味

oedema /ɪ'diːmə/ n.【BrE】*see* edema 浮肿;水肿

oesophagus /ɪ'sɒfəgəs; *AmE* ɪ'sɑː-/ n.【BrE】*see* esophagus 食管,食道

oestrogen /'iːstrədʒən/ n.【BrE】*see* estrogen 雌激素

oestrous /'iːstrəs/ adj.【BrE】*see* estrous 发情的:oestrous circle 发情周期

oestrus /'iːstrəs/ n.【BrE】*see* estrus 发情;发情期

off /ɒf; *AmE* ɔːf, ɑːf/ adv. **1**【racing】starting a race; leaving 起跑,开赛:The horses are off. 赛马起跑了。**2** decreased or cancelled 降低;取消:The horse is off its feed after catching an influenza. 这匹马得了流感后,就一直食欲不振。

adj. **1** (of a horse) passing a certain age (指马)过几岁的:This horse is five years off. 这匹马五岁多。**2** (of a horse) lame(指马)瘸的:The horse is off in the right front. 这匹马右前肢是瘸的。

n.【racing】the start of a race 开赛,起跑:All horses and riders are ready for the off. 所有骑手和马匹都准备好起跑。

vt. to begin a race 开赛

off center n.【driving】the horse on the right of a pair of centers 外侧中马 — *see also* **near center**

off horse n.【driving】the horse on the right of a pair 外侧马 compare near horse

off leader n.【driving】the horse on the right of a pair of leaders 外侧头马 compare near leader

off wheeler n.【driving】the horse on the right of a pair of wheelers 外侧辕马 compare near wheeler

off-course bet n. (also **off course bet**) *see* off-track bet 场外投注 compare on-course bet

offense /ə'fens/ n.【polo】an attempt made to

score 进攻 [opp] defense

offensive /əˈfensɪv/ adj. 【polo】(of a player) making an attack (球员)进攻的:an offensive player 进攻球员 [opp] defensive

official /əˈfɪʃl/ n. an individual authorized to work at an equestrian event 赛事官员

offset knees n. (also **bench knees**) a conformation defect of the horse's front legs in which the cannon bone is outside the direct line from the radius to the hoof (马前肢管骨朝外偏离桡骨至马蹄垂线的体格缺陷)管骨外偏,膝外偏 [syn] lateral deviation of the metacarpal bones

offside[1] /ˌɒfˈsaɪd; *AmE* ˌɔːf-/ adj. 【polo】(of a player) illegally ahead of the ball (球员)越位的

n. an offside position or play 越位

offside[2] /ˌɒfˈsaɪd; *AmE* ˌɔːf-/ adj. on the right side 右侧的,外侧的

n. the right side of a horse when viewed from behind (后观时马匹的右侧)外侧 [syn] far side

offside horse n. the right-hand horse in a pair 外侧马,右侧马

off-track /ˌɒfˈtræk; *AmE* ˌɔːf-/ adj. done or conducted away from a racetrack 场外的

off-track bet n. any wager made at a site other than the racetrack (在赛马场以外的地点进行的投注)场外投注 [syn] off-course bet

off-track betting n.【racing】(abbr. **OTB**) the practice of betting at legalized sites other than the racetrack (在赛马场以外的地点进行押注)场外投注:off-track betting sites 场外投注点 ◇ The money wagered on off-track betting is usually commingled with on-track betting pools. 场外押注金额通常添入场内赌金池中。 [compare] on-track betting

oil /ɔil/ vt. **1** to apply oil onto leather tack 上油,涂油 **2** to administer oil to a horse through a nasogastric tube to relieve gas or alleviate intestinal blockage (采用鼻胃管给马灌油以消除胃胀气或肠道阻塞)灌油 | **oiling** /ˈɔɪlɪŋ/ n.

oil cake n. the solid residue of oilseeds after extraction of oil (油籽榨油后的残渣)油饼,油渣饼

oilseed /ˈɔɪlsiːd/ n. any seed of plants grown for oil extraction (种植用于榨油的籽实)油籽:oilseed grains 油籽作物 ◇ oilseed meals 油籽粕

old scent n. *see* stale scent 变淡的踪迹

Oldenburg /ˈəʊldənˌbɜːɡ; *AmE* ˈoʊldənˌbɜːrɡ/ n. a heavy German warmblood developed from the Friesian in the 17th century as a strong carriage horse; stands 16.2 to 17.2 hands and usually has a bay, brown, black, or gray coat; has pronounced withers, straight back, a deep chest, broad croup, high-set tail, and short legs; used mainly for riding and dressage (德国在17世纪由弗里斯兰马育成作驭马使用的温血马品种。体高16.2~17.2掌,毛色多为骝、褐、黑或青,鬐甲明显,腰背平直,胸身宽广,尻宽,尾础较高,四肢较短,主要用于骑乘和盛装舞步)奥尔登堡马:The Oldenburg has been refined with infusions of Thoroughbred blood since the early 20th century and is now excelling in dressage, eventing and show jumping. 奥尔登堡马自20世纪初开始引入纯血马血统进行品种改良,目前在盛装舞步、三日赛和障碍赛中都有出色表现。

oleander /ˌəʊliˈændə(r); *AmE* ˌoʊli-/ n. a poisonous Eurasian evergreen shrub with fragrant white or purple flowers (一种生长在欧亚大陆的有毒长青灌木,开白色或紫色花朵)夹竹桃 [syn] rosebay

olfactory /ɒlˈfæktəri; *AmE* ɑːl-/ adj. of or relating to the sense of smell 嗅觉的

olfactory reflex n. *see* Flehmen response 性嗅反射

olive dun n. *see* olive grullo 青兔褐

olive grullo n.【color】(also **olive dun**) a color of horse with yellowish grullo coat and black points (被毛黄青色而末梢为黑色的毛色)青兔褐

Omnibus /ˈɒmnɪbəs; *AmE* ˈɑːm-/ n. a four-wheeled, horse-drawn public service carriage first appeared on the streets of Paris in the second half of the 17th century（于17世纪下半叶出现在巴黎街头的一种公用四轮马车）公共马车

omnivore /ˈɒmnɪvɔː(r); *AmE* ˈɑːm-/ n. an animal that eats both plants and meat（植物和肉类兼食的动物）杂食动物 compare carnivore, herbivore

onager /ˈɒnədʒə(r); *AmE* ˈɑːn-/ n. a fast-running wild ass of equine family that lives in central Asia and has an erect mane and a broad black stripe along its back（生活在中亚地区的马科野驴，善奔跑，鬃毛直立，黑色背线较宽）中亚野驴：Persian onager 波斯野驴

onchocerciasis/ˌɒŋkəʊsəːˈsaɪəsɪs; *AmE* -koʊ-/ n. a disease caused by infestation with filarial worms of the genus *Onchocerca*, often characterized by dermatitis and lesion of the eyes（由盘尾丝虫属丝虫大量寄生引发的疾病，其症状常表现为皮炎和眼部病变）盘尾丝虫病

on-course bet n. *see* on-track bet 场内押注 compare off-course bet

One Thousand Guineas n.【racing】(also **1,000 Guineas**) a one-mile race held annually for three-year-old fillies since 1814 at Newmarket, England（自1814年起每年在英国纽马基特为3岁雌驹举行的平地赛，赛程1.6千米）一千几尼赛

one-day event n. an equestrian competition consisting of dressage, show jumping, and cross country and conducted in one day with all three phases performed by the same rider and horse（人和马在一天之内完成盛装舞步、场地障碍赛和越野赛的马术项目）一日赛 compare two-day event, three-day event

one-horse adj. (of a carriage) pulled or drawn by a single horse（指车）单马拉的：a one-horse sleigh 单马拉的雪橇

onset /ˈɒnset; *AmE* ˈɑːn-/ n. the beginning of sth; start 起始，开端：The best time to cut hay is just at the onset of flowering. 牧草的最佳收割时间在开花初期。

one-sided /ˌwʌnˈsaɪdɪd/ adj. **1** (of a horse) bending more easily to one side than the other（指马）单向的 **2** (of a horse) responsive only to the rein aids on one side（指马）单侧灵敏的：one-sided mouth 单侧口角灵敏

on-site diagnostic test n. *see* stall-side diagnostic test 现场检测

on-track adj.【racing】(of betting) conducted on the racetrack（投注）场内的 opp off-track

on-track bet n.【racing】*see* track bet 场内投注 opp off-track bet

oocyte /ˈəʊəsaɪt/ n. a cell from which an egg develops by meiosis（可通过减数分裂产生卵子的细胞）卵母细胞：primary oocyte 初级卵母细胞 ◇ secondary oocyte 次级卵母细胞

open[1] /ˈəʊpən; *AmE* ˈoʊ-/ adj. **1** (of sports) not restricted in its entry（赛事）公开的：an open event 公开赛 **2**【hunting】(of a covert) not stopped（指洞穴）未堵的 **3** (of a carriage) having no covering or canopy（指马车）敞篷的：an open cart 一辆敞篷马车

open[2] /ˈəʊpən; *AmE* ˈoʊ-/ vi.【hunting】(of the hounds) to speak on the line of a fox in a covert for the first time（指猎犬追踪过程中首次发现狐狸踪迹时叫嚣）叫哮，狂吠 IDM **open on a fox** (of a hound) to bark at a fox（猎犬）朝狐狸狂吠

open bridle n. a bridle without blinkers covering the eyes of the horse 无眼罩的马勒

open fracture n. *see* compound fracture 开放性骨折，复合性骨折

open mare n. (also **yeld mare**) a mare who was not bred during the previous breeding season（上个繁育季节没有配种的母马）空怀母马

open race n.【racing】a race with little restriction on the eligibility of horses（对马匹参赛资格限制较少的比赛）公开赛

open top collar n. a harness collar that can be

spread open at the top to facilitate slipping it over the head of horse (顶部中段带开口的颈圈,可从下滑入马匹颈部)上开口套包,上开口颈圈

open-hocked /ˈəʊpənˌhɒkt/ adj. (of a horse) wide at the hocks and narrow between the hoofs (马的飞节)内弧的,外张的

opening meet n.【hunting】the first meet of a hunting season 首场狩猎会,开场狩猎会

Opera Bus n. a small, private horse-drawn vehicle first used by theater-goers in 1870s (19 世纪 70 年代出现的一种小型私人马车,最早由剧院看客使用)剧院马车

ophthalmia /ɒfˈθælmiə; *AmE* ɑːfˈθ-/ n. inflammation of the eye characterized by heat and pain (眼球出现发热、疼痛的炎症)眼炎:periodic ophthalmia 复发性眼炎

opposite rein vt. *see* neck rein 反向控缰,即颈部压缰

optic /ˈɒptɪk; *AmE* ˈɑːp-/ adj. of or relating to the eye or vision 视觉的:optic nerve 视神经 ◇ optic fiber 光纤

optimal /ˈɒptɪməl; *AmE* ˈɑːp-/ adj. most favorable or desirable; optimum 最佳的,最适宜的:The optimal time for mating is estimated at about 6 to 24 hours prior to ovulation. 排卵前 6～24 小时为最佳配种时期。

optimum /ˈɒptɪməm; *AmE* ˈɑːp-/ n. an amount or condition that is most favorable 最佳条件;最适用量

adj. most favorable or advantageous; best 最适合的;最佳的

optimum take-off zone n.【jumping】the ideal area where a horse should take off to clear an obstacle (可顺利跨过障碍的最佳区域)最佳起跳区:Size of the optimum take-off zone varies according to the height and type of jump to be cleared as well as the stride and jumping ability of the horse. 最佳起跳区的大小依障碍高度、类型以及马匹运步和跳跃能力而有所变化。

optional claimer n.【racing】a race for horses to be claimed for at a price within a limited range (马匹价格限定在特定范围的竞价赛)选择性竞价赛

oral /ˈɔːrəl/ adj. **1** of or relating to the mouth 口的 **2** (of medicine) taken through the mouth (药物)口服的:oral medicine 口服药物

orange dun n. a coat color of light red dun with red or light red points (红兔褐被毛上带浅红色斑点的毛色)橘兔褐

orchard grass n. (also **orchardgrass**) a Eurasian grass widely planted in pastures (一种欧亚大陆广泛种植的牧草)鸭茅:orchard grass hay 鸭茅干草

order /ˈɔːdə(r); *AmE* ˈɔːrd-/ n. a taxonomic category ranking above a family and below a class (分类学上高于科而低于纲的类群)目 compare kingdom, phylum, class, family, genus, species

ordinary hack n. *see* road hack 普通乘用马

ordinary walk n. *see* walk [普通]慢步

organ /ˈɔːgən; *AmE* ˈɔːrgən/ n.【physiology】any part of a horse that performs a specific function (发挥特定功能的机体部位)器官

organelle /ˌɔːgəˈnel; *AmE* ˌɔːrg-/ n.【biochemistry】a part inside a cell that performs a specific function (细胞内执行特定功能的构件)细胞器

orifice /ˈɒrɪfɪs; *AmE* ˈɔːr-/ n.【anatomy】an opening to a cavity or passage of the body (体腔或管道的)开口,外口

Orlov Trotter n. (also **Russian Trotter**) a Russian warmblood developed in the 1780s by putting the Arab stallion Smetanka to Danish and Dutch mares; stands 15.1 to 17 hands and usually has a gray coat, although black and bay do occur; has a light, powerful build, elegant conformation, long back, and a long, swan neck; is docile, but energetic; used mainly for farm work, harness, and riding (俄国在 18 世纪 80 年代采用阿拉伯公马斯米唐卡与丹麦和荷

O

兰母马杂交培育而来的温血马品种。体高 15.1～17 掌，毛色以青为主，黑和骝少见，体质轻，体格强健，体型优美，颈背长，鹤颈，性情温驯，富于活力，主要用于农田作业、驾车赛和骑乘）奥尔洛夫快步马；俄国快步马：The Orlov Trotter is commonly used to improve other breeds of horses in Russia. 在俄罗斯，奥尔洛夫快步马常用来改良其他的马匹品种。

orphan /ˈɔːfn; *AmE* ˈɔːrfn/ n. a young animal without a mother 无母幼兽

adj. (of a foal) deprived of parents（幼驹）无母的：orphan foal 无母幼驹

vt. to deprive an individual of one parent or both parents 使成孤儿：an orphaned foal 无母幼驹

orthopedic /ˌɔːθəˈpiːdɪk; *AmE* ˌɔːrθ-/ adj. (BrE **orthopaedic**) of or relating to the correction of deformities of bones 整形的，矫形的：orthopedic surgery 整形手术

orthopedics /ˌɔːθəˈpiːdɪks; *AmE* ˌɔːrθ-/ n. pl. (BrE **orthopaedics**) the branch of medicine dealing with the correction of deformities 整形科；矫形术

os pedis n.【Latin】the pedal bone 蹄骨

osselet /ˈɒsəlɪt/ n. a bony growth or swelling occurring on the fetlock or ankle joint（球节或踝关节上附生的小骨）赘骨，踝关节炎 syn arthritis of the fetlock joint

osseous /ˈɒsiəs; *AmE* ˈɑːs-/ adj. of or consisting of bone; bony 骨的，骨质的：osseous tissue 骨组织

ossification /ɒsɪfɪˈkeɪʃn; *AmE* ɑːsəf-/ n. the natural process of bone formation 成骨；骨化：endochondral ossification 软骨内成骨

ossify /ˈɒsɪfaɪ; *AmE* ˈɑːs-/ v. to form or convert cartilage into bone 成骨；骨化

oste(o)- pref. bone 骨

osteoarthritis /ˌɒstiəʊɑːˈθraɪtɪs; *AmE* ˌɑːstioʊɑːrˈθ-/ n. inflammation and degeneration of joints due to excessive wear or joint weakness（由于过度磨损或关节脆弱导致的骨关节发炎或退化的症状）骨关节炎 syn degenerative joint disease

osteoblast /ˈɒstɪəblæst/ n. a cell from which bone develops 成骨细胞：The osteoblasts synthesize and extrude the collagen and proteoglycan that form the extracellular bone matrix. 成骨细胞合成并分泌胶原和蛋白多糖，从而形成骨细胞间质。

osteochondrosis /ˌɒstɪəʊkɒnˈdrəʊsɪs/ n. (abbr. **OCD**) an abnormal development of cartilage and bone, often found in the hock and shoulders of horse and causing inflammation and pain of joints（由于软骨与骨的异常发育导致关节发炎、疼肿的症状，多见于马的跗关节和肩胛）骨软骨病

osteocyte /ˈɒstɪəsaɪt/ n. a bone cell embedded in the matrix of bone tissue（一种分布于骨组织中的分化细胞）骨细胞

osteodystrophia fibrosa n.【vet】a generalized bone disease primarily caused by dietary calcium deficiency with phosphorus excess resulting from diet or parathyroid problems, symptoms may include symmetric enlargement of the mandible and facial bones（由于日粮中钙不足但磷过量或甲状旁腺机能不全引起的全身性骨骼疾病，临床表现为下颌骨和面部骨骼增大等症）纤维性骨营养不良 syn miller's disease, bran disease, bighead disease, nutritional hyperparathyroidism

osteomalacia /ˌɒstɪəʊməˈleɪʃɪə/ n. a condition that often occurs in adult animals in which the bones become softer（一种常见于成年动物的病症，表现为骨质软化）骨软化，软骨病

osteomyelitis /ˌɒstɪəʊˌmaɪəˈlaɪtɪs/ n. an infection of the bone and bone marrow caused by microorganisms（骨和骨髓遭微生物感染发炎的症状）骨髓炎

osteon /ˈɒstɪɒn/ n.【anatomy】the structural unit of the dense bone 骨单位

osteoporosis /ˌɒstiəʊpəˈrəʊsɪs; *AmE* ˌɑːstioʊpəˈroʊ-/ n. a disease in which the bones become extremely porous due to demineralization（由于矿物质丢失

而使骨质出现孔洞的症状)骨质疏松

ostler /ˈɒslə(r); *AmE* ˈɑːs-/ n. a hostler(客栈的)马夫

OTB abbr. off-track betting 场外押注

out /aʊt/ adv. **1** away from sth 远离,不在 PHR **out of position**【cutting】(of a horse) losing the working advantage(指马)失去有利位置 **2** (of a horse) born by a mare 由母马所生:The foal was out of the mare Gillad. 这匹马驹为母马吉拉德所生。 compare sired

outbreak /ˈaʊtbreɪk/ n. a sudden increase or eruption; outburst 暴发,爆发

outclass /ˌaʊtˈklɑːs; *AmE* -ˈklæs/ vt. to surpass decisively 远超,远胜过

outcross /ˌaʊtˈkrɔːs/ vt. to breed by crossing individuals of different strains but usually of the same breed; crossbreed(种内不同品系间进行交配)杂交 | **outcrossing** /-sɪŋ/ n.

outdoor /ˈaʊtdɔː(r)/ adj. situated or done out of doors in the open air 室外的,户外的:outdoor sports 室外体育运动 compare indoor

outfit /ˈaʊtfɪt/ n. **1** a set of clothes worn for a particular occasion 套装:a riding outfit 骑马套装 **2** a group of people working together 群体;团体:Tom is the head of the ranch outfit. 汤姆是牧场工人的领班。**3** a set of equipment or articles needed for a particular activity(用于特定活动的所有装备和器具)用具,装备:a cowboy outfit 牛仔装备

vt. to provide sb/sth with a set of clothes or equipment for a particular purpose 提供装备

outlaw /ˈaʊtlɔː/ n. an intractable or unruly horse; rogue 桀骜不驯的马,难以制服的马

outlay /ˈaʊtleɪ/ n. the spending of money; expenditure 花费;开销,支出:capital outlay 资金支出

outlier /ˈaʊtlaɪə(r)/ n.【hunting】a fox found in the open rather than a covert 洞外活动的狐狸

outline /ˈaʊtlaɪn/ n. the profile of the horse(马匹的)面貌,面目:a horse with an elegant outline 一匹面貌清秀的马

out-of-bounds /ˈaʊtəvˈbaʊndz/ adv.【polo】(of a ball) crossing the side or end lines(指马球越过边线或底线)出界的

outrider /ˈaʊtraɪdə(r)/ n. **1**【racing】a mounted official who escorts racehorses to the track(护卫马匹至赛马场的赛事官员)护卫 **2** a mounted attendant who rides in front of or beside a carriage(车前或旁边的骑马随从)骑马侍卫,舆前侍卫

outside /ˌaʊtˈsaɪd/ n. the outer side of a horse(马的)外侧:When curving or turning to the left, the "outside" would be your right side. When curving or turning to the right, the "outside" would be the left side. 在左转弯时,外侧就是你的右面;而右转弯时,外侧就成了左面。 compare inside

adj. of, on or facing the outer side 外侧的,外方的:outside rein 外方缰

outside car n. *see* jaunting car 出游马车

outside mare n. a mare not owned by the stable which boards or breeds her(不为圈养马厩所有的母马)寄养母马

outsider /ˌaʊtˈsaɪdə(r)/ n.【racing】a racehorse having little chance of winning(比赛中)无望获胜的马:She always wagered on an outsider. 她总是在无望获胜的马上押注。

outspan /ˈaʊtspæn/ vt. to unharness a horse from a wagon 卸驾,卸套:The wagons were drawn up beside the road and the horses outspanned. 马车靠路停下,然后卸了马。

vi. to rest or camp at the roadside while travelling by wagon(车马旅行中)整顿休息

outspanner /ˌaʊtˈspænə(r)/ n.【driving】either one of the two horses put to the outside in a Russian troika(俄式三套车中置于外侧的马匹)外侧马

outstretched /ˈaʊtstretʃt/ adj. (of the limbs) stretched out 伸展的

ova /ˈəʊvə/ n. the plural form of ovum [复]卵子

ovarian /əʊˈveərɪən; *AmE* oʊˈ-/ adj. of or relating to the ovary 卵巢的 : ovarian artery 卵巢动脉 ◇ ovarian ligament 卵巢系膜

ovary /ˈəʊvəri; *AmE* oʊˈ-/ n.【anatomy】one of the pair of female reproductive glands that produce ova（雌性生殖腺,可产生卵细胞）卵巢

overalls /ˈəʊvərɔːls; *AmE* oʊˈ-/ n. pl. **1**（also **bib overalls**）a loose-fitting garment consisting of trousers with a front flap over the chest held up by shoulder straps（一种宽松工作裤,前襟由肩带吊着）背带裤 **2** a pair of tight riding pants with a stirrup through which the foot passes（一种紧身马裤,裤管下方带踏脚）马裤

overbite /ˈəʊvəbaɪt/ vi.（of a horse）have overshot jaw（指马）上颌突出

over-carted adj.（of a horse）harnessed to a vehicle too large for its size（指马）套车过大的

over-collected adj.（of a horse）moving with too much collection（指马）收步过大的

overface /ˈəʊvəfeɪs; *AmE* ˈoʊvərf-/ vt. **1**【jumping】to command a horse to jump a fence that is beyond its ability 跨越过高的障碍: Horse overfaced frequently would lose its confidence in jumping. 经常策马跨越过高的障碍,就会使马丧失信心。**2** to surfeit a horse with feed; overfeed 饲喂过度

overfeed /ˈəʊvəfiːd; *AmE* ˈoʊvərfiːd/ v. to feed or eat too often or too much 饲喂过度 [opp] underfeed

overfeeding /ˌəʊvəˈfiːdɪŋ; *AmE* ˌoʊvərˈf-/ n. the instance or state of being overfed 饲喂过度

overgraze /ˌəʊvəˈgreɪz; *AmE* ˌoʊvərˈg-/ vt. to allow too many stock to graze on grassland, especially to the detriment of the vegetation（使草场载畜量达到对植被有害的程度）过度放牧,过牧: Overgrazed pasture poaches more rapidly than well preserved grassland. 过度放牧的草场比维护良好的草地更易被踩坏。

over-horsed adj.（of a rider）too small for the mount in size（骑手）不相称的;人小马大的: A child would be over-horsed on a very large horse. 儿童骑高头大马,就会显得不相称。 [compare] under-horsed

overo /əʊˈverəʊ/ n. a coat pattern in pinto horses with irregular and jagged patches of white on the body（花马色型之一,被毛上的白色斑块呈不规则锯齿状）齿边白花毛,欧沃罗 [compare] sabino, tobiano, tovero

overreach /ˌəʊvəˈriːtʃ; *AmE* ˌoʊvərˈr-/ vi.（of a horse）to strike the heel of its forefoot with the toe of its hind hoof on the same side（马前肢蹄踵与同侧后肢蹄尖相撞）追突 | **overreaching** /-tʃɪŋ/ n. : Conformation, weakness or over-tracking when jumping are the main causes of overreaching. 马匹有肢体缺陷、体乏或跨栏时后肢前踏是追突发生的主要原因。 [compare] interfere, underreach

overreach boots n.（also **overreaching boots**）a bell-shaped protective covering for the front hoofs of a horse（套在马前蹄上的钟形护罩）护蹄碗: Overreach boots will help protect the heel and coronet area and tendon protectors, brushing boots or bandages will shield the leg between the knee and fetlock. 护蹄碗有助于保护马匹的蹄踵和蹄冠,而护腱绷带、防蹭绑腿和绷带则可以很好地保护前膝至球节的部分。 [syn] bell boots, racking boots

overreaching boots n. *see* over-reach boots 护蹄碗

override /ˌəʊvəˈraɪd; *AmE* ˌoʊvərˈr-/ vt. **1** to ride across or beyond 骑马经过 **2** to ride a horse too hard regardless of its physical condition（不顾马匹体况猛骑）过骑,猛骑

overrun /ˌəʊvəˈrʌn; *AmE* ˌoʊ-/ vt.【hunting】（of the hounds）to run past the line of the fox（猎犬）跑离踪迹: to overrun the line 跑离狐狸踪迹

overshot /ˌəʊvəˈʃɒt; *AmE* ˌoʊvərˈʃ-/ adj. having the upper jaw projecting beyond the lower 上颌突出的,天包地的

overshot jaw n.（also **overshot mouth**）a congenital malformation of the mouth in which the upper jaw projects beyond the bottom jaw（上

颌比下颌突出的先天口齿畸形）上颌突出，天包地 syn brachygnathia, parrot mouth, parrot jaw, salmon mouth compare undershot jaw

over-shot mouth n. *see* over-shot jaw 上颌突出，天包地

overstep /ˌəʊvəˈstep/ vi. (of the hind foot) to step beyond the corresponding forefoot print at the walk or trot（马慢步或快步行进时后蹄印跃过前蹄印）跨灶 syn overstride, over-track

overstock /ˌəʊvəˈstɒk/ vt. to have or allow more livestock than a pasture can hold（所养牲畜数目高于草场负荷）载畜过多：an overstocked grassland 载畜过多的草地 opp understock

overstrain /ˌəʊvəˈstreɪn; *AmE* ˌoʊvərˈs-/ vt. to exert or force beyond the natural limit 劳累过度

over-straining disease n. another term for exercise-related myopathy 运动相关性肌病

overstress /ˌəʊvəˈstres; *AmE* ˌoʊvərˈs-/ vt. to subject an individual to excessive physical or emotional stress（在体力和情感上使个体承受过多压力）压力过度，应急过度

overstride /ˌəʊvəˈstraɪd/ vi. *see* over-step 跨灶

over-track /ˈəʊvətræk; *AmE* ˈoʊvərt-/ vi. *see* overstep 跨灶

overtrain /ˌəʊvəˈtreɪn; *AmE* ˌoʊvərˈt-/ vt. to train a horse too much so as to incur injuries or loss of performance 训练过度 | **overtraining** /-ˈtreɪnɪŋ/ n. : To date studies have not been able to demonstrate if overtraining does exist in the horses. 目前的研究仍然不能说明马存在训练过度的情况。

overuse /ˌəʊvəˈjuːz; *AmE* ˌoʊvərˈj-/ n. excessive use 使用过度，滥用

vt. to use to excess 过度使用 compare underuse

overweight /ˌəʊvəˈweɪt; *AmE* ˌoʊvərˈw-/ adj. 【racing】(of a jockey) weighing more than the limit assigned to his mount（指骑师超过坐骑所限定的负重）超重的：According to the rules, a jockey may be a maximum of 5 pounds overweight, but it must be either posted on an information board or announced over the public address system prior to the race. 根据规则，骑师最多可超重5磅，但这必须要在赛前刊登在信息板上，或者通过公众广播进行通告。

n. the extra weight carried by a horse（马负载）超重

vi. to exceed the weight limit assigned to his mount 超重

oviduct /ˈəʊvɪdʌkt; *AmE* ˈoʊ-/ n.【anatomy】a tube through which the ova pass from the ovary to the uterus（卵子由卵巢运移至子宫所经过的管道）输卵管 syn Fallopian tube

ovulate /ˈɒvjuleɪt; *AmE* ˈɑːv-/ vi. to release ovum from the ovary 产卵，排卵 | **ovulation** /ˌɒvjʊˈleɪʃn; *AmE* ˌɑːv-/ n.

ovum /ˈəʊvəm; *AmE* ˈoʊ-/ n. (*pl.* **ova** /-və/) a female reproductive cell produced by the ovary; egg（由卵巢产生的雌性生殖细胞）卵子

owlhead /ˈaʊlhed/ n. an unruly horse 难调的马，难驾驭的马

owner /ˈəʊnə(r); *AmE* ˈoʊ-/ n. (also **horse owner**) one who owns a horse 马主

ownership /ˈəʊnəʃɪp; *AmE* ˈoʊnərʃɪp/ n. **1** the state of being a horse owner 马主身份 **2** the legal right to the possession of a horse（马匹）所有权

ownership brand n. *see* ranch brand 牧场烙印，所属烙印

ox /ɒks; *AmE* ɑːks/ n. **1** an adult castrated bull 阉公牛 **2** a suffix added to the name of any pure-bred Arab 标记纯种阿拉伯马的符号

ox fence n. *see* oxer 双横栏

oxbow /ˈɒksbəʊ; *AmE* -boʊ/ n. a U-shaped wooden piece fitted over the neck of an ox（驾牛用的U形木制挽具）套弓，牛轭

oxbow stirrup n. a stirrup with a round bottom and slightly curved sides, often used in Western saddles（一种弧边圆底的马镫，常用于西部牛仔鞍）环形马镫，圆弧马镫

oxer /ˈɒksə(r)/ n.【jumping】(also **ox fence**) an obstacle with two rails set aside（由两道横

O

杆组成的单道障碍)双横栏: ascending oxer 升坡双横栏 ◇ crossed oxer 交叉双横栏 ◇ descending oxer 降坡双横栏 ◇ parallel oxer 平行双横栏

Oxford cart n. *see* Oxford Dogcart 牛津马车

Oxford Dogcart n. (also **Oxford cart**) a light, horse-drawn vehicle of the dogcart type popular around the borough of Oxford during the 19th century (19 世纪流行于英国牛津地区的轻型马车)牛津马车

ox-kneed /ˈɒksniːd/ adj. (of knees) turning inwards when viewed from the front (马前膝前观时偏向内侧)膝内向的 syn knock-kneed

oxygenate /ˈɒksɪdʒəneɪ; *AmE* ˈɑːk-/ vt. to combine or infuse with oxygen 与氧气结合,充氧: oxygenated blood 富氧血 | **oxygenation** /ˌɒksɪdʒəˈneɪʃn; *AmE* ˌɑːk-/ n.

oxytocin /ˌɒksɪˈtəʊsɪn; *AmE* -ˈtoʊ-/ n.【physiology】a short polypeptide hormone released from the posterior lobe of the pituitary gland that stimulates the contraction of the uterus during labor and facilitates ejection of milk from the breast during nursing (一种由垂体后叶释放的短肽类激素,分娩时刺激子宫收缩,泌乳期促进乳汁排出)催产素

oxyuris equi n.【Latin】*see* pinworm 马尖尾线虫

oyster feet n. a horse's foot with horizontal ridges in the wall looking like an oyster shell, usually the result of chronic laminitis (马蹄壁出现环纹的情况,形似牡蛎壳,多由患慢性蹄叶炎引起)牡蛎蹄

P /piː/ abbr. pony 矮马

P Ⅰ n. (also **P 1**) *see* long pastern bone 长系骨，第一系骨

P Ⅱ n. (also **P 2**) *see* short pastern bone 短系骨，第二系骨

PⅢ n. (also **P 3**) *see* third pastern 第三系骨；蹄骨

pace /peɪs/ n. **1** the way or manner of moving 步法，步态：pace not true 步法混乱 **2** a two-beat lateral gait in which the foreleg and hind leg on the same side leave and strike the ground at the same time (一种两蹄音步法，其中同侧前后肢同时离地和落地)对侧步 **3**【racing】the rate of speed in a race (赛马中)步速

pace setter n.【racing】(also **pacemaker**) a horse who takes the lead and sets the pace for others in the field (开赛后领跑并为其他马匹确定步调的马)领跑马，带跑马

pacemaker /ˈpeɪsmeɪkə(r)/ n.【racing】*see* pace setter 领跑马，带跑马

pacer /ˈpeɪsə(r)/ n. a horse who moves with the two-beat lateral gait of pace 对侧步马 [syn] sidewheeler, wiggler

pachyderm /ˈpækidʒːm; *AmE* -dʒːrm/ n. a large, thick-skinned mammal, such as an elephant or rhinoceros (大型厚皮哺乳动物，如大象或犀牛)厚皮动物

pachyderm polo n. a ball game similar to polo but played on elephants 象上马球，象球 [syn] elephant polo

pacing /ˈpeɪsɪŋ/ n. (also **harness pacing**) a harness race in which the horses compete with the gait of pace (马以对侧步行进的驾车赛)对侧步驾车赛 [compare] trotting

pack /pæk/ n. **1**【hunting】a group of hounds that hunt together 猎犬群：pack of hounds 一群猎犬 **2** a bag carried on the back of an animal 驮载物：pack animal 驮畜

vt. to add loads on the back of an animal for conveyance 给驮畜加载

pack animal n. a farm animal used to carry heavy loads (用来驮运物资的农畜)驮畜

pack horse n. (also **packhorse**) a horse used to carry heavy loads (用来驮运物资的马)驮马 [syn] sumpter

pack pony n. a pony used to carry burdens on its back 驮用矮马

pack string n. (also **string**) a group of livestock tied one behind the other in a file to carry burdens and convey goods (前后相连驮运物资的牲畜纵队)驮队

packed cell volume n. (abbr. **PCV**) the percentage of total blood volume occupied by red blood cells (红细胞占血液总容积的百分比)红细胞比容：The PCV is often used as an indicator of dehydration, anemia, and other disorders. 红细胞比容常作为诊断脱水、贫血和其他疾病的指标使用。[syn] hematocrit

packer /ˈpækə(r)/ n. one who packs goods onto the backs of livestock (为驮畜加载货物的人)驮运工，装载工

packsaddle /ˈpæksædl/ n. (also **pack saddle**) a saddle on which loads can be secured (用于载物的马鞍)驮鞍

pad /pæd/ n. **1** the saddle pad 鞍垫 **2** the foot of a fox or hound (狐狸或猎犬的)趾垫，足垫 **3** part of the harness tack bridging the horse's back (置于马背上的一件挽具)搭腰 **4** the horseshoe pad 蹄铁垫 **5** the central part of a carriage step (马车的)踏垫 **6** a stirrup pad 镫垫

pad horse n.【dated】a road hack 出租用马

pad scent n. *see* stale line 变淡的踪迹

pad tree n. a wooden or metal frame to which harness pads are attached（用来放置挽鞍的木架或铁架）挽鞍架

paddling /ˈpædlɪŋ/ n. a gait defect in which the feet deviate outward from a straight line during flight and follow in an inner arc（马的一种运步缺陷，其中离地时蹄朝外偏离并形成内弧曲线）内向肢势 [syn] dishing, winging outward

paddock /ˈpædək/ n. **1** a small enclosure near the stable, chiefly used for grazing or exercising horses（厩舍旁边的小围场，供马吃草或运动）小牧场；运动栏 **2**【racing】(also **racecourse paddock**) the enclosure adjacent to a racecourse used for saddling, mounting, and checking of horses prior to the race（紧靠赛马场的围栏，用来备鞍、上马和进行赛前马匹检查）围场，围栏

paddock boots n. (also **jodhpur boots**) a kind of low boots made of leather or synthetic materials and available in laced, zipper, and pull-on styles（一种由皮革或合成材料制成的低帮马靴，有系带、拉链等几种样式）牧场马靴

paddock judge n.【racing】one responsible for all activity in the paddock（负责围场所有事务的人员）围场裁判

paddock sheet n. *see* summer sheet 牧场马衣

Pahlavan /pəˈlɑːvən/ n. a warmblood indigenous to Iran that descended from Plateau Persian crossed with Arab and Thoroughbred; stands 15.2 to 16 hands and may have a coat of any solid color; is strong in build and elegant in profile, and used mainly for riding（源于伊朗的温血马品种，由高原波斯马与阿拉伯马和纯血马杂交培育而来。体高15.2～16掌，各种单毛色兼有，体格强健，外貌俊秀，主要用于骑乘）帕拉瓦马

paint /peɪnt/ n. **1**【color】(also **pinto**) a coat color of horse with patches of white on any base color（被毛上带大块白斑的毛色）花毛 **2** a spotted cow 花毛牛

Paint Horse n. an American warmblood of a stock-type developed upon American Quarter Horses and Thoroughbreds; should have definite patches of white and markings required by its type for registration（美国采用夸特马和纯血马等育成的牧场用温血马品种，其白色花斑和别征需符合品种登记规定）花马 [syn] Pinto, Indian Pony

pair /peə; *AmE* per/ n. **1** a set of two things used together 一对 **2** a team of two horses driven or ridden side by side（并驾或并骑的两匹马）对马，双马

pair jumping n.【jumping】a show-jumping event in which two riders jump all the obstacles side by side（两名骑手并肩跨越所有障碍的比赛）双人障碍赛

palatability /ˌpælətəˈbɪləti/ n. the quality of being palatable 适口性

palatable /ˈpælətəbl/ adj. acceptable and agreeable to the taste or in flavor 合口的，适口的

palate /ˈpælət/ n. the roof of the inside of the mouth（口腔内侧的顶部）腭：hard palate 硬腭 ◇ soft palate 软腭 ◇ cleft plate 腭裂 | **palatal** /-lətl/ adj.

palatitis /ˌpæləˈtaɪtɪs/ n. *see* lampas 硬腭炎

palatum /ˈpælətəm/ n.【Latin】(*pl.* **palata** /-tə/) palate 腭：palatum durum/molle 硬/软腭

paleontologist /ˌpælɪɒnˈtɒlədʒɪst; *AmE* ˌpeliɑːnˈtɑːl-/ n. one who studies the fossils 古生物学家

paleontology /ˌpælɪɒnˈtɒlədʒi; *AmE* ˌpeliɑːnˈtɑːl-/ n. (BrE also **palaeontology**) the scientific study of living creature existing in prehistoric times by examining their fossils（通过生物化石研究史前生物体的学科门类）古生物学

palfrey /ˈpɔːlfrɪ/ n. (*pl.* **palfreys**) a light-weight horse used for ordinary riding in the Middle Ages（中世纪的）乘用马 [compare] sumpter, war horse

Palio /ˈpɑːlɪəʊ/ n. (also **Palio de Siena**) a racing meet held twice each year, on July 2 and August 16, in Siena, Italy, in which the competitors ride bareback on streets（意大利每年7

月2日和8月16日在锡耶纳举行的赛马比赛,选手在街上光背骑马进行比赛)锡耶纳赛马节

Palio de Siena n. *see* Palio 锡耶纳赛马节

palisade worms n. *see* large strongyles 大圆线虫

pallet /ˈpælɪt/ n. a rack or hard bed for stacking hay (堆放干草的支架或平台)托架,台架

pallium /ˈpælɪəm/ n.【anatomy】the mantle of gray matter forming the cerebral cortex (构成大脑皮层的灰质层)大脑皮质 [compare] medulla

palma /ˈpælmə/ n.【Latin】(*pl.* **palmae** /miː/) palm 掌

palmar /ˈpælmə(r)/ adj. of or relating to the lower part of horse's foreleg below the knee (马前肢)掌的,前掌的: palmar process 掌突 [syn] volar

palmar deviation of the carpal joints n. *see* calf knees 凹膝

palmar digital neurectomy n. the practice of cutting off the palmar digital nerve in the foreleg to relieve pain related to navicular disease (切断前肢掌神经以缓解舟骨疾病疼痛的做法)前掌神经切除术 [syn] heel nerving

palmaris /ˈpælmrɪs/ adj.【Latin】of or located near the palm 掌侧的 [compare] plantaris

Palmer, Lynwood n. (1868 – 1941) a British equestrian artist and leading figure on horsemanship, driving, and shoeing (英国马术绘画名家,在骑术、驾车和钉蹄方面也颇具名声)林恩伍德·帕尔玛

palomino /ˌpæləˈmiːnəʊ; *AmE* -noʊ/ n.【color】a color pattern in which the horse has a golden coat with flaxen or white mane and tail (被毛为金黄色而鬃、尾为亚麻黄或白色的毛色类型)淡黄色,银鬃: The coat color of palomino could be found in many horse and pony breeds, but it does not appear in purebred Thoroughbreds or Arabians. 银鬃在多个马与矮马品种都有,但在纯血马和阿拉伯马中却从未出现。

Palomino /ˌpæləˈmiːnəʊ; *AmE* -noʊ/ n. (also **Palomino horse**) a color breed developed in Spain during the 15th century; stands 14 to 17 hands and generally has a golden coat with flaxen or white mane and tail; physical characteristics vary and does not breed true to type; used for riding, parades, stock work, and driving (15世纪西班牙培育的一个毛色品种,体高14~17掌,被毛金黄色,鬃、尾为亚麻黄或白色,体貌特征变化大,品种类型不稳定,多用于骑乘、仪仗、牧场和挽驾)帕洛米诺马 [syn] golden horse, buttermilk horse

Palomino horse n. *see* Palomino 帕洛米诺马

Palomino Horse Association n. an organization founded in the United States in 1936 to promote the breeding and registry of Palomino horses (1936年在美国成立的组织机构,旨在促进帕洛米诺马的育种和登记工作)帕洛米诺马协会

Palomino Horse Breeders of America n. (abbr. **PHBA**) an organization founded in the United States in 1941 to preserve the blood purity and improve the breeding of Palomino horses (于1941年在美国成立的组织机构,旨在保护帕洛米诺马并对其进行育种改良)美国帕洛米诺马育种者协会

Palouse Pony n. *see* Appaloosa 阿帕卢萨马

Palousy /pəˈluːsi/ n. *see* Appaloosa 阿帕卢萨马

palpate /pælˈpeɪt/ vt. to examine by touch for medical purposes 触诊

palpation /pælˈpeɪʃn/ n. the act of examining by touch 触诊: rectal palpation 直肠检查

pancake /ˈpænkeɪk/ n.【slang】an English saddle 英式马鞍

pancreas /ˈpæŋkriəs/ n.【anatomy】a gland near the stomach that secretes digestive juices into the intestine and hormones that regulate the blood glucose level (位于胃附近的腺体,有向肠道分泌消化液并产生调节血糖水平的激素的功能)胰腺 | **pancreatic** /ˌpæŋkriˈætɪk/ adj.

pandemic /pænˈdemɪk/ adj. (of a disease)

P

spreading over the whole country（疾病）大流行的，流行性的
n. a widespread disease 流行性疾病

panel /ˈpænl/ n. **1**【racing, slang】one furlong 弗隆 **2** the saddle panel 鞍褥，鞍垫

pannier /ˈpæniə(r)/ n. one of a pair of large wicker baskets carried over the back of a pack animal（驮畜背上负载的两个藤篮）驮篮，驮筐

paper face n. *see* bald face 白脸

par /pɑː/ n. a normal amount or leve; average［平均］水平，［正常］水准：below-par performance 低水平发挥

parabola /pəˈræbələ/ n.【jumping】the figurative arc made by a horse when jumping over an obstacle in the air（马越过障碍时身体在空中划出的曲弧）弧线，抛物线

parade ring n.【racing】a ring on the racecourse where horses are showed before a race（赛马场开赛前展示马匹的地方）亮马圈，亮相圈

para-dressage /ˈpærəˌdresɑːʒ/ n.（also **para-equestrian dressage**）one of the two competitive events of para-equestrian sport, conducted under the same basic rules as conventional dressage, but with riders divided into different grades based on their functional abilities（残疾人马术运动的两个项目之一，与传统的盛装舞步规则基本相同，但根据骑手能力分成不同级别）残疾人舞步赛

para-driving /ˈpærəˌdraɪvɪŋ/ n.（also **para-equestrian driving**）one of the two competitive events of para-equestrian sport, conducted under the same basic rules as combined driving, but with competitors divided into different grades based on their functional abilities（残疾人马术运动的两个项目之一，与正式的驾车赛规则基本相同，但根据参赛者能力分成不同级别）残疾人驾车赛

para-equestrian /ˌpærəiˈkwestriən/ n. an equestrian sport governed by the FEI and consisting of para-dressage and para-driving with competitors divided into different grades based on their functional abilities（国际马术联合会监管的马术运动，分盛装舞步和驾车赛两个项目，根据参赛者能力分成不同级别）残疾人马术

para-equestrian dressage n. *see* para-dressage 残疾人舞步赛

para-equestrian driving n. *see* para-driving 残疾人驾车赛

parahippus /ˌpærəˈhiːpəs/ n. one of the ancestors of present-day horse that lived near the end of the Miocene epoch（现今马的远祖，生活于中新世末期）副马

parallel bars n.【jumping】a jumping obstacle consisting of two sets of parallel posts and rails jumped as a single element（由两道木杆与支架平行构成的单道障碍）平行跨栏，平行障碍

parallel oxer n.【jumping】an oxer with the front and back rails of the same height（前后跨杆高度相同的双横木跨栏）平行双横木

paralumbar /ˌpærəˈlʌmbə(r)/ adj. beside, near or alongside the lumbar vertebrae 腰椎旁的，腰椎附近的：paralumbar region 肷部

paralytic rabies n.（also **dumb rabies**）a rabies condition characterized by paralysis of body muscles（以身体肌肉麻痹为特征的狂犬病）麻痹性狂犬病

Parascaris equarum n.【Latin】equine ascarid 马蛔虫

parasite /ˈpærəsaɪt/ n. an organism that lives on or inside others and causes injury to the host（生活在其他生物表面或体内的生物体，常对宿主造成损害）寄生虫：Horses should receive regular deworming medicine to prevent parasites from forming. 马应当定期服用驱虫药以防止寄生虫滋生。

parasitic /ˌpærəˈsaɪtɪk/ adj. of or pertaining to parasites 寄生虫的，寄生的

parasiticide /ˌpærəˈsɪtɪsaɪd/ n. an agent used to destroy parasites（用于杀死寄生虫的制剂）寄生虫药，驱虫剂

parasympathetic /ˌpærəˌsɪmpəˈθetɪk/ adj. of, relating to, or affecting the parasympathetic nervous system 副交感神经的

n. the parasympathetic nervous system or any of the nerves of this system 副交感神经[系统]

parasympathetic nervous system n.【anatomy】the part of the autonomic nervous system originating from the brain stem and sacral regions of the body, responsible for stimulating digestive secretions, slowing the heart rate, constricting the pupils, dilating the blood vessels, and in general functions to inhibit and oppose the physiological effects of the sympathetic nervous system (源于脑干和荐部的自主性神经系统,有刺激消化液分泌、减缓心率、收缩瞳孔和扩张血管的作用,可总体上抑制和对抗交感神经系统的生理作用)副交感神经系统 [compare] sympathetic nervous system

paratyphoid /ˌpærəˈtaɪfɔɪd/ n. an infectious disease caused by *Salmonella*, characterized by septicemia and abortion in mares (由沙门氏菌引起的传染病,临床表现为败血症和妊娠母马流产)副伤寒 [syn] salmonellosis

parcel carter n. *see* vanner 拉货车的马

parcours /ˈpɑːkɔːz/ n.【French】a show-jumping or driving course 障碍赛场;驾车赛场

parentage /ˈpeərəntɪdʒ; *AmE* ˈper-/ n. the state or relationship of being a parent 父母身份;亲子关系

parentage testing n. (also **parentage verification**) the process of verifying the parentage of an individual by biological methods (采用生物学手段确定某一个体父母身份的方法)亲子鉴定,亲子检测

parentage verification n. *see* parentage testing 亲子鉴定,亲子检测

pari-mutuel /ˌpærɪˈmjuːtjʊəl; *AmE* -tʃʊəl/ n. (also **mutuel**)【racing】**1** a system of betting on a horse race in which the sum of money wagered are totalized and divided in proportion to the stake among those who placed bets on the winner after deducting a percentage for running cost and tax (赛马投注的一种机制,投注总额中扣除经营成本和税收后,赢家按股分得奖金)同注分彩 **2** (also **pari-mutuel machine**) a machine that records such bets and computes the payoffs; totalizator (用来记录赌金和计算赔率的机器)赌金计算器

pari-mutuel machine n. *see* pari-mutuel 赌金计算器

pari-mutuel pool n. (also **mutual pool**)【racing】the total amount bet on a certain wager 投注总额;赌金池

park Phaeton n. a light, open, horse-drawn vehicle of the Phaeton type adapted for park driving (一种轻便敞篷费顿马车,用于公园驾驶)公园费顿马车 [syn] lady's Phaeton

park trot n. a balanced, showy trot with collection and high action (一种步幅收缩的高蹄快步)游园快步

parlance /ˈpɑːləns; *AmE* ˈpɑːrl-/ n. a particular manner of speaking; jargon 行话;切口

parlay /ˈpɑːleɪ; *AmE* ˈpɑːrleɪ/ n.【racing】(also **parlay bet**) a multi-race bet in which the winnings earned on one race are wagered subsequently on its succeeding race (将一场比赛赢得的奖金押在随后比赛上的投注方式)累积投注,累积赌马 [syn] accumulator bet

vt. to bet the winnings earned in one race on its succeeding race 累积投注

parlay bet n. *see* parlay 累积押注,累积赌马

parotid /pəˈrɒtɪd/ n. a gland located under the cheek that secretes saliva (位于颊下方可分泌唾液的腺体)腮腺

adj. of or relating to the parotid gland 腮腺的: parotid gland 腮腺

parrot jaw n. *see* overshot mouth 鹦鹉嘴,上颌突出,天包地

parrot mouth n. a conformation defect of a horse's mouth in which the upper jaw overhangs the lower jaw (马匹上颌比下颌突出的体貌缺陷)上颌突出,鹦鹉嘴 [syn] overshot mouth

P

pars /pɑːz; *AmE* pɑːrz/ n.【Latin】(*pl.* **partes** /ˈpɑːtiːz; *AmE* ˈpɑːrt-/) a part 部,部分

Part /pɑːt; *AmE* pɑːrt/ n.【racing】a group of countries assorted by the International Cataloguing Standards Committee to categorize horses races held around the world (国际编目标准委员会为了划分世界各地的赛马比赛而对各个国家进行的分类)类,区: Part I countries 一区国家 ◇ Part Ⅱ countries 二区国家 ◇ Part Ⅲ countries 三区国家

Parthia /ˈpɑːθiə/ n. an ancient country of southwest Asia corresponding to modern northeast Iran from 250 BC to AD 226 (公元前250至公元226年地处西南亚的古国,位于现今伊朗的东北部)帕提亚,安息国

Parthian /ˈpɑːθiən/ adj. of or relating to Parthia 帕提亚的
n. the native or inhabitant of Parthia 帕提亚人,安息人

Parthian shot n. the skill of shooting arrows backward from the back of a galloping horse, named after the Parthians who used this tactic to win many victories in the 3rd century BC (在奔跑的马背上反身射箭的技艺,因公元前3世纪安息人多次以此取胜而得名)安息人射箭;反身射箭: The Parthian shot required superb equestrian skills, as the stirrups had not been invented at that time. 反身射箭需要高超的骑马技巧,因为那时还尚未发明马镫。

parturition /ˌpɑːtjʊˈrɪʃn; *AmE* ˌpɑːrt-/ n. the act or process of giving birth; delivery 分娩,生产: signs of impending parturition 立即分娩的迹象 [syn] foaling, labor

pas de deux n.【French】a vaulting competition performed by two vaulters (由双人进行马上体操表演)双人表演

paso /ˈpɑːsəʊ/ n. **1** a natural, four-beat gait unique to the Peruvian Paso (秘鲁帕索马特有的自然四蹄音)帕索步法 **2** (usu. **Paso**) *see* Peruvian Paso 秘鲁帕索马

Paso Fino n. a warmbood breed originating from Spain by crossing Barb, Andalusian and Spanish stocks; stands 13 to 15.2 hands and may have a coat of any color with bay, chestnut, gray and black occurring most common; is quite versatile in many disciplines and known for their smooth gaits and endurance (西班牙采用柏布马、安达卢西亚马和西班牙马杂交培育的温血马品种。体高13~15.2掌,各种毛色兼有,但以骝、栗、青和黑最为常见,可作多种用途,步态流畅,耐力持久)帕索菲诺马

Paso Fino Horse Association n. an organization founded in the United States in 1973 to promote and improve the Paso Fino (1973年在美国成立的组织机构,旨在推广并改良帕索菲诺马)帕索菲诺马协会

Paso Fino Owners and Breeders Association n. an organization founded in the United States in 1972 to promote, protect, and improve the Paso Fino; replaced by the Paso Fino Horse Association one year later (1972年在美国成立的组织机构,旨在推广、保护和改良帕索菲诺马,次年被帕索菲诺马协会取代)帕索菲诺马主与育种者协会

passade /pəˈsɑːd/ n.【dressage】a dressage movement in which the horse travels to and fro on two tracks (盛装舞步中马以横斜步侧向来回运步的动作)回转步,帕萨得

passage /pəˈsɑːʒ/ n.【dressage】a slow, cadenced movement in which the horse trots in an extremely collected and animated manner (马举蹄较高且动作活泼的慢快步)高蹄快步,帕萨基 [syn] elevated trot

passive immunity n. immunity acquired by the transfer of antibodies through injection or placental transfer to a fetus (通过注射或母源抗体获得的免疫力)被动免疫 [compare] active immunity

pastern /ˈpæstən; *AmE* -tərn/ n. the sloping part of the lower leg that connects the fetlock to the hoof (马四肢下方连接飞节与蹄的倾斜部分)蹄系: straight pastern 立系 ◇ sloping pastern 卧系 ◇ weak pastern 弱系 ◇ white pastern 系白

pastern bone n. **1** either of the two bones forming the fetlock and hoof（构成球节的两块骨）系骨：**long pastern bone** the long bone forming the bottom half of the fetlock joint（构成蹄系关节下半部的长骨）长系骨；第一系骨 [syn] P1, first phalanx, pastern bone **short paster bone** the small bone located between the long pastern bone and hoof（位于长系骨和蹄之间的短骨）短系骨；第二系骨 [syn] PⅡ, second phalanx, short pastern, coronary bone **2** the long pastern bone 长系骨

pastern joint n. the joint between the long and short pastern bones（介于长系骨与短系骨之间的关节）[蹄]系关节

pastime /ˈpɑːstaɪm; *AmE* ˈpæs-/ n. an activity done for enjoyment or recreation 娱乐[活动]：Nowadays, horse riding has become a national pastime in many countries. 目前，骑马已经成为许多国家的主要娱乐项目。

pasturage /ˈpɑːstʃərɪdʒ; *AmE* ˈpæs-/ n. *see* pastureland 草场

pasture /ˈpɑːstʃə(r); *AmE* ˈpæs-/ n. an expanse of land covered with grass and suitable for grazing farm animals（适合家畜放牧的一片草地）草场，牧场：high mountain pastures 高山牧场 ◇ The horses were turned out to pasture for free-choice exercise. 马散放在草场上任其自由活动。

vt. toput livestock in a pasture to graze 放牧

vi. to graze in a pasture 吃草

pastureland /ˈpɑːstʃəlænd; *AmE* ˈpæstʃərl-/ n. (also **pasturage**) land where farm animals feed on grass 草场

patella /pəˈtelə/ n.【anatomy】(*pl.* **patellae** /-liː/) a flat triangular bone located at the front of the knee joint（位于膝关节前部的一块扁状三角形骨）膝盖骨，髌骨 [syn] knee cap

patellar /pəˈtelə(r)/ adj.【anatomy】of or concerning patella 膝盖骨的；膝的：patellar ligament 膝韧带

patent leather n. black leather with a glossy surface, used chiefly for making saddlery（表面带亮漆的黑皮，用来制作鞍具）黑漆皮 [syn] japanned leather

paternal /pəˈtɜːnl; *AmE* -ˈtɜːrnl/ adj. received or inherited from a father 父本的；父系的：a paternal trait 父系遗传性状 [compare] maternal

paternal grandam n. the mother of a horse's sire 父系祖母马

pathogen /ˈpæθədʒən/ n. an agent that causes disease, especially a living microorganism such as a bacterium or fungus（引发疾病的细菌、真菌等微生物）病菌；病原体

pathogenic /ˌpæθəˈdʒenɪk/ adj. (capable of) causing disease 能致病的；病原的：pathogenic microorganism 病原微生物

pathological shoe n. *see* therapeutic shoe 矫正性蹄铁

pathological shoeing n. any specialized shoeing technique used to correct or remedy a disease or injury of the foot or leg（专门治疗或矫正马蹄伤病的装蹄技术）矫正性装蹄

pathology /pəˈθɒlədʒi; *AmE* -ˈθɑːl-/ n. the scientific study of the nature of diseases, their causes, symptoms, and consequences（研究疾病本质以及病因、症状和后果的学科）病理学

Pato /ˈpɑːtəʊ/ n.【Spanish】*see* El Pato 马上夺鸭

patrol judge n.【racing】any of the officials who monitor the progress of a horse race from different locations around the track（在赛场不同地点巡视比赛进程的专员）巡查裁判

pattern /ˈpætn; *AmE* -tərn/ n. the composition or arrangement of a coat color（毛色的）类型；构成：color pattern 毛色类型

paw /pɔː/ v. (of a horse) to strike or scrape ground with hoofs（马用蹄）刨扒

pawing /ˈpɔːɪŋ/ n. the vice of a horse who strikes ground with his hoofs（马用蹄刨地面的恶习）刨扒，刨地

P

payoff /ˈpeɪˌɒf/ n.【racing】the amount of money paid to a winning bettor（付给投注获胜者的金额）奖金，分红 syn dividend

PCr abbr. phosphocreatine 磷酸肌酸

PCV abbr. Packed Cell Volume 血细胞压积：As one of the blood test results，PCV indicates the degree of hydration of the horse. 在血液化验结果中，血细胞压积是马体脱水程度的指标。

peacock /ˈpiːkɒk；*AmE* -kɑːk/ n. a horse with an attractive profile（外貌俊秀的马）骏马

peacock neck n. a very high neck carriage of horse（马举颈过高的姿势）孔雀颈

peacocky /ˈpiːkɒki；*AmE* -kɑːki/ adj.（of horses）carrying head unusually high（指马）抬头过高的；雀颈的

peak /piːk/ n. the highest amount that comes in a sudden surge（突然出现的最高量或水平）高峰，峰值

peanut meal n.【nutrition】a high protein，high energy concentrate produced from the residue of peanuts after oil extraction（花生榨油后的残渣，是一种高能量、高蛋白精料）花生粕：hulled peanut meal 去壳花生粕 ◇ unhulled peanut meal 带壳花生粕

peanut skin n. the outer protective covering of a peanut（花生的外壳）花生壳：Although a good source of nitrogen，peanut skin is of limited nutritional value and palatability to most horses due to the high tannin content. 尽管是很好的氮源，但花生壳不仅营养价值低，而且由于单宁酸含量高而适口性较差。

peck /pek/ vi.【jumping】（of a horse）to touch or strike a jump slightly（指马）蹭到（跨栏） PHR V **peck at a jump** 蹭到跨栏

pecking order n. the social hierarchy among a herd of animals living together（动物群体内的社会等级）优胜序列；啄序

pectoral /ˈpektərəl/ adj. of or relating to chest 胸的：deep pectoral muscle 胸深肌 ◇ superficial pectoral muscle 胸浅肌

pedal /ˈpedl/ n.【slang】a stirrup 马镫

pedal bone n. *see* third pastern 蹄骨，即第三系骨

pedal joint n. a term for coffin joint 蹄骨关节，即系关节

pedigree /ˈpedɪgriː/ n. the record of a horse's parentage and ancestry kept in a stud book or registry（登记册中记录的有关马匹出生和血统的信息）系谱：pedigree verification 系谱鉴定 ◇ pedigree breeding 系谱选

peduncule /pɪˈdʌŋkl/ n.【anatomy】a stalklike bundle of nerve fibers connecting different parts of the brain（连接大脑各部的茎状神经纤维束）脚，茎：cerebral peduncule 大脑脚 ◇ olfactory peduncule 嗅脚 ◇ peduncule of flocculus 绒球脚 ◇ pineal peduncule 松果体脚

pedunculus /pɪˈdʌŋkjʊləs/ n.【Latin】(*pl.* **pedunculi** /-laɪ/) peduncule 脚

peel /piːl/ v. **1**（to cause）to lose or shed off（引起）剥落；脱落 **2**【cutting】（of a cow）to split away from the rider and return to the herd（指牛跑离骑手返回牛群）侧身跑开，侧身跑走：The calf peeled off to the herd. 小牛侧身跑回了牛群。

peg /peg/ n. *see* stud 防滑钉

Pegasus /ˈpegəsəs/ n. **1**【mythology】the winged horse in Greek mythology who caused the fountain Hippocrene to spring forth from Mount Helicon with a stroke of its hoof（希腊神话中的长翅飞马，用蹄踩赫利孔山后便使希波克里尼灵泉四季长涌）佩加索斯，飞马：According to Greek mythology，after Pegasus created the stream，he was captured and ridden off by Bellerophon with a golden bridle. Later，Pegasus threw Bellerophon and flew into outer space where he became the northern constellation that bears his name. 据希腊神话记载，在佩加索斯创造了灵泉后，他便被柏勒洛丰俘虏并用金勒驾驭。后来，佩加索斯将柏勒洛丰撂下马，逃到了九霄云外，化作北半星空的飞马座。**2** a constellation in the Northern Hemisphere near Aquarius and Andromeda（北半球星空靠近宝瓶座和仙女座的一个星座）飞马座

Pelham /ˈpeləm/ n. *see* Pelham bit 佩勒姆衔铁

Pelham Arabian n. *see* Alcock Arabian 佩勒姆·阿拉伯

Pelham bit n.（also **Pelham**） a curb bit with a single mouthpiece（带单根衔杆的链式口衔）佩勒姆衔铁：Pelham bit combines the action of a snaffle and a curb perfectly in one bit. 佩勒姆衔铁完美地将普通衔铁和链式衔铁的功用结合起来。

Pelham bridle n. any bridle fitted with a Pelham bit 佩勒姆勒

pellet /ˈpelɪt/ n. a small, compressed mass of feed（压缩颗粒化制成的饲料）颗粒，颗粒料：pellet feeds 颗粒饲料

vt. to make or form into pellet 颗粒化：pelleted forage 颗粒饲草 ◇ pelleted ration 颗粒料 ◇ pelleted grain mixture 颗粒化谷物混合料 | **pelleting** /-tɪŋ/ n.

pelleter /ˈpelɪtə/ n. a machine used to make pellet feeds 制粒机

Pellier, Jules Charles n. a Parisian riding master active in the 1830s（活跃在19世纪30年代的巴黎骑师）朱利叶斯·查尔斯·佩里埃

pelt /pelt/ n. the skin of an animal with the fur or hair still on it（动物的）毛皮

pelvic /ˈpelvɪk/ adj.【anatomy】of, near, or relating to the pelvis 骨盆的：pelvic cavity 骨盆腔 ◇ pelvic girdle 骨盆带

pelvic limb n. either of the two hind limbs 后肢

pelvis /ˈpelvɪs/ n.【anatomy】a basin-shaped structure of horse's skeleton that the femurs and spine are connected to（马后躯的盆状骨骼结构，髋骨和脊柱与之相连）骨盆

pen /pen/ n. a fenced enclosure for animals 畜栏

vt. to confine in a pen 圈起来，关起来

penalty /ˈpenəlti/ n. **1** a punishment or loss of advantage imposed for committing an offense（体育比赛中由于犯规而给予的惩罚或扣分）处罚，罚分：penalty point/score 罚分 **2**【polo】the polo penalty（马球比赛的）判罚

penalty goal n.【polo】a polo penalty awarded when a defending player commits a foul in the vicinity of his team's goal box to prevent a score by the opponent（防守球员在球门区为防止对方得分故意犯规后给予的判罚）罚球

penalty point n. any point deducted from a competitor's score for breaking rules（参赛者犯规后给予的减分）罚分

penalty zone n. the rectangular area surrounding a cross country obstacle in three-day event, within which faults are marked against the competitor（三日赛越野赛段中每道障碍周围的方形区域，参赛者在此出现失误后将给予罚分）罚分区

pendulous lip n. the lip of a horse when hanging low and lifeless 口唇松弛

Peneia /ˈpiːnɪə/ n.（also **Peneia pony**） a Greek pony breed developed in the district of Eleia in the Peloponnese; stands 10.1 to 14 hands and may have a bay, black, chestnut, or gray coat; is an easy keeper, frugal, hardy, and quiet; used for riding, light farm work, and packing（育成于希腊伯罗奔尼撒半岛伊莱亚地区的矮马品种。体高10.1～14掌，毛色多为骝、黑、栗或青，易饲养，耐粗饲，抗逆性强，性情安静，用于骑乘、农田轻役和驮运）皮尼亚矮马

penetrate /ˈpenɪtreɪt/ vt. to enter or force a way into; pierce 穿过，穿透：The larvae of *Strongylus vulgaris* penetrate the wall of the large intestine and migrate through the liver to the lining of the abdomen. 普通圆线虫的幼虫穿过大肠壁移行到肝脏，其后寄居在腹腔浆膜。

penile /ˈpiːnaɪl/ adj. of or relating to the penis 阴茎的：penile erection 阴茎勃起

penis /ˈpiːnɪs/ n.【anatomy】the male genital organ of copulation and urination（雄性交配和排尿的生殖器官）阴茎

pent /pent/ adj.（of an animal） penned or shut

up; closely confined(动物)关着的,圈着的

pent-up /ˌpentˈʌp/ adj. not given expression or release; repressed 抑制的;压抑的

pepsin /ˈpepsɪn/ n.【biochemistry】a protein-digesting enzyme found in gastric juice that catalyzes the breakdown of protein to peptides (一种由胃分泌的蛋白消化酶,可催化蛋白质分解为多肽)胃蛋白酶

pepsinogen /pepˈsɪnədʒən/ n.【biochemistry】the inactive precursor to pepsin (胃蛋白酶的前体,无生物活性)胃蛋白酶原

peptide /ˈpeptaɪd/ n.【biochemistry】an organic compound consisting of two or more amino acids linked together through chemical bonds (由两个以上氨基酸分子通过化学键构成的有机复合物)肽:peptide bond 肽键

perch[1] /pɜːtʃ; *AmE* pɜːrtʃ/ n. the main timber in the undercarriage of a horse-drawn vehicle (马车底盘的主体结构)底架

perch[2] /pɜːtʃ; *AmE* pɜːrtʃ/ vi.【racing】(of a rider) to squat on the saddle with his weight put on the pubis rather than the seat bones (指骑手蹲坐于马鞍,体重置于耻骨而非坐骨)蹲坐

Percheron /ˈpɜːʃərɒn/ n. a heavy coldblood breed originating in the Perche region of Normandy, France; stands 15.2 to 17 hands and often has a dappled gray coat; has a fine head, slightly arched neck, full mane and tail, short legs with no feathering, powerful hindquarters, and blue-horn hoofs; is quiet and docile, but energetic; originally used for pulling artillery and heavy coaches but now bred for general purposes (源于法国诺曼底佩尔什地区的重型冷血马品种。体高15.2~17掌,毛色多为斑点青,面貌俊秀,颈略呈弓形,鬃尾浓密,四肢短且无距毛附生,后躯发达,蹄质蓝色,性情安静而温驯,富有活力,过去用来拉军械和重型马车,现今用途较为广泛)佩尔什马:The highest horse in the world by now is a Percheron gelding named Firpon, who was foaled in 1959, stood 85 inches and weighed 2,976 pounds. 迄今为止,世界上最高的马是一匹名为"弗尔朋"的佩尔什骟马,生于1959年,体高2.16米,重达1 350千克。

Percheron Association n. an organization founded in the United States in 1877 to promote the breeding of Percheron horse and maintain its stud book (1877年在美国成立的组织机构,旨在促进佩尔什马的育种并对其进行品种登记)佩尔什马协会

Percheron Horse Association of America n. an organization founded in the United States to replace the Percheron Society of America in 1934, chartered to promote and preserve the Percheron breed and maintain a national stud book (1934年取代美国佩尔什马学会而成立的组织机构,旨在促进佩尔什马的保种工作并保存该品种的登记册)美国佩尔什马协会

Percheron Society of America n. an organization founded in 1905 in the United States to establish registration standards for the Percheron, was late replaced by the Percheron Horse Association of America in 1934 (1905年在美国成立的组织机构,建立了佩尔什马的登记准则,于1934年并入美国佩尔什马协会)美国佩尔什马学会

percussion /pəˈkʌʃn; *AmE* pərˈk-/ n. a method of medical diagnosis in which various areas of the body are tapped to determine the condition of internal organs by resonanc (医疗中通过敲击身体各处来诊断疾病的方法)叩诊

perennial /pəˈreniəl/ adj. lasting or existing through many years 多年生的

n. a perennial plant 多年生植物

perfecta /pəˈfektə/ n.【racing】*see* exacta 头二正序彩

performance /pəˈfɔːməns; *AmE* pərˈfɔːrm-/ n. the ability of a horse in work (马的)赛绩;性能:performance record 出赛记录

Performance Sale Internationale n.【French】(abbr. **PSI**) n. a top international auction held since 1980 for sports horses（自 1980 年起为竞技用马举行的国际顶级拍卖会）国际竞技马拍卖会

perfuse /pəˈfjuːz/ vt. to diffuse blood through an organ 使充血

perfusion /pəˈfjuːʒn/ n. the flow of blood into an organ or body part 肺充血

peri- pref. around; enclosing 外围的；包裹的

pericardiac /ˌperɪˈkɑːdɪæk / adj. *see* pericardial 心包的

pericardial /ˌperɪˈkɑːdɪəl/ adj.（also **pericardiac**）of or relating to the pericardia 心包的

pericardium /ˌperɪˈkɑːdɪəm/ n.【anatomy】(*pl.* **pericardia** /-dɪə/) the membranous sac that encloses the heart（包裹心脏的囊膜）心包

perimetrium /ˌperɪˈmiːtrɪəm/ n. the outer fine membrane of uterine wall 子宫外膜

perimysium /ˌperɪˈmɪzɪəm/ n.【anatomy】the sheath of connective tissue enveloping bundles of muscle fibers（包裹在肌束外的结缔组织膜）肌束膜 compare endomysium, epimysium

perineum /ˌperɪˈniːəm/ n.【anatomy】(*pl.* **perinea** /-niː/) the region between the scrotum and the anus in males, and between the posterior vulva junction and the anus in females（雄性阴囊和肛门间的部分，雌性阴户与肛门间的部分）会阴 | **perineal** /-ˈniːl/ adj.

period /ˈpɪəriəd; *AmE* ˈpɪr-/ n. a unit of time longer than an epoch and shorter than an era（介于代与世的地质年代划分单位）纪 compare epoch, era

periosteal /ˌperɪˈɒstɪəl/ adj. of or relating to the periosteum 骨膜的

periosteum /ˌperɪˈɒstɪəm/ n.【anatomy】(*pl.* **periostea** /-tiː/) a dense fibrous membrane covering the surface of bones（包裹在骨表面的致密纤维膜）骨膜 | **periosteous** /-tiəs/ adj.

peripheral /pəˈrɪfərəl/ adj. of the surface or outer part of a body or organ; external（机体或器官）外围的；外部的

perissodactyl /pəˌrɪsəʊˈdæktɪl/ n. a hoofed mammal of the order *Perissodactyla*, such as horse and rhinoceros, that has an uneven number of toes（蹄的数目为奇数的有蹄类哺乳动物，如马和犀牛）奇蹄动物

Perissodactyla /pəˌrɪsəʊˈdæktɪlə/ n.【Latin】an order of non-ruminant, hoofed mammals 奇蹄目

peristalsis /ˌperɪˈstælsɪs/ n. the wave-like contractions of the smooth muscles of the alimentary canal（消化道平滑肌的波动性收缩）蠕动 | **peristaltic** /-tɪk/ adj.

peritoneal /ˌperɪtəˈniːəl/ adj. of or relating to peritonea 腹膜的

peritoneum /ˌperɪtəˈniːəm/ n.【anatomy】(*pl.* **peritonea** /-niː/) the serous membrane that lines the walls of the abdominal cavity and folds inward to enclose the viscera（内衬于腹腔壁的浆膜层，包裹内脏）腹膜

peritonitis /ˌperɪtəˈnaɪtɪs/ n.【vet】inflammation of the peritoneum 腹膜炎

perlino /pəˈliːnəʊ; *AmE* pərˈliːnoʊ/ n.【color】a coat color of horse who has a white to cream-colored coat with pink skin, blues eyes, and slightly red points（马的毛色类型之一，其被毛为乳白色，皮肤为粉红色，眼为蓝色，四肢末梢带有粉色调）珍珠白

permanent /ˈpɜːmənənt; *AmE* ˈpɜːrm-/ adj. lasting or remaining unchanged 永久的，恒定的：permanent teeth 永久齿，恒齿 opp deciduous

Peruvian Paso n.（also **Paso**）a warmblood horse breed originating in Peru in the mid-16th century; descended from Spanish stock put to Barb, Friesian, and Andalusian mares; stands 14.1 to 15.1 hands and may have a coat of any solid color with bay and chestnut most common; has a medium head, broad and deep chest, a fairly short, muscular neck, short back, and low-set tail; is possessed of great stamina and used mainly for riding, farm work, and harness driving（秘鲁在 16 世纪中叶采用西班牙种公马与雌性柏布马、弗里斯马和安达卢西亚马母

马繁育而来的温血马品种。体高 14.1 ~ 15.1 掌,各种单毛色兼有,但以骝和栗常见,耐力好,尾础低,胸身宽广,头大小适中,背短,颈粗短,以其特征性帕索步著称,用于骑乘、牧场工作和挽车)秘鲁帕索马:The Peruvian Paso is the national horse of Peru and is known for its comfortable ambling gait. 秘鲁帕索马是秘鲁的国马,以其舒适的溜花蹄步法而知名。 syn Peruvian Stepping Horse

pesade /pəˈzɑːd/ n.【dressage】a high school movement in which the horse rears on its hind legs with its forelegs in the air(高等骑术中马以后肢直立而前肢悬于空中的动作)后肢直立,派萨德

pet /pet/ n. an animal kept for amusement or companionship(娱乐或陪伴用的动物)宠物:Nowadays, more and more people keep pony as a pet. 现在,越来越多的人把矮马作为宠物豢养。

peytral /ˈpeɪtrəl/ n. a piece of armour used to protect the horse's chest(马用盔甲部件,用来保护马匹前胸)护肩 — *see also* **champron**, **croupiere**, **crinet**

Phaeton /ˈfeɪtn/ n. a light, four-wheeled open carriage, usually drawn by a pair of horses(一种轻型敞篷四轮马车,多为双马联驾)费顿马车

phagocyte /ˈfæɡəsaɪt/ n. a type of cell that engulfs harmful microorganisms and other foreign bodies in the bloodstream and tissues(可吞噬血液或组织中有害微生物及异物的一类细胞)吞噬细胞

phalange /ˈfælændʒ/ n.【anatomy】*see* phalanx 指骨;趾骨

phalangeal /fəˈlændʒɪəl/ adj. of or relating to a phalanx or phalanges 指骨的;趾骨的

phalanx /ˈfælæŋks/ n.【anatomy】(*pl.* **-anxes**, or **phalanges** /fəˈlændʒiːz/) a bone of toe 趾骨:first phalanx 第一趾骨(长系骨) ◇ second phalanx 第二趾骨(短系骨) ◇ third phalanx 第三趾骨(蹄骨) syn phalange

phantom /ˈfæntəm/ n. an inanimate object similar to a female horse used for a stallion to mount and collecting semen(一件外形似母马的假体,供种公马爬跨采集精液使用)假台畜,假母马 syn dummy mare

pharmacology /ˌfɑːməˈkɒlədʒi; *AmE* ˌfɑːrməˈkɑːl-/ n. the branch of medicine concerned with the uses, effects, and action of drugs(研究药物使用、效果和作用的医学分支)药物学;药理学 | **pharmacological** /ˌfɑːməkəˈlɒdʒɪkl; *AmE* ˌfɑːrməkəˈlɑːdʒ-/ adj. 药物的:pharmacological therapy 药物治疗

pharyngitis /ˌfærɪnˈdʒaɪtɪs/ n. inflammation of the pharynx 咽炎

pharynx /ˈfærɪŋks/ n.【anatomy】the upper part of the throat where the passages to the windpipe and esophagus meet(喉上部气管与食管的交接处)咽

phenotype /ˈfiːnətaɪp/ n. the observable characteristics of an individual(个体的可见特征)表型 compare genotype

phenylalanine /ˌfiːnɪlˈæləniːn, ˌfɛn-/ n.【nutrition】an essential amino acid(一种必需氨基酸)苯丙氨酸

phenylbutazone /ˌfiːnɪlˈbjuːtəzəʊn, ˌfɛn-/ n.【vet】an anti-inflammatory, analgesic drug used in the treatment of traumatic musculoskeletal injuries or disorders(用来治疗肌肉骨骼损伤的消炎镇痛药)苯基丁氮酮,保泰松 syn bute, butazone

pheomelanin /ˌfiːəʊˈmelənɪn/ n. any of a group red or yellow pigments found in hair and skin of animals(见于动物皮肤和毛发的红色或黄色色素)褐黑色素 compare melanin

pheromone /ˈferəməʊn; *AmE* -moʊn/ n. a chemical substance secreted by an animal that influences the behavior of others of its species(动物分泌的化学物质,可影响同种其他个体的行为)外激素

PHF abbr. Potomac horse fever 波托马可热

phosphocreatine /ˌfɒsfəʊˈkriːətiːn; *AmE* ˌfɒsfoʊˈk-/ n.【biochemistry】(abbr. **PCr**) a high energy phosphate found in muscle tissue (一种存在于肌肉中的高能磷酸化合物)磷酸肌酸

phosphorus /ˈfɒsfərəs; *AmE* ˈfɑ:s-/ n.【nutrition】a mineral element essential for the bone formation, growth, and maintenance of horses (马体骨骼形成、生长和维持所必需的一种矿物质元素)磷: It is vital that the ration fed to your horse has a calcium to phosphorus ratio of 2:1 in terms of available minerals. 马匹日粮中,可利用矿物质元素钙磷比为2:1至关重要。

phosphorylate /ˈfɒsfərɪleɪt/ vt. to add a phosphate group to an organic molecule (为有机分子增加磷酸基)磷酸化

phosphorylation /ˌfɒsfɒrɪˈleɪʃn/ n. the act or process of phosphorylating 磷酸化: oxidative phosphorylation 氧化磷酸化

photo finish n.【racing】a finish so close that it is necessary to review photograph shot by a real-time camera to determine the winner (赛马通过终点时彼此过于接近,必须通过回放实时摄像机拍摄画面方可决定胜负)摄像决胜

photophobia /ˌfəʊtəˈfəʊbiə; *AmE* ˌfoʊtəˈfoʊ-/ n. an abnormal sensitivity to or intolerance of sunlight by the eyes (眼对于光线异常敏感或不可忍受)畏光,羞明

phylum /ˈfaɪləm/ n. (*pl.* **phyla** /-lə/) a principal taxonomic category that ranks below kingdom and above class (介于界和纲之间的主要分类单位)门 [compare] kingdom, class, order, family, genus, species

physiological dead space n.【physiology】(abbr. **$V_{D\ phys}$**) the volume of air that is brought in contact with a respiratory surface but gas exchange does not occur because the region is not currently being perfused by blood (指吸入后可与呼吸面接触但由于该处血液供应不足而不能发生气体交换的空气量)生理无效腔 [compare] anatomical dead space

physiological saline n. (also **saline**) the 0.9% sodium chloride solution that is isotonic with blood and is used in medicine and surgery (浓度为0.9%的氯化钠溶液,与血液等渗,常用于医药和外科手术)生理盐水

physiology /ˌfɪziˈɒlədʒi; *AmE* -ˈɑ:lə-/ n. **1** the functions of a living organism or any of its parts (生物有机体及其部位的功能)生理机能 **2** the biological study of the functions of living organisms and their parts (对有机体及其部位功能的生物学研究)生理学 | **physiological** /ˌfɪziəˈlɒdʒɪkl; *AmE* -ˈlɑ:dʒ-/ adj.

physitis /faɪˈsaɪtɪs/ n. *see* epiphysitis 骨骺炎,生长板发育不全

pia mater /ˈpaɪə ˈmeɪtə(r)/ n.【anatomy】the fine vascular membrane that closely envelops the brain and spinal cord (包裹大脑和脊髓的微血管膜)软膜: pia mater encephali 软脑膜 ◇ pia mater spinalis 软脊膜 [compare] dura mater, arachnoid

piaffe /pɪˈæf/ n.【dressage】a dressage movement in which the horse trots in place (盛装舞步中马以快步在原地踏步)原地快步,皮亚夫

vi. to perform piaffe 踏原地快步

pic /pɪk/ n. a wooden-handled lance used by a mounted picador in Spanish bullfight (西班牙斗牛表演中刺牛的骑者手中所持的木柄长矛)长矛,刺矛

pic six n.【racing】*see* pick six 六连赢

picador /ˈpɪkəˌdɔ:/ n. a rider in Spanish bullfight who pokes the bull's shoulder with a lance (西班牙斗牛表演中用矛刺牛的骑手)刺牛者;斗牛士 — *see also* **matador**

pick /pɪk/ vt. to choose or take from a group 选择;挑拣 PHR V **pick up** to choose a horse for use in competitions 挑选: to pick up a ride 挑了一匹马 **pick up a lead** (of a horse) to lead with either the right or left foreleg (指马)以某前肢起步

n.【racing】a multi-race bet in which the bettor tries to select the winners of all included races (一种多场次赛马投注方式,赌马者须选出

所有比赛的头马)连赢,连彩:pick six 六连赢

pick nine n.【racing】a wager in which the bettor tries to select the winners of nine consecutive races(赌马者选出连续九场比赛获胜头马的押注方式)九连赢

pick six n.【racing】(also **pic six**) a wager in which the bettor tries to select the winners of six consecutive races(赌马者选出连续六场比赛获胜头马的投注方式)六连赢

pick three n.【racing】a wager in which the bettor tries to select the winners of three consecutive races(赌马者须选出连续三场比赛获胜头马的投注方式)三连赢 [syn] daily triple

pickaxe /ˈpɪkæks/ n. *see* pick axe team 镐形联驾

pickaxe team n. (also **pickaxe**) a team of horses consisting of three leaders and two wheelers or two leaders and one wheeler to pull a carriage(由三匹头马与两匹辕马或两匹头马与一匹辕马组成的联驾队伍)镐形联驾,倒锥式联驾

pickup man n.【rodeo】a mounted official who assists bronc riders to dismount when the required ride time ends(超过规定时间后协助骑野马的选手下马的人员)接应人员,场地接应

picnic race n. a race meeting held in the outback of Australia in which amateur riders compete for small prizes on primitive bushland racetracks(澳大利亚内地由业余赛骑手在荒郊赛道上进行的一种奖金很少的比赛)野营赛,野餐赛:picnic race club 野营赛俱乐部

piebald /ˈpaɪbɔːld/ n. a horse with large white patches on black coat(在黑色被毛上有大块白斑的马)黑白花马 [compare] skewbald

pig sticking n. the mounted sport of hunting wild boar with a spear(骑马持矛追捕野猪的运动)猎野猪

pigeon-toed /ˈpɪdʒənˌtəʊd/ adj. (of the hoofs) turning abnormally inward(指蹄)内翻的,内八字的 [syn] toed-in

pig-eyed /ˈpɪɡˌaɪd/ adj. having eyes like that of a pig 像猪眼的:Pig-eyed horses have a smaller field of vision than others. 似猪眼的马视野较其他马要狭窄些。

pigging string n.【roping】(also **piggin' string**) a short length of rope used to immobilize an animal by tying its legs(用来捆绑动物四肢的短绳)绑腿绳

piggy /ˈpɪɡi/ adj. of or like pigs (似)猪的:A bold eye is a sign of intelligence and courage while a small piggy eye is an indication of meanness and wickedness. 马大而明澈的眼睛是智慧和勇敢的象征,而似猪的小眼则是卑微与邪恶的代名词。

pigment /ˈpɪɡmənt/ n. the natural coloring organic matter in animal or plant tissue(动植物组织内可产生颜色的有机化合物)色素:pigment cell 色素细胞

pigskin /ˈpɪɡskɪn/ n. the skin of a pig or leather made from this 猪皮;猪革

piker /ˈpaɪkə(r)/ n.【slang】a driver who tried to bypass a toll-gate without paying 绕路不缴路费的车夫

pill /pɪl/ n. **1** a small, rounded mass of medicine designed to be swallowed 药丸,药片 **2**【racing】*see* number ball 闸位球

Pill Box n.【slang】a light, horse-drawn vehicle of the Phaeton type used formerly by a doctor on his daily visits(过去医生出诊时使用的小型费顿马车)医药马车,出诊马车

pillars of the Stud Book n. the three founding sires, Byerley Turk, Godolphin Arabian, and Darley Arabian to which all Thoroughbreds can trace their lineage(育成纯血马的三匹奠基阿拉伯马公马,包括达利·阿拉伯、比艾尔力·土耳其和哥德芬·阿拉伯)纯血马登记册的奠基公马

pillion /ˈpɪliən/ n. (also **pillion saddle**) a padded seat placed behind the saddle on a horse for an extra rider(置于马鞍后面的坐垫,可供额外的骑者乘坐)附座,副鞍

vi. to ride on a pillion 乘附座,坐副鞍

pillion post n.【dated】a mounting block 上马台

pillion saddle n. *see* pillion 附座,副鞍

pin firing n. (also **point firing**) a method of treating chronic leg problems by inserting hot pins or needles to produce scar tissue and accelerate healing(通过火针穿刺生成烧伤组织,以加速治疗四肢慢性炎症的方法)火针刺疗,火针疗法

pin toes n. *see* pigeon toed 狭蹄

pincers /ˈpɪnsərz/ n. pl. **1** (also **pinchers**) a farrier tool having a pair of jaws, used to pull off nails or remove horseshoes(蹄铁匠用来拔蹄钉或卸蹄铁的工具)夹钉钳;夹蹄钳 **2** the incisors 切齿

pinch /pɪntʃ/ vt. **1** to squeeze or grip between finger and thumb(用拇指和指头捏)挤,捏;**2** to squeeze with the jaws of a tool(用钳)捏,夹:a pair of pinching boots 一双夹脚的靴子 ◇ Sam was pinched back on the homestretch. 山姆在终点直道被挤到了后面。

pinchers /ˈpɪntʃərz/ n. pl. *see* pincers 夹钉钳;夹蹄钳

Pindos /ˈpɪndəus/ n. a native pony breed indigenous to Greece; stands 12 to 13 hands and generally has a dark gray coat, although bay, black, and brown also occur; is strong, hardy, frugal, and possessed of exceptional stamina; used for light farm work, packing, and riding(希腊本土矮马品种。体高12~13掌,毛色多为暗青,但骝、黑和褐亦有之,体格健壮,抗逆性强,耐粗饲,耐力好,多用于轻挽、农田作业、驮运和骑乘)品德斯矮马

pineal /paɪˈniːəl/ n.【anatomy】(also **pineal body, pineal gland**) a small gland situated behind the third ventricle of the brain and responsible for melatonin synthesis(位于第三脑室下方的小腺体,与褪黑素的合成有关)松果体,松果腺

pineal body n. *see* pineal 松果体

pineal gland n. *see* pineal 松果体

pinhooker /ˈpɪnˌhʊkə(r)/ n.【racing】one who buys and sells racehorses for profit(为获利而倒卖赛马的人)赛马贩子

pinna /ˈpɪnə/ n. (*pl.* **pinnae** /-niː/)【anatomy】auricle 耳郭;外耳

pinto /ˈpɪntəʊ; *AmE* -toʊ/ n. **1** the coat color of paint 花毛 **2** (usu. **Pinto**) a Paint Horse 花马:Stock-type Pinto 牧场用花马 ◇ Hunter-type Pinto 狩猎用花马 ◇ Pleasure-type Pinto 娱乐用花马 ◇ Saddle-type Pinto 乘用型花马 ◇ Pinto is not a specific breed, but rather a color. Both piebald and skewbald are often referred to as "coloured" horses. 花马并非马品种,而属毛色。黑白花马和杂色花马都经常被称为花马。 compare piebald, skewbald

Pinto Horse Association of America n. (abbr. **PtHA**) an organization founded in Connecticut, USA in 1947 to maintain color registries for Pinto horses, improve their breeding, and promote the public interest in them(1947年在美国康涅狄格州成立的组织机构,旨在对花马进行毛色登记、品种改良并促进公众的兴趣)美国花马协会

pinworm /ˈpɪnwɜːm; *AmE* -wɜːrm/ n. a white round worm of the genus *Oxyuris equi* that infests the large intestine of horses and might cause itching and rubbing of the anus(一种寄生在马大肠内的尖尾属线虫,常引起马臀部瘙痒)蛲虫 syn seatworm

Pinzgauer Noriker n. *see* Noriker 诺立克马

pipe /paɪp/ n. **1**【hunting】a branch or hole of a fox's den(狐狸洞穴的分支)岔口 **2**【slang】the wind-pipe 气管

pipe stall n. an enclosure constructed of metal pipe rails 钢管围栏

pipe-opener n. a short gallop given to a horse to clear his wind-pipe(为打开马匹胸肺而进行的短途疾驰)开腔疾驰,开腔奔驰

piroplasmosis /ˌpɪrəplæzˈməʊsɪs; *AmE* -ˈmoʊ-/ n. *see* equine piroplasmosis 马梨形虫病

pirouette /ˌpɪruˈet/ n.【French】a dressage movement in which the horse circles on its hindlegs (盛装舞步中马以后肢为轴旋转的动作)旋转:pirouette on the center 中心旋转 ◇ The pirouette is often performed at the canter. 定后肢旋转通常在慢跑中进行。

vi. to perform such a movement 旋转

pirouette sur le centre n.【French】turn on the center 中心旋转

pirouette sur les haunches n.【French】a movement in which the horse turns around his inner hindleg(马)定后肢旋转

pisiform /ˈpɪsəfɔːm; *AmE* -fɔːrm/ n.【anatomy】the accessory carpal bone 副腕骨

pit /pɪt/ n. an underground coal mine 地下煤矿;矿井

pit pony n.【Brit】(also **pitter**) a pony used formerly in coal mines to haul coal (过去在煤矿用来驮运煤炭的矮马)矿井矮马 syn collier

pitch /pɪtʃ/ vt. (of a horse) to throw the rider off its back; unseat (指马)撂下骑手

pitchfork /ˈpɪtʃfɔːk; *AmE* -fɔːrk/ n. a long-handled fork used for lifting and pitching hay (用来挑取干草的一种长叉)干草叉

pitter /ˈpɪtə(r)/ n. *see* pit pony 矿井矮马

pituitary /pɪˈtjuːɪtəri; *AmE* -ˈtuːəteri/ n.【anatomy】a neuroendocrine gland located at the base of the brain of vertebrates that secrets a number of protein hormones, including the LH and FSH (位于脊椎动物大脑基底部的神经内分泌腺体,主要分泌包括促黄体素和促卵泡素在内的多种蛋白类激素)垂体: anterior pituitary gland 垂体前叶 ◇ posterior pituitary gland 垂体后叶

pivot /ˈpɪvət/ n. a half turn on the haunches performed by a horse (马以后躯为轴旋转半圈)旋转,转圈: to pivot on the hindquarters 定后肢旋转

place /pleɪs/ vi.【racing】**1** (of a horse) to finish second in a race (指赛马)位列第二 compare win, show **2** (of a horse) to finish a race in a ceratin position(指赛马)得名次: Against many bettors' expectation, Rock of Gibraltar only placed second in the Breeders' Cup Mile. 出乎许多赌马者预料的是,"直布罗陀岩石"在育马者杯一哩赛中仅得了第二名。**3** to put an amount of money on a racehorse; bet 押注,投注

place bet n.【racing】a wager in which the bettor wins if the selected horse finishes first or second in a race (赛马投注方式之一,赌马者所选赛马跑入前两名即可赢奖)头二彩,位置彩 compare win bet

place pool n.【racing】the sum of money wagered in a horse race for second place finishers (押在第二名赛马上的赌金总额)位置赌金池

placenta /pləˈsentə/ n.【anatomy】(*pl.* **placentae** /-tiː/ or **placentas**) a membranous vascular organ that develops in the uterus of female mammals to nurture the fetus during pregnancy (雌性哺乳动物子宫内壁上形成的富含脉管的膜性器官,在妊娠期可为胎儿提供养分)胎盘:The placenta consists of the chorionic and the allantoic membranes which fuse in early development of the embryo, it is ideally suited for facilitating material exchange between the maternal and fetal blood stream. 胎盘由绒毛膜和尿囊膜在胚胎发育的早期融合而成,其结构非常适合母体与胎儿间的物质交换。

placental /pləˈsentl/ adj. of or relating to the placenta 胎盘的: placental barrier 胎盘屏障 ◇ placental gonadotropin 胎盘促性腺激素

placentation /ˌplæsənˈteɪʃn/ n. the formation and development of a placenta in the uterus during pregnancy (妊娠期间胎盘在子宫的形成及发育过程)胎盘形成

placing judge n.【racing】a racing official who judges the order in which horses finish in a race (判决赛马冲过终点名次的人员)终点裁判,名次裁判

plain /pleɪn/ adj. ordinary or common 普通的,简单的: plain jointed snaffle 单关节衔铁

n. a large area of land with few trees (仅有稀疏树木的大片平地)平原

plain horseshoe n. a horseshoe without studs or other special features (不带防滑钉和其他特征的蹄铁)普通蹄铁

plains zebra n. *see* common zebra 普通斑马

plait /plæt/ n. a braid of hair 辫子

vt. to braid 编成辫子:A badly plaited tail is worse than no plait at all. 马尾辫糟的话还不如散着雅观。

plaited rein n. a rein made of braided leather strands 革编缰绳

plaiting[1] /ˈplætɪŋ/ n. the act of braiding the mane or tail 编辫

plaiting[2] /ˈplætɪŋ/ n. *see* rope walking 拧绳肢势

plane /pleɪn/ n.【anatomy】a flat or level surface 面:transverse plane 横切面 ◇ vertical plane 垂直面

plank /plæŋk/ n.【jumping】a flat piece of wood composing a jumping obstacle (搭建障碍的木质宽板)木板

plank jump n. *see* board fence 木板障碍

planta /ˈplæntə/ n.【Latin】the sole of foot 蹄底;跖

plantain /ˈplæntɪn/ n. any of various plants of the genus *Plantago* that produces dense spikes of small greenish flowers (车前草属植物,开淡绿色小花,穗状花序)车前草 [syn] ribwort, ribgrass

plantar /ˈplæntə(r)/ adj.【anatomy】of or relating to the sole of the hoof 蹄底的;趾的

plantar cushion n. (also **digital cushion**) a mass of fibro-fatty tissue that forms the bulb of horse's heel (构成马蹄球的纤维脂肪组织)趾垫:The plantar cushion functions as a shock absorber for horse's feet. 马的趾垫有减震作用。

plantaris /ˌplænˈteərɪs/ adj.【Latin】plantar 跖的 [compare] palmaris

Plantation Walking Horse n. *see* Tennessee Walking Horse 田纳西走马

plasma /ˈplæzmə/ n. **1** the clear, yellowish fluid portion of blood or lymph (血液或淋巴液中透明、淡黄色的成分)浆液 **2** the blood plasma 血浆:Plasma proteins consist of albumins, globulins, and fibrinogen. 血浆蛋白由白蛋白、球蛋白和纤维蛋白原组成。

plate /pleɪt/ n. **1** a silver or gold dish awarded as a prize in a race or competition (比赛中作为奖品的金或银盘)奖盘 **2** a light horseshoe used for racing (用于比赛的轻便)蹄铁:racing plate 赛用蹄铁 ◇ training plate 训练蹄铁

plate-and-screws stabilization n. a method of stabilizing a fractured leg bone (固定四肢骨折的方法)夹板螺丝固定法

Plateau Persian n. an Arab-type horse breed indigenous to the plateau regions of Iran; stands about 15 hands and generally has a gray, bay, or chestnut coat ; is strong and sure-footed, and used mainly for riding (伊朗高地地区的本土马品种,体貌似阿拉伯马。体高约 15 掌,毛色多为青、骝或栗,体格强健,步态稳健,主要用于骑乘)高原波斯马

platelet /ˈpleɪtlət/ n.【physiology】a minute, disk-like cytoplasmic body found in the blood plasma of mammals that promotes blood clotting when blood vessels are damaged (存在于哺乳动物血浆中的圆盘状胞质体,血管破裂时有促进凝血的作用)血小板 [syn] thrombocyte

plater /ˈpleɪtə(r)/ n.【racing】a farrier who specializes in shoeing racehorses 赛马蹄铁匠,赛马蹄师

platers Union n. *see* International Union of Journeyman Horseshoers of the United Stated and Canada 北美蹄铁匠联盟

play /pleɪ/ vi.【racing】to bet on horses in a race 赌马,押注:to play horses 赌马

player /ˈpleɪə(r)/ n. **1** one who plays in a mounted sports 球员,参赛者:a polo player 马球球员 **2**【racing】one who bets on a horse race 赌马者,投注者 **3** (usu. **players**) *see* keys 衔缀

pleasure /ˈpleʒə(r)/ n. an activity or event from which one gets enjoyment 娱乐,休闲:pleasure

P

riding 休闲骑乘

pleasure class n. a show class in which horses are shown at the walk, trot, canter, and backing (马匹展览赛的级别之一,马匹主要展示其慢步、快步、跑步及后退)娱乐级

Pleasure-Type Pinto n. one of four Pinto conformation types developed for both pleasure riding and Western disciplines; is medium in size and has a high tail carriage, a well-muscled thigh and gaskin (四个花马类型之一,适合娱乐骑乘和西部牛仔项目,体型大小适中,尾础高,四肢肌肉发达)娱乐用花马

pleura /ˈplʊərə; *AmE* ˈplʊrə/ n. 【anatomy】(*pl.* **pleurae** /-riː/) a thin serous membrane in mammals that envelops each lung and folds back to make a lining for the chest cavity (哺乳动物体内的薄层浆膜,包裹肺叶并回折形成胸腔内壁)胸膜: pulmonary pleura 肺胸膜

pleural /ˈplʊərəl; *AmE* ˈplʊrəl/ adj. of or relating to pleurae 胸膜的: pleural fluids 胸腔液 ◇ pleural membranes 胸膜 ◇ pleural cavity 胸膜腔 ◇ There are two pleural membranes, the parietal membrane and the visceral membrane. 胸膜有壁层胸膜和脏层胸膜两层。

pleurisy /ˈplʊərəsi; *AmE* ˈplʊr-/ n. (also **pleuritis**) inflammation of the pleura usually occurring as a complication of pneumonia, characterized by fever, chills, painful breathing and coughing accompanied by accumulation of fluid in the pleural cavity (由肺炎并发引起胸膜发炎的病症,其症状多表现为发热、风寒、呼吸困难和咳嗽并伴有胸膜腔积液)胸膜炎

pleuritis /plʊəˈraɪtɪs/ n. *see* pleurisy 胸膜炎

pleur(o)- pref. of or relating to pleura 胸膜的

pleuropneumonia /ˌplʊərəʊnjuːˈməʊnɪə/ n. pneumonia aggravated by pleurisy (由胸膜炎恶化所致的肺炎)胸膜肺炎: equine contagious pleuropneumonia 马传染性胸膜肺炎

Pleven /ˈplevən/ n. a warmblood breed developed in the late 19th century in Pleven, Bulgaria by crossing Russian Anglo-Arab stallions with purebred Arabs or local halfbred mares; stands about 15.2 hands and generally has a chestnut coat and Arab-like features; is a good all-round horse equally suitable for light farm work, riding, and competitive sports (19 世纪末保加利亚普列文地区采用俄国盎格鲁阿拉伯马公马与纯种阿拉伯马和当地半血马母马杂交培育而来的温血马品种。体高约 15.2 掌,毛色为栗,外貌与阿拉伯马相似,是优秀的兼用马种,也适用于农田轻役、骑乘和马术竞技)普列文马

pliable /plaɪəbl/ adj. easily bent or shaped; flexible 柔韧的;有弹性的: loose and pliable skin 柔韧而有弹性的皮肤

plica /ˈplaɪkə/ n. 【Latin】(*pl.* **plicae** /-kiː/) a fold or ridge of skin or membrane (皮肤或膜上的褶)皱襞;褶

Pliohippus /ˌplaɪəʊˈhɪpəs/ n. one of the early ancestors of the modern horse, which is the first true monodactyl (现代马的远祖之一,为真正的单蹄兽)上新马— *see also* **Eohippus**, **Hipparion**, **Merychippus**, **Mesohippus**

plough /plaʊ/ n. 【BrE】*see* plow 犁

vt. to plow 犁地

ploughland /ˈplaʊlænd/ n. 【BrE】 *see* plowland 耕地;狩猎乡野

plow /plaʊ/ n. (BrE **plough**) a farm implement consisting of a heavy blade at the end of a beam, used for breaking up soil and cutting furrows (一种农业用具,构架末端带钝刃,用来翻土和划沟)犁

vt. to break and turn over earth or make furrows with a plow 犁地

plow rein n. the harness rein used to draw the horse in the direction of the desired turn in plowing (犁地时用来操控马转向的挽缰)犁地缰 compare direct rein

plowland /ˈplaʊlænd/ n. (BrE **ploughland**) **1** land plowed for growing crops 耕地 **2**【hunting】the field over which a hunt is conducted 狩猎乡野

plowing harness n. the harness for horses plowing the field 犁地挽具

pluck[1] /plʌk/ vt. to pull hairs from the tail to make its look neater 拔毛,拔尾

pluck[2] /plʌk/ n. spirited courage 勇气;魄力

plucked tail n. *see* pulled tail 拔过的尾

plug /plʌg/ n. **1** a horse of common breed and profile 驽马 **2** a tired horse moving slowly 乏力的马,疲惫的马

PMSG abbr. pregnant mare serum gonadotropin 孕马血清促性腺激素

PMU abbr. pregnant mare urine 妊娠母马尿液

PMU farm n. a horse farm where the urine of pregnant mares are collected for the production of estrogen(收集母马尿液分离制备雌激素的马场)PMU 马场

pneumatic /njuːˈmætɪk; *AmE* nuː-/ n.【anatomy】having cavities filled with air(腔体内含有气体)含气的:pneumatic bone 含气骨

pneumo- /ˈnjuːməʊ; *AmE* nuːmoʊ/ pref. **1** of or relating to the lungs 肺的 **2** of or relating to air or gas 气体的,气的

pneumonia /njuːˈməʊnɪə; *AmE* nuːˈmoʊ-/ n. inflammation of the lungs caused by viruses, bacteria, or other microorganisms(由病毒、细菌或其他微生物引发的肺部炎症)肺炎

pneumonitis /ˌnjuːməʊˈnaɪtɪs; *AmE* ˌnuːmoʊ-/ n. inflammation of lung tissue 局部性肺炎 | **pneumonitic** /-tɪk/ adj.

pneumo-puncture /ˌnjuːməʊˈpʌŋktʃə(r); *AmE* ˌnuːmoʊ-/ n. a type of acupuncture in which air is injected into puncture points to control pain, treat internal disorders, and anesthetize(在穴位充气的针灸方法,可用来镇痛、治疗消化紊乱或麻醉)充气针灸

pneumotach / njuːˈmɒtæk; *AmE* nuː-/ n. a pneumotachometer 肺流速仪

pneumotachograph /ˌnjuːməˈtækəgrɑːf; *AmE* ˌnuː-/ n. *see* pneumotachometer 肺流速仪

pneumotachometer /ˌnjuːməʊtəˈkɒmɪtə(r); *AmE* ˌnuːmoʊ-/ n. (abbr. **pneumotach**) a medical device used for measuring the air flow in the lung(测定肺内气流的医学仪器)肺流速仪 [syn] pneumotachograph

pneumothorax /ˌnjuːməˈθɔːræks; *AmE* ˌnuː-/ n. a condition in which lung collapses inward resulting from the entering and accumulation of air into the pleural cavity(由于空气侵入胸膜腔致使肺内壁塌陷的症状)气胸

POA abbr. Pony of the Americas 美国矮马

poach /pəʊtʃ; *AmE* poʊtʃ/ vt. **1** to hunt illegally on sb else's property(在他人地界非法狩猎)偷猎,盗猎:The stags are poached for their velvet. 为获取鹿茸而偷猎雄鹿。**2** to disrupt land muddy by trampling 踏烂,踩烂

vi. (of land) to become muddy or sodden 变泥泞

poached /pəʊtʃt; *AmE* poʊtʃt/ adj.【hunting】(of the footing) sloppy and muddy(地面)泥泞的,稀烂的

poacher /ˈpəʊtʃə(r); *AmE* ˈpoʊtʃ-/ n. one who hunts illegally on the property of others(在他人地界非法狩猎的人)偷猎者,盗猎者

pocket /ˈpɒkɪt; *AmE* ˈpɑːk-/ n.【barrel racing】the turning area between the horse and the barrel(马与桶间的转弯区)绕弯处

pocketed /ˈpɒkɪtɪd; *AmE* ˈpɑːk-/ adj.【racing】(of a horse) surrounded by other horses and unable to speed up(指赛马被其他马匹包围而无法提速)被围的,被困的

podotrochilitis /pɒdəˌtrɒkˈlaɪtɪs/ n. *see* navicular disease 舟骨病

point /pɔɪnt/ n. **1** (usu. **points**) the ear tips, mane, tail, and lower legs of the horse grouped on the basis of color(毛色划分中对马匹耳梢、鬃、尾以及四肢下端的总称)末梢:points of the horse 马的末梢部位 ◇ a horse with black points 一匹末梢为黑色的马 **2** (usu. **points**) the knitted tips for a whip 鞭梢 **3** any protruding area on a horse's body(马体上的突出部位)端,角:point of the croup 尻端

vt. **1**【cutting】(of a cutting horse) to direct or

aim at the cow being worked（截牛中马）盯住，瞄准：to point a cow 盯住拦截的牛 **2** to indicate the direction or position 指明，标明

point firing n. *see* pin firing 火针刺疗

point of the hip n. the bony prominence on the croup of the horse（马体尻部突起）尻端，腰角 [syn] point of the croup

point of the shoulder n. the bony prominence formed by the scapula and humerus of the forearm（由前臂肩胛骨和肱骨形成的突起）肩端

point team n. a pair of horses positioned behind the leaders in an eight-horse hitch（八马联驾中位于头马之后的两匹马）端队，端马

pointing /ˈpɔɪntɪŋ/ n. **1** a condition in which a resting horse positions one affected foreleg ahead of another in an effort to relieve the weight on this side（马站立时将患病前肢伸向前方以减轻该肢负重的情况）点地：Pointing is usually a sign of navicular disease. 前肢点地通常说明马患有籽骨炎。**2** the action in which a horse places the toes of its forefeet on the ground before the heels when trotting（马快步行进时蹄尖先于蹄踵着地）蹄尖点地，蹄尖触地

point-to-point n. a mounted competition held in Britain centuries ago in which riders raced from one place to another over open fields（几个世纪前首次在英国流行的马上项目，骑手在荒野的各个分站进行比赛）分站越野赛 [syn] hunt race

poison /ˈpɔɪzn/ n. a substance that causes harm or death if introduced into the body（进入体内可造成损伤或死亡的物质）毒药，毒物：The hound was killed by rat poison. 猎犬被老鼠药毒死了。

vt. **1** to harm or kill by using poison 毒死；毒害 **2** to put poison in or on sth 投毒，下毒：The feed has been poisoned. 饲料已被人投毒。

poisonous /ˈpɔɪzənəs/ adj. capable of harming by poison or containing a poison；toxic 有毒的：poisonous plants 有毒植物

Poitevin /ˈpwɑːtvɪn/ n. a heavy draft breed originating from the province of Poitou，France from which the name derived；stands 15 to 17 hands and may have a dun，bay，gray，black，or palomino coat；has a heavy head，a short，straight neck，long，broad back，sloping quarters，thick，short legs with some feather，and broad feet；is calm，well balanced，and strong with good endurance；used for heavy draft，farming，and in the production of mules（源于法国普瓦图省的重挽马品种并由此得名。体高 15 ~ 17 掌，毛色多为兔褐、骝、青、黑或银鬃，头重，颈短而直，背长而宽，斜尻，四肢粗短且着生距毛，蹄大，性情安静，体格匀称，四肢强健，耐力较好，主要用于重挽、农田作业和骡的繁育）普瓦图马：Once in danger of extinction during the 1950s，the numbers of Poitevin are now on the increase. 普瓦图马在 20 世纪 50 年代曾一度面临灭种的危险，目前其数量则日趋增加。[syn] Cheval du Poitou

Poitou ass n.（also **Poitou donkey**）a large ass indigenous to the province of Poitou，France from which the name derived，often crossed with Poitevin mares to produce mules（产于法国普瓦图省的本土驴种并因此得名，常用来与普瓦图马杂交以繁育普瓦图骡）普瓦图驴 [syn] baudet de Poitou

Poitou mule n. a large mule developed by crossing the Poitou jackass with Poitevin mares；usually stands above 16 hands and has short，stout legs and broad hoofs；historically used for heavy agricultural work（采用普瓦图公驴和普瓦图母马杂交培育出的骡子，体型较大且多在 16 掌以上，四肢粗短，蹄大，过去主要用于农田作业）普瓦图骡

pole /pəʊl/ n. **1**【jumping】a wooden rail of a jumping obstacle（障碍的）横杆，栏杆 **2**【racing】one of the marking posts placed around the racetrack to indicate the distance to the finish（赛道上用来标记与终点间距离的木杆）标

杆:The quarter pole is located 1/4 mile from the finish. 四分杆置于与终点相距 1/4 英里的位置。**3**【driving】(also **coach pole**) a long, wooden bar fitted to the front of a horse-drawn vehicle on either side of which the wheelers are harnessed (马车前端的一根长木杆,一对辕马驾于两侧)驾杆,辀: pole chains 辀链 ◇ pole head 辀头 ◇ pole hook 辀钩 ◇ pole straps 辀带

pole bending n.【rodeo】(also **pole race**) a timed event in which the mounted rider travels in and out of a number of six vertically placed poles set in a straight line (计时竞技马术项目,其中参赛者按特定路线绕行直线排列的 6 根立杆)绕杆,穿杆: In pole bending, the rider who completes the course in the fastest time without knocking down any poles is the winner. 在穿杆比赛中,所用时间最少且没撞倒立杆的骑手即可获胜。

pole gear n.【driving】the parts and pieces of the harness used to attach the horses to the pole (用来将马套在辀上的挽具部件总成)辀件,驾杆部件

pole pin n.【driving】a steel pin that secures the pole between the futchells on a horse-drawn vehicle (用来将马车的驾杆固定在通柱之间的铁件)辀辖

pole race n. *see* pole bending 穿杆

polish /ˈpɒlɪʃ; *AmE* ˈpɑːl-/ vt. to make the surface smooth and shiny by rubbing or chemical action (通过擦拭或化学作用使表面变得光滑)打磨,擦亮

n. **1** the shine of the horse's coat when cleaned and groomed (马被毛刷拭后的亮光)光泽 **2** a substance used to shine a surface or impart a gloss 亮漆:hoof polish 蹄漆

Polish Warmblood n. *see* Wielkopolski 波兰温血马

poll /pəʊl; *AmE* poʊl/ n. the top of the horse's head between two ears (马头位于两耳之间的部位)头顶,项顶 [syn] nape, occipital crest

poll evil n. a soft swelling on the horse's poll caused by a blow to the head (马头顶遭撞击后留下的软肿)项顶肿包

poll guard n. *see* head bumper 护头

Pollard, James n. (1772 – 1867) a British painter noted for his hunting and sporting scenes (英国著名画家,擅长狩猎和运动场景写生)詹姆士・波拉德

polo /ˈpəʊləʊ; *AmE* ˈpoʊloʊ/ n. (also **pulu**) a ball game played by two teams of four mounted players on a grass field of 300 × 200 yards in size; consisting of six chukkers during which the players attempt to score by striking the ball into the goal with a mallet (一种马上球类运动,比赛由两队人马在 274 米 × 183 米的草场进行,每队四名球员;比赛分六局,骑手手持球杆将球击入对方球门方可得分)马球: Polo originated in Persia, and was soon popular in China. 马球起源于波斯,此后不久就流行于中国。◇ arena polo 室内马球 ◇ pachyderm polo 象球

polo ball n. a white wooden or plastic ball used in polo (马球比赛用的白色木质或塑料球)马球

polo boots n. a pair of knee-height leather boots worn by a polo player (马球运动员穿的齐膝皮靴)马球靴

polo cart n. *see* polo gig 马球车

polo club n. a club established to organize polo as sport 马球俱乐部: The first polo club in the Western world was officially established in London in 1873 and was controlled by the Hurlingham Polo Association. 西欧首个正式的马球俱乐部于 1873 年在伦敦成立,由赫林汉姆马球协会掌管。

polo field n. (also **polo ground**) a rectangular grass field upon which a polo game is played (举行马球比赛的长方形草地)马球场

polo game n. a polo match 马球比赛

polo gig n. (also **polo cart**) a horse-drawn sporting gig used in the late 19th century to exercise

P

a polo pony or carry the mallets to a match (19 世纪末使用的一种轻型运动马车,用来训练马球马或运送马球装备)马球马车

polo ground n. *see* polo field 马球场

polo handicap n. a rating awarded to each polo player on the basis of their mastery, horsemanship, sense of strategy and conduct, and pony quality (根据球员技术、骑术、策略运用和坐骑性能而进行的评分体系)马球分级[体系] syn polo rating

polo jacket n. a jacket worn by a polo player 马球衫 syn polo jersey

polo jersey n. *see* polo jacket 马球衫

polo mallet n. (also **mallet**) a long-handled wooden stick with a head, used by polo players to strike the ball (马球运动员用来击球的长柄木杆)马球杆 syn polo stick

polo match n. (also **polo game**) a game consisting of six chukkers with each chukker lasting 7.5 minutes and a 5 minutes of mid-term break (每场比赛共六局,每局7.5分钟,中场休息5分钟)马球比赛

polo mount n. *see* polo pony 马球马

polo penalty n. a penalty imposed for breaking the rules in the game of polo (马球比赛中因犯规给予的罚分)马球罚分

polo player n. one who plays polo; poloist 马球运动员: a professional polo player 一名职业马球运动员

polo pony n. (also **polo mount**) a horse used in the game of polo 马球马: Argentine Polo Pony 阿根廷马球马 ◇ The polo pony is a specific type of horse bred for the sport, and usually larger in size than pony. 马球马是专门培育的运动用马,其体型要比矮马大。

polo rating n. *see* polo handicap 马球分级体系

polo spur n. a blunt spur used by polo players (马球运动员使用的一种钝头马刺)马球马刺,马球靴刺

polo stick n. *see* polo mallet 马球杆

polo stroke n. (also **stroke**) an act or technique of striking the polo ball with the mallet (用球杆撞击马球)击球

polo team n. one of the two opposing groups in a polo game, with each consisting of four players (马球比赛中对抗的两组人马,每组由四名球员组成)马球队

polo widow n. 【slang】 the wife of a dedicated polo player (一位敬业马球运动员的妻子)马球寡妇

polocrosse /ˈpəʊləʊkrɒs; *AmE* ˈpoʊloʊ-/ n. a mounted team sport originating in Australia in which the riders scoop up a ball in a small net attached to a long stick, pass or carry it down the field to score by throwing the ball between the goal posts (一种源于澳大利亚的马上团体项目,球员手持带网兜的球杆,传球后通过将球掷入对方球门得分)网兜马球: Polocrosse is a combination of polo and lacrosse. 网兜马球是马球与长曲棍球结合的项目。

poloist /ˈpəʊləʊɪst; *AmE* ˈpoʊloʊ-/n. a polo player 马球运动员

polycythemia /ˌpɒlɪsaɪˈθiːmiə/ n. (BrE also **polycythaemia**) a condition marked by abnormal overproduction of red blood cells in the circulatory system (血液中红细胞异常增多的疾病)红细胞增多症

polyestrous /ˌpɒlɪˈestrəs/ adj. having several estrous cycles during a single breeding season (在一个繁殖季节有多个发情周期)多次发情的: Horses are seasonally polyestrous animals, the normal mares generally exhibit 9-10 estrous periods during the course of the covering season if they do not conceive. 马是季节性多次发情的动物,母马在交配期内如果没有受孕,一般可出现9~10次发情。

polygenic /ˌpɒlɪˈdʒenɪk; *AmE* ˌpɑːl-/ adj. 【genetics】 of or relating to more than one gene 多基因的

polygenic trait n. 【genetics】 a trait influenced by more than one gene 多基因性状 syn multiple gene trait

pommel /ˈpɒml; *AmE* ˈpɑːml/ n. **1** the upper front part of a saddle that rises up from the seat（马鞍前突出的部分）前鞍桥 [syn] head of the saddle **2** the knob on the hilt of a sword or dagger（剑柄末端的把手）圆头

pommel blanket n. *see* pommel pad 前鞍桥垫

pommel pad n.（also **pommel blanket**）a small pad placed under the pommel of the saddle to protect the withers from chafing（衬在前鞍桥下的软垫，可防止鬐甲被蹭伤）前鞍桥垫，鬐甲垫：Pommel pad is often used on horses with prominent withers. 前鞍桥垫常用于鬐甲过高的马匹。[syn] wither pad

pony /ˈpəʊni; *AmE* ˈpoʊni/ n. **1** a horse standing below 14.2 hands when fully grown（成年后体高低于14.2掌的马匹）矮马：Debao pony 德保矮马 ◇ Shetland pony 设特兰矮马 **2**【polo】a polo pony 马球马：For competition, polo ponies have their manes roached and tails braided so that there is no danger of being tangled in the mallet. 比赛时，马球马的鬃要剪短，尾要辫起来，以免与球杆发生缠结。

Pony Break n. a light, horse-drawn cart used formerly in Australia for exercise purposes, usually drawn by four ponies（澳大利亚过去使用的轻型调教马车，多由四匹矮马拉行）调教马车

Pony Club n. **1** a youth organization founded in 1929 in England to encourage young people to ride and care for horses and ponies while promoting the ideals of sportsmanship（1929年在英国成立的一个青年组织机构，旨在鼓励年轻人骑乘和料理马与矮马，并培养他们的体育精神）英国矮马俱乐部，英国青少年骑士会：The Pony Club has now become an international organization with branches in more than 20 countries dedicated to teaching young people to ride and learn all aspects of horsemanship. 英国青少年骑士会现已发展成为国际性组织，在全世界20多个国家设有分支机构，旨在教育青少年学习骑马和各方面的马术知识。**2** an organization founded in any country or region to encourage young people to ride and care for horses and ponies（任何国家或地区成立的组织机构，旨在鼓励年轻人骑乘和料理马与矮马）矮马俱乐部，青少年骑士会

Pony Club Mounted Games n. a mounted game first organized by Prince Philip in 1957 in UK to provide the Pony Club members with an opportunity to improve their equestrian skills through various competitions（1957年首次由英国菲利普亲王组织的马上竞技比赛，旨在通过各种竞技项目来提高矮马俱乐部成员的马术水平）矮马俱乐部马上运动会：The Pony Club Mounted Games are limited to riders under the age of 15 mounted on ponies at least four years old. 矮马俱乐部马上运动会限定选手年龄须在15岁以下，参赛矮马至少在4岁以上。[syn] Prince Philip Mounted Games

Pony Express n. a postal service operated through the West in the United States during the 1860s in which mail was carried by mounted relay riders（在19世纪60年代美国西部开展的骑马邮递业务）驿马快递，驿马快运

pony goal n.【polo】a goal scored when the polo ball is struck across the goal line by the feet of a polo pony（马球比赛中球被马蹄撞入球门得分的情况）马蹄射门，马匹射门

Pony of the Americas n.（abbr. **POA**）a pony breed developed in the 1950s in the United States by crossing Appaloosa with Shetland ponies; stands 11.1 to 13.1 hands, and usually has a coat pattern of an Appaloosa; has a nice head, prominent withers, short and straight back, well-muscled croup and loins, wide and deep chest, strong legs, and hard, vertically striped hoofs; is versatile, docile, and fast; often used as an all-purpose mount for children（美国在20世纪50年代采用阿帕卢萨马和设特兰矮马杂交繁育而来的矮马品种。体高11.1～13.1掌，毛色与阿帕卢萨马类似，面目俊秀，鬐甲突出，背短而平直，后躯发

P

达,胸身宽广,四肢粗壮,蹄质坚实且带竖纹,性情温驯,反应敏快,为多用途矮马,步速较快,是适合儿童策骑的全能马种)美国矮马

pony speed test n. an Australian horse race in which ponies ridden by boys are raced around a distance of 1/4 mile(澳大利亚举行的一项赛马比赛,由男孩策骑矮马参加,赛程400米)矮马速度赛:Frank Reys also rode in pony speed-tests, and was generally regarded as one of the top boy-riders in those events. 弗兰克·雷兹过去也参加过矮马速度赛,且被公认是该项目中最优秀的少年骑手之一。

pony tail n. **1** the tail of a pony or horse 马尾 **2** a hairstyle in which a bunch of hair is tied at the back of the head like a horse's tail(将头发束于脑后、形似马尾的发型)马尾辫

pool /puːl/ n.【racing】the total amount of money bet on a race by all bettors(赛马中赌马者投注的资金总额)投注总额;投注池 syn stake money

pop-eyed /ˈpɔpaɪd/ adj. having prominent or bulging eyes 眼球突出的

popped a splint n. *see* splint 小掌骨增生,赘骨

popped knee n. *see* carpitis 膝关节肿大,膝关节炎

popped sesamoid n. *see* sesamoiditis 籽骨炎

Porcellus /pɔːˈtʃeləs/ n. the original name for Julius Caesar's racing stallion(朱利叶斯·凯撒赛马的原名)小猪:Porcellus was later changed to Incitatus when the horse began to win races. "小猪"在比赛中获胜后,便易名为"迅驰"。

port[1] /pɔːt; *AmE* ˈpɔːrt/ n. the curved section in the center of a bit's mouthpiece under which the tongue fits(衔铁中段凸起的部分,马的舌位于其下)衔凸:low port 低衔凸 ◇ high port 高衔凸 ◇ The taller the port, the tighter the curb. 衔凸越高,衔链越紧。

port[2] /pɔːt; *AmE* ˈpɔːrt/ v. (of a horse) to paw or scrape the ground with its front legs(指马用前蹄)刨地,刨趴

ported /ˈpɔːtɪd; *AmE* ˈpɔːrt-/ adj. (of a bit) designed with a port(指衔铁)带衔凸的

Portland Cutter n. a light, horse-drawn sleigh having a single seat board with very high S-shaped dash in the front to ward off snow thrown up by horse's hoofs(一种单座马拉轻型雪橇,前端装有挡雪板,可挡住马蹄扬起的雪花)波特兰雪橇

position /pəˈzɪʃn/ n. **1** the arrangement of a horse's body 胎位;肢势:ventral position 腹胎位 ◇ dorsal position 背胎位 ◇ correct position 正肢势 ◇ incorrect position 不正肢势 ◇ ideal position 理想肢势 **2** the way a rider sits in the saddle; seat(骑马的)乘姿:two-point position 两点式乘姿

positive reinforcement n. a positive reward for certain behavior(行为的)正向强化,正面强化

post[1] /pəʊst; *AmE* poʊst/ n. **1** a long wooden or metal stick set upright as a marker(用作标记的竖立木杆或铁杆)立杆 **2**【racing】the starting or ending place on a racetrack(赛道的)起点;终点 **3** *see* post position 闸位

vt. **1**【racing】(of a rider or horse) to finish first in a race(指马在比赛中跑第一)获得,获胜:He posted 12 wins in 13 starts. 他出赛13次,获胜12次。**2** (of a rider) to rise from and fall into the saddle with the rhythm of the trot(指骑手随快步节奏上下起伏)起浪

PHR V **post trot** 起浪:When a rider is posting the trot, he rises out of the saddle as the outside foreleg moves forward, and sits down in the saddle as the inside foreleg moves forward. 当骑手快步起浪时,马右前肢起步时他要离开马鞍,左前肢起步时他应落入鞍座。syn rise trot

post[2] /pəʊst; *AmE* poʊst/ n. **1** the official service that delivers mails or parcels(官方运送邮件和包裹的服务)邮递,邮政 **2** one who rode to deliver mails between fixed stages(过去骑马

在驿站间投递信件的人）驿差，信使：post house 驿站，驿馆 ◇ post rider 驿差

post betting n.【racing】a betting that does not begin until the post positions of all competing horses are determined（赛马闸位确定后方才开始投注的赌马方式）先定闸位后投注，闸后投注

post chaise n. a four-wheeled, enclosed, horse-drawn vehicle used for public hire between inns and post houses in England in 1740s（18 世纪 40 年代出现在英国的四轮厢式马车，供客栈或驿站往返出租使用）驿站马车

post entry n. the nomination of a horse to run in a race on the day prior to the event（开赛前一天马匹报名参加比赛）闸位参赛，闸位报名

post master n. the owner of a post house 驿站长；驿馆老板

post parade n.【racing】a procession of racehorses moving from the paddock to the starting gate in front of the grandstand（赛马从围场步入看台前起跑闸的仪式）入场仪式，入闸亮相

post position n.【racing】（also **post**）the position of the horse in the starting gate（马在起跑闸中的位置）闸位，档位：The post position is usually drawn by ballot or selected by using a number ball at the close of entries the day prior to the race. 闸位通常在赛前报名结束当天通过抽签或数字球来决定。

post time n.【racing】the time at which all racehorses enter the starting gate to start a race（赛马进入起跑闸开赛的时间）入闸时间

post-and-rail n.【jumping】（also **post and rail**）a jumping obstacle consisting of upright posts with horizontal rails laid between them（由立杆和横木构成的障碍）单横栏 compare oxer

postboy /ˈpəʊstbɔɪ；*AmE* ˈpoʊ-/ n. a postilion 左马驭者

postboy's coat n. *see* great coat 车夫大衣

posterior /pɒˈstɪərɪə(r)；*AmE* pɑːˈstɪr-/ adj.【anatomy】situated behind or towards the rear 后面的，后部的：posterior digital nerve 趾后神经

post-estrus /ˌpəʊstˈestrəs；*AmE* ˌpoʊ-/ adj. after the estrus 发情后的：post-estrus ovulation 发情后排卵

n. the period of time after estrus 发情后期

posthouse /ˈpəʊsthaʊs；*AmE* ˈpoʊ-/ n.（also **post house**）**1** a house where post riders could rest and exchange horses（过去供驿差休息和更换马匹的客栈）驿站，驿馆 **2** a place for distributing the mails；a post office 投信处；邮局

Postier /ˈpəʊstjə(r)；*AmE* ˈpoʊ-/ n. *see* Postier-Breton 波斯提亚布雷顿马

Postier-Breton n.（also **Postier**）a coldblood developed in Brittany，France by crossing native stock with Norfolk Trotter during the Middle Ages；stands 15 to 16 hands and generally has a chestnut coat，although bay，gray，and roan also occur；has a well-proportioned head，broad forehead，short ears，flared nostrils，a short，broad，muscular neck，short and straight back，sloping croup，and short，powerful legs with heavy joints；used mainly for farm work and driving（中世纪时法国布列塔尼地区采用当地马与诺福克快步马杂交培育的冷血马品种。体高 15 ~ 16 掌，毛色以栗为主，但骝、青和沙毛也常见，头型比例适中，额宽，耳短，鼻孔大，颈粗短，背短而平直，斜尻，四肢粗短，主要用于农田作业和挽车）波斯提亚布雷顿马 syn medium Breton

postilion /pəˈstɪljən/ n. one who rides the near horse of the leaders to guide a team of horses pulling a coach（多马联驾中策骑左侧头马以协助驾车的人）左马驭者 syn postboy

posting trot n. the rise-and-fall movement of a rider's body with the rhythm of the trot（马快步行进时骑手身体随着节律上下起伏的动作）起浪：At the posting trot，the rider rises out of the saddle and sits back down in the saddle once in each stride. 快步起浪时，骑手要在马快步行进的每步中上、下起坐一次。

post-mortem /ˌpəʊstˈmɔːtəm; *AmE* ˌpoʊstˈmɔːrtəm/ n. (also **post-mortem examination**) a medical examination of the body after death to find the possible causes (为了寻找死因而对尸体进行的检验)死后尸检:to conduct/carry out a post-mortem 进行死后尸检

postnatal /ˌpəʊstˈneɪtl; *AmE* ˌpoʊ-/ adj. of or occurring after birth 产后的,出生后的:Postnatal routes of microbial entry include the mouth, airways and umbilicus. 产后病菌可通过幼驹的口、气管和脐带侵入身体。 opp prenatal

post-operative myopathy n. (abbr. **POM**) a condition in which the muscle is repeatedly injured by its surrounding tough fibrous tissue in constant growth, often associated with reduced muscle blood flow and malignant hyperthermia (肌肉组织被周围增生的硬纤维组织反复损伤的症状,常表现为患处血流降低和异常高热)术后肌病,术后肌纤维瘤

postpartum /ˌpəʊstˈpɑːtəm adj. 【Latin】postparturient 产后的

postpartum complications n. the effects of pregnancy, parturition and heavy lactation following foaling, such as debilitation, paralysis, and founder of mares (母马妊娠、分娩及产后大量泌乳所引发的症状,如体质虚弱、瘫痪和蹄叶炎等)产后并发症

postparturient /ˌpəʊstˈpɑːtjʊərɪənt; *AmE* -ˈpɑːrtʊr-/ adj. of or occurring in the period shortly after foaling 产后的:postparturient debilitation 产后虚弱

postparturient founder n. laminitis of a mare's hoof caused by retention of the placental membranes in the uterus following foaling (母马产后胎衣滞留所致的蹄叶炎)产后蹄叶炎

post-training /ˌpəʊstˈtreɪnɪŋ; *AmE* ˌpoʊ-/ adj. of the time after the training of horses (马)训练后的:post-training care of horses 训练后的马匹料理

posture /ˈpɒstʃə(r); *AmE* ˈpɑːs-/ n. the position of the head, neck and extremities to the fetal body in foaling (分娩前胎儿的头、颈及四肢的位置)胎势

potash /ˈpɒtæʃ; *AmE* ˈpɑːt-/ n. any of several compounds containing potassium 钾盐:potash alum 明矾

potassium /pəˈtæsiəm/ n. 【nutrition】a natural, metallic element that is essential for acid-base balance, body fluid regulation, nerve and muscle function and carbohydrate metabolism (一种天然金属元素,为机体酸碱平衡、体液调节、神经和肌肉发挥功能以及糖代谢所必需)钾

pot-bellied /ˈpɒtˌbelɪd; *AmE* ˈpɑːt-/ adj. (of a horse) having a large, protruding stomach (指马)草腹的:a potbellied horse 一匹草腹马

potbelly /ˈpɒtbeli; *AmE* ˈpɑːt-/ n. (also **pot-belly**) a large protruding stomach 草腹,草包肚:A large pot-belly could falsely represent a fat horse in good condition and may be an indication of worms, quite common in young stock. 草腹往往造成马匹膘情良好的假象,而其实马匹有可能正遭受虫害,这在青年马中较为常见。

potential /pəˈtenʃl/ n. the inherent ability or capacity 潜力,潜能:The key to unlock the potential of a horse really comes down to training. 真正挖掘马匹潜力的关键在于训练。

adj. having the capacity for further development; latent 有潜力的;潜在的

Potomac horse fever n. (abbr. **PHF**) a fatal disease of horses resulting from infection by *Ehrlichia risticii*, recognized first in 1979 around the Potomac River Valley from which the name derived; symptoms may include fever, diarrhea, colic, and laminitis (马感染立氏埃利希体后所患的致死性疾病,于1979年首次在美国的波托马可河谷发现并因此得名,症状表现为高热、腹泻、腹痛和蹄叶炎等)波托马可马瘟 syn equine monocytic ehrlichiosis

Pottok /pəˈtɔːk/ n. (also **Pottoka**) *see* Basque 巴斯克矮马

poultice /ˈpəʊltɪs; *AmE* ˈpoʊ-/ n. any moist substance containing antiseptics that applied to the skin for healing purposes (一种含有抗菌剂的膏药,敷在皮肤上以达到治愈效果)敷剂;膏药

vt. to apply a poultice to a wound or an inflamed area 敷药 | **poulticing** /-tɪsɪŋ/ n.

poultice boot n. a plastic or leather boot filled with poultice and used to treat certain hoof problems (内填膏药的塑料或皮套,用来治疗某些肢蹄疾病)药靴

pound /paʊnd/ n. (abbr. **lb**) a unit for measuring weight equal to 454 grams (重量测量单位,相当于454 克)磅

powder /ˈpaʊdə(r)/ n.【racing】slight body contact between horses during a race (赛马间轻微的身体接触)摩蹭,碰触

Powys Cob n. an ancient Welsh pony breed developed in the 12th century by putting Welsh Mountain Ponies to imported Spanish stock (源于威尔士的古老矮马品种,于 12 世纪采用威尔士山地矮马与西班牙马杂交繁育而来)普沃斯柯柏马

prad /præd/ n.【BrE,dated】a horse 马

prairie /ˈpreəri; *AmE* ˈpreri/ n. a wide area of grassland in North America (北美)大草原

prairie schooner n. (also **schooner**) a covered horse-drawn wagon used by American pioneers moving westward across the prairie and plains during the 1820s (19 世纪 20 年代北美拓荒者西进穿越大草原时所驾的带篷马车)草原篷车 [syn] ship-of-the-plains, covered wagon

PRCA abbr. Professional Rodeo Cowboys Association 职业牛仔竞技协会

Preakness /ˈpriːknɪs/ n. *see* Preakness Stakes 普瑞克尼斯大赛

Preakness Stakes n. (also **Preakness**) a 9.5-furlong stakes race for three-year-olds run annually since 1873 on the Pimlico course in Baltimore, Maryland, USA (自 1873 年起每年在美国马里兰州巴尔的摩赛马场举行的比赛,由 3 岁驹参加,赛程 1.9 千米)普瑞克尼斯大赛

pregnancy /ˈpregnənsi/ n. the period of time or state of being pregnant 怀孕;妊娠:pregnancy hormone 孕激素 ◇ There are mainly three methods of pregnancy diagnosis: rectal palpation; blood test for the presence of PMSG; and urine analysis for estrogens. 妊娠诊断的主要方法有三种:直肠检查、孕马血清促性腺激素血液检测和尿液雌激素分析。

pregnancy hormone n. *see* progesterone 孕激素

pregnancy rate n. the percentage of the total number of animals which become pregnant after insemination (授精后妊娠个体占全体母畜的百分比)妊娠率

pregnant /ˈpregnənt/ adj. carrying fetus within the body 怀孕的,妊娠的:pregnant broodmare 妊娠母马

pregnant mare serum gonadotropin n. (abbr. **PMSG**) a hormone secreted by the endometrial cups of the equine placenta (由母马胎盘上的子宫内膜杯分泌的一种激素)孕马血清促性腺激素

pregnant mare urine n. (abbr. **PMU**) urine collected from pregnant mares from which the estrogen is extracted (用于分离雌激素而收集的妊娠母马的尿液)妊娠母马尿液

preliminary canter n.【racing】a short warm-up canter executed shortly before a horse race (为了赛前热身而进行的短跑)热身预跑,预热短跑

premature /ˈpremətʃə(r); *AmE* ˌpriːməˈtʃʊr/ adj. born earlier before gestation period (在妊娠期满之前出生)早产的

prematurity /ˌpriːməˈtjʊrəti/ n. the condition of being born earlier 早产,早熟

premolar /priːˈməʊlə(r); *AmE* -ˈmoʊ-/ n.【anatomy】one of the 12 teeth of horse located between the incisors and molars on each side of the upper and lower jaws (马上、下颌两侧位于切齿与臼齿之间的 12 颗齿)前臼齿

prenatal /ˌpriːˈneɪtl/ adj. existing or occurring before birth 出生前的；产前的 [opp] postnatal

pre-novice /ˌpriːˈnɒvɪs; *AmE* -ˈnɑː v-/ n. (also **pre-novice class**) a level of equestrian competition ranking between the intro class and the novice class（介于入门级和初级之间的赛事级别）预备级 —*see also* **intro**, **novice**, **intermediate**, **advanced**

prep race n.【racing】a workout or race in which a horse is trained or prepared for a future competition（为了训练马匹或备战大赛而举行的比赛）预赛，预备赛

prepotent /prɪˈpəʊtənt/ adj.【genetics】(of parents) having greater potential in transmitting their inheritable characteristics to their progeny or offspring（指亲本有将遗传性状传递给子代的优势）具有遗传优势的 | **prepotency** /-tənsi/ n.

P

prepuce /ˈpriːpjuːs/ n. the loose fold of skin covering the glans penis; sheath（包裹在阴茎头上的软褶）包皮

presence /ˈprezns/ n. the bearing and appearance of a horse 外貌；姿态：A lovely presence is sought in showing horses and ponies because it is eye-catching. 参加展览赛的马和矮马要有秀丽的外貌，因为这样才会引人注目。

presentation /ˌpreznˈteɪʃn/ n. **1** the way how a horse is presented 外观；姿态 **2** the way in which the fetal body presents to the pelvic inlet in delivery（分娩时胎儿身体从母体骨盆产出的方式）胎向：anterior presentation 顺胎 ◇ posterior presentation 倒胎 ◇ Normal presentation is for the foal to come head and forelimbs first. 在正常胎向中，马驹的头和前肢先出产道。

presternal /priːˈstɜːnl; *AmE* -ˈstɜːrnl/ adj. of or located at the presternum 胸骨的，前胸的：presternal region 前胸部

presternum /priːˈstɜːnəm; *AmE* -ˈstɜːrn-/ n.【anatomy】the handle-shaped part of sternum 胸骨柄 [syn] manubrium

pre-training /priˈtreɪnɪŋ/ adj. of the time before the training of horses 训练前的

prevent /priˈvent/ vt. to keep sth from happening 预防；阻止

preventive /priˈventɪv/ adj. preventing or slowing the course of an illness or a disease（阻止或延缓疾病进程的）预防性的：preventive measures 预防措施

prick /prɪk/ vt. (of an animal) to make the ears stand erect when on alert（动物警觉时竖立耳朵）竖立，竖起：The horse pricked its ears at the sight of his owner. 马看到主人便竖起了耳朵。

pricket /ˈprɪkɪt/ n. a young buck in its second year before the antlers branch（鹿角尚未分叉的青年雄鹿）两岁雄鹿

primary /ˈpraɪməri; *AmE* -meri-/ adj. occurring first in time or sequence; primitive 最初的；原始的，初级的：primary corpus luteum 初级黄体 ◇ primary follicle 初级卵泡

primigravida /ˌpraɪmɪˈgrævɪdə/ n. a mare in foal for the first time; maiden mare（初次怀孕的母马）初孕母马

primitive /ˈprɪmətɪv/ adj. existing in its original form in ancient times 古老的；原始的：primitive breeds 原始马种

n. the early sub-species of *Equus caballus*, including the Asian Wild Horse, the Tarpan, the Forest Horse and the Tundra Horse（早期家马的亚种，包括亚洲野马、西藏野马和森林马以及冻原马）原始马种

primitive marking n. (also **primitive mark**) any of the coat markings seen in certain horses and asses, including the dark stripes along the spine and over the withers, knees, and hocks that can be traced to ancient breeds（某些马和驴脊背、鬐甲、前膝和飞节等处的深色条纹，多可追溯至相应的古老品种）原始别征：Primitive markings are most commonly seen on dun-colored horses. 原始别征在兔褐马中最为常见。— *see also* **dorsal stripe**, **zebra**

stripes, wither stripe

Prince of Wales Stakes n. a 9. 5-furlong Canadian stakes race held for 3-year-old Thoroughbreds in July at Fort Erie racetrack, Ontario (加拿大每年7月在安大略省伊利堡赛马场为3岁纯血马举行的奖金赛,赛程1 900米) 威尔士亲王大赛

Prince Philip Cup n. a cup awarded to the champion who wins in the Horse of the Year Show held every October in UK (英国每年10月举行的马匹展览赛上颁给冠军的奖杯)菲利普亲王杯

Prince Philip Mounted Games n. *see* Pony Club Mounted Games 菲利普亲王马术运动会

prince seat n.【vaulting】a freestyle movement in which the vaulter kneels on one leg with both arms outstretched from his sides (马上杂技动作之一,表演者单膝跪于马背,双臂外张)王子跪姿

principal horse n. *see* cast horse 主角马

pritchel /ˈprɪtʃəl/ n.【farriery】a pointed tool used to make nail holes in a hot horseshoe (用于在蹄铁上冲孔的工具)铳子

vt. to make nail holes in a horseshoe 冲孔

private auction n. an auction exclusive to invited participants (只有受邀者方可参加的拍卖会)私人拍卖会 compare public auction

private pack n.【hunting】a pack of hounds owned solely by the Hunt Master (狩猎主个人拥有的犬队)私人犬队

Prix des Nations n.【French】an international team show-jumping with four riders from each participating country (一项国际团体障碍赛,每个参赛国家有4名骑手参与)民族杯国际障碍赛: It was not until in 1953 that the Prix des Nations opened its gate to female competitors. 民族杯国际障碍赛直到1953年才允许女骑手参赛。 syn Nations Cup, Prize of Nations

Prix St. Georges n.【dressage】a level of international dressage competition below the Intermediare I (国际盛装舞步赛中低于I号中级赛的级别)圣·乔治奖[赛] — *see also* **Intermediare I, Intermediare II, Grand Prix**

prize money n. a sum of money awarded to a competitor for victory (奖给比赛获胜者的金额)奖金

Prize of Nations n. *see* Prix des Nations 民族杯国际障碍赛

prize winner n. the horse or rider that wins a prize in a race (比赛中赢得奖金的马匹或骑手)获奖马;获奖者

PRL abbr. prolactin 催乳素

probiotic /ˌprəʊbaɪˈɒtɪk; *AmE* ˌproʊ-/ n.【nutrition】any nutritional additive or bacterium that supplements normal gastrointestinal flora (对胃肠道菌群有补充作用的营养添加剂或菌种)益生素; 益生菌: Nowadays, the lactic acid bacteria is generally used as a probiotic that can stimulate the immune system of host and to expel the invasion of pathogen by antagonism, adhesion resistance and competition. 目前,乳酸菌被普遍当作益生菌使用,它可激活寄主的免疫系统并通过颉颃、黏附阻断及竞争来防止病原微生物的侵袭。

probstmayria vivipara n.【Latin】*see* small pinworm 小尖尾线虫

produce *verb, noun*

vt. /prəˈdjuːs; *AmE* -ˈduːs/ to give birth 生产;繁育

n. /ˈprɒdjuːs; *AmE* ˈprɑːduːs/ **1** the progeny or offspring of a mare (母马的)后代: A foal is the produce of a mare and the get of a stallion. 幼驹是母马和公马共同的后代。 compare get **2** the fodder for horses (喂马的)饲草

produce race n.【racing】a flat race for the progeny of selected mares and/or stallions (由特定马匹繁育的后代参加的平地赛)子代赛

producer /prəˈdjuːsə(r); *AmE* -ˈduːs-/ n.【racing】a mare whose progeny has won at least one race (所产后代在比赛中获胜的母马)冠军母马

proestrus /prəʊ'estrəs; *AmE* proʊ-/ n.【physiology】the period proceeding estrus in which the mare becomes increasingly receptive to mating (发情期之前的阶段,此时母马有接受交配的意愿)前情期,发情前期

professional /prə'feʃənl/ n. one engaged in an activity or sport as profession (作为职业从事某种活动或体育项目的人)职业人;职业选手 [compare] non-professional

adj. of or relating to a profession 职业的: professional horse trainer 职业调马师 ◇ professional jockey 职业骑师

proffer /'prɒfə(r); *AmE* 'prɑːf-/ vt. to offer or provide; tender 提供;提出: If the advices proffered by nutritionists and vets are reasonably applied to horse breeding, the productivity would be enhanced greatly without doubt. 如果营养学家和兽医的建议被合理地应用于马匹育种中的话,生产率毫无疑问会提高很多。

profile /'prəʊfaɪl; *AmE* 'proʊ-/ n. **1** a side view or outline of an object or structure 外形;轮廓: blood profile 血象 **2** the outline of the face of a horse (马的)面貌: a horse with fine profile 一匹面貌俊秀的马

profundus /prə'fʌndəs/ adj.【Latin】deep 深的 [compare] superficialis

profuse /prə'fjuːs/ adj. plentiful; copious 大量的;丰富的: A horse who work hard in winter coat will sweat profusely soon and be difficult to clean and dry. 长有冬季被毛的马负重役时,一会儿就出很多汗,导致清洁和干燥比较困难。

progenitor /prəʊ'dʒenɪtə(r); *AmE* proʊ-/ n. the sire of a group of descendants 祖先;父辈

progeny /'prɒdʒəni/ n. the offspring or descendants of either a mare or stallion (一匹母马或公马繁衍的个体)后代,后裔: Despite extensive chromosomal differences, the various equine species generally can be successfully crossed to produce viable progeny. 尽管染色体差异较大,但是各种马属动物之间杂交都可产出可存活的后代。

progesterone /prə'dʒestərəʊn; *AmE* -roʊn/ n.【physiology】a steroid hormone secreted by the corpus luteum of ovaries that is essential for maintaining pregnancy (由卵巢黄体分泌的类固醇激素,对于妊娠的维持有重要作用)孕酮,黄体酮 [syn] pregnancy hormone

prognathic /prɒg'næθɪk/ adj. (also **prognathous**) having the lower jaw that projects forward obviously 下颌突出的,凸颌的

prognathism /'prɒgnəθɪzəm/ n. (also **prognathia**) a congenital malformation in which the lower jaw projects beyond the upper jaw to a marked degree (下颌比上颌明显突出的情况的口齿畸形)下颌突出,地包天 [syn] undershot jaw

prognathous /'prɒgnəθəs/ adj. *see* prognathic 下颌突出的,凸颌的

progressive color pattern n. a coat color that changes as the horse ages, such as the gray (随着马匹年龄增长而发生变化的毛色,如青毛)渐进式毛色类型 [compare] nonprogressive color pattern

prolactin /prə'læktɪn; *AmE* proʊ'l-/ n.【physiology】(abbr. **PRL**) a pituitary hormone that stimulates and maintains the secretion of milk (垂体所分泌的激素,可刺激并维持母体泌乳)催乳素

prolapse /'prəʊlæps; *AmE* 'proʊ-/ vi. to fall or slip out of place 脱垂,脱出

n. the falling down or slipping out of place of an organ or part (器官)脱垂,脱出: uterine prolapse 子宫脱出

prolapsed sole n. *see* dropped sole 蹄底脱出,蹄底下坠

prolong /prə'lɒŋ; *AmE* -'lɔːŋ/ vt. to extend the duration; protract 延长;持久: prolonged diestrus 间情期延长

promethazine /prəʊ'meθəziːn; *AmE* proʊ-/ n. *see* acetylpromazine 异丙嗪,乙酰丙嗪

prominence /'prɒmɪnəns; *AmE* 'prɑːm-/ n.【anatomy】a small projection or protuberance 隆

凸;突起

prominent /ˈprɒmɪnənt; *AmE* ˈprɑːm-/ adj. projecting outward from surface 明显的,突出的:a horse with prominent withers 一匹鬐甲明显的马

prominentia /ˌprɒmɪˈnenʃɪə/ n.【Latin】(*pl.* **prominentiae** /-ʃiː/) prominence 隆凸;突起

pronate /ˈprəʊneɪt; *AmE* ˈproʊ-/ vi. to turn or rotate inward 内转

prong /prɒŋ; *AmE* ˈprɔːŋ/ n. a thin, pointed, projecting part, such as of a fork; tine 尖端,叉齿: three-pronged fork 三股叉

prop /prɒp; *AmE* prɑːp/ n. **1**【racing】a horse who refuses to break at the start of the race 拒出闸的马 **2**【racing】a horse who stops suddenly by planting its front feet (前肢蹬地突然止步的马)猛停的马

Prophet Strain n. a line of horses selected, according to legend, by the Prophet Mohammed as the foundation stock of Arab (传说中穆罕默德为培育阿拉伯马所选的奠基马群)先知家系 compare Mohammed's Ten Horses

Prophet's thumb n. (also **Prophet's thumb mark**) a small indentation or dimple that often appears in the shoulder muscles of Arabs (常见于阿拉伯马肩部的凹痕或浅窝)先知拇指印:The Prophet's thumb is often regarded as a sign of good luck as legend goes that the mark was made by the thumb of the Prophet Mohammed. 阿拉伯马肩上的先知拇指印常被认为是吉兆,因为传说这个标记乃先知穆罕默德用拇指所按。

Prophet's thumb mark n. *see* Prophet's thumb (见于阿拉伯马的)先知拇指印

prophylactic /ˌprɒfɪˈlæktɪk; *AmE* ˌprɑːf-/ adj. acting to defend against or prevent disease; preventive 预防疾病的

prophylaxis /ˌprɒfɪˈlæksɪs; *AmE* ˌprɑːf-/ n. action taken to prevent disease 疾病预防

proppy /ˈprɒpi; *AmE* ˈprɑːpi/ adj. (of a horse) stiff in gait and action (指马)步法僵硬的,步法呆滞的

proprietor /prəˈpraɪətə(r)/ n.【driving】one who has his own team of horses and coach (自己拥有车马的人)车主

prostaglandin /ˌprɒstəˈglændɪŋ/ n.【physiology】(abbr. **PG**) any of a group of hormones derived from fatty-acids in various tissues and active in many physiological functions (在多种组织产生的脂肪酸衍生类激素,参与机体内许多生理机能活动)前列腺素

prostate /ˈprɒsteɪt; *AmE* ˈprɑːs-/ n.【anatomy】(also **prostate gland**) a gland surrounding the base of the bladder in male mammals (位于雄性哺乳动物体内膀胱基部的腺体)前列腺

prostate gland n. *see* prostate 前列腺

prostatic /ˈprɒsteɪtɪk/ adj. of or relating to the prostate gland 前列腺的:The prostate gland consists of two lateral lobes that are connected by a thin isthmus. Approximately 15-20 prostatic ducts perforate the pelvic urethra. 前列腺由两侧腺叶和连接两腺叶的峡部组成,有15~20个导管开口于骨盆处的尿生殖道。

prostatitis /ˌprɒstəˈtaɪtɪs/ n. inflammation of the prostate gland 前列腺炎

prostrate /ˈprɒstreɪt; *AmE* ˈprɑːs-/ adj. physically or emotionally exhausted 衰竭的,乏力的

prostration /prɒˈstreɪʃn; *AmE* prɑːˈs-/ n. extreme physical weakness or exhaustion 虚脱,衰竭:heat prostration 热虚脱,热性衰竭

protein /ˈprəʊtiːn; *AmE* ˈproʊ-/ n. a group of complex organic macromolecules consisting of one or more chains of amino acids (由单条或多条肽链构成的有机大分子化合物)蛋白质:protein feeds 蛋白质饲料 ◇ protein supplements 蛋白质补充料 ◇ Proteins are essential in the diet of animals for the growth and repair of tissue and can be obtained from foods such as meat, fish, eggs, milk, and legumes. 蛋白质是动物膳食所必需的成分,对生长发育和组织损伤修复都至关重要,可从肉、鱼、鸡蛋、牛奶和豆类食品中摄取。

P

protein bumps n. a skin condition caused by excessive protein content in feed (由于饲料中蛋白质含量过高而导致皮肤出现肿块)蛋白疹

protest *verb*, *noun*
vt. /prəˈtest; *AmE* also ˈproʊ-/ to express objection in a formal statement 申诉,抗议
n. /ˈprəʊtest; *AmE* ˈproʊ-/ an objection or complaint submitted to the organizing body (向赛事组织机构提出的异议)申诉,抗议

protrude /prəˈtruːd; *AmE* proʊˈtr-/ vi. to push or thrust outward 突出;伸出

protrusion /prəˈtruːʒn; *AmE* proʊˈtr-/ n. the act of protruding 伸出;突出

proud flesh n. the swollen flesh that surrounds a healing wound (伤口愈合后周围形成的肿块)肉肿

prowess /ˈpraʊəs/ n. **1** superior strength or ability 能力,本领 **2** great skill at doing sth 才干,才能

proximal /ˈprɒksɪməl; *AmE* ˈprɑːk-/ adj.【anatomy】near to a joint or organ 近侧的: proximal sesamoid bones 近侧籽骨

proximalis /ˌprɒkˈsɪməlɪs; *AmE* ˌprɑːk-/ adj.【Latin】proximal 近侧的 [compare] distalis

pruritus /prʊˈraɪtəs/ n. severe itching of the skin (皮肤的)瘙痒

Przewalski's Horse n. a primitive breed of wild horse that once lived in the Gobi desert and steppes of Mongolia in large herds but became extinct in the mid-20th century; first discovered in 1876 by Russian explorer Przewalski for whom the breed was named; generally had a dun coat with short, erect mane, and stripes on their hind legs, having no forelock (过去大量生活在蒙古戈壁沙滩的原始马种,于20世纪中叶在野外绝种,首次于1876年由俄国探险家普泽瓦尔斯基发现并由此得名。毛色多为兔褐,鬃短而直立,后肢生有条纹,不生额发)普氏野马: Nowadays, the total number of Przewalski's horse in captivity is around 1,000, and some of them have been reintroduced into the wild in Mongolia and China. 目前,全世界圈养普氏野马的数量约为1 000匹,部分已被引入蒙古国和中国营野生生活。[syn] *Equus ferus przewalskii*

pseudomonad /sjuːˈdɒmənæd; *AmE* ˌsuːdəˈmoʊ-/ n.【vet】a pathogenic bacterium of the genus *Pseudomonas* that often causes contagious equine metritis in horses (假单胞菌属致病菌,可引发母马传染性子宫炎)假单胞菌

pseudomonas /sjuːˈdɒmənəs; *AmE* ˌsuːdəˈmoʊnəs/ n. the genus including several phyla of gram-negative, rod-shaped, mostly aerobic flagellated bacteria (包括多个革兰氏阴性厌氧菌的属,该属细菌呈杆状且多带鞭毛)假单胞菌属

PSI abbr. performance sale internationale 国际竞技马拍卖会

Psoroptes ovis n. a species of mite 羊痒螨

psoroptic scabies n. a contagious skin disease caused by a species of mite called *Psoroptes ovis*, characterized by lesions in sheltered body areas such as the forelock, mane, root of the tail, under the chin, between the hind legs, and sometimes in the ears (由羊痒螨所致的传染性皮肤病,症状表现为额发、鬃、尾根、下颌、两后肢间以及耳内等处糜烂)痒癣

PTH abbr. parathyroid hormone 甲状旁腺激素

PTR abbr. Performance Thoroughbred Registry 纯血马性能登记

puberty /ˈpjuːbəti; *AmE* -bərti/ n. the stage in which an individual becomes physiologically capable of sexual reproduction (个体在生理上达到生殖繁育的时期)初情期;性成熟期

pubic /ˈpjuːbɪk/ adj. of or relating to the region of the pubis or the pubes 阴部的,耻骨的: pubic region 耻骨部

pubic bone n. *see* pubis 耻骨

pubis /ˈpjuːbɪs/ n.【anatomy】(*pl.* **pubes** /-iːz/) a bone forming the front arch of the pelvis (位于骨盆前弓处的骨)耻骨 [syn] pubic bone, os pubis

public auction n. an open auction requiring no invitation (不需要邀请即可参加的拍卖)公开

拍卖 [compare] private auction

public stable n.【racing】a stable where horses belonging to more than one owner are boarded (寄宿多个马主马匹的厩房)公共马厩,公用马厩

public trainer n.【racing】one who trains horses for more than one owner or stable (为多个马主或厩房训练马匹的人)公共练马师

puddle /ˈpʌdl/ vi. (of a horse) to shuffle its body in motion(指马)摇摆,摇晃

puffer /ˈpʌfə(r)/ n. *see* shill 马托

puffy /ˈpʌfi/ adj. (of a joint) swollen or bloated (关节)肿胀的

Puissance /ˈpwiːsɒ̃s; *AmE* ˈpwɪsəns/ n. a non-timed show jumping competition in which competitors jump a course consisting of six to eight obstacles ranging from 1.4 to 1.6 meters in the first round, the height and width being progressively increased in each successive round until only one competitor remains (一种不计时的障碍赛,场内设6～8道障碍,首轮障碍高度在1.4～1.6米,在随后各轮中障碍高度和宽度逐渐增加,直到只剩1名参赛者获胜为止)超越障碍赛

puissance wall n.【jumping】*see* brick wall 砖墙

pull /pʊl/ vt. **1** to draw back 扯,拽 PHR V **pull leather**【rodeo】(of a bronc rider) to hold part of the saddle with his free hand during the eight-second ride in competition (指骑野马者在8秒内)手拽马鞍皮革 **pull slack**【rodeo】(of a roper) to pull the rope tight after the loop has set on the target (套牛者)拽紧套绳 **2** make tail or mane tidy and neat by combing or brushing 梳理:To pull the mane and tail, a mane comb will do the job. 梳理马匹鬃、尾时,用鬃梳就可以了。**3** to pluck hair on the upper edges of dock to achieve a neat, slim tail carriage 拔尾根毛发:a pulled tail 被拔过毛的尾巴 **4**【racing】to hold back a horse to avoid winning a race 拽马,勒马 **5** to draw blood by using a syring 采血,抽血:The vet pulled three cc of blood form the horse for bloodtyping. 兽医从马体采3毫升血样用于血液分型。

pulled tail n. a tail style in which the hair on the upper edges of dock is pulled 拔过的尾 [syn] plucked tail

puller /ˈpʊlə(r)/ n. a horse who draws back the reins 拽缰的马,扯缰的马

pulling comb n. a metal comb used to trim the mane or tail 理鬃梳 [syn] mane drag

pulmonary /ˈpʌlmənəri; *AmE* -neri/ adj. of, relating to or affecting the lung 肺的:pulmonary artery 肺动脉 ◇ pulmonary circulation 肺循环 ◇ pulmonary emphysema 肺气肿 ◇ pulmonary hemorrhage 肺出血

pulmonary oedema n. the swelling of lung resulting from excessive accumulation of interstitial fluids in the alveoli or airways (由于肺泡或肺导管内积液过多导致的肺部肿胀)肺水肿

pulmonitis /ˌpʌlməˈnaɪtɪs/ n. *see* pneumonia 肺炎

pulse /pʌls/ n. the rhythmic beating or throbbing of arteries pumped by the regular contractions of the heart (由于心脏有规律的收缩产生的动脉节律性跳动)脉搏:The normal pulse of a horse at rest is between thirty-six and forty beats per minute. 马静息时的脉搏为每分钟36～40次。

pulu /ˈpuːluː/ n.【dated】polo 马球

pulverize /ˈpʌlvəraɪz/ vt. to pound or crush to powder or dust 捣烂;碾碎

pumiced foot n. *see* dropped sole 蹄底脱出,蹄底下坠

punch[1] /pʌntʃ/ n.【BrE】a horse having short legs and a sturdy body 矮壮马

punch[2] /pʌntʃ/ vt. **1** to brand cattle or horses 打烙印 **2** to pierce holes on a horseshoe (蹄铁)打孔:a set of finely punched horseshoes 一副钉孔整齐的蹄铁

puncture /ˈpʌŋktʃə(r)/ n. **1** a small hole 孔,洞 **2** a deep cut made by a sharp object 刺伤,扎伤:puncture wound 扎伤

P

vt. to make a puncture 扎孔，扎伤

punt /pʌnt/ vi. 【BrE】to bet or gamble 押注；赌博

punter /ˈpʌntə(r)/ n. 【racing, slang】a bettor 博彩者，赌马者

puppy show n. a competition for young hounds judged on the basis of conformation, gait, and training（由幼犬参加的比赛，评分依据体貌、步态和受训程度进行）幼犬展览，幼犬选美

puppy walker n. one hired to take care of hound puppies（负责料理幼年猎犬的人员）遛犬员

Pure Blood Lusitano n. *see* Lusitano 卢西塔尼亚马，卢西塔诺马

purebred /ˈpjʊəbred; *AmE* ˈpjʊrbred/ adj.（of animals）born from parents of the same breed（指动物由同种父母本繁育而来）纯种的：purebred Arab 纯种阿拉伯马 ◇ a purebred Orlov Trotter 一匹纯种奥尔洛夫快步马

n. a horse with parents of the same breed 纯种马

purebred percentage n. the percentage of pure blood in a horse born from a purebred stallion and an unregistered mare or mare of another breed（一匹纯种公马与非登记母马或其他品种母马所生后代的纯血比例）纯种比例

purgative /ˈpɜːgətɪv; *AmE* ˈpɜːrg-/ adj. tending to cause evacuation of the bowels; laxative 通便的，轻泻的

n. a purgative agent or medicine 泻药 syn cathartic

purge /pɜːdʒ; *AmE* pɜːrdʒ/ vt. to cause or induce evacuation of the bowels 使腹泻，通便

purple roan n. 【color】a coat color consisting of a uniform mixture of red, white, and black hairs（由红、白、黑三种毛发混杂而成的毛色）紫沙毛

purse /pɜːs; *AmE* pɜːrs/ n. **1** a small bag used for carrying money 钱包，钱袋 **2**【racing】a sum of money awarded to the first three finishers in a race（奖给比赛前三名的金额）奖金：Prize money was originally contained in a purse hung on the finish line, a system from which the present name derived. 过去奖金常装在钱袋挂在终点线上，故而沿用了这个名称。 PHR V **take down the purse** to win the prize money in a horse race（比赛中）获得奖金

purulent /ˈpjʊərələnt; *AmE* ˈpjʊr-/ adj. containing or discharging pus 化脓的，脓性的：a purulent infection 化脓性感染

pus /pʌs/ n. a viscous, yellowish fluid formed in infected tissue（感染组织产生的淡黄色黏液）脓，脓液：pus pocket 脓包

push-button horse n. a well-trained horse who responds quickly to the rider's signals 训练有素的马，反应敏快的马：A push-button horse can be used a good mount for a novice rider. 一匹训练有素的马非常适合新手骑乘。

put /pʊt/ vt. **1** to place or move sth in a certain place or position 处置，放置 PHR V **put down** to kill a horse humanely 安乐处死：Many horses were put down after being detected with glanders. 许多马在检测患有鼻疽后，就被施以安乐死。**put him on his head**（of a rider）to be bucked off a horse（骑手）被撂下马，撂在地上：That horse really put him on his head. 那马果真将他撂在地上。**2**（of a stallion）to be bred to a mare（指公马）配种，与配：Beberbeck is a horse breed descended from Thoroughbred stallions put to local mares. 波波贝克马是由纯血马与配当地母马繁育而来的马品种。**3** to hitch a horse; harness 套马，套驾：put a pony to a cart 用矮马套车

pyemia /paɪˈiːmiə/ n. septicemia caused by the spread of pus-forming bacteria in the blood stream（由血液中的化脓性细菌引起的败血症）脓毒症，脓血症

pyogenesis /ˌpaɪəˈdʒenəsɪs/ n. the formation of pus 生脓，化脓

pyogenic /ˌpaɪəˈdʒenɪk/ adj. producing pus 化脓的：pyogenic microorganisms 化脓性微生物

pyramidal disease n. a condition of horse's foot

involving new bone growth in the region of the extensor process of the distal phalanx; symptoms may include pointing with the affected foot, a shortened stride with a tendency to land heavily on the heel, heat, pain, swelling, lameness, and arthritis (马蹄远侧趾骨伸肌附近着生骨刺的病症，表现为患肢蹄尖着地、步幅缩短、蹄踵负重、肿痛、跛行和关节炎)锥蹄病 syn buttress foot, extensor process disease

pyridoxine /ˌpɪrɪˈdɒksiːn/ n.【biochemistry】a water-soluble vitamin essential in amino acid synthesis (一种水溶性维生素，为氨基酸合成代谢所必需)吡哆胺；维生素 B_6 syn Vitamin B_6

P

Q /kjuː/ abbr. quinella 头二组合彩

QH abbr. Quarter horse 夸特马

quadrella /kwɒˈdrelə/ n.【racing】a wager in which the bettor tries to select the winners of four specific races（赌马者选出特定四场比赛的头马方可赢奖的投注方式）四连赢

quadriceps /ˈkwɒdrɪseps; *AmE* ˈkwɑːd-/ n.【anatomy】the large four-part extensor muscle at the front of the thigh（后股前端的四方形肌）四头肌

quadriga /kwɒˈdriːɡə, kwɒˈdraɪɡə/ n.（*pl.* **quadrigae** /-ɡiː/）an ancient Roman chariot drawn by a team of four horses（古罗马时期四马联驾的战车）驷马战车，驷驾马车 — *see also* **biga**, **triga**

quadrille /kwəˈdrɪl/ n. an orchestrated performance given by teams of four riders who execute dressage or High School figures on horseback（由四名骑手策马跟随音乐行进的高等骑术表演）四对方舞，方阵舞

Quagga /ˈkwæɡə/ n. an extinct South African animal of the the horse family resembling zebra but only having stripes on the head, neck, and shoulders（现已灭绝的南非马属动物，体貌与斑马相似，但仅头、颈和肩部带斑纹）非洲斑驴：Quagga became extinct in Africa since the late 19th century. 斑驴在 19 世纪末就从非洲绝迹了。 [syn] *Equus quagga*

qualifying race n.【racing】a horse race held to evaluate the ability, manners, and eligibility of horses for which there is neither purse nor betting（为了衡量马匹体能和赛事资格而举行的赛马，比赛既不设奖金也不进行投注）资格赛

qualitative trait n.【genetics】a trait controlled by a pair of alleles（由一对等位基因控制的性状）质量性状 [compare] quantitative trait

quality /ˈkwɒləti; *AmE* ˈkwɑːl-/ n. the essential characters or traits of certain horse breeds（某些马种的主要特征）品性，特质

quantitative trait n.【genetics】a trait controlled by many pairs of genes functioning together（由多个基因共同作用控制的性状）数量性状：quantitative trait loci 数量性状位点 [compare] qualitative trait

Quarab /kwɔːˈrɑːb/ n. a horse breed developed in the United States in the 1960s by crossing Quarter and Arabian horses; stands 14 to 16 hands and usually has a bay or gray coat; has a noble head with large bold eyes, long neck, wide chest, and hard hoofs; is intelligent and easy to train, and used mainly for riding, showing and driving（美国在 20 世纪 60 年代采用夸特马与阿拉伯马杂交培育的马品种。体高 14 ~ 16 掌，毛色多为青或骝，面貌俊秀，眼大而亮，颈长，胸身宽广，蹄质结实，反应敏捷，易于调教，主要用于骑乘、展览赛和驾车）夸拉伯马

quarantinable /ˌkwɒrənˈtiːnəbl/ adj.（of diseased animals）that should be isolated（病畜）应隔离的

quarantine /ˈkwɒrəntiːn; *AmE* ˈkwɔːr-, ˈkwɑːr-/ n. a period or practice of isolation to prevent contagious disease from spreading（为预防传染病传播而施行的隔离或时段）隔离；隔离期：Large breeding establishments should maintain a separate barn for isolating sick animals or for keeping newly arriving horses for a short quarantine period. 大型育种机构应当预留一个厩舍，以便隔离病畜或在隔离期饲养新进马匹。

vt. to put an animal into quarantine（使动物）

隔离

quarantine barn n. a building or stall used to isolate infected horses from others (用来隔离患病马匹以防传染其他个体的厩舍)隔离栏,隔离马房

quarry /ˈkwɒri; *AmE* ˈkwɔːri, ˈkwɑːri/ n. 【hunting】 an animal pursued by a hound or hunter; prey (被猎犬或猎人追捕的动物)猎物

quarter[1] /ˈkwɔːtə(r); *AmE* ˈkwɔːrt-/ n. **1** either side of a horse's hoof between the toe and the heel (马蹄两边介于蹄尖与蹄踵的部分)蹄侧 **2** (usu. **quarters**) the hindquarters of a horse (马体)后躯 PHR **quarters falling out** (of a dressage horse) with quarters falling out (舞步马)后躯外撇 **quarters leading** (of a dressage horse) with quarters leading its forehand in the lateral movement (舞步马)后躯起先 ◇ **quarters not engaged** (of a dressage horse) with hind legs not sufficiently brought under its body (舞步马)后躯没收拢的 **quarters trailing** (of a dressage horse) with quarters leaving behind the movement in half pass (舞步马以横斜步行进时)后躯滞后 PHR V **range the quarters** (of a horse) to move the haunches either to the left or right (指马)挪动后躯,调整后躯 **3**【racing】one fourth of a mile 1/4 英里

quarter[2] /ˈkwɔːtə(r); *AmE* ˈkwɔːrt-/ vt. to groom a horse briefly before exercise in the morning (马匹晨练前)简单刷拭 | **quartering** /-tərɪŋ/ n. : Quartering is usually carried out at early morning stables after mucking out. 马体简单刷拭多在清早打扫厩舍后进行。

quarter blanket n. *see* quarter sheet 后躯马衣

quarter clip n.【farriery】a V-shaped extension of the horseshoe on the left and/or right sides between the toe and heel to fit the shoe more securely to the hoof (蹄铁两侧位于铁尖和铁尾的V形突起,可使蹄铁装得更牢固)[侧面]蹄铁唇 compare toe clip

quarter crack n.【farriery】a crack on the lateral sides of the hoof wall 蹄侧开裂

Quarter Horse n. (abbr. **QH**) a versatile horse breed developed in the 17th century in Virginia and Carolina by crossing Spanish-type mares with imported English Thoroughbred stallions; stands 14.1 to 16 hands and may have a coat of any solid color, although chestnut is most common; has a short-coupled body, a short and wide head, a long neck, and well-defined withers; is compact, agile, fast, well-balanced, and possessed of quick reflexes; a versatile horse used in rodeo, driving, and many other equestrian disciplines (17 世纪殖民者在弗吉尼亚和卡罗莱纳州采用西班牙马母马与雄性纯血马杂交育成的兼用型马品种。体高 14.1 ~ 16 掌,各种单毛色兼有,但以栗色最为常见,后腰短,头短而宽,颈长,鬐甲突出,体质结实,运步稳健,反应敏快,用于牛仔竞技、驾车和诸多马术比赛项目)夸特马

quarter marks n. a pattern of square marks made on the horse's hindquarters to improve its appearance (出于美观而在马体后躯被毛上修剪出的方格图案)后躯方格

quarter pole n.【racing】a post located adjacent to the infield rail exactly 2 furlongs from the finish (立于赛道内栏的标杆,与终点相距 400 米) 1/4 英里杆,四分杆

quarter sheet n. **1** (also **quarter blanket**) a covering placed over the quarters of a horse for warmth (为了保暖而披在马后躯的薄毯)后躯罩衣,后躯披毯 **2** *see* summer sheet 夏季马衣

quarter strap n.【driving】a strap of the harness passing over the quarters of horse and connecting to the breeching (挽具中搭在马体后躯且与坐鞧相连的革带)后鞧革

quarters-in n.【dressage】*see* travers 腰向内斜横步

quarters-out n.【dressage】*see* renvers 腰向外斜横步

Quashquai /ˈkwɒʃkaɪ; *AmE* ˈkwɑːʃ-/ n. a horse breed indigenous to Iran; stands around 15 hands and usually has a bay, chestnut, or gray coat; used mainly for riding and farm work (伊朗本土的马品种,体高约15掌,毛色多为骝、栗或青,主要用于骑乘和农田作业)夸什凯马

queen of saddles n. another term for sidesaddle (侧鞍的别名)淑女鞍,侧鞍

Queen's Plate n. a 1.25-mile Canadian stakes race held for 3-year-old Thoroughbred horses each June since 1860 in Toronto (自1860年起加拿大每年6月在多伦多举行的奖金赛,由3岁纯血马参加,赛程2千米)女王杯[大赛]: The Queen's Plate is North America's oldest Thoroughbred horse race. 女王杯是北美历史最悠久的纯血马比赛。

quick /kwɪk/ n.【anatomy】the sensitive tissue or structure within the hoof of a horse (马蹄)敏感组织

vt. to penetrate the sensitive structures of the hoof with nails while shoeing a horse (打掌时蹄钉扎伤马蹄敏感组织)钉伤 [syn] nail quick, nail prick | **quicking** /ˈkwɪkɪŋ/ n.: Quicking generally results in lameness. 钉伤通常可导致跛行。

quick-response diagnostic test n. *see* stall-side diagnostic test 快速诊断检测,现场诊断检测

quick-stop n. a Western training bridle used to train a horse to stop quickly on its haunches (美国西部马勒,用来训练马匹速停)速停勒

quid /kwɪd/ n. partially chewed fodder 嚼过的草料

quiddor /ˈkwɪdə(r)/ n. a horse who drops feed from its mouth while chewing (咀嚼时从口中掉落饲料的马)漏嘴的马

quiescent /kwiˈesnt/ adj. being quiet or still; inactive 安静的,静息的

quinella /kwɪˈnelə/ n.【racing】(abbr. **Q**) a wager in which the bettor selects the first two winners in a race regardless of the order (比赛中投注者仅选出前两名马而不考虑次序即可获奖的押注方式)头二彩,中二元 [compare] exacta

quinidine /ˈkwɪnɪdiːn/ n. a colorless crystalline alkaloid resembling quinine and used mainly in treatment of atrial fibrillation in horses (类似奎宁的无色、晶体状生物碱,多用于马匹心脏房颤的治疗)奎尼丁

quinsy /ˈkwɪnzi/ n.【vet】an abscess in the throat resulting from acute inflammation of the tonsils and the surrounding tissue (由于扁桃体及周围组织的急性炎症而导致咽喉脓肿)咽门炎;扁桃体周脓肿

quintain /ˈkwɪntɪn/ n. a post used as a target in the medieval exercise of tilting (中世纪刺枪练习时当靶使用的柱杆)矛靶,枪靶

quirt /kwɜːt; *AmE* kwɜːrt/ n. a short riding whip (赛用)短鞭

vt. to strike with a quirt 鞭抽,鞭打

quit /kwɪt/ vt.【cutting】(of a horse) to stop working a cow (指马)停止拦截 PHR V **quit a calf /cow** to stop working a cow (指马)停止拦截: In National Cutting Horse Association competitions, to quit a cow is permissible without penalty if the cow has obviously stopped, turned away from the cutter, or if it is behind the time line. 在全国截牛协会举办的比赛中,如果所拦截的牛已停下、远离截牛者或跑到计时线外时,马停止拦截不进行罚分处理。

quittor /ˈkwɪtə(r)/ n. a chronic inflammation of the hoof cartilages resulting from injury to the coronet or pastern or puncture to the sole of the foot, characterized by swelling and abscessation around the coronary band, degeneration of hoof tissue, and ultimately lameness (马蹄冠、系部或蹄底受伤造成蹄软骨的慢性炎症,症状表现为蹄冠周围肿胀化脓、蹄组织腐烂以及最终跛行)马蹄疽 [syn] coronary sinus

R

rabicano /ˌræbɪˈkɑːnəʊ/ n. 【color】 a coat color pattern with white hairs on the flanks and at the base of the tail (马肋侧与尾础被毛中散生白毛的毛色类型)散沙毛,拉比卡诺:Rabicano generally occurs in Quarter Horse, Arab, Noriker, Brabant, and North American Spanish. 散沙毛多见于夸特马、阿拉伯马、诺里克马、布拉邦特马和北美西班牙马。

rabid /ˈræbɪd/ adj. (of animals) affected with rabies (动物)患狂犬病的

rabies /ˈreɪbiːz/ n. 【vet】 a contagious fatal viral disease of mammals that attacks the central nervous system and causes madness and convulsions; is transmitted by the bite of infected animals (哺乳动物所患的一种致死性病毒传染病,能破坏中枢神经系统并导致疯癫和抽搐,可通过患畜咬伤传播)狂犬病:furious rabies 狂躁型狂犬病 ◇ rabies vaccine 狂犬病疫苗

race /reɪs/ n. **1** (also **horse race**) a competition or contest of speed between horses 比赛;赛马:allowance race 让磅赛 ◇ claiming race 竞价赛 ◇ endurance race 耐力赛 ◇ handicap race 让磅赛 ◇ normal race 常规赛 ◇ stakes race 奖金赛 ◇ race course 赛道 ◇ welter race 负重赛 ◇ Stakes races declared off will generally be rescheduled at a later date, while normal races will not be rerun. 一般来说,奖金赛宣布取消后会重新安排比赛,而常规赛取消后不予重赛。**2**【genetics】 a population of organisms differing from others of the same species in their hereditary traits; a subspecies (遗传特征与种内其他个体不同的生物群体)亚种 **3** a facial white marking of horses which runs from the forehead to one nostril (从前额延伸到鼻孔一侧的面部别征)弯流星

race card n. 【racing】(also **race program**) the printed program of a race meeting, including the name and time of each race, the names of all competing horses, their owners, trainers and riders, and the weights to be carried (为比赛而印制的日程表,包括赛事名称与时间、参赛马匹、马主、练马师和骑手的名字以及马匹的负重)比赛日程,赛程表

race meeting n. **1** a place where horse races are held 赛马场 **2** a festival during which racing events are conducted 赛马节,赛马会

Race of Champions n. 【racing】(abbr. **ROC**) an annual endurance race held in the United States in which the competing team with the lowest average time wins (美国每年举行的最为重要的一项耐力赛,参赛代表队所用平均时间最少的获胜)冠军联赛:To qualify for the Race of Champions, horses must have completed five or more 100-mile races with at least two top-ten finishes, while qualifying riders must have completed at least one 100-mile race. 要参加冠军联赛的马必须参加过5场以上160千米的比赛且至少两次跑入前10名,而骑手则至少要参加过一次160千米的长途赛。

race position n. 【racing】 the position of a horse competing in a race (比赛中马匹的位置)跑位

race program n. *see* race card 比赛日程,赛程表

race tracker n. one who frequents a racetrack; racegoer 赛马迷

racecourse /ˈreɪskɔːs; *AmE* -kɔːrs/ n. **1** a track used for horse racing 赛道 [compare] raceway **2**【racing】 an area where horse races are held 赛马场,跑马场

racecourse paddock n. *see* paddock (赛场上的)围场,围栏

racegoer /ˈreɪsɡəʊə(r); *AmE* -ɡoʊ-/ n. (also **race goer**) one who attends a race meeting; race tracker (赛马场)看客,赛马迷

racehorse /ˈreɪshɔːs; *AmE* -hɔːrs/ n. (also **race horse**) a horse bred or trained for racing 赛马: Racehorse generally refers to the Thoroughbred in flat racing, but sometimes it is also extended to trotters, pacers and sprinters. 赛马多指参加平地赛的纯血马,但有时也泛指快步马、对侧步马和短途赛马。 syn racer, bangtail

racer /ˈreɪsə(r)/ n. **1** a horse running in a race; racehorse 赛马 **2** a riding boot worn by jockeys (骑师穿的)赛马靴

racetrack /ˈreɪstræk/ n. (also **race track**) an oval course on which horse races are held (举行赛马的椭圆形场地)赛道;赛场

raceway /ˈreɪsweɪ/ n. a track upon which harness racing is held 轻驾车赛场

rachitic /rəˈkɪtɪk/ adj. of or relating to rachitis 佝偻病的

rachitis /rəˈkaɪtɪs/ n.【vet, dated】*see* rickets 佝偻病,脊柱炎

racing /ˈreɪsɪŋ/ n. a competition of speed between horses in which the fastest wins (马匹间的速度赛,最快者获胜)比赛,赛马: barrel racing 绕桶赛 ◇ flat racing 平地赛 ◇ harness racing 轻驾车赛 ◇ horse racing 赛马

adj. of or relating toa race or horse races 比赛的;赛马的: racing breastplate 竞赛胸带 ◇ racing girth 竞赛胸带 ◇ racing lad 骑手 ◇ racing officials 赛事官员 ◇ racing secretary 赛事秘书 ◇ racing steward 赛事监管 ◇ racing yard 赛马场

Racing Biga n. an ancient Roman racing chariot drawn by a pair of horses harnessed abreast (古罗马使用的双轮赛用战车,多由双马并肩套驾)赛用战车

racing blinkers n. a pair of blinkers consisting of a hood fitted over the head of a racehorse to restrict its vision to the sides and rear (戴在赛马头上的眼罩套兜,用来遮挡两侧和后方的视线)竞赛眼罩: The use of racing blinkers must be approved by the racing stewards. 赛用眼罩的使用必须经过赛事主管的许可。

racing breastplate n. *see* Aintree girth 赛用胸带

racing calendar n. a periodical publication in which upcoming races, advertisements for sales breeding, weights and acceptances for handicaps, racing colors, etc. are noted (赛马业定期刊发的出版物,上面有赛事预告、育种广告、评磅重量和赛服颜色等信息)比赛日历,赛马日历 syn sheet calendar

racing cap n. *see* racing helmet 竞赛头盔

Racing Clearance Notification n. (abbr. **RCN**) an official document issued by the racing authority to qualify a horse to compete in races (赛马监管机构出具的文件,马匹凭此方可参赛)出赛许可证,赛马许可证

racing colors n.【racing】(also **colors**) a uniquely patterned jockey jacket and cap as selected by the owner of the mount and worn by the jockey during a race (骑师比赛所穿的夹克和头盔的特殊颜色)赛服颜色,骑师彩衣 syn racing silks

racing form n.【racing】a sheet containing the information of a racehorse's past performance that can be used by bettors as reference for wagering (详细记录赛马以往成绩的表单,博彩者以此作为参考进行押注)出赛记录表

racing gallop n. an extended gallop with speed ranging from 30 to 45 mph (一种伸长袭步,时速在48~72千米)竞赛袭步

racing girth n. *see* Aintree breast girth 赛用胸带

racing helmet n. (also **helmet**) a hard head covering worn by a jockey in racing (比赛中骑师戴的头盔)竞赛头盔 syn crash skull, racing cap

racing iron n. a light, sturdy aluminum-made stirrup designed for racing (赛马使用的铝制马镫)赛用马镫

racing on the flat n. *see* flat racing 平地赛

racing plate n. (also **plate**) a light horseshoe, usually made of aluminium, used on racehorses (赛马用的一种铝制轻型蹄铁)竞赛蹄铁,赛用蹄铁

racing saddle n. a saddle used on racehorse 赛用鞍: There are two types of racing saddles, the light used for flat racing generally weigh less than 2 pounds, and the heavy are mainly used for hurdling and steeplechasing. 赛用鞍分两种,轻型鞍用于平地赛,重量通常少于1千克;重型鞍则主要用于跨越障碍和越野障碍赛。

racing seat n. 【racing】 the position assumed by a jockey who crouches in the saddle with stirrup leathers adjusted extremely short when riding a racehorse in competition (骑师在赛马中所采用的坐姿,其中骑手蹲坐于马鞍,镫革调的非常短)赛马骑姿

racing secretary n. 【racing】 one responsible for establishing track conditions and sometimes for assigning weights to riders in handicap races (负责设立赛场规则并在让磅赛中为骑手确定负重的人员)赛事秘书

racing silks n. 【racing】 *see* racing colors 赛服颜色,骑师彩衣

racing snaffle n. a Dee-ring snaffle used in race [D环]赛用衔铁

racing sound adj. 【racing】 (of a horse) able to compete in a race (指马)可参赛的

racing strip n. 【racing】 the part of a racecourse between the inside and outside rails; racetrack (赛场内、外栏杆之间的区域)赛道区

racing trot n. a trot that has a longer stride than extended trot and is common to racing Thoroughbreds (一种步幅比伸长快步更大的快步,在赛马中较为常见)竞赛快步

rack[1] /ræk/ n. **1** a wooden or iron framework used to hold hay, etc. (用来盛干草等物的木质或铁制框架)支架: hay rack 干草架 **2** a pair of antlers of a stag 鹿角 **3** 【racing】 the framework on the racecourse 赛场栏架 PHR **by the rack** (of a bettor) purchasing every possible daily-double or other combination ticket (指赌马者)投注组合彩: to bet by the rack 投注所有组合彩 **4** 【hunting】 a break in the hedge or bush resulting from repeated passage of animals (由于动物经常穿梭而在篱笆或灌木丛中形成的缺口)裂口,开口

vt. **1** to add fodder for livestock in a hay rack 添草料 **2** to tie a horse to a post or a ring in the wall 拴马: to rack up a horse 拴马 **3** 【racing】 (of a jockey) to collide into other horses running in a race (骑师)挤堆,冲撞 PHR V **rack up** (of jockeys) to collide into one another (指骑师)挤堆,冲撞

rack[2] /ræk/ n. a fast, artificial 4-beat gait in which each foot touches the ground separately and at equal intervals, often seen in American Saddlebred horses (一种四蹄音人工步法,四蹄触地间隔均等,多见于美国骑乘马)单蹄快步 syn single foot

vi. to move or ride at a rack 以单蹄快步行进

rack chain n. a chain formerly used to lead and tie a horse before lead rope was introduced (在采用缰绳之前用来牵行和拴马的索链)拴索,拴链

racking boots n. *see* overreach boots 护蹄碗

Racking Horse n. a light horse breed recently developed in the United States out of the Tennessee Walking Horse; stands about 15.2 hands and may have a black, bay, sorrel, chestnut, brown, or gray coat; has an attractive and graceful build with a long sloping neck, good bone, and smooth legs; used for riding, Western tack, and driving (美国采用田纳西走马培育而来的轻型马品种。体高约15.2掌,毛色多为黑、骝、红褐、栗、褐或青,体态优雅,颈长而斜,管骨强健,四肢修长,用于骑乘、西部牛仔项目和驾车)单蹄快步马

Racking Horse Breeders' Association of America n. (abbr. **RHBAA**) an organization founded in 1971 in Alabama, USA to maintain the registry for the Racking Horse breed (1971年在美

国亚拉巴马州成立的组织机构,旨在对单蹄快步马进行品种登记)单蹄快步马育种者协会

racy /ˈreɪsi/ adj. (of a horse) suitable for racing (指马)适于比赛的:There is no argument that the Thoroughbreds are more racy than any other breeds. 毋庸置疑,纯血马比其他任何马品种都更适合赛马比赛。

radial /ˈreɪdiəl/ adj.【anatomy】of, relating to, or near the radius or forearm 桡骨的,桡侧的:radial carpal joint 桡侧掌关节 [compare] ulnar

radialis /reɪdɪˈeɪlɪs/ adj.【Latin】radial 桡骨的,桡侧的

radiograph /ˈreɪdɪəʊgrɑːf; *AmE* -dɪoʊ-/ n. an image produced on a radiosensitive film by radiation of X-rays passed through an object (采用X线照射物体而在感光胶片上留下的图像) X片

vt. to make a radiograph of sth 拍X片

radiography /ˌreɪdiˈɒgrəfi; *AmE* -ˈɑːgrəfi/ n. a technique in which the image is produced on a radiosensitive film by x-rays passed through an object (通过X线扫描物体而在感光胶片表面形成图像的技术)放射显影,X线显影

radius /ˈreɪdɪəs/ n.【anatomy】(*pl.* **radii** /-diaɪ/) the shorter one of the two forearm bones (前臂两骨中较短的一支)桡骨:lateral tuberosity of radius 桡骨外侧隆起 [compare] ulna

radix /ˈreɪdɪks/ n.【Latin】(*pl.* **radixes**, or **radices**) a root or point of origin (根部或生长点) 根:radix penis 阴茎根

rag /ræg/ n. **1** a piece of old or worn cloth used for cleaning or dusting 破布;抹布: The blacksmith wiped his oily hands on a rag. 铁匠用一块破布擦自己油污的双手。**2** a herd of young horses 马群

ragged /ˈrægɪd/ adj. (of a horse) poorly fleshed or muscled (指马)瘦弱的:ragged hips 腰角突兀

ragwort /ˈrægwɜːt; *AmE* -wɜːrt/ n. any of several plants of the genus *Senecio* having yellow flower heads (千里光属草本植物,有黄色花冠)千里光

rail /reɪl/ n. **1**【jumping】a round wooden bar used to create a jump (用来搭建障碍的圆木杆)横杆:ground rail 地杆 [syn] pole **2**【racing】a metal barrier set on either side of the racetrack (置于赛道两侧的栏杆)围栏,护栏:inside rail 内栏 ◇ outside rail 外栏 [syn] fence [PHR] **on the rail** (of a horse) running close to the infield rail (指马)靠内栏杆跑

rail runner n.【racing】a horse who prefers to run along the inside rail 靠内栏跑的赛马

rain rot n. *see* rain scald 雨斑病

rain scald n. (also **rain rot**) a painful, infectious skin inflammation caused by the organism *Dermatophilus congolensis* which infects the horse's coat and gains entry into the skin when it is saturated by prolonged rain (一种产生痛感的传染性皮炎,由寄生在马被毛中的刚果嗜皮菌所致,被毛遭雨水长时浸湿后可侵入皮肤) 雨斑病 [syn] streptothricosis

rake[1] /reɪk/ n. **1** a long-handled agricultural implement with a row of teeth or tines at its head (一种头部带排齿或细股的长柄农具)耙:horse rake 马拉耙 **2**【BrE】a number of carriages or coaches coupled together (连接在一起的多辆马车)车队

rake[2] /reɪk/ vi.【rodeo】(of a rider) to urge or incite the bronc by swinging the spur form its shoulders to the flanks (指骑手用马刺从肩部向肋部滑动)侧踢:to rake up a bronc 侧踢马刺以激野马

Ralli car n. (also **Ralli cart**) a small, light, two-wheeled horse-drawn carriage with back-to-back seating for four passengers, first introduced at the end of the 19th century and named after the family by whom it was originally designed (一种轻型双轮英式客用马车,背靠背可坐四名乘客,首次出现在19世纪末并以设计者的姓氏命名)拉利马车

Ralli cart n. *see* Ralli car 拉利马车

ram-headed adj. (of a horse) having a convex head profile (指马)羊头的

ramulus /ˈræmjʊləs/ n.【Latin】(*pl.* **ramuli** /-laɪ/) a small branch 小支

ramus /ˈreɪməs/ n.【anatomy】(*pl.* **rami** /-maɪ/) a branch of a bone, nerve, or blood vessel (骨、神经或血管的分支)支

ranch /rɑːntʃ; *AmE* ræntʃ/ n. a large farm on which livestock are raised (饲养家畜的大型农场)牧场

ranch brand n. an identifying mark branded on the hip of livestock (烫在家畜臀部的记号)牧场印号 syn vanity brand, ownership brand

ranch house n. the building or house built on a ranch 牧场房屋,牧场建筑

ranch man n. one who works on a ranch 牧场工人

ranch sorting n. a timed Western event in which two riders cooperate to cut out and drive the ten numbered cattle one by one from one pen to another (西部牛仔竞技项目之一,两名骑手合作将编号的10头牛逐个从一个围栏隔离出来并转移至另一个围栏)牧场分栏

rancher /ˈrɑːntʃə(r); *AmE* ˈræntʃər/ n. **1** one that owns or manages a ranch 牧场主 **2** one who is employed to work on a ranch 牧场工人

ranchero /rɑːnˈtʃeərəʊ; *AmE* rænˈtʃeroʊ/ n.【Spanish】(*pl.* **rancheros**) one who owns or works on a farm; rancher 牧场主;牧场工人

ranchette /ˌrɑːnˈtʃet; *AmE* ˌræn-/ n. a little farm or ranch 小牧场;小农场

ranching /ˈrɑːntʃɪŋ; *AmE* ˈræntʃɪŋ/ n. the activity of running a ranch 农牧经营: cattle ranching 牧牛经营

rancho /ˈræntʃəʊ; *AmE* -tʃoʊ/ n. (*pl.* **ranchos**) **1** a large ranch 大牧场 **2** a hut or building for housing farm workers on a ranch 牧场棚屋

randem /ˈrændəm/ n.【driving】a team of three horses hitched in single file to pulled a cart (前后以三匹马纵列驾车)三马纵驾 compare tandem

adv. driving in such way 三马纵驾车: to drive randem 三马纵驾

range /reɪndʒ/ n. a large, open area or land for keeping livestock (饲养家畜的开阔场地)牧场,草场

vt. to keep livestock on open land for grazing 放牧

v. to wander freely; roam 游荡,驰骋

range horse n. **1** any horse kept on a range 牧场马 **2** *see* cow pony 牛仔马

ranger /ˈreɪndʒə(r)/ n. one employed to maintain and protect a forest or other natural reserves (看护森林和自然保护区的人员)护林员,林管员

rangy /ˈreɪndʒi/ adj. (of a horse) having long legs; slender (指马)四肢修长的: a rangy horse 一匹四肢修长的马

rank /ræŋk/ vi.【racing】(of a horse) to disobey the rider and run in a headstrong manner in a race (指赛马不听骑手驾驭而狂奔)乱冲;拒从

n. an instance of running in a headstrong manner 乱冲,狂奔: to run rank 横冲,狂奔

RAO abbr. recurrent airway obstruction 复发性呼吸道阻塞

rape /reɪp/ n. a plant grown as forage crop and for its seeds that yield oil (一种饲料栽培作物,其籽实可榨油)油菜

rapeseed /ˈreɪpsiːd/ n. the seeds of the rape plant 油菜籽

rapeseed meal n.【nutrition】a protein supplement acquired from the residue of rapeseeds after extraction of the oil (油菜籽榨油后的残渣,可作为蛋白质补充料)菜籽粕

raphe /ˈreɪfiː/ n.【anatomy】a seam-like line or ridge between two similar parts of a body organ, as in the scrotum (机体器官两部分之间的缝合线或嵴,如阴囊缝)缝: raphe of pharynx 咽缝 ◇ raphe of pons 脑桥缝 ◇ raphe of scrotum 阴囊缝

rapping /ˈræpɪŋ/ n.【jumping】a training practice of using a rapping pole to encourage a horse to lift his feet higher when jumping（跨越障碍时用叩杆训练马往高收蹄的方法）收蹄训练

rapping pole n.【jumping】a wooden pole supported by jump stands and getting raised as a horse jumps so that its hind feet hit the pole（用架杆支撑的木杆，马跨越时其高度会有所增加，故马后肢会撞击到杆上）收蹄杆：People use rapping pole to encourage the horse to lift its feet higher on subsequent jumps to avoid the pain of striking. 人们采用收蹄杆来鼓励马匹在以后跨越障碍时举蹄更高从而避免撞蹄。

Rarey, John Solomon n.（1827 – 1866）an American who used the technique of whispering to calm and train horses during the mid-19th century（美国马语者，于19世纪中叶推广使用口哨声来安抚和训练马匹）约翰·所罗门·罗瑞

rasp /rɑːsp; *AmE* ræsp/ n. a long-handled, coarse file with sharp, raised ridges for smoothing uneven surfaces（一种带粗棱的长柄锉，用来将表面打磨平整）锉刀：rasp handle 锉刀柄

vt. to file or scrape with a rasp; float 锉平 | **rasping** /ˈrɑːspɪŋ; *AmE* ˈræs-/ n.

rasper /ˈrɑːspə(r); *AmE* ˈræs-/ n.【hunting】a large, untrimmed hedge that scratches the horses and riders when jumped over（一种尚未修剪的篱笆，跨越时易刮伤人马）刮人树篱

rat tail n. **1**【slang】a horse's tail sparsely covered with hair 毛稀的马尾 **2** *see* mule ear 提靴带

ratcatcher /rætˈkætʃə(r)/ n. **1**【dated】one who catches mice and rats for sale 捕鼠者 **2**【hunting】a riding outfit worn during show jumping and cub-hunting（障碍赛或猎幼狐时穿的套装）骑装：Ratcatcher usually comprises a tweed coat and dark breeches with a bowler hat and long boots. 全套骑装通常包括一件花呢大衣、一条深色马裤、一顶常礼帽和一双长筒靴。

rate /reɪt/ n. a measurement of the speed at which sth moves or happens（物体移动或行进的速度）速率，率：heart rate 心率

vt. **1**【rodeo】(of a rider) to adjust the strides of horse to turn around a barrel（骑手）调整（步伐）**2**【racing】(of a jockey) to restrain a horse early in a race to save energy for a dash at the finish（指骑手在起跑后限制步速以节省体力以备终点冲刺之需）限速 **3**【hunting】to scold or rebuke a hound 责骂，训斥（猎犬）

rated /ˈreɪtɪd/ adj.【racing】(of a horse) restrained by the jockey early in a race（指马）被限速的，被遏制的

ratio /ˈreɪʃɪəʊ; *AmE* -oʊ/ n. (*pl.* **-os**) the relation between two quantities expressed as the quotient of one divided by the other（两个数量之间的关系，通过二者的商来表示）比例：ratio of forage to concentrate 粗精料比例

ration /ˈræʃn/ n.【nutrition】a fixed amount of fodder allotted to domestic livestock（定量供给家畜的饲料）日粮：balanced ration 平衡日粮 ◇ total ration 全价料 ◇ ration formulation 日粮配合

vt. to supply with rations 定量供给

rattle /ˈrætl/ v. **1**【racing】(of a horse) to make a quick succession of percussive sounds when running on a firm track（马在硬赛道上奔跑时发出的声音）咯嗒作响 **2**【hunting】PHR V **rattle a fox** (of hounds) to press hard on the trail of a fox（指猎犬）紧追狐狸

rattler /ˈrætlə(r)/ n. a plastic or wooden ball tied around the fetlock of a horse's forefeet to encourage him to step higher at the trot（拴在马前肢系部的塑料或木质球，用来提高快步举蹄高度）响铃，响球

Rau, Gustav n.（1880 – 1954）a famous German hippologist（德国著名马学专家）格斯塔夫·劳

raw /rɔː/ adj. **1** not cooked or processed; crude 天生的；天然的 **2** recently finished; fresh 新

鲜的:raw semen 鲜精

rawhide /ˈrɔːhaɪd/ n. **1** untanned hide of cattle or other animals (未鞣制的兽皮)生牛皮;生皮 **2** a whip or rope made of rawhide 生皮鞭

ray /reɪ/ n. a dorsal stripe 背线

razor-backed adj. (of a horse) having a prominent and sharp backbone (指马)脊背突兀的,瘦骨嶙峋的

RBC abbr. Red Blood Cells 红细胞:When reading a blood test results, the RBC count indicates the oxygen carrying capacity of the blood. 在血液化验结果中,红细胞数是血液运氧能力的指标。 compare WBC

RCN abbr. Racing Clearance Notification 出赛许可证,赛马许可证

RCVS abbr. Royal College of Veterinary Surgeons [英国]皇家兽医学会

RDAs abbr. Recommended Dietary Allowances 推荐饲养标准

ready /ˈredi/ adj. (of a horse) prepared to participate in an event (指赛马)准备就绪的

ready-made shoe n.【farriery】a machine-made shoe with rounded heels and nails holes punched perpendicular to the shoe instead of corresponding to the slope of the hoof (一种由机器铸造的蹄铁,蹄铁尾圆滑,钉孔竖直而非与蹄角度平行)预制蹄铁 syn cold shoe 冷蹄铁

reagent /riˈeɪdʒənt/ n. a substance used in chemical reaction to detect, measure, or analyze other substances (通过化学反应来检测分析其他物质的化学物质)试剂

rear[1] /rɪə/ vt. to bring up or nurture; raise 抚养,养育;饲养

rear[2] /rɪə/ adj. of or located at the hind part; hind 后面的;后部的:rear leg 后肢

n. the hind part of horse's body (马体)后部

vi. (a horse) to stand up on its hind legs (马用后腿)直立,起仰

rear boot n. an enclosed compartment at the back a carriage for storing luggage 车后行李箱

rearing /ˈrɪərɪŋ/ n. the vice of a horse who rises on its hind legs for attack (马以后肢直立袭击人的恶习)直立,起扬

reata /rɪ(ː)ˈɑːtə/ n.【Spanish】a lasso or lariat 套索

vt. to catch with reata 用套索捕捉

reata strap n. a short strap used to secure a lariat to a Western saddle (西部鞍上的)系套索皮带

receiving barn n.【racing】a barn where horses entering for a race are stabled (为参赛马匹提供的畜舍)接待马棚

receptacle /rɪˈseptəkl/ n. a container for holding water or fodder (放水或饲料的)容器;饲槽

receptive /rɪˈseptɪv/ adj. (of a mare) ready or willing to accept a stallion for mating (指母马)接受交配的;有求配欲的 | **receptivity** /ˌriːsepˈtɪvəti/ n.

recess /rɪˈses, ˈriːses/ n.【anatomy】an indentation or small hollow 隐窝:laryngopharyngeal recess 喉咽隐窝 ◇ optic recess 视隐窝 ◇ piriform recess 梨状隐窝

recessive /rɪˈsesɪv/ adj.【genetics】of, relating to an allele that does not produce a characteristic effect when present with a dominant allele (与显性基因共同出现时不产生表型)隐性的:recessive character 隐性性状 ◇ recessive allele 隐性基因 opp dominant

n. a recessive allele or trait 隐性基因;隐性性状

recessus /rɪˈsesəs/ n.【Latin】(*pl.* **recessus**) recess 隐窝

recipe /ˈresəpi/ n. a set of directions for making or preparing food or feed (配制食物或饲料的说明)食谱;处方

recipient /rɪˈsɪpiənt/ adj. functioning as a receiver; receptive 接受的

n. one that receives or is receptive 接受者,受体:recipient mare 受体母马 compare donor

recognized hunt n.【hunting】a Hunt having permanent status from the association governing hunting activity in the area (永久地位被地区狩猎协会所认可的猎场)知名猎场,认证猎场 compare registered hunt

recombination /ˌriːkɒmbəˈneɪʃn/ n.【genetics】the natural process in which the pair of homologous maternal- and paternal-derived chromosomes exchange segments (来自母本和父本的同源染色体之间发生片段交换的过程)重组 | **recombinational** /-nl/ adj.

Recommended Dietary Allowances n.【nutrition】(abbr. **RDAs**) quantities of nutrients in the diet that are required to maintain good health of an individual (个体维持健康所需的营养成分含量)推荐日粮供给量

recover /rɪˈkʌvə(r)/ vt. **1** to get back or regain strength, etc.; restore (体力)恢复 **2**【hunting】(of the hounds) to pick up the scent again after a check (指猎犬在短暂停顿后)重新找到狐狸踪迹

recovery /rɪˈkʌvəri/ n. the act, process, duration of recovering 恢复;复原

rectal /ˈrektəl/ adj. of or relating to rectum 直肠的:rectal temperature 直肠温度,肛温

rectal palpation n. an examination of the reproductive tract or organ to determine the pregnancy in mare by feeling with the hand through the rectal wall (用手隔着直肠壁触摸诊断母马生殖道来判定其妊娠状况)直肠检查

rectum /ˈrektəm/ n.【anatomy】the terminal portion of the large intestine (大肠末段的部分)直肠

recur /rɪˈkɜː(r)/ vi. to happen or show up again or repeatedly 复发;重现

recurrent /rɪˈkʌrənt; *AmE* -ˈkɜːr-/ adj. occurring or appearing again or repeatedly 重现的;复发的:The large strongyles are generally thought to be the commonest cause of recurrent bouts of spasmic colic. 大圆线虫通常被认为是痉挛性腹痛复发的主要病因。

recurrent uveitis n. *see* moon blindness 复发性葡萄膜炎,夜盲症

red bay n.【color】a light shade of bay color 红骝毛

red blood cell n. (abbr. **RBC**) *see* erythrocyte 红细胞:The resting normal mean of a horse's red blood cell count is approximately 8×10^{12} to 15×10^{12} cells/litre of blood. 马在静息时的平均红细胞数为$(8 \sim 15) \times 10^{12}$个细胞/升。compare WBC

red chestnut n.【color】a color of horse who has a bright, shiny reddish coat (马被毛鲜红闪亮的毛色)红栗毛:Red chestnuts always have red highlights that stand out. 红栗毛色泽艳丽,因而非常显眼。syn red sorrel

red deer n. a deer with a rich red-brown summer coat that turns dull in winter (一种夏季被毛为红棕色的鹿,其被毛在冬季变得暗淡)赤鹿 —*see also* **hind**, **hart**

red dun n.【color】a light red or yellow coat color with brown, red, or flaxen points, and commonly primitive marks (被毛淡褐而末梢为褐红或褐黄的毛色,多带原始别征)红兔褐 syn claybank dun

red flag n.【jumping】a red marker used to denote the right side of a jumping obstacle (用来标识障碍右侧的旗子)红旗 compare white flag

red fox n. a carnivorous mammal of the genus *Vulpes* that has a reddish fur, pointed muzzle, erect ears, and a long bushy tail (狐属肉食哺乳动物,被毛红色,口鼻突出,耳朵直立,尾巴较大)红狐,赤狐:Red fox is noted for its cunning and alertness. 赤狐以其狡猾和机警而出名。

red hot scent n. the burning scent of prey (猎物)浓烈的气味

red ribbon n. a red strip of cloth tied around the tail of a horse known to kick (系在有踢蹴恶习马匹尾巴上的红色丝带)红带

red roan n. *see* bay roan 沙骝毛

red urine n. *see* hemoglobinuria 血红蛋白尿,血尿

redboard /ˈredbɔːd; *AmE* -bɔːrd/ n. **1**【rac-

R

ing】a red board used for declaring a race in the past (过去用来宣布比赛开始的)红色公告板 **2**【racing】one who claims to have selected the winning horse in a race following it (赛后声称选中头马的人)事后先生

reducible hernia n. a kind of hernia characterized by a non-inflammatory, painless, soft swelling that may vary from time to time where the protruding organ can be pushed back into the correct body cavity (一种无痛、非炎性疝气,软肿大小可变,且脱出部分可以推回到原腔体内)可复性疝,易复性疝 compare irreducible hernia

redworm /ˈredwɜːm; *AmE* -wɜːrm/ n. *see* large strongyles 大圆线虫

referee /ˌrefəˈriː/ n.【polo】*see* third man 仲裁

refit /ˌriːˈfɪt/ vt. to fit the horseshoes again 重装蹄铁

reflex /ˈriːfleks/ n.【physiology】an involuntary response to a stimulus (对刺激的非自主性反应)反射作用:nervous reflex 神经反射

refusal /rɪˈfjuːzl/ n. **1**【jumping】act or instance of refusing to jump an obstacle 拒跳:The horse had one refusal on the course. 这匹马在跨越过程中拒跳一次。**2**【racing】instance of failing to break the start in a race 拒闸

refuse /rɪˈfjuːz/ vi. **1**【jumping】(of a horse) decline to jump an obstacle (指马)拒跳 **2**【racing】(of a racehorse) fall to break the start in a race (赛马)拒绝出闸,拒闸

regime /reɪˈʒiːm/ n. *see* regimen 养生法

regimen /ˈredʒiːmən/ n.【nutrition】(also **regime**) a regulated system, as of diet, therapy, or exercise, intended to restore and promote health (以恢复和促进健康而制定的膳食结构、治疗方案或锻炼计划)养生法:Proper feeding should be considered as an integral part of the sick horse nursing and the therapeutic regime. 患病马匹的合理饲喂应看作是其护理和养生治疗的一种必要手段。

regional anesthesia n. partial loss of sensation induced by anesthetics to an area of the body 局部麻醉

regional anesthetic n. an agent that causes loss of sensation to an area of body (可让身体某个部位丧失知觉的制剂)局部麻醉剂

register /ˈredʒɪstə(r)/ n. a book for recording of items, entries, etc. 登记册,登记簿
vt. to record in an official register 登记;注册

registered /ˈredʒɪstəd/ adj. (of a horse) having the pedigree recorded and verified by an authorized association of breeders (指马匹血统经相关育种机构登记认可)注册的,登记的

registered hunt n.【hunting】a Hunt which meets the Association standards but has not been granted permanent status (指达到了狩猎协会标准但尚未获得永久性地位的猎场)注册猎场:The status of registered hunt is only provisional and the hunt is expected to improve its facilities and performance before being identified as a recognized hunt. 注册猎场的地位只是暂时性的,只有在改善设施并提高业绩后才有望被认定为知名猎场。compare recognized hunt

Registered Shoeing Smith n. (abbr. **RSS**) one of three levels of certification formerly awarded to practicing farriers in Britain by the Worshipful Company of Farriers on the basis of written, oral, and practical examinations (由英国蹄铁匠公会经过笔试、口试以及专业技能考核而颁发给从业蹄铁匠的培训资格证书)注册蹄铁匠 compare Associate of the Worshipful Company of Farriers, Fellow of the Worshipful Company of Farriers

registration /ˌredʒɪˈstreɪʃn/ n. **1** the act of registering 注册;登记 **2** the number registered; enrollment 登记数;注册数目

registry /ˈredʒɪstri/ n. **1** the act of registering; registration 注册 **2** a book for official records 记录簿;登记册

regular /ˈregjələ(r)/ adj. (of the horse's gait) even in length and rhythmic in steps 运步均

R

一的

regurgitate /rɪ'gɜːdʒɪteɪt; *AmE* -gɜːrdʒ-/ vi. to rush or surge back 返流，回涌：The cleft plate causes the foal to regurgitate milk down its nostrils after sucking. 腭裂使幼驹吮吸的乳液又从鼻孔返流出来。

rehabilitate /ˌriːə'bɪlɪteɪt/ vt. to restore to normal condition from fatigue or illness 使恢复，复原 | **rehabilitation** /ˌriːəˌbɪlɪ'teɪʃn/ n.

rein /reɪn/ n. (*pl.* **reins**) a long, narrow strap of leather or synthetic material attached on one end to the bit of a bridle and held in the rider's or driver's hands on the other to control a horse (一条细长的皮带或纤维绳，其一端系在勒衔铁环上，另一端为骑手或车夫把持以驾驭马匹) 缰绳：auxiliary rein 辅助缰 ◇ bearing rein 姿态缰 ◇ driving rein 驾车缰 ◇ guiding rein 控马缰

vt. to control a horse by the reins (用缰来驾驭马匹) 驾驭，控缰：to rein a horse back 扯缰使马后退

rein aids n. the use of the reins to communicate instruction to the horse (用缰绳向马传达指令的方法) 缰绳扶助

rein back n. **1**【dressage】a movement in which the horse steps backwards at a two-beat gait with the diagonal feet touching the ground simultaneously (舞步马对角线两肢同时着地后退的动作) 倒退，后退 **2** a movement in which the horse backs up with four distinct hoofbeats (马以四蹄分别向后移动) 后退，倒退

rein billet n. the part of the rein that passes around the bit ring and back to itself where it is buckled or attached (缰绳绕过衔铁环并反扣的部分) 缰绳扣带

reiner /'reɪn(ə)/ n. one that participates in the sport of reining 驭马者

reining /'reɪnɪŋ/ n. a Western riding discipline in which the athletic ability of a horse is shown through execution of several stunting movements (西部马上竞技项目，骑手策马表演多个特技动作) 驭马赛，雷宁

reining horse n. any horse, usually a Quarter, trained for reining maneuvers 驭马赛用马；雷宁马

reining pattern n. any of the stunting movements a horse performs in reining competition 驭马动作：The reining patterns include small circles, large fast circles, flying lead changes, rollbacks over the hocks, 360 degree spins in place to the left and to the right, and the sliding stop. 驭马动作包括转小圈、转大圈、空中换腿、急速回转、360°左右原地旋转和滑停。

reining point n.【cutting】a one-point penalty given to a rider who picks up his hand and the reins while working a cow (截牛比赛中因骑手抬手扯缰而给予的罚分) 扯缰罚分

reject /rɪ'dʒekt/ vt. (of a mare) to refuse to accept the stallion during the period of diestrus (指间情期母马) 拒绝交配：In diestrus, the mare rejects the stallion by kicking, laying the ears back and biting. 处于间情期的母马会通过踢蹴、两耳后拢和撕咬来拒绝公马交配。

rejection /rɪ'dʒekʃn/ n. the act or instance of rejecting the stallion 拒绝交配

rejon /reɪ'həʊ/ n.【Spanish】(*pl.* **-jones** /-'həʊnes/) the long lance used by a bullfighter who works the bull on horseback (马背斗牛士使用的长矛) 斗牛长矛

rejoneador /reɪˌhəʊnɪə'dɔːr/ n.【Spanish】a Portuguese bullfighter who works the bull on horseback (葡萄牙) 马背斗牛士

rejoneo /ˌreɪhəʊ'neɪəʊ/ n.【Spanish】(*pl.* **rejoneos**) a form of horseback bullfighting originated in Portugal in which the bull is not killed (源于葡萄牙的马上斗牛，牛不被杀死) 马背斗牛

relax /rɪ'læks/ vt. to relieve from tension or strain; slacken 放松，松弛

vi. to become less restrained or tense; ease up 休息；放松

relaxant /rɪ'læksənt/ adj. relieving muscular or

nervous tension （肌肉或神经）弛缓的，缓解的
n. a drug or therapeutic treatment that relaxes or relieves muscular or nervous tension （缓解肌肉或神经紧张的药物或疗法）弛缓剂，弛缓疗法

relaxin /rɪˈlæksɪn/ n.【physiology】a female hormone secreted by the ovary that helps soften the cervix and relax the pelvic ligaments in parturition （卵巢分泌的一种雌激素，分娩时可舒张子宫颈和松弛骨盆关节韧带）松弛素

remedy /ˈremədi/ n. a medicine or therapy that relieves pain, cures disease, or corrects a disorder （能缓解疼痛、治疗疾病或消除紊乱的药物或疗法）治疗；药剂

remission /rɪˈmɪʃn/ n. a diminution of the intensity of disease or pain （病症或疼痛的）缓解，缓和

remit /rɪˈmɪt/ vi. (of a disease) to diminish or subside; abate （疾病）缓解，缓和

remount
noun, *verb*/ ˌn. riːˈmaʊnt/ an extra horse used to replace tired or lame horses （用来代替乏马或瘸马的其他马匹）备用马
vt. /riːˈmaunt/ to mount a horse again 重骑，再骑

remouth /rɪˈmaʊθ/ vt. to resensitize a horse's spoiled mouth by use of a gentler bit and soft hands （通过使用较为轻柔的衔铁和手法矫正马口角失灵）矫口 | **remouthing**/rɪˈmaʊðɪŋ/n.

remuda /rɪˈmuːdə/ n. a herd of horses turned out on a ranch （牧场上放养的一群马）马群

ren /ren/ n.【Latin】(*pl.* **renes** /-iːz/) kidney 肾，肾脏

rental horse n. a horse for hire 出租马，租用马 [syn] job horse

renvers /ˈrenvɜːs; *AmE* ˈrenvɜːrs/ n.【dressage】a dressage movement performed by a horse on two tracks with quarters out（马）腰[向]外斜横步 [syn] haunches out, quarters out;

renvers half volte n.【dressage】a dressage movement in which the horse scribes one half of a 6-meter circle haunches out, and returns to the track traveling in the opposite direction （盛装舞步中马臀朝外在6米的圆圈中转半圈后又转回原处的动作）腰向外半圈乘，腰向外转半圈

renvers-volte n.【dressage】a dressage movement in which the horse performs a 6-meter circle with the haunches out （盛装舞步中马臀朝外在6米的圆圈中旋转）腰向外圈乘，腰向外转圈

rep /rep/ n.【slang】a cowboy hired to search and round up stray cattle from a ranch 寻找离群牛的牛仔

repercussion /ˌriːpəˈkʌʃn; *AmE* -pərˈk-/ n. an indirect or reciprocal effect or influence; side effect 负面影响；副作用：It is important for vets to recognize both the beneficial consequences and the untoward repercussion in applying drugs. 兽医用药时对药物正、副两方面的作用须心中有数，这一点是非常重要的。

replace /rɪˈpleɪs/ vt. to take or fill the place of sb/sth 取代，代替

replacer /rɪˈpleɪsə(r)/ n. one used to replace another 代替者；代用品：milk replacer 代乳料

replenish /rɪˈplenɪʃ/ vt. to fill or make complete again; inspire 补充，恢复：Recovery from strenuous work requires several days for the horse to replenish its energy reserves. 马匹负重役后一般需要好几天来恢复体力和储存能量。

repository /rɪˈpɒzətri; *AmE* rɪˈpɑːzətɔːri/ n. a yard where horses are kept in an auction 马匹交易栏，马匹拍卖栏

reproduce /ˌriːprəˈdjuːs; *AmE* -duːs/ vt. to give birth to offspring 生殖；繁殖

reproduction /ˌriːprəˈdʌkʃn/ n. the act of reproducing or process of being reproduced 繁殖；生殖：sexual reproduction 有性生殖

reproductive /ˌriːprəˈdʌktɪv/ adj. of or relating

to reproduction 繁殖的，生殖的：reproductive efficiency 繁殖力 ◇ reproductive performance 繁殖性能 ◇ reproductive rate 繁殖率 ◇ reproductive system 生殖系统 ◇ reproductive tract 生殖道

reproductivity /ˌriːprədʌkˈtɪvəti/ n. the capability of reproducing offspring 繁殖力

RER abbr. respiratory exchange ratio 呼吸交换率

re-ride /riːˈraɪd/ n.【rodeo】a second ride awarded to a bronc or bull rider when he was not afforded a fair opportunity to perform（在选手没有得到公平竞争机会时给予的二次骑乘）重骑

vt. to ride a horse again; remount 再次骑马，重骑

reserve[1] /rɪˈzɜːv; *AmE* rɪˈzɜːrv/ vt. to keep for future use or for a special purpose 保留；储备

n. a supply of sth kept or saved for future use 储备物，贮备：body fat reserves 体脂储备 ◇ energy reserves 能贮 ◇ glucose reserves 葡萄糖储备

reserve[2] /rɪˈzɜːv; *AmE* rɪˈzɜːrv/ n.【BrE】(also **reserve price**) the minimum price that an owner will accept to sell a horse for at a public auction（拍卖会上马主可接受的最低出价）底价 PHR V **put a reserve on sth** to set the lowest price for sth at an auction 订底价：On a public auction, the consignor can either register the reserve with the sales company which will not let the horse be sold unless the price exceeds that amount, or the consignor can bid on the horse himself until the reserve is reached. 公开拍卖会上，委托人即可以与拍卖公司拟定马匹的底价等出价高于底价时拍卖，也可以自己叫价直到超过底价为止。IDM **reserve not achieved**（**RNA**）a condition when the reserve price is not reached in a public auction（指公开拍卖会上）底价未到

reserved /rɪˈzɜːvd; *AmE* rɪˈzɜːrvd/ adj. **1**（of a horse）held or kept for a particular race（指马）保留的，备用的 **2**【racing】（of a horse）held off in a race（指赛马）收步的，有保留的

reset /riːˈset/ vt.【farriery】to remove a horseshoe from the hoof and fit it again; refit 重装蹄铁

resin /ˈrezɪn/ n. a sticky substance exuded by certain trees and used in varnishes and adhesives（某些树木分泌的黏性物质，常用于制造清漆和黏剂）树脂；松香

resin back n.【vaulting】*see* rosinback 背上涂松香的马

resist /rɪˈzɪst/ vt.（of a horse）to evade or act against the rider's aids（指马）抵抗，拒从

resistance /rɪˈzɪstəns/ n. the act or an instance of resisting or the capacity to resist 抵抗，抗性；耐药性

resistant /rɪˈzɪstənt/ adj.（also **resistent**）of or possessing resistance 抵抗的，有抗性的

resorb /rɪˈsɔːb; *AmE* -ˈsɔːrb/ vt.【physiology】to dissolve and assimilate 消溶；吸收

resorption /rɪˈsɔːpʃn; *AmE* -ˈsɔːrp-/ n. the act or process of resorbing 消溶；吸收：Resorption describes the death of a fetus or embryo in the early period of pregnancy, the fetus and its membranes may become absorbed rather than expelled. 消溶是胎儿或胚胎在妊娠早期死亡之后，胎儿和胎膜被子宫吸收而非排出的情况。

respiration /ˌrespəˈreɪʃn/ n. the act of breathing 呼吸：artificial respiration 人工呼吸

respiratory /rəˈspɪrətri, ˈrespərətri; *AmE* ˈrespərətɔːri/ adj. connected with breathing 呼吸的：respiratory diseases 呼吸性疾病 ◇ respiratory frequency 呼吸频率 ◇ respiratory muscle 呼吸肌

respiratory exchange ratio n.（abbr. **RER**）the ratio of carbon dioxide production to oxygen consumption during metabolic process（机体代谢过程中产生的二氧化碳与所耗氧气之间的比率）呼吸交换率

respiratory system n. the integrated system of organs including the nostrils, pharynx, larynx, trachea, bronchi, bronchioles, and alveoli that involved in breathing and exchange of oxygen and carbon dioxide in the lung (参与呼吸活动和肺内氧气与二氧化碳交换的器官所组成的系统,其中包括鼻孔、咽、喉、气管、支气管、细支气管以及肺泡)呼吸系统

respire /rɪˈspaɪə(r)/ vi. to inhale and exhale; breathe 呼吸

respite /ˈrespaɪt/ n. a short interval of rest or relief 暂歇;休息: Nowadays, most studs wean foals at about six months of age, giving the mare a respite from lactation during the winter months and facilitating young stock management. 目前,大多育马场都在幼驹6月龄时断奶,这样既可缓解母马冬季泌乳的负担,又有助于幼驹管理。

response /rɪˈspɒns; *AmE* rɪˈspɑːns/ n. a reaction made by an organism to a specific stimulus (有机体对外界刺激作出的反应)应答: immune response 免疫应答

responsive /rɪˈspɒnsɪv; *AmE* -ˈspɑːn-/ adj. (of a horse) readily responding to the commands or aids of its rider (指马积极回应骑手的指令或扶助)反应敏快的;听从指令的 | **responsiveness** /-vnəs/ n.

rest[1] /rest/ n. **1** a cessation of motion, exertion, or work 间歇,休息 **2** a metal projection on a suit of armor located above the waist, upon which a jouster rests his lance while tilting (马上刺枪者盔甲齐腰处的铁板,用来支撑长矛)铁托

rest[2] /rest/ v. to stop growing crops on land in order to restore the fertility of soil (为恢复土壤肥力而停止耕种)休耕

rest horse n. a spare horse 备用马

restiform /ˈrestɪfɔːm; *AmE* -fɔːrm/ adj.【anatomy】similar that of a rope or cord 绳状的,索状的: restiform body 绳状体

resting /ˈrestɪŋ/ adj. (of a leg) not bearing weight when changing leads (指某肢)不负重的,暂歇的: resting leg 暂歇肢,非负重肢

restless /ˈrestləs/ adj. unable to rest or relax due to anxiety to boredom; restive 不安的,烦躁的 | **restlessness** n.: the restlessness of a tied horse 马被拴后的烦躁不安

restore /rɪˈstɔː(r)/ vt. to bring back to an original condition 恢复,复原 | **restoration** /ˌrestəˈreɪʃn/ n.

restrain /rɪˈstreɪn/ vt. to limit or restrict the activity or liberty of an individual (限制个体活动或自由)束缚;保定

restraint /rɪˈstreɪnt/ n. any method or device used to restrain an animal 保定措施;保定器具: In handling with a fidgeting and less patient horse, many methods of restraints, such as a lip twitch or hobbles could bring him under control. 如果碰到一匹急躁不安的马,采用鼻捻、蹄拌等保定措施就可以将其制服。

restrict /rɪˈstrɪkt/ vt. to keep or confine within limits 限制,限定: restricted ration 日粮限制 ◇ restricted feeding 限饲

result chart n.【racing】(also **chart**) a form upon which the results of a given race are posted (一张刊登比赛结果的表单)比赛结果单,赛事结果表

retain /rɪˈteɪn/ vt. to keep or hold; maintain 滞留;保持

retained afterbirth n. *see* retained placenta 胎衣不下,胎盘滞留

retained placenta n. (also **retained afterbirth**) a case in which the afterbirth is not expelled within six hours of foaling (母马分娩6小时后胎衣尚未排出的情况)胎衣不下,胎盘滞留

retained sole n. the sole of a horse's hoof that does not exfoliates normally (指马蹄底不正常剥落的情况)蹄底滞留 [syn] false sole

retainer /rɪˈteɪnə(r)/ n. **1** the fee paid in advance to retain a jockey for competing in horse races (雇佣骑手参赛而预先支付的费用)预聘金 **2** one who retains the riding services of a professional rider 预聘人

rete /ˈriːti/ n.【anatomy】(*pl.* **retia** /ˈriːtɪə/) an anatomical mesh or network of blood vessels or nerves（由血管或神经构成的网状结构）网：dorsal rete of carpus 腕背侧动脉网 ◇ ethmoid rete 筛网 | **retial** /ˈretiəl/ adj.

retention /rɪˈtenʃn/ n. **1** the act of retaining 保留，保持 **2** the condition of being retained 滞留，迟滞：In late-foaling mares, feeding it with fresh spring grass of high quality will reduce the problems of meconium retention. 在围产期前，给母马饲喂优质的嫩草可减少幼驹胎粪滞留的问题。

reticulocyte /rɪˈtɪkjʊləsaɪt/ n.【physiology】an immature red blood cell that contains a network of basophilic filaments（一种尚未成熟的红细胞，内含嗜碱性细丝网）网织红细胞

retina /ˈretɪnə; *AmE* ˈretənə/ n.【anatomy】a delicate, multilayered, light-sensitive membrane lining the inner eyeball and connected by the optic nerve to the brain（衬在眼球内侧的复层光敏膜，通过视神经与大脑相连）视网膜

retinaculum /ˌretɪˈnækjʊləm/ n.【anatomy】(*pl.* **-la** /-lə/) a band or bandlike structure that holds an organ or a part in place（支持机体器官或组成部分的带或带状结构）支持带，系带：retinaculum of extensor muscle 伸肌支持带 ◇ retinaculum of flexor muscle 展肌支持带

retinal[1] /ˈretɪnl/ adj. of or relating to retina 视网膜的：retinal scan 视网膜扫描

retinal[2] /ˈretɪnl/ n.（also **retinene** /ˈretɪniːn/）the oxidized form of retinol（视黄醇的氧化形式）视黄醛

retinol /ˈretɪnɒl/ n. a fat-soluble vitamin essential for vision and the health of mucous membranes（一种脂溶性维生素，为视觉与黏膜健康所必需）视黄醇；维生素 A [syn] vitamin A

reverse shoe n. *see* backwards shoe 反向蹄铁

reverse wedge shoe n.【farriery】a horseshoe with thickness reducing gradually from the toe the heel, used to elevate the hoof toe and thus lower the hoof angle（一种厚度从头至尾递减的蹄铁，可垫高蹄尖从而降低蹄角度）反向楔形蹄铁

RH abbr. relative humidity 相对湿度

rhabdomyolysis /ˌræbdəʊmaɪˈəʊlɪsɪs; *AmE* -doʊmaɪˈoʊ-/ n. disintegration of striated muscle fibers with excretion of myoglobin in the urine（横纹肌纤维降解导致尿液中出现肌球蛋白）横纹肌溶解症：acute rhabdomyolysis 急性横纹肌溶解症 ◇ equine rhabdomyolysis 马横纹肌溶解症

RHBAA abbr. Racking Horse Breeders' Association of America 美国单蹄快步马育种者协会

Rhenish /ˈriːnɪʃ, ˈrɛn-/ n. *see* Rhineland Heavy Draft 莱茵兰德重挽马

Rhineland Heavy Draft n. a coldblood breed developed in Rhineland, Germany during the 19th century; stands 16 to 17 hands and may have a bay, chestnut or red roan coat; has a strong build, massive quarters and shoulders, a deep, broad back, and short, and strong legs with heavy feathering; used for heavy draft and farm work（19 世纪在德国莱茵兰德地区育成的冷血马品种。体高 16～17 掌，毛色多为骝、栗或沙栗，体格强健，后躯和肩部肌肉发达，背腰宽广，四肢粗短且距毛浓密，主要用于重挽和农田作业）莱茵兰德重挽马：The stud book for the Rhineland Heavy Draft was established in 1876. 莱茵兰德重挽马的品种登记始于 1876 年。[syn] Rhenish, Rhinelander, German Coldblood

Rhinelander /ˈraɪnˌlændə(r)/ n. *see* Rhineland Heavy Draft 莱茵兰德重挽马

rhinencephalon /ˌraɪnenˈsefəlɒn/ n.【anatomy】(*pl.* **-la** /-lə/) the olfactory region of the brain（大脑的嗅觉控制区）嗅脑

rhinopneumonitis /ˈraɪnəʊˌnjuːməˈnaɪtɪs/ n. an infectious disease caused by the equine herpes virus Ⅰ, characterized by nasal and pulmonary inflammation of foals and abortion in pregnant mares（由马疱疹病毒Ⅰ型引起的马属动物传染病，幼驹表现鼻肺炎症状，妊娠母马发生流产）鼻肺炎：equine infectious rhinopneumonitis 马传染性鼻肺炎

rhodopsin /rəʊˈdɒpsɪn; *AmE* roʊ-/ n. the pigment sensitive to light in the retinal rods of the eyes, consisting of opsin and retinene（存在于眼视网膜视杆细胞内的光敏色素，由视蛋白和视黄醛构成）视紫红质 syn visual purple

rhombencephalon /ˌrɒmbenˈsefəlɒn/ n.【anatomy】the portion of the embryonic brain from which the metencephalon and myelencephalon develop（胚胎内发育为后脑及终脑的部分）菱脑，后脑 syn hindbrain

rhomboid /ˈrɒmbɔɪd; *AmE* ˈrɑːm-/ adj. shaped like a rhombus 菱形的：cervical part of rhomboid muscle 颈菱形肌

rhythm /ˈrɪðəm/ n. the regularity and evenness of the hoof beats in movement（马运行中马蹄落地的规律性）节奏，节律 PHR V **change rhythm**（of a horse）to alter its steps by breaking the length of stride for one or more steps（指马）改变运步节奏，调整运步 **lack rhythm**（of a horse）to lack regular, even steps in gait（指马）运步缺乏节奏

rhythmic /ˈrɪðmɪk/ adj. of, relating to, or having rhythm 有节律的

rhythmicity /rɪðˈmɪsəti/ n. the character of quality of being rhythmic 节律性：inherent rhythmicity 固有节律，自动节律

riata /rɪ(ː)ˈɑːtə/ n. *see* reata 套索

rib /rɪb/ n. each of the 18 curved bones attached in pairs to the vertebral column and enclosing the thoracic cavity of a horse（与脊柱相连的18根成对弧状扁骨，构成马体的胸腔）肋骨 PHR **short of (a) rib**（of a horse）with the distance between the last rib and the point of the hip greater than normal（指马）肋胁过短，后腰过长

rib bar n. a primitive marking of dark stripes appearing over the ribs of the horse（马肋部出现的原始深色条纹）肋纹

ribbon /ˈrɪbən/ n.【racing】a strip of colored silk signifying a place won by a horse in a competition（标志马匹获胜名次的彩色丝带）彩带：The blue ribbon is often used to identify the first place in a race. 蓝带在赛马中通常用来表示第一名。

ribcage /ˈrɪbkeɪdʒ/ n.（also **rib cage**）the structure formed by the curved ribs and the bones to which they are attached（由肋骨及附属骨骼形成的结构）胸腔，胸廓

ribgrass /ˈrɪbgrɑːs/ n.（also **ribwort**）plantain 车前草

riboflavin /ˌraɪbəʊˈfleɪvɪn/ n. a water-soluble vitamin that acts as an essential component of many enzymes involved in protein and carbohydrate metabolism（一种水溶性维生素，是蛋白质和糖类代谢中多种酶的组成成分）核黄素 syn vitamin B_2

ribonucleic acid n.【genetics】(abbr. **RNA**) a biochemical substance found in the cytoplasm and nuclei of cells that promotes the synthesis of cell proteins（见于细胞质和细胞核中的生化物质，参与细胞蛋白质的合成）核糖核苷酸 compare deoxyribonucleic acid

ribosome /ˈraɪbəsəʊm; *AmE* -soʊm/ n.【biochemistry】the tiny particle within a cell where protein synthesis takes place（细胞中的颗粒结构，蛋白质在此合成）核糖体

ribs /rɪbs/ n. the chest or barrel of a horse's body（马的）肋部

rice bran n. the outer layer of the rice kernel removed during milling（稻谷碾米所剩的外壳碎片）稻糠：Rice bran has a shelf life of up to one year. 稻糠的存放期约为一年。

rice hull n. the outer covering of a rice grain; husk 稻壳：As a bedding material by many horsemen, rice hulls are highly absorbent, dust free, and attractive. 作为马厩的垫料，稻壳吸水性强、粉尘少、外观整洁。

rickets /ˈrɪkɪts/ n. a metabolic disease resulting from deficiency of vitamin D or calcium, characterized by deformities of bone growth and occurring chiefly in foals（由于维生素D或钙缺乏而导致的代谢性疾病，其症状表现为骨骼发

R

育畸形,多发于马驹)佝偻病 syn rachitis

ridable /ˈraɪdəbl/ adj. **1** (of a horse) able or suitable to be ridden (指马)可骑的,适合骑乘的 **2** (of a wood or brook) capable of being ridden through or over (指树林或小溪)可骑马通过的

ride /raɪd/ n. **1** the act of riding on a horse 骑马,策骑 **2**【hunting】a path or trail made for riding through woodlands (骑马穿过林地的)骑道,小径

v. to sit and travel on a horse, etc; hack 骑马,策骑 PHR V **ride for a fall** to lose a race by falling intentionally from the horse (赛马中)假装落马 **ride for hire** to receive payment for riding in a race 以骑马为业: Kino was an excellent jockey, and has ridden for hire since 1997. 奇诺是一名出色的骑师,从1997年起就以骑马为业。**ride hard** (of a rider) to push the horse to his physical limits 策马疾驰: The hunters all ride hard to the hounds. 猎手们全都策马紧随猎犬。**ride hell for leather** to push the horse to the extreme and ride in a reckless speed regardless of the consequences 策马狂奔 **ride into the ground** (of a rider) to fall on the ground (骑手)摔倒在地 **ride off**【polo】(of a player) to knock against his opponent (球员)策马撞击: In polo match, to ride off an opponent at an angle greater than about 45 degrees is ruled as a foul. 在马球比赛中,以大于45°的角度侧面撞击对手将被视为违例。**ride short**【racing】to ride with short stirrups 短镫骑马 IDM **ride sb down** to direct one's horse at ab to knock him down 策马撞倒某人 **ride to a blue** to win the first place in race 赢得蓝带,获得第一 **ride on horse of ten toes** to walk on foot 步行 **ride in the Master's pocket**【hunting】(of a rider) to follow closely on the heels of the Field Master so as not to lag behind the pack (指骑手)紧追猎主,紧随猎主 **ride to hounds**【hunting】to go fox-hunting on horseback 骑马猎狐

ride and tie n. (also **ride & tie**) a cross-country race developed in the United States in the 1970s in which two partners alternate turns to run and ride a horse at least six times (20世纪70年代起源于美国的一项越野赛,有很强的策略性。团队两名成员轮流跑步和骑马至少6次以上)骑拴[运动]: In ride and tie, when the first runner reaches the tied horse, he mounts, and rides to catch his partner running in front of him, and this alternative process of riding and tying continues until all three team members have crossed the finish line. 在骑拴运动中,当首名跑步队员遇到拴着的马,便上马追赶跑在前面的伙伴;这种骑、拴交替的过程持续至人、马三组队员都通过终点为止。

Ride and Tie Association n. (abbr. **RTA**) an association founded in the United States in 1989 to organize and promote the sport of Ride & Tie (1989年成立于美国的组织,旨在组办并推广骑拴运动项目)美国骑拴[运动]协会

rider /ˈraɪdə(r)/ n. one who rides a horse, mule or ass 骑者,骑手

ridgeling /ˈrɪdʒəlɪŋ/ n. (also **ridgling**) a male horse with one or two undescended testicles; cryptorchid (单侧或两侧睾丸尚未下降的马匹)隐睾马

ridgling /ˈrɪdʒlɪŋ/ n. *see* ridgeling 隐睾马

riding /ˈraɪdɪŋ/ n. (BrE also **horse riding**; AmE also **horseback riding**) the sport or activity of riding horses 骑马; 骑乘; 乘骑: riding camp 乘马野营 ◇ riding instructor 骑术教练 ◇ distance riding 长途赛马 ◇ endurance riding 耐力赛 ◇ pleasure riding 休闲骑乘 ◇ riding horse 乘用马 ◇ riding resort 骑马胜地 ◇ rough riding 野骑

riding coat n. a dense, rainproof coat worn by a rider in bad weather (骑手在天气恶劣时穿的防水外套)骑乘大衣

riding crupper n. *see* saddle crupper 鞦褡

riding fee n. a sum of money paid to a professional rider（付给职业骑手的酬金）策骑费

riding horse n. any horse suitable for riding 乘用马

riding instructor n.（also **instructor**）one who teaches others the art of horsemanship 骑术指导，骑术教练

riding pony n. a pony ridden for pleasure or competition 乘用矮马

riding school n. an institution where people are taught the art of horsemanship by professional instructors（向人们传授骑术的机构）骑术学校

rig /rɪg/ n. a cryptorchid 隐睾马：Rigs are often a nuisance because they still look for mares and often have an inconsistent temperament. 隐睾马经常制造麻烦，因为它们依然寻觅母马且性情不稳。

right diagonal adj.（of a horse）with the right foreleg moving in unison with the left hind at the trot（指马快步中右前肢与左后肢同时落地的）右对角的：the right diagonal pairs 右对角两肢

right of way n.【polo】a governing rule in polo that entitles the player having hit the ball to follow the line of that ball and to take a further shot（马球比赛规则，击球球员有权跟随球行进路线再次击球）进攻权，行进权

right rein n. the rein on the right ride of a horse 右方缰

right-lead adj.（of canter or gallop）with the right foreleg extending farther forward（跑步或袭步中右前肢靠前）右起的，右领的：right-lead canter 右领跑步◇ right-lead gallop 右起袭步 [opp] left-lead

rigor mortis n.【Latin】the muscular stiffening of the body following death（身体）死后强直

rim shoe n.【farriery】any horseshoe with a long cleat on the outer edge, used for horses with bad tendons（外侧带铁棱的蹄铁，用于筋腱有问题的马匹）镶边蹄铁

ring /rɪŋ/ n. **1** an enclosed area where equestrian activities are carried out（举行马术运动的场地）赛场；竞技场：indoor ring 室内赛场 ◇ outdoor ring 室外赛场◇ show ring 展览场 ◇ rectangular ring 方形赛场 ◇ rounded-ends ring 圆角赛场 ◇ oval ring 椭圆形赛场 ◇ For both indoor and outdoor rings, good footing is the single essential ingredient for success. 无论是室内还是室外比赛，场地基脚是赛事成功的一个关键因素。**2**【racing】（usu. **the Ring**）the bookmakers collectively（总称）赌马经济人

ring bit n. *see* tattersall ring bit 环形衔铁

ring boots n. *see* fetlock ring boot 护蹄圈

ring sour adj.（of a horse）bored with the training and exercise in an enclosed area（指马厌倦围栏训练）厌圈的

ring steward n.（also **ringmaster**）an official responsible for overseeing competitive activity held in the riding or show ring（在赛场内负责巡视比赛的人员）场内监管

ringbone /ˈrɪŋbəʊn/ n.【vet】a bony growth occurring within the pastern and coffin joints of a horse that may result in lameness（系骨和蹄骨间附生赘骨的症状，可导致马跛行）环骨瘤：high ringbone 上位环骨瘤 ◇low ringbone 下位环骨瘤 [syn] interphalangeal arthritis

ringer /ˈrɪŋə(r)/ n. **1**【AmE, racing】a horse entered in a race under a false name to obtain more favorable betting odds（为了获取高赔率而以假名参赛的马）冒名参赛马：It is a dishonest practice to enter the ringer in a race below its class where it is almost certain to win. 让马冒名参赛而稳中取胜是一种欺诈行为。**2**【hunting】（also **ringing fox**）a fox who runs in circles rather than in a straight line 绕圈跑的狐狸

ringing fox n. *see* ringer 绕圈跑的狐狸

ringmaster /ˈrɪŋmɑːstə(r)；*AmE* -mæs-/ n. *see* ring steward 场内监管

ringworm /ˈrɪŋwɜːm；*AmE* -wɜːrm/ n. a contagious skin disease caused by several related fungi, characterized by hair loss and lesions of

R

skin around infected area(由多种真菌所致的皮肤传染病,症状表现为患处脱毛和皮肤溃烂)皮癣 syn dermatophytosis

rinse /rɪns/ vt. **1** to wash lightly with water or liquid 冲洗,清洗 **2** to remove soap, etc by washing lightly in water 漂洗: Once the horse is washed he can be rinsed off and scarped with the sweat-scraper. 给马匹洗浴后,冲去肥皂沫,再用汗刮刮擦。

ripe /raɪp/ adj. **1** mature 成熟的 **2** thoroughly matured or ready for mating 适合交配的

RMHA abbr. Rocky Mountain Horse Association 落基山马协会

RNA abbr. ribonucleic acid 核糖核酸 compare DNA

roach /rəʊtʃ; *AmE* roʊtʃ/ vt. to clip a horse's mane close to its roots; hog 剪鬃: roached mane 被剪过的鬃 | **roaching** /ˈrəʊtʃɪŋ; *AmE* ˈroʊ-/ n.

roach back n. a conformation defect in which the horse's back is convex between the withers and loins(马鬐甲与后腰之间脊背向上凸起的体格缺陷)凸背 syn hog back

roach backed adj. (of a horse) having an arched back(指马)凸背的 syn hog backed

road cart n. a two-wheeled, horse-drawn passenger vehicle drawn by a single horse(一种客用双轮马车,由单马拉行)公路双轮马车

road coach n. a four-wheeled, horse-drawn, public transportation vehicle popular in the late 19th century; was brightly colored and could hold 12 passengers(19 世纪末使用的一种公用四轮马车,车身颜色鲜艳,可坐 12 名乘客)公路马车

road founder n. inflammation of the horse's laminae caused by excessive concussion over hard surfaces(马蹄与硬基路面相撞导致的蹄叶发炎)公路蹄叶炎

road hack n. a horse suitable for pleasure riding rather than competition 普通乘用马 syn ordinary hack

road hunter n.【hunting】a hound good at tracing the line of a fox along a hard road 路面寻踪猎犬

road puff n. *see* windgall 关节软肿

road wagon n. a four-wheeled, horse-drawn open wagon used for transporting goods(一种用于货物运输的敞篷四轮马车)货运马车

roadcraft /ˈrəʊdkrɑːft; *AmE* ˈroʊ-/ n. the skills of driving a horse-drawn vehicle 驾车技巧

roadster /ˈrəʊdstə(r); *AmE* ˈroʊ-/ n. a horse used for riding on a road 公路乘用马

road-worthy /ˈrəodwəːði/ adj. (of a horse or carriage) suitable for long-distance traveling(指马或车)适合长途行进的

roam /rəʊm; *AmE* roʊm/ vi. to move about without aim or plan; wander(无目的地到处移动)游荡,驰骋: Whenever possible the horse should be allowed to roam freely at pasture to break up the monotony of the day once he has worked. 应尽可能让马在草场上自由驰骋,以此打破每天沉闷的劳作。

roan /rəʊn; *AmE* roʊn/ n.【color】a black, bay or chestnut coat interspersed with white hairs(黑、骝或栗色被毛中散生白毛的毛色)沙毛: bay roan 沙骝 ◇ blue roan 沙青 ◇ chestnut roan 沙栗

roaring /ˈrɔːrɪŋ/ n. the sound made by a horse with paralysis of vocal cords when air is inhaled; whistling(马由于声带瘫痪而在吸气时发出声响)喘鸣

ROC abbr. Race of Champions 冠军联赛

Rockaway /ˈrɒkəweɪ/ n. a four-wheeled, horse-drawn American passenger vehicle with two seats and a top, used throughout the 19th century(19 世纪使用的一种美式四轮马车,设两个座位,带顶篷)洛克维马车

Rockaway Coupé n. a smaller version of the Rockaway 小型洛克维马车

Rockaway Landau n. a horse-drawn vehicle popular during late 19th century, the features of which is a combination of the Rockaway and the

Landau（流行于19世纪末的四轮马车，结合了洛克维马车和兰道马车的设计）洛克维兰道马车

rocker /ˈrɒkə(r); *AmE* ˈrɑːk-/ n. **1** a rocking horse 摇摆木马 **2** one of the two curved pieces upon which a wooden horse rocks（摇摆木马立地的两个弧板）摇板，摇脚

rocker shoe n. *see* rocker-toe shoe 卷头蹄铁

rocker-toe shoe n.（also **rocker shoe**）a therapeutic horseshoe with rolled toe 弯头蹄铁，卷头蹄铁 [syn] roller motion shoe, set-toe shoe

rocking horse n. a wooden toy horse set on curved planks or rollers for a child to ride（固定在弧形板或滑轮上供儿童骑乘的木马玩具）摇摆木马 [syn] hobby horse

rocking-chair canter n. a smooth, collected canter of the Tennessee Walking Horse with a rolling movement（田纳西走马的一种跑步，运步流畅，步幅缩短，伴有摇摆动作）摇摆跑步，摇椅跑步

Rocky Mountain Horse n. a breed of horse originating in the late 1800s in the foothills of the Appalachian Mountains of eastern Kentucky, United States; stands from 14.2 to 16 hands and may have any solid coat color; has a fine profile, bold eyes, and wide chest; is sure-footed, easy-gaited in ambling, docile and easy to manage; used mainly for farm work, herding cattle, riding, and driving（19世纪末源于美国肯塔基州阿巴拉契亚山部地区的马品种。体高14.1～16掌，各种单毛色兼有，面貌俊秀，眼大而明，胸身宽广，步法稳健，善走溜花蹄，性情温驯，易于驾驭，主要用于农田作业、牧牛、骑乘和驾车）落基山马

Rocky Mountain Horse Association n.（abbr. **RMHA**）an organization founded in the United States in 1986 to promote the breeding of the Rocky Mountain Horse and to maintain its registry（1986在美国成立的组织机构，旨在推动落基山马的育种工作并进行品种登记）落基山马协会

rodeo /ˈrəʊdiəʊ, rəʊˈdeɪəʊ; *AmE* ˈroʊdioʊ, roʊˈdeɪoʊ/ n.（*pl.* **rodeos**）**1** any of the competitions including roping, bronc riding, steer wrestling, etc. originally derived from the daily chores of cowboys working on the ranches（源于牧场牛仔日常工作的多项竞技比赛，包括套牛、骑野马、摔牛等）牛仔竞技：rodeo events 牛仔竞技项目 [syn] Cattle Men's Carnivals **2** an enclosure for keeping cattle that have been rounded up（用来圈住赶拢的牛群的地方）围场，围栏

vi. to compete in a rodeo 参加牛仔竞技

rodeo rider n. one who performs in rodeo exhibitions 牛仔竞技骑手

Roentgen ray n. *see* X-ray 伦琴射线，X射线

rogue /rəʊɡ; *AmE* roʊɡ/ n. a horse having a fierce disposition 性情暴烈的马；难驾驭的马

roll /rəʊl; *AmE* roʊl/ v. **1** to turn over its back on the ground 打滚 **2** to turn inward 卷，翘：rolled toe 蹄尖上翘，蹄尖上卷 **3** to spread or compress by applying pressure with a roller（用辊子）滚压，辗：Rolling helps to incorporate the nitrogen granules into the ground, consolidating and flattening out the areas damaged by poaching. Besides, rolling also encourages the grass to tiller. 草地滚压可以把氮肥颗粒挤压入土壤中，同时也可紧实、平整马踩烂的区域。此外，滚压有助于牧草分蘖。**4** PHR V **roll back**（of a horse）to turn back to gallop in the reverse direction by lifting its forelegs and turning 180 degrees（指马直立回转180°后继续奔驰）急转，回转

rollback /ˈrəʊlbæk; *AmE* ˈroʊl-/ n. a reining movement in which a horse turns back to gallop in the reverse direction 急转，回转 [syn] set-and-turn

roller /ˈrəʊlə(r); *AmE* ˈroʊ-/ n. **1** a heavy, revolving cylinder that is used to level, crush, or smooth land（用来平整土地或压碎土块的圆辊）轧辊，铁辊 **2**（also **body roller**）a padded leather band that buckled over withers, fitted with rings to tie a rug or blanket in place（一种系于马髻甲处的带衬里皮带，上面的扣环可以用来系马衣）马衣固定带 **3** *see* bit roller 口衔滚珠

roller bolt n.【driving】an upright, metal projection fitted at the end of the singletree to which the traces of a pair or coach wheelers attach（挂杆末端装的直立铁件，用来拴系辕马的挽绳）挂杆闩，挂杆桩 syn bollard

roller motion shoe n. *see* rocker-toe shoe 卷头蹄铁

rolling /ˈrəʊlɪŋ; *AmE* ˈroʊ-/ n. an excessive side-to-side shoulder motion in a horse（马匹）肩部摇摆：Horses wide between the forelegs and lacking muscles in that area are prone to rolling their shoulders. 两前肢之间过宽且此处肌肉不发达的马匹容易出现肩部摇摆的情况。

rolltop /ˈrəʊltɒp; *AmE* ˈroʊ-/ n.【jumping】a fence or jumping obstacle that is arched in a half-circle（外形呈半弧状的障碍）圆顶障碍

Roman nose n. a profile of horse in which the bridge of nose between poll and muzzle is prominently convex when viewed sideways（马鼻梁侧观时显著外凸的面貌特征）罗马鼻：Although Roman nose is a feature often seen in mustangs and several draft breeds such as the Shire, Lippizaner and Saddlebred, some people deem it a mark of stupidity. 尽管罗马鼻在美国野马和多个挽马品种如夏尔马、利比扎马以及乘骑马中都很普遍，但有些人认为这是马驽钝的标志。syn Roman profile

Roman profile n. *see* Roman nose 罗马鼻

Romanic style n. a highly collected riding style derived from the Roman period on the basis of lightness and dexterity of the rider's hands（源于古罗马的骑乘方式，马以高蹄缩步行进，强调骑者手法的轻柔与娴熟）罗马式乘姿，罗马式骑乘

romp /rɒmp; *AmE* rɑːmp/ n. **1** a rapid or easy pace 轻快的步伐 **2**【racing】an easy win 轻而易举取得的胜利

vi. **1** to move in a rapid or easy manner 轻快地奔跑 **2**【racing】to win a race easily 轻易取胜：Kelvin romped in the long endurance race, mainly depending on the excellent stamina of his mount. 在这次长途耐力赛中，凯文凭借坐骑出色的耐力轻而易举地取得了胜利。

root /ruːt/ n. **1** the part of a plant that grows under the ground（植物生长于地下的部分）根茎 **2**（usu. **roots**）【hunting】a field of tuberous plants, such as carrots or turnips（种植块茎作物的田地）萝卜地

root crops n. any of the tuberous vegetables fed to livestock a dietary supplement including carrots, sugar beets, and turnips（作为补充料喂给家畜的块茎作物，如胡萝卜、甜菜和芜菁）块茎作物

rope /rəʊp; *AmE* roʊp/ n. a lasso or lariat used to catch a calf or horse（用于捕捉牛或马的）套索

vt. to catch a horse or cattle as with a rope 捕捉；套捕（牛或马）

rope bag n.【rodeo】（also **rope can**）a leather bag in which a lariat is packed or stored for travelling 套索包

rope horse n. *see* roping horse 套牛用马

rope walk vi. to walk or move in a twisted way 拧绳运步，绞步行进：Aged horses often rope walk. 老年马常绞步行进。

rope walking n. a gait defect of horse in which the feet move in a twisted way（马运步呈交错状态的一种缺陷）拧绳肢势，绞步 syn lacing, plaiting, winding

roping /ˈrəʊpɪŋ; *AmE* ˈroʊ-/ n. **1** the act of catching horses or cattle with a rope 套捉，套捕 **2** the rodeo event of calf roping（西部牛仔竞技项目）套小牛

roping dummy n.【roping】（also **dummy calf**）any target that a roper uses to practice throwing a loop（套牛者做掷套练习用的靶）套牛靶

roping horse n.（also **rope horse**）a horse trained for calf-roping 套牛用马

rose gray n.【color】the coat color of a horse who has black skin and a medium gray coat lightly mixed with reddish or chestnut hairs, thus giving the coat a rosy tint（皮肤为黑色且在青色被毛中散生红毛的毛色）玫瑰青

rosebay /ˈrəʊzbeɪ; *AmE* ˈroʊ-/ n. *see* oleander 夹竹桃

rosette /rəʊˈzet; *AmE* roʊ-/ n. an ornament made of ribbon or silk to resemble a rose, used to award the winners in an equestrian event（用丝带叠成的玫瑰形饰品，用来奖励马术竞赛中的获胜者）玫瑰结

rosin /ˈrɒzɪn; *AmE* ˈrɑːzn/ n. a translucent yellowish resin applied to increase traction（黄色半透明树脂，可用来增加摩擦力）松香，松脂：In vaulting, rosin is often applied to the the back of horse, mainly around the quarters, to prevent performers from slipping. 在进行马上杂技表演时，经常在马背的后躯涂上松香，以防止表演者滑落下马。

vt. to rub with rosin 涂松香

Rosinante /ˌrɒzɪˈnæntɪ/ n. **1** the mount of Don Quixote, the protagonist in Miguel de Cervantes' book of the same name（塞万提斯小说《堂吉诃德》中同名主人公的坐骑）罗西南特 **2** a useless horse 无用的马；驽马

rosinback /ˌrɒzɪnˈbæk/ n.【vaulting】a wide or flat-backed horse on which one or several vaulters perform（马上杂技表演用马，脊背宽而平坦）杂技表演用马 syn resin back

rota /ˈrəʊtə; *AmE* ˈroʊtə/ n. a list showing the duties of each member of a team 轮班表；出勤表

rotate /rəʊˈteɪt; *AmE* ˈroʊteɪt/ vt. **1**（cause to）to turn around on an axis or center 转动，旋转 **2** to grow crops in a fixed order of sequence; alternate 轮流；轮种，轮牧

rotation /rəʊˈteɪʃn; *AmE* roʊ-/ n. **1** the act of turning around a center 旋转 **2** the process of growing or grazing grass alternatively 轮种，轮牧

rotational /rəʊˈteɪʃənl; *AmE* roʊˈ-/ adj. of or relating to rotation 旋转的，转动的；轮种的，轮牧的：rotational crops 轮种作物 ◇ Rotational grazing is an efficient way of reducing worm infestation on the pasture, and also gives paddock-owner the chance to do weed control and fertilize the field. 轮牧是牧场有效控制虫害的一种措施，同时也使牧场主有时间进行杂草除防和施肥。

rough[1] /rʌf/ adj.（of a terrain）difficult to travel over or through（地形）崎岖的，难行的：rough country 崎岖乡野

rough[2] /rʌf/ v. to feed a horse with little or no grains and concentrates（用很少的谷物或精料喂马）粗饲 PHR V **rough off/up** to reduce grains form a horse's diet 粗饲

rough stock n.【rodeo】an untrained horse or bull used in certain rodeo events（某些西部牛仔竞技项目中使用的未调教的马或牛）生牛；生马

rough stock rider n.【rodeo】a rough rider 驯生马者；骑生畜人

roughage /ˈrʌfɪdʒ/ n. any feed with a high fiber content and low digestible nutrients, such as hay and straw（富含纤维且可消化养分含量少的饲料，比如干草和秸秆）粗饲料

rough-out /ˈrʌf aʊt/ adj.（of tanned hide）with the rough side worn outside（指鞣革粗面在外）翻毛的：The chaps and some types of cowboy boots are usually made of rough-out leather. 皮套裤和有些牛仔靴通常由翻毛皮制成。

roughride /ˈrʌfraɪd/ vt.【rodeo】to train or ride rough stock, such as broncs or bulls 骑生畜，野骑；驯生畜

roughrider /ˈrʌfˌraɪdə/ n.【rodeo】one who rides and trains rough stocks 骑野畜者；驯野畜者 syn rough stock rider

round /raʊnd/ n. **1** a stage in equestrian competitions（马术比赛中的）轮，场 **2**【jumping】a complete way around the course in show jumping（障碍赛的）一轮：the first horse to jump a clear round 第一匹跨栏无罚分的马

vt. **1**【hunting】to cut off the tips of puppy's ears for a smart appearance（为美观而剪去幼犬耳梢）剪耳梢：To round off the tips of dogs has been censured severely by many animal protectionists. 给犬剪耳梢已遭到了动物保护主义者的强烈谴责。**2** PHR V **round up** to gather cattle or animals together; herd（将牲口）驱拢，赶拢

round pen n. an enclosed, round arena where a horse is trained 圆形围场；练马圈

rounded /ˈraʊndɪd/ adj.（of horse's back）shaped like an arc; arched（马背）弧形的，弓形的：rounded back 背部拱起

roundup /ˈraʊndˌʌp/ n.（also **round-up**）**1** the herding together of cattle or horses for inspection or branding purposes 赶拢，聚拢（牲畜）**2** the cattle or herd of horses rounded up 牲畜群

round-up wagon n. *see* chuck wagon 野营四轮马车

roundworm /ˈraʊndwɜːm; *AmE* -wɜːrm/ n. *see* ascarid 蛔虫

rount /rʌnt/ n.【BrE】a horse with a roan coat mixed with white or peach 沙灰毛马 syn grizzle, grissel

rouse /raʊz/ vt. to startle game from its cover or lair（将猎物从藏身处赶出）惊起，突袭

vi.（of animals）to cease to rest or be inactive; stir（动物）惊觉

route /ruːt, raʊt/ n.【racing】a horse race over 1-1/8 miles in length（赛程超过1 800米的比赛）中途赛

router /ˈraʊtə(r)/ n.【racing】a horse who excels in races over 1-1/8 miles（在1 800米以上的比赛中表现出色的马）中途赛马

routine /ruːˈtiːn/ n. a fixed and regular way or procedure of doing sth 规程；常规

adj. usual; regular 常规的；定期的：routine checks 常规检查

rowel /ˈraʊəl/ n. a sharp-toothed wheel attached to the shank of a spur（马刺杆上所带的带齿小轮）刺轮

Royal Canadian Mounted Police n. the famous mounted police force of Canada founded in 1873（加拿大在1873年成立的骑警队）加拿大皇家骑警 syn Mountie, Mounty

Royal College of Veterinary Surgeons n.（abbr. **RCVS**）an institution founded in UK in 1844 to train veterinary surgeons and regulate the profession（英国于1844年创立的组织机构，旨在培养兽医人员并对该行业进行监管）皇家兽医学会：Anyone wishing to practice as a vet in the United Kingdom must be registered with the RCVS. 任何人如果想在英国当兽医，都必须经过皇家兽医学会认证登记。

Royal International Horse Show n. an annual international horse show first held in 1907 in London, England, formerly known as the International Horse Show and granted the prefix of Royal in 1957（自1907年起每年在英国伦敦举行的国际赛事，原名国际马匹展览赛，于1957年冠名"皇家"字样）皇家国际马匹展览赛 syn London International, Royal International

Royal Society for the Prevention of Cruelty to Animals n.（abbr. **RSPCA**）an organization founded in Britain in 1824 to encourage kindness and prevent cruelty to animals（于1824年在英国成立的机构，旨在鼓励人们善待动物并禁止虐待动物）英国皇家动物反虐协会

Royal Veterinary College n.（abbr. **RVC**）a veterinary school established in 1791 in London, UK which provides undergraduate and postgraduate programs in veterinary science, clinical practice and education（1791年在英国伦敦成立的兽医学校，为兽医教学、临床研究开展本科与研究生教育）皇家兽医学院

RSS abbr. Registered Shoeing Smith 注册蹄铁匠，注册蹄师

rub /rʌb/ v. to move back and forth over a surface with firm pressure 揉按,擦拭 PHR V **rub down** to rub a horse with a rough towel for cleaning purpose 擦拭干净:The mounted police stopped to rub down his horse with a cloth. 这名骑警停了下来,用布将马擦拭干净。

rubber band n. an elastic rubber band used to hold braids of the mane and tail in showing horse (马匹展览赛中用来扎鬃和尾的)橡皮筋

rubber boot n. 【AmE】*see* wellington [长筒]胶鞋

rubber curry n. (also **rubber curry comb**) an oval-shaped grooming tool having rows of soft rubber teeth, mainly used to remove dirt settled under the horse's hair (一种带橡胶齿的椭圆形刷拭工具,用来除去积蓄在马被毛中的污垢)橡胶梳 compare metal curry comb

rubber shoe n. a kind of horseshoe made of rubber with a steel center (一种带铁芯的)橡胶蹄铁

rubefacient /ˌruːbɪˈfeɪʃjənt; *AmE* -ʃənt/ n. an agent applied externally to the skin, causing mild heat and irritation to reduce pain and heal trauma (一种涂在皮肤表面使之轻微红肿的药剂,可减轻疼痛并促进治愈)发红剂

rudimentary /ˌruːdɪˈmentri; *AmE* -təri/ adj. being in the earliest stages of development or incompletely developed; incipient 原始的;发育不全的: rudimentary teeth 原始齿(即犬齿)

rug /rʌg/ n. a heavy blanket or fabric used as covering for a horse 披毯,马衣: jute rug 黄麻披毯 ◇ flax rug 亚麻披毯 ◇ day rug 日间马衣 ◇ New Zealand rug 新西兰马衣 ◇ sweat rug 防汗马衣

vt. to provide a horse with rug 穿马衣:Horses should be rugged up at night in winter. 冬季夜间最好给马穿上马衣。

rug fillet n. *see* fillet string (马衣)系带

ruga /ˈruːgə/ n.【Latin】(*pl.* **rugae** /-dʒiː/) a fold or wrinkle, as in the lining of the stomach (胃等脏器内壁的)皱,皱褶:palatine rugae 腭褶

rugged /ˈrʌgɪd/ adj. **1** (of a horse) wearing a rug (指马)穿马衣的 **2** (of a horse) having a sturdy build or strong conformation (指马)体格粗壮的

ruined /ˈruːɪnd/ adj. (of a horse) spoiled due to improper training or breaking (指马)调坏的,毁了的

rule /ruːl/ n. any regulation observed by the players in a game or competition (竞技运动的)规则

vt. to decide or declare authoritatively 判罚,裁决 PHR V **rule off** to prohibit a horse or rider from competing in a race 禁赛:Red Storm was ruled off for one year. "红色风暴"被禁赛一年。

Rum Pony n. an ancient strain of the Highland Pony living on the Isle of Rum, Scotland; stands around 13 hands and generally has a mouse dun or chestnut coat; is extremely hardy and sure-footed, and used mainly for riding and light draft (生活在苏格兰兰姆岛上的高地矮马品种。体高约13掌,被毛多为鼠灰或栗色,抗逆性强,步伐稳健,主要用于骑乘和轻挽)兰姆矮马:Due to the isolation of the ponies on the island, the Rum ponies have remained in a form close to their original endemic type. 由于兰姆岛的地理隔离,兰姆矮马因而保留了最初的本土特征。

rumen /ˈruːmen/ n.【anatomy】the first division of the stomach of a ruminant animal where most food collects immediately after being swallowed and from which it is later retuned to mouth as cuds for thorough chewing (反刍动物的第一胃室,采食后食物储存于此并形成食糜供以后反刍咀嚼)瘤胃 syn paunch

ruminant /ˈruːmɪnənt/ n. any of the animals having rumen 反刍动物

ruminate /ˈruːmɪneɪt/ v. to chew the returned cud from rumen 反刍 | **rumination** /ruː-

mɪˈneɪʃn/ n.

rump /rʌmp/ n. the croup of horse（马的）尻部，臀部

run /rʌn/ n. **1** a fast gallop of horse（马的）袭步；奔跑 PHR V **make a run**【racing】(of a horse) to increase its speed and moves up in the field（指马）加速 **2**【hunting】the trail traveled by a fox 狐狸的行踪：the run of a fox 狐狸的行踪 **3**【cutting】the period of 2.5 minutes allotted to each cutting contestant participating in a competition during which the cutter attempts to work and separate as many calves as possible from the herd（截牛比赛两分半的时间，期间每位参赛者从牛群尽可能多地拦截小牛）场，回合，轮

vi. **1** to move at a fast speed; gallop 飞奔，疾驰 PHR V **run down**【racing】(of a horse) to scrape the flesh off its heels on the track surface while racing（指赛马）蹭破蹄踵 **run in**【racing】(of a horse) to win unexpectedly（指赛马）意外获胜 **run loose**【racing】(of a horse) to run unbacked in a race（指马）空跑：Big Apple ran loose over the homestretch after its rider falling off the saddle. 在骑手落马后，'大苹果'空跑完了终点直道。**run rank**【racing】(of a horse) to run in a headstrong manner（指马）横冲，狂奔：Most jockeys don't wish their mount to run rank early in the race so as to conserve its energies for a push later on. 大多骑手都不太希望坐骑在起跑不久就狂奔，以便保存体力为后段冲刺做准备。**run out** ①【racing】(of a horse) to finish out of the money（指马）未跑入三甲 ②【racing】(of a horse) prefer to run along the outside rail（指马）靠外栏奔跑 ③【jumping】(of a horse) to avoid jumping an obstacle by running aside of it（指马）跑离障碍：An uncertain fencer sometimes would choose to run out from obstacles. 怯跳的马有时会选择跑离障碍。**run through**【cutting】(of a cow) to run under a horse in an attempt to return to the herd（指牛）从马腹下穿过 **run to ground**【hunting】(of a hound) to chase the fox into its earth（指猎犬）追入狐洞 **run wide**【racing】(of a horse) to run away from the inside rail and cover extra ground（指赛马离开内栏）跑大圈 IDM **run a hole through the wind** (of a horse) to run very fast in a race（指马）疾驰 **run downhill**【racing】(of a horse) to run in quick, hurried strides with little balance and too much weight on the forehand（指马在奔跑中步法匆乱，缺乏平衡且前肢负重过多）下山跑 **run hot and cold** (also **run in and out**)【racing】(of a horse) to run inconsistently in races（指赛马）发挥不稳：To bet on a horse who runs hot and cold definitely incurs a greater risk of losing money. 在出赛发挥不稳定的马身上投注无疑会冒更大的赌输风险。**2**【racing】to finish a horse race in a ceratin position 获得名次：Jack ran second in the Hong Kong Sprint. 杰克在香港短途锦标赛中获得了第二名。

runaway /ˈrʌnəweɪ/ adj. (of a horse) having escaped restraint or control（指马）脱缰的；失控的：runaway horses 脱缰的马

n. a horse that has escaped control or restraint 脱缰的马

run-down bandage n. a bandage wrapped around the lower leg and fetlock to provide support and to avoid abrasive injury of heels in races or workouts（比赛或训练中裹在马下肢球节上的绷带，起支撑作用并可防止蹄踵蹭破）弱系绷带

run-down boots n. *see* tendon boots 弱系绑腿

runners /ˈrʌnəːrz/ n. pl. *see* keepers（固定挽具皮带的）皮圈，鞘环

running double n.【racing】a wager in which the bettor must select the winners in two successive races in order to win（赛马押注方式之一，赌马者只有选出连续两场比赛的头马才能赢奖）二连赢 compare daily double

running martingale n. a piece of tack used in

English riding to prevent the horse from raising its head too high; consisting of a looped strap which buckles around the horse's neck and another strap which attaches at one end to the girth and divides into two branches at another with each branch ending with a ring through which the left and right reins pass（英式骑乘中所用的马具，可防止马抬头过高，由扣在马颈上的环革和另一条皮带构成，此皮带一端系于肚带，另一端分为两股，左、右方缰绳从两股末端的环孔中穿过）跑步鞅，低头革

running rail n.【racing】the inside rail 内侧栏杆，内栏

running rein n. an auxiliary rein used to control the head position of horse, attached on one end to the girth, passing between the horse's forearms, and then dividing into two branches with each connecting to the left and right bit rings（一种用来控制马头姿的辅助缰，其一端系于肚带，穿过两前肢之间后分为两股，然后各自与衔铁环相连）低头缰，跑步缰 compare draw reins

running walk n. a four-beat gait of certain breeds in which each foot strikes the ground separately at regular intervals（某些步态马走的四蹄音步法，每个蹄以均等间隔落地）大走步，疾走步：The running walk is similar to the trot, but the difference is that the diagonal pairs do not strike the ground that the same time. 疾走步与快步较为相似，其差别在于马对角线两蹄并不同时落地。

running-quick mount n. a method of mounting in which the rider runs towards a standing horse, jumps with the left foot into the left stirrup, and swing the right leg over the saddle（上马方式之一，骑手跑向站立马匹，起跳时左脚伸入左侧马镫，然后迅速侧身跨入马鞍中）飞身上马 syn flying mount

runny /ˈrʌni/ adj.（of nose or eyes）tending to exude mucus 流鼻涕的；流泪的：a runny nose 流鼻涕

n. a horse that has escaped control or proper confinement; runaway 脱缰之马

run-out bit n.【racing】a racing bit that exerts extra leverage on one side and thus prevents the horse from bearing out to the left or right（一种可单侧额外施力的赛用衔铁，可防止马向一侧奔跑）防侧跑衔铁

run-under heels n. *see* underrun heels 蹄踵着地

rupture /ˈrʌptʃə(r)/ n.【vet】a tear or breaking in body tissue; hernia 破裂；疝

vi. to break or suffer a rupture 破裂；患疝：ruptured bladder 膀胱破裂 ◇ The rupture of the arteries running in the uterine broad ligament at the time of foaling may give rise to serious, and possibly fatal, hemorrhage. 分娩时，分布于子宫扩韧带上的动脉破裂会导致严重失血，有时甚至是致死的。

rushing /ˈrʌʃɪŋ/ adj.【dressage】（of a horse）lacking balance and rhythm in pace（指马）步法匆乱的

Russian cab n. *see* Drosky 俄式马车，卓斯基马车

Russian Heavy Draft n. a heavy draft breed developed in the mid-19th century in Russia by putting local mares to a variety of heavy draft breeds; stands 14.1 to 15 hands and may have a chestnut, bay, or roan coat; has a well-balanced and muscular build, a short, straight back, full mane, tail and forelock, and short, lightly feathered legs; used for heavy draft and farm work（俄国在19世纪中叶采用当地母马与多个重挽品种公马杂交育成的重挽马品种。体高14.1～15掌，毛色多为栗、骝或沙，体格匀称，肌肉发达，鬃尾及额发浓密，背短而平直，四肢粗短，距毛稀疏，主要用于重挽和农田作业）俄罗斯重挽马

Russian Saddle Horse n. a breed of horse descended from the Orlov Trotter crossed with Thoroughbreds and Arabs（由奥尔洛夫快步马和纯血马及阿拉伯马杂交培育的马品种）俄国乘用马

R

Russian Steppe Horse n. *see* Kustanair 俄国高原马,库斯塔奈马

Russian style n. *see* troika【俄式】三套车

Russian Trotter n. *see* Orlov trotter 俄国快步马,奥尔洛夫快步马

rustle /rʌsl/ vi. to make or move with a soft, crackling sound 沙沙响:The leaves of the trees rustled in the breeze. 树叶在微风中沙沙作响。

vt. to steal farm animals 偷牲口

rustler /ˈrʌslə(r)/ n. one who steals farm animals 偷牲口的人

rut[1] /rʌt/ n. a deep track or groove made by the passage of the wheels of vehicles on road (车轮在路上行驶留下的)车辙:The coach sunk in a deep rut, and just could not move. 马车陷到了深深的车辙中,怎么也动不了。

rut[2] /rʌt/ n. **1** the sound made by a stag in mating season 雄鹿发情时的叫声 **2** the time of year when male deer become sexually active (雄鹿的)发情,发情期:be in rut 处于发情期

vi. (of deer) be in rut (鹿)发情 | **rutting** /ˈrʌtɪŋ/ n.

rutting season n. the period during which deer experience sexual arousal and mate (鹿的)发情季节,发情期

RVC abbr. Royal Veterinary College 皇家兽医学院

rye /raɪ/ n. an annual cereal plant widely grown for its grain as forage (一年生禾本科作物,因谷粒可作饲料而广泛种植)黑麦

rye straw n. the dried rye grass highly prized as a stuffing for horse collars (黑麦干草常作马匹套包的填料使用)黑麦草,黑麦秸

ryegrass /ˈraɪgrɑːs/ n. (also **rye grass**) the grass of rye 黑麦草:rye grass hay 黑麦干草

S

S-A node abbr. sinoatrial node 窦房结

saber /ˈseɪbə(r)/ n. (BrE also **sabre**) a heavy cavalry sword with a curved blade (骑兵用的)马刀,军刀

saber hocks n. *see* sickle hocks 镰状飞节

saber-legged /ˈseɪbə(r)ˌlegɪd/ adj. *see* sickle-hocked (飞节)镰状的

sabino /səˈbiːnəʊ/ n.【color】a coat color pattern in pinto horses in which the white areas range from distinct, irregular, sharp patches to small white spots on the base body color (花马毛色类型之一,其被毛基色中分布的白斑大小、形态差异变化较大)散沙毛,萨比诺: a sabino horse 一匹萨比诺花毛马 [syn] freckled roan [compare] overo, tobiano, tovero

Sable Island horse n. (also **Sable Island Pony**) a small-sized feral horse breed found on the Sable Island, Canada; descended from French stock released during the 18th century; stands 13 to 14 hands and usually has a coat color of bay, brown, black, or chestnut; has a heavy head, large ears with turned in tips, a short, broad neck, short back, full girth, strong legs, and small hoofs; is extremely hardy, and frugal, and noted for its ambling gait (加拿大塞布岛上的小型野生马品种,由 18 世纪放生的法国马繁衍而来。体高 13 ~ 14 掌,毛色多为骝、褐、黑或栗,头重,耳大且稍内卷,颈短而粗,背短,胸围宽,四肢粗壮,蹄小,抗逆性极强,耐粗饲,善走溜花蹄)塞布岛马 : The Sable Island horse is often pony-sized, but with a horse phenotype and usually dark in color. 塞布岛马多为矮马大小,但却具有马的形体,且毛色通常为深色。

Sable Island Pony n. *see* Sable Island horse 塞布岛[矮]马

sabre /ˈseɪbə(r)/ n.【BrE】a saber 军刀,马刀

sac /sæk/ n.【anatomy】a pouch or pouch-like structure in an organ or body part, sometimes filled with fluid (机体器官内的囊状结构,有时内含液体)囊;液囊: air sacs 肺泡 ◇ conjunctival sac 结膜囊 ◇ lacrimal sac 泪囊

sacral /ˈseɪkrəl, ˈsæ-/ adj. of or situated near the sacrum 荐骨的: sacral region 荐部 ◇ sacral vertebrae 荐椎

sacrococcygeal /ˌsækrəˈkɒksɪdʒiːl/ adj. of or relating to the sacra and the coccyges 荐尾的

sacroiliac /ˌsækrəʊˈɪlɪæk; *AmE* -kroʊ-/ adj. of or relating to the sacrum and ilium 荐髂的

n.【anatomy】the joint between the sacrum and ilium 荐髂关节

sacrum /ˈseɪkrəm, ˈsæ-/ n.【anatomy】(*pl.* **sacra** /-krə/) a triangular bone made up of five fused vertebrae and forming the posterior section of the pelvis (由五块椎骨愈合而成的三角形骨,构成脊柱后段)荐骨

saddle /ˈsædl/ n. a padded leather seat secured on the horse's back by a girth for the rider to sit comfortably on (一种带衬垫的皮革座,以肚带系于马背供骑者乘坐)马鞍: military saddle 军鞍 ◇ dressage saddle 盛装舞步鞍 ◇ English saddle 英式马鞍 ◇ saddle rack 鞍架 ◇ saddle mark 鞍迹 ◇ stock saddle 牛仔鞍 ◇ Western saddle 西部牛仔鞍 ◇ The common saddle colors are black, brown, chocolate, and tan. 常见的鞍色有黑、褐、黑褐和黄褐。 IDM **under saddle** (of a horse) equipped with the saddle and bridles (指马)备好鞍的: a horse under saddle 一匹备好鞍的马

vt. to put saddle on the back of a horse 备鞍: saddle up a horse 给马备鞍

saddle airer n. (also **saddle stand**) a wooden or

metal frame upon which a saddle is placed to be aired when not in use（用来晾置马鞍的木质或金属支架）晾鞍架

saddle bag n. either one of a pair of bags laid across the back of a pack animal（搭在驮畜背上的袋囊）鞍褡，鞍袋

saddle blanket n.（also **saddle cloth**）a square cotton covering placed between the back of horse and saddle to absorb sweat, provide padding, and prevent galling（衬在马背和鞍之间的方形棉毯，用来吸收汗液并防止马背蹭伤）鞍毯，鞯

saddle bow n. the arched front part of a saddle 鞍弓

saddle bracket n. *see* saddle rack 鞍架

saddle bronc riding n. a rodeo event in which one rides a wild bucking horse on a regulation saddle for at least ten seconds（牛仔竞技项目之一，骑手策骑弓背跳跃的野马至少 10 秒以上）骑野马比赛

saddle channel n.（also **channel**）the open portion of the saddle beneath the seat which runs from the pommel to the cantle（马鞍底侧从前鞍桥至后鞍桥的空档）鞍沟

S

saddle cloth n. **1** *see* saddle blanket 鞍毯，鞯 **2**【racing】*see* number cloth 号码布

saddle crupper n.（also **riding crupper**）a riding tack consisting of a strap attached on one end to the saddle by means of metal dees and a loop fitted on the other around the dock of horse, used to prevent the saddle from slipping forward onto the withers（一件乘用皮革马具，其前端通过挂钩与马鞍相连，后端则绕于马尾根下方，可防止马鞍向前滑动）马鞍尻带

saddle dee n. *see* D-ring（马鞍上的）D 环

saddle fall n. an act performed by a stunt man who falls from a galloping horse on cue in movie（电影中特技演员依提示从马背摔落）落马，跌落

saddle flap n.（also **flap**）part of the English saddle on either side of the seat upon which the rider's leg rests（英式马鞍的侧翼部分，骑者两腿靠于上面）鞍翼

saddle fork n.（also **fork**）the open part of the underside of the saddle that rests upon the horse's withers（马鞍底部搭在马匹鬐甲处的空档部位）鞍叉

saddle gaits n. the gaits of the horse including the pace, stepping pace, and the rack（包括对侧步、高蹄对侧步和单蹄快步在内的步法）乘用步法

saddle horn n.（also **horn**）a horn-shaped structure located in front of the Western saddle, used to tie the lasso in roping（位于西部牛仔鞍前侧的角形结构，套牛时用来拴系套绳）角柱，鞍角，鞍头

saddle horse n. a horse suited or used for riding 乘用马

saddle leather n. tanned leather used in saddlery（制鞍用的皮革）鞍革

saddle mark n. an acquired mark on the back of a horse resulting from abrasion by a poorly fitted saddle（由于马鞍装配不当而在马背压出的印迹）鞍迹，鞍印

saddle pad n. a saddle-shaped cotton pad placed under a saddle（置于马鞍下面的棉垫）鞍垫 syn numnah

saddle panel n.（also **panel**）the padded part of the saddle below the saddle tree（马鞍位于鞍芯之下的部分）鞍褥

saddle rack n. a metal or wooden support attached to a wall upon which a saddle rests when not in use（安装在墙上的金属或木制架，马鞍不用时置于其上）鞍架 syn saddle bracket

saddle skirt n. the part of the Western saddle on either side of the seat which covers the stirrup bars（西部牛仔鞍座两侧盖在镫杆之上的部分）鞍裙

saddle soap n. a mild soap used for cleansing saddle（清洗马鞍用的肥皂）鞍皂

saddle sore n.（also **saddlesore**）**1** a gall or open

sore that develops on the back of the horse from an ill-fitting or ill-adjusted saddle（由于装鞍不当而在马背造成的硬块或伤疤）鞍伤 **2** a sore on the skin of the rider chafed by the saddle（骑手被马鞍蹭出的伤疤）鞍伤

saddle stand n. *see* saddle airer 晾鞍架

saddle strings n. the narrow strips of leather attached to the front and rear skirts of a Western saddle to tie objects to the saddle（西部牛仔鞍的鞍襟上用来拴系物品的细皮条）鞍绳

saddle tree n. a wooden or metal frame around which the saddle is built（制作马鞍的木质或金属骨架）鞍芯，鞍骨

Saddlebred /ˈsædlbred/ n. *see* American Saddlebred 美国骑乘马

saddler /ˈsædlə(r)/ n. one who makes or sells saddles and other equipment for horses（制造或贩卖马鞍和马具的人）鞍匠；鞍具商

saddlery /ˈsædləri/ n. **1** saddles, bridles and other equipment for horses; tack 鞍具；马具 **2** a shop that sells tack 鞍具店，马具店 **3** the craft or trade of making or selling tack 鞍具业；马具业

saddlesore /ˈsædlsɔː; *AmE* ˈsædlsɔːr/ n. *see* saddle sore 鞍伤

adj. **1** (of a rider) chafed from riding on a saddle（指骑手被鞍）蹭伤的 **2** feeling sore and stiff after riding a horse（骑马后）酸痛的

Saddle-type Pinto n. one conformation type of Pinto with high head carriage and animated, high action for standard gaited and parade events（花马体貌类型之一，头姿高昂，步法灵活，举蹄较高，主要用于步态和仪仗项目）乘用型花马 compare Stock-type Pinto

Sado /ˈseɪdəʊ/ n. a two-wheeled, horse-drawn cart used on the island of Java, Indonesia to transport people and goods（印度尼西亚爪哇岛上过去使用的双轮马车，主要用来运送乘客和货物）爪哇马车

safe /seɪf/ n. the leather lining of some tack that prevents the buckles or bars from chafing the skin of the horse（马具的皮革衬里，可防止带扣等蹭伤马的皮肤）皮衬里

safety helmet n. (also **helmet**) a hard headpiece worn by riders for safety purposes 安全帽，头盔

safety stirrup n. (also **safety iron**) a stirrup designed to release the rider's foot easily in case of a fall（人落马时脚可轻易脱开的马镫）安全马镫

saffron /ˈsæfrən/ n. a corn-producing plant having purple or white flowers with orange stigmas（一种番红花属球茎植物，开紫色或白色花朵，花柱为橘黄色）藏红花

sainfoin /ˈseɪnfɔɪn/ n. a Eurasian legume plant having pinnately compound leaves and pink or white flowers, grown widely as a forage plant（一种生长在欧亚大陆的豆科植物，羽状复叶，花多为粉红色或白色，因用作饲料而广泛种植）驴食豆，红豆草

Saint Christopher n. the patron saint of all horsemen（骑马人的守护神）圣・克里斯托弗

Saint Hubert n. the patron saint of the Hunt（猎场的守护神）圣・休伯特

sais /seɪz/ n. **1** *see* syce 印度马夫 **2** a Russian mounted game played by herdsmen in which each rider attempts to grab his opponent by the hands or arms and pull him off his mount（一项俄国牧民的马上运动，某骑手抓住对方手臂而将其拽下马）马上角力

Salerno /səˈlɛrnəu; *AmE*-nou/ n. a warmblood breed native to the Maremma and Salerno districts of Italy; descended from the Neapolitan breed crossed with Spanish and Oriental stock; stands 16 to 16.2 hands and may have a coat of any solid color, with bay, black, and chestnut occurring most common; has a light head, long neck, pronounced and muscular withers, short loins, and a deep girth; is quiet, balanced, and energetic; historically a popular cavalry mount, now used mainly for riding and jumping（源于意大利的温血马品种，由那不勒斯马与西班

牙马及东方马杂交繁育而来。体高 16 ~ 16.2 掌,各种单毛色兼有,但以骝、黑和栗最为常见,头轻,颈长,鬐甲突出,腰短而胸深,性情安静,体格匀称,富有活力,过去主要用作军马,现在主要用于骑乘和障碍赛)萨莱诺马

saline /ˈseɪlaɪn; *AmE* -liːn/ adj. of, relating to, or containing salt; salty 咸的;含盐的: saline solution 盐水溶液

n. *see* physiological saline 生理盐水

saliva /səˈlaɪvə/ n. the watery fluid secreted from the oral mucous glands that lubricates chewed food and aids digestion (口腔腺体分泌的水样液体,可润滑咀嚼的食物并协助消化)唾液

salivary /səˈlaɪvəri; *AmE* ˈsæləveri/ adj. of or relating to saliva 唾液的: salivary gland 唾液腺 ◇ salivary amylase 唾液淀粉酶

Salix /ˈsælɪks/ n. another registered trade name for furosemide (利尿磺胺的注册商标名)速尿灵,呋塞米

salt /sɔːlt/ n. the chemical compound of sodium chloride 食盐

salt block n. *see* salt lick 舔砖,盐块

salt lick n. (also **salt block**) a chunk of natural or man-made salt upon which horses lick to balance the mineral elements (一种天然或人造盐块,供马匹舔食以补充矿物质元素)舔砖,盐块

salve /sælv/ n. a soothing ointment or balm applied to wounds or sores (在伤痛处涂的)软膏,药膏

vt. to soothe or heal with salve 敷药膏

SAN abbr. sinoatrial node 窦房结

San Fratello n. a horse breed indigenous to Sicily where it is bred in the wild; stands 15 to 16 hands and usually has a bay, brown, or black coat; has a slightly heavy head, sloping croup, wide and deep chest, and prominent withers; is hardy and used mainly for riding and packing (西西里岛的本土马品种,在野生条件下繁育。体高 15 ~ 16 掌,毛色多为骝、褐或黑,头偏重,斜尻,胸身宽广,鬐甲高,抗逆性强,多用于骑乘和驮运)圣弗兰特罗马

sand bath n. **1** a bath taken by a horse rolling in sand (马在沙地打滚)沙浴 **2** an artificial sand pit where horses roll 沙浴场

vi. (of a horse) to roll in sand(指马)沙浴

sand colic n. a digestive distress resulting from the casual consumption of sand while grazing or eating hay rations directly from the ground (由于马匹吃草时误食沙土而导致的消化性疾病)积沙性腹痛

Sandalwood /ˈsændlwʊd/ n. (also **Sandalwood pony**) a pony breed indigenous to the islands of Sumba and Sumbawa, Indonesia; stands 12 to 13 hands and may have a coat of any color; has a small, well-shaped head, a full forelock, mane, and tail, wide chest, deep girth, and strong, hard legs; is quiet and energetic, quite fast for its size, and possessed of good endurance; used for bareback racing, riding, packing, light draft, and farm work (印度尼西亚松巴岛、松巴哇岛上的本土矮马品种。体高 12 ~ 13 掌,各种毛色兼有,头小,面貌俊秀,额发、鬃和尾毛浓密,胸腔宽广,四肢粗壮,性情活泼、温驯,步速较快,耐力好,主要用于光背赛马、骑乘、驮运、轻挽以及农田作业)檀香木矮马

sandwort /ˈsændwɜːt; *AmE* -wɜːrt/ n. any of numerous low-growing herbs of the genus *Arenaria*, having small, usually white flowers often grouped in cymose clusters (一种蚤缀属草本植物,植株矮小,开白色小花,聚伞状花簇)蚤缀

sandy bay n.【color】a light red coat with black points (被毛淡红末梢为黑的毛色)沙骝毛: Sandy bay is the lightest shade of bay. 沙骝是色调最浅的骝毛类型。

Sanhe n.【Chinese】(also **Sanhe horse**) a horse breed indigenous to Inner Mongolia, China; later improved with Orlov Trotters, Anglo-Arabs, and Thoroughbreds; stands about 15 hands and generally have a bay or chestnut coat; has an elegant profile, a strong build, prominent withers, broad chest and deep girth, and strong legs; is hardy, frugal, and good-natured; used as an excellent riding and draft horse (源于中国内蒙古三河地区的马品种,后经奥尔洛夫马、盎格鲁阿拉伯马和纯血马改良。体高约15掌,毛色以骝、栗为主,面貌俊秀,体质结实,鬐甲明显,胸身宽广,四肢强健,抗逆性强,耐粗饲,性格温驯,是优良的乘挽兼用马)三河马

sanitary /ˈsænətri; *AmE* -teri/ adj. of or relating to health; hygienic 健康的,卫生的

sanitation /ˌsænɪˈteɪʃn/ n. the equipment and systems that keep places clean and protect health 卫生[体系]

saponin /ˈsæpənɪn/ n. any of various plant glucosides that form soapy lathers when mixed and stirred with water, used in detergents, foaming agents, and emulsifiers (与水混合后能产生泡沫的多种植物糖苷,可用作除垢剂、发泡剂和乳化剂)皂角苷

sarc(o)- pref. muscle 肌

sarcoid /ˈsɑːkɔɪd/ n. a viral tumor composed mainly of connective tissue that appears on the skin of horse (马皮肤上由结缔组织形成的病毒性肿瘤)肉样瘤:Sarcoid is the most common tumor in horses. 肉样瘤是马最常患的肿瘤。

sarcolemma /ˌsɑːkəˈlemə/ n. the outer cell membrane of muscle fiber (肌纤维的外层细胞膜)肌膜

sarcomere /ˈsɑːkəmɪə/ n.【physiology】one of the segments into which a myofibril of striated muscle is divided (横纹肌的肌原纤维分成的节段)肌节:The sarcomere is the functional unit of contraction and is made up largely of special contractile proteins. 肌节是肌肉收缩的功能单位,主要由一些特殊的收缩蛋白组成。

sarcoplasma /ˌsɑːkəʊˈplæzmə; *AmE* ˌsɑːkoʊ-/ n. the cytoplasm of muscle fibre (肌纤维的胞质成分)肌质 | **sarcoplasmic** /-mɪk/ adj.

sarcoplasmic reticulum n.【physiology】(abbr. **SR**) a reticular structure within the muscle fibres that serves to store intracellular calcium (肌纤维内贮存钙离子的网状结构)肌质网:The arrival of the electrical impulse at the sarcoplasmic reticulum (SR) causes calcium ion channels to open within the SR membrane. 到达肌质网的电脉冲信号导致肌质网膜上的钙离子通道被打开。

Sardinian /sɑːˈdɪnɪən; *AmE* sɑːrˈdɪnɪən/ n. **1** the Sardinian pony 撒丁尼亚矮马 **2** the Sardinian donkey 撒丁尼亚驴

Sardinian Anglo-Arab n. a horse breed indigenous to Sardinia, Italy; developed by putting Arab stallions to native mares and revived by infusion of Thoroughbred blood in the early 20th century; stands 15.1 to 16 hands and usually has a gray, chestnut, or bay coat; has a light square head, pronounced withers, slightly hollow back, short loins, broad chest, full girth, and strong legs; is hardy, swift, and possessed of good endurance; used for riding and show jumping (意大利撒丁岛的本土马品种,主要采用阿拉伯马公马与当地母马杂交培育而成,20世纪初又引入纯血马加以改良。体高15.1~16掌,毛色多为青、栗或骝,头轻且呈方形,鬐甲突出,背稍凹,腰短,胸身宽广,四肢强健,抗逆性强,步伐敏捷,耐力好,主要用于骑乘和障碍赛)撒丁尼亚盎格鲁阿拉伯马

Sardinian Donkey n. (also **Miniature Sardinian donkey**) a miniature donkey breed originated in the Mediterranean island of Sardinia, Italy; stands below 12 hands and usually has gray coat; is hardy, frugal, and resistant to weather (源于意大利撒丁岛的微型驴品种。体高12掌以下,毛色以青为主,抗逆性强,耐粗饲)撒丁尼亚微型驴

Sardinian pony n. an ancient pony breed indigenous to Sardinia, Italy; stands 12.1 to 13 hands and generally has a bay, brown, black, or liver chestnut coat; has a heavy head, long neck, full mane, tail, and forelock, slightly hollow back, slender legs, and small feet; is hardy, frugal, lively and agile; mainly used for farm work and riding (意大利撒丁岛的本土矮马品种。体高12.1~13掌,毛色多为骝、褐、黑或深栗,头重,颈长,鬃、尾和额发浓密,背稍凹,四肢纤细,蹄小,抗逆性好,耐粗饲,性情活泼,反应敏捷,主要用于农田作业和骑乘)撒丁尼亚矮马

Saumur /ˈsəʊmjʊə(r)/ n. a town in western France where the Cavalry School of Saumur was established in the early 18th century (法国西部城镇,18世纪初在此建立了索米尔骑兵学校)索米尔

sausage boot n. a leather or rubber ring with an adjustable strap, placed around the pastern of a horse prone to capped elbow (一种可调节式皮革或橡胶圈,戴在系部可防止马患肘关节肿胀)护腿圈 [syn] doughnut boot

savage /ˈsævɪdʒ/ adj. (of a horse) aggressive and violent; vicious (指马)具攻击性的;袭人的

n. a mean horse prone to bite and attack other horses or person (易于攻击其他马和人的马匹)恶马

vt. (of a horse) to bite or attack other horses or person (指马)攻击;袭人

savaging /ˈsævɪdʒɪŋ/ n. a stable vice of horse who attempts to bite and attack horses or people (咬其他马匹和人的恶习)攻击;袭人

Savanilla Phaeton n. a light, horse-drawn carriage widely used in Bangkok, Thailand in the 19th century; usually drawn by a single horse between shafts (19世纪泰国曼谷使用较为广泛的轻型马车,由单马驾辕拉行)撒瓦尼拉费顿马车

sawdust bedding n. a bedding consisting of wood sawdust 锯末垫料:Sawdust bedding has a high dust content which makes it unsuitable for horses with allergies or other respiratory ailments. 锯末垫料含尘量较高,因而对那些有过敏症状和呼吸道疾病的马来说并不合适。

sawhorse /ˈsɔːhɔːs; *AmE* -hɔːrs/ n. (also **saw horse**) a wooden frame with four extended legs that supports wood to be sawed (用来支撑锯木且带四条腿的木架)锯木架,马凳

sawhorse stance n. the position assumed by a horse afflicted with tetanus in which both its fore and hind legs spread and extend outward (马患破伤风后前后肢外展的肢势)马凳肢势

scab /skæb/ n. **1** a crust that forms over a a cut or wound (伤口痊愈时形成的硬壳)痂 **2** a skin disease of animals; mange 疥癣

vi. to become covered with scabs or a scab 结痂

scald /skɔːld/ vt. to burn with or as if with hot liquid or steam (用高温液体或蒸气)烫伤;蹭伤:In hunter clip, the unclipped saddle patch will offer some protection from sores and scalding. 在狩猎型修剪中,鞍坐处未修剪的被毛可保护皮肤免受疼痛和蹭伤。

scapula /ˈskæpjʊlə/ n.【anatomy】(*pl.* **scapulae** /-liː/ or **scapulas**) either of two large, flat, triangular bones forming the back part of the shoulder (构成肩背的两块大而扁平的三角形骨)肩胛骨 [syn] shoulder blade

scapular /ˈskæpjʊlə(r)/ adj. of or relating to the shoulder blade or scapula 肩胛骨的:scapular cartilage 肩胛软骨 ◇ scapular spine 肩胛冈

n. the shoulder bone; scapula 肩胛骨

scapulary /ˈskæpjʊləri/ adj. scapular 肩胛骨的

scar /skɑː(r)/ n. a mark left on the skin after a surface injury or wound has healed (表皮损伤或伤口愈合后留下的痕迹)伤痕,疤痕

scarlet /ˈskɑːlət; *AmE* ˈskɑːrlət/ n. *see* scarlet coat 红色狩猎服

scarlet coat n.【hunting】(also **scarlet**) a red hunting coat 红色狩猎服

scent /sent/ n. the smell left by a fox or other

quarry on the ground（狐狸或其他猎物留在地面的）气味：A dry, hard ground usually does not hold a scent well, while vegetation such as grass and shrubs does. 干燥而坚硬的地表一般留下的猎物气味较淡，而有植被的地带如草地和灌木丛中气味则较浓烈。

scent hound n. a hound who pursues the line of a quarry by scent instead of sight 靠气味捕猎的猎犬：The scent hounds such as stag hounds, foxhounds, and coon hounds all have very good sense of smell. 许多靠气味捕猎的猎犬，如用来追捕公鹿、猎狐和浣熊的猎犬嗅觉都十分灵敏。

Schleswig Heavy Draft n.（also **Schleswig horse**）a coldblood developed in Schleswig-Holstein, Germany in 19th century from which the name derived; stands 15.1 to 16.1 hands and may have any of the solid colors; has a well-proportioned head, small eyes, a short, arched neck, low withers, a short back, sturdy, short legs with feathering, and large hoofs; used for draft, farm work, and transportation（德国什勒斯威格－霍斯坦地区在 19 世纪培育的冷血马品种并由此得名。体高 15.1～16.1 掌，各种单毛色兼有，头型适中，眼小，颈短且呈弓形，低鬐甲，背短，四肢粗短且附生距毛，蹄大，主要用于重挽、农田作业和运输）什勒斯威格重挽马

Schleswig horse n. *see* Schleswig Heavy Draft 什勒斯威格重挽马

school /skuːl/ n. **1** an institution for the instruction of riding, dressage, etc. 骑术学校：Spanish riding school at Vienna 维也纳西班牙骑术学校 ◇ the cavalry school at Saumur 法国索米尔骑兵学校 **2** the movements or exercises performed in the classical equitation 古典骑术动作：high schools 空中动作，高等骑术

vt. to train a horse 训练，调教：a well-schooled horse 一匹调教好的马

school horse n. any horse used as mount to teach people how to ride（用来教人骑乘的马）教学用马

school movements n.【dressage】any movements or figures performed to improve the suppleness and versatility of a horse（用来提高马匹柔韧性和技能的各种动作）高等骑术动作

School of Saumur n. *see* French Cavalry School at Saumur 法国索米尔骑兵学校

school riding n. the riding exercised from classical equitation 高等骑术

schooling /ˈskuːlɪŋ/ n. the act or process of training a horse 训练，调教，驯致：schooling principles 调教原则

schools above the ground n. *see* airs above the ground 空中动作

schooner /ˈskuːnə(r)/ n. a prairie schooner 草原篷车

sciatic /saɪˈætɪk/ adj.【anatomy】of or relating to the ischium or the region where it is located 坐骨的：sciatic nerve 坐骨神经

scissors /ˈsɪzəz; *AmE* ˈsɪzərz/ n.【vaulting】a compulsory exercise in individual vaulting in which the performer swings both his legs in a handstand around the body of the horse（马上体操规定动作，表演者以手撑在马背，双腿前后绕过马体一周）剪式动作 —*see also* **basic seat, flag, mill, stand, flank**

sclera /ˈsklɪərə/ n.【anatomy】the tough, white, fibrous outer envelope of tissue covering all the eyeball except the cornea（包裹在眼球外的一层坚韧的白色纤维膜）巩膜

scoop /skuːp/ n. a shovel-like utensil, usually having a deep, curved dish and a short handle（一种底深、柄短的铲形用具）铲勺：a heaped scoop 一满勺 ◇ a level scoop 一平勺

vt. to pick up and move with a scoop 铲：Philip began to scoop barley grain into the bag. 菲利普开始往袋子里铲大麦。

scope /skəʊp; *AmE* skoʊp/ n. the athletic ability or freedom of movement of a horse（马的）运动体能；灵活程度：a horse having scope 一匹行动灵活的马

scopy /ˈskəʊpi; *AmE* ˈskoʊpi/ adj. (of a horse) having freedom of movement（指马）行动灵活的

scorpion /ˈskɔːpɪən/ n.【cutting】an athletic, agile, quick-moving horse 动作敏捷的马

scour /ˈskaʊə/ vt. **1** to scrub sth in order to clean or polish it 擦洗(亮) **2** (of livestock) to have diarrhea（家畜）腹泻；痢疾

n. diarrhea 腹泻；痢疾：infectious white scour 传染性白痢

scraper /ˈskreɪpə(r)/ n. a plastic or metal grooming tool used to remove sweat or water from a horse's coat and to gently massage the muscle masses（用来刮除马体被毛上的汗液或水渍的塑料或金属用具，同时也可起到按摩肌肉的作用）刮子：sweat scraper 汗刮 ◇ water scraper 水刮

scratch /skrætʃ/ vt. **1**【racing】to withdraw a horse from an event or competition; forfeit 使退出比赛；取消参赛资格 **2**【rodeo】to spur a horse vigorously in bronc riding competition（骑野马时）使劲踢，猛踢

scratched /skrætʃt/ adj. (of a horse) formally withdrawn form an event or competition（指马）退出比赛的；取消参赛资格的

scratches /ˈskrætʃɪz/ n. inflammation around the fetlock and heel of horses, caused by infection of *Dermatophilus congolensis* which gains entry into the skin when dirty or saturated by rain water（马匹球节或蹄踵发炎的症状，常由于皮肤受污泥或雨水浸泡导致刚果嗜皮菌侵入所致）蹄踵炎，蹄踵裂 syn greasy heel, greased heels, cracked heels

screw /skruː/ n. an inferior, unsound, or worn-out horse 驽马；乏马

screwworm /ˈskruːwɜːm/ n. the larva of the screwworm fly 螺旋蝇蛆，螺旋锥蝇蛆

screwworm fly n. one type of blowfly that lays eggs on wounds or open cuts（在伤口下蛆的一种丽蝇）螺旋蝇，螺旋锥蝇

scrotal /ˈskrəʊtəl/ adj. of or pertaining to scrotum 阴囊的

scrotal hernia n. a congenital condition in which the intra-abdominal tissue descends through the inguinal canal into the scrotum（一种先天病症，其中腹腔组织通过腹股沟落入阴囊）阴囊疝

scrotum /ˈskrəʊtəm/ n.【anatomy】a skin pouch containing the testes 阴囊

scrub /skrʌb/ vt. to clean or wash by hard rubbing 清洗，擦洗：The manager and water trough should be scrubbed out regularly. 食槽和饮水池应当定期进行彻底擦洗。

scruffy /ˈskrʌfi/ adj. (of horse's coat) shabby and untidy（马被毛）脏乱的，不整洁的

scurry /ˈskʌri/ n.【racing】a short-distance race; sprint 短途赛

scurvy /ˈskɜːvi; *AmE* ˈ-ɜːrvi/ n. a disease caused by deficiency of vitamin C, often characterized by spongy and bleeding gums, bleeding under the skin, and extreme weakness（由维生素 C 缺乏引起的一种疾病，症状为牙龈多孔性出血，皮下出现血斑和体质异常虚弱）坏血病

scut /skʌt/ n. the short tail of a hare or deer（野兔或鹿的）短尾

SDF abbr. synchronous diaphragmatic flutter 横膈膜同步颤动

SDFT abbr. superficial digital flexor tendon 趾浅屈肌腱

sea horse n. a small marine fish（一种小型海洋鱼类）海马

seal /siːl/ n. a substance used to fill cracks 密封剂

vt. to close by sticking the opening together 密封

seal brown n.【color】a coat color of dark brown close to black, with brown or yellow areas on the muzzle, flanks, and inner legs（被毛为暗褐色而口鼻、腹侧和四肢内侧呈黄褐色的毛色）暗褐毛

sealant /ˈsiːlənt/ n. *see* hoof sealant 封蹄胶

season /ˈsiːzn/ n. **1** the estrus 发情期: to be in season 处于发情期 **2**【racing】the period of time during which racing is held on a track or course 赛季 **3**【hunting】the hunting season 狩猎季节

seat /siːt/ n. **1** the part of a saddle that a rider sits on 鞍座 **2** the posture of a rider in the saddle 坐姿,乘姿,骑姿: correct seat 正确乘姿 ◇ hunting seat 狩猎乘姿 ◇ incorrect seat 不正骑姿 ◇ hunch seat 弓背骑姿 **3** a sitting place for a passenger in a horse-drawn vehicle (马车的)座位: seat box 座下行李箱

seat bones n. the two prominent points of the rider's bottom upon which his body weight is distributed in riding (骑手臀部的两个骨突,骑马时承载体重)坐骨

sebaceous /sɪˈbeɪʃəs/ adj. of or relating to secreting fat or sebum; fatty 皮脂的;脂肪的: sebaceous gland 皮脂腺

sebum /ˈsiːbəm/ n. the semifluid secretion of the sebaceous glands (皮脂腺产生的分泌物)皮脂

second dam n. *see* granddam 祖母马

second incisors n. *see* lateral incisors 第二切齿,边切齿

second leg n.【racing】the second half of a double event (连赛中的下半场)后场,第二场

second phalanx n.【anatomy】the portion of the horse's leg including the short pastern 第二趾骨

second pommel n. *see* top pommel 上鞍桥

second thigh n. *see* gaskin 小腿,胫

second whipper-in n.【hunting】one of two principal assistants to the Hunt Master from whom instructions are received 猎狐主副助理

second wind n. the renewed energy and strength of a horse to move forward again (指马)第二股劲,再次发力

secondary /ˈsekəndri; *AmE* ˈsekənderi/ adj. occurring second in time, sequence or derived from what is primary or original 二级的,次级的;继发性的: secondary corpus luteum 次级黄体 ◇ secondary follicle 次级卵泡 ◇ secondary yellow body 次级黄体,副黄体

secondary corpus luteum n.【physiology】a mass of endocrine cells formed in follicle about one month after conception (母马受孕约1个月后由卵泡形成的内分泌细胞团块)次级黄体

secondary joint disease n. *see* osteoarthritis 骨关节炎

Secretariat /ˌsekrəˈteəriət; *AmE* -ˈteriət/ n. (1970 – 1989) an American chestnut Thoroughbred with three white socks who won the U. S. Triple Crown in 1973 and was hailed as one of the greatest racehorses of all time (一匹栗色的美国纯血马,其三肢带管白,于1973年夺得美国三冠王,被认为是有史以来最杰出的赛马)秘书处: Secretariat ranked second behind Man O'War in *The Blood-Horse*'s list of the top 100 US racehorses of the 20th century. 在《纯血马》杂志列出的20世纪美国100名最佳赛马榜单上,"秘书处"屈居"战神"之后位列第二。— *see also* **Big Red**

sedate /sɪˈdeɪt/ vt. **1** to administer a sedative to 给服镇静剂: The veterinary surgeon must be called in to sedate a horse by administering an appropriate tranquilizer. 必须请外科兽医采用合适的镇静剂给马匹镇静。**2** to calm or relieve by means of a sedative drug (用药物)使镇静 | **sedation** /-ˈdeɪʃn/ n.

sedative /ˈsedətɪv/ adj. having a calming or tranquilizing effect, and thus reducing or relieving anxiety, stress, irritability, or excitement (具有镇定并可缓解疼痛的)镇静的

n. any drug having a soothing, calming, or tranquilizing effect (具有镇定并可缓解疼痛的药物)镇静剂

seed /siːd/ n. **1** the small hard part of a plant, from which a new plant can grow 种子,籽实 **2** the sperm cells; semen 精子;精液

vt. to plant seeds in land; sow 播种: seeding

rate 播种率

seed hay n. *see* cereal grass hay 禾本科牧草

seedy toe n. a condition characterized by a separation of the hoof wall from the sensitive laminae, commonly associated with chronic laminitis（蹄壁与蹄叶分离的症状，多伴有慢性蹄叶炎）蹄匣分离 [syn] hollow wall, dystrophia ungulae

select /sɪˈlekt/ vt. to choose or pick out superior ones for breeding 选育

selection /sɪˈlekʃn/ n. the act of selecting or the fact of having been selected 选择，选育：selection coefficient 选择系数 ◇ selection index 选择指数

selective /sɪˈlektɪv/ adj. of or relating to selection 选择的：selective breeding 选育 ◇ selective mating 选配

selenium /səˈliːniəm/ n.【nutrition】a trace mineral essential to muscle formation and function（一种微量矿物质元素，为肌肉形成及机能活动所必需）硒

Selle Français n. a breed of horse descended form Norman horse with heavy introduction of Arab and Thoroughbred blood; stands 15.2 to 16.3 hands and usually has a chestnut coat; has a small head, robust frame, powerful shoulders, strong and long back, and deep girth; used for riding, show jumping and eventing（由法国诺曼马繁育而来的马品种，受阿拉伯马和纯血马影响较大。体高 15.2 ~ 16.3 掌，毛色以栗为主，头小，体格健壮，肩部发达，背长而有力，胸深，主要用于骑乘、障碍赛和三日赛）法兰西乘用马，塞拉法兰西马 [syn] French Saddle Horse

semen /ˈsiːmən/ n.【physiology】a thick, whitish fluid secreted by the testes and the accessory gland of a stallion that contains the sperm for fertilization of the egg（由公马睾丸和副性腺分泌的乳白色黏液，其中的精子可使卵子受精）精液：fresh semen 鲜精 ◇ raw semen 鲜精 ◇ Each ejaculation of semen is estimated to contain 4 to 18 billion sperm with 500 million sperm per ejaculation required for fertilization of one egg. 公马单次射精量中大概含有 40 亿 ~ 180 亿个精子，而要保证卵子受精需要 5 亿个精子。

semilunar /ˌsemɪˈljuːnə(r); *AmE* -ˈluːnə(r)/ adj. shaped like a half-moon; crescent-shaped 半月形的

semilunar valve n.【anatomy】one of the valves found between the ventricles and the arteries that prevent the blood backflow（心室与动脉之间的瓣膜，可阻止血液回流）半月瓣：nodule of semilunar valve 半月瓣小结

semimembranous /ˌsemɪˈmembrənəs/ adj.（of muscle）partly membranous（指肌肉）半膜的：semimembranous muscle 半膜肌

seminal /ˈsemɪnl/ adj. of, relating to, or containing semen or seed 精液的，精子的：seminal vesicles 精囊 ◇ seminal characteristics 精子特性 ◇ seminal plasma 精清 ◇ seminal quality 精液品质

seminiferous /ˌsemɪˈnɪfərəs/ adj. conveying, containing, or producing semen 输送精液的，产生精子的：seminiferous tubules 生精小管

semitendinous /ˌsemɪˈtendɪnəs/ adj. partly tendinous 半腱的：semitendinous muscle 半腱肌

send /send/ vt.【racing】to put or enter a horse in a race 报名参赛 PHR V **send away** to open the starting gate and begin a horse race 开闸起跑

sensitive /ˈsensətɪv/ adj. **1** easily hurt or irritated 敏感的，易受伤的：a sensitive skin 易受伤的皮肤 **2** capable of perceiving with sense 敏感的：Always start clipping at the shoulders which is a less sensitive area and less likely to alarm a nervous horse. 修剪首先要从马的肩部开始，因为此处皮肤敏感性差，不易令情绪紧张的马受惊。

sensitive laminae n. the sensitive area of the hoof between the hoof wall and coffin bone（介于蹄壁与蹄骨之间的敏感层）敏感性蹄叶，蹄叶

敏感层 compare insensitive laminae

sensitive sole n. part of the hoof below the coffin bone that nourishes the horn-producing layer（位于蹄骨下的组织层，可为蹄底角质层细胞提供营养）蹄底敏感层

sensory /ˈsensəri/ adj. of or relating to the senses or sensation 感觉的；感官的

sensory nerve n.【anatomy】any of the nerver fibers that mediate sensations of touch, smell, taste, hearing, and sight when stimulated（机体受刺激引发触觉、嗅觉、味觉、听觉和视觉的神经纤维）感觉神经[元]

sepsis /ˈsepsɪs/ n. a poisoned condition of the system resulting from the spread of infection throughout the blood（由于血液感染而导致全身中毒的症状）脓毒症

septicemia /ˌseptɪˈsiːmiə/ n.（BrE also **septicaemia**）a systemic infection of the blood by harmful microorganisms（由致病微生物造成全身血液感染的症状）败血症 syn blood poisoning

sequela /sɪˈkwiːlə/ n.（*pl.* **sequelae** /-liː/）a pathological condition resulting from a disease（由一种疾病导致的另一病症）继发症

serine /ˈseriːn/ n.【nutrition】a non-essential amino acid（一种非必需氨基酸）丝氨酸

serology /sɪˈrɒlədʒi; *AmE* -ˈrɑːl-/ n. the scientific study of blood serum 血清学 | **serological** /ˌsɪərəʊˈlɒdgɪkl/ adj. : serological test 血清学检测

serous /ˈsɪərəs; *AmE* ˈserəs/ adj. containing, secreting, or resembling serum 浆液的: serous gland 浆液腺

serous arthritis n. inflammation of a joint resulting from trauma and characterized by excessive accumulation of serous fluid in the joint space（关节受伤后出现的炎症，表现为关节腔内浆液大量淤积）浆液性关节炎

serpentine /ˈsɜːpəntaɪn; *AmE* ˈsɜːrpəntiːn/ n.【dressage】a schooling exercise in which the horse travels around the arena in a series of S-shaped loops（训练马匹的一种方法，马在场地内以S形曲线行进）S形运步，蛇形行进：A serpentine can be done in any gait and increases suppleness and control in horse and rider. 蛇形运步适用于任何步法，这种行进方式在增加马匹柔韧性的同时提高了骑手的驾驭能力。

serrate /ˈserɪt/ adj. resembling the teeth of a saw 锯齿状的: cervical part of ventral serrate muscle 颈下锯肌 ◇ thoracic part of ventral serrate muscle 胸下锯肌

serum /ˈsɪərəm; *AmE* ˈsɪrəm/ n.【physiology】（*pl.* **sera** /-rə/ or **serums**）**1** the clear yellowish fluid exuded from the blood when it clots（血液形成凝块时析出的浅黄色清液）血清 **2** watery fluid from animal tissue（组织中的清液）浆液

serve /sɜːv; *AmE* sɜːrv/ vt.（of male animals）to mate with female animal（雄性动物）与配，配种

service /ˈsɜːvɪs; *AmE* ˈsɜːrv-/ n.（of a stallion）mating with a mare（种公马与母马的）交配

service fee n.（also **stud fee**）a sum of money paid to the owner of a stallion for its breeding services（公马为母马配种后付给公马马主的钱款）[公马]配种费

serving hobbles n. *see* breeding hobbles 配种蹄绊

sesamoid /ˈsesəmɔɪd/ n.【anatomy】（also **sesamoid bone**）a small bone or bony nodule that develops in a tendon（在肌腱中出现的小骨或骨状结节）籽骨: the proximal sesamoid 近端籽骨 ◇ the distal sesamoid 远端籽骨

adj. of or relating to a sesamoid bone 籽骨的

sesamoid fracture n. a break in one or all of the sesamoid bones 籽骨骨折

sesamoidian /ˌsesəˈmɔɪdɪən/ adj. of or relating to sesamoid bone 籽骨的

sesamoiditis /ˌsesəmɔɪˈdaɪtɪs/ n. inflammation of the sesamoid bones or sesamoidian ligaments caused by injuries and tearing due to excessive stress placed on the fetlock during fast exercise; symptoms include swelling and varying degrees of lameness (剧烈运动时球节受力过大造成籽骨和籽骨韧带受伤发炎,症状表现为患处肿胀和不同程度的跛行)籽骨炎

set[1] /set/ n. the four horseshoes nailed to a horse 一副蹄铁

vt. to attach a horseshoe to the foot; fit 钉蹄,装蹄

set[2] /set/ vt. to cut the tendon and muscles below the tail of horse to achieve a high tail carriage when its heals (割断马尾下的肌腱,愈合后使尾巴高翘)翘尾,垫高尾巴:to set tail 垫高尾式 | **setting** /ˈsetɪŋ/ n.

set-and-turn n. *see* rollback 急转,回转

setfast /ˈsetfɑːst/ n. *see* sitfast 背部硬瘤,坐�散瘤

setter /ˈsetə(r)/ n. a breed of long-haired hunting dog (一种长毛猎犬)塞特犬:red setter 爱尔兰塞特犬

settle /ˈsetl/ vi. **1** (of a mare) to conceive as a result of mating with a stallion (母马)受孕,怀孕 PHR V **settle to service** to become pregnant after mating 配种后受孕 **2**【cutting】(of a rider) to calm a herd of cattle by moving back and forth in front of them prior to cutting (骑手在拦截牛时前后移动以平息牛群)安抚,平息:to settle cattle 平息牛群

settled /ˈsetld/ adj.【cutting】(of cattle) composed and not spooking away when cut from the herd (指牛被拦截后沉着而不惊逃)镇静的,冷静的

set-toe shoe n. *see* rocker-toe shoe 卷头蹄铁

severe /sɪˈvɪə(r); *AmE* -ˈvɪr/ adj. (of a bit) causing great pain or applying lots of pressure (指口衔)严厉的,厉害的:a severe bit 一副厉害的衔铁

severity /sɪˈverəti/ n. the amount of pain inflicted by using a bit 严厉,厉害

sex allowance n.【racing】a weight reduction given to fillies competing in races against males (雌驹与雄驹同场竞技时给予的减重)性别让磅

sex chromosome n.【genetics】the chromosome that determines the sex of a horse, i. e. the X or Y chromosome (决定马匹性别的染色体,即X或Y染色体)性染色体 compare autosome

sexual /ˈsekʃuəl/ adj. of, relating to, involving, or characteristic of sex and sexuality 性的:sexual stimulation 性刺激

SH abbr.【racing】Southern Hemisphere 南半球

Sha Tin Racecourse n. a racecourse built in 1978 on reclaimed land in Hong Kong for horse racing meets (香港于1978年填海建造的赛马场)沙田马场 — *see also* **Happy Valley Racecourse**

shabrack /ˈʃæbræk/ n.【dated】(also **shabraque**) a piece of cloth or leather covering for the saddle and horse (棉质或皮质的)鞍毯,鞍垫

shabraque /ˈʃæbræk/ n. *see* shabrack 鞍毯,鞍垫

shadbelly /ˈʃædˌbeli/ n. *see* tail coat 燕尾服

shadow roll n.【racing】a piece of sheepskin or synthetic material secured around the noseband of a bridle to prevent the horse from seeing the shadows on the ground (缠在马勒鼻革上的一块羊皮或合成材料,可防止马匹看到地面上的阴影)鼻革卷

shaft /ʃɑːft; *AmE* ˈʃæt/ n. **1**【driving】each of the pair of wooden bars between which a horse is harnessed to pull a vehicle; thill (车身的两根长木杆,马套在中间拉车)辕杆 **2**【anatomy】the midsection of a long bone; diaphysis 骨干

shaft horse n.【driving】a horse harnessed between the two shafts of a vehicle; wheeler (套在辕杆中间拉车的马)辕马 syn thiller, thill horse

Shagya Arab n. (also **Hungarian Shagya**) a relatively rare horse breed developed in the 1830s at Austro-Hungarian military stud farms by using desert-bred Arabs as foundation stock and introducing Thoroughbreds and Lipizzaners later to increase its size and improve movement;

stands 15 to 16 hands and generally has a gray coat; has a tall and big frame, long hip, and better riding qualities than a purebred Arab; mainly used for riding and driving (19 世纪 30 年代奥匈帝国军马场采用沙漠型阿拉伯马作为奠基群育成的稀有马品种,后引入纯血马和利比扎马血统以增加体格并改善步法。体高 15~16 掌,毛色多为青,体型高大,臀端长,乘用性能优于纯种阿拉伯马,主要用于骑乘和驾车)沙迦阿拉伯马:The name of Shagya Arab is derived from one foundation sire, Shagya, an Arab horse born in Syria in 1830 and imported to the Austro-Hungarian military stud in 1836 for breeding purposes. 沙迦阿拉伯马的名称源于一匹名为"沙迦"的奠基阿拉伯马公马,该马 1830 年生于叙利亚,并于 1836 年输入奥匈帝国军马场作种用。 syn Hungarian Arabian。

shake /ʃeɪk/ vt. **1** to tremble or shiver 摇摆;战栗 **2** to excite or agitate 使兴奋,使振奋 PHR V **shake up**【racing】(of a rider) to strike a horse with a whip in order to make him run faster (骑手用鞭抽马使之加速)使抖擞精神,使奋力奔驰

shaker foal syndrome n. another term for botulism (肉毒杆菌中毒症的别名)幼驹颤抖综合征

Shan pony n. *see* Burmese 缅甸矮马,掸邦矮马

Shandan n.【Chinese】(also **Shandan horse**) a breed of horse developed in the 1970s at the Shandan Military Stud Farm in Gansu, China by crossing native mares with Yili stallions and later improved with Don; stands about 14 hands and generally has a bay or chestnut coat; has a light head, prominent withers, deep chest, straight back, short loins, sinewy legs with hard hoofs; used for riding, light draft and farm work (20 世纪 70 年代中国甘肃省山丹军马场育成的马品种,主要由当地母马与伊犁马等品种杂交,后经顿河马改良而成。体高约 14 掌,毛色多为骝或栗,头轻,鬐甲突出,胸深,背腰平直,腰短,四肢强健,蹄质结实,主要用于骑乘、轻挽和农田作业)山丹马

shandrydan /ˈʃændrɪdæn/ n. **1** an Irish two-wheeled cart with a hood (爱尔兰的一种带篷双轮马车)带篷马车 **2** any decrepit old-fashioned vehicle 破旧马车

shank /ʃæŋk/ n. (also **shank bone**) the bone of horse's lower hindleg between the hock and fetlock (马后肢飞节至球节之间的长骨)胫骨 syn shannon, hindcannon

shank bone n. *see* shank 胫骨

shannon /ˈʃænən/ n. (also **shannon bone**) the shank bone of a horse's hindleg 胫骨

shape /ʃeɪp/ n. the conformation or condition of a horse (马的)体况 IDM **in good shape** (of a horse) healthy and well-conditioned (指马)体况良好的,健康的

vt.【cutting】to move a cow or a herd towards certain direction 赶牛 PHR V **shape cattle** to drive a herd of cattle in a specific direction 赶拢牛群 **shape a cow** to separate a cow from the herd to the middle of the arena to begin work (将牛从群中赶到场地中央)隔牛拦截

shaps /ˈʃæps/ n. *see* chaps 皮套裤

shaving /ˈʃeɪvɪŋ/ n. a thin slice of wood that is shaved off (木料的)刨花:wood shavings 刨花

shd abbr.【BrE, racing】short-head 短头长

she-ass /ˈʃiːæs/ n. a female donkey; jenny, jennet 母驴

sheath /ʃiːθ/ n. **1** the case for a blade 刀鞘 **2** the outer covering of penis; prepuce 阴鞘,包皮

shed[1] /ʃed/ n. a stall or shelter for horses or cattle 马棚;牛棚

shed[2] /ʃed/ v. (of hairs) to fall off by natural process; molt (毛发)脱落 PHR **to shed coat** to lose its coat due to season change (因季节变化)脱毛,换毛 ◇ The winter coat will begin to shed off in early spring, horses are usually clipped providing the coat is not changing rapidly at this time. 早春季节马身上的冬毛开

S

始脱落,这时要对换毛太慢的马进行被毛修剪。

shedding blade n. a grooming tool used in the spring to help loosen and get rid of the extra hair that the horse may shed off due to the season change (一种刷拭工具,在换季时用来疏松并除去春季马体尚未脱落的多余被毛)毛刮子

sheep knees n. *see* calf knees 凹膝

sheet /ʃiːt/ n. a thin clothing for horses 马衣:fly sheet 驱蝇马衣 ◇ summer sheet 夏季马衣 ◇ sweat sheet 防汗马衣 ◇ exercise sheet 训练马衣

Shelborne Landau n. *see* Shelburne Landau 谢尔本兰道马

Shelburne Landau n. a heavy, four-wheeled, horse-drawn carriage with a square box, two doors, and a double folding hood, seated four vis-à-vis; was usually pulled by a pair of horses (一种车厢呈方形的重型四轮马车,两边双车门,带前后折叠顶篷,可对坐四名乘客,多由双马联驾拉行)谢尔本兰道马车 [syn] Square Landau, Angular Landau

shelly /ˈʃeli/ adj. (of horse's hoofs) small, brittle and thin-soled (马蹄)干裂的:shelly feet 干裂蹄

Sheltie /ˈʃelti/ n. 【colloq】the Shetland pony 设特兰矮马

Shetland Pony n. (also **Sheltie**) a breed of pony originating from the Shetland Isles, north of Scotland; stands below 11.2 hands and may have a paint, brown, chestnut, gray, or black coat; has a well-shaped but small head, broad chest, short back with heavily muscled loins, deep girth, broad quarters, short legs with sharply defined joints, light feathering, and an exceptionally thick mane, forelock, and tail; is hardy, strong, lively, and possessed of a quick, free action; historically to haul and carry peat in mining; now used for riding and light draft (产于英国苏格兰北部设特兰岛的矮马品种。体高低于11.2掌,毛色多为花、褐、栗、青或黑,面貌俊秀,头小,胸宽,背短,腰肌发达,胸较深,后躯宽广,四肢短,关节突出,距毛稀疏,鬃尾和额发浓密,抗逆行强,体格健壮,性情活泼,步法快而顺畅,过去常被煤矿用来运送泥炭,现在主要用于骑乘和轻挽)设特兰矮马:Shetland pony is the most famous and smallest of the British pony breeds. 设特兰矮马是英国矮马品种中最为知名且体型最小的矮马。

Shetland Pony Stud-Book Society n. (abbr. **SPSBS**) an organization founded in Great Britain in 1891 to promote the breeding and maintain the registry of Shetland Ponies (1891年在英国成立的组织机构,旨在促进设特兰矮马的育种和登记工作)设特兰矮马登记委员会

Shift Seat Wagon n. *see* Jump Seat Wagon 弹座马车

shigellosis /ˌʃɪgəˈləʊsɪs/ n. an intestinal infection caused by any of various species of *Shigellae*, occurring most frequently in areas where poor sanitation and malnutrition are prevalent and commonly affecting newborn animals (由多种志贺菌引起的肠道感染,常发生在卫生和营养条件差的地方,新生幼畜易受感染)志贺菌病

shill /ʃɪl/ n. one hired to act as a horse buyer to run up the price; bonnet, puffer (受雇假装买马以抬高价格的人)马托

vi. to act as a shill 当托

vt. to deceive one into a swindle; decoy 诱骗,拐骗

shin /ʃiːn/ n. (also **shin bone**) *see* cannon bone 胫骨,管骨

shinbuck /ˈʃiːnbʌk/ n. *see* bucked shin 胫凸

ship /ʃip/ vt. to transport a horse, as by truck, rail, ship, or plane 运输

ship-of-the-plains n. *see* prairie schooner 草原篷车

shipping boots n. padded boots wrapped around the legs of a horse during shipping to provide protection and support (马运输时所缠的护

腿,可保护四肢避免受伤)运马护腿 [syn] traveling boots

shipping fever n. a severe respiratory disorder often found in young animals following shipping (青年马匹在运输后出现的严重呼吸道疾病)船运热,运输热 [syn] transit fever

shire[1] /ˈʃaɪə/ n. a former administrative division of Great Britain (英国以前的行政区划)郡

Shire[2] /ˈʃaɪə/ n. a heavy draft breed originating from England; stands 16.1 to 17.3 hands or even higher, and may have a bay, brown, black, chestnut or gray coat with white markings; has a small head relative to its body size, a long, arched and muscular neck, a short back, powerful croup, clean, short strong legs with heavy feathering; is docile and good natured; historically used as a carriage horse and for agriculture, is now used for heavy draft and farm work (源于英格兰的重挽马品种。体高 16.1 ~ 17.3 掌或更高,毛色多为骝、褐、黑、栗或青,带白色别征,头相对其体型较小,颈长且发达、呈弓形,背短,后躯发达,四肢粗短,距毛浓密,管围较粗,性情温驯,过去主要用于驾车和役用,现在用于重挽和农田作业)夏尔马:The Shires are generally raised in the Midland regions of Great Britain. 夏尔马多饲养在英国中部地区。

shiver /ˈʃɪvə(r)/ vi. to tremble involuntarily with cold, fear, or excitement (出于寒冷、惊恐或兴奋而颤抖)战栗,颤抖

shivers /ˈʃɪvəz; *AmE* -vərz/ n. a condition in which the muscle in horse's hind legs shivers and flexes (马后肢肌肉颤抖抽搐的症状)后肢战栗

shock /ʃɒk; *AmE* ʃɑːk/ n. a condition in which the vital functioning fails due to circulatory collapse of the body (由血液循环障碍引起的机体主要功能衰退)休克:cardiac shock 心脏休克 ◇ hypovolemic shock 血量减少性休克 ◇ shock condition 休克症状

shoe /ʃuː/ n. a horseshoe 蹄铁 vt. (**shod**, **shod** /ʃɒd; *AmE* ʃɑːd/) to fit horseshoe to the hoof of a horse 钉蹄铁,钉掌;装蹄

shoe board n. **1** a wooden board used to display all kinds of horseshoes in a farrier's workshop (铁匠铺用来陈列各种蹄铁的木板)蹄铁板 **2** a sign that identifies the type of horseshoes each racehorse wears (发布赛马所用蹄铁类型的公告栏)蹄铁公告栏

shoe pad n. *see* horseshoe pad 蹄铁垫

shoeing /ˈʃuːɪŋ/ n. the act or process of fitting horseshoes 钉蹄,装蹄:hot shoeing 热装蹄 ◇ cold shoeing 冷装蹄

shoeing block n. *see* foot stool 装蹄凳

shoeing chaps n. *see* apron 装蹄围裙

shoeing forge n. *see* forge 锻铁炉

shoeing hammer n. *see* driving hammer 钉蹄锤

shoer /ˈʃʊə(r)/ n. a horseshoer; farrier 蹄铁匠,装蹄工

shoot /ʃuːt/ v. to hunt and kill game using a gun (用枪)打猎

shooter /ˈʃuːtə(r)/ n.【cutting】a cow who bolts from the herd and interferes with the performance of a cutter (猛然冲出畜群而干扰拦截的牛)横冲的牛,猛冲的牛

shooting /ˈʃuːtɪŋ/ n. the sport or practice of killing games with a gun (用枪)打猎

short horse n.【racing】a horse inept in long distance races; sprinter 短途赛马

short price n.【racing】a small mutual payoff in pari-mutuel betting 低额赔率,低额奖金

short side n. one of the two short ends of a rectangular arena (长方形赛场两端的底边)短边

short-coupled adj. (of a horse) short in back (指马)腰背短的 [syn] close-coupled

shortfall /ˈʃɔːtfɔːl; *AmE* ˈʃɔːrt-/ n. **1** a failure to attain a specified amount or level; a shortage (没有达到要求的数量或水平)不足,缺少 **2** the amount by which a supply falls short of need or demand 缺少量,不足量:If the horse is fed with hay of poor quality, more concen-

S

trates should be added to make up the protein shortfall of hay. 如果饲喂马匹的干草质量不高,就应当增加精料用量来弥补干草中蛋白质的不足。

short-head n.【BrE, racing】(abbr. **shd**) a winning margin equal to the length of a horse's muzzle (赛马胜出的距离,大致相当于马口鼻长度)短头长 compare dead heat, nose, short-neck, neck, head, length

short-neck /ˈʃɔːtnek; *AmE* ˈʃɔːrt-/ n.【BrE, racing】(abbr. **snk**) a winning margin shorter than the neck but longer than the head (赛马胜出的距离,小于颈长而大于头长)短颈长 compare dead heat, short-head, head, neck, length

shot /ʃɒt; *AmE* ʃɑːt/ n.【polo】an attempt to score by striking the ball 射门

vi.【polo】to attempt to score by striking the ball with the mallet (用球杆击球得分)射门 PHR V **shot on goal** to strike the ball in an attempt to make a goal 射门

shoulder /ˈʃəʊldə(r); *AmE* ˈʃoʊ-/ n. **1**【anatomy】the joint of certain vertebrates that connects the forelimb to the chest (脊椎动物连接前肢与胸腔的关节)肩胛 **2** the body part of an animal near this joint (此关节附近的部位)肩部: point of shoulder 肩端 ◇ straight shoulder 直肩 ◇sloping shoulder 斜肩

shoulder atrophy n. *see* sweeney 肩肌萎缩

shoulder blade n. *see* scapula 肩胛骨

shoulder hang n.【vaulting】an exercise performed by a vaulter who hangs upside down against the shoulder of the horse by holding onto the handles of the vaulting roller (马上体操动作之一,表演者双手握住肚带上的把手倒立于马匹肩部)肩部倒立

shoulder stripe n. *see* wither stripe 鹰斑

shoulder-in /ˈʃəʊldərˌɪn/ n.【dressage】a two-track in which the horse travels forward while bending the length of its spine away from the direction in which he is traveling (一种斜横步法,其中马向前运行时脊柱偏离行进方向)肩[向]内斜横步,肩[向]内行进 compare shoulder-out

vi. (of a horse) to travel such a movement (指马)肩向内行进

shoulder-out /ˈʃəʊldərˌɔːt/ n.【dressage】a two-track in which the horse travels forward while bending the length of its spine towards the direction in which he is traveling (一种斜横步法,马向前运行时脊柱偏向行进方向)肩[向]外斜横步,肩[向]外行进 compare shoulder-in

vi. (of a horse) to travel such a movement (指马)肩向外行进

show[1] /ʃəʊ/ vi.【racing】(of a horse) to finish third in a race (赛马)获得第三名,位列第三 compare win, place

show[2] /ʃəʊ/ n. a horse show 马匹展览赛,马匹选美

vi. to present or compete in a horse show 参展

show buggy n. an extremely light, four-wheeled, horse-drawn American buggy with a single seat used for showing trotting horse or ponies (一种四轮单座美式轻型马车,主要用于快步马或矮马的展览赛)展览赛马车

show horse n. a horse that competes in a show 选美马匹

show jumper n.【jumping】a horse competing in the event of show jumping 障碍马: Many top show jumpers are tall warmblood types bred specifically for the sport. 许多顶级障碍马都是体格高大的温血马,且专为此项运动而培育。

show jumping n. (also **jumping**) a discipline of the Olympic equestrian sports in which the riding horse is required to clear a succession of obstacles in order (奥运会马术运动项目之一,选手骑马须按规定依次跨越多道障碍)[场地]障碍赛

show-ring n. an arena or ring used for horse show 马匹展览圈,马匹展览场

SHPF abbr. Société Hippique Percheronne de

France 法国佩尔什马协会

shred /ʃred/ n. (*pl.* **shreds**) chopped or smashed feed of animals 铡料,碎料

vt. to cut or tear into shreds 切碎,撕碎

shut /ʃʌt/ vt. to close a gate, etc 关闭;阻碍 PHR V **shut off** 【racing】(of a jockey) to cross in front of another rider during a race (骑手)切道,阻挡 **shut out** 【racing】 to close the betting window and stop accepting wager of a bettor on a race (赌马窗口关闭后停止投注) 关窗,停投 IDM **shut the stable door after the horse has bolted** to lock the stable door after the horse is stolen 亡马锁厩,为时已晚

shy /ʃaɪ/ vi. to startle or jump aside in fright 惊跳,闪躲 PHR V **shy at/from sth**; **shy away** (of a horse) to jump sideways when startled by a real or imaginary object (指马)惊跳,闪躲

shy feeder n. a horse or hound who has a poor appetite 食欲不振的马;胃口差的猎犬

Sicilian /sɪˈsɪljən/ n. an Anglo-Arab horse bred on the Island of Sicily, Italy; stands 14.3 to 15.2 hands and may have a bay, black, chestnut or gray coat; has long legs, a short back, prominent withers, and small ears; is spirited and possessed of good stamina; used for riding and light draft (西西里岛上繁育的盎格鲁阿拉伯马,体高14.3~15.2 掌,毛色多为骝、黑、栗或青,四肢修长,背短,鬐甲突出,耳小,性情活泼,耐力好,主要用于骑乘和轻挽)西西里马

Sicilian donkey n. (also **Miniature Sicilian donkey**) a rare miniature donkey breed native to the Mediterranean island of Sicily; stands 9 to 10 hands and usually has a chestnut or gray coat with dorsal and wither stripes; is intelligent and affectionate (源于地中海西西里岛上的微型驴品种。体高9~10 掌,毛色多为栗或青,带背线和鹰斑,反应灵敏,性情温驯)西西里微型驴

sick /sɪk/ adj. suffering from or affected with illness; unhealthy 患病的;体弱的: sick horses 患病马匹

sickle hocks n. (also **saber hocks**) a conformation defect of the horse's lower hind legs in which the angle of the hock between the gaskin and cannon is too acute, thus giving the shape of a sickle when viewed sideways (马的一种后肢缺陷,胫部与管部形成的角度过小,而使飞节侧观时似呈镰刀状)镰状飞节 syn curby hocks, excessive angulation of the hock joints, sabre hocks

sickled /ˈsɪkld/ adj. sickle-hocked (飞节)镰状的: sickled hocks 镰状飞节

sickle-hocked /ˌsɪklˈhɒkt/ adj. (also **sickle-legged**) (of a horse) having the conformation defect of sickle hocks (飞节)镰状的 syn saber-legged

sickle-legged /ˌsɪklˈlegɪd/ adj. *see* sickle-hocked (飞节)镰状的

side pass n. 【dressage】 a horizontal movement of the horse without forward or rearward movement (马沿水平面而非前后的运动)横向行进,侧向运步 compare half pass

side puller n. a horse who pulls harder on one side of the bit than the other 侧拽的马

side reins n. a pair of training reins attached on one end to the bit rings and to the girth or saddle rings on the other to teach the horse head carriage in lungeing (一端连衔铁环上,另一端连在肚带或鞍环的辅助绳,长缰训练中用来调教马匹头姿)侧缰

side saddler n. one who rides sidesaddle 侧鞍骑者

side step n. (also **sidestep**) *see* traverse 横步

side wheeler n. *see* pacer 对侧步马

side wheeling n. the side-to-side rolling movement of a pacer (对侧步马行进中的)侧摆

sidebone /ˈsaɪdbəʊn/ n. 【vet】 (usu. **sidebones**) ossification of the lateral cartilage on either side of the coffin bone; symptoms may include heat and the appearance of hard lumps on the coronet (蹄骨两侧软骨硬化的症状,表现为患处

S

发热及蹄冠附生硬块)蹄骨骨瘤,赘骨

sideline /ˈsaɪdlaɪn/ n.【polo】(also **side line**) one of two marking lines along the two sides of a polo field that define the polo area in conjunction with the endlines (马球赛场两边的界线,与底线共同确定球场边界)边线 compare backline /endline

sidepiece /ˈsaɪdpiːs/ n. *see* cheekpiece 颊革

sides /saɪdz/ n. the ribs or chest of horse 肋部,胸侧

sidesaddle /ˈsaɪdˌsædl/ n. (also **side-saddle**) a saddle formerly designed for women to ride with both legs on the near side of the horse (过去为女性骑乘设计的马鞍,双腿同时放在马的左侧)侧鞍 syn lady's saddle 淑女鞍

adv. with both legs on the same side of the horse 侧鞍骑:to ride a horse side-saddle 侧鞍骑马,侧骑

sifting /ˈsɪftɪŋ/ n. a method of separating the top performers from a group of horses in large show classes (马匹展览赛中从群中选出优胜马匹的方法)筛选,海选

sight /saɪt/ n. **1** the ability or faculty of seeing 视力:to hunt by sight 靠视觉捕猎 **2** the field within which one can see 视野:The deer ran out of his sight. 鹿跑出了他的视野范围。

sight hound n. a hound who pursues the quarry by sight 靠视觉追捕的猎犬

Siglavy /ˈsɪgləvi/ n. one of the three primary subbreeds of the Assil (阿西利阿拉伯马的三个亚种之一)西格拉维阿拉伯马 — *see aslo* **Kuhailan, Muniqi**

sign /saɪn/ n. any indication of a disease 病征

silage /ˈsaɪlɪdʒ/ n. a brown colored fodder prepared by storing and fermenting moist green forage plants such as grasses and legumes under anaerobic conditions (青草和苜蓿等绿色饲草在无氧条件下发酵制成的褐色饲料)青贮料:bag silage 袋装青贮料 ◇ corn silage 青贮玉米

vt. to store and ferment green fodder in a silo; ensile, ensilage 青贮

silent /ˈsaɪlənt/ adj. (of estrus) without marked signs; quiet (发情)不明显的;安静的

silent estrus n. *see* silent heat 静息发情

silent heat n. (also **silent estrus**) a condition in which the mare undergoes physiological estrus and ovulates but shows no behavioral signs (指处于生理发情期的母马虽排卵但不表现发情行为的情况)静息发情

silo /ˈsaɪləʊ; *AmE* -loʊ/ n. a tall cylindrical structure used for storing and fermenting fodder (用来贮藏和发酵饲料的柱状容器)青贮窖,青贮塔:vinyl-vaccum silo 聚乙烯真空青贮窖

silver buckskin n.【color】a creamy yellow coat with black points (被毛乳黄而末梢为黑色的毛色)银沙黄,银鹿皮

silver dapple n.【color】a dark brown coat with light dapples (棕色被毛中带淡色斑点)斑褐毛

silver dun n.【color】a sandy coat with primitive marks and black points (被毛沙黄色而末梢为黑色)银兔褐

silver grullo n.【color】a dun coat with slate blue points(被毛兔褐而末梢为石板青的毛色)淡青灰

simple change n. (also **simple lead change**) the change of a horse's leading leg when cantering (马在跑步中更换领先肢)简单换腿 syn simple change of leg at the canter

simple change of leg at the canter n. *see* simple change 简单换腿

simple dismount n.【vaulting】a compulsory vaulting exercise in which the performer swings down from the horse's back (马上体操的规定动作,其中表演者绕过马背下马)简单下马;下马动作

simple fracture n. a bone fracture along a single line that causes little or no damage to the surrounding soft tissues (一种单道开裂的骨折,不对周围软组织造成损伤)单纯骨折

simple lead change n. *see* simple change 简单换步

simulcast /ˈsɪmlkɑːst; *AmE* -kæst/ vt. to broadcast an equestrian event or horse race simultaneously on radio or television (通过广播或电视即时直播马术项目或赛马比赛)同步直播:The Grand National Steeplechase has been simulcast on television in the United Kingdom since 1960. 全英越野障碍赛自1960年起就开始电视同步直播。

n. a broadcast so transmitted 同步直播

sinew /ˈsɪnjuː/ n. **1** a tendon or ligament 筋腱 **2** vigorous strength; muscular power 精力,体力

sinewy /ˈsɪnjuːi/ adj. **1** consisting of or having many sinews 肌腱的;多腱的 **2** muscular and strong 强健有力的

singe /sɪndʒ/ vt. to burn off the ends of long hair by subjecting briefly to flame (快速掠过火焰烧掉长毛的末梢)燎,烧掉:Before electric clippers were available, horsemen would clip their animals with razors, singe the long hairs with a candle or gas lamp, or use hand-cranked clippers. 在没有电动剪刀之前,养马者会用剃刀来进行修剪,或者用烛焰、油灯来燎掉长毛,还有的采用手动型剪刀。

single bank n.【jumping】a jumping obstacle consisting of a bank with a ditch on one side 单侧水沟跨堤

single foot n. *see* rack 单蹄快步

singleton /ˈsɪŋgltən/ n. an animal born singly or alone 单胎,独生:Twins born alive full term are usually below the normal singleton size. 妊娠期满生下来的双胞胎通常比单胎个体要小。

singletree /ˈsɪŋgltriː/ n. the pivoted horizontal crossbar to which the harness traces of a draft animal are attached and which is in turn attached to a vehicle or an implement such as a plough (前端连接挽马套绳,后端挂接在马车或犁等农具上的枢纽横杆)横杠,挂杆 [syn] splinter bar, swing bar, swingletree, whiffletree, whippletree

Sini horse n. *see* Xinihe horse 锡尼河马

sinister /ˈsɪnɪstə(r)/ adj.【Latin】left 左的

sinistra /ˈsɪnɪstrə/ adj.【Latin】sinister 左的

sinker /ˈsɪŋkə(r)/ n.【vet】a severe case of founder in which much of the laminae became degenerated and the bone begins to sink into the foot (蹄叶大部分腐烂导致蹄骨内陷的情况)蹄骨内陷

sinking fox n.【hunting】a fox who tires after a hard hunt (猛追后体乏的狐狸)力衰的狐狸,疲惫的狐狸

sinoatrial /ˌsaɪnəʊˈeɪtrɪəl/ adj. of or relating to the sinoatrial node 窦房结的

sinoatrial node n.【anatomy】(abbr. **S-A node**, or **SAN**) a specialized cardiac tissue near the conjunction of superior vena cave and the right atrium that initiates the heart beat (心脏位于前腔静脉与右心房结合处的特化组织,是心脏的起搏点)窦房结

sinuatrial /ˌsaɪnjʊˈeɪtrɪəl/ adj.【anatomy】sinoatrial 窦房的

sinuatrial nodus n. sinoatrial node 窦房结

sinus /ˈsaɪnəs/ n.【anatomy】a dilated channel containing chiefly venous blood or air-filled cavity in the bones (充盈静脉血的膨大部位或骨中充气的空腔)窦:accessory sinus 副鼻窦,鼻旁窦 ◇ aortic sinus 主动脉窦 ◇ carotid sinus 颈动脉窦 ◇ frontal sinus 额窦

sire /ˈsaɪə/ n. the male parent of a foal; stallion 父本;种公马:a national sire 国有种公马 [compare] dam

sire stakes n.【racing】a stakes race restricted to the progeny of stallions standing at stud in a certain region (一种仅限特定地区内种公马的后代参加的奖金赛)父系奖金赛:the New York sire stakes 纽约州父系奖金赛

sister /ˈsɪstər/ n. a relationship of female horses out of the same mare (同匹母马生的雌驹之间的亲缘关系)姐妹 — *see also* **full sister**, **half sister**

S

sitfast /ˈsɪtfɑːst/ n.（also **setfast**）a hard and painful swelling on the back of a riding horse; gall（乘用马背部出现的硬质肿块）硬瘤，鞍伤

sit-still vi.【racing】（of a jockey）to lose a race due to disuse of the whip（骑师因不用马鞭）坐失比赛

Six Steeds of Zhao Mausoleum n. the six war horses Emperor Taizong（599 – 649）of the Tang dynasty rode in his life time, whose sculptures appear on the stone relief of his tomb Zhao Mausoleum situated in Liquan County of Shaanxi province, China（唐太宗李世民生前所骑的六匹战马，其浮雕见于中国陕西省礼泉县昭陵甬道的石壁上）昭陵六骏

sixteenth pole n.【racing】a distance-marking post located on the infield rail exactly 100 meters from the finish line（立于赛道内栏的标杆，与终点相距 100 米）1/16 杆，百米杆

size /saɪz/ n. the physical dimensions of a horse（马的）体型大小：a horse of medium size 一匹体型大小适中的马

skate /skeɪt/ n.【slang】a horse of poor condition or quality 驽马，劣马

skeletal /ˈskelɪtl/ adj. of, relating to, or forming skeleton 骨骼的：skeletal system 骨骼系统

skeletal muscle n.【anatomy】a type of muscle that is attached to the bones by tendons and contracts on stimuli from nerve impulses to exert forces on joints（通过肌腱与骨骼相连的肌肉，在接受神经冲动收缩后可施力于关节）骨骼肌 syn striated muscle, voluntary muscle

skeleton /ˈskelɪtn/ n. the bony framework that protects and supports the soft organs, tissues and other body parts of the body（保护和支持身体软组织、器官和其他部分的骨结构）骨架，骨骼：skeleton of horse 马体骨骼

skeleton bridle n. a harness bridle without blinkers 不带眼罩的挽用马勒

skep /skep/ n.（also **skip**）an open, round container used to collect droppings of horses in a stable（厩房用来清理马粪的圆形敞口容器）清粪桶，拾粪桶：rubber skep 橡胶拾粪桶

vt. to collect droppings form the stall with a skep 清粪 PHR V **skep out** muck out with a skep 清粪：To skep out properly the groom should not take out any clean bedding unnecessarily. 在用粪桶清理厩舍内的马粪时，马夫无须将干净的垫草也带走。

skewbald /ˈskjuːbɔːld/ adj.（of a horse）having patches of chestnut or brown other than black on a body color of white（在马的白色被毛上有栗色或褐色花斑）杂色的，花色的

n. a horse having such a coat color pattern 杂色花马：Both skewbald and piebald are often referred to as "coloured" horses. 杂色花马和黑白花马又经常统称为花马。 compare piebald

skiboy /ˈskiːbɔi/ n. one who competes in a skijoring event 马拉滑雪选手

skid /skɪd/ v. **1**（of a vehicle）to slide sideways（车辆）侧滑，打滑 **2** to fasten a skid to a wheel as a brake 安止滑器

n. **1** the movement of a vehicle that slides sideways in an uncontrolled way 打滑，侧滑 **2**（also **skid-shoe**）a curved iron plate placed under wheels of a heavy horse-drawn vehicle before descending a slope（重型马车下坡时，置于车轮下的弧形铁板）防滑板，止轮器：skid chain 防滑板固定链

skid shoe n. *see* skid 防滑板，止轮器

skidproof /ˈskɪdˌpruːf/ adj.（of a surface）resisting slide or skid; non-slippery 防滑的：The floor in the aisle must be skidproof to prevent slipping and easy to sweep clean. 厩舍走廊的地面必须防滑以防马滑倒，而且要易于清理。

skijoring /ˈskiːdʒɔːrɪŋ, skiːˈdʒɔː-/ n. a winter sport in which a skier is pulled over snow by seizing a long rope attached to the saddle of a horse（冬季运动项目，其中滑雪者由系在马鞍上的长绳牵拉滑行）马拉滑雪：Skijoring is derived from the Finnish sport of the same name in which competitors ski behind reindeer or dogs. 马拉滑雪源于芬兰的同名运动,那里的滑雪者由驯鹿或雪橇犬牵拉。

skim /skim/ vt. to remove the floating fat from milk, etc.(从乳中除去悬浮的脂肪)脱脂

skimmed /skɪmd/ adj.(of milk) with fat removed(乳)脱脂的:skimmed milk 脱脂奶,脱脂乳

skin[1] /skɪn/ n. **1** a thin layer of membranous tissue consisting of epidermis and dermis and forming the outer covering of the body(体表的膜状组织层,由表皮和真皮构成)皮肤:The skin may constitute 12 to 24 percent of the horse's total body weight depending on age. 根据不同的年龄,马的皮肤可占到其体重的12% ~24%。**2** the pelt of an animal 毛皮:calf skin 小牛皮 ◇ goat skin 山羊皮

vi. to remove skin from an animal; peel off 剥皮 [saying] There is more than one way to skin a cat. <谚>凡事都不止一种做法。

skin[2] /skɪn/ vt. **1**【racing】to harden the surface of a racetrack by rolling(用辊子将赛道)压整,压实 **2**【slang】to take money from one by cheating; swindle 诈骗,诈取

skirt[1] /skɜːt; *AmE* skɜːrt/ n. **1** saddle skirt 鞍裙 **2** apron 围裙

skirt[2] /skɜːt; *AmE* skɜːrt/ vi.【hunting】(of a hound) to cut corners when pursing the scent of the fox(猎犬追踪狐狸时)抄近道

skirter /ˈskɜːtə(r); *AmE* ˈskɜːrt-/ n.【hunting】a hound who cuts corners in pursuing the scent of a fox(猎狐中)抄近道的猎犬

skittish /ˈskɪtɪʃ/ adj.(of a horse) easily frightened, nervous or shy(指马)易受惊的,紧张的

skive /skaɪv/ vt. to cut prepared hide into thin layers; pare(将皮革)剥层,分层 | **skiving** /ˈskaɪvɪŋ/ n.

Skyros /ˈskaɪrəs/ n.(also **Skyros pony**) an ancient pony breed indigenous to the Greek island of Skyros from which the name derived; stands 9.1 to 11 hands and may have a bay, brown, dun, or gray coat; has a small head, short back, poorly developed croup and chest, and slender legs; is quiet and friendly; used principally for packing and riding(源于希腊斯基罗斯岛的矮马品种并由此得名。体高9.1 ~11 掌,被毛为骝、褐、兔褐或青,头小且背短,尻部和胸部肌肉不太发达,性情温驯而友好,主要用于驮运和骑乘)斯基罗斯矮马:The Skyros pony is believed to be descended from horses brought to the island of Skyros during the 5th to 8th centuries BC by Athenian colonists. 斯基罗斯矮马据信于公元前5 ~8世纪由雅典人带到斯基罗斯岛上的马匹繁衍而来。

SL valve abbr.【anatomy】the semilunar valve 半月瓣

slab fracture n. a fracture of bone in a joint that extends from one articular surface to the other(从一个骨关节延伸至另一个关节面的骨折)板层骨折

slab-sided /ˌslæbˈsaɪdid/ adj.(a horse) having a narrow ribcage(指马)胸腔狭窄的

slack[1] /slæk/ adj.(of rope or reins) not tight or taut; loose; relaxed or flaccid(绳索或套绳)不结实的;松弛的 IDM **be slack in loins**(of a horse) having weak loins(指马)腰部乏力的 PHR V **pull slack**【roping】(of a roper) to take hold of the lariat with the hand after the loop has settled on the target so as to pull the loop tight(在套索套住牛颈后)拽紧套绳 opp taut

slack[2] /slæk/ n. **1**【BrE】a depression between hills in a hillside or on the surface of ground 山凹;峡谷 **2** a boggy or wet area 沼泽,洼地

slate grullo n. *see* grullo 石板青

slat-sided /ˈslætˈsaidid/ adj.(of a horse) flat-sided(指马)狭胸的

slaughter /ˈslɔːtə(r)/ n. the killing of animals for meat 屠宰

S

vt. to kill animals for meat; butcher 屠宰,屠杀

slaughterhouse /ˈslɔːtərˌhaus/ n. a place where farm animals are butchered for meat; abattoir (屠宰家畜的场所)屠宰场

SLE abbr. systemic lupus erythematosus 全身性红斑狼疮

sleeper /ˈsliːpə(r)/ n.【racing】an underrated horse who wins in a race unexpectedly 出人意料的马,一鸣惊人的马

sleeping sickness n. another term for encephalomyelitis (脑脊髓炎的别名)昏睡症

sleepy /ˈsliːpi/ adj. ready for or needing sleep; sluggish 困乏的 | **sleepiness** n.

slice /slaɪs/ n. a thin piece of sth 片;条

vt.【rodeo】(of a horse) to turn a barrel too close (指马转弯时离桶过近)削桶,蹭桶

slick[1] /slɪk/ adj. (of coat) smooth and glossy; sleek (被毛)光滑的,光亮的

vt. to make smooth, glossy or neat; trim or tidy 擦光,擦亮 PHR V **slick up** to make smooth and neat by cleaning and trimming 刷拭光亮,打扮:Most owners slick up their horses before up a show. 很多马主在展览赛之前都会将马打扮一番。

slick[2] /slɪk/ n. an unbranded range animal; maverick 没有烙印的牲畜

slicker /ˈslɪkə(r)/ n.【AmE】a long waterproof coat made of oilskin; a raincoat 雨衣

slide[1] /slaɪd/ n. **1** an image on a transparent film for projection on a screen (在屏幕上投影用的透明胶片)幻灯片 **2** a small glass plate for placing specimens to be examined under a microscope (放置样本用于显微镜观察的小块玻片)载玻片

slide[2] /slaɪd/ vi. to move smoothly over a surface 滑动:slide to a stop 滑停下来

sliding stop n.【rodeo】an abrupt stop performed by a galloping horse who slides to a stop on his hind feet (奔马以后肢蹬地急速滑动停下)滑停:The horse needs to melt into the ground when performing a sliding stop. 马在完成滑停时,需要后肢跐入地面方能完成。

slime /slaɪm/ n. a thick, sticky substance, as accumulated inside the water receptacles (饮水设备内壁沉积的黏稠物)黏污,污垢:Buckets and automatic drinking bowels will need wiping out with a sponge or cloth regularly to ensure that any built-up slime is removed. 水桶和自动引水设备要定期用海绵或抹布彻底进行清洁,以除去沉积的污垢。

slip[1] /slɪp/ vt. (of animals) to bring forth young prematurely; abort, miscarry (动物)早产;流产

slip[2] /slɪp/ n.【racing】a betting ticket 彩票

slippery /ˈslɪpəri/ adj. causing or tending to cause sliding or slipping 易滑的;打滑的

slip-straw n. a thin layer of straw used to cover the floor after the stall was mucked out (清扫厩房后铺在地上的薄层垫草)防滑垫草

slit /slɪt/ n. a narrow cut or opening 切口,狭缝

vt. to make a slit; cut 切开,割开 PHR V **slit the nostrils** to cut the nostrils of a horse wide to make it breathe more freely (将马匹鼻孔割大使其呼吸顺畅)割宽鼻孔:Dating to ancient Egypt around 1350 BC, the method of slitting the nostrils is still practiced even today. 割宽鼻孔最早可追溯到公元前1350年的古埃及,这种做法甚至今天仍有人效仿。

Sloan, Tod n. (1874 – 1933) an American jockey who popularized the crouched-style of flat racing in which the rider sits well over the shoulders of the horse with very short stirrup leathers (推广了猴蹲式策骑姿势的美国骑手,骑手坐于赛马肩部,镫革调得很短)泰德·斯隆

slop /slɒp; *AmE* slɑːp/ n. waste food used to feed pigs or other animals; swill (用来喂猪或其他动物的残羹)泔水

sloping shoulders n. (also **oblique shoulders**) a conformation defect of a horse whose shoulders slope sharply from the withers to the point of the shoulder (马肩胛从鬐甲至肩端过于倾斜的

体格缺陷)斜肩 compare straight shoulders

sloppy /ˈslɒpi; *AmE* ˈslɑːpi/ adj. containing too much liquid; watery 稀薄的;积水的

sloppy track n.【racing】a racing track covered with puddles but not muddy (有水坑但不泥泞的赛道)水洼赛道,积水赛道

slot /slɒt; *AmE* slɑːt/ n. **1**【racing】a post position 闸位 **2** the track or trail of a deer 鹿的蹄迹

slow /sləʊ/ adj. moving at low speed 缓慢的 vt. to make slow in speed 减速,缓速 PHR V **slow down** (of a horse) to decrease in speed; decelerate (指马)减速,放慢

slow pace n. *see* stepping pace 慢对侧步,破对侧步

slow track n.【racing】a slightly wet track in which the speed of horses is to be slowed (由于潮湿而降低赛马速度的赛道)慢赛道

slug /slʌg/ n. a slow, lazy horse 行动懒散的马,动作迟缓的马

small intestine n.【anatomy】the section of a horse's alimentary tract including the duodenum, jejunum, and ileum (马匹消化道包括十二指肠、空肠和回肠的部分)小肠 compare large intestine

small metacarpals n. *see* splint bone 小掌骨,赘骨

small metacarpal bones n. *see* splint bone 小掌骨,赘骨

small pinworm n. a type of pinworm found in the large intestine of horse (寄居在马大肠中的蛲虫)小蛲虫 syn *Probstmayria vivipara*

small red worms n. *see* small strongyles 小圆线虫

small station wagon n. *see* Brougham 小型驿站马车;布鲁厄姆马车

small strongyles n.【vet】one type of parasite of the genus *Triodontophorus* that infests the cecum or colon and often causes severe damage to a horse's gut, thus resulting in weight loss and diarrhea (寄生在马的盲肠或结肠的三齿属寄生虫,可导致严重的肠道损伤,从而造成马体重下降和腹泻)小圆线虫

smart money n.【racing】an insider's bet 知情押注,精明投注

smegma /ˈsmegmə/ n. a sebaceous secretion that collects under the prepuce or around the clitoris (蓄积在包皮之下或阴蒂窝内的皮脂分泌物)阴垢;包皮垢 | **smegmatic** /ˌsmegˈmætɪk/ adj.

smithery /ˈsmɪθəri/ n. the craft or occupation of a smith or farrier 铁匠行当;蹄匠行业

smithy /ˈsmɪði; *AmE* ˈsmɪθi/ n. **1** a farrier 蹄铁匠 **2** the workshop of a farrier or blacksmith 铁匠铺 syn smithery

smock /smɒk; *AmE* smɑːk/ n. a loose-fitting outer garment worn to protect the clothes while working; overalls (工作时穿在外面起保护作用的宽松外衣)工作服,罩衣

smoky black n. the coat color of black with the body color lighter than the points (体色较末梢颜色浅的黑毛色)灰黑,淡黑

smoky eye n. *see* wall eye 玻璃眼,玉石眼

smoky gray n. a coat color of horse with black skin and light shades of gray hair all over the body (肤色为黑、周身被毛呈淡青色的毛色)淡青色,青灰色

smooth /smuːð/ adj. *see* unshod 未钉蹄铁的,未挂掌的

smooth muscle n. any unstriated muscle that is controlled by the autonomic nervous system and found in the walls of the internal organs, such as the stomach, intestine, bladder, and blood vessels (受自主神经系统调控的无纹肌,多见于胃、肠道、膀胱和血管等内脏器官)平滑肌 syn involuntary muscle

smooth-mouthed adj. (of the teeth) worn smooth due to long-time use 牙口磨光的,牙磨平的

smutty /ˈsmʌti/ adj. dirty or smeared 泥灰的,污浊的

smutty buckskin n. a yellow coat of buckskin mixed with black hairs (兔褐毛中混生黑色毛

S

发的毛色）灰沙黄，暗兔褐 syn dark buckskin

smutty palomino n. a coat color of palomino evenly mixed with black hairs, thus giving a dirty appearance（淡黄色被毛中混有黑色毛发，看似呈土灰色）土黄色，灰银鬃 syn sooty palomino

snaffle /ˈsnæfl/ n. *see* snaffle bit 小衔铁，斯奈夫：Snaffle is the simplest and most commonly bit for horses. 小衔铁是马用口衔中最简单也最常用的。◇ jointed snaffle 单关节衔铁 ◇ full-cheek snaffle 全口衔铁 ◇ half-cheek snaffle 半口衔铁 ◇ D-ring race snaffle D 环赛用衔铁

snaffle bit n.（also **snaffle**）a kind of bit consisting of jointed bars with a ring on each end to which one pair of reins is attached（由单关节衔杆与两侧穿缰的圆环构成的衔铁）小衔铁，斯奈夫

snaffle bridle n. a bridle used in conjunction with a snaffle bit 普通马勒，水勒

sniff /snɪf/ v. to smell in a short, audible breath 嗅，闻

snip[1] /snɪp/ vt. to cut or clip with short, quick strokes 剪开，剪断

snip[2] /snɪp/ n. a white mark over the horse's muzzle between the nostrils（位于马口鼻上的白斑）鼻端白— *see also* **bald face**, **blaze**, **snip**, **stripe**, **white lip**

snk abbr.【racing】short-neck 短颈长

snort /snɔːt; *AmE* snɔːrt/ vi.（of a horse）to force air out through the nostrils and make a loud, fluttering noise（马从鼻孔猛然喷气而发出响声）打响鼻

n. the act or the vibrating sound of snorting 打响鼻；响鼻声

snorter /ˈsnɔːtə(r); *AmE* snɔːrt-/ n. an excitable horse 悍性强的马

snotty /ˈsnɒti; *AmE* ˈsnɑːti/ adj. covered with nasal mucus; runny 流鼻涕的：snotty nose 流鼻涕

snowball hammer n.【farriery】a tool used by a farrier to break up and remove packed snow and ice from the underside of a horse's hoofs（用来刨除蹄底冰雪的工具）除雪锤

snowball pad n. a pad inserted between the hoof and the horseshoe to prevent snow balling on the hoofs（置于蹄铁和马蹄间的衬垫，用来防止马蹄刨雪）防雪垫

snowflake /ˈsnəʊfleɪk; *AmE* ˈsnoʊ-/ n.【color】one of the color patterns of the Appaloosa breed with small white spots on a dark base coat（阿帕卢萨马的毛色类型之一，其深色被毛上散布有许多白色斑点）雪花斑 compare blanket, frost, leopard, marble

snub /snʌb/ vt. to tie a horse to a post to restrict its movement 保定；拴马：The groom snubbed the disobedient colt to the post. 马夫将这匹桀骜不驯的雄驹拴在桩上。

soak /səʊk/ vt. to immense in liquid to make sth thoroughly wet; saturate（将某物放入溶液使之湿透）浸湿，浸透

soapwort /ˈsəʊpwɜːt/ n. a perennial Eurasian herb having dense clusters of pink to whitish flowers（一种多年生欧亚草本植物，花簇生，花色介于粉红色至白色之间）肥皂草 syn bouncing Bet

Sociable Barouche n.（also **Barouche**）a light, four-wheeled passenger carriage popular in the 19th century with seats for four passengers facing each other（流行于 19 世纪的轻型四轮客用马车，可对坐四名乘客）对坐马车，巴洛齐马车

Société Hippique Percheronne de France n.【French】（abbr. **SHPF**）an organization founded in 1883 in France to maintain the stud book for the Percheron（1883 年在法国成立的组织机构，旨在对佩尔什马进行品种登记）法国佩尔什马协会

sock /sɒk; *AmE* sɑːk/ n. a leg marking which extends from the coronet to the middle of the cannon bone（马四肢别征之一，其白色区域

由蹄冠延伸至管骨中段)半管白 [syn] half cannon

soft /sɒft; *AmE* sɔːft/ adj. **1** (of a horse) getting fatigued easily (指马)体乏的,体虚的 **2**【roping】(of a cow) getting tired easily and having little play (指牛)易疲劳的,体乏的

soft bone n. *see* cartilage 软骨

soft mouth n. the mouth of a horse who requires the least bit and rein action to achieve the desired response (以口衔和缰绳轻微施力后,马就对指令作出反应)口角灵敏 [compare] hard mouth

soft palate n.【anatomy】the muscular tissue at the posterior part of the roof of the mouth, separating the nasal and oral cavities in conjunction with the hard palate (口腔顶部后缘的肌性组织,与硬腭相连而将鼻腔和口腔隔开)软腭 [compare] hard palate

soft track n.【racing】a racetrack with high moisture content into which the hoofs sink easily (水分过多致使马蹄容易下陷的赛道)软赛道

soften /ˈsɒfn; *AmE* ˈsɔːfn/ v. to make or become less hard and tight 变软;软化 PHR **softening of the bones** the expansion of a mare's pelvic girdle during foaling (母马产驹时骨盆带扩张)盆骨松弛

soft-mouthed /ˌsɔːftˈmauðd/ adj. **1** (of a horse) responsive to the bit and rein action (指马)口角灵敏的 **2** (of a hound) having a low voice (指猎犬)叫声低的:a soft-mouthed hound 叫声低的猎犬

soil[1] /sɔɪl/ n. the top layer of the earth's surface, consisting of rock and mineral particles mixed with organic matter (地壳的最外层,由岩石、矿物质和有机质构成)土壤: sandy soil 沙土 ◇ clay soil 黏土 ◇ soil nutrient testing 土壤养分测定

soil[2] /sɔɪl/ vt. to feed cattle on fresh-cut fodder to fatten them (用优质青草喂牛使之长膘)养膘,育肥

Sokolsky /səˈkɒlski/ n. a Polish warmblood developed in the late 19th century with heavy influence of Belgian Heavy Draft, Ardennes, and Norfolk blood; stands 15 to 16 hands and usually has a chestnut, bay, or gray coat; has a heavy head, broad neck, short and straight back, strong legs and large feet; used mainly for heavy draft and farm work (波兰在 19 世纪末育成的温血马品种,受比利时重挽马、阿尔登马和诺福克马影响较大。体高 15 ~ 16 掌,毛色多为栗、骝或青,头重颈宽,背直而短,四肢强健,蹄大,主要用于重挽和农田作业)索科尔斯基马

solar /ˈsəʊlə(r); *AmE* ˈsoʊ-/ adj. of or situated under the horse's hoof 蹄底的

sole /səʊl; *AmE* soʊl/ n. the bottom of a horse's hoof 蹄底

soleus /ˈsəʊlɪəs; *AmE* ˈsoʊ-/ n.【anatomy】(*pl.* **solei** /-laɪ/) a broad, flat muscle of the calf of the leg, situated under the gastrocnemius (小腿位于腓肠肌下的扁平状肌肉)比目鱼肌

solid color n.【color】a coat consisting of only one color without white markings (被毛仅由一种颜色构成且无白色别征)单毛色 [compare] compound color

soluble /ˈsɒljəbl; *AmE* ˈsɑːl-/ adj. (of a substance) able to be dissolved in a liquid (指物质)可溶的:Glucose is soluble in water. 葡萄糖可溶于水。[opp] insoluble

soma /ˈsəʊmə; *AmE* ˈsoʊ-/ n. **1** the entire body of an organism 体,躯体 **2** the body of nerve cell (神经细胞的)胞体

Somali wild ass n. a subspecies of the African wild ass found in Somali, Eritrea, and Ethiopia, having distinctive zebra stripes on their legs (非洲野驴的亚种,分布于索马里、厄立特里亚与埃塞俄比亚,四肢带有明显的斑马纹)索马里野驴:The Somali wild ass is critically endangered and there are likely less than 1,000 animals in the wild. 索马里野驴属极度濒危动物,野外存活数量可能不到 1 000 头。

S

somatic /səʊˈmætɪk/ adj. of or relating to the body, exclusive of the germ cells 身体的：somatic cell 体细胞

sooty /ˈsʊti/ adj. dirty or smeared; smutty 污色的，污浊的

sooty Palomino n. *see* smutty palomino 土黄色，灰银鬃

sophomore /ˈsɒfəmɔː(r); *AmE* ˈsɑːf-/ n.【racing】a horse in its second year of race 出赛两年的马

sore /sɔː(r)/ adj. feeling or causing pain 疼痛的：He was sore from the ride. 骑马后他浑身酸痛。

n. a wound or ulcer causing pain 伤口，伤痛

sore shins n. *see* bucked shins 胫凸，胫痛

sorghum /ˈsɔːgəm; *AmE* ˈsɔːrgəm/ n. a plant of the genus *Sorghum bicolor* widely cultivated for its grain as forage（一种两色蜀黍属植物，因籽实可作饲料而广泛种植）高粱：Overfeeding of sorghum can result in founder or severe digestive disturbances in horses. 给马饲喂过量的高粱，可能诱发蹄叶炎或严重的消化紊乱。

Sorraia /səˈraɪə/ n. *see* Sorraia pony 索拉亚矮马

Sorraia pony n.（also **Sorraia**）an ancient pony breed originating in western Spain along the Sorraia river for which the breed was named; stands 12.2 to 13 hands and may have a dun, gray, or palomino coat with zebra markings on the legs and a dorsal stripe; has a large head, a slender and long neck, high withers, low-set tail, and long, solid legs; is fugal, hardy and possessed of good endurance; historically used for farm work but nowadays for riding and packing（源于西班牙西部索拉亚河流域的古老矮马品种并由此得名。体高 12.2～13 掌，毛色多为兔褐、青或银鬃，四肢有斑马纹，有背线，头大，颈细长，鬐甲高，尾础低，四肢粗壮，耐粗饲，抗逆性强，耐力好，过去常用于农田作业，现在用于骑乘和驮运）索拉亚矮马：As Spain's sole native pony breed, Sorraia pony is thought to have descended from the Tarpan which it resembles. 作为西班牙本土唯一的矮马品种，一般认为索拉亚矮马由欧洲野马繁衍而来，因而体貌与之较为相似。

sorrel /ˈsɒrəl; *AmE* ˈsɔːr-/ n. **1**【color】a bright reddish coat color 红栗毛：Sorrel is always appealing since it is usually bright and shiny, and very saturated. 红栗毛由于光彩绚丽、色泽丰满而非常抢眼。 syn red chestnut **2** a horse having such a coat 红栗毛马

sorrel roan n. *see* strawberry roan 沙栗毛

sound /saʊnd/ adj.（of a horse）free from any illness, blemish or conformation defect that may affect its performance（指马）健康的，体况良好的 opp unsound

soundness /ˈsaʊndnəs/ n. the state or condition of being sound and healthy 体况良好

soundness examination n. *see* veterinary inspection 兽医体况检查

soup /suːp/ n.【jumping】the muddy water pool in a course（障碍赛中的）水坑 PHR **in the soup**（of a rider）falling from the horseback into water（骑手）落入水沟，落入水坑

sour /ˈsaʊə(r)/ adj. becoming inactive or bored due to overuse or overtraining（由于使用或训练过度变得迟钝或厌烦）迟钝的；懒散的

sour cattle n.【cutting】（also **stale cattle**）an inactive cow that does not respond well to the cutting horse due to overuse（指在截牛中使用过频而反应迟滞的牛）用酸的牛，麻木的牛

South German Coldblood n. *see* Noriker 诺里克马

Southern and Eastern Gidran n. one of the two types of Gidran Arabian that is lighter than the other and thus used mainly for all-purpose competition（两个基特兰阿拉伯马类型中体型较轻的一种，主要用于马术竞技比赛）东南欧基特兰阿拉伯马 — *see also* **Middle European Gidran**

Soviet Heavy Draft n. a heavy draft breed devel-

S

oped in Russia from 1890 to 1930 by putting natives mares to imported Belgian stallions; stands about 15 hands and generally has a bay, chestnut, or roan coat; has a massive build, a well-proportioned head with relatively short neck, low withers, deep chest, and short legs with moderate feathering; is quiet, energetic, strong, and sure-footed; mainly used for heavy draft and farming (俄国在 1890—1930 年采用当地母马与进口比利时马公马杂交繁育的重挽马品种。体高约 15 掌,毛色多为骝、栗或沙,体格强健,头型适中,颈短,低鬐甲,胸身宽广,四肢粗短,距毛适中,性情安静,爆发力好,步伐稳健,主要用于重挽和农田作业)苏维埃重挽马,苏联重挽马

sow mouth n. *see* undershot jaw 下颌突出,地包天

sowar /səʊˈwɑː(r)/ n.【dated】a member of the cavalry troop of British India (印度殖民时期的)骑兵

soybean /ˈsɔɪbiːn/ n. an annual leguminous plant widely cultivated as forage for its nutritious seeds (一年生豆科植物,因籽实富含养分可作饲料而广泛种植)大豆

soybean meal n.【nutrition】a high energy, high protein supplement produced from the residue of the soybean following oil extraction (大豆榨油后产生的残渣,是一种高能、高蛋白的补充料)大豆粕

span /spæn/ n. a period of time 时段: life span 寿命

Spanish Anglo-Arab n. *see* Hispano 西班牙盎格鲁阿拉伯马,即西班牙马

Spanish Arab n. *see* Hispano 西班牙阿拉伯马,即西班牙马

Spanish Riding School (of Vienna) n. a riding school founded in Vienna, Austria in 1572 to instruct nobility in classical equitation (1572 年在奥地利维也纳创建的骑术学校,主要向贵族传授古典骑术)[维也纳]皇家骑术学校 syn Imperial Riding School of Vienna

Spanish snaffle bit n. *see* Kimblewick 西班牙普通衔铁

Spanish walk n. an artificial, exaggerated walk often seen in Spanish horses who extend their straightened forelegs up to about chest height (一种举蹄较高的人工步法,常见于西班牙马,马伸直前肢并举至齐胸的高度)西班牙慢步

Spanish-Norman n. a warmblood horse breed developed from a cross between the Andalusian and Percheron breeds; stands 15.3 to 17 hands and usually has a gray coat; has a refined, convex head, long neck, broad chest, short back, well-muscled hindquarters; used mainly in show ring and exhibition (由安达卢西亚马与佩尔什马杂交培育的温血马品种。体高 15.3 ~ 17 掌,毛色以青色为主,面貌清秀,颈长胸宽,背短,后躯发达,多用于展览赛)西班牙诺曼马

SPAOPD abbr. summer pasture-associated obstructive pulmonary disease 夏季牧场阻塞性肺病

spare /speə(r); *AmE* sper/ adj. additional to what is required for ordinary use 备用的,多余的: a spare horse 一匹备用马

n.【driving】an extra horse kept at the post station along the road to replace a tired horse in the team (养在沿途驿站的多余马匹,用来替换联驾中的乏马)备用马 syn rest horse

spasm /ˈspæzəm/ n. a sudden, involuntary contraction of a muscle or group of muscles (肌肉或肌肉群突然的、不自主性收缩)痉挛: spasms of the gastrointestinal tract 肠胃痉挛 | **spasmic** /-zmɪk/ adj.

spatium /ˈspeɪʃɪəm/ n.【Latin】(*pl.* **spatia** /-ʃə/) a small space between two cavities, tissues, or body parts (身体内腔、组织或部位之间的空隙)隙,间隙: spatium intercostale 肋间隙 ◇ spatium interosseum cruris 小腿骨间隙 ◇ spatium mandibulare 下颌间隙

spavin /ˈspævɪn/ n. (also **hock spavin**) a swelling around the hock of a horse 飞节肿胀,跗关节肿 — *see also* **bog spavin**, **bone spavin**, **jack spavin**, **occult spavin**

spavin shoe n. a horseshoe used on spavined horses to relieve stress on the hock joint 飞节肿胀用蹄铁

spavined /ˈspævɪnd/ adj. (of a horse) suffering from a spavin (指马)飞节肿胀的

spay /speɪ/ vt. to remove surgically the ovaries of a female animal (手术移除雌性动物的卵巢)摘除卵巢

spayed mare n. a mare whose ovaries have been removed to render her incapable of conception 摘除卵巢的母马

speak /spiːk/ vi.【hunting】(of a hound) to cry or bark (猎犬)咆哮,吠叫

speciation /ˌspiːsiːˈeɪʃn/ n.【genetics】the evolutionary formation of new biological species, usually by the division of a single species into two or more genetically distinct ones (生物进化中新物种的形成过程,常由单个物种分化为遗传特征不同的多个物种)物种形成: Numerical as well as morphological chromosomal differences show that the speciation process in equines involved exchanges of DNA sequence blocks in many chromosomes. 染色体数目和形态的差异说明,在马属动物的物种形成过程中发生了许多DNA序列片段的交换。

species /ˈspiːʃiːz/ n. a fundamental category of taxonomic classification, ranking below a genus or subgenus and consisting of related organisms capable of interbreeding (低于属或亚属的分类学基本单位,其个体之间可以繁育后代)物种 [compare] kingdom, phylum, class, order, family

speculum /ˈspekjələm/ n.【vet】(*pl.* **specula** /-lə/) an instrument for opening the vagina of female stock for estral examination (一种用来开张母畜阴道进行发情检查的器具)开膣器

speedy-cutting n. any limb interference occurring at a fast gait; interfering (马行进中发生的四肢交突)撞蹄;交突

speedy-cutting shoe n. *see* feather-edged shoe 削边蹄铁,缺边蹄铁

spent grains n. the brewer's dried grains 酒糟

sperm /spɜːm; *AmE* spɜːrm/ n. (*pl.* **sperm** or **sperms**) a male gamete or reproductive cell; spermatozoon (雄性配子或生殖细胞)精子

spermatic /spɜːˈmætɪk/ adj. of, relating to, or containing sperm 精子的: spermatic cord 精索 ◇ sperm ducts 输精管 ◇ sperm penetrating assay 精子穿透试验

spermatocyte /spəˈmætəsaɪt/ n. a germ cell formed by cell division of spermatogonia 精母细胞: primary spermatocyte 初级精母细胞 ◇ secondary spermatocyte 次级精母细胞

spermatogenesis /ˌspɜːmətəˈdʒenəsɪs; *AmE* ˌspɜːrm-/ n. the process of producing sperm 精子发生,精子形成

spermatogonium /ˌspɜːmətəˈgəʊnɪəm; *AmE* ˌspɜːrmətəˈgoʊ-/ n. (*pl.* **spermatogonia** /-nɪə/) the stem germ cells of males 精原细胞

spermatozoon /ˌspɜːmətəˈzəʊən; *AmE* ˌspɜːrmətəˈzoʊən/ n. (*pl.* **spermatozoa** /-ˈzəʊə; *AmE* -ˈzoʊə/) a sperm cell 精子,精细胞

sphenoid /ˈsfiːnɔɪd/ adj. of or relating to the sphenoid bone 蝶骨的
n. the sphenoid bone 蝶骨

sphenoid bone n.【anatomy】a compound bone with winglike processes, situated at the base of the skull (颅腔底部有翼状突起的复合骨)蝶骨

sphincter /ˈsfɪŋktə(r)/ n.【anatomy】a ringlike muscle that maintains constriction of a body passage or orifice (维持身体通道或开口处收缩的环状肌)括约肌: anal sphincter 肛门括约肌 ◇ cardiac sphincter 贲门括约肌 ◇ pyloric sphincter 幽门括约肌 ◇ vaginal sphincter 阴道括约肌

Spider Phaeton n. a light, four-wheeled, horse-drawn carriage of the Phaeton type popular in the 1860s with a covered seat in front and a footman's seat behind; pulled by a single horse in shafts or a pair in pole gear (流行于19世纪60年代的一种轻型四轮费顿马车,前侧座位带篷,后侧为脚夫座位,由一匹辕马或两匹马套驾杆拉行)蛛形费顿马车

spike /spaɪk/ n. a heavy nail 大头钉

spike team n. *see* unicorn[2] 锥形三马联驾

spill /spɪl/ vt. to cause to fall 使跌落,摞倒:The rider was spilled by his horse. 骑手被马摞倒在地。

n. a fall, as from a horse 落马,摞倒

spin /spɪn/ vi. (of a horse) to turn around on its hind legs 旋转

n. the act of spinning 旋转

spinal /ˈspaɪnl/ adj. of or relating to the spine 脊柱的:spinal column 脊柱 ◇ spinal cord 脊髓

spine /spaɪn/ n. **1**【anatomy】the vertebral column of an animal 脊柱 **2**【anatomy】any of various pointed projections or processes of animals (动物的各种突出)棘

spinny /ˈspɪni/ n.【hunting】a small covert or grove 矮树林,灌木丛

spinous /ˈspaɪnəs/ adj. resembling a spine or thorn 刺状的,棘状的:spinous process 棘突

spiral fracture n. a break in bone that spirals around the bone circumference, often caused by a sudden twist of the bone (绕骨外周开裂的骨折,多由骨突然扭折所致)螺旋形骨折

spit box n.【racing】**1** a receptacle used to hold urine, saliva, or blood sample taken from a horse for post-race drug testing (赛后药物检测时收集马匹尿液、唾液或血液的容器)药检盒 **2** a barn where horses are brought for drug testing; test barn 药检室,药检棚

Spiti horse n. *see* Spiti pony 斯比提矮马

Spiti pony n. (also **Spiti horse**) a pony breed originating from the Himalayan mountain region of India; stands around 12 hands and usually has a brown or gray coat; has a heavy neck, a strong, short back, short legs, round feet, and a full mane and tail; is tough, sturdy, sure-footed, and vigorous; mainly used for packing in mountain areas (印度靠近喜马拉雅山脉的本土矮马品种。体高约12掌,毛色多为褐或青,颈粗,背短而有力,四肢粗短,蹄圆,鬃、尾浓密,外貌粗糙,步伐稳健,体力充沛,主要用于山区驮运)斯比提矮马

spiv /spɪv/ n. **1**【BrE, slang】one who makes a living by cheating or disreputable dealings 诈骗贩,奸商 **2**【dated】a part-time groom who does odd jobs 兼职马夫,临时马夫

splashboard /ˈsplæʃbɔːd/ n.【driving】a panel that protects the upper part of a horse-drawn vehicle from splashes of mud kicked up from the road (马车前安装的挡板,可防止车身被路面溅起泥水弄脏)挡泥板 [syn] fender

splashed white n. a coat color pattern of horse with large marked patches on a white body (白色被毛上有大片斑块的毛色类型)斑块白

splatter /ˈsplætə(r)/ vi.【cutting】(of a horse) to drop down in fornt of a cow (马截牛时)跌倒

splay-footed /ˌspleɪˈfʊtɪd/ adj. (of hoofs) turning abnormally outward 蹄外翻的,外八字的 [syn] toed-out

spleen /spliːn/ n.【anatomy】a large lymphoid organ near the stomach that serves to store blood, filter foreign substances from the blood, and produce lymphocytes (位于胃下方的一个淋巴器官,有贮存血液、过滤血液中异物和生成淋巴细胞的功能)脾

spleen(o)- pref. of spleen 脾

splenic /ˈsplenɪk/ adj. of or relating to the spleen 脾脏的

splenic fever n. another term for anthrax 脾热,炭疽

splenius /ˈspliːnɪəs/ n.【anatomy】(*pl.* **-nii** /-nɪaɪ/) either of two superficial muscles of the back of the neck extending from the upper vertebrae to the base of the skull that rotate and extend the head and neck (后颈部从颈椎到颅骨底部的两块浅层肌,控制头颈的转动与伸张)夹肌

S

| **splenial** /-niəl/ adj. :splenius muscle 夹肌

splint /splɪnt/ n. a bony swelling of the splint bones resulting from a proliferation of fibrous tissue and osteoperiostitis, usually characterized by swelling, heat, and lameness (小掌骨纤维组织增生或骨膜发炎导致的骨质增生,其症状表现为肿胀、发热和跛行)小掌骨增生,赘骨: interosseous splints 骨间赘骨 ◇ edge splints 边缘赘骨 ◇ creeping splints 潜生赘骨 ◇ knee splints 前膝赘骨 [syn] popping a splint

splint bone n.【anatomy】(also **splint-bone**) either of the two small metacarpal bones that lie along the posterior parts of the cannon bone in the fore or hind legs of horse or other related animals (马属动物四肢管骨后侧两块小掌骨中的一块)小掌骨: lateral splintbone 外侧小掌骨 ◇ medial splintbone 内侧小掌骨 [syn] small metacarpas, small metacarpal bones (foreleg), metatarsus bone (hind leg)

splinter bar n. *see* singletree 挂杆,衡杆

split /splɪt/ vi.【hunting】(of hounds) to separate into two packs and move in different directions (指猎犬)分道,分群

split-up behind adj. (of a horse) with the thighs dividing too high when viewed from behind (马后观时两股开叉过高)分叉过高的: The Akhal-Teké is often long-backed with a tendency to be split-up behind. 阿哈捷金马通常腰背长,而且有后躯分叉过高的倾向。

split-up quarters n. a conformation defect in which the horse's thighs divide too high when viewed from behind (马的一种肢体缺陷,后观时两股开叉过高)后躯分叉过高

spoiled mouth n. a horse's mouth that turns insensitive to the bit and rein action due to mishandling (由于操控不当而使马匹嘴角对衔铁和缰绳的辅助反应变得迟钝)口角不灵

spoke /spəʊk; *AmE* spoʊk/ n. any of the metal or wooden rods connecting the hub of a wheel to its rim (连接轮缘与轮毂的铁杆或木条)辐条,轮辐

spoke brush n. a brush with short, hard bristles used to clean wheels of a horse-drawn carriage (用来清洁马车车轮的硬毛刷)辐条刷

sponge /spʌndʒ/ n. **1** a piece of porous, synthetic material used to wipe and clean the body of a horse (用来擦拭清洁马体的泡沫材料)海绵 **2** an act of wiping or cleaning with a sponge 用海绵清洁,用海绵擦拭

vt. **1** to clean or wipe with a sponge 用海绵擦拭 **2**【racing】to insert a piece of sponge into the nostrils of a horse to impede its ability to breath, thus impacting his performance (用海绵等材料堵塞马匹鼻孔以阻碍其呼吸,从而影响其赛绩的做法)堵塞鼻孔

spongiosa /ˌspʌndʒɪˈəʊsə/ n.【anatomy】the cavernous structure of the deeper layer of bone under the substantia compacta (骨深层的海绵样结构,位于骨密质下面)骨松质: substantia spongiosa 骨松质 [compare] compacta

spongy /ˈspʌndʒi/ adj. resembling a sponge in elasticity, absorbency, or porousness 海绵状的

spongy bone n.【anatomy】the tissue on the inside of the bone shaft (骨干的内部组织)骨松质 [compare] compact bone

spontaneous /spɒnˈteɪniəs; *AmE* spɑːn-/ adj. happening or arising without apparent external cause; self-generated (不受外力引发而产生)自发的

spook /spuːk/ v. (to cause) to become frightened and nervous; stampede 受惊,逃窜;使受惊: The horse spooked at the siren. 警报一响,马就惊了。

spooky /ˈspuːki/ adj. (of a horse) easily frightened by sounds, moving objects or changes of light and color; jumpy (指马对声音、运动物体以及光线和颜色变化反应过度的)易受惊的;神经质的: a spooky horse 容易受惊的马 [opp] traffic-proof

sporadic /spəˈrædɪk/ adj. appearing or occurring singly 散发的;散在的,零星的

spore /spɔː(r)/ n. a small, single-celled reproductive body produced by certain bacteria, fungi, and algae; is highly resistant to adverse environment and capable of developing into a new organism under favorable conditions (由某些细菌、真菌和藻类产生的单细胞再生体,有较强的抗逆性,在适宜条件下可发育为新的个体)孢子

sport of kings n.【racing】*see* horse racing 王者的运动;赛马

spot /spɒt; *AmE* spɑ:t/ n. **1**【color】a small white mark on the horse's forehead (马前额的)白斑,白点 **2**【jumping】the most suitable place from which a horse leaves the ground to clear an obstacle with ease (马离开地面跨越障碍的理想地点)起跳点: The spot in front of an obstacle is usually determined by obstacle height and type as well as the jumping ability of the horse. 障碍前的起跳点主要由障碍的高度、类型及马匹的跨越能力决定。

spotted blanket n.【color】one type of coat color patterns of the Appaloosa with colored spots on the white over the hips (阿帕卢萨马的被毛类型之一,臀部白色区域上着生其他颜色的斑点)斑点披毯型

spotted horse n. **1** any horse with a spotted coat pattern 斑毛马 **2** the Appaloosa horse 阿帕卢萨马

spotty /ˈspɒti; *AmE* ˈspɑ:ti/ adj.【hunting】(of a fox scent) uneven and periodic (指狐狸气味)若有若无的,时断时续的: a spotty scent 若有若无的气味

sprain /spreɪn/ n. a painful hurt of the ligaments of a joint (关节韧带的)扭伤,拉伤

vt. to cause a sprain to a joint or ligament 扭伤,拉伤

spray /spreɪ/ n. a mass of dispersed droplets released from a pressurized container (从压力容器中向外喷出的水雾)喷雾: spray irrigation 喷灌

vt. to disperse or discharge spray 喷雾: Spraying should be carried out twice a year, it is necessary to remove the stock and make alternative arrangements for grazing for up to 14 days after spraying. 喷雾除草一年两次,将牲畜撤离喷雾区域并采用分区轮牧,喷雾后 14 天以上可重新放牧。

sprayer /ˈspreɪə(r)/ n. a device with a pressurized container for spraying 喷雾器: knapsack sprayer 背桶式喷雾器

spread /spred/ n. **1** the communication and transmission of a disease (疾病的)传播 **2**【jumping】the width of an obstacle (跨栏的)横宽 **3**【farriery】the distance between the two heels of a horseshoe 蹄铁尾间距

spread fence n.【jumping】an obstacle that is wide with wings (侧翼较宽的)横展障碍

spring[1] /sprɪŋ/ n. an elastic device underlying the frame of a carriage, used as a cushion to absorb the impact in traveling due to unevenness of road (支撑在马车底部的弹性装置,用来缓冲路面不平带来的颠簸)[减震]弹簧: C-springs C 形弹簧 ◇ semi-elliptic springs 半椭圆形弹簧

spring[2] /sprɪŋ/ vi.【driving】to gallop 狂奔

spring bar n. *see* stirrup bar 蹄铁杆

spring tree n. a saddle tree with a strip of metal at the waist that provides increased flexibility (中腰连有铁条的鞍芯,可借此增加其弹性)弹力鞍芯

Spring Wagon n. a four-wheeled, horse-drawn American passenger carriage with its body hung on shallow semi-elliptical springs and drawn by a single horse in shafts (一种美式四轮客用马车,车身悬于半椭圆形弹簧上,由单马驾辕拉行)四轮弹簧马车

springhalt /ˈsprɪŋhɔ:lt/ n. *see* stringhalt 后肢痉挛

sprint /sprɪnt/ n.【racing】a short race at top speed for a brief period, usually 7 furlongs or less (急速短途赛,赛程多在 1 400 米以内)短途赛: a sprint race 短途赛

v. to run or race at top speed for a brief period

over short distance（短时间高速奔跑）冲刺，疾驰：Horses are born with the capacity for sprinting. 马天生具有疾驰的能力。

sprinter /ˈsprɪntə(r)/ n.【racing】a horse trained to compete in short races 短途赛马，短跑赛马 compare middle distance horse, stayer

SPSBS abbr. Shetland Pony Stud-Book Society 设特兰矮马登记委员会

spur /spɜː(r)/ n. a metal device with a small spike or spiked wheel, worn on the heel of a rider's boot to urge a horse forward（骑手马靴后跟上的金属刺轮，用来催马前进）马刺，靴刺：dressage spurs 舞步赛马刺 ◇ English spurs 英式马刺 ◇ hunt spurs 狩猎马刺 ◇ Texan spurs 德州马刺 ◇ Western spurs 西部牛仔马刺 syn gad

vt. to activate or urge a horse on by use of spurs（用马刺踢马使之前进）刺激，踢马：Bronc riders are required to spur throughout the duration of their ride. 在骑野马比赛中，骑手可在策骑时间之内用马刺踢马。

spur rest n. a small, square piece of leather attached to the heel of a boot upon which a spur rests（马靴后跟上突出的皮革，马刺靠于其上）马刺托

spur shield n. a leather lining fitted beneath the spur strap to prevent excessive buckle wear of the boots（位于靴刺带下的皮革衬里，可防止带扣磨损马靴）靴刺衬皮

spur strap n. a leather strap used to fasten the spur to the boots of the rider（用来将马刺系到骑手靴子上的皮带）靴刺带

spurgall /ˈspɜːɡɔːl/ n.（also **spur gall**）a skin sore on the belly of horse made by friction with spurs（靴刺在马腹部磨出的伤疤）靴刺伤

spurred /spɜːd/ adj.（of a rider）wearing spurs（骑手）戴马刺的

spurrier /ˈspɜːrɪə/ n. one who makes spurs 靴刺匠，马刺制造匠

spurt /spɜːt; *AmE* spɜːrt/ n. a sudden burst or increase 喷发；猛增：growth spurt 快速生长期 ◇ On the homestretch the horse put on a spurt and reached second place. 这匹马在终点直道上突然发力，并最终夺得了第二名。

squad /skwɒd; *AmE* skwɑːd/ n. an athletic team（体育）代表队：the Chinese Olympic squad 中国奥运会代表队

squat /skwɒt; *AmE* skwɑːt/ vi. to crouch down 蹲坐，下蹲

squeal /skwiːl/ n. a loud, shrill cry or sound of horse; neigh（马的）高声嘶叫声，尖叫声

vi. to make such a loud, shrill cry or noise 高声鸣叫，尖叫

squire /ˈskwaɪə/ n. a country gentleman who owns an estate in the countryside（在乡下拥有田产的绅士）乡绅

squirearchy /ˈskwaɪərɑːki; *AmE* -ɑːrki/ n. the landowners considered collectively as a class having political or social influence（具有特定政治和社会影响的地主阶级）乡绅阶层

SSI abbr. Standard Starts Index 标准出赛索引

st abbr. stone 英石

ST abbr. Standardbred 美国标准马

St. Leger Stakes n.（also **St. Leger**）a one 1-13/16 mile flat race held annually for three-year-old Thoroughbreds since 1776 at Doncaster, England（英国自1778年起每年在唐克斯特为三岁纯血马举行的平地赛，赛程2 937米）圣莱切奖金赛，圣莱切大赛：Established in 1776, the St. Leger is the oldest of Britain's five Classic Races. 创立于1776年的圣莱切奖金赛在英国五项经典大赛中历史最为悠久。

stabilizer bar n. *see* anti-sway bar 稳定杆，防摆杆

stable /ˈsteɪbl/ n. a building used for sheltering and feeding of horses 马厩，马房：stable construction 厩舍搭建 ◇ stable hygiene 厩舍卫生 ◇ stable management 厩舍管理 ◇ stable routines 厩舍日常工作 ◇ stable diary 厩舍日志 ◇ stable tools 厩舍用具 IDM **lock the stable door after the horse is bolted** to try to prevent or avoid loss or damage when it is already too late 亡马锁厩，为时晚矣。

vt. to keep (horses) in a stable (马)圈养，厩饲

stable bandage n. another term for standing bandage (站立绷带的别名)马房绷带

stable blanket n. *see* night rug 夜用马衣，厩用马衣

stable boots n. **1** (also **stable wraps**) a padded covering strapped around the lower legs of a horse for protection (系在马下肢起防护作用的衬里护套)[厩用]蹄靴 **2** strong leather boots worn by people working in a stable 厩用马靴，厩靴

stable fly n. a type of feeding fly that inflicts painful bites and discomfort on farm animals (一种叮咬家畜的蝇类)厩蝇

stable fork n. a steel fork used for pitching hay or mucking out stables (用来挑干草或清除马粪的钢叉)厩用钢叉

stable rubber n. a piece of linen or cotton cloth used to wipe over the horse to remove any dust (用来擦去马匹被毛上灰尘的亚麻布或棉布)厩用擦布

stable rug n. (also **stable blanket**) *see* night rug 夜用马衣

stable sheet n. *see* summer sheet 厩用马衣，夏用马衣

stable vice n. *see* vice 恶习，恶癖

stable wraps n. *see* stable boots [厩用]蹄靴

stable yard n. (also **yard**) an enclosed area besides the stable for horses to exercise 运动场；围场

stabled /ˈsteɪbld/ adj. confined or fed in a stable 厩饲的：stabled horses 厩饲马匹

stablehand /ˈsteɪblhæd/ n. (also **stable hand**) one who works in a stable 马厩工人

stableman /ˈsteɪblmən/ n. (*pl.* **stablemen**) one employed to manage a stable 马厩管理员

stablemate /ˈsteɪblmeɪt/ n. any of the horses kept in the same stable 同厩的马

stack /stæk/ n. a pile of straw or fodder; heap 草垛，草堆

vt. to arrange in a stack; pile 堆垛

stacker /ˈstækə(r)/ n. a machine used to stack hay (干草)堆垛机

stag /stæg/ n. **1** a male deer over four years of age (四岁以上的)雄鹿 — *see also* **deer**, **doe**, **fawn** **2** a castrated horse; gelding 骟马

stag face n. *see* dished face 面部凹陷

stag hunting n. a mounted sport of hunting stags with a pack of hounds (用猎犬来捕猎雄鹿的运动)猎雄鹿

stagecoach /ˈsteɪdʒˌkəʊtʃ; *AmE* -koʊtʃ/ n. (also **stage coach**) a large, closed, horse-drawn vehicle formerly used to carry passengers and mail along a regular route; usually pulled by a team of four horses (过去在固定路线上载客和送信的厢式四轮马车，常由驷马联驾拉行)驿站马车

staggard /ˈstægəd/ n. a four-year-old male red deer 四岁雄赤鹿，成年公赤鹿

staggers /ˈstægərz/ n. (also **blind staggers**) a nervous disorder of livestock caused by worm infestation, impaired circulation, or poor digestion; characterized by sudden reeling, staggering gait, dizziness, and frequent falling (家畜由于寄生虫滋生、血液循环受阻或营养不良等导致的神经系统疾病，表现为患畜突然转圈、站立不稳、眩晕和跌跤)打摆子，眩晕症

staghound /ˈstægˌhaʊnd/ n. a breed of dog used for hunting deer and stag 猎鹿犬

stake[1] /steɪk/ n. a wooden post pointed at one end, driven into ground as fence pole, or to fasten a horse (嵌入地下的尖头棒，用作栏柱或用来拴马)木桩

vt. **1** to support a tree or plant with a stake or stakes 用木桩加固 **2** to fasten or tether a horse to a stake 拴马于桩上

stake[2] /steɪk/ n. (usu. **stakes**) **1**【racing】a commission paid to a winning jockey, trainer, or groom (付给骑师、练马师和马夫的)佣金：The jockey acquired a 10% stake of the prize

S

money. 骑师从奖金中获得了10%的佣金。 **2** (**stake**) a sum of money gambled on the outcome of a game or race (在游戏或比赛上投注的金额)赌金:to bet on a race for high stakes 在比赛上投高额赌金 **3** (**stake**) the prize money in horse racing (赛马的)奖金**4** (also **stakes race**) a horse race in which all the owners of the racehorses contribute to the prize money (赛马主之间的)奖金赛,大奖赛:The horse is to run in the Belmont Stakes 这匹马要参加贝尔蒙特大赛。

vt. to gamble money on the outcome of a game or race (在游戏或比赛上)投注,押注:One gambler staked everything he had got and lost. 他把所有家当都押上,结果却输了个精光。

stake and bound n.【jumping】(also **stake-and-bound**) an obstacle consisting of a hedge made of vertical stakes interlaced horizontally with supple saplings and bound between strong upright poles (由木桩横向编上藤条构成的篱笆障碍,两侧以立柱固定)栅栏障碍:This painting shows a hunter jumping a stake-and-bound fence in great style. 在这幅油画中,一匹狩猎马正以优美的姿态跨越栅栏障碍。

stakes race n. *see* stakes 奖金赛,大奖赛

stakes-placed adj.【racing】(of a horse) finished second or third in a stakes race (指赛马跑入二、三名)赢奖金的

stakes-producer n.【racing】a mare who has produced winners in stakes races 产获奖驹的母马

stale[1] /steɪl/ adj. **1** (of food) smelling or tasting bad or no longer fresh (饲料)变味的,发霉的:Never put fresh feeds on top of old otherwise the old will eventually become stale and may go mouldy making it both unpalatable and dangerous to feed. 切忌不要将新鲜饲料堆在原有饲料的上面,否则原有饲料就会变味并逐渐发霉,使之不仅适口性差,而且容易引起饲喂事故。**2** (of a horse) sour or bored (指马)厌倦的,厌烦的:A horse tends to get sour when stabled all day. 整天圈在厩房,马很容易生厌。

stale[2] /steɪl/ n. the urine of a horse 马尿

vi. (of horses) to urinate (指马)排尿:The tail of a horse should not be raised for any length of time unless he is staling or posing dung. 除了排粪尿以外,马尾不应在其他时间竖立。

stale cattle n. *see* sour cattle 用酸的牛,麻木的牛

stale line n.【hunting】the scent of a prey that is old and difficult to follow (因时间久而难以跟踪的猎物气味)变淡的踪迹 [syn] cold line, cold scent, pad scent

stalk[1] /stɔːk/ n.【anatomy】a stem-like structure to which the embryo is attached to the amnionic cavity (连接胚胎至羊膜腔壁的柄状结构)蒂,体蒂:body stalk 体蒂

stalk[2] /stɔːk/ vt. to track or pursue a prey or quarry stealthily;跟踪;追踪:a lioness stalking a gazelle 一头跟踪瞪羚的母狮

stall /stɔːl/ n. a compartment for a horse in a stable (马厩中的隔间)厩舍,马房 [syn] loose box

v. **1** to stop running or doing sth 熄火;停止 **2** to put or keep an animal in a stall (动物)圈养

stall gate n.【racing】a starting gate in which each racehorse has its own compartment (带隔间的)起跑闸

stall walker n. (also **box walker**) a horse who walks in the stall 在厩舍绕行的马,走厩的马

stall walking n. (also **box walking**) a stable vice of a horse who walks restlessly in its stall (马的一种恶习,表现为绕厩舍慢走)厩舍绕行;走厩

stall-bound adj. (of a horse) tied and confined in a stall (指马)圈在厩内的;厩养的

stall-feed /ˈstɔːlfiːd/ vt. to tie and feed horses in a stall (马)圈养,厩饲

stallion /ˈstæliən/ n. an adult male horse capable of reproducing (有繁殖能力的成年雄性马匹)公马:stallion station 种公马站 ◇ lead stallion 领头公马

stallion breeding report n. a form listing the stallion's name, registration number, owner, and the information concerning all mares bred during the calendar year (记载公马名称、登记号码、马主以及年度内与配母马相关信息的表格)公马配种报告,公马繁育报告

stallion cage n. (also **stallion support**) a frame used to support a strong stallion during breeding 公马配种架

stallion donkey n.【BrE】a male donkey; jackass 公驴

stallion groom n. one who takes care of stallions 公马饲养员

stallion hound n.【hunting】a male hound used for breeding purposes 种[用]公犬

stallion ring n. *see* stud ring 种马环

stallion season n. the period in a year during which a stallion stands at stud for breeding 公马配种季

stalljack /ˈstɔːldʒæk/ n.【farriery】a small anvil on which farriers shape horseshoes (用来锤锻蹄铁的)小型铁砧

stall-side diagnostic test n. a quick examination developed from up-to-date biotechnologies to detect reproductive concerns, a variety of diseases, and other physiological problems of horses on site by using tests strips and kits (一种针对马匹繁育状况、特定疾病和生理指标的快速检测手段,借助现代生物技术开发的检测条和试剂盒现场进行)快速诊断检测,现场诊断检测 [syn] on-site diagnostic test, quick-response diagnostic test

stamina /ˈstæmɪnə/ n. the physical or mental strength and ability to conduct sth difficult for long periods of time; endurance (长时间从事繁重作业的能力)毅力;耐力:If we select on the basis of racecourse test for speed and stamina, taking no consideration of factors such as fertility and good mothering quality, we may inadvertently increase the risk of infertility, poor growth rate and unsoundness. 如果我们仅依靠赛场测试记录对速度和耐力性状进行选育,而对繁殖力和母性性状不加考虑,那么我们无意中便增加了不育、生长速度慢和形体失格的危险。

stampede /stæmˈpiːd/ n. a sudden rush or flight of panic-stricken animals (动物受惊后)惊逃,逃窜

vi. to run or flee in a rush 惊逃,逃窜:On seeing the hidden lion in the jungle, the herd of zebras stampeded away frenzily. 看到隐藏在丛林中的狮子后,斑马群四处惊慌逃窜。

stance /stæns/ n. the attitude or position of a standing horse (马站立时的)肢势,站姿:regular stance 正肢势 ◇ sawhorse stance 马凳肢势

stanch /stɔːntʃ/ vt. (also **staunch**) to stop or check bleeding 止血

stand /stænd/ *noun*, *verb*

n. **1**【jumping】the jump stand (障碍)支架,标杆 **2**【vaulting】a compulsory exercise in individual vaulting in which the performer moves from an astride position onto both feet (马上体操个人项目的规定动作,表演者从坐姿直立于马背)马背站立 — *see also* **basic seat, flag, flank, mill, scissors**

vi. **1** (of a horse) to be on its feet in a certain manner (指马)站立 PHR V | **stand close** (of a horse) to have the left and right feet set too close (马左右两蹄间距过小)狭踏 [syn] base narrow | **stand wide** n. (of a horse) to have the left and right feet set wide apart (马左右两蹄间距过大)广踏 [syn] base wide | **stand near the ground** (of a horse) having a deep body with short legs (指马)四肢过短 | **stand under behind** (of a horse) to place the hind limbs too far beneath its body when viewed from the side (侧观时马后肢姿势过于内敛)后肢内踏 [compare] camped behind | **stand under in the front** (of a horse) to place the front limbs too far beneath its body when viewed

S

from the side（侧观时马前肢肢势过于内敛）前肢内踏 [compare] camped in the front **2**（of a stallion）to be available for breeding（指公马）可配种 IDM **stand at stud** 可参加配种 **3**【racing】(of a bet) can not be cancelled（投注）不予撤回，不退：Once wagered，the bet stands. 投注后将不予撤回。

vt.（of a horse）be of a certain height at withers when standing straight（马匹站立时）体高为：A Thoroughbred usually stands 14.3 to 17 hands. 一匹纯血马体高通常在 14.3～17 掌。

standard /ˈstændəd；*AmE* -dərd/ n. **1** a required or agreed level 标准 **2**【jumping】*see* jump stand（跨栏）支架，标杆：Each standard sits upon a base for stability. 每个跨栏支架底部都有基座来增加稳固性。

Standard Starts Index n.【racing】(abbr. **SSI**) an American index calculated on the basis of average earnings per start of a racehorse（表示赛马每场比赛平均赢得奖金的指数）标准出赛指数 [compare] Average-Earnings Index

Standardbred /ˈstændədˌbred/ n.（also **American Standardbred**）an American warmblood breed developed from the English Thoroughbred stallion "Messenger" imported to the United States in 1788；stands 14.1 to 16 hands and generally has a bay，brown，black，or chestnut coat；has a long back，short legs，and powerful shoulders，and tremendous speed and stamina；used mainly used for harness racing，trotting，and pacing（美国温血马品种，由 1788 年进口的雄性英纯血马"信差"繁育而来。体高 14.1～16 掌，毛色多为骝、褐、黑或栗，背长，四肢短，肩部有力，速度和耐力极佳，在轻驾车赛的快步赛和侧步赛中表现非凡）美国标准马 [syn] American Trotter

Standardbred racing n. *see* Harness racing 轻驾车赛

standing bandage n. a thick，cotton leg-wrap used to protect the horse's lower leg during shipping or while stabled（缠在马匹下肢的棉质绑腿，在运输或圈养时可起保护作用）站立绷带 [syn] stable bandage

standing martingale n. a piece of tack consisting of a strap which buckles around the horse's neck on one end and connects to the girth on the other，used to prevent a horse from raising his head too high（一端套在马颈、另一端系在肚带上的马具，用来阻止马抬头过高）站立鞅，低头革 [syn] tiedown martingale，fast martingale

Stanhope Gig n. a horse-drawn vehicle of the Gig type designed by British captain Fitzroy Stanhope around 1814 for whom it was named（1814 年左右由英国军官菲茨罗伊·斯坦诺普设计的一种双轮马车，故而得名）斯坦诺普马车

Stanhope Phaeton n. a light-weight，horse-drawn vehicle of Phaeton type designed by British captain Fiezroy Stanhope and becoming popular in the early 19th century（19 世纪初流行的一种轻型费顿马车，由英国军官菲茨罗伊·斯坦诺普设计）斯坦诺普费顿马车

stapes /ˈsteɪpiːz/ n.【anatomy】(*pl.* **stapes**) the innermost of the three small bones of the middle ear，shaped somewhat like a stirrup（中耳三个小骨中最里面的一块，形似马镫）镫骨 [syn] stirrup bone

staphylococcus /ˌstæfɪləˈkɒkəs；*AmE* -ˈkɑːkəs/ n.（*pl.* **-cocci** /-ˈkɒksaɪ；*AmE* -ˈkɑːksaɪ/）a spherical gram-positive parasitic bacterium of the genus *Staphylococcus*，usually occurring in grapelike clusters and causing septicemia and other infections（一种球形革兰氏阳性葡萄球菌属寄生菌，常呈葡萄串状，可导致败血症和其他传染性疾病）葡萄球菌

star /stɑː(r)/ n. **1** a white marking on a horse's forehead（马前额上的白色别征）额星— *see also* **bald face，blaze，snip，stripe，white lip** **2** *see* dental star 齿星

star gazer n. a horse who holds its head too high 看星头，探星头：By no means should a star gazer be selected for showing jumping. 无论如何不能选择看星头马匹参加障碍赛。

starch /stɑːtʃ; *AmE* stɑːrtʃ/ n.【nutrition】a natural carbohydrate found chiefly in the seeds, fruits, roots, and tubers of plants（一种天然碳水化合物，主要见于植物的种子、果实、根以及块茎中）淀粉

staring /ˈsteərɪŋ; *AmE* ˈsterɪŋ/ adj.（of hairs）dull, scraggly and unhealthy due to malnutrition, parasite infestation, or illness（指被毛由于营养不良、虫害或疾病而无光泽且杂乱的）杂乱的，无光泽的：a staring coat 被毛杂乱无光

start /stɑːt; *AmE* stɑːrt/ n. **1** a signal to begin a race 起跑，开始 **2** an instance of beginning a race 开赛，出赛：to win three times in five starts 出赛5次获胜3次

vt. **1** to begin an event or process 开始 PHR V **start a horse** to begin training a young horse 开始调马 **2**（of a horse）to move suddenly as from surprise or pain（指马由于受惊或疼痛而突然移动）起动，启动

starter /ˈstɑːtə(r); *AmE* ˈstɑːrt-/ n. **1**【racing】one responsible for opening the starting gate and ensuring the fair start of a horse race（赛马中负责打开起跑闸确保开赛公平的人员）开闸员，司闸员：deputy starter 副司闸员 **2**【racing】any horse in the starting gate ready to run a race 待跑赛马 **3** an animal who starts as a competitor 参赛起跑者

starter's list n.【racing】a list of horses entered to run in a race 出赛名单

starter's orders n.【racing】a call made by a racing official to begin a horse race 出闸口令，起跑口令 IDM **under starter's orders**（of racehorses）standing in the starting gate and ready to run（指赛马）准备待发的：After hearing the starter's orders, all racehorses should break the starting gate and gallop off as soon as possible. 在听到出闸口令后，所有赛马都要尽快出闸疾驰。

starting gate n.【racing】（also **gate**）a partitioned mechanical device consisting of stalls with front and back doors used to ensure a fair and equal start to a horse race（由前后带门栅栏构成的机械设备，可确保起跑的公正性）起跑闸，马闸：The starting gate was first used in London, England in 1900. 起跑闸于1900年在英国伦敦首次使用。

starting order n. *see* jumping order 起跳次序

starting price n.【racing】the betting odds on a horse at the beginning of a race（赛马的）开赛赔率

startle /ˈstɑːtl; *AmE* ˈstɑːrtl/ vt. to alarm, frighten, or surprise suddenly 使受惊

State coach n.（also **coronation coach**）an elegant, enclosed coach used by the royalty on formal state occasions; usually drawn by a team of four horses（王公贵族在国事庆典上使用的豪华四轮马车，多为驷马联驾）仪仗马车，加冕马车 compare Town coach

state-bred adj.（of a horse）bred in a particular state of the United States 州培育的

static exercises n.【vaulting】any one of the three figures, including the basic seat, flare, and stand, performed in vaulting competition（马上体操选手表演的坐立、展臂和站立三个动作）静立动作

station wagon n. *see* Depot Wagon 车站马车

stature /ˈstætʃə(r)/ n. the height of an individual in an upright position（个体直立时的高度）身高；体高

staunch /stɔːtʃ/ vt.【BrE】to stanch 止血

stayer /ˈsteɪə(r)/ n.【racing】a horse who performs well over long-distance races（在长途赛中表现出色的马）长跑马 compare middle-distance horse, sprinter

steamflaking /ˌstiːmˈfleɪkɪŋ/ n.【nutrition】a feed processing method in which the grain is steamed and flaked 蒸汽压片

steed /stiːd/ n. a literary term for a horse［文学用语］骏马，战马

S

steel gray n. a dark, silvery gray color of horse who has a black base coat with lightly mixed white/gray hairs（在黑色被毛基础上混有白色或灰色毛发而形成的银黑色）铁青色：Many horses who are born steel gray, will turn into a dapple gray or light gray with age. 许多生来就是铁青毛的马随年岁的增长，被毛逐渐变为斑点青或白青毛。

steeplechase /ˈstiːpltʃeɪs/ n.（also **steeplechase race**）a cross-country jumping race over natural terrain with obstacles（在带障碍的野外进行的跨栏赛马）越野障碍赛：Steeplechase was originated around 1750s, at which time races were run from church to church, the steeples serving as markers. 越野障碍赛源于18世纪50年代，当时人们以尖顶作为标示在教堂之间开展比赛。

steeplechase jockey n. one who rides in steeplechases; timber rider 越野障碍赛骑手

steeplechase meeting n. the location at which a steeplechase is held 越野障碍赛场

steeplechase race n. *see* steeplechase 越野障碍赛

steeplechaser /ˈstiːpltʃeɪsə(r)/ n. **1** a horse competing in a steeplechase race 越野障碍赛马 **2** one who rides in a steeplechase race 越野障碍赛骑手

steer /stɪə(r); *AmE* stɪr/ n. a young bull castrated and raised for its meat（阉割作肉用饲养的青年公牛）阉公牛，肉牛

steer roping n. a rodeo event in which two horsemen endeavor to rope a steer, with one rider roping the steer's head while the other roping the heels（牛仔竞技项目之一，两名骑手一人套牛头，另一人套牛腿）套牛比赛

steer wrestler n.（also **wrestler**）one who competes in steer wrestling events 摔牛者 syn bulldogger

steer wrestling n.【rodeo】(also **wrestling**) a timed rodeo event in which a mounted rider gallops along a running steer, drops from the horseback onto the horns of the steer and attempts to throw the steer unto the ground, and the contestant completing the event in the shortest time wins（牛仔竞技项目之一，骑手策马与牛并列奔驰，从马背上侧身抓住牛角并将其摔倒在地，比赛所用时间最少者获胜）摔牛比赛 syn bulldogging

stem /stem/ n. the main body or stalk of a plant（植物的）茎，秆

stem cell n.【genetics】an unspecialized cell capable of differentiating into specific cells（能分化成特定细胞类型的非特化细胞）干细胞：embryo stem cell 胚胎干细胞 ◇ multipotential stem cell 多能干细胞

stemmy /ˈstemi/ adj. having too much thick stems 茎秆粗的；多茎秆的

step /step/ n. the act of lifting its foot in movement 运步；举蹄

v. **1**（of a horse）to lift its foot and move forward（指马）运步，举蹄 **2** to move in a certain direction 运动，运行 PHR V **step up**【racing】（of a horse）to move up into a higher class（指马进入高级别赛事）晋级

Steppe Bashkir n.（also **Steppe Bashkir Curly**）one type of Bashir Curly originating in Bashkiria around the southern foothills of the Ural Mountains in Russia; stands 13.1 to 14 hands and usually has a bay, chestnut or palomino coat; has a thick, curly long coat, thick mane, tail, and forelock, a short neck, low withers, elongated and sometimes hollow back, a wide and deep chest, short and strong legs, and small hard hoofs; is docile, strong, quiet, and hardy and generally used for driving, packing, and riding（源于俄国乌拉尔山脉南面巴什基尔地区的矮马品种。体高13.1～14掌，毛色多为骝、栗或银鬃，被毛长而卷曲，鬃、尾以及额发浓密，颈短，鬐甲较低，背长且呈凹形，胸身宽广，四肢粗壮，蹄小，性情温驯、安静，抗逆性强，主要用于驾车、驮运和骑乘）高原巴什基尔卷毛马：Due to the introduction of

Ardennais and Trotter blood, Steppe Bashkir usually is heavier than the Mountain Bashkir. 由于引入了阿尔登马和快步马外血，高原巴什基尔卷毛马的体躯通常较山地巴什基尔卷毛马重些。 compare Mountain Bashkir

Steppe Bashkir Curly n. *see* Steppe Bashkir 高原巴什基尔卷毛马

stepper /ˈstepə(r)/ n. a horse with a high knee action in gait 举蹄高的马

stepping pace n. (also **four-beat stepping pace**) a variation of the pace in which the horse's hind foot touches the ground slightly before the fore foot on the same side (一种对侧步的变体步法，同侧后肢着地较前肢稍早) 破对侧步 syn slow pace

sterile /ˈsteraɪl/ adj. **1** incapable of producing offspring 不育的 **2** free from bacteria or other microorganisms 无菌的；消过毒的

sterility /stəˈrɪləti/ n. the condition of being sterile 不育：male sterility 雄性不育

sternal /ˈstɜːnl; *AmE* ˈstɜːrnl/ adj. of, relating to, or near the sternum 胸骨的：sternal region 胸部

stern(o)- pref. sternum 胸[骨]

sternocephalic /ˌstɜːnəʊˈsefəlɪk; *AmE* ˌstɜːrnoʊ-/ adj. of or relating to the sternum and head 胸头的：sternocephalic muscle 胸头肌

sternocostal /ˌstɜːnəʊˈkɒstəl; *AmE* ˌstɜːrnoʊ-/ adj. of or relating to the sternum and ribs 胸肋的

sternomandibular /ˌstɜːnəʊˈmædɪbjələ(r); *AmE* ˌstɜːrnoʊ-/ adj. of or relating to the sternum and the mandible 胸下颌的

sternum /ˈstɜːnəm; *AmE* ˈstɜːrnəm/ n. 【anatomy】(*pl.* **sterna** /-nə/) a long flat bone situated along the ventral midline of vertebrate with which the coastal cartilages articulate (位于脊椎动物胸腔中间、与肋软骨相连的长扁骨) 胸骨 syn breastbone

steroid /ˈsterɔɪd/ n. 【physiology】a group of compounds resembling cholesterol chemically, including the adrenal hormone and sex hormones such as androgens, estrogens and progestagens, etc. (一类结构与胆固醇相似的化合物，包括肾上腺皮质激素与性激素，如雄激素、雌激素和孕激素等) 类固醇：steroid hormone 类固醇激素

stethoscope /ˈsteθəskəʊp; *AmE* -skoʊp/ n. an instrument used for listening to sounds produced within the body (用来听诊身体内声音的医疗器械) 听诊器

steward /ˈstjuːəd; *AmE* ˈstuːərd/ n. (also **stipendiary steward**) an official appointed by the equestrian sports' governing body to ensure that a competition is conducted according to the rules (马术监管机构委派官员，负责确保比赛根据规则进行) 赛事监管

stick /stɪk/ n. **1** a riding crop; a jockey's whip 马鞭 **2**【polo】the polo mallet 马球杆 **3**【jumping】(usu. **sticks**) an obstacle 障碍：to jump over the sticks 跨越障碍

vt. **1** to prod with a spur 用马刺踢 **2**【racing】to whip a horse 鞭打

stick horse n.【racing】a horse who runs better when whipped 加鞭出成绩的马

stiff /stɪf/ adj. **1** not easily bent; rigid 难曲的；僵硬的 **2** (of muscles) lacing suppleness (肌肉)僵硬的，无弹性的

stifle /ˈstaɪfl/ n.【anatomy】the part of a horse's hind leg analogous to the human knee (马后肢中与人的膝盖对等的部分) 后膝：stifle joint 膝关节

stile /staɪl/ n.【jumping】a jumping obstacle consisting of a fence or wall elongated by wings on either side 带侧翼的障碍

stillbirth /ˈstɪlbɜːθ; *AmE* -bɜːrθ/ n. the birth of an offspring that has died in the womb (产出前已在子宫内死亡) 死胎，死产

stillborn /ˈstɪlbɔːn; *AmE* -bɔːrn/ adj. dead at birth 死产的，死胎的：stillborn delivery 死胎分娩

stimulant /ˈstɪmjələnt/ n. any drug or chemical

S

agent that temporarily arouses or accelerates physiological or organic activity in an organism or some part of it such as the circulatory, respiratory, or central nervous system (能瞬间激发并加快机体血液循环、呼吸或中枢神经系统等生理活动的物质或药剂)兴奋剂:stimulant testing 兴奋剂检测

stimulate /ˈstɪmjuleɪt/ vt. to rouse to action as by spurring or goading; provoke 激发;刺激

stimulation /ˌstɪmjuˈleɪʃn/ n. the action of stimulating or being stimulated 刺激

stimulus /ˈstɪmjələs/ n. (*pl.* **stimuli** /-laɪ/) a thing or event that incites one to action; an incentive 刺激

stint[1] /stɪnt/ n. **1** *see* agistment 寄养放牧 **2** the right to pasture horses on public land (在公共草地上的)放牧权

stint[2] /stɪnt/ vt. 【dated】(of a mare) be mated or covered by a stallion (母马)受配:a stinted mare 一匹配过种的母马

stipe /staɪp/ n. 【racing, slang】the office of the Stipendiary Steward 赛事监管之职;赛事监管办公室

Stipendiary Steward n. *see* Steward 赛事监管

stir /stɜː(r)/ vt. to move an object through a liquid in circular motion so as to mix the contents (用物件搅动液体以混合其中的内容物)搅拌,搅动

stirrup /ˈstɪrəp/ n. (also **stirrup iron**) one of the D-shaped metal ring hung on each side of a horse's saddle to support the rider's foot in mounting and riding (悬在马鞍两侧的D形金属环,供骑手上马或骑乘中撑脚)马镫:According to archeological findings, stirrups were first invented in ancient China around AD 300. 根据考古学发现,中国最早在公元300年左右发明了马镫。◇ Stirrups too narrow will trap the rider's foot and cause a frightful accident. 马镫过于狭小容易卡脚,并会酿成严重事故。 PHR V **run up the stirrup iron** to slide the stirrup iron to the top of the stirrup leather 收起马镫

stirrup bar n. either one of the two metal bars built into the front of saddle beneath the skirt to which the stirrup leathers attached (马鞍前襟下面用来拴系镫带的铁杆)镫带杆

stirrup bone n.【anatomy】*see* stapes 镫骨

stirrup cup n. a cup of wine offered to a mounted traveler ready to go on his journey (献给骑马远行者的告别酒)上路酒,上马酒 syn one for the road

stirrup iron n. *see* stirrup 马镫

stirrup leather n. (also **stirrup strap**) an adjustable leather strap by which the stirrup is attached to a saddle (连接马镫与马鞍的一根长度可调的皮带)镫带

stirrup pad n. (also **pad**) a rubber piece fitted into the slit of a stirrup to provide grip for the rider's foot (装在马镫狭缝的橡胶,用来增加骑手鞋底的摩擦)镫垫

stirrup strap n. *see* stirrup leather 镫带

stitch /stɪtʃ/ n. a loop of thread that sews the edges of a wound together in surgery (外科缝合伤口的)缝针,缝线:The cut needed ten stitches. 这道伤口需要缝十针。

vt. to join or suture with stitches 缝合

stock[1] /stɒk; *AmE* stɑːk/ n. **1** farm animals kept or raised for their meat or milk; livestock 家畜,牲畜:stock farm 畜牧场 ◇ stock feed 家畜饲料 ◇ stock font 牲畜饮水器 ◇ stock market 牲畜市场 ◇ stock owner 畜主 **2** the ancestry or lineage of a horse breed (马的)家系,血统:a horse of Arabian stock 一匹阿拉伯马血统的马

adj. **1** of or relating to the raising of livestock 家畜(饲养)的 **2** used for breeding 种用的:a stock mare 种用母马

vt. **1** to keep or raise farm animals on a farm; herd 饲养;放牧 **2** to supply a farm with livestock 提供家畜

stock[2] /stɒk; *AmE* stɑːk/ n. **1** the handle of a whip 鞭杆 **2** *see* barstock 蹄铁条 **3** a wooden or metal framework used to immobilize a horse for branding, shoeing, etc. (打号或装蹄时用来固定马匹的木质或铁质栏)保定栏

stock car n. a railroad car for transporting livestock 运载牲畜的列车

stock saddle n. *see* Western saddle 西部牛仔鞍

stock whip n. a short-handled whip with a long lash used for driving cattle (一种长皮条的短柄鞭,多用来驱赶牲畜)牧鞭

stockbreeder /ˈstɒkˌbriːdə(r); *AmE* ˈstɑːk-/ n. one who raises and breeds livestock 家畜繁育者;家畜养殖者

stockbreeding /ˈstɒkˌbriːdɪŋ; *AmE* ˈstɑːk-/ n. the breeding of livestock 家畜育种,家畜繁育

stocker /ˈstɒkə(r); *AmE* ˈstɑːk-/ n. **1** a farm animal kept for meat until matured or fattened 育肥家畜;育肥牛 **2** a stock car 运载牲畜的列车

stockhorse /ˈstɒkhɔːs; *AmE* ˈstɑːkhɔːrs/ n. (also **stock horse**) **1** a horse used to herd livestock in a ranch 牧用马 **2** a stallion used for breeding 种公马

stocking[1] /ˈstɒkɪŋ; *AmE* ˈstɑːk-/ n. the raising and herding of livestock 家畜饲养;放牧: mixed stocking 混合放牧 ◇ stocking density 载畜量 ◇ stocking intensity 放牧强度 ◇ stocking level 放牧水平

stocking[2] /ˈstɒkɪŋ; *AmE* ˈstɑːk-/ n. (also **stocking white**) a leg marking consisting of white that extends up to or near the horse's knee or hock (马四肢别征之一,其白色区域延伸至前膝或飞节)管白

stocking white n. *see* stocking 管白: half stocking white 半管白 ◇ quarter stocking white 1/4 管白 ◇ three-quarters stocking white 3/4 管白 ◇ full white stocking 全管白

stockman /ˈstɒkmən; *AmE* ˈstɑːk-/ n. **1** one who owns or raises livestock 养殖户;牧场主 **2** one who takes care of livestock or works on a stock farm 牧场饲养员;牧场工人

Stock-type Pinto n. a Pinto horse of Quarter breed that is suitable for work on a farm or Western disciplines (花色夸特马品种,适用于牧场工作和西部牛仔项目)牧用型花马 [compare] Saddle-type Pinto

stock-up /ˈstɒkʌp; *AmE* ˈstɑːk-/ n. 【vet】 a swelling of the horse's leg resulting from accumulation of excessive fluid in the soft tissues, usually caused by stabling a horse immediately following strenuous exercise or prolonged standing (由软组织积液造成的下肢水肿,多由于马高强度训练后立即被圈在马厩或站立时间过长所致)下肢水肿 [syn] Monday morning leg, Monday morning evil

stockyard /ˈstɒkjɑːd; *AmE* ˈstɑːkjɑːrd/ n. a large enclosed yard with pens or stables, where livestock are temporarily kept until slaughtered, sold, or shipped elsewhere (一个带栅栏和厩舍的围场,用来临时圈养有待屠宰、出售或运输的牲畜)牲畜栏,牲畜圈

stomach /ˈstʌmək/ n. a saclike enlargement of the alimentary canal located in vertebrates between the esophagus and the small intestine (脊椎动物消化道介于食管与小肠间的囊状膨大部位)胃: stomach tube 胃管

stomach bot n. *see* bot 马胃蝇蛆

stomach staggers n. *see* colic 腹痛

stomach worm n. any of various parasitic nematode worms that infest the stomachs of animals (寄生在动物胃内的多种圆线虫)胃线虫

stone /stəʊn; *AmE* stoʊn/ n. (abbr. **st**) a British unit of weight equal to 14 pounds (英制重量单位,相当于6.35千克)英石

stone bruise n. a bruise on the sole of a horse's hoof resulting from impact with a hard or sharp object (马蹄底与硬物撞击留下的瘀伤)蹄底撞伤,蹄底瘀伤

stone horse n. *see* stallion 公马

stone wall n. 【jumping】 a solid jumping obstacle consisting of faux stone blocks; puissance wall (一道由人造石块筑成的坚实障碍)石墙

S

stool /stuːl/ n. a piece of feces 粪便

stoop /stuːp/ v. to bend one's body downward 弯腰,弯身: Mark stooped to pick up the bottles. 马克弯腰捡起瓶子。

stooper /ˈstuːpə(r)/ n.【racing, slang】one who stoops to collect winning tickets carelessly discarded by others at a racetrack (在赛马场捡拾获奖彩票的人,多因他人疏忽而丢失)捡彩票者

stop /stɒp; *AmE* stɑːp/ vt.【hunting】to block the entrance of a fox den 堵塞(狐狸等洞穴): to stop earth 堵塞狐狸洞穴

stopper /ˈstɒpə(r)/ n.【hunting】one hired to block the entrance of a fox den 堵洞人

storage /ˈstɔːrɪdʒ/ n. the act of storing goods or the state of being stored 贮藏;储存: storage areas 储藏地点 ◇ Hygienic, low temperature and dry areas are essential for successful long-term storage. 清洁、低温、干燥的环境是饲料长期成功储藏的关键。

store /stɔː(r)/ vt. to keep or reserve sth for future use 贮藏; 储存

n. a large quantity of resource or food kept for later use 储物,储备: glycogen stores 糖原储备

STPD abbr. standard temperature (0℃), pressure (101.3 kPa), and dryness (0% relative humidity) 标准温度气压干燥度

straddle /ˈstrædl/ vi. to stand or sit astride; bestride 叉腿站着;跨骑: In presence of the stallion, the mare in estrus will straddle her hind legs, raises the tail, and urinates quantities of yellowish fluid with a characteristic odour. 在有公马的情况下,发情母马会叉开后肢翘起尾巴,并排出有特殊气味的淡黄色尿液。

straggle /ˈstrægl/ vi. to stray or fall behind 离群,走失;掉队

straggler /ˈstræglə(r)/ n. one who straggles 离群者

straight-neck fox n.【hunting】(also **straight-necked fox**) a fox who runs in a straight line 直行的狐狸

straight shoulders n. a conformation fault in which the horse's shoulder is too upright from the withers to the point of the shoulder and thus lacks sufficient angulation (马肩胛由于从鬐甲至肩端过直而角度太小的体格缺陷)直肩 compare oblique/sloping shoulders

straight Six n.【racing】a wager in which the bettor tries to select the winners of six consecutive nominated races in order to win; pick six (投注者选出连续六场比赛的头马方可赢得奖金的押注方法)六连赢

straight trifecta n.【racing】a wager in which the bettor selects the first, second, and third place horses in the exact order (赌马者以正确次序选出前三名赛马的投注方式)头三正序彩 compare boxed trifecta

straightway /ˈstreɪtweɪ/ n.【racing】the straight section of a racecourse (赛道的)直道: straightway race 直道赛 syn stretch

strain[1] /streɪn/ vi. to stretch one's muscles or nerves to the utmost 拉伤;(分娩)努责

n. a force tending to pull or stretch sth to its extreme 拉力;应力

strain[2] /streɪn/ n. a breed or variety of an animal or plant (动植物的)品系

strangle /ˈstræŋgl/ vt. to kill by squeezing the throat so as to choke or suffocate; throttle (掐住喉咙使其窒息而死)扼死;掐死

strangles /ˈstræŋglz/ n.【vet】a highly contagious disease of horses and related animals caused by the bacterium *Streptococcus equi*, characterized by inflammation of the nasal mucous membrane and abscesses under the jaw and around the throat that cause a strangling or choking sensation (由马腺疫链球菌引发的马和其他动物的急性传染病,症状表现为鼻黏膜发炎、下颌和咽喉周围脓肿,可引起马匹窒息)马腺疫 syn equine adenitis

strap /stræp/ n. a long, narrow piece of leather or other synthetic materials 皮带;皮条

vt. to beat with a strap 鞭打,抽打

strapping /ˈstræpɪŋ/ n. the practice of banging the horse's body for healing purposes (为治疗目的而用皮带轻抽马体)捶打(按摩)

stratum /ˈstrɑːtəm; *AmE* streɪtəm/ n.【anatomy】(*pl.* **strata** /-tə/) a layer of tissue in an organ or body part (机体器官组织的)层

stratum germinativum n.【Latin】a layer of cells that produce the horny sole 角质生成层

straw /strɔː/ n. the dried stalks of grain used as fodder or bedding for animals (作饲草或垫料用的)秸秆,草料 IDM **travel in straw** (of groom) to travel in a van with the horses during transport (指马夫)随马出行

strawberry roan n. (also **sorrel roan**) the color of a horse who has a chestnut coat interspersed with white hairs (栗色被毛中散生白色毛发)栗沙毛 compare roan, blue roan, red roan

strawyard /ˈstrɔːjɑːd; *AmE* -jɑːrd/ n. an outdoor area bedded with straw where horses lie down for rest or recovery (室外铺有草垫的区域,供马躺在上面休息或康复)稻草铺

streak /striːk/ n. **1** a long line or mark 条纹 **2**【color】a stripe (马面部的)流星
vt. to cover a surface with streaks 使带条纹

Strelets /ˈstreleɪts/ n. (also **Strelets horse**) a horse breed developed at the Strelets Stud in the Ukraine by putting Orlov mares to Anglo-Arab and purebred Arab stallions in the early 20th century; stands 15 to 16 hands and usually has a gray coat; mainly used for riding and eventing (20世纪初乌克兰斯特雷勒茨种马场采用奥尔洛夫马母马与盎格鲁阿拉伯马和纯种阿拉伯马公马繁育而来的马品种。体高15~16掌,毛色以青毛为主,主要用于骑乘和三日赛)斯特雷勒茨马:By the 1920s the Strelets had nearly died out. Only two stallions and a few mares remained. 至20世纪20年代,斯特雷勒茨马几近灭绝,仅存两匹公马和几匹母马。

strength /streŋθ/ n. the power or quality of being strong 力量

strenuous /ˈstrenjuəs/ adj. requiring great effort, energy, or exertion; laborious 艰辛的;费力的:You shouldn't expect a horse over the age of twenty to perform strenuous tasks. 不要指望一匹20岁以上的老马去完成艰巨的任务。

streptococcus /ˌstreptəˈkɒkəs; *AmE* -ˈkɑːkəs/ n. (*pl.* **streptococci** /-ˈkɒkaɪ; *AmE* -ˈkɑːkaɪ/) a gram-positive, pathogenic bacterium of the genus *Streptococcus* that causes various diseases, including erysipelas and scarlet fever (一种链球菌属革兰氏阳性致病菌,可引发多种疾病,包括丹毒和猩红热)链球菌

streptothricosis /ˌstreptəʊθrɪˈkəʊsɪs/ n. the alias for the disease of rain rot (雨斑病的别名)链丝菌病

stress /stres/ n. a mental or emotional disruption occurring in response to adverse external influences (由外界不利影响导致的精神或情感上的烦乱)应激:A damaged placenta is a potent cause of fetal stress. 胎盘受损是胎儿应激的潜在因素。

stress fracture n. a fracture of bone caused by repeated concussion on a hard surface (反复与坚硬表面撞击造成的骨折)应力性骨折:Stress fracture generally occurs in the front of the cannon bone and tibia. 应力性骨折常见于前肢的管骨和胫骨。

stressful /ˈstresfl/ adj. causing mental stress 紧张的,胁迫的

stretch[1] /stretʃ/ vt. to extend one's limbs to full length 伸张,舒展 PHR-V **stretch the topline** (of a horse) to extend its head and neck out and downward to a loose rein (指马)伸直脖颈
n. the act of stretching or the state, extent of being stretched 伸展

stretch[2] /stretʃ/ n. the straight part of a racecourse or track, especially the homestretch (跑道的)直道:the head of the stretch 直道前段 compare homestretch, back stretch

stretch call n.【racing】a call made at the eighth

pole from the finish for the horses to charge (离终点200米八分杆处给予的铃声，提示马匹冲刺)直道响铃

stretch runner n.【racing】a horse who runs fastest near the finish 直道冲刺马

stretch turn n.【racing】the final turn on the racetrack before the homestretch 直道转弯处

stria /ˈstraɪə/ n.【anatomy】(*pl.* **striae** /ˈstraiː/) a thin, narrow line or band, especially one of several that are parallel or close together (互相平行或紧挨着的窄线或带)条纹；纹：terminal stria 终纹 ◇ habenular stria thalamus 丘脑缰纹

striate /ˈstraɪeɪt/ vt. to mark with striae or striations 条纹化

adj. marked with striae; striped 有条纹的

striated muscle n. another term for skeletal muscle (骨骼肌别名)横纹肌

stride /straɪd/ n. a single step of a horse measured from where one hoof leaves the ground to the spot where the same hoof touches the ground again (马行进中同肢的离地点与落地点之间的距离)步子，步幅：stride length 步长

strike /straɪk/ vi. **1** (of a horse) to take its first step of the canter (指马)起步 PHR V **strike off early/late**【dressage】to take the first step of the canter before/after the appropriate arena marker (指马在盛装舞步中)起步过早/晚 **2** to knock or scrape one hoof with another (马)撞蹄 **3** (of a horse) to hit with a foreleg (马)前踢 **4** PHR V **strike a fox**【hunting】(of hound) to pursue and find a fox (指猎犬)追击狐狸

striking /ˈstraɪkɪŋ/ n. a stable vice of horse who strikes with its forelegs (马以前肢踢人的恶习)踢蹴

string /strɪŋ/ n. **1** a group of racehorses kept and trained at one stable 同厩的赛马 **2** *see* pack string 驮队

stringhalt /ˈstrɪŋhɔːlt/ n. (also **springhalt**) a nervous disease or disorder of horse characterized by involuntary, jerky flexion of the hind legs toward its abdomen (一种神经紊乱性疾病，表现为后肢不自觉地朝腹部痉挛性屈伸)后肢痉挛

stripe /straɪp/ n. **1** a dark line found on the horse's coat 暗线，条纹：dorsal stripe 背线 ◇ zebra stripe 斑马纹 ◇ stripe across the withers 鹰膀 **2**【color】(also **streak**) a narrow band of white marking running from the horse's forehead to the muzzle (从马前额延伸至口鼻的白色条纹)流星：narrow stripe 窄流星 ◇ wide stripe 宽流星 ◇ interrupted stripe 断流星 — *see also* **blaze, snip, star, bald face, white lip**

striped /straɪpt/ adj. having lines or stripes of different color 有条纹的：striped legs 四肢带条纹

striped hoof n. a hoof having vertical stripes 带直纹的蹄

stroke /strəʊk; *AmE* stroʊk/ n. **1** an act of hitting or striking 撞击，敲击 **2** *see* polo stroke 击球

stroke volume n.【physiology】(abbr. **SV**) the volume of blood pumped out of the left ventricle at each beat (左心室一次搏动的射血量)搏出量

strongyle /ˈstrɒŋgaɪl; *AmE* ˈstrɑːndʒɪl/ n. any of various nematode worms that infest the gastrointestinal tract of horses and cause extensive damage to the blood vessels and the mucous membrane (附生在马胃肠道内的圆线虫科寄生虫，可造成血管和黏膜损伤)圆线虫：small strongyles 小圆线虫 ◇ large strongyles 大圆线虫

strongyloide /ˈstrɒndʒɪlɔɪd/ n. a parasitic roundworm found in the small intestine of young foals (一类附生在幼驹小肠内的寄生虫)类圆线虫

strongylosis /ˌstrɒndʒɪˈləʊsɪs; *AmE* -ˈloʊsɪs/ n. a disease caused by infestation of strongyles and characterized by weakness and anemia in serious cases (由寄生圆线虫导致的病症，严重

时表现为体虚和贫血)圆线虫病

strongylus /strɒnˈgaɪləs; *AmE* ˈstrɒndʒɪləs/ n.【Latin】strongyle 圆线虫

Strongylus edentatus *n.*【Latin】a type of strongyle that lives mainly in the colon of horse and measures 2 to 5 cm long(一种圆形寄生虫,成虫体长 2 ~5 厘米,主要寄生于马的结肠内)齿圆线虫

Strongylus equinus *n.*【Latin】a type of large strongyle measuring up to 5 cm in length(一种大圆线虫,体长在 5 厘米以上)马圆线虫

Strongylus vulgaris *n.*【Latin】a round worm that mainly infests the cecum of horse and measures 2 to 8 cm long(一种圆形寄生虫,成虫体长 2 ~8 厘米,主要寄生在马的盲肠内)普通圆线虫

stud[1] /stʌd/ n. **1** a stallion kept for breeding 种公马:stud book 配种登记册 ◇ stud fee 配种费 ◇ stud farm 种马场 ◇ stud horse 种公马 ◇ stud season 配种季节 **2** a farm that breeds and raises horses; studfarm 马场:military stud 军马场 ◇ national stud 国有种马场

stud[2] /stʌd/ n. (also **road stud**) a metal spike threaded in the hole of horseshoe to increase traction on slippery roads(通过螺纹拧在蹄铁孔上的铁钉,用于增加附着摩擦力,以防打滑)防滑钉: stud hole 钉孔 ◇ competition stud 竞赛用防滑钉 ◇ Roads studs are mostly fitted in the hind shoes only but in some cases it may be necessary to fit front studs. 防滑钉大多只在后肢蹄铁上安装,但在有些情况下,前肢蹄铁也有必要安装。 syn peg

stud book n. **1** a registry book recording the pedigree of a ceratin animal breed(记载家畜品种谱系的)良种登记册 **2** (usu. **Stud Book**) the registry book recording the pedigree of the English Thoroughbreds 纯血马登记册;纯血马登记处: the International Stud Book Committee 国际纯血马登记委员会 syn Weatherby's

stud farm n. (also **studfarm**) a farm used for breeding horses 种马场,育马场:studfarm routine 种马场日常事务

stud fee n. *see* service fee [公马]配种费:Northern Dancer was the world's leading Thoroughbred sire from 1965 to 1990, with his stud fee reaching as high as $1,000,000. 在 1965 - 1990 年期间,“北方舞蹈家”可谓是世界上最优秀的种公马,其配种费最高可达 100 万美元。

stud groom n. the head groom, usually in a stud(种马场的)领班马夫

stud horse n. (also **stud**) a stallion kept for breeding 种公马

stud mule n. an uncastrated male mule(未阉的)公骡

stud ring n. (also **stallion ring**) a rubber ring slipped over the head of a stallion's penis to discourage erection and ejaculation when not being bred(套在种公马阴茎头上的橡胶环,在非配种期可抑制公马勃起和射精)种马环

stud-breed n. a breed of horse which maintains its own stud book 有登记册的马品种

studman /ˈstʌdmən/ n. one who works on a stud farm 马场工人

stumble /ˈstʌmbl/ vi. to move unsteadily and almost fall; trip 跌绊,绊倒: The sick horse stumbled forward in a stupor, then fell off the ground. 这匹病弱的马跌绊着向前走了几步,然后就倒在了地上。

stumer /ˈstjuːmə(r)/ n. **1** a failure 失败,失利 **2**【racing, slang】a horse who loses in a race 跑输的马,失利的马

stump /stʌmp/ n. a bottom part left after the main part has been cut down(主体被截去后残留的部分)断头,残桩:After the umbilical cord ruptures naturally, the stump may be dressed with antibiotic or antiseptic power. 脐带自然断裂后,可在其断头处敷上抗生素或防腐剂。

stump sucking n. *see* cribbing 咬槽癖

stunt[1] /stʌnt/ vt. to prevent the growth or development 阻碍,抑制:stunted growth 生长受阻

S

stunt[2] /stʌnt/ n. a dangerous performance displaying unusual skill or daring 特技表演：a stunt rider 一名特技骑手
vi. to perform stunts 表演特技

sturdy /ˈstɜːdi; *AmE* ˈstɜːrdi/ adj. physically strong or firmly built 强健的，结实的：sturdy fence 结实的围栏 ◇ a sturdy horse 强健的马匹

subacute /ˌsʌbəˈkjuːt/ adj. (of a disease) between acute and chronic (指疾病介于急性与慢性之间)亚急性的

subacute laminitis n. inflammation of the laminae characterized by a rapid onset and a brief duration (发病快但持续期短的蹄叶炎症)亚急性蹄叶炎

subclinical /sʌbˈklɪnɪkl/ adj. (of a disease) not manifesting characteristic clinical symptoms (指疾病未表现临床特征症状)亚临床的：subclinical symptoms 亚临床症状 ◇ subclinical illness 亚临床疾病

subcutaneous /ˌsʌbkjuˈteɪnɪəs/ adj. located or placed beneath the skin 皮下的：subcutaneous implant 皮下包埋 ◇subcutaneous injection 皮下注射 ◇ subcutaneous fat 皮下脂肪

subdue /səbˈdjuː; *AmE* -ˈduː/ vt. to bring sb/sth under control; submit 制服：A lip twitch is an effective means to subdue a horse. 唇捻是制服马匹的有效手段。

subfertile /sʌbˈfɜːtaɪl; *AmE* -ˈfɜːrtaɪl/ adj. having a low fertility 生育力低的

sublingual /sʌbˈlɪŋgwəl/ adj. situated beneath or on the underside of the tongue 舌下的

sublingual gland n. one of the salivary glands locating beneath the tongue 舌下腺

subluxate /sʌbˈlʌkseɪt/ v. (to cause) to dislocate partially or incompletely 使半脱位

subluxation /ˌsʌblʌkˈseɪʃn/ n. the incomplete or partial dislocation of a joint (关节不完全或部分脱位)半脱位

submaxillary gland n. *see* mandibular gland 颌下腺

subminimum /sʌbˈmɪnɪməm/ adj. below the minimum 低于最小量的

submission /səbˈmɪʃn/ n. the obedience of a horse to its rider or handler 屈服，温驯

submit /səbˈmɪt/ vt. (of a horse) to accept the control and commands of the rider (指马)听命，屈服：A noble horse would submit himself to nobody but his owner. 一匹高贵的马只会听命于自己的主人。

submucosa /ˌsʌbmjuːˈkəʊsə; *AmE* -ˈkoʊsə/ n. 【anatomy】a layer of loose connective tissue beneath a mucous membrane (位于黏膜下的松散结缔组织层)黏膜下层 | **submucosal** /-sl/ adj.

subscribe /səbˈskraɪb/ vt. to contribute or offer 捐助，提供：The stallion and the mare subscribe fifty percent of the genetic material respectively of the foal at conception. 受孕时公马和母马各自为幼驹提供50%的遗传物质。

subscription /səbˈskrɪpʃn/ n.【racing】the fees paid by the owner of a horse to nominate or maintain eligibility for the horse in a stakes race (马主为了让自己的马匹参赛而交的费用)参赛费，报名费

subsidiary /səbˈsɪdiəri; *AmE* -dieri/ adj. serving to assist or supplement; auxiliary 辅助的：The teaser plays an important subsidiary role to determine the sexual state of mares by showing marked interest in them that are in estrus and a relative lack of interest in those that are in diestrus. 试情马在判断母马发情状况中起着重要的辅助性作用。如果它对母马表现出强烈的兴趣，说明母马正处于发情期；如果兴趣不足，则说明母马处于间情期。

substance /ˈsʌbstəns/ n. **1** a chemical or drug with a certain effect 物质；药物：forbidden substance 违禁药物 **2** the physical quality or build of a horse's body 体格，体质：a horse having substance 一匹体格结实的马

suck /sʌk/ v. (of a foal) to draw milk from the mare's udder (幼驹)吮吸，吮乳

n. the act or sound of sucking; suction 吮吸: If a orphan foal is premature or lacks a normal suck reflex, it must be fed through a stomach tube until it learned to suckle. 如果一匹无母幼驹尚未成熟或缺乏正常的吮吸反射,那么在它学会吮乳之前,必须用胃管来喂养。

sucked-up adj. *see* tucked-up (指马)后腰凹陷的

sucker /ˈsʌkə(r)/ n. *see* suckling 吮乳幼驹,哺乳幼驹

sucking louse n. an external blood-sucking parasite that lives on the body of a horse (寄居在马体表的吸血寄生虫)吮虱

suckle /sʌkl/ vi. (of a foal) to suck the udder of the mare (指驹)吮乳

vt. to nourish or nurse with the milk of breast 哺乳;哺育

suckling /ˈsʌklɪŋ/ adj. unweaned 未断奶的

n. (also **sucker**) a foal that has not been weaned (尚未断奶的马驹)吮乳幼驹

sucrose /ˈsuːkrəʊz; *AmE* -kroʊz/ n.【biochemistry】a disaccharide carbohydrate that is the chief component of the cane or beet sugar (一种存在于甘蔗和甜菜中的主要双糖)蔗糖

suction /ˈsʌkʃn/ n. the act or process of sucking 吮吸

Sudan grass n. a grass indigenous to Sudan and widely grown for hay 苏丹草: Sudan grass hay 苏丹干草

sudden death n.【polo】an overtime period played until one team scores when a match is tied at the end of the sixth chukker (马球比赛第六局结束出现平局时,进行加时至一方得分获胜)突然死亡法,一球制胜

sufficient /səˈfɪʃnt/ adj. more than what is needed; enough; abundant 足够的;充分的

suffocate /ˈsʌfəkeɪt/ v. (cause to) die from lack of oxygen; asphyxiate (使)窒息 | **suffocation** /ˌsʌfəˈkeɪʃn/ n.: It may be wise to peel the amnion from the foal's muzzle and lift its head out of the amniotic fluids immediately after the foal has been delivered in case of suffocation. 为避免窒息,在母马分娩后立即剥离幼驹口鼻上的羊膜并将它的头部移出羊水是明智之举。

Suffolk /ˈsʌfək/ n. (also **Suffolk Punch**) a heavy draft breed native to Suffolk, Britain; stands 15.3 to 16.2 hands and is exclusively chestnut in coat color; has a compact, yet large body set on short, powerful, clean legs with little feather; is frugal, gentle, and possessed of exceptional pulling power; used mainly for heavy draft and farm work (源于英国萨福克郡的挽马品种。体高15.3～16.2掌,被毛仅有栗色,体格结实,四肢粗短,距毛稀疏,耐粗饲,性情温驯,挽力强劲,主要用于重挽和农田作业)萨福克马: The Suffolk is the smallest and oldest of the British heavy draft breeds. 萨福克马在英国重挽马品种中体型最小也最为古老。

Suffolk Horse Society n. an organization founded in Great Britain in 1878 to maintain the breed registry for the Suffolk Punch (1878年在英国成立的组织机构,旨在对萨福克马进行品种登记工作)萨福克马匹协会

Suffolk Punch n. *see* Suffolk 萨福克马

sugar beet n. (also **beet**) a plant with a large round root from which sugar is extracted (一种根茎粗大的植物,主要用来榨糖)甜菜

sugar beet pulp n. *see* beet pulp 甜菜渣

sugar cane n. a tall tropical plant having thick stems from which sugar is extracted (一种茎秆可榨糖的热带植物)甘蔗

sulfonamide /sʌlˈfɒnəmaɪd/ n. (also **sulphonamide**) a group of organic sulfur compounds used as antibacterial drugs (作抗菌药使用的一类有机硫化物)磺胺药物

sulk /sʌlk/ vi.【racing】(of a horse) to refuse to run or respond to the urging of the rider (指马)愠怒,倔强

sulky /ˈsʌlki/ n. a light, two-wheeled vehicle used in harness racing, having curved shafts and mounted on wire-spoked wheels with pneumatic tires (用于驾车赛的轻型双轮马车,车辕弧形,带充气轮胎)赛用马车

S

sulky stirrup n. an adjustable foot rest fitted on a sulky 驾车赛脚镫

Sullivan, Con n. a famous Irish horse-whisperer of the l8th century（18 世纪爱尔兰著名的马语者）考·沙利文

sulphonamide /sʌlˈfɒnəmaɪd/ n. *see* sulfonamide 磺胺药物

Sumba /ˈsuːmbə/ n.（also **Sumba pony**）a pony breed originating on the island of Sumba, Indonesia; stands around 12 hands and usually has dun coat with a dorsal stripe, dark mane, tail and points; has a slightly heavy head, a short, broad neck, long back, and short, strong legs; is very tough, willing, intelligent, and docile, and possessed of good endurance; used mainly for riding and packing（源于印度尼西亚松巴岛的矮马品种。体高约 12 掌，被毛多为兔褐且带背线，鬃、尾和末梢为黑色，头重，颈粗短，背长，四肢粗短，抗逆性好，性情温驯，反应灵敏，耐力好，主要用于骑乘和驮运）松巴矮马

Sumbawa /suːmˈbɑːwə/ n.（also **Sumbawa pony**）a pony breed originating on the island of Sumbawa, Indonesia; is quite similar in appearance to Sumba pony（源于印度尼西亚松巴哇岛的矮马品种，体貌与松巴矮马较为相似）松巴哇矮马

summer /ˈsʌmə(r)/ vt. to herd or pasture cattle or horses for the summer 夏牧

summer pasture-associated obstructive pulmonary disease n.（abbr. **SPAOPD**）a chronic allergic pulmonary disease of horses caused by hypersensitivity to soil- or grass-borne fungus in summer; symptoms may include difficulty exhaling, flared nostrils, elevated heart and respiratory rates, abnormal respiratory sounds, and coughing not associated with exercise（马在夏季由于对土壤或牧草真菌过敏而引发的慢性肺病，症状主要表现为呼吸困难、鼻孔开张、心率和呼吸频率加快、异常呼吸声响以及非运动性干咳）夏季牧场阻塞性肺病

summer sheet n. a light blanket used to keep a horse clean in summer（夏季用来保持马体清洁用的薄单）夏季马衣 [syn] stable sheet, paddock sheet

summer sores n. *see* cutaneous habronemiasis 皮肤线虫病，夏季皮肤溃烂

summering /ˈsʌmərɪŋ/ n. the turning out of horses to pasture during the summer months（夏季将马匹放逐在牧场吃草的措施）夏季放牧

sumpter /ˈsʌmptə(r)/ n. a pack animal, such as a horse or mule 驮畜；驮马 [compare] palfrey, war horse

Sun Yang n.（c. 680 – 610 BC）a Chinese who lived in the state of Qin during the Spring and Autumn Period, known often as "Bo Le" for his keen judgments on horses and his work *On Evaluating Horses* which was lamentably lost（公元前 680—前 610 年春秋时期秦国著名的相马名家，世称"伯乐"，所著《相马经》不幸失传）孙阳

sunburn /ˈsʌnbɜːn; *AmE* -bɜːrn/ n. skin inflammation caused by overexposure to the sunlight（皮肤经日光暴晒后发炎的症状）日光灼伤，晒伤

v. to affect or be affected with sunburn 晒伤，日光灼伤：The skin may appear red and blisters may develop after being sunburned. 皮肤被日光晒伤之后，常出现红肿并伴有水疱。

sunflash /ˈsʌnflæʃ/ n.（also **sun circle**）a brass decoration in the pattern of the sun, placed on the forehead of a horse to ward off the evil eye（形似太阳的铜饰，常佩戴在马的前额用来避邪）日形铜饰

sunflower meal n.【nutrition】a high-energy protein supplement produced from the residue of the sunflower seed following oil extraction（葵花籽榨油后产生的残渣，可作为马匹的高能蛋白质补充料）葵花籽粕

sunstroke /ˈsʌnstrəʊk; *AmE* -troʊk/ n. *see* heat stroke 日射病，中暑

superfecta /ˈsjuːpəˌfektə/ n.【racing】a wager in

which the bettor tries to select perfectas in two nominated races in order to win (赌马者须选出指定两场比赛的头二正序赛马方可赢得奖金的投注方式)超级正序彩

superficial /ˌsuːpəˈfɪʃl, ˌsjuː-; *AmE* ˌsuːpərˈf-/ adj. of or near the surface of an organ or body part (器官或机体)表面的,浅层的

superficial crack n. a shallow crack in the hoof wall 浅层蹄裂 [syn] surface crack

superficial digital flexor tendon n.【anatomy】a tendon that runs down the back of knee in the foreleg and hock in the hindleg of horse (马前膝及飞节后侧的肌腱)趾浅屈肌腱 [syn] superficial flexor tendon

superficial flexor tendon n. *see* superficial digital flexor tendon 趾浅屈肌腱

superficial laceration n. an injury penetrating only the upper layers of the skin 表层拉伤

superficial wound n. an injury that only damages the upper layers of the skin (仅造成上层皮肤组织损伤的)浅层伤口

superior check ligament n.【anatomy】a ligament situated above the knee and attached to the superficial flexor tendon (位于前膝上方并与趾浅屈肌腱相连的韧带)上止韧带

supinate /ˈsuːpɪneɪt/ vt. to turn outward 外旋 | **supination** /ˌsuːpɪˈneɪʃn/ n.

supinator /ˌsuːpɪˈneɪtə(r)/ n. a muscle whose contraction assists in the supination of a limb 外旋肌

supple /ˈsʌpl/ adj. moving smoothly with ease; flexible 轻盈的;敏快的 | **suppleness** /-plnəs/ n.
vt. to make a horse supple through exercises 使柔韧;使敏快

supplement /ˈsʌplɪmənt/ n. any substance added to a horse's diet to balance it (为平衡马匹日粮而在其中添加的成分)补充料;添加剂:Supplements come in various forms: liquids, powders, meals and pellets and injections. 补充料有液状、粉剂、片剂、丸剂和针剂等多种形式。

supporting limb n. the leg that supports the weight of horse in movement 负重肢:supporting limb lameness 负重肢跛行

supraorbital /ˌsuːprəˈɔːbɪtl, ˌsjuː-; *AmE* ˌsuːprəˈɔːrb-/ adj.【anatomy】located above the orbit of the eye 眶上的:supraorbital foramen 眶上孔

suprarenal /ˌsuːprəˈriːnl, ˌsjuː-/ adj.【anatomy】located on or above the kidney 肾上的

suprarenal gland n. *see* adrenal gland 肾上腺

suprascapular /ˌsjuːprəˈskæpjələ/ adj. located above the scapula 肩胛上的: suprascapular nerve 肩胛上神经

supraspinatus /ˌsʌprɑːspɪˈneɪtəs/ n.【Latin】the shoulder muscle on the upper side of the scapular spine (位于肩胛冈上方的肌肉)冈上肌

surcingle /ˈsɜːsɪŋgl/ n. a wide band of leather circled around a horse's chest behind its forelegs to hold a blanket or pad in place (系在马胸腔紧靠前肢处的皮带,用来固定乘毯或坐垫)胸带,肚带

sure-footed /ˌʃʊə(r)ˈfʊtɪd/ adj. (of horse's gait) steady and firm (指马)步法稳健的 | **sure-footedness** /-tɪdnəs/ n.

surface crack n. *see* superficial crack 浅层蹄裂

surfeit /ˈsɜːfɪt; *AmE* ˈsɜːrfɪt/ vt. to feed or supply to excess or satiety; overfeed 饲喂过度,供给过剩
vi. to stuff with foods; gorge 暴食,吃饱

surgical leg brace n. a surgical device that holds or supports an injured leg of a horse (用来支持固定马匹受伤四肢的医疗装置)四肢夹板 [syn] leg brace

surgical shoe n. *see* therapeutic shoe 矫正性蹄铁,治疗性蹄铁

surrey /ˈsʌri/ n. a light, four-wheeled, horse-drawn pleasure carriage with a canopy or folding top, developed in Surrey, England from which the name derived (一种有华盖或顶篷的轻便四轮游览马车,源于英国的萨里郡并因此而得名)萨里马车

susceptibility /səˌseptəˈbɪləti/ n. the quality or

S

condition of being susceptible 易感性：susceptibility to endemic diseases 对流行病的易感性

susceptible /sə'septəbl/ adj. easily influenced or affected by disease, etc. 易受影响的；易感染的：susceptible to tetanus 对破伤风易感

suspend /sə'spend/ vt. **1** to hang or stay in air 悬挂，悬空 **2** to declare off a race 取消，撤销 **3**【racing】to prohibit a horse or jockey from competing in a race as a punishment; rule off（惩罚性地禁止马匹或骑手参加比赛）吊销，禁赛

suspension /sə'spenʃn/ n. **1** the act of suspending or the state or period of being suspended 悬空；撤销 **2**【racing】the temporary prohibition of a horse or jockey from participating in an event（马或骑手暂时的）禁赛

suspension time n.【jumping】the period of time during which the horse is in the air over the fence after take-off（马跨越障碍时起跳后在空中的时段）悬空时间，滞空时间

suspensory ligament n.【anatomy】a ligament that supports the fetlock and fetlock joint（用来支撑球节的韧带）悬韧带

S

sutura /'suːtʃərə/ n.【Latin】(*pl.* **suturae** /-riː/) the fibrous or cartilaginous juncture found in the skull; suture（存在于颅骨中的）缝：sutura coronalis 冠状缝

suture /'suːtʃə(r)/ n. **1** the surgical method used to sew a wound or join tissues（用来缝合伤口或连接组织的外科方法）缝合 **2** the fine thread used in surgery 缝线

SV abbr. stroke volume 搏出量

swab /swɒb; *AmE* swɑːb/ n. a piece of absorbent material, such as cotton, used for cleaning wounds or taking samples（用来清洁伤口或取样的吸水棉球）药签；棉棒：cotton swab 棉签

vt. to apply a swab on or clean with a swab（用棉签）涂药；消毒

swage /sweɪdʒ/ n. a tool with moulds for shaping cold metal 型铁，铁模

vt. to shape metal using a swage（金属）塑形

swage block n.（also **swedge block**）a moulding block used for shaping metal objects, such as barstock（用来轧制蹄铁块等铁件的模具）型砧，型铁

swamp fever n. another term for leptospirosis（钩端螺旋体病的别名）沼瘟，沼泽热

swan neck n. **1** a conformation defect in which the horse's neck is long and narrow（马颈又细又长的体貌缺陷）天鹅颈 **2** *see* pole hook 驾杆钩

sward /swɔːd; *AmE* swɔːrd/ n. the upper layer of soil covered with dense grass; turf（表层为浓密青草覆盖的土壤）草皮

sway back n. a conformation defect of horse when its back is hollow and arched downwards（马背部向下凹陷的体格缺陷）凹背：The sway back often occurs in older horses who have weak backs, have been ridden in a poor fitting saddle, or ridden by a rider too heavy for that particular horse over a period of time. 凹背常见于年岁较大的马匹，它们有的因脊背软弱，有的因马鞍不合适，还有的因骑者体重太大且骑乘时间过久所致。 syn dipped back, hollow back, bobby back

sway bar n. *see* anti-sway bar 防摆杆

sway-backed /ˌsweɪ'bækt/ adj. having the back hollow and arched downwards; sagged 背部凹陷的，凹背的：a sway-backed horse 一匹凹背马

swayed /sweɪd/ adj.（of a horse）sway-backed; dipped（指马）凹背的

sweat /swet/ n. drops of saline liquid excreted through the pores of the skin; perspiration（通过皮肤毛孔排出的含盐液滴）汗液；流汗 IDM **get a sweat on a horse** to exert a horse too much so that it sweats all over 使马劳累过度

vi. to excrete sweat through the pores in the skin; perspire（从皮肤毛孔分泌汗液）流汗；出汗 PHR V **sweat the brass**【racing】to overwork a horse 使马劳累过度：sweating rate 发汗率 ◇ Sweating involves the evaporation of fluid from pores in the surface of skin, taking with it excessive heat. 出汗是体液从皮肤表

面毛孔的蒸发,在此过程中带走大量的热。

sweat scraper n. a grooming tool used to scrape off the lather when a horse is bathed (洗浴马匹时用来刮去其体表汗沫的工具)汗刮

swedge block n. *see* swage block 型砧

Swedish Ardennes n. a Swedish heavy draft breed developed in the 19th century by crossing Belgian and French Ardennais horses with the Swedish Horse; stands 15 to 16 hands and may have a black, bay, chestnut, or brown coat; has a heavy head, short back, rounded croup and sturdy legs with light feathering; is docile and quiet but energetic in nature; used mainly for heavy draft and farm work (19 世纪在瑞典育成的重挽马品种,主要采用比利时和法国阿登马与瑞典马杂交培育而来。体高 15 ~ 16 掌,毛色多为黑、骝、栗或褐,头重,背短,尻部圆浑,四肢粗壮,距毛稀疏,性情温驯而活泼,主要用于重挽和农田作业)瑞典阿登马

Swedish Gotland n. *see* Gotland [瑞典]哥特兰马

Swedish Halfbred n. *see* Swedish warmblood 瑞典温血马,瑞典半血马

Swedish Warmblood n. (abbr. **SWB**) a Swedish warmblood breed developed in the 17th century by crossing native mares with imported Arab, Andalusian, Friesian, Hanoverian, Trakehner, and Thoroughbred stallions; stands 16.2 to 17 hands and may have a coat of any solid color; is a strong, sound riding horse of good temperament and used mainly as a dressage mount in equestrian competitions nowadays (16 世纪在瑞典育成的温血马品种,主要由当地母马和进口的优良雄性阿拉伯马、安达卢西亚马、弗里斯兰马、汉诺威马、特雷克纳马和纯血马交配繁育而来。体高 16.2 ~ 17 掌,各种毛色兼有,体格强健,性情温驯,是出色的乘用马品种,现在主要用于盛装舞步赛)瑞典温血马 [syn] Swedish Halfbred

Swedish Warmblood Association n. an organization founded in Sweden in 1928 to promote the breeding and maintain the stud book for the Swedish Warmblood (1928 年在瑞典建立的组织机构,旨在推动瑞典温血马的育种工作并进行品种登记)瑞典温血马协会

sweeney /ˈswiːni/ n. 【vet】 an atrophy of the shoulder muscles due to damage of the nerve supply (马肩部由于神经受损导致肌肉萎缩的症状)肩肌萎缩 [syn] shoulder atrophy

sweeps /swiːps/ n. *see* sweepstakes 奖金赛

sweepstakes /ˈswiːpsteɪks/ n. **1** a type of betting on horse races in which all the prize money are divided among the winners in previously agreed percentage of the total (获胜者按预定份额分配所有奖金)赌金全赢制 **2** (also **sweeps**) a race on which money is bet and divided among the winners (获胜者分得投注金额的比赛)奖金赛 [syn] stakes race

sweet feed n. any feed added with molasses to improve its flavor or palatability (为了增加适口性而添加糖蜜的饲料)甜味饲料

sweet itch n. a seasonal skin condition caused by an allergic reaction to certain blood-sucking insects; may result in intense irritation and produce patches of thick, scaly, and sometimes ulcerated skin (皮肤被吸血性昆虫叮咬后出现的过敏反应,可造成剧烈的瘙痒并导致皮肤结痂和溃烂)瘙痒症 [syn] Culicoides hypersensitivity, Queensland itch

swell /swel/ vi. to increase in size or grow larger 肿胀,肿大

n. **1** a full or gently rounded shape or form 圆形;突起 **2** (also **swell fork**) the part of a Western saddle below the horn in front of the pommel (西部牛仔鞍前鞍桥位于角柱的部分)前鞍桥突起 [syn] Western saddle fork

swell fork n. *see* swell 前鞍桥突起

swing /swɪŋ/ vi. **1** to move back and forth 摆动,晃动 **2** to move or turn laterally 侧向移动

PHR V **swing up** to jump on the back of horse by swinging one's right leg 纵身跨上马背: Most trick riders could swing up on the back of

S

a horse with no difficult. 大多特技骑手都可以毫不费力地纵身跨上马背。

swing bar n. *see* singletree 挂杆;横杠

swing team n. (also **swings**) the middle pair in a six-horse hitch or the team in front of the wheelers in an eight-horse hitch (六马联驾中的)中马;(八马联驾中的)副马

swingletree /ˈswɪŋgltriː/ n. (also **swingle tree**, **swingle bar**) *see* singletree 横杠;挂杠

swings /swɪŋz/ n. *see* swing team (六马联驾的)中马;(八马联驾的)副马

swing-up /ˈswɪŋ-ʌp/ n. *see* Indian style 印第安式上马

swipe /swaɪp/ n.【slang】a groom or stable hand 马夫

vt. to brush and clean a horse; groom, strap 刷拭,轻捶

swirl /swɜːl; *AmE* swɜːrl/ n. a twist or curl of hair on the head of a horse (马头上的)卷毛

swish /swɪʃ/ v. (to cause) to make or move with a hissing or whistling sound; rustle 嗖嗖响,唰唰响:The horse swished its tail to drive away the flies. 马唰唰摆动着尾巴赶走苍蝇。

Swiss Anglo-Norman n. *see* Einsiedler 瑞士盎格鲁诺曼马,艾因斯德勒马

Swiss Warmblood n. a warmblood breed developed in the 1960s in Switzerland by crossing Thoroughbred, Swedish and German horses with the Einsiedler; stands 15 to 16.2 hands and may have any solid coat color with chestnut and bay occurring most common; has a broad and deep chest, sloping shoulders, prominent withers, and strong legs; used mainly for riding (瑞士在20世纪60年代采用纯血马、瑞典和德国的公马与爱因斯德勒马母马交配繁育的马品种。体高15~16.2掌,各种单毛色兼有,但以栗、骝最为常见,胸深而宽,斜肩,鬐甲突出,四肢强健,主要用于骑乘)瑞士温血马

switch /swɪtʃ/ n. the hairy tip of the tail of a cow or pig (牛或猪的)尾尖

syce /saɪs/ n. an Indian stableman or groom (印度的)马夫,马倌 [syn] sais

symbion /ˈsɪmbɪɒn/ n. (also **symbiont**) an organism living on a symbiotic relationship with others (依靠共生关系存活的有机体)共生生物,共生体

symbiont /ˈsɪmbaɪɒnt/ n. *see* symbion 共生生物,共生体

symbiosis /ˌsɪmbaɪˈəʊsɪs; *AmE* -ˈoʊsɪs/ n. a close relationship between two different species that benefit from each other in a particular way (两个物种间互利互惠的依从关系)共生 | **symbiotic** /-ˈɒtɪk/ adj.

symmetry /ˈsɪmətri/ n. similar or exact correspondence in size and shape between two parts or sides (双方在大小和形态上对等)对称 | **symmetrical** /sɪˈmetrɪkl/ adj. symmetrically marked 别征对称的

sympathetic /ˌsɪmpəˈθetɪk/ adj.【anatomy】of, relating to, or acting on the sympathetic nervous system 交感神经的:sympathetic neuron 交感神经元

n. the sympathetic nervous system or any of the nerves of this system 交感神经

sympathetic nervous system n.【anatomy】the part of the autonomic nervous system originating in the thoracic and lumbar regions of the spinal cord that inhibits or opposes the physiological effects of the parasympathetic nervous system (始于胸椎和腰椎处的自主神经纤维,有对抗副交感神经生理作用的功能)交感神经系统 [compare] parasympathetic nervous system

sympathicus /sɪmˈpæθɪkəs/ n.【Latin】the sympathetic nervous system 交感神经系统

symphyseal /simˈfɪzɪəl/ adj. of or relating to symphysis or symphyses 骨联合的

symphysis /ˈsɪmfəsɪs/ n.【anatomy】(*pl.* **symphyses** /-siːz/) a type of joint in which the bone surfaces are firmly united by a plate of fibrocartilage (两关节面通过纤维软骨板紧密相连的关节)联合:ischiatic symphysis 坐骨联合 ◇ mandibular symphysis 下颌联合 ◇ pelvic symphysis 盆骨联合 ◇ pubic symphysis 耻骨联合 | **symphysial** /-zɪəl/ adj.

symptom /ˈsɪmptəm/ n. any of the characteristic signs or indications of a disease（疾病的）症状

symptomize /ˈsɪmptəmaɪz/ vt. to be a symptom or sign of a certain disease 症状为，表现为：The glanders is usually symptomized by swellings below the jaw and mucous discharge from the nostrils. 马腺疫通常表现为下颌肿大以及鼻腔流出黏液。

synchronous diaphragmatic flutter n.（abbr. **SDF**）an involuntary contraction of the diaphragm accompanied with abnormal heartbeat, resulting from electrolyte losses following physical exertion or hypocalcemic tetany（伴有心律不齐的横膈膜不自主收缩，多由运动过度或低钙性搐搦造成的电解质丢失引起）横膈膜同步颤动

syndicate /ˈsɪndɪkət/ n. a joint partnership of buying a horse for race, show, or breeding purposes（合伙买马参加比赛、展览或者育种）辛迪加

synovia /sɪˈnəʊvɪə; *AmE* -ˈnoʊ-/ n. the plural form of synovium［复］滑膜，滑液

synovial /sɪˈnəʊviəl; *AmE* -ˈnoʊ-/ adj. of or relating to synovia 滑膜的，滑液的：synovial bursa 滑膜囊 ◇ synovial layer 滑膜层

synovial fluid n. the sticky transparent lubricating fluid secreted by the synovial membrane lining joint cavities and tendon sheaths（由关节囊和腱鞘内衬的滑膜分泌的半透明润滑性黏液）滑液 [syn] synovia, joint fluid, tendon oil

synovial joint n. a movable joint with a cavity lined by synovial membrane and lubricated by synovial fluid（一种带腔的活动关节，内衬滑膜且润滑作用由滑液提供）滑膜关节

synovial membrane n. a thin membrane that lines the joint capsules and tendon sheaths and secretes synovial fluid to prevent friction（衬在关节囊和腱鞘内的薄膜，所分泌的滑液可降低磨损）滑膜

synovial sheath n. a thin membrane that lines a tendon sheath and produces synovial fluid（衬在腱鞘内并分泌滑液的薄膜）滑膜鞘

synovitis /ˌsɪnəˈvaɪtɪs/ n. inflammation of the synovial membrane 滑膜炎

synovium /sɪˈnəʊvɪəm; *AmE* -ˈnoʊ-/ n.【anatomy】(*pl.* **-via** /-vɪə/) the synovial membrane or fluid 滑膜；滑液

Syrian /ˈsɪrɪən/ n.（also **Syrian horse**）an ancient horse breed originating in Syria and closely related to the Arab; stands 14.2 to 15.2 hands and generally has a gray or chestnut coat; is fast, frugal, energetic, and possessed of good endurance and strength（源于叙利亚的古老的马品种，与阿拉伯马亲缘关系较近。体高14.2~15.2掌，毛色多为青或栗，步速快，耐粗饲，体能充足，耐力好）叙利亚马

Syrian horse n. *see* Syrian 叙利亚马

syringe /sɪˈrɪndʒ/ n. a medical device used to inject fluids into the body 注射器：a hypodermic syringe 皮下注射器

vt. to inject fluids with a syringe 注射

systemic /sɪˈstemɪk/ adj. **1** of or relating to the entire body 全身的：the systemic venous system 全身静脉系统 **2** of or relating to a whole system 系统的

systemic lupus erythematosus n.（abbr. **SLE**）an auto-immune disease rare in horses, symptomized by swollen eyelids, facial pain and swelling, visual impairment, photosensitivity, whitening of the hair in patches, and scabbing（马的一种罕见自体免疫性疾病，症状表现为眼睑红肿、面部疼痛浮肿、视力衰弱、光敏、被毛出现白斑和皮肤结痂）全身性红斑狼疮

systole /ˈsɪstəli/ n. the rhythmic contraction of heart 心脏收缩：atrial systole 心房收缩［期］◇ ventricular systole 心室收缩［期］| **systolic** /ˌsɪsˈtɒlɪk/ adj.

systolic pressure n. the arterial pressure in the systermic circulation when the heart contracts（心脏收缩时的动脉血压）收缩压

tabanid /ˈtæbənɪd/ n. (also **tabanid fly**) any of various bloodsucking flies of the family *Tabanidae*, such as horseflies and gadflies (一类虻科吸血性蝇虫,其中包括马蝇和牛虻)虻虫

table bank n.【jumping】an obstacle consisting of a mound of earth with a flat top and vertical takeoff and landing sides (一道由土堆筑成的障碍,带平顶且起跳与落地两侧与地面垂直)桌式跨堤

tack /tæk/ n. all the equipment such as bridle, saddle and halter used on horses; saddlery (马勒、鞍和缰等用具的总称)马具:horse tack 马具 ◇ tack thefts 马具失窃 ◇ tack trunk 马具箱

vt. to equip with tack 装勒,备鞍 PHR V **tack up** to equip a horse with saddle, bridle, etc. (给马)装勒,备鞍

vi.【racing】(of a rider) to weigh a specific amount (骑手)体重:The jockey tacks less than 50 kilograms. 这位骑师体重低于50千克。

tack room n. a room where tack and saddlery are kept 马具房

tack shop n. a shop where tack and saddlery are sold 马具店

tacking-up n. the process of saddling up a horse 戴勒,备鞍

tag /tæg/ n. **1**【racing】the claiming price asked of a horse (马的)要价,标价:tag price 标价 **2** the tip of a fox's tail (狐狸的)尾尖,尾梢

tail /teɪl/ n. **1** the posterior part of a horse that includes the dock (马体后端包括尾根在内的部分)尾巴:tail pulling 尾巴梳理 ◇ tail plaiting 辫尾巴 **2** (*pl.* **tails**) *see* tail coat 燕尾服

tail bandage n. (also **tail wrap**) a bandage wrapped around the tail of a horse either to keep the hair in place or to improve appearance (缠在马尾上用来保持尾毛整齐或增进美观的绷带)扎尾绷带

tail carriage n. the way or manner that a horse carries its tail 尾巴样式

tail coat n.【dressage】(also **tails**) a formal black coat cut to waist length in front and divided at the back, worn by riders in advanced equestrian competitions (一种衣襟齐腰而后摆开叉的黑色礼服,是骑手参加高级马术比赛的着装)燕尾服 syn shadbelly

tail guard n. a cloth or synthetic covering over the top of a horse's tail to protect the hair from being rubbed while stabled or during transport (一种套在马尾上端的布料筒,可防止尾毛在圈养或运输中蹭乱)尾套

tail hound n.【hunting】a hound who follows behind the pack rather than leading (猎狐时跟在犬群后面的猎犬)后梢猎犬,队尾猎犬

tail rubbing n. a stable vice of horse who rubs its tail against wall continuously due to boredom or parasite infestation (马由于厌烦或寄生虫刺痒而不停用尾巴蹭墙的恶习)蹭尾

tail seat n.【vaulting】a vaulting exercise in which the performer sits on the back of the horse with both legs crossed (马上体操中选手盘腿静坐于马背的动作)盘坐

tail set n. **1** a device used to lift a horse's tail to achieve a high carriage (可抬高尾础使马尾高翘的用具)垫尾用具,提尾带 **2** the carriage of the horse's tail 尾巴样式

tail shot n.【polo】a way of hitting the ball backhand across the rump of his mount (从坐骑胯下反向击球)尾后击球,胯下击球

tail style n. the way how a horse carries its tail 尾巴样式

tail wrap n. *see* tail bandage 扎尾绷带

tailer /ˈteɪlə(r)/ n. 【roping】*see* heeler 套后腿者

Taipa Racecourse n. a racecourse built in Macau for horse racing meets (澳门举行赛马比赛的场地)氹仔马场

take /teɪk/ vt. **1** to carry or move from one place to another 带走;离开 PHR V **take back/up** 【racing】(of a jockey) to pull back his mount to avoid collision with others (骑师)回拽,回扯 **take down**【racing】to disqualify a racehorse who has run in the money for an infraction 取消获胜资格:Any racehorse will be taken down if detected having been doped. 取消任何被检测服用兴奋剂赛马的获胜资格。**take off** 【jumping】(of a horse) to jump off ground to clear an obstacle (指马跨越障碍时)起跳:take off too soon 起跳过早 ◇ The horse will generally hit the jump with his forelegs if he takes off too late. 起跳过晚时,马的前肢通常会撞到障碍上。**2** to get possession of sth 拿到,获取 PHR V **take out**【racing】to deduct money from mutuel pool to cover track revenue and taxes (从投注总额中)抽款,提取:It is just unreasonable that 50 percent has been taken out from the mutuel pool. 赛马场从投注总额中抽取50%的款项实在是太不合理。

take-off /ˈteɪkˌɒf; *AmE* -ˌɑːf/ n.【jumping】the act of jumping off the ground (跨栏时)起跳

take-off point n. the point from which a horse jumps off the ground to clear a obstacle (跨越障碍的)起跳点

take-off side n. the side of an obstacle from which the rider jumps off the ground (骑手跨越障碍时的)起跳侧 compare landing side

take-off zone n.【jumping】the ideal area in which a horse must take off to clear a jump (马匹离地跨越障碍的理想区域)起跳区:The optimum take-off zone size varies according to the height and type of jump to be cleared as well as the stride and jumping ability of the horse. 最佳起跳区的大小依据障碍的高度、类型以及马匹的运步和跨障能力而有所改变。

tallow /ˈtæləʊ; *AmE* -loʊ/ n. any hard fat obtained from livestock (从家畜身上获取的硬性脂肪)硬脂;牛脂

talus /ˈteɪləs/ n.【anatomy】(*pl.* **tali** /-laɪ/) one of the tarsal bones articulating with the tibia and large metatarsal to form the tarsal joint (连接胫骨和大趾骨并构成跗关节的跗骨之一)距骨 syn ankle bone, astragalus

tame /teɪm/ adj. not wild or fierce; tractable 温驯的;驯化的 vt. to domesticate an animal 驯服;驯化:It is necessary to tame the horse in order to have some degree of control over him. 人类要想驾驭马,就必须驯服它。

tandem /ˈtændəm/ n.【driving】a carriage driven by two horses one hitched behind the other 双马纵驾马车 IDM **in tandem** 纵驾 compare randem

adv. with one behind the other 一前一后;纵列地 PHR V **to drive tandem** 纵列驾车:In tandem driving, the wheeler is placed between the shafts of the vehicle and the leader is out in front. 在双马纵驾中,辕马套于两辕之间,而首马则套在前面。

tandem cart n. a two-wheeled cart driven by two horses in tandem 纵驾双轮马车

Tanghan /ˈtæŋgən/ n. a horse breed originating in the Himalayan mountains of northern India; stands around 13.2 hands and usually has a gray coat; has a short neck, shaggy mane, straight shoulder, and short, strong legs; is frugal and possessed of good endurance; used for riding and packing (源于印度北部喜马拉雅山区的马品种。体高约13.2掌,毛色以青为主,颈短,鬃毛粗乱,直肩,四肢短而结实,耐粗饲,耐力较好,主要用于骑乘和驮运)汤甘马

tangle /ˈtæŋgl/ n. a confused, intertwined mass of hair or mane (毛发或鬃交错而成的乱团)缠结:A brush or comb through the mane and tail

T

is all that necessary to remove caked-in dirt and keep it free of tangles. 要除去马鬃、尾中沉积的灰尘并保持其不缠结,用一把刷子和一把梳子来梳理就足够了。

tapadera /ˌtæpəˈderə/ n.【Spanish】**1** the leather cover for the stirrup on a Western saddle (西部鞍马镫的皮革护罩)镫罩 **2** a stirrup with a leather cover 带罩马镫 syn hooded stirrup

taper /ˈteɪpə(r)/ vt. to gradually reduce the training intensity of a horse before competition for full restoration (赛前为恢复体力逐渐减小训练强度)降级训练:Tapering ensures that there is enough time for total tissue repair and full restoration of muscle glycogen stores. 赛前降级训练可保证马匹有足够的时间进行组织修复并存贮充足的肌糖原。

tapeworm /ˈteɪpwɜːm; *AmE* -wɜːrm/ n. a gastrointestinal parasite found in the cecum and small intestines of horses, heavy infestation may cause digestive disturbances, unthriftiness, colic, and anemia (寄生在马盲肠和小肠的肠道寄生虫,大量繁殖时可造成消化紊乱、体重下降、腹痛和贫血)绦虫 syn flat worm

taproot /ˈtæpˌruːt/ n. the direct female line of descent (血统的)母系主支,直系母本

taproot mare n. a foundation mare 奠基母马

Tarai /təˈraɪ/ n. (also **Tarai pony**) a pony breed indigenous to Nepal; stands about 11 hands and usually has a bay coat; used for riding and packing (尼泊尔本土矮马品种,体高约 11 掌,毛色以骝为主,用于骑乘和驮运)德赖矮马

Tarbenian /ˈtɑːbənɪən; *AmE* ˈtɑːrb-/ n. a French horse breed indigenous to Tarbes from which the name derived; stands about 15 hands and generally has bay, chestnut, or gray coat; is fast, intelligent and used mainly for riding and equestrian competition (源于法国塔波尼亚地区的马品种。体高约 15 掌,毛色多为骝、栗或青,速度快,反应敏捷,用于骑乘和马术竞技)塔波尼亚马

target /ˈtɑːgɪt; *AmE* ˈtɑːrgɪt/ n. an object or place selected as the aim of an activity 目标,靶位:target organ 靶器官 ◇ target tissue 靶组织 vt. to aim at an object 盯上,瞄准 PHR V **target a cow**【cutting】(of a rider) to identify a cow in a herd to work (指骑手)瞄准要截的牛

Tarpan /ˈtɑːpæn/ n. a primitive wild horse breed with a tan or mouse coat found formerly in eastern Europe (过去生活在东欧的古老原始野马品种,毛色多为黄褐或鼠灰)欧洲野马,鞑靼野马 syn *Equus ferus ferus*, Eurasian wild horse

tarsal /ˈtɑːsl; *AmE* ˈtɑːrsl/ adj. of or relating to tarsus 跗骨的:tarsal tendon 跗腱

tarsus /ˈtɑːsəs; *AmE* ˈtɑːrsəs/ n.【anatomy】(*pl.* **tarsi** /-saɪ/) the section of the vertebrate hindleg between the tibia and the metatarsus; hock (脊椎动物后肢位于胫骨与跖骨的部分)跗骨;跗关节

tarsus valgus n. *see* cow hocks 飞节内偏;X 形飞节

tattersall ring bit n. (also **ring bit**) a circular bit with the upper half of the circle placed in the horse's mouth while the lower half fitted under the horse's chin (一种环形衔铁,半环衔在马口中,另外半环绕在马下颌)环形口衔

tattoo /təˈtuː; *AmE* tæˈtuː/ n. (*pl.* **tattoos**) an artificial permanent mark made on the horse's skin by pricking and ingraining an indelible dye for identification or registration purpose (在马的皮肤上刺刻并涂上颜料后留有的永久性标记,主要用于身份认证和登记)刺纹,刺青:lip tattoo 唇刺青

vt. to mark the skin with a tattoo or tattoos (在皮肤上)刺纹,刺青:Thoroughbreds are tattooed on the inside of the lips. 纯血马在唇内侧进行刺青。| **tattooing** /-ˈtuːɪŋ/ n.

tattooer /təˈtuːə(r); *AmE* tæˈtuːə(r)/ n. one trained to tattoo horses (马匹)刺纹师

Tattu /təˈtuː; *AmE* tæˈtuː/ n. (also **Tattu po-**

ny) a pony breed originating in the Himalayan mountains of northern India; stands 11.2 hands and generally has a gray coat; has a short neck, shaggy mane, straight shoulder, and short, strong legs; is frugal and has good endurance; used for mountain packing (源于印度北部喜马拉雅山区的矮马品种。体高约 11.2 掌,毛色为青,颈短,鬃毛杂乱,直肩,四肢粗壮,耐粗饲,耐力好,主要用于山地驮运)塔图矮马

taut /tɔːt/ adj. (of muscles) pulled or drawn tight; tense (肌肉)拉紧的,绷紧的 opp slack

tautness /tɔːtnəs/ n. the condition or state of being taut 紧度

taxing /ˈtæksɪŋ/ adj. physically demanding; burdensome 苛刻的;繁重的:Never ask a horse to do anything too taxing beyond his capacity. 切忌不要让马做任何超过其体能的训练。

TB abbr. Thoroughbred 纯血马

TBA abbr. Thoroughbred Breeders Association 英纯血马育种者协会

T-Cart n. an open, horse-drawn wagon usually pulled by a single horse; had two seats and a T-shaped body from which the name derived (19 世纪下半叶使用的一种敞篷四轮马车,多由单匹马拉行,带两个座位,因车身呈 T 形而得名) T 形马车

Tchenarani /ˌtʃenəˈrɑːni/ n. (also **Tchenarani horse**) an old half-breed developed in present northern Iran by crossing Persian Arab stallions with Turkmen mares; stands 14.2 to 15.1 hands and usually has a coat of solid color; has an athletic, wiry build, sloping croup, and powerful hindquarters; is spirited and gentle, although historically used as a cavalry mount, now used for riding (一个古老的半血马品种,由现今伊朗北方地区采用阿拉伯马公马与土库曼马母马杂交培育而来。体高 14.2 ~ 15.1 掌,多为单毛色,体格健壮,斜尻,后躯发达,性情活泼而温驯,过去常用作军马,现在主要用于骑乘)契纳诺尼马

tea /tiː/ n. 【racing, slang】any of the illegal chemicals administered to a horse to improve its performance (为提高马匹赛绩而采用的非法药剂)违禁药物

team /tiːm/ n. **1** a group of horses harnessed together to pull a wagon or plow [联驾]马队 **2** a group of riders forming one side in an equestrian event (马术项目的)代表队,参赛队:In the years leading up to the 2008 Beijing Olympics, the Chinese equestrian team focused on show jumping and three-day eventing. 在 2008 年北京奥运会举办之前的几年,中国马术代表队将重心放在障碍赛和三日赛上。

team penner n. a rider who participates in team penning 团体圈牛者,团体截牛者

team penning n. (also **penning**) a mounted competition in which a team of three riders try to separate three specifically identified cattle from a herd of 30, and drive into an enclosed pen through an opening at the opposite end of the arena within 60 to 90 seconds depending on the class (一种马上团体竞技项目,3 名骑手策马从 30 头牛的牛群中拦截出指定的 3 头牛,并将其赶入赛场另一端的围栏中,所用时间依据级别在 60 ~ 90 秒之间)团体截牛,团体圈牛

team rope vi. 【rodeo】to participate in team roping 参加团体套牛

team roper n. 【rodeo】a rider, either a header or heeler, who participates in team roping 团体套牛者

team roping n. 【rodeo】a timed rodeo event in which two riders, a header and heeler, try to rope the head and one hind leg of a calf respectively (牛仔竞技项目之一,两名骑手一个套牛头,另一个套牛后腿)团体套牛,双人套牛

team sorting n. a mounted sport developed recently in California, USA, in which a team of three mounted riders attempt to separate ten calves from a herd in numerical order within two minutes（美国加利福尼亚州最近发展起来的一项马上运动，三名骑手组成的团队在2分钟之内依次从牛群中拦截10头小牛）团体分牛

tease /tiːz/ vt. to stimulate a mare sexually with a stallion 试情

teaser /ˈtiːzə(r)/ n.（also **teaser stallion**）a male horse used to bring a mare into heat or to test if the mare is ready for mating（用来刺激母马发情或判断母马发情状况的公马）试情公马：The teaser can be a stallion or gelding, but is not the horse of which the mare is actually bred to. 试情马可以是一匹公马或骟马，但一般并不是与母马配种的种公马。

teasing /ˈtiːzɪŋ/ n. the act of process of using a male horse to determine the stage of the estrus cycle of the mare（用公马来鉴定母马所处发情期的手段）试情 syn trial

teasing board n. *see* trying board 试情栏

teat /tiːt/ n. a nipple of a mare's mammary gland 乳头

technical delegate n. an official responsible for ensuring that an international equestrian event is held according to the FEI rules（负责核实国际马术比赛按照FEI规则举行的人员）技术代表

ted /ted/ vt. to spread sth out for drying 摊晒

tedder /ˈtedə(r)/ n.（also **hay tedder**）a machine that spreads newly mown hay for drying 摊草机；翻晒机 syn spreader

teeth /tiːθ/ n. the plural form of tooth［复］牙齿

telega /təˈlegə/ n. a rough, four-wheeled, springless Russian coach（一种俄式粗制四轮马车，无减震弹簧）俄式四轮马车

telencephalon /ˌtelenˈsefəlɒn/ n.【anatomy】the anterior portion of the forebrain（前脑靠前的部位）端脑 syn endbrain

teletheater /ˌtelɪˈθɪətə(r)/ n.【racing】a place where horse races are simulcast and off-track bets are placed（转播赛马实况并进行场外赌马的地方）赛马直播厅

teletimer /ˌtelɪˈtaɪmə(r)/ n.【racing】an electronic device that accurately measures the fractional and finish times of horses running in a race（可准确实时记录赛马分段用时和最终用时的电子设备）电子计时器

Tellington Method n. *see* Tellington-Jones Equine Awareness Method 特林顿训练法

Tellington-Jones Equine Awareness Method n.（abbr. **TTEAM**）a system of integrated horse training, healing, and communication developed by Linda Tellington-Jones to increase a horse's willingness and ability to perform in an anxiety-free environment（琳达·特林顿－琼斯开创的综合训练、治疗和沟通的方法，借此可使马匹在轻松氛围下自由表演发挥）特林顿－琼斯马匹意识训练法；【简称】特林顿训练法 syn Tellington Method

Tellington-Jones, Linda n.（1937－）an American horsewoman and trainer most noted for her holistic health care and training approach for horses（美国女骑手和练马师，以其综合性治疗训练方法而闻名）琳达·特林顿－琼斯

temperament /ˈtemprəmənt/ n. the nature of a horse as shown in its behavior or reaction（马在行为反应中表现出来的本性）性情：a horse of docile temperament 一匹性情温驯的马 ◇ The sex of a horse directly affects its temperament and, accordingly, its performance. 一匹马的性别直接影响着它的性情，进而也影响着它的出赛水平。

temperamental /ˌtemprəˈmentl/ adj. excessively sensitive or excitable（性情）暴烈的

temperature /ˈtemprətʃə(r); *AmE* -tʃʊr/ n.（also **body temperature**）the measurement or degree of heat present in the horse's body（马体内的热度）体温：The body temperature of a horse at work may rise by as much as five degrees. 马在作业时体温最多可升高5℃。

temporal /ˈtempərəl/ adj. of, relating to, or near the temples of the skull 颞的：temporal fossa 颞

窝 ◇ temporal muscle 颞肌

temporary /ˈtemprəri; *AmE* -pəreri/ adj. lasting for a limited period of time; not permanent 暂时的;不定的

temporary teeth n. *see* deciduous teeth 不定齿,脱落齿: In the young horses, all the incisors and the three premolars are temporary teeth, whereas the last three molars at the back of each jaw and the four canine teeth are permanent. 青年马的所有切齿与三个前臼齿都是不定齿,而上、下颌后方的三个臼齿与犬齿则是永久齿。 compare permanent teeth

temporomandibular /ˌtempərəʊmænˈdɪbjʊlə(r); *AmE* -bjələ/adj. of or relating to the temporal bone and the mandible 颞下颌的: temporomandibular joint 颞下颌关节

tendinitis /ˌtendɪˈnaɪtɪs/ n. (also **tendonitis**) inflammation of the tendon 肌腱炎

tendinous /ˈtendɪnəs/ adj. of, having, or resembling a tendon; sinewy 腱的;似腱的

tendinous windgall n. a soft, fluid-filled swelling of the digital flexor tendon sheath (屈肌腱鞘内积液导致的软肿)屈腱软肿 compare articular windgall

tendinous windpuff n. *see* tendinous windgall 屈腱软肿

tendo /ˈtendəʊ; *AmE* -doʊ/ n.【Latin】(*pl.* **tendines** /-dəniːz/) tendon 肌腱

tendon /ˈtendən/ n. a band of tough, fibrous tissue that connects a muscle to the bone (连接肌肉与骨的坚韧纤维组织)肌腱: calcaneal tendon 跟腱 ◇ common tendon 总腱 ◇ contracted tendons 屈腱炎 ◇ The tendons should be hard and sinewy and not in any way of soft or puffy. 肌腱应当坚韧有力,而不应有任何软肿迹象。

tendon boots n. a protective covering for the cannon bones of the horse's forelegs (套在马前肢管部的护套)肌腱护腿

tendonitis /ˌtendəˈnaɪtɪs/ n. *see* tendinitis 肌腱炎,屈腱炎

tendosynovitis /ˌtendəʊˌsɪnəˈvaɪtɪs; *AmE* ˌtendoʊ-/ n. *see* bowed tendon 腱鞘炎

tenectomy /tɪˈnektəmi/ n.【vet】the surgical removal of a tendon 肌腱切除术

ten-minute halt n.【eventing】a compulsory ten-minute rest of the horse and rider near the end of the Speed and Endurance phase, during which the second veterinary inspection is conducted and the horse cooled down and refreshed (三日赛中速度与耐力项结束前规定的10分钟休息,期间对马进行第二次兽医检查,人马也可借此稍作歇息)十分钟休停

Tennessee Walker n. *see* Tennessee Walking Horse 田纳西走马

Tennessee Walking Horse n. (also **Tennessee Walker**) a horse breed developed in the 19th-century in Tennessee, USA; stands 15 to 17 hands and usually has a bay, black, or chestnut coat; is large-boned, deep and short-coupled; has a steady and reliable temperament and is known for its comfortable gaits (美国田纳西州在19世纪培育的马品种。体高15~17掌,毛色多为骝、黑或栗,管围较粗,腰深而短,性情稳重可靠,以其步法舒适而闻名)田纳西走马 syn Plantation Walking Horse, Turn-Row Horse, Walking Horse, Walker

Tennessee Walking Horse Breeders' and Exhibitors' Association n. (abbr. **TWHBEA**) an organization founded in 1935 in Lewisberg, Tennessee, USA to register and preserve the pedigrees of the Tennessee Walking Horse (于1935年在美国田纳西州李维斯博格市建立的组织机构,旨在对田纳西走马进行品种登记和保护)美国田纳西走马育种者和展览者协会

tenocyte /ˈtenəsaɪt/ n.【physiology】a tendon cell 腱细胞

tenosynovitis /ˌtenəʊˌsɪnəˈvaɪtɪs; *AmE* ˌtenoʊ-/ n. inflammation of the tendon sheath 腱鞘炎

tense /tens/ adj. **1** (of muscles) stretched tight and rigid (肌肉)紧绷的,僵硬的: Body mas-

T

sage helps to relax the tense muscles. 身体按摩有助于放松紧绷的肌肉。**2** mentally anxious and nervous 焦躁的；紧张的

tension /ˈtenʃn/ n. **1** the state of being stretched tight 紧绷，紧张 **2** the mental state of being nervous［神经］紧张

tensor /ˈtensə(r)/ n.【anatomy】a muscle that stretches or tightens a body part（拉伸身体部位的肌肉）张肌：tensor muscle of lateral fascia 阔筋膜张肌

tent pegger n. one who participates in the mounted sport of tent pegging 挑帐桩骑手

tent pegging n. a mounted sport originating in India for military purposes in which mounted riders charge on galloping horses and try to pick out the tent pegs with long lances carried in hand（源于印度军队的一项马上运动，骑手策马疾驰冲向帐篷，手持长矛将帐桩挑起）挑帐桩比赛：In the sport of tent pegging, the size of the peg varies depending on the skills of the competitors, with larger pegs used by beginners and smaller ones by more advanced riders. 在挑帐桩比赛中，桩孔大小依参赛者的技术水平而定，一般初学者所用的桩孔较大而技术高的骑手使用的桩孔则较小。

teratogen /ˈterətədʒən; *AmE* təˈræt-/ n. any agent that causes malformation of an embryo or a fetus（导致胚胎或胎儿畸形的制剂）致畸因子 | **teratogenic** /ˌterətəˈdʒenɪk/ adj. 致畸的

Terek /ˈterək/ n. *see* Tersky 捷尔斯基马

terminology /ˌtɜːmɪˈnɒlədʒi; *AmE* ˌtɜːrməˈnɑːl-/ n. the vocabulary of technical terms used in a particular field of subject or science; nomenclature（用于特定学科领域的专业词汇）术语；学名

terms race n.【racing】*see* condition race 规格赛，条款赛

terrain /təˈreɪn/ n. the surface features of an area of land 地形；地势

terret /ˈterɪt/ n. one of the metal rings or loops on a harness through which the driving reins pass（挽具上供穿过缰绳的金属圆环）穿缰环，鞍环

terrier /ˈteriə(r)/ n. any of several small, active breeds of hunting dog originally developed for driving game from burrows（一种活泼的小型猎犬，最初用来狩猎穴居动物）㹴犬，㹴狗：Scotch terrier 苏格兰㹴犬

terrier man n.【hunting】one who takes care of the hunt terrier and is responsible for earth stopping（猎狐中照看㹴犬并负责堵狐穴的人）㹴犬员

territorial /ˌterəˈtɔːriəl/ adj. of or relating to territory 地域的

territory /ˈterətri; *AmE* ˈterətɔːri/ n. an area occupied by a single animal, mating pair, or group and defended vigorously against intruders, especially those of the same species（动物占据的一片地域，为捍卫此区常对入侵同类进行驱赶）领地，地域

Tersk horse n. *see* Tersky 捷尔斯基马

Tersky /ˈtɜːski; *AmE* ˈtɜːrs-/ n.（also **Tersk horse**）a Russian warmblood horse breed developed at the Tersk and Stavropol Studs from the 1920s to 1950; stands 14.3 to 15.1 hands and always has a gray coat; is good-natured, active, and possessed of good stamina; used for riding, racing, and circus（1921—1950年在俄国捷尔斯基和斯塔夫罗波尔种马场育成的温血马品种。体高14.3～15.1掌，毛色为青毛，性情温驯而活泼，耐力好，主要用于骑乘、赛马和马戏表演）捷尔斯基马 [syn] Terek

test /test/ n. a procedure intended to determine the presence or quality of sth; trial（测定某物存在与否或品质好坏的过程）测试；检测：drug test 药物检测 ◇ dope test 违禁药物检测

test barn n.【racing】a room where blood and urine samples of horses are taken and examined to detect the presence of illegal drugs following each race（赛后采集马匹血液和尿样以检测有无违禁药物的房间）检测棚，检测室

test cross n.【genetics】a method used to determine the genotype of an individual by crossing with a homozygous recessive one（通过与纯合隐性个体杂交来检测特定个体基因型组成的方法）测交

testes /ˈtestiːz/ n. the plural form of testis［复］睾丸 syn nuts, balls

testicle /ˈtestɪkl/ n. *see* testis 睾丸

testicular /tesˈtɪkjələ(r)/ adj. of or relating to testicles 睾丸的：testicular degeneration 睾丸退化

testis /ˈtestɪs/ n.【anatomy】(*pl.* **testes** /-tiːz/) either of the pair of oval-shaped reproductive glands in the scrotum of a male mammal that produce sperm and androgens（雄性哺乳动物阴囊内的一对卵圆形生殖腺，可产生精子并分泌雄激素）睾丸 syn testicle

testosterone /teˈstɒstərəʊn/ n.【physiology】one type of androgen produced primarily in the testes and responsible for the development and maintenance of male secondary sex characteristics（由睾丸分泌的雄激素，有促进机体发育和维持雄性第二性征的作用）睾酮

testy /ˈtesti/ adj.（of a horse）easily irritated or impatient（指马）暴躁的，暴戾的

tetanus /ˈtetənəs/ n. an acute, fatal disease caused by the toxin of the bacillus *Clostridium tetani* that infects the body through a deep wound, usually symptomized by spasmodic contraction of voluntary muscles of the neck and jaw（由破伤风杆菌通过伤口感染而引起的急性、致死性疾病，其症状常表现为颈、颌肌肉痉挛）破伤风 syn lockjaw

tetanus shot n. an injection that helps protect an animal from developing tetanus 破伤风疫苗

tether /ˈteðə(r)/ n. a long rope or chain by which an animal is tied to restrict its range of movement（一根拴系动物的长绳索，用来限定其范围活动）绳索，长绳

vt. to fasten with a tether 长绳拴系

Tevis Cup Ride n. a 100-mile endurance race held annually since 1955 in the United States from Tahoe City, Nevada to Auburn, California over the steep Sierra Mountain range（自1955年起每年在美国举行的一项耐力赛，赛程从内华达州塔霍湖市至加利福尼亚州奥本市，为160千米，沿途经过陡峭的内华达山脉）特维斯杯耐力赛，又名西部越野赛 syn Western States Trail Ride

thalamencephalon /ˌθæləmenˈsəfəlɒn/ n. *see* diencephalon 间脑，丘脑

thatch /θætʃ/ n. **1** dried straw used for a roof covering 茅草 **2** a roof made of this material 茅草屋顶：The thatch was badly damaged in the storm. 茅草屋顶在暴风雨中遭到严重破坏。

vt. to cover with thatch 加盖茅草

Thelazia lacrymalis n. the scientific name for eye worm［学名］眼虫

therapeutic /ˌθerəˈpjuːtɪk/ adj. having or exhibiting healing powers 医疗的，有疗效的：therapeutic riding 骑马疗养，治疗骑乘

therapeutic shoe n.（also **therapeutic horseshoe**）a horseshoe used to treat or correct conformation defect, disease, or injury of the foot or leg such as founder or bruised heels（为矫正四肢缺陷或治疗蹄叶炎、蹄底挫伤等肢蹄疾病而使用的蹄铁）治疗性蹄铁，矫正蹄铁 syn pathological shoe, surgical shoe, corrective shoe

therapeutic ultrasound n. the medical use of high frequency sound waves to break down unwanted tissues and promote healing by stimulating circulation（采用高频声波来破碎多余组织或通过促进血液循环来加速治愈的方法）超声波治疗

therapy /ˈθerəpi/ n. the treatment or healing of illness or disability 治疗

thermography /θɜːˈmɒgrəfi/ n. the technique of using the infrared radiation emitted from the skin surface to form a visual image for diagnostic analysis（根据体表发射的红外线成像进行诊断的技术）红外线成像术

T

thermoregulation /ˌθɜːməʊˌregjʊˈleɪʃn/ n. the maintenance and regulation of a constant internal body temperature independent from the environmental changes (在环境可变情况下维持并调节自身的体温)体温调节:During low to moderate intensity exercise a significant amount of cardiac output is also directed to the skin to aid thermoregulation. 在做低、中强度运动时,相当一部分心输出的血液流到皮肤表面协助体温调节功能。| **thermoregulatory** /-ˈleɪtəri/ adj.

Thessalian /ˌθeɪsəˈlɪən/ n. an ancient Greek pony breed extinct today (古希腊矮马品种,现已灭绝)塞萨利矮马

thiamine /ˈθaɪəmɪn/ n.【nutrition】a water-soluble vitamin essential for normal metabolism and nerve function of body, deficiency may result in loss of appetite, nervousness, reduced fertility, and a lack of coordination (一种水溶性维生素,为机体正常代谢和维持神经活动所必需,其缺乏症常表现为食欲下降、紧张、生育力下降及运动失调)硫胺素 syn vitamin B_1

thick wind n.【AmE】difficult breathing of a horse 呼吸受阻,呼吸不畅

thickset /θɪkˈset/ adj. (of a horse) having a solid, stocky body; stout (指马)体格结实的,粗壮的

thief /θiːf/ n. **1** one who steals sth from another 盗贼: cattle thief 盗牛贼 **2**【racing】a horse who fails to win a race against all expectations 发挥失常的赛马,辜负众望的赛马

thigh /θaɪ/ n.【anatomy】the part of a horse's hind leg between the hip and the gaskin (马后肢从臀至小腿的部分)股:thigh bone 股骨

thill /θil/ n. **1** either of the two long shafts between which a horse is hitched to pull a carriage (马车的两根长杆,挽马套于其间)辕杆: thill horse 辕马 **2** a shaft horse; thiller 辕马

thiller /ˈθilə(r)/ n. a shaft horse 辕马

third eyelid n. *see* nictitating membrane 第三眼睑;瞬膜

third incisors n. *see* corner incisors 第三切齿

third man n.【polo】a referee who arbitrates when the two field umpires disagree on a call (比赛中当两名场内裁判意见出现分歧时进行公断的人)第三人,仲裁

third pastern n. the largest bone in a horse's foot 第三系骨,蹄骨 syn os pedis, third phalanx, coffin bone, distal phalanx, pedal bone

third phalanx n. *see* third pastern 第三趾骨(即蹄骨)

third sire n. the paternal great-grandfather of a horse 曾祖公马

thong /θɒŋ; *AmE* θɔːŋ/ n. a strip of leather used as the lash of a whip (鞭上的)皮条
vt. to flog or lash with a whip 鞭打,抽打

thoracic /θɔːˈræsɪk/ adj. of, relating to, or situated in or near the thorax 胸的:thoracic cavity 胸腔 ◇ thoracic vertebrae 胸椎 ◇ thoracic inlet 胸腔口

thorax /ˈθɔːræks/ n. the part of a horse's body between the neck and the diaphragm; chest (马体位于颈与横膈膜之间的部分)胸部,胸廓

thoropin /ˈθʌrəpɪn/ n. *see* thoroughpin 鞘膜肿胀

thoroughbrace /ˈθʌrəbreɪs/ n. (also **thorough brace**) any of the strong leather straps upon which the body of a carriage is suspended in place of springs (减震弹簧普及之前用来悬挂马车车身的皮带)减震皮带,吊车皮带

Thoroughbred /ˈθʌrəbred; *AmE* ˈθɜːroʊbred/ n. a horse breed developed in England in the early 18th century and descended mainly from the three famous stallions-Darley Arabian, Byerly Turk and Godolphin Arabian; stands 14.3 to 17 hands and usually has a bay, chestnut or gray coat with markings; used mainly for racing and riding (英国在18世纪初育成的马品种,主要由达利·阿拉伯、比艾尔力·土耳其和哥德芬·阿拉伯三匹奠基公马繁育而来。体高14.3~17掌,毛色多为骝、栗或青,带白色别征,主要用于赛马和骑乘)纯血马:

The Thoroughbred is most commonly known as a racing horse, although Thoroughbreds excel in many disciplines. 尽管纯血马在许多项目上都表现不俗,但它们主要以赛马而著称于世。

Thoroughbred Breeders Association n. (abbr. **TBA**) an organization founded in 1917 in England to promote the breeding of Thoroughbreds by encouraging the cooperation between breeders (1917 年在英国成立的组织机构,旨在通过加强纯血马育种者的合作来提高纯血马的品质)英纯血马育种者协会

Thoroughbred Racing Association n. (abbr. **TRA**) an association founded in 1942 in North America by racetrack owners and managers to promote and maintain the integrity of racing business and provide related support (1942 年在北美由赛马场主和管理者成立的行业协会,旨在促进并维护赛马产业的公正性并提供相关支持)北美纯血马赛马协会

Thoroughbred Racing Protective Bureau n. (abbr. **TRPB**) a wholly owned subsidiary of the Thoroughbred Racing Association of North America founded in 1945 to operate as a national investigative agency in the racing business against corruption (北美纯血马赛马协会的独立分支机构,于 1945 年成立,主要负责调查赛马业腐败情况)美国赛马监管局

thoroughpin /ˈθʌrəpɪn/ n. (also **thoropin**) a fluid-filled swelling of the sheath of the deep flexor tendon at the hock joint (马跗关节鞘膜由于内液淤积造成的水肿)趾深屈腱鞘肿胀

Thracian /ˈθreɪʃn/ n. (also **Thracian pony**) an ancient Greek pony breed extinct today (古希腊矮马品种,现已灭绝)色雷斯矮马

threadworm /ˈθredwɜːm; *AmE* -wɜːrm/ n. (also **thread worm**) a round worm of the genus *Strongyloides westeri* about 0.5 ~ 1 cm long that lives in the duodenum of a foal (寄生在幼驹十二指肠内的类圆线虫属圆形虫,体长 0.5 ~ 1 厘米)韦氏类圆线虫

three-day event n. (also **eventing**) an equestrian competition conducted over a period of three days and consisting of dressage, show jumping, and speed and endurance phases (一项为期三天的马术比赛,分盛装舞步、障碍赛和速度耐力三个部分)三日赛 [syn] combined training

three-eighths pole n. 【racing】 a marking pole located on the inside rail three furlongs from the finish line (立于赛道内栏的标杆,距终点 600 米) 3/8 英里杆,600 米杆

three-gaited /ˌθriːˈgeɪtɪd/ adj. (of a horse) capable of performing the three natural gaits of walk, trot, and canter (指马能够采用慢步、快步和跑步三个自然步法行进)三步调的: three-gaited saddler 三步调乘用马

three-quarter pole n. 【racing】a distance marking pole located on the inside rail six furlongs from the finish line (立于赛道内栏的标杆,距终点 1 200 米)3/4 英里杆,1 200 米杆

three-quarters /ˌθriːˈkwɔːtəz/ adj. 【genetics】 (of horses) genetically having very close relationship (指马)有 3/4 血缘的: three-quarters brothers 有 3/4 血缘的兄弟 ◇ three-quarters sisters 有 3/4 血缘的姐妹 ◇ three-quartes in blood 共有 3/4 的血缘

threonine /ˈθriːəniːn/ n. 【nutrition】 a colorless crystalline essential amino acid (一种无色晶体状必需氨基酸)苏氨酸

thrifty /ˈθrɪfti/ adj. (of animals) growing vigorously on small rations; thriving (动物)壮实的,耐粗饲的: a thrifty doer 一匹壮实的马 [opp] unthrifty | **thriftiness** /-tɪnəs/ n.

thrive /θraɪv/ vi. to grow vigorously; gain condition 茁壮生长;长膘: In good condition the horse will be seen to be not only surviving but also thriving. 体况好的马不但能存活,还能长膘。

throat /θrəʊt; *AmE* θroʊt/ n. **1** the front part of the neck 咽部 **2** 【anatomy】the portion of the digestive tract between the rear of the mouth

T

and the esophagus（消化道位于口腔和食管之间的部分）咽喉 syn throttle

throatlash /ˈθrəʊtlæʃ; *AmE* ˈθroʊ-/ n.（also **throat lash**）a narrow strap of the bridle buckled under the throat to prevent the bridle from slipping over the horse's head（马勒系于喉部的一条皮带，可防止马勒滑脱）喉革 syn throatlatch

throatlatch /ˈθrəʊtlætʃ; *AmE* ˈθroʊ-/ n.（also **throat latch**）*see* throatlash 喉革

Thro-bred /ˌθrəʊˈbred; *AmE* ˌθroʊ-/ n.【colloq】*see* Thoroughbred 纯血马

thrombin /ˈθrɒmbɪn; *AmE* ˈθrɑːm-/ n.【physiology】an enzyme in the blood partially responsible for the process of clotting（在血液凝固中发挥功用的一种酶）凝血酶

thrombocyte /ˈθrɒmbəsaɪt; *AmE* ˈθrɑːm-/ n. *see* platelet 血小板

thrombosis /θrɒmˈbəʊsɪs; *AmE* θrɑːmˈboʊ-/ n. the formation or development of a thrombus 血栓形成；凝血作用

thrombus /ˈθrɒmbəs; *AmE* ˈθrɑːm-/ n.（*pl.* **thrombi** /-baɪ/）a fibrous clot formed in a blood vessel or in a chamber of the heart（血管或心脏腔室中形成的纤维状凝块）血栓

throttle /ˈθrɒtl; *AmE* ˈθrɑːtl/ n. *see* throat 咽喉

throw /θrəʊ; *AmE* θroʊ/ vt. **1**【roping】to toss or fling in the air 扔起，抛起：to throw a calf to the ground 将小牛放倒在地 **2**（of a horse）buck a rider off its back（指马将骑手）撂下马 **3** PHR V **throw a race**【racing】（of a rider）to intentionally prevent the horse from winning a race（指骑手）有意输掉比赛，假输 **throw a shoe**（of a horse）to lose a horseshoe（指马）跑掉蹄铁 **throw tongue**【hunting】（of hounds）to give tongue（猎犬）伸舌头

throw-in /ˈθrəuin/ n.【polo】an act of throwing the polo ball into play from the midfield（从中场）掷球开赛，掷界外球

throwing hobbles n. hobbles attached to the pasterns to throw a horse to the ground（系在马蹄系处将其放倒在地的绳索）绊马绳

thrown /θrəʊn/ adj. **1**（of a rider）bucked off the back of the horse（骑手）被撂下马的 **2**（of livestock）cast or tossed to the ground for medical treatment, etc.（牲畜）被放倒的 **3**（of a horseshoe）detached from the hoof for any reason other than intentional removal（蹄铁）脱落的

thrush /θrʌʃ/ n. inflammation of the frog of a horse's foot caused by bacterial or fungal infection, often characterized by rotting and a foul smell（由于细菌或真菌感染导致的蹄叉发炎，症状表现为马蹄腐烂和恶臭）蹄叉腐烂 syn frush

thrust[1] /θrʌst/ vi.（of a stallion）to push its hindquarters forward and pull back after intromission into the vagina（公马阴茎插入母马阴道后臀部前后运动）抽送

n. such movement of penis after intromission in mating（交配时公马阴茎插入后的动作）抽送：pelvic thrust 胯部抽送

thrust[2] /θrʌst/ vi.【hunting】to ride hard into the hounds or other hunting staff 突入，撞击

thruster /ˈθrʌstə(r)/ n.【hunting】a member who rides hard into the hounds or hunting staff 突入猎犬的骑手；撞上他人的骑手

thumb of the Prophet n. *see* Prophet's thumb 先知拇指印，肩部凹痕

thumps /θʌmps/ n. *see* synchronous diaphragmatic flutter 膈膜同步颤动

thymus /ˈθaɪməs/ n.【anatomy】（*pl.* **thymuses** /-siːz/）a small glandular organ situated behind the top of the breastbone, consisting mainly of lymphatic tissue and serving as the site of T cell differentiation（位于胸骨后上部的小腺体，由淋巴组织构成，是T细胞分化的场所）胸腺：The thymus increases gradually in size and activity until puberty, undergoing involution thereafter. 在动情期之前胸腺不断变大且逐渐活跃，而后开始逐渐退化。

thyrocalcitonin /ˌθaɪrəʊˌkælsɪˈtəʊnɪn; *AmE* -roʊ

ˌkælsɪˈtoʊ-/ n.【physiology】*see* calcitonin 降钙素

thyroid /ˈθaɪrɔɪd/ n.【anatomy】the thyroid gland 甲状腺

adj. of or relating to the thyroid gland 甲状腺的

thyroid gland n.【anatomy】(also **thyroid**) a two-lobed, largest endocrine gland situated on either side of the larynx, which produces thyroid hormones that control the rate of basic body function (体内最大的内分泌腺,位于喉两侧,所分泌的甲状腺激素控制着机体的基础代谢率)甲状腺

thyroid hormone n.【physiology】any of the hormones secreted by the thyroid gland 甲状腺激素

thyroid stimulating hormone n.【physiology】(abbr. **TSH**) a hormone secreted by the pituitary gland that stimulates the thyroid gland to increase production of thyroxin (垂体分泌的激素,可刺激甲状腺素的分泌)甲状腺刺激素,促甲状腺素

thyroxin /θaɪˈrɒksɪn/ n.【physiology】(also **thyroxine**) an iodine-containing hormone produced by the thyroid gland that increases the rate of cell metabolism and regulates growth (甲状腺分泌的含碘类激素,能加速细胞新陈代谢并调节生长)甲状腺素

thyroxine /θaɪˈrɒksiːn/ n. *see* thyroxin 甲状腺素

Tibetan /tɪˈbetn/ n. (also **Tibetan horse**) a horse breed originating in the mountain areas of Tibet, China; stands about 13 hands and may have a coat of any color; has a straight profile, broad forehead, full forelock, mane and tail, small ears, a short, muscular neck, short back and short, sturdy legs; used for riding, packing, and farm work (源于中国西藏山区的矮马品种。体高约 13 掌,各种毛色兼有,面貌普通,额宽,额发与鬃尾浓密,耳小,颈粗短,背短,四肢粗短,用于骑乘、驮运和农田作业)藏马

Tibetan horse n. *see* Tibetan 藏马

Tibetan wild ass n. *see* kiang 藏野驴

tibia /ˈtɪbiə/ n.【anatomy】(*pl.* **tibiae** /-biiː/) a larger bone of the hind legs that extends from the thigh to the hock (后肢从股到跗的骨)胫骨 | **tibial** /-biəl/ adj.: tibial crest 胫骨脊 ◇ tibial tarsal 距骨 ◇ tibial tuberosity 胫骨隆起 ◇ lateral tibial condyle 胫骨外踝

tick /tɪk/ n. **1**【jumping】a slight touch of the obstacle without knocking down any part of it (触到障碍而没有撞掉任何部分)轻触,碰触 **2** any of various external, bloodsucking parasites that lives on the skin of horse (寄生于马体表的多种吸血性寄生虫)扁虱: Heavy infestation of ticks may cause anemia, unthriftiness of horses, and can transmit a large variety of diseases including encephalomyelitis and swamp fever. 扁虱大量寄生可造成马匹贫血、体况下降等并可传播多种疾病,如脑脊髓炎和马传染性贫血。

ticket /ˈtɪkɪt/ n. **1** a paper slip or card indicating that the holder is entitled to a certain service (证明持有者享有某种服务的单据)票,券: entrance ticket 门票,入场券 **2**【racing】(also **betting ticket**) a slip of paper serving as evidence of the bet placed on a racehorse (证明赛马投注的票据)彩券,彩票: pari-mutuel ticket 同注分利彩券

tie[1] /taɪ/ vt. **1** to fasten or secure with a cord, rope, or strap (用绳索或皮革系紧)系,拴 PHR V **tie up** 拴马: A horse should first be taught to accept being tied up in a stable and not pull back. 首先应当让马习惯被拴在厩舍而不扯缰。**2**【roping】to wrap a string around the three legs of a calf to immobilize it (将小牛的三条腿用绳子捆绑起来)捆绑 PHR V **tie off** 拴紧,系牢

tie[2] /taɪ/ vi. to equal or achieve an equal score in a contest 平局,平分

tie stall n. a partitioned box in which a horse is tied 拴马圈,拴马舍

tiedown martingale n. (also **tiedown**) *see* standing martingale［站立］低头革

tie-down roping n.【rodeo】*see* calf roping 套小牛

Tieling Draft n.【Chinese】a draft breed developed in Liaoning province of China by crossing the Anglo-Norman with Ardennais in the 1950s; stands about 15 hands and generally has a bay or black coat; has a medium head, large eyes, broad forehead, long, arched neck, deep girth, study legs with little feathering, and hard hoofs; used for draft and farm work（20 世纪 50 年代育成于中国辽宁省的挽马品种，主要由盎格鲁诺曼马与阿尔登马杂交育成。体高约 15 掌，毛色多为骝或黑，头适中，眼大，额宽，颈长且呈弧形，胸深，四肢强健，距毛少，蹄质结实，主要用于重挽或农田作业）铁岭挽马

tie-ring /ˈtaɪrɪŋ/ n. a metal ring fixed into the wall for tying up a horse（固定在墙上用来拴马的金属环）拴马环

tiger stripes n. *see* zebra stripes 斑马纹；虎斑

tight /taɪt/ adj. **1** keeping close to sth 紧跟的；紧追的 PHR **tight on a cow**【cutting】(of a horse) working a cow in close confrontation（指马）紧逼所截的牛 **2**【racing】(of a horse) ready to race（指马）准备出赛的

tightener /ˈtaɪtənə(r)/ n. **1**【racing】a race intended to bring a horse to the peak of its physical fitness（为提高马匹竞技状态组织的比赛）备战赛 **2** a leg brace or bandage 夹板；绑腿

Tilbury /ˈtɪlbərɪ/ n. *see* Tilbury gig 提尔堡马车

Tilbury gig n.（also **Tilbury**）a light, open, two-wheeled, horse-drawn vehicle developed by the London firm of Tilbury in the early 19th century（19 世纪初由伦敦提尔堡公司推出的轻型敞开式双轮马车）提尔堡马车

till /tɪl/ vt. to prepare land for the raising of crops, as by plowing and harrowing; cultivate（平整土地以种植庄稼，如犁地和耙地等过程）翻耕，耕种

tillage /ˈtɪlɪdʒ/ n. the preparation of land for growing crops 耕种，耕地

tiller /ˈtɪlə(r)/ n. one who tills land and grows crops 耕夫；种田人

tilt[1] /tɪlt/ vi. ~（**at sb/sth**）to charge or thrust with a lance in jousting 骑马刺枪：The medieval armoured knights used to tilt on horseback. 中世纪的骑士们常骑在马背上用长矛刺杀。

tilt[2] /tɪlt/ n. a covering or canopy of a horse-drawn wagon or cart（马车的）顶篷，顶盖

vt. to cover a vehicle with a covering or canopy 给马车搭顶篷

tilter /ˈtɪltə(r)/ n. one who tilts on horseback 骑马刺枪者

tilting /ˈtɪltɪŋ/ n. a mounted medieval sport or exercises in which the riders galloped to strike target post, tree truck, etc. by a long lance（一种中世纪的马上项目，骑士手持长矛策马刺向木杆或树桩等物）马上刺物

timber /ˈtɪmbə(r)/ n. **1**（AmE **lumber**）trees or wood used in building 树木；木料 **2**【jumping】a jumping obstacle constructed of wood 圆木障碍

timber rider n.【racing, slang】a steeplechase rider 越野赛骑手

timber topper n.【racing, slang】a horse who runs in jump races 障碍赛马

time /taɪm/ n. an interval during which an event occurs 时间：time faults/penalty 时间罚分

vt. to record the time of an event 计时：a timed event 计时项目

time allowed n.【jumping】a period of time in which a rider must complete a show jumping course without incurring time faults（骑手为了避免时间罚分而跨越所有障碍的）规定时间：A rider incurs time faults at the rate of one-quarter-fault for every one second over the time allowed. 骑手超过规定时间后，罚分以每秒 0.25 分累计。compare time limit

time limit n.【jumping】the maximum time in which a rider may complete a show jumping course（骑手跨越场地障碍所容许的最长时间）时限：A rider exceeding the time limit is eliminated. 骑手超过时限将遭淘汰。

Timor /ˈtiːmɔː(r)/ n.（also **Timor pony**）a pony breed indigenous to the island of Timor, Indonesia; stands 9 to 11 hands and usually has a black, bay, or brown coat; has a heavy head, short ears, flared nostrils, short neck and back, and strong quarters; is sure-footed, docile, strong, and willing; used for riding, farm work, and packing（源于印度尼西亚帝汶岛的矮马品种。体高 9 ~ 11 掌，毛色多为黑、骝或褐，头重，耳短，鼻孔大，颈背短，后躯发达，步伐稳健，性情温驯、顽强，主要用于骑乘、农田作业和驮运）帝汶矮马

timothy /ˈtɪməθi/ n. *see* timothy grass 梯牧草，猫尾草

timothy grass n.（also **timothy**）a perennial Eurasian coarse grass with dense cylindrical spikes or bristly spikelets, widely cultivated for hay（一种产于欧亚大陆的多年生草本植物，锥状花序，小穗单花，大量种植以生产干草）梯牧草

timothy hay n. the dried hay made from timothy grass 梯牧干草

tincture /ˈtɪŋktʃə(r)/ n. an alcohol solution of a nonvolatile medicine（非挥发药剂的酒精溶液）酊剂：mother tincture 酊剂母液 ◇ iodine tincture 碘酒

tine /taɪn/ n. a prong a fork or pitchfork 叉齿，叉股：a four-tine fork 四股叉

tip[1] /tɪp/ n. the inside information as in a race or event（赛马中的内部信息）消息，机密
vi. to provide private or secret information 泄密

tip[2] /tɪp/ vt. **1** to give a small sum of money for some service 付小费 **2** to strike lightly and sharply; to tap 轻敲，轻击

tipster /ˈtɪpstə(r)/ n.【racing】one who makes a business by providing betting information about horse races to bettors 兜售赛马机密的人

tit /tɪt/ n.【slang】a small, undersized horse 短小的马，瘦小的马

toad eyes n.【BrE】a condition typical of Exmoor Pony in which the eyes bulge prominently with fat eyelids（马眼高突且眼睑过厚的情况，多见于埃克斯穆尔矮马）蛤蟆眼 [syn] buck eye

tobiano /ˌtəʊbɪˈɑːnəʊ/ n. a color pattern in pinto horses with white legs and regular and smooth-edged patches of white on the body（花马毛色类型之一，其四肢为白色，被毛上的白色斑块较为规则且边缘整齐）齐边白花毛，托比阿诺 [compare] overo, sabino, tovero

tocopherol /təʊˈkɒfərɒl; *AmE* toʊ-/ n.【nutrition】any of a group of closely related, fat-soluble alcohols that function as a biological antioxidant（一组结构类似的脂溶性醇类，是生物体内的抗氧化剂）生育酚 [syn] vitamin E

toe /təʊ; *AmE* toʊ/ n. **1** the forepart of a horse's hoof 蹄尖 IDM **on its toes**（of a horse）eager or anxious to move（指马）不安的；急躁的 **keep sb on their toes** to make sb alert or ready for any eventuality 使人保持戒备；使人常备不懈：He carries out random checks to keep the staff on their toes. 他进行随机检查，以便让员工都常备不懈。**make sb's toes curl** to make sb feel embarrassed or uncomfortable 使人尴尬，使人难为情 **2**【farriery】the forepart of a horseshoe 蹄铁头，蹄铁尖
vi.（of a horse）to stand with the toes pointed in a certain direction（指马站立时）蹄尖朝向 PHR V **toe in**（of the toes）to point towards each other（蹄尖）朝内，内翻 **toe out**（of the toes）to point away from each other（蹄尖）朝外，外翻

toe clip n.【farriery】a V-shaped extension at the toe of certain horseshoe（蹄铁尖的 V 形突起）蹄铁唇 [compare] quarter clip

toed-in /ˌtəʊdˈɪn/ adj. *see* pigeon-toed 蹄内翻的，内八字的 [compare] toed-out

toed-out /ˌtəʊdˈaʊt/ adj. *see* splay-footed 蹄朝外的，外八字的 [compare] toed-in

toeing knife n.【farriery】a mallet-driven blade

T

used by a farrier to trim the hoof wall (用来削蹄壁)削蹄刀

tölt /tølt/ n. (also **tølt**) a natural four-beat gait similar to rack and typical of the Icelandic horse (一种类似单蹄快步的四蹄音步法,为冰岛马所特有)托特步

tone /təʊn; *AmE* toʊn/ n. **1** the normal elastic tension or baseline contraction in resting muscles (处于静息状态的肌肉正常的紧张度和收缩性)张力: Tone exists to prepare muscle for contraction and to maintain posture. 肌肉张力可以使肌肉做好收缩准备,同时也可维持身体姿势。**2** the normal firmness of a tissue or an organ (组织或器官具有的正常紧缩度)紧缩性;弹性: muscle tone 肌肉弹性

tonga /ˈtɔŋgə; *AmE* ˈtɑːŋgə/ n. (also **tonga cart**) a two-wheeled, hooded, horse-drawn vehicle formerly used in India, usually pulled by a single horse (印度过去使用的带篷双轮马车,多由单马拉行)印度双轮马车

tongue /tʌŋ/ n. a freely movable, muscular organ attached to the floor of an animal's mouth that aids in grazing, chewing, and drinking (附着在动物口腔底部可自由活动的肌性器官,主要协助采食、咀嚼与饮水)舌 IDM **with tongue over the bit** 舌位于衔铁之上 PHR V **give tongue**【hunting】(of a hound) to bark on the line of the quarry (指猎犬)狂吠,吠叫 syn throw tongue, throw his tongue

tonic /ˈtɒnɪk; *AmE* ˈtɑːn-/ adj.【nutrition】producing or stimulating physical, mental, or emotional vigor 滋补的: tonic feeds 滋补性饲料

n. (usu. **tonics**) any agent or medication that restores or increases vigor 滋补品

tonsil /ˈtɒnsl; *AmE* ˈtɑːnsl/ n.【anatomy】a small mass of lymphoid tissue embedded around pharynx and acting to protect the body from respiratory infections (位于咽喉处的一块淋巴组织,有防止上呼吸道感染的作用)扁桃体: lingual tonsil 舌扁桃体 ◇ palatine tonsil 腭扁桃体

tonsilla /ˈtɒnsɪlə; *AmE* ˈtɑːn-/ n.【Latin】(*pl.* **tonsillae** /-liː/) tonsil 扁桃体

tonsillar /ˈtɒnsɪlə(r); *AmE* ˈtɑːn-/ adj. of or relating to the tonsil 扁桃体的

tool[1] /tuːl/ n. a device used to perform a certain task 用具,工具: farrier's tools 装蹄工具

tool[2] /tuːl/ vt.【driving】to drive a horse-drawn vehicle 驾驶马车: tooled the cart at 10 miles an hour. 驾马车以 16 千米每小时的速度行进。

tool budget n.【driving】a leather box attached to the forepart of a horse-drawn vehicle for keeping tools (马车前面安装的小皮箱,用来放置相关工具)工具箱

tooth /tuːθ/ n. (*pl.* **teeth** /tiːθ/) any of a set of hard bones in the mouth for biting and chewing food (口腔中用以撕咬或咀嚼食物的一排硬骨)牙齿 PHR **long in the tooth** (of a horse) old in age (指马)年老的,牙口老的

tooth decay n. a decay of tooth resulting from bacterial infection (牙齿因细菌感染导致的腐烂)龋齿,蛀牙 syn caries

tooth rasp n. a rasp used to level off sharp edges on the molars of horses (用来磨平马臼齿锋棱的锉)牙锉

tooth rasping n. (also **rasping**) the practice of filing off the sharp edges of the upper molars (锉平臼齿锋棱的做法)锉牙 syn floating, filing

tooth star n. *see* dental star 齿星

top /tɒp; *AmE* tɑːp/ n. **1** the upper part of sth 顶端 PHR **on top**【racing】(of a horse) running in the lead (指马)领先的: The horse kept on top all through the course. 这匹马在比赛中始终保持了领先。**2** *see* carriage hood (马车的)顶篷,车盖

vt. to pinch or cut the top of a plant (剪掉植物的顶端)去顶

adj. of first quality; first-rate 顶级的,一流的

top horse n.【racing】the first horse listed in a race program (列在赛事程序首位的马)首马

top line n. **1** the line from the withers to the end of the croup (马背从鬐甲到尻的线条)背部轮廓线 [syn] backline **2** the breeding line on the sire's side of a pedigree chart (育种系谱中的)父系系谱

top pommel n. (also **near pommel**) the higher of the two padded pommels on the side-saddle (侧鞍两鞍桥中位置较高的前桥)上鞍桥 [syn] fixed head, hunting head, near head [compare] lower pommel

top rider n. an accomplished rider 顶级骑师,顶级骑手

topping /ˈtɒpɪŋ; *AmE* ˈtɑːp-/ n. the act of cutting off the top of grass or plants; mowing (割去草本或植物的顶端)去顶;刈割

top-quality /ˌtɒpˈkwɒlətɪ/ adj. of high quality 优质的:top-quality hay 优质干草

toreador /ˈtɒrɪədɔː/ n. a bullfighter 斗牛士

torero /təˈrerəʊ/ n.【Spanish】(*pl.* **toreros**) a bullfighter 斗牛士

Toric /ˈtɔːrɪk/ n. (also **Torisky**) an Estonian warmblood developed recently by crossing Arab, Ardennais, Hackney, East Friesian, Hanoverian, Orlov, Thoroughbred and the Trakehner with the Klepper; stands about 15 hands and may have a bay or chestnut coat; has a long neck, short back, deep and wide chest, and tough hoofs; used for draft and farm work (爱沙尼亚育成的温血马品种,主要采用阿拉伯马、阿尔登马、哈克尼马、东弗里斯兰马、汉诺威马、奥尔洛夫快步马、纯血马以及特雷克纳马与克莱波马杂交培育而成。体高约15掌,毛色多为骝或栗,颈长,背短,胸深,蹄质结实,主要用于重挽和农田作业)托利马 [syn] Estonian Klepper

Torisky /ˈtɔːrɪski/ n. *see* Toric 托里斯基马,托利马

torsion /ˈtɔːʃn; *AmE* ˈtɔːrʃn/ n. the act of twisting or condition of being twisted 扭曲,缠结

torsion colic n. an intense abdominal colic resulting from the twisting of digestive tract or blocked passage of feed through the intestines (由消化道缠结或食物阻塞引起的剧烈腹痛)肠绞痛,肠扭转 [syn] twisted gut, twisted bowel

torso /ˈtɔːsəʊ; *AmE* ˈtɔːrsoʊ/ n. (*pl.* **-os**) the trunk of the body 躯干

total digestible nutrients n.【nutrition】(abbr. **TDN**) the total amount of fat, protein, and carbohydrate in the feed that indicates its energy density (饲料中脂肪、蛋白质和碳水化合物的总含量,是饲料能量密度的指标)总消化养分

total lung capacity n.【physiology】the total volume of air that can be contained by the lung (肺部可容纳的气体总量)肺活量

totalizator /ˈtəʊtəlaɪzeɪtə(r); *AmE* ˈtoʊt-/ n.【racing】(also **totalizer**) a pari-mutuel machine showing the total number and amounts of bets at a racetrack (一种同注分彩电子仪器,可显示投注场次和金额)赌金计算器

totalizator board n.【racing】(also **tote board**, **odd board**) an electronic board located at the racecourse upon which the betting odds for each horse and the amounts wagered in each mutual pool are displayed (赛场上的电子公告板,可显示每场比赛中每匹赛马的投注赔率以及每种投注类型的赌金总额)投注显示屏,赌金显示屏

totalizer /ˈtəʊtəlaɪzə(r); *AmE* ˈtoʊt-/ n. *see* totalizator 赌金计算器

tote /təʊt; *AmE* toʊt/ n.【colloq】the totalizator 赌金计算器

tote board n.【colloq】the totalizator board 投注显示屏,赌金显示屏

touch-and-out n.【jumping】a jumping competition in which the rider will be disqualified once the obstacle is touched (骑手碰触障碍就被淘汰的比赛)触障淘汰赛

tournament /ˈtʊənəmənt; *AmE* ˈtʊrn-/ n. **1** a series of events in which teams, individuals, or horses compete against one another (团队、个

体或马匹之间的系列比赛)锦标赛;联赛 **2** a medieval competition in which two knights jousted on horseback with long lances; tilting (中世纪两名骑士策马以长枪挑刺的比赛)马上刺枪

tout /taʊt/ n. 【racing】 one who peddles betting information to the racegoers in advance of a race (开赛前)兜售赛马信息者

vi. **tout around** to obtain and peddle information on racehorses 兜售赛马信息

tovero /təʊˈverəʊ/ n. a color pattern in pinto horses that commonly comes from crossing a Tobiano horse with an Overo colored horse (花马毛色类型之一,多由齐边白花马和齿边白花马杂交繁育而成)杂花毛,托沃罗 compare overo, sabino, tobiano

towel /ˈtaʊəl/ n. a piece of thick absorbent cloth used for wiping the body of a horse (用来擦拭马体的吸水性厚织物)毛巾

vi.【driving】to flog a coach horse 抽打

Town coach n. an elegant, heavy coach with a box-seat covered with fringed hammer cloth and a rear platform for two footmen; often drawn by a pair of matching bay horses (一种外观优雅的定制四轮马车,车厢缀有流苏,车身后台可立乘两门随从,由两匹体貌相称的骝毛马拉行)市镇仪仗马车 compare State coach

Town Plate (at Newmarket) n. *see* Newmarket Town Plate 纽马基特奖杯赛

toxemia /tɒkˈsiːmiə; *AmE* tɑːk-/ n. (BrE also **toxaemia**) a condition in which the blood contains toxins produced by body cells at a localized source of infection or derived from the growth of microorganisms (血液中含有毒素的症状,由病灶感染处的细胞或入侵的微生物产生)毒血症 syn blood poisoning | **toxemic** /-mɪk/ adj.

toxic /ˈtɒksɪk; *AmE* ˈtɑːk-/ adj. of, relating to, or caused by a toxin or other poison 有毒的;中毒的

toxicity /tɒkˈsɪsəti; *AmE* tɑːk-/ n. the quality or state of being toxic 毒性

toxicosis /ˌtɒksɪˈkəʊsɪs; *AmE* ˌtɑːksɪˈkoʊ-/ n. an abnormal or diseased condition resulting from poisoning 中毒症状

toxin /ˈtɒksɪn; *AmE* ˈtɑːk-/ n. any of the poisonous substances of plant or animal origin that are capable of causing disease when introduced into the body tissues (源于动植物的有毒物质,进入身体组织后能够导致疾病)毒素

toxoid /ˈtɒksɔɪd; *AmE* ˈtɑːk-/ n. a substance that has been treated to destroy its toxic properties but retains the capacity to stimulate production of antitoxins in immunization (经过处理毒性已被破坏,但仍能刺激机体免疫系统产生抗毒素的一种物质)类毒素

trace /treɪs/ n. **1** an extremely small amount or quantity 微量,痕量:trace element 微量元素 **2**【driving】one of two side straps or chains connecting a harnessed draft horse to a vehicle (左右两条用来将挽马和马车连接起来的皮带或铁链)挽绳,靷绳

trace-mineralized salt n. salt containing a mixture of the microminerals 含微量矿物质的盐

trachea /trəˈkiːə; *AmE* ˈtreɪkiə/ n.【anatomy】a thin-walled tube of cartilaginous and membranous tissue descending from the larynx to the bronchi and carrying air to the lungs (一个由软骨和膜组织构成的薄壁管道,从喉部向下进入支气管,并携带氧气进入肺部)气管 syn windpipe | **tracheal**/-kiːl/ adj.

trache(o)- pref. trachea 气管

tracheobronchial /ˌtreɪkɪəʊˈbrɒŋkɪəl; *AmE* -kɪoʊ-/ adj. of or relating to the trachea and the bronchi 气管的;支气管的

tracheotomy /ˌtrækiˈɒtəmi; *AmE* ˌtreɪkiˈɑːt-/ n. the practice of cutting into the trachea and inserting a breathing tube to enable a horse to breathe in emergent situations where the trachea is obstructed or swollen (在马气管阻塞或肿胀等紧急情况下将其切开并插入细管辅助呼吸的手术)气管切开术 syn tubing

track /træk/ n. **1**【racing】an area or a course where horses run during training or racing (马匹进行训练或赛跑的场地)赛道,跑道:dirt track 泥土赛道 ◇ fast tract 高速赛道 ◇ hard track 硬赛道 ◇ muddy track 泥赛道 ◇ sand track 沙土赛道 ◇ soft track 软赛道 ◇ track announcer 赛场播音员 ◇ turf track 草地赛道 [syn] racetrack, race track, racecourse, course, racing strip, strip **2** a path or trail left by repeated passage of persons, animals, or vehicles (行人、动物或车辆留下的行径)路径 **3** a succession of footprints or other marks left on the ground; trace (地面上留下的)踪迹,行踪

vt. to follow the track of a fox, etc.; trail, pursue 跟踪,追踪

track bandage n. *see* exercise bandage 训练绷带

track bet n.【racing】any bet placed at the racetrack 场内投注 [compare] off-track bet

track condition n.【racing】the state or nature of a racing surface 赛场地况

track master n.【racing】one responsible for maintaining the condition of the race track (负责维护赛道的人员)赛道管理员

tract /trækt/ n.【anatomy】a passage or canal in an organ or body part (机体器官内的通道)管道

tractable /ˈtræktəbl/ adj. (of a horse) easily managed or controlled; obedient (指马)易于驾驭的;温驯的

traffic jam n.【racing】*see* jam 扎堆,拥挤

traffic-proof /ˈtræfɪkˌpruːf/ adj. (of a horse) not frightened by running vehicles (指马)不怕车辆的;镇定的

trail /treɪl/ n. **1** a mark or trace left by sth that has moved or been dragged by; trace, track (物体移动或拖曳留下的痕迹)痕迹;踪迹 **2** a marked or beaten path through woods or countryside (树林或乡野的)小径,小道:trail riding 越野骑乘

v. **1** to drag or be dragged heavily on the ground 拖曳;拖行 **2** to follow the course taken by; pursue 跟踪,追踪 **3** to lag behind 落后 [PHR V] **trail up** (of a racehorse) to lose in a race (指赛马)失利,跑输

trail horse n. a horse used for cross-country riding 越野乘用马

trail ride n. a cross-country riding or endurance race 越野骑乘;越野赛马

trailer /treɪlə(r)/ n. (also **horse trailer**) a vehicle or van towed behind a truck or tractor to transport horses (连接在卡车后面用来运输马匹的车辆)拖车

vt. to transport by a trailer 用拖车运

trailing foreleg n. *see* nonleading leg 非起步前肢,滞后前肢

train /treɪn/ vt. to teach a horse with instruction and practice to complete certain performance (以口令训练马匹完成特定动作)训练,调教:a well-trained horse 一匹调好的马 [PHR V] **train off** to overtrain a horse to the point that his performance deteriorates 训练过度 | **training** /-nɪŋ/ n.

trainable /ˈtreɪnəbl/ adj. easily trained 易于调教的,可调教的:The superior mentality of horse makes him more trainable than others animal except the dog. 除犬以外,马杰出的悟性使它比其他任何动物都易于调教。

trained /ˈtreɪnd/ adj. gained or acquired by training; artificial 训练获得的;人工的

trained gait n. a gait trained to the horse; artificial gait (马经调教后具有的步法)人工步法

trainer /ˈtreɪnə(r)/ n. **1** one who trains athletes or horses 教练;驯马师:certified horse trainer 注册练马师 ◇ An excellent trainer is one who can balance the horse's work and feed to produce an athlete ready to give it's best. 一名出色的练马师应当根据马匹工作量合理搭配饲料,让马发挥最大的运动潜能。**2** an apparatus used in training horses 练马机

training /ˈtreɪnɪŋ/ n. the process or state of being trained 训练,调教:interval training 间断性训练 ◇ racehorse training 赛马训练 ◇ training

T

duration 训练时间 ◇ training facilities 训练设备 ◇ training frequency 训练频率 ◇ training intensity 训练强度 ◇ training objectives 训练目的 ◇ training principles 训练原则 ◇ training programme 训练计划 ◇ training volume 训练量 ◇ combined training 综合训练 ◇ Training leads to an increase in fitness and an improvement in performance. 训练可提高马匹的体能和出赛成绩。

training cavesson n. *see* lungeing cavesson 调教索,调教笼头

training plate n. a light horseshoe made of steel and used on racehorses in training (一种轻型钢制蹄铁,赛马训练时使用)训练蹄铁

training response n. a long-term physiological adaptation to repeated bouts of increase muscular activity (机体肌肉在强度逐渐增加的训练过程中所形成的长期生理性适应)训练性应答

training track n.【racing】a racetrack used for exercising horses rather than for competition 训练赛道,训练场

trait /treɪt/ n. **1** a distinguishing feature or quality (事物的显著特点)特征 **2** a genetically determined characteristic or quality (由基因决定的)遗传特征;性状:qualitative trait 质量性状 ◇ quantitative trait 数量性状

Trakehner /ˈtrækənə(r)/ n. a warmblood breed developed at the stud of Trakehnen in East Prussia during the 18th century; stands around 16 hands and may have a chestnut, bay, or black coat; has a well-proportioned head, long neck, pronounced withers, and deep chest; is intelligent, lively and elegant; often used in dressage and combined training (18 世纪育成于东普鲁士特雷克纳种马场的温血马品种。体高约 16 掌,毛色多为栗、骝或黑,头型匀称,颈长,鬐甲突出,胸深,反应灵敏,性情活泼,主要用于盛装舞步赛和三日赛等马术竞技项目)特雷克纳马 [syn] East Prussian horse

trandem /ˈtrædəm/ n. a team of three horses driven abreast 三马并驾 [syn] Manchester team

tranquil /ˈtræŋkwɪl/ adj. free from tension, anxiety or disturbance; calm 安静的

tranquilize /ˈtræŋkwəlaɪz/ vt. to sedate or relieve of anxiety or tension by the administration of a drug (采用药剂来镇静或缓解焦虑及紧张)使镇静

tranquilizer /ˈtræŋkwəlaɪzə(r)/ n. a drug used to reduce tension or anxiety; sedative (用来缓解紧张或焦虑的药物)镇定剂,镇静剂

Transcaspian wild ass n. *see* kulan 土库曼野驴,库兰野驴

transit fever n. *see* shipping fever 船运热,运输热

transition /trænˈzɪʃn/ n. a change of pace or speed from one gait to another 步法转换,换步:upward transition 上行换步,向上转换 ◇ downward transition 下行换步,向下转换 ◇ rough transition 换步过猛

transition not defined n. a change from one gait to another without distinct difference between the speed or pace (步法转换中速度与步幅没有明显变化)换步不明显

transtracheal aspiration n. (abbr. **TTA**) a medical technique used to collect a sample of bronchial secretions or to remove fluid from the lungs by inserting a needle into the windpipe through the skin (用注射器经皮肤扎入气管采集支气管分泌液样品或抽吸肺内积液的医疗技术)气管穿刺抽吸

trap /træp/ n. a light two-wheeled carriage with springs, mainly designed for country driving (一种装有弹簧的双轮轻便马车,主要用于野外驾乘)双轮轻便马车

trapezius /trəˈpiːzɪəs/ n.【anatomy】either of the two large, flat, triangular muscles running from the second cervical vertebrate to the middle of the back that support and make it possible to raise the head and shoulders (自第二颈椎到脊背中段的两大块三角形肌,有支撑并提举头与肩的作用)斜方肌:cervical part of trapezius muscle 颈斜方肌 ◇ thoracic part of trapezius muscle 胸斜方肌

trapper /ˈtræpə(r)/ n.【slang】a trotting harness horse 轻驾车赛马

trappings /ˈtræpɪŋz/ n. the ceremonial harness and decorations for a parade horse (仪仗马所用的挽具和饰物)仪仗挽具

trappy /ˈtræpi/ adj. (of a surface) uneven and rough (路面)崎岖的,艰险的:Hunting horses must be able to negotiate trappy terrain for many miles and hours. 狩猎马须能穿越艰险的地段,长途跋涉几个小时并行进上百公里。

trauma /ˈtrɔːmə; *AmE* ˈtraʊmə/ n. a serious physical injury resulting from an accident (意外事故给身体造成的严重伤害)创伤,外伤 | **traumatic** /trɔːˈmætɪk/ adj.

travers /ˈtrævəs/ n.【dressage】a movement performed on two tracks 腰[向]内斜横步 syn haunches-in, quarters-in

traverse /ˈtrævɜːs; *AmE* -vɜːrs/ n.【dressage】a lateral movement of the horse to the left or right without moving forward or backward (马向左或右横向运行的步法)横步 syn side step

travois /trəˈvɔi/ n. (also **travoy** or **travoise**) an A-shaped, horse-drawn wooden frame widely used by native Americans for conveying goods and belongings (一种A字形马拉托运架,北美印第安人主要用来运送物资)马拉拖架

travoise /trəˈvɔiz/ n. *see* travois 马拉拖架

travoy /trəˈvɔi/ n. **1** *see* travois 马拉拖架 **2** a military stretcher used during the First World War with one end attached on to a horse, while the rear end supported and steered by a medical orderly on foot (第一次世界大战中使用的担架,一端搭在马背,另一端由医疗勤务兵步行抬着)马拉担架

tread /tred/ v. to press or crush beneath the feet; trample 踩,踏

treadmill /ˈtredmɪl/ n. a machine consisting of a wide moving belt upon which the horse can trot or canter for exercising purposes (一台带有转动皮带的器械,马匹可以在上面进行快步和跑步练习)跑步机

treble /ˈtrebl/ n.【jumping】*see* triple combination 三连组合障碍

tree /triː/ n. *see* saddle tree 鞍芯

trefoil /ˈtrefɔɪl, ˈtriːfɔɪl/ n. any of various plants of the genus *Trifolium* having compound trifoliate leaves (一种车轴属植物,生有三个复合小叶)车轴草,三叶草

trek /trek/ n. **1** a long journey made on foot; hike 徒步旅行 **2** a long journey travelled on horseback 野外骑乘

vi. **1** to go on a long journey 长途旅行 **2** to travel on horseback 野外骑乘 **3** to travel by pony pulled wagon 驾车旅行 | **trekking** /ˈtrekɪŋ/ n. 徒步旅行;野外骑乘

trekking center n. a commercial establishment that provides horses for holiday trekking (出租马匹供游客野外骑乘的商业机构)野外骑乘中心

trencher /ˈtrentʃə(r)/ n. a wooden board on which food is cut or served (切割食物的)案板;(盛餐的)木盘

trial /ˈtraɪəl/ n.【racing】a preparatory race or workout held shortly before an important race to improve its speed (大赛前为了提高马匹速度而进行的预赛或训练)试闸,试跑:the Derby trial 德比赛试闸

Tribus /ˈtraɪbəs/ n. a horse-drawn carriage first introduced in the 1840s that could accommodate three passengers (19世纪40年代出现的一种出租马车,可容纳三名乘客)三座马车

triceps /ˈtraɪseps/ n.【anatomy】a large three-headed muscle running along the upper arm and serving to extend the forearm (位于上臂的一块大三头肌,用于伸展前臂)三头肌:triceps muscle 三头肌

tri-color hound n. a hound with a mixed black, tan, and white coat (混有黑、褐、白三种毛色的猎犬)三色犬

trifecta /traɪˈfektə/ n.【racing】a wager in which the bettor selects the first three winners of a

T

race in any order（投注者选出前三名而不考虑次序的押注方式）头三彩：The trifecta has two types, straight trifecta and boxed trifecta. 头三彩有头三正序彩和头三组合彩两类。

trifecta box n. a trifecta wager in which all possible combinations for a specific number of horses are included（一定数目的马匹构成的各种可能的头三彩）头三彩组合

triga /ˈtriːgə, ˈtraɪgə/ n.（*pl.* **trigae** /ˈtriːgaɪ, ˈtraɪdʒiː/）a two-wheeled Roman chariot pulled by three horses harnessed abreast（古罗马时期由三匹马并驾的双轮战车）三驾战车 —*see also* **biga, quadriga**

trim /trɪm/ vt. to make neat or tidy by clipping or cutting 修剪：When using clippers to trim the ears of horse, you should hold the ear flat in one hand. 修剪马耳朵的毛发时，要用一只手将马的耳朵弄平展后再修剪。

trimmer /ˈtrɪmə(r)/ n. **1** one who trims 修剪工 **2** one who decorates the interior of horse-drawn vehicles with materials such as silk and lace（用丝绸和彩带装潢马车内部的人）马车装潢工

T

trip[1] /trɪp/ n.【racing】the course taken by a horse in a race from the start to finish（赛马从起点到终点所行进的路程）路线，里程

trip[2] /trɪp/ n. a stubble or fall 摔倒，绊倒
v.（to cause）to stumble or fall in strides 摔倒，绊倒

triple bar n.【jumping】a jumping obstacle consisting of three rails set up at a ramped slant（由三条横杆搭建的梯形坡面障碍）三横栏 compare oxer, post-and-rail

triple buckboard n. a four-wheeled, horse-drawn carriage having three seats in a row（一种四轮马车，单排座位可坐三人）三座平板马车

triple combination n.【jumping】an obstacle consisting of three consecutive jumps with one or two strides between each jump（由相继三个跨栏组成的障碍，其间马匹可跨1～2步）三连组合障碍 syn treble

Triple Crown n. **1** the three classic races for three-year-olds run in the United States since 1867, which include the Kentucky Derby, the Preakness Stakes, and the Belmont Stakes（美国自1867年起为3岁驹举行的三项经典比赛，包括肯塔基德比赛、普瑞克尼斯大赛和贝尔蒙特大赛）三冠王：The equivalent races of Triple Crown are also run in England（the 2,000 Guineas, Epsom Derby, and St. Leger Stakes）and Canada（the Queen's Plate, Prince of Wales Stakes, and Breeders' Stakes）. 与美国三冠王等同的赛事在英国有两千几尼赛、艾普森德比赛和圣莱切奖金赛，在加拿大有女王杯、威尔士王子大赛和育马者大赛。**2** a racing series consisting of three Thoroughbred races（由三项重要赛事构成的系列赛）三冠王[系列赛]：Japanese Triple Crown 日本三冠王系列赛

Triple Crown Winner n. a three-year-old horse winning all three races of the Triple Crown（同年获得三项系列赛的马匹）三冠王

triple oxer n.【jumping】a combination jumping obstacle consisting of three oxers in a line, each separated by one to two strides（一种组合障碍，由三道双横木构成，其间可跨1～2步）三道双横木跨栏

tristeza /trɪsˈteɪzə/ n. *see* equine piroplasmosis 马焦虫病，马梨形虫病

triticale /ˈtraɪtɪkəl; *AmE* ˌtrɪtɪˈkeɪliː/ n. **1** a hybrid of wheat and rye having a high yield and rich protein content（小麦和黑麦的杂交种，产量高且富含蛋白质）黑小麦 **2** the grains of this hybrid 黑小麦

trochanter /trəʊˈkæntə(r); *AmE* troʊ-/ n.【anatomy】any of several bony processes on the upper part of the femur of many vertebrates（脊椎动物股骨上端着生的骨质突起）转子：major trochanter of femur 股骨大转子 ◇ third trochanter of femur 股骨第三转子 ◇ lesser trochanter 小转子 | **trochanteric** /-rɪk/ adj.

trochlea /ˈtrɒklɪə/ n.【anatomy】(*pl.* **trochleae** /-liiː/) an anatomical structure that resembles a pulley (类似滑轮的解剖结构)滑车:metarcarpal trochlea 掌骨滑车 ◇ humeral trochlea 肱骨滑车

troika /ˈtrɔɪkə/ n. **1** a Russian method of hitching three horses abreast with the two outside horses harnessed by side reins and the center horse put between shafts under an arched douga (一种俄式三马并驾,其中外侧两匹马由侧缰套挽,驾辕的中马则套在辕弓下)[俄式]三套车,俄式三马并驾 [syn] Russian style **2** a Russian vehicle pulled by three horses hitched abreast [俄式]三套车

Trojan Horse n. **1** 【mythology】a large, hollow wooden horse inside which Greek soldiers hid in order to enter the city of Troy (古希腊神话中的巨型中空木马,希腊士兵藏于其内借此侵入特洛伊城)特洛伊木马 **2** a device used to trick an opponent (用来迷惑对手的)阴谋,诡计

trooper /ˈtruːpə(r)/ n. **1** a member of a unit of cavalry 骑兵 **2** a cavalry horse 骑兵马 **3** a mounted police officer 骑警

trot /trɒt; *AmE* trɑːt/ n. a two-beat gait in which the horse's legs move in diagonal pairs (马对角线两肢同时移动的双蹄音步法)快步:collected trot 缩短快步 ◇ extended trot 伸长快步 **PHR V** **rise trot** (of a rider) to post trot (骑手)起浪,打浪 **sit trot** (of a rider) to ride the trot without rising from the saddle to post (骑手)坐浪 — *see also* **walk**, **canter**, **gallop**
vi. move with this gait 以快步行进 **PHR V** **trot level** to move at the trot with ease 以快步平稳行进 **trot up** (of a horse) to move at the trot (指马)快步行进

trot pole n. *see* ground pole 地杆,快步杆

trotter /ˈtrɒtə(r); *AmE* ˈtrɑːt-/ n. **1** a horse bred to trot in harness racing 快步马:French Trotter 法国快步马 ◇ Orlov Trotter 奥尔洛夫快步马 **2** the foot of a pig prepared as food (食用的)猪蹄

trotting /ˈtrɒtɪŋ; *AmE* ˈtrɑːt-/ n. (also **harness trotting**) a harness race in which the horses compete in the gait of trot (马以快步行进的驾车赛)快步赛 [compare] pacing

trotting breed n. a breed of horse gaited to trot in harness racing 快步马种

TRPB abbr. Thoroughbred Racing Protective Bureau 美国纯血马赛马监管局

TRTA abbr. The Ride and Tie Association 骑拴协会

true skin n. *see* dermis 真皮

truncal /ˈtrʌŋkl/ adj. of or relating to the trunk of body 躯干的

truncus /ˈtrʌŋkəs/ n. 【Latin】(*pl.* **trunci** /-saɪ/) the trunk of body 躯干

trunk /trʌŋk/ n. **1** the main body or stem of an organ or nerve apart from the branches (器官、神经等除分支以外的主体部分)躯干;主干:cerebral trunk 脑干 **2** *see* boot (马车的)行李箱

try /traɪ/ vt. to subject a mare to the stallion to determine its sexual state; tease (让母马靠近公马以断定其发情状态)试情

trying board n. (also **teasing board**) a fence or padded board used to separate a mare from the teaser or stallion to determine if she is ready to be mated (用来隔开母马与试情公马的隔栏,以此判别母马是否适合配种)试情栏

tryptophan /ˈtrɪptəˌfæn/ n.【nutrition】an essential amino acid (一种必需氨基酸)色氨酸

Tubal Cain n. the person mentioned in the *Holy Bible* as the first smith (《圣经》中记述的第一个铁匠)土巴·该隐:Zillah bore Tubal-Cain, who made all kinds of bronze and iron tools (Genesis,4:22). 洗拉又生了土八该隐,他打造各种铜铁利器(创世纪,4 :22)。

tuck jump n.【vaulting】a leap performed by a vaulter on the back of a horse 马背跳跃

tucked-up adj. (of a horse) with loins depressed behind the ribs due to illness or undernutrition (马由于疾病或营养不良导致腰部凹陷)后腰凹陷的 [syn] sucked up

tug /tʌg/ n.【driving】one of the oval-shaped leather loops connected to the harness back-band through which the shafts pass when a horse is hitched to a vehicle（与挽具背带相连的椭圆形革圈，套车时辕从中穿入）搭腰圈，搭腰环

tumbleweed /ˈtʌmblwiːd/ n. *see* horse devil 风滚草

tumbrel /ˈtʌmbrəl/ n.（also **tumbril**）**1** a two-wheeled farm cart formerly used in western Europe that can be tilted to dump a load（西欧过去使用的双轮农用马车，车身可翻斗卸物）翻斗马车 **2** a two-wheeled cart used mainly to carry prisoners in the 18th century, drawn by a single horse in shafts（在 18 世纪使用的单马拉双轮车，主要用来押送囚犯）囚车

tumbril /ˈtʌmbrəl/ n. *see* tumbrel 翻斗马车；囚车

tundra /ˈtʌndrə/ n. a vast, flat, treeless arctic region of northern Europe where the subsoil is permanently frozen（北欧广阔的无树平原，地下为永久冷冻土层）苔原，冻原

Tundra horse n. a primitive horse breed that originally lived in the frigid region of northern Europe（过去栖息于北欧寒冷地区的原始马种）冻原马

tune /tjuːn; *AmE* tuːn/ vt. to adjust or bring into perfection or harmony 调整，调节 PHR V **tune up** to improve the performance of a horse by conducting exercises prior to a competition（赛前通过锻炼来提高马匹竞技水平）上调竞技状态

tunic /ˈtjuːnɪk; *AmE* ˈtuː-/ n.【anatomy】a coat or layer enveloping an organ or a part（包裹器官或部位的膜层）被膜

tunica /ˈtjuːnɪkə; *AmE* ˈtuː-/ n.【Latin】(*pl.* **tunicae** /-kiː/) tunic 膜，被膜

tunicary /ˈtjuːnɪkəri; *AmE* ˈtuː-/ adj. having or covered with coat 有膜的

turf /tɜːf; *AmE* tɜːrf/ n. **1** a dense growth of short grass upon the upper layer of earth（地表生长稠密的矮草）草地，草皮：newly laid truf 新铺的草皮 **2**【racing】(usu. **the turf**) the sport of horse racing 赛马业：He spent his money gambling on the turf. 他把钱都花在赌马上了。

turf course n.【racing】*see* turf track 草地赛道

turf track n.【racing】(also **turf course**) a grass track used for horse racing 草地赛道

turfman /ˈtɜːfmən; *AmE* ˈtɜːrf-/ n.【racing】a person interested in horse racing 赛马迷

turgid /ˈtɜːdʒɪd; *AmE* ˈtɜːrd-/ adj. (of a body part) swollen or bloated（身体部位）肿胀的

Turinsky /tuːˈrɪŋski/ n. a native horse breed of Russia（俄国的本土马品种）图林斯基马

Turk /tɜːk; *AmE* tɜːrk/ n. *see* Turkish horse 土耳其马

Turkish horse n. (also **Turk**) a horse breed indigenous to Turkey that has predominantly Persian and Arab blood（土耳其的本土马匹品种，带有波斯马和阿拉伯马血统）土耳其马

Turkman /ˈtɜːkmən; *AmE* ˈtɜːrk-/ n. (also **Turkman horse**) a warmblood bred in Turkmenistan for centuries; stands 15 to 16 hands and may have any of the coat colors; has a slender body, straight profile, long neck, and sloping shoulders; is possessed of exceptional speed and endurance and used mainly for riding and endurance races（土库曼斯坦经过几个世纪育成的温血马品种。体高 15 ~ 16 掌，各种毛色兼有，身体细长，面貌平直，颈长，斜肩，速力惊人，主要用于骑乘和耐力赛）土库曼马：The Turkman horse was noted for its endurance. 土库曼马以耐力著称于世。

Turkman horse n. *see* Turkman 土库曼马

Turkmenian kulan n. *see* kulan 土库曼野驴，库兰野驴

turn /tɜːn; *AmE* tɜːrn/ v. **1** to bring the undersoil to the surface by plowing（将底层土壤用犁翻上来）翻耕，翻土 **2** to change direction of moving 转向，变向 PHR V **turn on the center**【dressage】(of a horse) to turn around the central axis of its body（指马）以中心为轴旋转 **turn on the forehand**【dressage】to turn around its inside foreleg 定前肢旋转 **turn on the haunches/quarters**【dressage】to turn around its inner hind leg 定后肢旋转 **turn out** ① to set horses loose on a pasture 放逐，放牧；② to groom a horse for show competition 刷拭，装扮：a well turned out horse 一匹装扮漂亮的马 **turn tail**【cutting】(of a horse) to turn and stop working a cow（指马）掉头停止截牛

turnback help n. *see* turnback rider 截牛帮手

turnback rider n.【cutting】(also **turnback help**) either one of the mounted riders responsible for turning the cow being worked back toward the cutter in competition（比赛中负责将所拦截的牛赶回到截牛者面前的骑手）截牛帮手

turnout /ˈtɜːnaʊt; *AmE* ˈtɜːrn-/ n. **1** the act of turning a horse loose in a pasture or paddock 外放，放牧 **2** the dress and appearance of horse and rider（马匹和骑手的）装束，装扮

Turn-Row Horse n. *see* Tennessee Walking Horse 田纳西走马

turtle boot n. (also **turtle-back boot**) the foreboot located beneath the box seat of some horse-drawn carriages（马车前驾驶座位下面的工具箱）杂物箱，龟背箱

turtle-back boot n. *see* turtle boot 杂物箱，龟背箱

tush /tʌʃ/ n. a canine tooth of a male horse（公马的）犬齿 [syn] tusk

tusk /tʌsk/ n. a canine tooth; tush 犬齿

TWHBEA abbr. Tennessee Walking Horse Breeders' and Exhibitors' Association of America 美国田纳西走马育种者和展览者协会

twin /twɪn/ n. one of two offspring born at the same birth 双胞胎，双生：identical twins 同卵双生 ◇ twin conception 双胎妊娠

vi. to give birth to twins 生双胞胎，孪生：twinning rate 双胎率

twin trifecta n.【racing】a betting option in which the bettor wins two successive trifecta（投注者连赢两场头三彩的押注方式）双连头三彩

twist /twɪst/ vt. to break a horse 调教：twist a stallion 调教一匹公马

twisted bowel n. *see* torsion colic 肠绞痛，肠扭转

twisted gut n. *see* torsion colic 肠绞痛，肠扭转

twitch /twɪtʃ/ n. a metal or leather chain looped on a stick, used to restrain a horse by tightening it around its upper lip（拴在鞭杆末端的铁链或皮环，可拧在马的上唇将其制服）鼻捻：muzzle twitch 鼻捻

two hole position n【harness racing】the position immediately behind the leading horse（驾车赛）第二名 [syn] win hole

Two Thousand Guineas n.【racing】(also **2,000 Guineas**) a one-mile flat race held annually for three-year olds since 1809 at Newmarket, England（自1809年起每年在英国纽马基特为3岁驹举行的平地赛，赛程1.6千米）两千几尼赛

two track n.【dressage】(also **two track movement**) any movement in which the horse's hind legs follow in a paralle track to that made by the forelegs（马前、后肢蹄迹平行的运步方式）横斜步，双蹄迹 [syn] sidestep, appuyer **PHR** **move on two tracks** (of a horse) to move in parallel tracks（指马）双蹄迹运步，走横斜步

two-day event n. a three-phase eventing competition consisting of dressage, show jumping, and speed and endurance phase conducted by the same rider and horse over a period of two days（人马在两天内参加盛装舞步、跨越障碍以及速度耐力赛三个项目的比赛）两日赛 [compare] one-day event, three-day event

two-wheeler n. a two-wheeled, horse-drawn carriage 双轮马车

two-year-old adj. (of a horse) being at age of two years old（指马）两岁的：a two-year-old colt 两岁雄驹

n. (abbr. **T-Y-O**, or **2-Y-O**) a horse of two

T

years old 两岁驹

tying-up n. *see* tying-up syndrome 僵直症，强拘综合征

tying-up disease n. *see* tying-up syndrome 僵直症，强拘综合征

tying-up syndrome n.（also **typing up**, or **typing-up disease**）a disease that occurs in racehorses under heavy exercise, symptoms include profuse sweating, rapid pulse, lameness, and rigidity of the muscles（赛马和轻型马由于训练过度所致的病症，临床表现发汗、心率加快、跛行和肌肉僵直等）僵直症，强拘综合征：Some authorities diagnose tying-up as a mild form of azoturia. 一些专家将僵直症诊断为轻度形式的氮尿症。[syn] cording up

T-Y-O abbr. two-year-old 两岁驹：A T-Y-O filly 一匹两岁雌驹

type /taɪp/ n. **1** a breed type of horse 马种类型 **2** a group of horses classified according to the particular purpose they serve, such as hunter, cart horse, and hack（根据马匹用途进行的分类，如狩猎马、驾车马和乘用马）马匹类型 [PHR] **true to type**（of a horse）having typical characteristics of its breed（指马）品种特征明显

tyrosine /ˈtaɪrəsiːn, ˈtɪr-/ n.【nutrition】a nonessential amino acid（一种非必需氨基酸）酪氨酸

U /ju:/ abbr. unit 单位

udder /ˈʌdə(r)/ n. the mammary gland of a mare (母马的)乳房:udder disturbance 乳房机能紊乱 ◇ udder edema 乳房水肿

Ukrainian Riding Horse n. a horse breed developed in Ukraine in the late 1940s by crossing Trakehner, Hanoverian, and Thoroughbred stallions with local mares; stands 15.1 to 16.1 hands and usually has a chestnut, bay, or black coat; has a long neck, deep chest, and long croup; used for riding, light draft, farm work, and competition (乌克兰20世纪40年代末采用特雷克纳马、汉诺威马和纯血马与当地母马杂交培育而成的马品种。体高15.1~16.1掌,毛色多为栗、骝或黑,颈长,胸深,尻长,可用于骑乘、轻挽、农田作业和体育竞技)乌克兰乘用马

ulcer /ˈʌlsə(r)/ n. a lesion of the skin or tissue (皮肤或组织的损伤)溃疡,溃烂:gastric ulcer 胃溃疡 ◇ intestinal ulcer 肠溃疡
vi. to ulcerate 溃疡,溃烂

ulcerate /ˈʌlsəreɪt/ vi. to form an ulcer 溃疡,溃烂 | **ulceration** /ˌʌlsəˈreɪʃn/ n.

ulcerative /ˈʌlsərətɪv/ adj. or relating to ulcer or ulceration 溃疡的,溃疡性的:ulcerative lymphangitis 溃疡性淋巴管炎

ulna /ˈʌlnə/ n.【anatomy】(*pl.* **ulnae** /-ni:/) the inner bone in the forearm of a horse (马前臂内侧的骨)尺骨 compare radius | **ulnar** /ˈʌlnə(r)/ adj.

ultrasonography /ˌʌltrəsəˈnɒgrəfɪ/ n. a medical technique using ultrasonic waves for diagnostic or therapeutic purposes (采用超声波进行诊断或治疗的医学技术)超声诊断,超声检查 | **ultrasonographic** /ˌʌltrəsəʊnəˈgræfɪk/ adj.:ultrasonographic examination 超声检查

ultrasound /ˈʌltrəsaʊnd/ n. ultrasonic waves used for diagnostic purposes 超声波:therapeutic ultrasound 超声治疗 ◇ diagnostic ultrasound 超声诊断

umbilical /ʌmˈbɪlɪkəl/ adj. of or relating to the umbilicus 脐的:umbilical cord 脐带 ◇ umbilical artery 脐动脉 ◇ umbilical region 脐部 ◇ umbilical veins 脐静脉

umbilical hernia n. the protrusion of the intestines through the umbilical opening shortly after birth (新生个体肠道从脐孔脱出的情况)脐疝

umbilicus /ʌmˈbɪlɪkəs/ n.【anatomy】(*pl.* **umbilici** /-lɪsaɪ/) the navel of a horse 脐

umpire /ˈʌmpaɪə(r)/ n. an official appointed to enforce the rules of play in mounted sports (马术项目中执行比赛规则的人员)裁判

unassisted /ˌʌnəˈsɪstɪd/ adj. (of foaling) without help or aids (产驹)无人辅助的:unassisted foaling 无人辅助产驹

unbacked /ʌnˈbækt/ adj. **1** (of a horse) having never been ridden (马)未被骑过的 **2**【racing】(of a horse) with no bets placed upon (赛马)无人押注的

unbalanced /ˌʌnˈbælənst/ adj. **1** not in balance or proportion 非平衡的:unbalanced ration 非平衡日粮 **2** (of weights) not equally or evenly distributed (负重)不均的;失调的:unbalanced gait 步态失衡

unbridled /ʌnˈbraɪdld/ adj. **1** (of a horse) not wearing or being fitted with a bridle (指马)未戴勒的,不羁的:an unbridled horse 无羁之马 **2** unrestrained; uncontrolled 不受约束的;放纵的:Unbridled power breeds corruption. 不受制约的权力会滋生腐败。

unbroken /ʌnˈbrəʊkən; *AmE* -ˈbroʊ-/ adj. (of a horse) not broken and trained yet (指马)未驯

服的,未调教的

uncastrated /ʌnˈkɑːstreɪtɪd/ adj. (of male horses) not castrated (公马)未去势的,未骟的

uncertain /ʌnˈsɜːtn; *AmE* ʌnˈsɜːrtn/ adj. 【jumping】(of a horse) not determined or showing no confidence(指马)不坚定的,胆怯的:uncertain fencer 怯跳之马

uncontaminated /ˌʌnkənˈtæmɪneɪtɪd/ adj. not contaminated; unstained 未污染的;清洁的:uncontaminated water 未污染的水

uncouple /ʌnˈkʌpl/ vt. **1** to detach a horse from the vehicle; unharness 卸开,卸车 **2** to release dogs from a pair of joined collars 卸套,解套

undemanding /ˌʌndɪˈmɑːndɪŋ/ adj. (of a horse) requiring little ration and care for maintenance (饲喂)要求不高的,耐粗饲的

undercarriage /ˈʌndəkærɪdʒ; *AmE* -dərk-/ n. 【driving】 the supporting structure under the body of a horse-drawn vehicle (马车底端的支承结构)底架,底盘

under-conditoned /ˈʌndəkənˌdɪʃnd/ adj. (of a horse) in poor conditions (指马)体况差的

underfed /ˌʌndəˈfed *AmE* -dərˈf-/ adj. not fed well or sufficiently 饲喂不足的,没喂饱的 [opp] overfed

underfeed /ˌʌndəˈfiːd; *AmE* -dərˈf-/ vt. not to give enough feed 饲喂不足:Horses will run into less problems from underfeeding than overfeeding. 马匹饲喂不足出现的问题要比饲喂过度少些。[opp] overfeed

under-horsed /ˌʌndəˈhɔːst; *AmE* -dərˈhɔːrst/ adj. (of a rider) too big for the mount in size 人大马小的,人马不相称的:An adult would be under-horsed on a pony. 成年人骑矮马会显得人马不相称。[compare] over-horsed

undernourish /ˌʌndəˈnʌrɪʃ; *AmE* -dərˈnɜːr-/ vt. to provide with insufficient food for good health and condition (所供食物不足以维持健康和体况)使营养不足 | **undernourishment** /-rɪʃmənt/ n.

undernourished /ˌʌndəˈnʌrɪʃt; *AmE* -dərˈnɜːr-/ adj. not providing enough food; underfed 营养不足的

undernutrition /ˌʌndənuːˈtrɪʃn/ n. the condition of being undernourished 营养不足,营养不良

underpinning /ˈʌndəˌpɪnɪŋ; *AmE* -dərˌp-/ n. the legs and feet of a horse (马的)肢蹄

underreach /ˈʌndəriːtʃ/ n. (also **underreaching**) a gait defect of a horse in which the toe of a front horseshoe strikes the toe of the hind shoe on the same side (马前肢蹄铁尖撞击同侧后肢蹄铁前端的运步缺陷)追突,追蹄

vi. (of a horse) to strike the toe of a front horseshoe with the toe of the hind hoof at trot (马快步行进时前肢蹄铁尖撞击后蹄前端)追突:A horse may underreach either due to conformation defect or simply long toes. 马匹肢体缺陷或蹄尖过长,都可能引起追突。| **underreaching** /ˌʌndəˈriːtʃɪŋ/ n. [compare] interfere, overreach

underrun heels n. a gait defect of a horse in which the hoofs touch ground with the heels (马以蹄踵着地的运步缺陷)蹄踵着地 [syn] run-under heels, underslung heels

undershot /ˌʌndəˈʃɒt/ adj. (of an animal) having the lower jaw or teeth projecting beyond the upper 下颌突出的,地包天的 [compare] overshot

undershot jaw n. (also **undershot mouth**) a congenital malformation of the mouth in which the lower jaw protrudes beyond the upper and the incisors do not meet properly (一种先天口齿畸形,下颌比上颌突出故而切齿咬合不齐)下颌突出,地包天 [syn] monkey mouth, hog mouth, sow mouth, prognathism [compare] overshot jaw

undershot mouth n. *see* undershot jaw 下颌突出,地包天

undershrub /ˈʌndəʃrʌb/ n. a very low-growing shrub 小灌木

undersized /ˌʌndəˈsaɪzd; *AmE* -dərˈs-/ adj. (of an animal) less than normal or sufficient size

(动物)体格弱小的

underslung heels n. *see* underrun heels 蹄踵着地

understock /ˌʌndəˈstɒk/ vt. to have or allow less number of livestock than a pasture can hold (所养牲畜数目低于草场负荷)载畜不足 [opp] overstock

understocked /ˌʌndəˈstɒkt/ adj. (of a farm) having less number of livestock than a pasture can hold (农场)载畜不足的:an understocked ranch 载畜不足的牧场 [opp] overstocked

underuse /ˌʌndəˈjuːs/ vt. to use sth insufficiently 使用不足

n. insufficient use 使用不足,使用不全

underused /ˌʌndəˈjuːst/ adj. not fully used 未充分利用的,使用不足的

underweight /ˌʌndəˈweɪt; *AmE* -dərˈw-/ adj. (of an animal) weighing less than normal or required (动物)体重不足的

n. insufficiency of weight 重量不足,体重不足

undescended /ˌʌndɪˈsendɪd/ adj. (of testicles) not descended into the scrotum (睾丸)未下降的

undesirable /ˌʌndɪˈzaɪərəbl/ adj. not wanted; objectionable 不需要的,不良的 : undesirable traits 不良特征

undigested /ˌʌndɪˈdʒestɪd/ adj. (of feed) not digested (饲料)未消化的: undigested residue 未消化残渣

undo /ʌnˈduː/ vt. to untie or loosen 解开,松开: The plaits should be undone carefully as not to damage the mane. 松解辫子时要格外细心,以防损伤马鬃。

undulant fever n. another term for brucellois (布鲁氏菌病的别名)波状热

unentered /ʌnˈentəd; *AmE* -tərd/ adj. 【hunting】(of hounds) not having been put into a pack yet (指猎犬尚未加入猎队)未出猎的

uneven /ʌnˈiːvn/ adj. not smooth or level; rough 不平的,崎岖的:Work over uneven and hilly ground is much more arduous than on flat terrain. 在崎岖的山地行进比在平坦地段要费力的多。

unfertilized /ʌnˈfɜːtɪlaɪzd/ adj. **1** (of land) supplied with no fertilizers (土地)未施肥的 **2** (of ovum) not fertilized (卵子)未受精的:unfertilized egg 未受精卵

ungula /ˈʌŋgjʊlə/ n.【Latin】hoof 蹄

ungulate /ˈʌŋgjʊlət, -leɪt/ adj. (of animals) having hoofs (动物)有蹄的

n. a hoofed mammal 有蹄动物: even-toed ungulate 偶蹄动物 ◇ odd-toed ungulate 奇蹄动物

unharness /ʌnˈhɑːnɪs/ vt. **1** to remove the harness from a horse 卸马具 [opp] harness **2** to release or liberate 释放,解开

unhealthy /ʌnˈhelθi/ adj. **1** being ill or sick 病弱的 **2** unwholesome or harmful to health 不健康的

unherded /ˈʌnhədɪd/ adj. (of livestock) grazing on pasture without a herder; loose-grazed (家畜)无人放牧的,无人看管的

unhitch /ʌnˈhɪtʃ/ vt. to release a horse from a hitch; unfasten 卸马;解开 [opp] hitch

unhooked /ʌnˈhʊkt/ adj.【cutting】(of a horse) losing the attention of the cow being worked (指马)盯防不紧的

unhulled /ʌnˈhʌld/ adj. not having the hull removed 未去壳的,带壳的: unhulled peanut meal 带壳花生粕 [opp] hulled

unicellular /ˌjuːnɪˈseljələ(r)/ adj. having or consisting of one cell; one-celled 单细胞的

unicorn[1] /ˈjuːnɪkɔːn; *AmE* -kɔːrn/ n.【mythology】a mythical beast like a white horse with a long straight horn on its forehead (希腊神话中体貌似马的野兽,前额长有犄角)独角兽

unicorn[2] /ˈjuːnɪkɔːn; *AmE* -kɔːrn/ n.【driving】a hitch of three horses consisting of one leader and two wheelers (由一匹头马和两匹辕马构成的)锥形三马联驾 [syn] spike team

unidentified /ˌʌnaɪˈdentɪfaɪd/ adj. not identified; unclear 未知的,不明的: unidentified growth factor 未知生长因子

U

unilateral /ˌjuːnɪˈlætrəl; *AmE* -tərəl/ adj. of or affecting only one side 单侧的：unilateral cryptorchid horse 单侧隐睾马

uniparous /juːˈnɪpərəs/ adj. producing only one offspring at a time 单胎的：uniparous mammals 单胎哺乳动物

unisexual /ˌjuːnɪˈseksjʊəl; *AmE* -kʃʊrl/ adj. **1** of or relating to only one sex 单性的 **2** having only one type of sexual organ 雌雄异体的

united /juˈnaɪtɪd/ adj. (of gait) properly coordinated and executed 步法协调的，步调均匀的

United States Calf Ropers Association n. (abbr. **USCRA**) an organization founded in the United States in 1996 to organize calf roping events and ensure the fairness of the sport (1996 年在美国成立的组织机构，负责组办国内套牛比赛并维护赛事的公正性)美国套牛协会

United States Combined Training Association n. (abbr. **USCTA**) an organization founded in the United States in 1959 to sponsor and facilitate horse trials, combined tests, two-day events, and three-day events (1959 年在美国成立的组织机构，负责协调组办两日赛和三日赛等各种综合全能比赛)美国综合全能赛协会

United States Equestrian Federation n. (abbr. **USEF**) the governing body for the equestrian sports in the United States, which first began as the American Horse Shows Association in 1917 and merged with United States Equestrian Team in 2003 to form the present organization (美国的马术监管机构，最早源于 1917 年成立的美国马术展览协会，2003 年与美国马术队合并后更为现名)美国马术联合会

United States Equestrian Team n. (abbr. **USET**) an American organization founded in 1950 to represent the United States in international show jumping, eventing, dressage, and driving competitions (美国在 1950 年成立的组织机构，负责代表美国参加国际性障碍赛、三日赛、盛装舞步以及轻驾车赛)美国马术队

United States Polo Association n. (abbr. **USPA**) an organization founded in 1890 to organize and regulate polo matches in the United States (1890 年在美国成立的组织机构，主要负责组织并协调国内马球比赛)美国马球协会

United States Team Penning Association n. (abbr. **USTPA**) a non-profit American organization founded in 1993 to promote the sport of team penning and ranch sorting (1993 年在美国成立的非营利组织，旨在促进团体圈牛和牧场分栏比赛)美国团体截牛协会

uniungulate /ˌjuːniˈʌŋgjʊlət/ adj. (of animals) having one hoof (动物)单蹄的

n. an animal having one hoof 单蹄动物

unkennel /ˌʌnˈkenl/ vt. 【hunting】to drive a fox from its covert 赶出窝

unknown /ˌʌnˈnəʊn; *AmE* -ˈnoʊn/ adj. not known or identified 不明的，不详的：unknown pedigree 谱系不明

unlevel /ʌnˈlevl/ adj. 【dressage】(of a horse) taking uneven strides (指马)步态不稳的，步态失衡的

unmade /ʌnˈmeɪd/ adj. (of a horse) not broken or schooled yet (指马)未调教的

unmarked /ʌnˈmɑːkt; *AmE* -ˈmɑːrkt/ adj. (of a horse) having no white markings on its body (指马)无别征的：an unmarked horse 无别征的马

Unmol /ˈʌnməl/ n. (also **Unmol horse**) a rare horse breed indigenous to the north-western Punjab, Pakistan; stands around 15 hands and usually has a bay or gray coat; has a compact body, long mane and tail; is hardy, elegant and used mainly for light draft and riding (源于巴基斯坦旁遮普西北部地区的稀有马品种。体高约 15 掌，毛色以骝、青为主，体质结实，鬃尾较长，抗逆性强，体貌俊秀，主要用于轻挽和骑乘)昂穆尔马

unnerve /ʌnˈnɜːv; *AmE* -ˈnɜːrv/ vt. to cut off or remove the nerver; denerve 割断神经

unpatterned /ʌnˈpætənd/ adj. not having the typical patterns 未成形的，类型不明显的：unpatterned leopard 非典型豹纹花斑

unpick /ʌnˈpɪk/ vt. to undo the sewing of stitches or plaiting of braids 拆散，解开：After all the plaits have been unpicked the mane should be brushed or combed out and damped down with a water brush to encourage it to lie flat. 解开所有的辫子后，要对马鬃进行刷拭或梳理，接着再用水刷将鬃毛梳理平展。

unraced /ʌnˈreɪst/ adj.【racing】(of a horse) not having competed in a race yet（指马）未出赛的

unrecorded /ˌʌnrɪˈkɔːdɪd; *AmE* -ˈkɔːrd-/ adj. (of horses) not recorded or registered in a stud book; unregistered（指马）未登记的，未注册的

unruly /ʌnˈruːli/ adj. (of a horse) difficult to control or rule（指马）难以驾驭的

unsaddle /ʌnˈsædl/ vt. **1** to remove the saddle from a horse 卸鞍 **2** (of a rider) to be unseated from the saddle（骑手）落马，坠马

unseat /ʌnˈsiːt/ vt. to throw the rider from the saddle; buck off 撂下马：an unseated rider 落马的骑手

unshod /ʌnˈʃɒd/ adj. (of horses) not having or wearing horseshoes; barefoot（指马）未钉蹄的；未钉掌的：an unshod horse 一匹未钉掌的马

unshoe /ʌnˈʃuː/ vt.【farriery】to remove the horseshoes 卸蹄铁，卸掌

unskid /ʌnˈskɪd/ vt. to remove the skid from the wheel of a horse-drawn vehicle 去掉制轮器

unsound /ʌnˈsaʊnd/ adj. (of horses) not physically ideal or having conformation faults 体型欠佳的，有缺陷的；失格的 [opp] sound

unsoundness /ʌnˌsaʊndˈnɪs/ n. the state of being unsound or faulty in conformation 缺陷，失格：unsoundness of feet 四肢缺陷

unstained /ʌnˈsteɪnd/ adj. not contaminated; clear 未污染的；清洁的：unstained water sources 清洁水源

unsteady /ʌnˈstedi/ adj. not firm or solid, unstable; wavering 不稳的；摇摆的：unsteady gait 步伐不稳 ◇ unsteady head 头姿不稳 ◇ unsteady halt 停步不稳 ◇ The horse cantered to an unsteady halt, and still shifting his legs and head. 那马跑着停了下来，不时地挪步摆头，站立不稳。

untack /ʌnˈtæk/ vt. to remove the tack from a horse after work（马匹训练后卸下马具）卸马具

unthrifty /ʌnˈθrɪfti/ adj. (of horses) not healthy or thriving 不壮实的，体况差的 [opp] thrifty | **unthriftiness** n.

untried /ˌʌnˈtraɪd/ adj. **1**【racing】(of a horse) not previously raced or tested for speed; unraced（指马）未出赛的，未测速的 **2** (of a stallion) not having been used for breeding（公马）未配过种的

unwind /ˌʌnˈwaɪnd/ v. **1**【racing】to release a horse from intense training gradually; relax 使放松，使松懈 **2** (of a horse) start to buck（指马）弓背跳

unyoke /ʌnˈjəʊk; *AmE* ʌnˈjoʊk/ vt. to release a farm animal from harness; untack 卸挽具；卸轭 [opp] yoke

upgrade /ˌʌpˈgreɪd/ vt.【racing】to raise a race to a higher grade or level 升级，提级 [opp] downgrade

upper aids n. the rider's hands employed to communicate instructions to the horse by acting on the bit through the reins（骑手持缰操控衔铁向马传达指令的双手扶助）上肢扶助 [compare] lower aids, voice aids

upright /ʌpˈraɪt/ adj. being in a vertical position or direction 挺直的，竖立的：upright mane 鬃毛直立 ◇ upright pastern 立系 ◇ upright shoulder 直肩

upset[1] /ʌpˈset/ adj. having disturbed digestion 消化不良的，失调的：an upset stomach 消化不良的胃

n. the condition of being disturbed 消化不良；

失调

upset² /ʌpˈset/ adj. (of price) fixed at a lowest amount 固定的,保底的

upset price n.【AmE】the lowest acceptable price offered for a horse at an auction (马匹拍卖时可接受的最低价)起拍价 compare reserve price

upside-down neck n. *see* ewe neck 羊颈;颈峰内凹

uptake /ˈʌpteɪk/ n. the absorption of nutrients 吸收,摄取:Insulin acts to lower the blood glucose level by increasing the uptake of glucose into the muscle and liver. 胰岛素通过增加肌肉和肝脏对葡萄糖的吸收来降低血糖水平。

upward transition n. a transition from a slower gait to a faster one (步法由慢变快的)上行换步,向上转换 compare downward transition

upwind /ˌʌpˈwɪnd/ adv. in the direction from which the wind is blowing 逆风的,戗风的 opp downwind

unweaned /ʌnˈwiːnd/ adj. not weaned; suckling 未断奶的,未断乳的:an unweaned foal 未断乳的幼驹

urachus /ˈjʊrəkəs/ n.【anatomy】a tube in the umbilical cord that connects the fetal bladder to the allantoic cavity (脐带中连接胎儿膀胱和尿囊腔的导管)脐尿管:The allantoic fluid is formed by the placenta and from urine passed through the urachus. 尿囊液由胎盘代谢物和经脐尿管汇集的尿液形成。

ureter /jʊˈriːtə(r)/ n.【anatomy】one of the two long, narrow ducts by which urine passes from the kidney to the bladder (尿液从肾脏输至膀胱的两根细长导管)输尿管

urethra /jʊˈriːθrə/ n.【anatomy】the canal through which urine is discharged from the bladder (膀胱贮存尿液排泄的通道)尿道

urethral /jʊˈriːθrəl/ adj. of or relating to urethra 尿道的:urethral orifice 尿道外口 ◇ urethral fossa 尿道隐窝

uric /ˈjʊərɪk; *AmE* ˈjʊrɪk/ adj. of or relating to urine 尿的:uric acid 尿酸

urinalysis /ˌjʊərɪˈnælɪsɪs/ n. a laboratory analysis of urine 尿液分析,尿液化验

urinary /ˈjʊərɪnəri; *AmE* ˈjʊrəneri/ adj. of or relating to urine or urination 尿的;泌尿的:urinary bladder 膀胱 ◇urinary calculi 尿结石 ◇ urinary energy 尿能 ◇ urinary system 泌尿系统

urinate /ˈjʊərɪneɪt; *AmE* ˈjʊrən-/ vi. to discharge urine 排尿 | **urination** /ˌjʊrɪˈneɪʃn/ n.

urine /ˈjʊərɪn; *AmE* ˈjʊrən/ n. a deep yellow to brown liquid discharged through the urethra (从泌尿道排出的黄褐色液体)尿液:A full-grown horse will discharge six liters of urine per day on average. 成年马每天平均排出大约6升的尿液。

urine test n. a chemical analysis of a horse's urine to detect if illegal drugs have been administered to improve its performance (对马匹尿液进行的化学分析,以检测是否服用违禁药物来提高赛绩)尿检

urogenital /ˌjʊərəʊˈdʒenɪtl/ adj. of or relating to both the urinary and genital structures or functions 泌尿生殖的:urogenital tract 泌尿生殖道 ◇ urogenital sinus 泌尿生殖窦 ◇ urogenital system 泌尿生殖系统

urolith /ˈjʊərəlɪθ; *AmE* ˈjʊrəlɪθ/ n. a hard mass of mineral salts formed in the urinary tract (矿物质盐在尿道形成的硬块)尿[结]石 syn urinary calculus

urticant /ˈɜːtikənt/ n. any substance causing an itching or stinging sensation 刺痒物

urticaria /ˌɜːtɪˈkeəriə; *AmE* ˌɜːrtɪˈkeriə/ n. *see* hives 荨麻疹

urticate /ˈɜːtɪkeɪt/ vi. to cause a stinging or prickling sensation 刺痒 | **urtication** /ˌɜːtɪˈkeɪʃn/ n.

USCTA abbr. United States Combined Training Association 美国综合全能比赛协会

USDA abbr. United States Department of Agriculture 美国农业部

USEF abbr. United States Equestrian Federation

美国马术联合会

USET abbr. United States Equestrian Team 美国马术队

USPA abbr. United States Polo Association 美国马球协会

USTPA abbr. United States Team Penning Association 美国团体截牛协会

uterectomy /ˌjuːtəˈrɛktəmi/ n. the surgical removal of the uterus; hysterectomy 子宫切除术

uterine /ˈjuːtəraɪn/ adj. of or relating to the uterus 子宫的:uterine artery 子宫动脉 ◇ uterine atony 子宫弛缓 ◇uterine body 子宫体 ◇ uterine broad ligament 子宫阔韧带 ◇ uterine cavity 子宫腔 ◇ uterine cervix 子宫颈 ◇ uterine horn 子宫角

uterine prolapse n. the falling or sliding of the uterus from its normal position in the pelvic cavity (子宫滑离骨盆腔内正常位置的情况)子宫脱出

uteritis /ˌjuːtəˈraɪtɪs/ n. inflammation of the uterus; metritis, hysteritis 子宫炎

uter(o)- pref. uterine 子宫的

uterus /ˈjuːtərəs/ n. 【anatomy】(*pl.* **uteri** /-təraɪ/) the reproductive organ of a mare where the fertilized egg implants and develops into a foal (母马体内供受精卵着床和发育的生殖器官)子宫 [syn] womb

utility /juːˈtɪləti/ adj. used or designed for various purposes 多用途的;兼用的:utility vehicle 兼用型车辆

utility cart n. a two-wheeled, low-slung cart used for breaking young horses (一种低底双轮马车,多用于调教青年马匹)兼用马车,调教马车

utility saddle n. a saddle designed for general purpose 兼用鞍,通用鞍

UV abbr. ultraviolet 紫外线

uvea /ˈjuːvɪə/ n. the vascular middle layer of the eye (眼的)葡萄膜

uveitis /ˌjuːvɪˈaɪtɪs/ n. inflammation of the uvea 葡萄膜炎:recurrent equine uveitis 马复发性葡萄膜炎

U

V /viː/ n.【roping】the V-shaped figure created by tying a calf's fore and hind legs together in calf roping(套牛时将小牛前后肢绑起来形成的字形)V 形

vaccinate /ˈvæksɪˌneɪt/ vt. to treat with vaccine to stimulate immune response against certain disease; inoculate(注射疫苗以刺激机体对疾病产生免疫应答)接种,注射疫苗 | **vaccination** /ˌvæksɪˈneɪʃn/ n. : vaccination program 接种计划

vaccine /ˈvæksiːn/ n. a substance administered into body to stimulate immune response against certain diseases(注射后可刺激机体对特定疾病产生免疫应答的物质)疫苗

vagina /vəˈdʒaɪnə/ n.【anatomy】the passage leading from the opening of the vulva to the uterus of a mare(母马由阴门开口到子宫颈的通道)阴道

vaginal /vəˈdʒaɪnl/ adj. of or relating to vagina 阴道的: vaginal artery 阴道动脉 ◇ vaginal atresia 阴道闭锁 ◇ vaginal inspection 阴道检查 ◇ vaginal vestibule 阴道前庭

vaginitis /ˌvædʒəˈnaɪtɪs/ n. inflammation of the vagina 阴道炎 [syn] colpitis

valet /ˈvæleɪ/ n.【racing】one who takes care of a jockey's clothing, tack, and the horse (料理骑师衣物、马具和坐骑的人)骑师助理

valine /ˈvæliːn/ n.【nutrition】(abbr. **Val**) one kind of essential amino acid(一种必需氨基酸)缬氨酸

van /væn/ n. a covered wagon used for transporting goods or livestock(用于运载货物或牲畜的厢式车辆)篷车;货车

vt. to transport horses by van 用货车运输马匹: to van the horses to the racetrack 用货车把马运到赛马场

vanity brand n. *see* ranch brand 牧场烙印

vanner /ˈvænə/ n. a horse used to pull a tradesman's van 拉货车的马 [syn] parcel carter

vaquero /vɑːˈkerəʊ/ n.【Spanish】a cowboy or cattle-driver; buckaroo 牛仔;驯马师

varicose /ˈværɪkəʊs/ adj. (of veins) abnormally swollen or knotted (静脉) 曲张的: varicose veins 静脉曲张

varmint /ˈvɑːmɪnt; *AmE* ˈvɑːrm-/ n. **1** a troublesome child 顽童 **2**【slang】a fox 狐狸

varnish roan n.【color】a coat color pattern that varies from a white to roan blanket, often seen in the Appaloosa horse(由白至沙毛的毛色类型,常见于阿帕卢萨马)淡沙毛

vas /væs/ n.【Latin】(*pl.* **vasa** /ˈveɪsə/) a vessel or duct 脉管,导管 | **vasal** /væsl/ adj.

vas deferens n.【Latin】(*pl.* **vasa deferentia**) one of the two muscular tube that carry semen from the epididymis to the ejaculatory duct(将精液从附睾运送到射精管的肌质细管)输精管 [syn] ductus deferens

vascular /ˈvæskjələ(r)/ adj. of or relating to vessels that carry blood or lymph through the body (有关体内运送血液或淋巴的管道)脉管的;血管的

vascularize /ˈvæskjələˌraɪz/ v. to make or become vascular(使)脉管化

vascularized /ˈvæskjələraɪzd/ adj. having or containing vessels 脉管化的,脉管丰富的: The pituitary is a highly vascularized neuroendocrine gland, thus it can monitor levels of certain hormones in the blood stream with a high degree of sensitivity. 垂体是个脉管分布非常丰富的神经性内分泌腺,可灵敏地监控血液中特定激素水平的变化。

vasculogenic /væskjʊləʊˈʒenɪk/ adj. concerning or

related to blood vessels; vasogenic 血源性的，血管性的

vasculogenic shock n. (also **vasogenic shock**) a collapse resulting from the dramatic reduction of circulating blood volume in the vascular system (由于循环系统中血量锐减而引发的虚脱) 血源性休克

vas(o)- pref. of blood vessel 血管的

vasoactive /ˌveɪzəʊˈæktɪv/ adj. (of a drug) causing constriction or dilation of blood vessels(指药物可引起血管收缩或扩张)作用于血管的

vasoconstricting /veɪzəʊkənˈstrɪktɪŋ/ adj. tending to constrict the blood vessels 血管收缩的

vasodilating /væsəʊdaɪˈleɪtɪŋ/ adj. tending to dilate the blood vessels 血管扩张的 | **vasodilation** /-ˈleɪʃn/ n.

vasogenic /ˌvæsəˈʒenɪk/ adj. originating in the blood vessels 血源性的，血管性的

vasogenic shock n. *see* vasculogenic shock 血源性休克

vasopressin /ˌvæsəʊˈpresɪn/ n. 【physiology】 a hormone secreted by the posterior lobe of the pituitary gland that constricts blood vessels, raises blood pressure, and reduces excretion of urine(脑垂体后叶分泌的一种激素，有促进血管收缩、升高血压并能减少排尿的作用)血管升压素 [syn] antidiuretic hormone

vaulter /ˈvɔːltə/ n. one who performs vaulting on horseback 马上体操选手：In competition, vaulters compete as individuals, pairs (pas-de-deux), and teams. 比赛中，马上体操选手以单人、双人或团体进行角逐。

vaulting /ˈvɔːltɪŋ/ n. (also **equestrian vaulting**) an equestrian sport in which one performs gymnastics on the back of a moving horse(马术运动项目之一，其中选手在马背上进行体操表演) 马上体操，马上技巧：In individual vaulting competition, the six compulsory exercises — basic seat, flag, mill, scissors, stand, and flank — must be performed in order. 在马上体操个人赛中，基本坐姿、迎风展臂、回旋、剪式、站立和侧摆六个规定动作必须按顺序完成。

vaulting barrel n. a padded barrel with four legs and handles used by vaulters to practice movements in place of horses(有四条腿和把手的带衬垫铁桶，马上体操选手借此代替真马来进行动作练习)马上体操练习桶 [syn] metal horse

vaulting horse n. a heavy horse used in vaulting 马上体操用马

vaulting roller n. *see* vaulting surcingle 马上体操胸带

vaulting surcingle n. (also **vaulting roller**) a wide leather band buckled around the girth of the vaulting horse, with two handles on either side to aid vaulters in performing certain movements(系在马上体操用马前胸的宽皮带，两侧把手可协助表演者完成特定动作)马上体操胸带

veal /viːl/ n. **1** the meat of a calf 小牛肉 **2** a vealer 肉用小牛

vealer /ˈviːlər/ n. a calf raised for meat 肉用小牛

vealers /ˈviːlərz/ n. any riding boot made with veal hide 小牛皮靴

VEE abbr. Venezuelan Equine Encephalomyelitis 委内瑞拉马脑脊髓炎：VEE is an emerging infectious disease in Latin America. 委内瑞拉马脑脊髓炎是拉丁美洲出现的传染病。

vehicle /ˈviːəkl; *AmE* ˈviːhɪkl/ n. a thing used for transporting goods or passengers 交通工具；车辆：Horse was used as a vehicle of nomadic herdsmen before his introduction to the chariot and to the cavalry. 在用来驾战车和作为骑兵坐骑之前，马已是游牧民族的交通工具。

vein /veɪn/ n. 【anatomy】 any of the membranous tubes that carry blood back to the heart(将血液运回到心脏的脉管)静脉：The blood flowing through veins is deoxygenated and darker in color than arterial blood. 静脉中血液由于脱氧颜色要比动脉中血液深一些。[compare] artery

velvet /ˈvelvɪt/ n. the soft, furry covering on the developing antlers of deer(生长期鹿角上覆盖的绒毛层)鹿茸 PHR **in velvet**(of a stag) with antlers covered by velvet 长鹿茸的

vena /ˈviːnə/ n.【Latin】a vein 静脉

vena cava n.【anatomy】(*pl.* **venae cavae**) either of the two large veins entering the heart 腔静脉: inferior vena cava 下腔静脉,后腔静脉 ◇ superior vena cava 上腔静脉,前腔静脉 syn large vein

venereal /vəˈnɪəriəl; *AmE* -ˈnɪr-/ adj. of or relating to sexually transmitted disease or sexual intercourse 性传染的;性的: venereal diseases 性病

venery /ˈvenəri/ n.【dated】the sport of hunting 打猎

venezuelan equine encephalomyelitis n.(abbr. **VEE**) a mosquito-borne viral disease of all equine species that causes inflammation of the white matter in the brain and spinal cord; often symptomized by high fever, aggression, muscular stiffness, convulsions, coma, and even death(马科动物所患的病毒性疾病,以蚊虫为媒介,可造成大脑白质和脊柱发炎,患畜常表现高热、肌肉僵直、痉挛、昏迷甚至死亡)委内瑞拉马脑脊髓炎 syn sleeping sickness

venison /ˈvenɪsn, -zn/ n. meat from a deer 鹿肉: Historically, venison was considered to be a status symbol among many Europeans. 在过去,鹿肉在许多欧洲人看来是社会身份的标志。

venous /ˈviːnəs/ adj. of or relating to the veins 静脉的: venous blood 静脉血 ◇ venous bleeding 静脉出血 ◇ venous circulation 静脉循环 ◇ venous injection 静脉注射

ventilate /ˈventɪleɪt/ vt. to allow air to enter and circulate freely in a building 通风,换气 | **ventilation** /ˌventɪˈleɪʃn/ n. 换气: minute ventilation volume 每分钟换气量

ventral /ˈventrəl/ adj.【anatomy】**1** of, situated on or near the abdomen 腹侧的: ventral arteries 腹动脉 **2** inferior 下部的: ventral serrate muscle 下锯肌

ventral hernia n. a protrusion of the intestines through the weakened abdominal wall(肠道从腹腔壁突出的症状)腹壁疝

ventralis /ˈventrəlɪs/ adj.【Latin】ventral 腹(侧)的 compare dorsalis

ventricle /ˈventrɪkl/ n.【anatomy】**1** a small cavity or chamber within a body or an organ(身体或器官内的)腔;室 **2** either of the two chambers of the heart 心室: left/right ventricle 左/右心室

ventricle stripping n. *see* laryngeal ventriculotomy 喉室切开术

ventricular /venˈtrɪkjʊlə/ adj. of or relating to a ventricle or ventriculus 心室的;腔室的: ventricular ejection 心室射血[期]

ventriculus /venˈtrɪkjʊləs/ n.【Latin】(*pl.* **ventriculi** /-laɪ/) a ventricle 室;腔: ventriculus dexter 右心室 ◇ ventriculus sinister 左心室

venule /ˈvenjuːl/ n.【anatomy】a small vein 微静脉 | **venular** /ˈvenjuːlə(r)/ adj.

vermin /ˈvɜːmɪn; *AmE* ˈvɜːrm-/ n. any of the pests or insects 害虫: vermin control 虫害防治

vermin-proof /ˈvɜːmɪnˌpruːf/ adj. designed to prevent vermin and pests from intruding 防虫害的

versatile /ˈvɜːsətaɪl; *AmE* ˈvɜːrsətl/ adj.(of a horse) able to be adapted to many different works or purposes(指马)多用途的,兼用的: a versatile horse breed 兼用马种 | **versatility** /ˌvɜːsəˈtɪləti; *AmE* ˌvɜːrs-/ n.: The Morgan horse is best known for its versatility. 摩根马尤以用途广泛而著名。

vertebra /ˈvɜːtɪbrə; *AmE* ˈvɜːrt-/ n.【anatomy】(*pl.* **vertebrae** /-beɪ; -briː/) any of the irregular bones forming the spinal column(构成脊柱的不规则骨)椎骨: cervical vertebrae 颈椎 ◇ coccygeal vertebrae 尾椎 ◇ lumbar vertebrae 腰椎 ◇ sacral vertebrae 荐椎 ◇ thoracic vertebrae 胸椎

vertebral /ˈvɜːtɪbrəl/ adj. of or relating to the vertebrae 脊椎的:vertebral arch 椎弓 ◇ vertebral body 椎体 ◇ vertebral canal 椎管 ◇ vertebral column 脊柱 ◇ vertebral foramen 椎孔

vertebral column n. the backbone of a horse 脊柱 syn spinal column

vertebrate /ˈvɜːtɪbrət; *AmE* ˈvɜːrt-/ n. any animal having a backbone 脊椎动物

adj. of or characteristic of vertebrates 脊椎动物的

vertical /ˈvɜːtɪkl; *AmE* ˈvɜːrt-/ adj. being or situated at right angle to the horizon; upright 垂直的

n. **1** the imaginary vertical line 垂线;垂轴 PHR **behind the vertical** (of head) carried behind the imaginary line drawn perpendicular to the ground(指头姿位于垂线后方)低头过度 **in front of the vertical**(of head) carried in front of the imaginary line drawn perpendicular to the ground(指头姿位于垂线前方)抬头过高 **2**【jumping】*see* vertical jump 垂直障碍

vertical jump n.【jumping】(also **vertical**) a jumping obstacle consisting of a fence built vertical to the ground with no spread(与地面垂直且无侧翼的障碍)垂直障碍:Verticals are commonly designed from gates, planks, walls, straights posts or rails. 垂直障碍多由门板、木板、矮墙、竖杆或横杆搭建。syn upright jump

vesical /ˈvesɪkəl/ adj. **1** of or relating to vesicles 囊的 **2** of or relating to bladder 膀胱的:vesical calculus 膀胱结石

vesicant /ˈvesɪkənt/ adj. (of an agent) causing to form blisters 发疱的

n. a blistering agent applied to the skin toincrease blood supply and flow and thus promote healing(涂在皮肤表面的一种制剂,可通过促进血液循环加快修复过程)发疱剂

vesicle /ˈvesɪkl/ n. a small sac or cavity 囊;疱:seminal vesicles 精囊腺

vesicular /vəˈsɪkjʊlə/ adj. of or relating to vesicles; vesical 囊的;疱的

vesicular stomatitis n. a contagious viral disease characterized by inflammation of the mucous tissues of mouth, tongue, coronary band, and occasionally other parts of the body(一种病毒性传染病,症状表现为口、舌、蹄冠等处的黏膜组织发炎)疱疹性口炎

vessel /ˈvesl; *AmE* ˈvɛsəl/ n. a duct or canal carrying or circulating blood or lymph through the body(在体内运送血液或淋巴液的管道)脉管;血管:blood vessels 血管 ◇ lymphatic vessels 淋巴管

vestibular /vesˈtɪbjʊlə/ adj. of or relating to vestibule 前庭的:vestibular glands 前庭腺

vestibule /ˈvestɪbjuːl/ n.【anatomy】a cavity or chamber that leads to another cavity 前庭:buccal vestibule 颊前庭 ◇ labial vestibule 唇前庭 ◇ laryngeal vestibule 喉前庭 ◇ nasal vestibule 鼻腔前庭 ◇ oral vestibule 口腔前庭 ◇ vaginal vestibule 阴道前庭

vestibulitis /ˈvestɪbjʊlɪtɪz/ n. inflammation of the vaginal vestibule 前庭炎

vet /vet/ abbr.【BrE】veterinarian 兽医

vet check n. *see* veterinary inspection 兽医检查

vet clean adj. (of a horse) sound and free form contagious disease, conformation fault, injury or blemishes when inspected by a veterinarian(马匹经兽医检查尚无传染病、肢体缺陷、外伤或瑕疵)健康的

vetch /vetʃ/ n. any of various legume plants cultivated as fodder for farm animals(一种野生豆科植物,可作家畜饲草种植)野豌豆

veterinarian /ˌvetərɪˈneəriən; *AmE* -ˈner-/ n. (abbr. **vet**) one who practices veterinary medicine; animal doctor 兽医 syn veterinary medical doctor

veterinary /ˈvetrənəri; *AmE* ˈvetərəneri/ adj. of or concerned with the medical or surgical treatment of animals 兽医[学]的:traditional Chinese veterinary medicine 中兽医学 ◇ veterinary acupuncture 兽医针灸 ◇ veterinary medi-

V

cine 兽医内科 ◇ veterinary surgery 兽医外科

veterinary certificate n. a certificate of soundness issued by a veterinarian following evaluation of the horse's physical condition(兽医对马匹体况评估后出示的健康证明)兽医证明

veterinary examination n. *see* veterinary inspection 兽医检查

veterinary inspection n. (also **vet check**, **veterinary examination**) an examination of the physical condition and soundness of a horse(对马匹体况进行的检查)兽医检查: Veterinary inspection is compulsory in endurance riding and eventing. 在耐力赛和三日赛中,兽医检验是例行项目。 syn soundness examination

veterinary medical doctor n. (abbr. **VMD**) a veterinarian 兽医

veterinary medicine n. the branch of medical science concerned with the study of animal diseases (研究动物疾病的医学分支)兽医学

veterinary surgeon n. *see* veterinarian 兽医

VFAs abbr. volatile fatty acids 挥发性脂肪酸

viability /ˌvaɪəˈbɪləti/ n. the capability of being viable 成活力,生存力

viable /ˈvaɪəbl/ adj. capable of surviving or living 能活的,能生存的

Viatka /ˈvjɑːtkə/ n. (also **Vyatka**) an endangered horse breed indigenous to the Viatka regions of the former Soviet Union; stands 13 to 14 hands and may have a bay, chestnut, roan, or mouse dun coat; has a muscular, long neck, long back, wide and deep chest, and solid legs; is sturdy, lively, energetic and quiet; used for pulling troikas and light farm work(源于苏联维亚特卡地区的濒危马品种。体高 13～14 掌,毛色可为骝、栗、沙或鼠灰,颈背长,胸身宽广,四肢强健,体格强健,性情活泼而安静,主要用于挽车和农田轻役)维亚特卡马

vibrio /ˈvɪbrɪəʊ/ n.【vet】any of various short, S-shaped bacteria that causes cholera(引发霍乱的S形细菌)弧菌: *Vibrio cholerae* 霍乱弧菌

vibriosis /ˈvaɪbraɪəsɪs/ n. a venereal infection in cattle and sheep caused by the bacterium *Vibrio fetus*, often producing infertility or spontaneous abortion(见于牛羊的一种性病,由胎儿弧菌引起,常造成家畜不孕或自然流产)弧菌病

vice /vaɪs/ n. (also **stable vice**) any bad habit or abnormal behavior of a horse(马的不良习惯或异常行为)恶习,恶癖: The common stable vices are wood chewing, biting, wind sucking, pawing, kicking and rocking, etc. 常见的恶习有嚼癖、咬癖、咽气、扒刨、踢癖和摇摆等。

vice-breaker n. a device used to prevent the developing of vices(用来阻止马匹恶习发生的器件)恶习矫正器

Viceroy /ˈvaɪsrɔɪ/ n. a light, elegant, four-wheeled, air-tired, single-horse vehicle used for show harness classes(一种轻型单马四轮马车,外形美观,采用充气轮胎,多用于挽车展览)维瑟罗马车

vicious /ˈvɪʃəs/ adj. cruel, evil or immoral 邪恶的;恶毒的: vicious circle 恶性循环

Victoria /vikˈtɔːriə/ n. a half-hooded, four-wheeled, horse-drawn carriage with one forward-facing seat for two passengers and a raised driver's seat; was named after Queen Victoria and quite popular in the late 19th century(一种半敞式四轮马车,朝前的单座可容两名乘客,车夫驾驶座较高,流行于 19 世纪末并以维多利亚女王命名)维多利亚马车

view /vjuː/ vt.【hunting】to see a fox 看见(狐狸)

view halloo interj.【hunting】an exclamation uttered by the huntsman when viewing a fox(狩猎者看到狐狸时发出的叫喊)看到了!出现了!

vigilance /ˈvɪdʒɪləns/ n. the state of being watchful; alertness 警惕(性)

vigilant /ˈvɪdʒɪlənt/ adj. extremely careful to possible danger or trouble; alert, watchful 警觉的,警惕的

vigor /ˈvɪɡə/ n. physical or mental strength 体力,活力:vigor at birth 出生活力

vigorous /ˈvɪɡərəs/ adj. energetic and active in mind or body; robust 有活力的,精力充沛的

villus /ˈvɪləs/ n.【anatomy】(*pl.* **villi** /-laɪ/) any of the minute projections set closely on a mucous membrane(黏膜上紧密排列的微小突起)微绒毛

vinegary /ˈvɪnɪɡəri/ adj. having the taste, smell, or nature of vinegar 有酸味的

viral /ˈvaɪrəl/ adj. of or relating to viruses 病毒的:viral diseases 病毒性疾病 ◇ equine viral arteritis 马病毒性动脉炎

viral abortion n. abortion caused by viral infection 病毒性流产

virgin /ˈvɜːdʒɪn; *AmE* ˈvɜːrdʒ-/ n. a female animal that has not copulated 未配种的雌性动物:a virgin mare 处女母马

virosis /vaɪəˈrəʊsɪs; *AmE* vaɪˈroʊsɪs/ n. (*pl.* **viroses** /-siːz/) a disease caused by viruses; viral disease 病毒病

virulent /ˈvɪrələnt; *AmE* -rjəl-/ adj. extremely malignant or poisonous; lethal 剧毒的,致命的

vis-à-vis /ˌviːz ɑː ˈviː/ adv.【French】(of passengers) sitting face-to-face on opposite seats (乘客)面对面的:to sit vis-à-vis in a coach 对坐在四轮马车中

viscera /ˈvɪsərə/ n. pl.【anatomy】the internal organs of the body 内脏,脏腑

visceral /ˈvɪsərəl/ adj. of, situated in, or affecting the viscera 内脏的:visceral fat 内脏脂肪 ◇ visceral muscle 肠道平滑肌

viscerate /ˈvɪsəreɪt/ vt. to remove the bowels or intestines from the abdominal cavity; disembowel; eviscerate(从腹腔将内脏移除)取内脏,掏内脏

vision /ˈvɪʒn/ n. the faculty of sight; eyesight 视力;视觉:vision defects 视力缺陷

visual /ˈvɪʒuəl/ adj. of or relating to sight 视觉的:visual examination 视诊 ◇ visual inspection 肉眼检查 ◇ visual stimulation 视觉刺激

vital /ˈvaɪtl/ adj. **1** of, relating to, or characteristic of life 生命的:vital capacity 肺活量 **2** necessary to existence; essential 重要的;本质的:vital function 重要功能 ◇ vital organs 重要器官

vital signs n. any of the indicators of body health including heart rate, body temperature, and respiration rate(表示机体健康状况的参数,如心率、体温和呼吸频率等)生命指征,生命体征

vitality /vaɪˈtæləti/ n. the capacity to live, grow, or develop; vigor 生命力;活力

vitamin /ˈvɪtəmɪn; *AmE* ˈvaɪtəmɪn/ n.【nutrition】any of various organic substances essential in small amounts for normal growth and activity of the body(机体生长和活动所必需的多种微量有机物)维生素:fat-soluble vitamins 脂溶性维生素 ◇ water-soluble vitamins 水溶性维生素

vitamin B complex n. a compound consisting of ten water soluble vitamins(由十种水溶性维生素构成的复合物)维生素 B 复合物,复合维生素 B

vitamin B_1 n. *see* thiamin 维生素 B_1

vitamin B_2 n. *see* riboflavin 维生素 B_2

vitamin B_6 n. *see* pyridoxine 维生素 B_6

vitamin B_{12} n. a complex, cobalt containing water-soluble vitamin essential to normal blood formation, neural function, and growth(一种结构复杂的含钴水溶性维生素,为正常血液形成、神经活动和机体生长所必需)维生素 B_{12} [syn] cobalamin, cyanocobalamin

vitamin C n. a water-soluble vitamin required for production of certain essential amino acids and building of intracellular material(一种水溶性维生素,为某些必需氨基酸合成和细胞间质形成所必需)维生素 C [syn] ascorbic acid

vitamin E n. a fat-soluble vitamin associated with muscle development and fertility(一种脂溶性维生素,与肌肉生长和繁育紧密相关)维生素 E [syn] tocopherol

vitamin H n. *see* biotin 维生素 H

vitamin K n. a fat-soluble vitamin essential in blood clotting(一种脂溶性维生素,为血液凝固所必需)维生素 K

vitelline /ˈvaɪtəlɪn/ adj.【anatomy】of or relating to the yolk of an egg 卵黄的:vitelline membrane 卵黄膜 ◇ vitelline reaction 卵黄膜反应

vitiligo /ˌvɪtɪˈlaɪgəʊ/ n. a skin disorder characterized by the loss of pigmentation of the hair or skin(一种毛发或皮肤色素缺乏的皮肤病)白癜风

vitreous /ˈvɪtriəs/ adj.【anatomy】of, relating to or resembling glass 玻璃的,玻璃状的:vitreous body 玻璃体

viviparous /vɪˈvɪpərəs/ adj. giving birth to living offspring that develop within the mother's body (子代从母体中发育出生)胎生的

vixen /ˈvɪksn/ n. a female fox 雌狐:Once the egg is fertilized, the vixen enters a period of gestation that can last from 52 to 53 days. 在卵子受精后,雌狐妊娠期通常为 52 ~ 53 天。

Vizir /vɪˈzɪə; *AmE* vɪˈzɪr/ n. one of Napoleon's grey Arab stallions(拿破仑所骑的一匹青毛阿拉伯公马)维兹尔

Vladimir Heavy Draft n. a heavy draft breed developed in the second half of the 19th century in Vladimir, Russia; stands 15.1 to 16.1 hand, and may have a chestnut, bay, brown, or black coat; has a medium head, arched neck, prominent withers, short back, deep chest, and short powerful, well-feathered legs; is energetic, and willing; used for heavy draft and farm work(俄国弗拉基米尔地区于 19 世纪下半叶育成的重挽马品种。体高 15.1 ~ 16.1 掌,毛色可为栗、骝、褐或黑,头型适中,颈弓形,鬐甲突出,背短,胸深,四肢粗短,距毛浓密,精力充沛,性情顽强,主要用于重挽和农田作业)弗拉基米尔重挽马

VMD abbr. Veterinary Medical Doctor 兽医学博士

vocal /ˈvəʊkl; *AmE* ˈvoʊkl/ adj. of or relating to the voice or larynx 声音的;喉的

vocal cord n. the folds of membrane in the larynx which produce sound when vibrated(喉腔内的皱褶,振动时可发出声音)声带

voice aids n. any means to communicate instructions to the horse by using orders or raising voice(通过声音向马匹传达指令)声音扶助 compare lower aids, upper aids

voice box n. *see* larynx 喉

void /vɔɪd/ vt. to excrete body wastes; discharge 排泄

adj. useless; invalid 无用的;无效的

void bet n.【racing】a wager declared off in which the money bet is returned to the bettors (投注取消后赌金返还给博彩者的赌马)无效投注

volar /ˈvəʊlə; *AmE* ˈvoʊlə/ adj. *see* palmar 掌的

volatile /ˈvɒlətaɪl; *AmE* ˈvɑːlətl/ adj. evaporating readily at normal temperatures and pressures 挥发性的:volatile fatty acids 挥发性脂肪酸

volte /vəʊlt/ n.【dressage】a circular movement executed by a horse 环骑,圈乘:half volte 半圈乘 ◇ double volte 双圈乘

voluntary /ˈvɒləntri; *AmE* ˈvɑːlənteri/ adj. normally controlled by or subject to individual volition(由个体意志控制的)自主的,随意的:voluntary muscle 随意肌

vomer /ˈvəʊmə; *AmE* ˈvoʊmə/ n.【anatomy】a thin flat bone dividing the nostrils in most vertebrates(一块位于鼻中隔的扁平骨)犁骨

vomeronasal /vəməˈrəʊnəsl; *AmE* -ˈroʊn-/ adj. of the vomer and nasal bones 犁鼻[骨]的:vomeronasal cartilage 犁鼻软骨

von Osten, Wilhelm n. (1838 – 1909) a Russian psychologist who developed the theory of equine intelligence in 1890s by training horses to finish intellectual tasks(俄国心理学家,于 19 世纪 90 年代通过训练马匹完成智力技巧发展了马匹心智理论)威廉·冯·奥斯汀

Vulpes fulva n.【Latin】the scientific name for red fox [学名]赤狐

vulpicide /ˈvʌlpɪˌsaɪd/ n.【hunting】**1** the action

of killing a fox without the use of hounds(不采用猎犬)捕杀狐狸 **2** a person who kills a fox by shooting 枪杀猎狐者,捕杀狐狸者

vulpine /ˈvʌlpaɪn/ adj. **1** of or like a fox 狐狸的 **2** crafty,cunning 狡猾的,狡黠的:Karl gave a vulpine smile. 卡尔露出了狡黠的笑容。

vulva /ˈvʌlvə/ n.【anatomy】(*pl.* **vulvae** /-viː/) the external genital organs of the female(雌性动物的外生殖器)外阴;阴门

vulval /ˈvʌlvəl/ adj. of or relating to vulva 阴门的:vulval aperture 阴门裂 ◇ vulval lips 阴唇

vulvitis /vʌlˈvaɪtɪs/ n. inflammation of the vulva 外阴炎

Vyatka /ˈvjɑːtkə/ n. *see* Viatka 维亚特卡马

W /ˈdʌbljuː/ abbr. weight 重量

wafer /ˈweɪfə(r)/ n. a flat block of hay 干草块
vt. to prepare in the form of wafers 制块,压块

waferer /ˈweɪfərə/ n. (also **hay waferer**) a machine used for making hay bales 干草压块机 [syn] wafering machine

wager /ˈweɪdʒə(r)/ n. an act of risking money on the result of a race; bet 投注,押注
vt. to bet on horse races 押注: to wager $ 50 on a horse 一匹马上投注 50 美元

wagon /ˈwægən/ n. a four-wheeled, horse-drawn vehicle with a box-shaped body, formerly used for transporting loads and passengers(过去用来运送货物和乘客的厢式四轮马车)四轮马车 [compare] cart

wagon train n. a long line of wagons traveling together 马车队

wagoner /ˈwægənə/ n. one who drives a wagon; driver 车夫

wagonette /ˌwægəˈnet/ n. a light, four-wheeled, open passenger vehicle pulled by a single horse (一种轻型客用四轮马车,由单马拉行)小型四轮马车: wagonette Break 布里克四轮马车 ◇ wagonette Omnibus 公共四轮马车 ◇ wagonette Phaeton 费顿四轮马车

wagonload /ˈwægənləʊd/ n. an amount of goods that can be carried in one wagon 马车载重;一车之量

WAHO abbr. World Arabian Horse Organization 世界阿拉伯马协会

wain /weɪn/ n.【dated】**1** a wagon 四轮马车 **2** an open farm cart drawn by a single horse in shafts(由单马驾辕拉行的农用双轮马车)农用马车

wainwright /ˈweɪnraɪt/ n. a wagon-builder 造车匠

Waler /ˈweɪlə(r)/ n. (also **Australian Waler**) a light saddle horse of mixed breed developed in New South Wales, Australia by crossing Thoroughbreds with Arabs and some pony breeds(源于澳大利亚新南威尔士的乘用马品种,主要采用纯血马与阿拉伯马及某些矮马品种杂交育成)澳洲威尔士马

walk /wɔːk/ n. a natural four-beat gait of horse in which each foot leaves and strikes the ground at a separate interval(一种四蹄音自然步法,马的四肢分别离地和落地)慢步: collected walk 收缩慢步 ◇ extended walk 伸长慢步 ◇ lazy walk 舒缓慢步 ◇ ordinary walk 普通慢步—*see also* **trot**, **canter**, **gallop**
vi. to move at the gait of walk 以慢步行进: walk on a loose rein 自由行进
vt. **1** to go or pass over by walking 走遍 PHR V **walk the course**【jumping】(of a rider) to walk along the jumping course to evaluate the footing and obstacles prior to competition(赛前骑手沿障碍路线行走以熟悉场地和障碍情况)走场 **2** to let a horse or hound walk for exercising purposes 遛马;遛犬

walk ring n.【racing】an enclosure beside the racetrack where horses are walked and jockeys mount before the start of the post parade(位于赛道旁的围场,入场仪式前赛马被牵至此处以便骑师上马)入场栏

walker /ˈwɔːkə(r)/ n. **1** a horse breed good at walking(善以慢步行进的马种)走马 [syn] walking horse **2** (usu. **Walker**) *see* Tennessee walking horse 田纳西走马

Walking Horse n. *see* Tennessee Walking Horse 田纳西走马

walkover /wɔːkˈəuvə/ n. **1**【racing】a race in which the sole qualified horse wins the race by walking(唯一有资格的赛马慢走即可获胜的比赛)走步获胜 **2** an easy or uncontested win in a competition; breeze 轻易获胜

walk-up start n.【racing】a start of a race in which the horses walk toward the starting line and gallop off at the starter's command(参赛马匹走到起跑线并在听到发令后起跑的比赛)走步式起跑:The walk-up start is sometimes conducted without a starting gate. 在没有起跑闸时,有时也采用走步式起跑。

wall /wɔːl/ n. **1**【jumping】(also **brick wall**) an upright jumping obstacle made of wooden blocks painted and stacked to look like a wall of bricks(由木块堆积并漆成砖墙造型的直立障碍)砖墙 **2** a cross-country obstacle made of bricks or stones(越野赛中由砖块或石头搭建的障碍)跨墙 **3** the hoof wall 蹄壁: wall of the hoof 蹄壁

wall eye n. the eye of a horse showing more white than usual around the pupil due to a lack of iris pigmentation(由于马的眼虹膜缺乏色素导致瞳孔周围泛白的情况)玻璃眼,玉石眼:Horses with wall eyes see exactly the same as horses who have the normal eye pigment, and they are born with it. 眼白过多的马属于天生,其视力与眼内色素正常的马一样。 syn china eye, fish eye, glass eye, smoky eye | **wall-eyed** /ˌwɔːlˈaɪd/ adj.

wallop /ˈwɒləp; *AmE* ˈwɑːl-/ vt. to beat or strike hard 痛打,猛击

vi. (of a vehicle) to move in quick, rolling manner(指车辆)颠簸

wander /ˈwɒndə; *AmE* ˈwɑːn-/ vi. to move about aimlessly; amble (无目的)游荡,漫步

wanderer /ˈwɒndərə(r); *AmE* ˈwɑːn-/ n. *see* barker 患出生适应不良综合征的幼驹;游走幼驹

war bonnet paint n. a Paint horse with colored head and neck but white body(头颈为花色而体色为白色的花马)战帽花马: Some American Indian tribes thought war bonnet paints were endowed with supernatural powers. 一些美洲印第安部落相信战帽花马拥有超自然的力量。

war horse n. (also **war-horse**) a powerful horse used as mount in battle; charger, destrier 战马,戎马 compare palfrey, sumpter

warble /ˈwɔːbl; *AmE* ˈwɔːrbl/ n. **1** a hard lump of swelling on a horse's back caused by rubbing against the saddle(由马鞍磨蹭马背形成的硬肿块)鞍瘤 **2**【vet】an abscess or swelling on the back of an animal formed by infestation of warble fly(动物背部寄生皮蝇后形成的脓肿)皮蝇瘤,皮蝇脓胀

warble fly n. any of several flies of the family *Oestridae* which lays eggs under the skin of cattle and horses and causes subdermal warbles on the back when hatched(狂蝇科的多种蝇类,在牛马背部皮下产卵孵化后可形成皮瘤)皮蝇

Warde, John (1753 – 1839) n. a British sportsman who pioneered the sport of foxhunting and is often regarded as the father of foxhunting(英国现代猎狐运动的先驱,常被人称为猎狐之父)约翰·瓦德 syn Father of Foxhunting

ware /weə(r)/ interj.【hunting】a call made by huntsman to the field to arouse attention to sth (狩猎时提示大家留意的叫喊)注意!当心!:Ware hound! 注意猎犬! ◇ Ware ditch! 小心水渠!

warehouse /ˈweəhaʊs; *AmE* ˈwerh-/ n. a building for storing grain or fodder; barn(贮藏草料的房屋)仓库

warm /ˈwɔːm; *AmE* ˈwɔːrm/ vi. to do gentle exercises 热身 PHR V **warm down** to cool down after competition 赛后放松 **warm out of it** to exercise until the muscles warmed up 活动筋骨,完成热身 **warm up** to prepare for competition beforehand by doing gentle exercises 赛前热身

warmblood /ˈwɔːmˌblʌd/ n. (also **halfblood**) a crossbreed developed by crossing coldbloods with hotbloods(由冷血马和热血马杂交育成

的马匹类型)温血马;半血马 compare coldblood, hotblood

warmblooded /ˌwɔːmˈblʌdɪd/ adj. of or relating to a warmblood 温血马的:a warmblooded horse 一匹温血马

warm-down /ˈwɔːmˌdaʊn/ n. the act or process of cooling down 赛后放松 compare warm-up

warm-up /ˈwɔːmˌʌp/ n. an act of preparing for a race or competition(赛前)热身:warm-up arena 热身场 compare warm-down

warrantable /ˈwɒrəntəbl/ adj.【hunting】(of deer) reaching a legal age for hunting(指鹿)到捕猎年龄的,适猎的

n. a deer of five years' age or older(5 岁以上的鹿)成年鹿

warranty /ˈwɒrənti; *AmE* ˈwɔːr-/ n. a guarantee given to a horse sold as sound in conformation and capable of performing certain task(买卖中对马匹体貌和性能给予的保证)担保,保证

washed-out /ˌwɒʃtˈaʊt/ adj. **1** faded in color 色淡的 **2** (of a horse) exhausted or tired out(指马)疲惫的;乏力的

washy /ˈwɔʃi; *AmE* ˈwɑːʃi/ adj. **1** (of color) light or faded(颜色)淡的,浅的;弱的:An old saying suggests 'washy in colour, washy in constitution.'常言道,"色淡则体弱"。**2**【racing】(of a horse) sweated all over body prior to race(指马)赛前出汗的

water[1] /ˈwɔːtə(r)/ vt. to give water to an animal to drink 饮水:to water horses 饮马 | **watering** n. :watering place 饮水点

water[2] /ˈwɔːtə(r)/ n.【hunting】**1** a ditch or stream in a jumping course 水沟,水渠 **2** *see* water jump 水沟障碍

water bag n. *see* amnionic sac 羊膜

water brush n. a grooming tool used to brush the coat, mane and tail of horse(被毛刷拭时使用的工具)水刷

water bush n.【jumping】a small hedge or row of bushes positioned on the takeoff side of a water jump(水沟障碍起跳侧放置的低矮灌木)水沟树篱,水沟灌木

water founder n. inflammation of the laminae of horse's hoof caused by drinking too much cold water following strenuous exercise(马剧烈运动后饮大量冷水导致蹄叶发炎)水性蹄叶炎

water jump n.【jumping】a cross-country jumping obstacle consisting of a shallow, water-filled ditch(越野障碍赛中的浅水渠沟)水渠障碍:The water jump is always preceded by a water bush. 水沟障碍前面通常设有树篱。syn water[2]

water scraper n. *see* scraper 水刮

water-soluble /ˈwɔːtərˌsɒljʊbl/ adj. able to be dissolved in water 水溶性的:water-soluble vitamins 水溶性维生素 compare fat-soluble

wattle /ˈwɒtl; *AmE* ˈwɑːtl/ n. **1** a structure consisting of poles interlaced with branches, twigs, or reeds for making fences(由木竿与枝条或芦苇编织的围栏结构)树篱,篱笆 **2**【hunting】a hurdle or jumping obstacle 跨栏;障碍

wax /wæks/ n. a layer of dried colostrum formed on the teats of a pregnant mare ready to foal(待产母马乳头上形成的初乳干层)乳蜡

vi. to secrete colostrum on the teats before foaling 分泌乳蜡:waxed teats 泌乳蜡的乳头 | **waxing** /ˈwæksɪŋ/ n.

way of going n. the way how a horse moves; gait 步态,步法

WB abbr. warmblood 温血马

WBC abbr. White Blood Cells 白细胞

WCF abbr. Worshipful Company of Farriers 蹄铁匠公会

WCTPA abbr. World Champion Team Penning Association 国际团体截牛锦标赛协会

WD abbr. withdrawal 退赛

weak /wiːk/ adj. lacking physical strength or vigor; feeble 体弱的,虚弱的:weak heat 微弱发情 ◇ weak pastern 弱系 | **weakness** /ˈwiːknəs/ n.

wean /wiːn/ vt. to stop feeding a foal with the mare's milk 断奶,断乳:A foal is often weaned at age of six months. 幼驹通常在 6 月龄断奶。

weaning /ˈwiːnɪŋ/ n. the gradual process to stop a foal sucking its dam 断奶：artificial weaning 人工断乳 ◇weaning weight 断奶重

weanling /ˈwiːnlɪŋ/ n. a foal weaned from the mare 断乳幼驹：weanling colt 断乳雄驹 ◇ weanling filly 断乳雌驹

wear /weə(r)/ v. **1** (of teeth) to diminish or erode by use or rubbing(牙齿)磨损 **2** (of a horse) to carry or move in a certain way(指马)走姿 PHR V **wear itself well** (of a horse) to move with its head and tail carried high(指马)头尾高翘；趾高气扬 syn carry both ends

n. the process or state of being worn off 磨损(程度) PHR **in wear**(of teeth) used in chewing process(指牙齿)磨损的

Weatherby /ˈweðəbi/ n. a British family who maintained and published the first volume of the *General Stud Book* for the English Thoroughbred in 1793(登记并于 1793 年出版首册《英国纯血马登记册》的家族)韦瑟比

Weatherby's n. the Stud Book [英国]纯血马登记册

weave /wiːv/ vi. **1** (of a horse) to sway or wander from side to side (指马)摇晃，摇摆 **2** to move in a zigzag pattern 蛇形移动，窜行

weaver /wiːvə(r)/ n. a horse having the vice of weaving 患摇摆症的马

weaving /ˈwiːvɪŋ/ n. a vice of horse who sways his shoulders, neck or head from side to side relentlessly(马左右前后不停地晃动肩部、颈和头的恶习)摇摆症：Weaving usually results from boredom of confinement in stall box, and is generally corrected when the horse is turned out to pasture. 摇摆症多半是由于马厌倦厩舍束缚所致，所以，马被放到草场上后便可矫正。

web /web/ n. **1** the area between the left and right sides of a horseshoe(位于蹄铁左右两支之间的区域)蹄铁掌面 **2** the width of the barstock from which a horseshoe is made(制作蹄铁的)铁条宽度

web martingale n. *see* bib martingale 围嘴鞅

wedge /wedʒ/ n. **1** a piece of wood, rubber, or metal with thick end and one thin end, placed between two objects to secure or separate them (一端厚而另一端薄的木质、橡胶或金属件，置于两物体间起加固或分离目的)楔子 **2**【farriery】a piece of horseshoe-shaped rubber with increased thickness from toe to heels, placed between the horseshoe and the hoof to elevate the heel thus change the hoof angle(形似蹄铁且从头至尾厚度渐增的橡胶楔，衬在马蹄和蹄铁间用来增加蹄踵高度并借此改变蹄角度)蹄楔

vt. **1** to fix in position by using a wedge 加楔固定：The door was wedged open. 门用楔子撑开着。**2** to force into a narrow space 塞入

wedge pad n. a saddle pad with the thickness gradually increasing towards the rear(后端厚度逐渐增加的鞍垫)楔形鞍垫，楔式鞍垫

wedge shoe n. (also **wedge-heeled shoe**) a horseshoe with increased thickness from toe to heel (从头至尾厚度逐渐增加的蹄铁)楔形蹄铁

wedge-heeled shoe n. *see* wedge shoe 楔形蹄铁

weed /wiːd/ n. **1** a plant or grass considered undesirable or troublesome 杂草：weed control 杂草防除 **2** a small, poorly built horse 瘦弱的马

weedicide /ˈwiːdɪsaɪd/ n. (also **weedkiller**) any chemical used to destroy or inhibit the growth of weeds(用于除灭或阻止杂草生长的化学制剂)除草剂，除莠剂：weedicide sprays 除草喷雾剂 syn herbicide

weedkiller /ˈwiːdkɪlə(r)/ n. *see* weedicide 除草剂

weedless /ˈwiːdlɪs/ adj. (of pasture) without weeds(牧场)无杂草的

weedy /ˈwiːdi/ adj. **1** (of grassland) full of weeds (草场)杂草丛生的：a weedy pasture 杂草丛生的牧场 **2** (of a horse) weak and poorly built (指马)瘦弱的

WEG abbr. World Equestrian Games 世界马术

大赛

weigh /weɪ/ vt.【racing】to measure the weight of sb/sth 称重 PHR V **weigh in/out** to weigh the rider and saddle before or after a race 赛前/后称重：The riders are required to weigh in and out twice during a race, and the ending weight of him and gear should be equal to beginning weight. 赛马规定骑手在赛前、赛后都要进行称重，而且赛后和赛前重量要相等。

weighbridge /ˈweɪbrɪdʒ/ n. a machine with a platform for weighing heavy loads or large livestock（用来称量载重卡车和牲畜的仪器）地秤，地磅

weighing room n.【racing】a room where the jockeys are weighed 称重室

weight /weɪt/ n. **1** the amount of heaviness 重量；体重：beginning weight 赛前体重 ◇ ending weight 赛后体重 ◇ body weight 体重 ◇ weight loss 体重下降 **2**【racing】the extra loads carried by a horse（马匹）负重 PHR **weight for age**【racing】（abbr. **WFA**）a condition in which the weight a horse carries depends on its age with younger horses carrying less than older ones（赛马中根据马匹年龄确定负重的方法，青年马较年长者负重少）按龄负重

weight aid n. the rider's body weight employed to give instructions to the horse by shifting the center of gravity（骑手通过改变身体重心传达指令的方式）体重辅助

weight allowance n.【racing】a reduction in the amount of weight carried by a racehorse（给予一匹赛马的负重减少量）负重让磅：Weight allowance is usually granted to apprentice jockeys or fillies racing against colts. 负重让磅多给予见习骑师，或那些与雄驹同场竞技的雌驹。

weight cloth n.【racing】*see* lead pad 负重袋，铅块袋

weight-for-age race n.【racing】a race in which the weight carried by a horse is determined by its age（依照马的年龄确定负重磅数的比赛）按龄负重赛

weightape /ˈweɪtæp/ n. a tape that uses a measurement of the horse's heart girth to give an approximate body weight（根据马匹胸围测量值给出相应体重的带尺）体重尺

well-conditioned /ˌwelkənˈdɪʃənd/ adj.（of livestock）in good condition（牲畜）体况良好的

well-drained /ˌwelˈdreɪnd/ adj. that can be drained efficiently 排水良好的

well-fed /ˌwelˈfed/ adj.（of animals）having good ration or feeds（动物）营养好的，饲喂好的

well-formed /ˌwelˈfɔːmd; *AmE* -ˈfɔːrmd/ adj.（of a horse）having a good conformation（指马）体型良好的

well-groomed /ˌwelˈgruːmd/ adj. carefully groomed and cleaned 精心刷拭的：well-groomed coat 精心刷拭的被毛

wellington /ˈwelɪŋtən/ n.【BrE】（also **wellington boot**）a knee-length water-proof rubber boot（一种齐膝高的防水胶靴）长筒胶靴 syn rubber boot

well-muscled /ˌwelˈmʌsld/ adj. having strong muscles 肌肉发达的，肌肉良好的

well-schooled /ˌwelˈskuːld/ adj.（of a horse）well broken and trained（指马）调教好的

well-sprung /ˌwelˈsprʌŋ/ adj.（of ribs）long and well arched（肋骨）开张良好的：The well-sprung ribs of a horse will allow ample room for the function of the heart and lungs. 一匹马的肋骨开张良好，心肺活动就有充足的空间。

Welsh /welʃ/ n. *see* Welsh Pony 威尔士矮马

Welsh Cart Horse n. an ancient horse breed developed in Wales around the 12th century by crossing Welsh ponies with Spanish horses; was hardy, quiet, powerful and used mainly for driving cart and farm work（12 世纪左右威尔士采用当地矮马与西班牙马杂交培育的马品种，抗逆性强，性情温驯，速力较好，主要用于挽车和农田作业）威尔士辕马

Welsh Cob n. an ancient pony breed native to

Wales, developed in the early 11th century by crossing Welsh Mountain ponies with Spanish horses and later improved with Norfolk Roadsters and Yorkshire Coach Horses in the 18th century; stands 14 to 15.2 hands and may have a coat of any solid color; has a compact body, strong shoulders, powerful back, and slight, silky feathering; historically used as a carriage horse, but now also used for equestrian competitions and sports(源于威尔士的古老矮马品种,于11世纪初采用威尔士山地矮马与西班牙马培育而成,后于18世纪经诺福克挽马和约克夏挽马改良。体高14~15.2掌,各种单毛色兼有,体格结实,肩部发达,后躯有力,四肢附生白色距毛;过去多用于拉车,现在也用于马术竞技比赛)威尔士柯柏马

Welsh Mountain n. *see* Welsh Mountain Pony 威尔士山地矮马

Welsh Mountain Pony n. (also **Welsh Mountain**) a native pony breed indigenous to the mountain regions of Wales; stands under 12 hands and has usually has a gray coat; has a small head with bold eyes, a compact body with deep girth, powerful loins and hind legs, and exceptionally hard hoofs; is hardy, spirited, and intelligent; formerly used to haul coal in the underground mines, now used for both riding and harness(源于威尔士山区的本土矮马品种。体高在12掌以下,毛色以青为主,头小而眼大,体型短小精悍,胸深,腰和后肢肌肉发达,蹄质结实,抗逆性好,性情活泼,反应敏快,过去常用于煤矿运输,现在主要用于骑乘和驾车)威尔士山地矮马

Welsh Pony n. a pony breed derived from the Welsh Mountain pony and improved with Hackney blood; stands 12 to 13.2 hands and may have a coat of any solid color; has deep girth, a full mane and tail, long, well-proportioned limbs, and strong quarters and hocks; is hardy, quiet, and energetic; historically used for shepherding, now used mainly for riding and harness(由威尔士山地矮马经哈克尼马改良育成的矮马品种。体高12~13.2掌,各种单毛色兼有,胸身宽广,鬃尾浓密,四肢修长,后躯发达,抗逆性强,性情温驯,富有活力,过去主要用于放牧,现在主要用于骑乘和驾车)威尔士矮马

Welsh Pony and Cob Society n. (abbr. **WPCS**) an organization founded in UK in 1901 to maintain and promote the breeding and registries of Welsh Cobs and Ponies(1901年在英国成立的组织机构,旨在促进威尔士矮马和矮壮马的育种和品种登记)威尔士矮马与柯柏马协会:The Welsh Pony and Cob Society classifies the native ponies into four sections – A, B, C, and D. Section A is the Welsh Mountain Pony, Section B is the Welsh Pony, Section C is the Welsh Pony of Cob Type, and Section D is the Welsh Cob. 威尔士矮马与柯柏马协会将本地矮马分为A、B、C、D四类。A类是威尔士山地矮马,B类是威尔士矮马,C类是柯柏型威尔士矮马,D类是威尔士柯柏马。

Welsh Pony of Cob Type n. a pony breed indigenous to Wales descended from the Welsh Mountain mares put to smaller Norfolk Trotters and Hackneys; stands 13.2 to 14 hands and may have a coat of any solid color; has powerful quarters, short back, deep chest, and relatively short legs; originally used for shepherding and general farm work, but now primarily for light draft and riding(威尔士的本土矮马品种,主要采用威尔士山地马母马与体型较小的雄性诺福克马及哈克尼挽马交配繁育而来。体高13.2~14掌,各种单毛色兼有,后躯发达,背短,四肢相对较短,过去常用于放牧和农田作业,现在主要用于轻挽和骑乘)柯柏型威尔士矮马

welt /welt/ n. a red, swollen mark left on the skin by a blow(皮肤被抽打后出现的红肿印记)红肿

vt. to strike or lash hard 猛击;抽打

welter /ˈweltə(r)/ adj. (of a racehorse) weighted 负重的,加磅的

n. (also **welterweight**) the extra weight carried by a racehorse according to its previous performance and earnings(根据以往赛绩和奖金而额外给予的负重)负重;加磅

welter race n.【racing】a race in which weights are carried by some horses according to their previous winnings and purse earnings(根据马匹以往获胜次数和奖金决定负重的比赛)负重赛,加磅赛 compare allowance race

welterweight /ˈweltəweɪt/ n. *see* welter 磅数,负重

Western /ˈwestən; *AmE* -tərn/ adj. concerning the life of American pioneers and cowboys(涉及美国西进运动或牛仔生活的)西部的;西部牛仔的:Western movies 西部片

western bit n. one kind of bit mainly used in western America (美国)西部口衔

western equine encephalomyelitis n. (abbr. **WEE**) one of three primary strains of encephalomyelitis 西部马脑脊髓炎

Western horse n. a type of horse suitable for cutting, calf roping, etc. (适用于截牛、套牛等牧场工作的马匹)西部马,牛仔马

Western Isles Highland Pony n. one type of Highland Pony 西岛高地矮马

Western riding n. a riding style popularized by the American cowboys in which a Western saddle and related tacks are used(美国牛仔采用的骑马方式,使用西部牛仔鞍和相应马具)西部骑术,西部骑乘 syn Western style

Western saddle n. a saddle with a large, deep seat and saddle horn used mainly by cowboys (牛仔常用的一种鞍座很深且带角柱的马鞍)西部牛仔鞍 syn stock saddle, working saddle

Western States Trail Ride n. *see* Tevis Cup Ride 特维斯杯耐力赛,又名西部越野赛

Western stirrup n. a stirrup attached to a Western saddle 西部马镫

Western style n. *see* Western riding 西部骑术,西部骑乘

Westlands pony n. *see* Fjord Pony 挪威峡湾马

Westphalian /ˈwestfeɪljən/ n. a Hanoverian horse bred in Westphalia, Germany; stands 16 to 17 hands and may have a coat of any solid color; has a solid build and is quiet, powerful, and athletic; used as a competition and riding horse (育成于德国威斯特伐利亚地区的汉诺威马,体高16~17掌,各种单毛色兼有,体格强健,性情温驯,挽力较大,运动能力强,主要用于竞技和骑乘)威斯特伐利亚马

wet mare n. a mare who nurses a foal 哺乳母马

wetland /ˈwetlənd/ n. a lowland area that is saturated with water, especially the natural habitat of wildlife(持水性较好的洼地,是野生动物的自然栖息地)湿地

Weymouth bit n. a kind of curb bit 魏茂斯衔铁

WFA abbr. weight for age 按龄负重

whang /wæŋ/ n. **1** a thong or whip of hide or leather 皮带,皮鞭 **2** a lashing blow, as of a whip 抽打

vt. to beat or whip with a thong 抽打;鞭笞

wheal /wiːl/ n. a red swelling on the skin caused by lashing or allergic reaction; welt(皮肤上因鞭打或过敏反应而产生的红肿)鞭痕;红肿

wheat /wiːt/ n. an annual cereal plant widely cultivated for its edible grain(一年生禾本科农作物,因谷粒可食而广泛种植)小麦:wheat bran 麦麸 ◇ wheat flour 面粉 ◇ wheat middling 粗小麦粉 ◇ wheat straw 麦秸

wheat bran n. the outer coating of the wheat grain separated from the flour in milling(小麦磨面时脱落的种皮)麸皮

wheat middlings n. the coarser particles of ground wheat mingled with bran(含麸皮的面粉)粗粉

wheatgrass /ˈwɪtgræs/ n. a kind of perennial grass(一种多年生草本植物)冰草:western wheatgrass 蓝茎冰草

wheatgrass hay n. the cut and dried hay made from wheatgrass 晒干的冰草

wheel[1] /wiːl/ n. a circular wooden or iron frame that turns around the axle of a vehicle(绕车轴旋转的木质或铁制圆形框架)车轮

wheel[2] /wiːl/ vi. **1**【racing】(of a horse) to turn sharply (指马)猛然转向 **2**【racing】(of a bettor) to wager on every type of combination bets for a specific horse(针对某匹马的所有组合票都投注)组合通押

wheel bar n. *see* swingle tree 挂杠;挂杆

wheel horse n.【driving】the horse in a team that is harnessed nearest the front wheels(马队中套在前车轮旁的马匹)副马

wheel hub n. the hub of a wheel 轮毂

wheelbarrow /ˈwiːlbærəʊ/ n. a one-wheeled vehicle with handles at the rear, mainly used to carry small loads(一种带把手的单轮推车,用来运送少量物资)手推车

wheelbase /ˈwiːlbeɪs/ n. the distance from the center of the front wheel to that of the rear wheel in a vehicle(车辆从前轮中心至后轮中心间的距离)前后轴距

wheeled /wiːld/ adj. (of vehicle) having wheels 带轮的: a four-wheeled carriage 一辆四轮马车

wheeler /ˈwiːlə(r)/ n. **1** a wheelwright 车轮匠 **2** a wheel horse 副马 **3**【driving】either one of the pair of horses hitched directly to the shafts of a vehicle(直接套在车辕上的马匹)辕马: near wheeler 内侧辕马 ◇ off wheeler 外侧辕马 [compare] center pairs, leaders

wheelwright /ˈwiːlraɪt/ n. (also **wheeler**) one who makes and repairs wheels(制造并修理车轮的人)车轮匠

wheeze /wiːz/ vi. to breathe with a rattling sound in the chest due to obstruction of the air passages(由于呼吸道阻塞导致呼吸时胸腔发出呼哧声响)呼哧,喘息

n. a sound made by a horse who wheezes 呼哧声,喘息声

whelp /welp/ n. an unweaned puppy of a hound or wolf(未断奶的)幼兽,幼崽 [PHR] (**be**) **in whelp** (of a bitch or vixen) carrying young or litter(母犬或雌狐)怀崽的

v. to give birth to the young 下崽,产仔

whey /weɪ/ n.【nutrition】the watery part of milk that separates from the curds, as in the process of making cheese(在奶酪制作过程中,从乳块中分离出来的水样液体)乳清: whey meal 乳清粉

whicker /ˈwɪkə(r)/ n. a gentle sound a horse makes; whinny 嘶鸣声

vi. to whinny, nicker 轻声嘶鸣

whiffletree /ˈwɪfltriː/ n. *see* singletree 衡杆;挂杠

whinny /ˈwɪni/ vi. (of a horse) to make a gentle sound; nicker(马)轻声嘶鸣 [compare] neigh

n. a gentle sound made by a horse (马的)嘶鸣声

whip /wɪp/ n. **1** (also **horsewhip**) a thin piece of lash attached to a handle or rod, used to drive, control or punish a horse(末端连有皮条的木棍,用来驱赶或责罚马匹)鞭子,马鞭: schooling whip 调教鞭 ◇ whip hand 持鞭的手 [syn] crop, fiddle **2** one who drives a carriage with a whip 车夫

vt. to lash orstrike with a whip 鞭打

whipcord /ˈwɪpkɔːd; *AmE* -kɔːrd/ n. **1** a twisted or braided cord used in making whiplashes 鞭条,鞭绳 **2** a strong fabric with a distinct diagonal rib, often used for making riding habits or breeches(一种带斜纹织物,质地结实,常用来制作骑装和马裤)马裤呢

whiplash /ˈwɪplæʃ/ n. **1** the lash of a whip 马鞭的皮条 **2** a hit with a whip 鞭打

whipper-in /ˈwɪpərˈɪn/ n.【hunting】one who assists the huntsman in handling a pack of hounds in the field(野外协助猎犬主看管猎犬的人)猎主助理

whippletree /ˈwɪpltriː/ n. (also **whipple tree**) *see* singletree 挂杠;挂杆

whirlicote /ˈwɜːlɪkəʊt/ n. a heavy and luxurious carriage 豪华马车

whisk /wɪsk/ n.【driving】a whip with a small bunch of hair attached to its end(末梢束毛的鞭杆)毛刷鞭

whiskers /ˈwɪskəz/ n. long stiff bristles growing around the muzzle of a horse(马口鼻周围生长的硬毛)触须:Some owners would leave the whiskers around the horse's muzzle unclipped, but still others prefer them trimmed off for cosmetic reasons. 一些马主对马口鼻周围的触须不予修剪,而有的马主出于美观喜欢将其剪掉。

Whiskey /ˈwɪski/ n. a light, one-horse carriage popular in the late 18th and early 19th centuries(在18世纪末至19世纪初流行的一种轻型马车,多由单马拉行)威斯奇马车

whisperer /ˈwɪspərə/ n. *see* horse whisperer 马语者

whistle /ˈwɪsl/ n. the sound made by blowing sth 哨声

vi. **1** to make a high sound by blowing sth 吹哨 PHR V **whistle on the play**【polo】to blow a whistle to stop the play 吹哨暂停 **2** (of a horse) to make a heavy inhaling sound while running(指马奔跑时)喘气,喘息 | **whistling** /ˈwɪslɪŋ/ n.

white[1] /waɪt/ n. **1** the color of white 白色 PHR **in the white** (of a carriage) not yet painted after construction(马车)未上漆的 **2** a white horse 白马:dominant white 显性白

white[2] /waɪt/ n. a condition in which the cornea of the eye is short of pigment(眼睛角膜缺少色素导致的症状)白眼:Horses which show the white of their eye are regarded as being bad tempered. 普遍认为带白眼的马脾性不好。

white blanket n. a coat color pattern of the Appaloosa who has a solid colored body with white over the hips and an absence of colored spots(阿帕卢萨马花色类型之一,马被毛为其他单色,臀部毛色为白且不混生花色斑点)白毯型

white blood cell n. (abbr. **WBC**) a colorless blood cell active in defense against infection(血液中起防御作用的无色细胞)白细胞 syn leukocyte, leucocyte

white body n.【physiology】the white fibrous tissue in an ovary that forms after the involution and regression of the yellow body(卵巢内黄体退化萎缩后形成的白色纤维化组织)白体 syn corpus albicans

white brass n. *see* German silver 德国银

white coronet n. (also **coronet**) a leg marking referred when the coronet is white 蹄冠白 syn coronet white

white face n. *see* bald face 白脸

white fetlock n. (also **fetlock**) a leg marking referred when the white extends to the fetlock 球节白 syn fetlock white

white flag n.【jumping】a white marker used to denote the left side of a jumping obstacle(用来标识障碍左侧的旗)白旗 compare red flag

white heel n. a leg marking referred when the heel of hoof is white 蹄踵白

white leg n. a leg marking consisting of white that extends above knees on the forelegs and hock on the hindlegs(马四肢别征之一,其白色区域延伸至前膝或飞节以上)白腿

white lethal n.【genetics】an inherited genetic condition of a foal with a lethal dominant white gene(幼驹的一种遗传疾病,由显性白化基因控制)白化致死基因

white line n. a band of pale yellow-colored juncture between the wall and the sole of the foot (介于马蹄壁与蹄底之间的浅黄色条带)白线:white line disease 白线病 ◇ The white line indicates to the farrier the amount of wall he has in which to place the nails. 对蹄铁匠来说,白线是蹄钉镶入蹄壁分寸的标记。

white lip n. a facial marking referred when the lips of horse are white 唇白 — *see also* **bald face, blaze, snip, star, stripe**

W

white marking n. (also **marking**) any white mark on the body of a horse 别征,白彰:White markings on the head and legs are prominent characteristics of domestic animals. 头部和四肢上的白彰是家畜的明显特征。

white muscle disease n. (abbr. **WMD**) a nutritional disease caused by deficiency of the trace mineral element selenium and vitamin E(由于微量元素硒和维生素 E 缺乏引起的一种营养疾病)白肌病

white muzzle n.【color】a facial marking that covers both lips and extends to the nostrils(一种白色区域从唇延伸至鼻的面部别征)口唇白,粉口

white of the eye n. part of the white sclera showed between partially closed eye lids(眼睛半闭时巩膜的白色部分)眼白

white pastern n. (also **pastern**) a leg marking referred when the pastern is white(一种蹄系被毛为白的四肢别征)系白:half pastern white 1/2 系白 ◇ three-quarter pastern white 3/4 系白

White Rose (**International**) n. a 100-mile endurance ride held annually in Yorkshire, UK (每年在英国约克郡举行的长途耐力赛,行程 160 千米)白玫瑰耐力赛

white sock n. (also **sock**) a leg marking referred when the white extends above the fetlock and includes part of the cannon bone(马四肢别征之一,其白色区域从球节延伸至管骨某处)管白

white spot n. **1**(also **white mark**) a leg marking consisting of a small tuft of white hairs on the foot of a horse(由一小撮白毛构成的四肢别征)白点 **2** a white mark on the eye of the horse caused by past injury that may interfere with vision(马眼由于受伤留有的白点,可能会影响马匹视力)白斑,白点

white stocking n. (also **stocking white**) a leg marking consisting of white that extends above the fetlock(马四肢别征之一,白色区域延伸至球节以上)管白:quarter white stocking 1/4 管白 ◇ half white stocking 1/2 管白 ◇ three-quarters white stocking 3/4 管白 ◇ full white stocking 全管白

whoa /wəʊ; *AmE* woʊ/ interj. a call used to command a horse to stop or stand still(用来吆喝马停下的口令)吁! 唷!

whole colored adj. (of coat) composed of hairs of one color(被毛)单色的,纯色的

whorl /wɜːl; *AmE* wɜːrl/ n. a small area where two tufts of hairs turn or meet at opposite direction(两撮毛发反向生长的区域)旋毛:tufted whorl 簇旋毛 ◇ simple whorl 单旋毛 [syn] hair whorl, cowlick

wid /wɪd/ adj.【dated】(of a horse) unsound in breathing(马)呼吸不畅的,阻气的

wide-chested /waɪdˈtʃestɪd/ adj. (of a horse) having a broad chest(指马)胸身宽广的 [opp] narrow-chested

Wielkopolski /ˌwiːlkəˈpɒlski/ n. a warmblood horse breed recently developed in Poland; stands 15.1 to 16 hands and may have any of the solid colors; has a compact body and a high, deep chest; is sturdy, well-balanced, and hard-working and used mainly for riding and driving(波兰最近培育的温血马品种。体高 15.1 ~16 掌,各种单毛色兼有,体格结实,胸身宽广,体质粗壮匀称,工作能力强,主要用于骑乘和驾车)波兰温血马 [syn] Polish Warmblood

wiggler /ˈwɪɡlə/ n.【slang】a pacer 对侧步马

wild /waɪld/ adj. (of animals) living or growing in a natural state; undomesticated 天然的;野生的:wild ass 野驴 ◇ wild horse 野马 ◇ wild oat 野燕麦 ◇ wild grassland 天然草场 ◇ wild species 野生品种

wild burro n. an ass living wild in the western part of the United States(在美国西部野生的驴)野驴

wild game n. wild animals hunted for food or sport 野味,猎物

W

wild horse n. **1** any horse living in the wild 野马 **2** *see* Mustang 北美野马

Wild Horses of America Registry n. an organization founded in the United States in 1974 to register and protect the wild horses and burros in North America, subsumed later into the International Society for the Protection of Mustangs and Burros(1974 年在美国成立的组织机构，旨在登记并保护北美的野马和野驴，后并入国际野马野驴保护协会)美国野马登记委员会

wildlife /ˈwaɪldlaɪf/ n. wild animals living in a natural, undomesticated state 野生动物: the wildlife of the Tibetan steppe 青藏高原上的野生动物

willing /ˈwɪlɪŋ/ adj. tame and compliant 温驯的，听话的: a willing horse 一匹温驯的马

wilt /wɪlt/ vi. (of plants) wither or droop because of heat (植物)枯萎；发蔫: semi-wilted forage 半干饲料

win /wɪn/ v. **1** to gain victory in competition 获胜，夺冠: to win by a nose 以微弱优势获胜 compare place, show **2** (of a bettor) to wager on a horse to finish first in a race (投注者)押中头马；赌赢

win and place n.【racing】a bet in which the player will win if he selects the winner and the second place horses(博彩者选对前两名赛马方可赢奖的投注方式)头二彩

win bet n.【racing】(also **win only**) a bet in which the player picks the horse who finishes first(博彩者选出头马赢奖的投注)头彩，独赢 compare place bet

win hole n.【harness racing】*see* two hole position (驾车赛)第二名，位列第二

win only n.【racing】*see* win bet 独赢，头彩

wind[1] /wɪnd/ n. **1** the breathing capacity of a horse (马的)呼吸性能: wind broke 呼吸沙哑 PHR **touched in the wind**(of a horse) slightly unsound in breathing (指马)呼吸不畅 **2** *see* flatulence 胃肠胀气

wind[2] /waɪnd/ v. to have or make many twists and turns 曲折；蜿蜒 PHR V **wind a fox**【hunting】(of hounds) to catch the scent of a fox (猎犬)嗅到狐狸踪迹

wind testing n. a test of horse's breathing ability (马)呼吸性能测试: The wind testing is usually checked with a stethoscope after demanding work. 呼吸性能测试多在马匹高强度运动后用听诊器进行测定。

windgall /ˈwɪndgɔːl/ n. (also **windpuff**) a painless swelling near the fetlock joint or elsewhere due to enlargement of the synovial sacs(由于滑液囊增大导致的下肢球节肿胀)软肿: articular windgall 关节软肿 ◇ tendinous windgall 屈腱软肿 syn road puff

winding /ˈwaɪndɪŋ/ n. *see* rope walking 拧绳肢势: Winding increases the likelihood of interference and stumbling. 拧绳肢势增加了交突和跌绊发生的可能。

windpipe /ˈwɪndpaɪp/ n.【anatomy】the trachea 气管

windpuff /wɪndˈpʌf/ n. *see* windgall 软肿

wind-sucker /ˈwɪndˌsʌkə(r)/ n. a horse having the vice of wind-sucking 有咽气癖的马

wind-sucking /ˈwɪndˌsʌkɪŋ/ n. a vice of horse who arches its neck as he sucks in air(马伸颈吸气的恶习)咽气癖 syn aerophagia

windy /ˈwɪndi/ adj. *see* roaring 呼吸沙哑的，喘鸣的

wing /wɪŋ/ n.【jumping】each of two extensions to the side of a fence(障碍两侧的延伸部分)侧翼

winging outward n. *see* paddling 内向肢势

wink /wɪŋk/ vt. (of a mare) to open and expose its genitalia and assume a mating position(母马接受交配时阴门开张)外张，张合 | **winking** /ˈwɪŋkɪŋ/ n.

winkers /ˈwɪŋkə(r)z/ n. pl. *see* blinkers 眼罩

winner /ˈwɪnə(r)/ n. a rider or horse who wins in a race or competition 获胜者；头马

winner's circle n. *see* winner's enclosure 获胜

栏,冠军栏

winner's enclosure n.【racing】(also **winner's circle**) an area on a racecourse where the winning horses and jockeys are awarded and photographed(赛马场上为获胜马匹和骑手颁奖并进行拍照的区域)获胜栏,冠军栏

winning post n.【racing】the finish post or finish line 终点,终线

wiry /ˈwaɪəri/ adj. (of a horse) sinewy and lean (指马)瘦而结实的;精悍的: a wiry looking horse 一匹短小精悍的马

withdraw /wɪðˈdrɔː, wɪθˈd-/ *verb* (**withdrew** /-ˈdruː/, **withdrawn** /-ˈdrɔːrn/) vi. (of a rider or horse) to stop taking part in a race or an equestrian event(指骑手或马)退出,退赛: In the 2016 Rio Olympic Games, US equestrienne Elizabeth Madden withdrew from team show jumping final due to a tendon injury of her mount. 在2016年里约热内卢奥运会上,美国马术选手伊丽莎白·马登由于坐骑肌腱受伤,只得退出团体障碍赛的决赛。

vt. to remove a rider or horse from competitor due to an injury or accident (骑手或马因受伤或意外)使退赛;取消参赛 compare disqualify

withdrawal /wɪðˈdrɔːəl, wɪθˈd-/ n. (abbr. **WD**) the act of withdrawing or the state of being withdrawn from a competition(在比赛中)退赛;取消参赛: voluntary withdrawal 主动退赛

wither pad n. *see* pommel pad 前桥垫,鬐甲垫

wither stripe n. (also **shoulder stripe**) a dark stripe that runs across the withers and extends down the shoulders of certain horse breeds(有些马种中横穿鬐甲延伸至肩部的深色条纹)鹰斑 — *see also* **dorsal stripe**, **zebra stripes**

withers /ˈwɪðəz; *AmE* -ðərz/ n. the highest part of the horse's back between the shoulders(马背与肩胛交界处最高的部分)鬐甲,肩隆

withers height n. (also **body height**) *see* height [鬐甲]体高

WMD abbr. white muscle disease 白肌病

wobble /ˈwɒbl; *AmE* ˈwɑːbl/ vi to move unsteadily from side to side; tremble, waver 摇晃;颤抖

n. the act or an instance of wobbling 摇晃;颤抖

wobbler /ˈwɒblə/ n. a horse who wobbles 患摇晃症的马: wobbler disease 摇晃症 ◇ wobbler syndrome 摇晃症

wobbles /wɒblz/ n. a syndrome causing uncoordinated gaits and even lameness(导致马步态失调甚至跛行的综合征)摇晃症,蹒跚病 syn ataxia

wolf /wʊlf/ vt. (of a horse) to eat extremely fast or greedily; bolt (指马)吞食,抢食: to wolf feed 吞食饲草

wolf teeth n. *see* canine teeth 犬齿,犬牙

womb /wuːm/ n.【anatomy】the female reproductive tract where the embryo implants and develops; uterus(雌性生殖道内供胚胎着床和发育的)子宫

wood shavings n. the thin curly slices shaved off wood(刨推木料时产生的卷曲薄片)刨花: Sometimes wood shavings can also be used as bedding material in stables. 有时刨花也可以作厩房垫料使用。

wood-chewing /ˈwuːdˌtʃuːɪŋ/ n. a vice of horse who chews wooden board without good reasons (马匹无故啃咬木板的恶习)啃木癖: Application of creosote to the chewed boards will help prevent woodchewing. 在啃咬处涂抹木焦油可制止啃木癖的发生。

work /wɜːk; *AmE* wɜːrk/ vt. **1** to exercise a horse 训练: to work on the long reins 长缰训练 ◇ to work out of it 活动开筋骨 **2**【cutting】to separate a cow from a herd; cut 拦截: working advantage 拦截优势 PHR V **lose the working advantage** (of a horse) to lose the attention of the cow being worked(指马)拦截失利

n.【cutting】the action of separating a cow from a herd 截牛,拦截

work level n. the intensity of work that a horse undertakes 运动强度,运动量: light work level

轻役 ◇ medium work level 中役 ◇ hard work level 重役

work on two tracks n. *see* lateral work 侧向运步

work tab n.【racing】a list of morning workouts recorded according to distance and time(记录赛马晨练行程和时间的清单)训练记录单

worked up adj. (of a horse) tense or nervous(指马)紧张的

workhorse /ˈwɜːkhɔːs; *AmE* ˈwɜːrkhɔːrs/ n. a horse that is used for farm work rather than for racing or riding(用于农田作业而非竞赛或骑乘的马匹)役用马,工作马 [syn] working horse

working canter n. a type of canter 工作跑步

working horse n. *see* workhorse 役用马

working saddle n. *see* stock saddle 牧用鞍,西部牛仔鞍

working trot n. a type of trot 工作快步

working walk n. a type of walk 工作慢步

workload /ˈwɜːkləʊd; *AmE* ˈwɜːrkloʊd/ n. the amount of work to be finished 工作量

workman /ˈwɜːkmən; *AmE* ˈwɜːrk-/ n.【slang】a skilled coachman 技术娴熟的车夫

workout /ˈwɜːkaʊt; *AmE* ˈwɜːrk-/ n. a session of exercises to improve a horse's fitness for competition(为提高竞赛水平而进行的训练)训练,热身

workstock /ˈwɜːkstɒk/ n. a domestic animal used for farm work 农畜,役畜

World Arabian Horse Organization n. (abbr. **WAHO**) an international organization founded in 1970 in Uk for the preservation, improvement, and registration of purebred Arabian Horses(1970 年在英国成立的国际组织,旨在对纯种阿拉伯马进行保种、改良和登记)世界阿拉伯马协会

World Champion Racer n.【racing】a horse who wins the highest number of races in the World Racing Championships in one year(一年中在世界赛马锦标赛中获奖数目最多的马匹)世界马王

World Champion Team Penning Association n. (abbr. **WCTPA**) an organization founded in the United States in 1978 to promote, organize, and regulate the sport of team penning(1978 年在美国成立的组织机构,旨在促进、组织和监管团体截牛运动)世界团体截牛冠军联赛协会

World Elephant Polo Association Games n. an invitational elephant polo tournament held every December in southern Nepal since 1982 (自 1982 年起每年 12 月在尼泊尔南部举行的象球邀请赛)世界象球协会邀请赛

World Equestrian Games n. (abbr. **WEG**) a group of modern international equestrian sports and competitions held every four years since 1990 in a different city(一项现代国际性马术盛会,自 1990 年起每四年一届在不同城市举行)世界马术运动会,世界马术大赛: Administered by the Fédération Equestre Internationale(FEI), the World Equestrian games have been held every four years, halfway between sets of consecutive Summer Olympic Games. 世界马术运动会由国际马术联合会监管,这场四年一届的盛会正好介于两届夏季奥运会中间。

World Racing Championships n. a series of horse racing consisting of 13 major international races around the world to encourage intercontinental racing and find the world Champion racer (包括全球 13 项重大赛马比赛的系列赛,旨在增进洲际间的竞争并决出世界马王)世界赛马锦标赛: The World Racing Championships was inaugurated in 1999. 世界赛马锦标赛始于 1999 年。

worm /wɜːm; *AmE* wɜːrm/ n. any parasite hosting the body of a horse 寄生虫: worm burden 寄生虫负荷 ◇ worm control 寄生虫防治 ◇ worm disease 寄生虫病

vt. to administer drugs designed to expel parasitic worms from the body; deworm(采用药物驱除体内的寄生虫)驱虫: worming programme 驱虫计划

wormer /ˈwɜːmə(r); *AmE* ˈwɜːrm-/ n. any drug or chemical administered to expel parasitic worms(用来驱除体内寄生虫的药剂)驱虫药

wormfree /wɜːmˈfriː/ adj. (of animals) haboring no parasitic worms (动物)无虫的

Worshipful Company of Farriers n. (abrr. **WCF**) *see* Worshipful Company of Farriers of London 伦敦蹄铁匠公会

Worshipful Company of Farriers of London n. a guild established in London, England in 1356 to train, conduct exams, and certify farrier members(1356年在英国伦敦成立的行会，负责从业蹄铁匠的培训、考试并颁发相应的证书)伦敦蹄铁匠公会

wound /wuːnd/ n. an injury caused by a cut, blow, or impact 伤口，创伤

vt. to inflict an injury on the body 使受伤；伤害

WPCS abbr. Welsh Pony and Cob Society 威尔士矮马协会

wrangle /ˈræŋgl/ vt. to herd or round up horses or other livestock 放养，赶拢 | **wrangling** /-lɪŋ/ n.

wrangler /ˈræŋglə(r)/ n.【AmE】one who takes care of horses or other livestock on a ranch; cowboy(牧场上照料马匹或其他牲畜的人员)牧工；牛仔

wrap /wræp/ vt. to apply a piece of fabric around a body part or wound 缠裹；包扎：to wrap a bandage around the foreleg of a horse 在马匹前肢缠裹绷带

n. a piece of cloth used for wrapping 缠布，绷带：tail wrap 扎尾绷带 PHR V **run under wraps**【racing】(of a racehorse) to run under restraint from the rider(指赛马)跑步受束

wrestler /ˈreslə(r)/ n. *see* steer wrestler 摔牛者

wrestling /ˈreslɪŋ/ n. *see* steer wrestling 摔牛

wring /rɪŋ/ vt. to wrench or twist forcibly by using a clencher or pair of pincers(用弯钉钳)拧弯，夹断：wring off the nails 拧断蹄钉

wrong bend n.【dressage】a condition in which the horse fails to achieve the correct degree of bend when turning corners or circles(马在拐角或转弯时弧度不符要求)转弯失误，转弯失利

wrong leg n.【dressage】a condition in which the horse canters on the wrong lead(舞步马跑步行进时起步肢出错的情况)起步错误：The dressage rider found that it is too late to correct the wrong leg of his mount. 等骑手发现坐骑起步错误，意欲矫正却为时已晚。

wrung withers n. a horse's withers bruised by an ill-fitting saddle 鬐甲挫伤

wry /raɪ/ adj. abnormally twisted or bent to one side 扭曲的，歪斜的：wry muzzle 歪鼻 ◇ wry neck 歪颈 ◇ wry nose 歪鼻

Württemberg /ˈwɜːtəmˌbɜːg/ n. a German warmblood developed in the late 16th century by putting native mares to Arab and Suffolk Punch stallions; stands about 16 hands and may have a bay, brown, black, or chestnut coat; is strong, hardy, and used mainly for riding and driving (德国在16世纪末期育成的温血马品种，主要采用当地母马与阿拉伯马和萨福克马公马杂交培育而来。体高约16掌，毛色以骝、褐、黑或栗多见，体格强健，抗逆性好，主要用于骑乘与驾挽)符腾堡马

X /eks/ n.【dressage】the center point of a dressage arena(盛装舞步赛场的中心)中央,中心

X chromosome n.【genetics】a sex chromosome that determines the sex of females(决定雌性个体性别的染色体)X 染色体 — *see also* **Y chromosome**

Xanthus /ˈzænθəs/ n.【mythology】one of two immortal horses who pulled the chariot of Achilles and was granted human speech by Hera and foretold his death(传说中为阿基里斯挽战车的两匹神马之一,此马被赫拉赋予人的言语并预告了主人的死亡)桑索斯 — *see also* **Balius**

xenogenous /ˌzenəˈdʒenəs/ adj. of or between different species 种间的,异种的:xenogenous fertilization 异种受精

Xenophon /ˈzenəfən/ n. (c. 430 – 354 BC) a Greek calvary officer and historian of ancient Athens, often regarded as the founder of the classical equitation and most notably famous for his work *On the Art of Horsemanship*(古希腊雅典骑兵将领与历史学家,被誉为古典骑术之父,因著有《论骑术》而闻名于世)色诺芬

xeroderma /ˌzɪərəʊˈdɜːmə/ n. a condition characterized by excessive or abnormal dryness of the skin(皮肤过度或异常干燥的病症)干皮症

xerophthalmia /ˌzɪərɒfˈθælmiə/ n. abnormal dryness of eye resulting from a deficiency of vitamin A(由于缺乏维生素 A 引起的眼异常干燥)干眼病,干眼症

xeroradiography /ˌzɪrəreɪdɪˈɒgrəfi/ n. the use of X-rays to produce an image X 射线显影,X 射线照相

xerosis /zɪəˈrəʊsɪs; *AmE* zɪˈroʊ-/ n. abnormal dryness of the skin, eyes, or mucous membranes (皮肤、眼睛和黏膜异常干燥的病症)干燥症

Xinihe horse n. (also **Sini Horse**) a Chinese horse breed originating along the Xini river of Inner Mongolia and closely related to Sanhe horses; stands 14.2 to 14.6 hands and generally has a bay, chestnut, or black coat; used mainly for light draft and farm work(产于中国内蒙古锡尼河流域的马品种,与三河马亲缘关系较近,体高 14.2 ~ 14.6 掌,毛色多为骝、栗或黑,主要用于轻挽和农田作业)锡尼河马

xiphoid /ˈzɪfɔɪd/ adj.【anatomy】shaped liked a sword 剑状的:xiphoid cartilage of sternum 剑状软骨 ◇ xiphoid process 剑突

X-ray photograph n. a picture produced by an X-ray X 片

X-ray therapy n. the treatment of a disease by use of X-rays X 射线治疗;放射治疗

Xu Beihong n. (1895 – 1953) a renowned Chinese artist known for his ink paintings of horses and birds and one of the first artists to create monumental oil paintings with epic Chinese themes(中国著名画家,以水墨奔马和飞鸟作品而闻名,是较早采用巨幅油画表达宏大中国主题的艺术家)徐悲鸿

xylazine /ˌzaɪləˈziːn/ n.【vet】a drug administered as tranquilizer that relieves pain and relaxes muscles(一种能缓解疼痛并松弛肌肉的镇静药)塞拉嗪,甲苯噻嗪

Y chromosome n.【genetics】a chromosome that determines the sex of males(决定雄性个体性别的染色体)Y 染色体 — *see also* **X chromosome**

yaboo /ˈjɑːbuː/ n. **1** a saddle horse 乘用马 **2** an aged, worthless horse; nag 驽马

yabusame /ˌjɑːbuːˈsæm/ n. a traditional Japanese equitation in which a mounted archer shoots arrows successively at three targets on a running horse(日本传统骑术项目,其中骑士策马奔驰,拉弓连射三个箭靶)日式骑射,弓马道:a yabusame archer 弓马道骑手

Yanqi n.【Chinese】(also **Yanqi horse**) an all-purpose horse breed developed in Xinjiang, China; stands about 14 hands and generally has a bay, chestnut, black, or gray coat; is strongly built and mainly used for farm work and driving (育成于中国新疆的兼用型马品种。体高约 14 掌,毛色多为骝、栗、黑或青,体格健壮,适于农田作业和驾车运输)焉耆马

yap /jæp/ vi. (of a hound) to bark sharply; yelp(猎犬)叫啸,狂吠:The pack yaps vehemently at a brown bear. 犬群对着棕熊狂吠。n. a sharp, shrill bark; a yelp 狂吠

yard /jɑːd; *AmE* jɑːrd/ n. an enclosed area used to exercise or train horses(用来训练或调教马的围场)运动场;训练场:yard feeding 圈饲 ◇ competition yard 赛场 ◇ training yard 训练场

yearling /ˈjɪəlɪŋ; *AmE* ˈjɪrlɪŋ/ n. a horse between one and two years of age 周岁驹:yearling colt 一岁雄驹 ◇ yearling filly 一岁雌驹 ◇ yearling sales 周岁驹拍卖会

yearling bit n. a bit used on a yearling colt, usually with players on mouthpiece(周岁雄驹用的衔铁,口衔上常带衔缀)岁驹衔铁 [syn] colt bit

yeld /jeld/ adj. (of a mare) infertile or producing no offspring in the previous breeding season(指母马在上个繁育季节不孕或未产驹)不孕的;空怀的:yeld mare 空怀母马

yellow body n.【physiology】*see* corpus luteum 黄体:functioning yellow body 功能性黄体 ◇ non-functioning yellow body 非功能性黄体 ◇ yellow body of pregnancy 妊娠黄体

yellow dun n.【color】a coat color of horse with yellow coat hair and brown points(体色为黄而末梢为褐的毛色)黄兔褐

yelp /jelp/ vi. (of hound or fox) to make or give a short, sharp cry; yap(猎犬或狐狸)吠,尖叫

yerk /jɜːk; *AmE* jɜːrk/ vi. **1** to strike with a stick or whip 鞭打 **2** to crack a whip 打响鞭

yielding /ˈjiːldɪŋ/ adj.【racing】(of the turf) having a high moisture content(赛场)渗水的,积水的:a yielding turf course 渗水的草地赛道

Yili n.【Chinese】(also **Yili horse**) a Chinese horse breed developed by crossing native Kazakhs with Don and Orlov Trotters in the Yili-Kazakh area from which the name derived; stands 14 to 15 hands and generally has a bay or chestnut coat; has an elegant profile, relatively prominent withers, deep girth, straight back, strong legs with hard hoofs; is possessed of good speed and stamina and used as an excellent riding horse(中国新疆伊犁哈萨克地区采用当地哈萨克马与顿河马和奥尔洛夫快步马杂交育成的马品种。体高 14 ~ 15 掌,毛色多为骝或栗,面貌俊秀,鬐甲中等,前胸宽广,背腰平直,四肢强健,蹄质结实,速力兼备,是优秀的乘用马)伊犁马

Yiwu n.【Chinese】(also **Yiwu horse**) a horse breed developed mainly by crossing Kazakhs with Yili horses in Xinjiang, China; stands 14 to 15 hands and usually has a bay, black, or chestnut coat; has a medium head with large nostrils, broad and wide chest, straight back, and strong legs with hard hooves; is hardy, frugal, and used for riding, packing and farm work (中国新疆采用哈萨克马与伊犁马杂交育成的马品种。体高 14 ~ 15 掌,毛色以骝、黑或栗为主,头型适中,鼻孔大,背腰平直,胸宽而深,四肢健壮,蹄质结实,抗逆性强,耐粗饲,用于骑乘、驮运和农田作业)伊吾马

yoick /jɔɪk/ interj. *see* joicks 唷!哟!

vi. to make or give such a cry 唷,哟

yoicks /jɔɪks/ interj. 【hunting】(also **yoick**) a cry used by hunters to urge hounds after a fox (狩猎时催促猎犬追赶狐狸的喊声)唷!哟!

yoke /jəʊk; *AmE* joʊk/ n. (*pl.* **yokes**) **1** a wooden frame or crossbar fitted around the necks of a pair of oxen or other draft animals for harnessing them together(套在双牛或其他挽畜颈上的横杆)轭 **2** (*pl.* **yoke**) a pair of draft animals harnessed together; couple(用轭联套使役的)共轭挽畜

vt. **1** to fit or join with a yoke 上轭,驾轭 **2** to harness a draft animal to a plow, etc. 套轭,套犁 [opp] unyoke

Yomud /ˈyəʊməd/ n. *see* Iomud 伊慕德马

Yorkshire boot n. a brushing boot used to protect the fetlock(保护球节用的防蹭绑腿)约克夏护腿

Yorkshire Coach Horse n. a British horse breed descended from Cleveland Bay with heavy infusion of Thoroughbred blood in Yorkshire; stands about 16 hands and may have a bay or brown coat; has a long body and relatively short legs, and deep girth; used for driving and show purposes(源于英国约克夏郡的挽马品种,由克利福兰骝马与纯血马杂交育成。体高约 16 掌,毛色多为骝或褐,体长,四肢较短,胸深,用于驾车和展览赛)约克夏挽马

Yorkshire Derby n. *see* Kiplingcotes Derby 约克夏德比赛,吉卜林科兹德比赛

Yorkshire gallop n.【BrE】a medium canter 中袭步

Yorkshire Packhorse n. another term for Cleveland Bay(克利福兰骝毛马的别名)约克夏驮马

young /jʌŋ/ adj. having lived or existed for only a short time; not fully developed 年轻的,幼小的:young stock 青年马;青年牛

n. young animals or offspring 幼畜,幼驹

Y'sabella /ˌɪzəˈbɛlə/ n. *see* Isabella 淡银鬃

Yuan-Heng's Therapeutic Treatise of Horses n. a book on the treatment of horses and cattle by traditional Chinese veterinary medicine, written by the Yu brothers — Yu Benyuan and Yu Benheng — and published in 1608 during the Ming dynasty(一本汇集中兽医治疗牛马疾病的手册,于明朝万历年间(1608 年)由喻本元和喻本亨兄弟编写出版)《元亨疗马集》

Yunnan pony n.【Chinese】a pony breed originating in Yunnan province of China; stands 11 to 12 hands and may have a bay, chestnut, black or gray coat; used mainly as pack animals in mountain areas(源于中国云南省的矮马品种,体高 11 ~ 12 掌,毛色可为骝、栗、黑或青,是山区常用的驮畜)云南矮马

Z

zebra /ˈzebrə, ˈziːbrə/ n. an African wild animal with alternating white and black stripes on its body(生活于非洲的野生马属动物，身上有黑白相间的条纹)斑马

zebra dun n.【color】the coat color of dun with a yellow coat, black points, and zebra marks(被毛沙黄、末梢为黑且带斑马纹的毛色)斑马兔褐

zebra marks n. *see* zebra stripes 斑马纹；虎斑

zebra stripes n.(also **zebra marks**) the primitive stripes found on the limbs and around the withers of certain horse breeds(在有些马种的四肢和鬐甲上出现的原始条纹)斑马纹 [syn] tiger stripes, leg barring — *see also* **dorsal stripe, wither stripe**

zebrine /ˈziːbraɪn/ adj. of, relating to, or resembling a zebra 斑马的；像斑马的

zebroid /ˈziːbrɔɪd/ n. the hybrid offspring of a male zebra and a female ass(公斑马与母驴交配产生的杂交后代)杂交斑马

Zeeland horse n. an ancient horse breed originating in the Netherlands(源于荷兰的古老马种)泽兰马

Zemaituka /ziːˈmaɪtuːkə/ n.(also **Zemaituka pony**) an ancient pony breed indigenous to Lithuania and generally thought to have descended from the Asiatic wild horse; stands 13.2 to 14.2 hands and may have a bay, brown, black, mouse dun, or palomino coat; has a medium head and a coarse profile, low withers, short back, and short strong legs; is hardy, frugal, and used for riding, light draft, and farm work(源于立陶宛的古老矮马品种，普遍认为由亚洲野马繁衍而来。体高13.2～14.2掌，毛色可为骝、褐、黑、鼠灰或银鬃，头型适中，面貌粗糙，鬐甲低平，背短，四肢粗短，抗逆性强，耐粗饲，用于骑乘、轻挽和农田作业)泽迈图卡矮马 [syn] Zhmud

Zhmud /ˈziːmuːd/ n. *see* Zemaituka 泽慕德矮马，泽迈图卡矮马

zigzag /ˈzɪgzæg/ n. **1** a jumping obstacle consisting of timbers placed at overlapping angles to each other(栏杆相互交叠成Z形的障碍)Z形跨栏 **2**【dressage】a line or course of movement having abrupt alternate turns Z字行进；Z形运步

vi. to move along in a zigzag course Z字行进

zinc /zɪŋk/ n. a metallic element essential in the feeds of livestock(一种金属元素，为家畜饲料所必需)锌：Zinc deficiency in horses is rare, but it can cause loss of appetite, skin lesions, and stunted growth of foals. 马匹锌缺乏十分少见，但有可能引起幼驹食欲下降、皮肤溃烂和发育迟滞等症。

zona pellucida n.(abbr. **ZP**) the transparent outer layer of membrane covering the developed mammalian oocyte(哺乳动物成熟卵母细胞的外层半透明膜)透明带：zona reaction 透明带反应

zony /ˈzəʊni; *AmE* ˈzoʊni/ n. a hybrid cross between a pony and a zebra(矮马与斑马的杂交后代)矮斑马

zoonosis /zəʊˈɒnəsɪs/ n.(*pl.* **zoonoses** /-siːz/) a disease that can be transmitted from animals to humans or vice versa(可在人与动物之间交叉传染的疾病)人畜共患病，人兽共患病

zoophile /ˈzəʊəfaɪl/ n. one who is devoted to the welfare and protection of animals(致力于动物福利和动物保护的人)动物爱好者

zoophilia /zəʊəˈfɪliə/ n. *see* zoophilism 酷爱动物；恋兽欲，兽色情

zoophilism /zəʊˈɒfɪlɪzəm/ n.(also **zoophilia**)

1 the tendency to be emotionally attached to animals 嗜爱动物,酷爱动物 2 erotic attachment to or sexual contact with animals 恋兽癖,兽色情:Even today,the play *Equus* written by Peter Shaffer in 1973 is still regarded highly provocative due to its theme of zoophilism. 彼得·谢弗于1973年创作的剧本《恋马狂》,由于主题涉及恋兽癖至今依然颇具争议。

zoophilist /zəʊˈɒfɪlɪst/ n. 1 a lover of animals 动物爱好者 2 one who has sexual contact with animals 恋兽狂

zoophobia /ˌzəʊəˈfəʊbiə/ n. an abnormal fear of animals 动物恐惧症

zorse /zɔːs; *AmE* zɔːrs/ n. a hybrid cross between a male horse and a female zebra(公马与母斑马交配产生的杂交后代)杂交斑马

ZP n. abbr. zona pellucida 透明带

zygoma /zaɪˈgəʊmə; *AmE* -ˈgoʊmə/ n.【anatomy】(*pl.* **zygomata** /-mətə/) a paired bone on each side of the cheek(构成马面颊两侧的一对骨)颧骨 [syn] cheekbone

zygomatic /ˌzaɪgəˈmætɪk/ adj.【anatomy】of, relating to or locating in the area of zygoma 颧骨的:zygomatic arch 颧弓 ◇ zygomatic bone 颧骨 ◇ zygomatic muscle 颧肌

zygote /ˈzaɪgəʊt; *AmE* -goʊt/ n. a cell formed by the union of the male and female gametes(雌雄配子结合形成的细胞)合子 | **zygotic** /zaɪˈgɒtɪk/ adj.

Z

Appendices

附　　录

Appendix 1 附录1

Fahrenheit to Celsius Temperature Conversion

华氏—摄氏温度换算表

Formula 换算公式：$°F = (°C \times 9/5) + 32$ or $°C = (°F - 32) \times 5/9$

°F	°C	°F	°C	°F	°C
100	37.8	77	25.0	54	12.2
99	37.2	76	24.4	53	11.7
98	36.7	75	23.9	52	11.1
97	36.1	74	23.3	51	10.6
96	35.6	73	22.8	50	10.0
95	35.0	72	22.2	49	9.4
94	34.4	71	21.7	48	8.9
93	39.9	70	21.1	47	8.3
92	33.3	69	20.6	46	7.8
91	32.8	68	20.0	45	7.2
90	32.2	67	19.4	44	6.7
89	31.7	66	18.9	43	6.1
88	31.1	65	18.3	42	5.6
87	30.6	64	17.8	41	5.0
86	30.0	63	17.2	40	4.4
85	29.4	62	16.7	39	3.9
84	28.9	61	16.1	38	3.3
83	28.3	60	15.6	37	2.8
82	27.8	59	15.0	36	2.2
81	27.2	58	14.4	35	1.7
80	26.7	57	13.9	34	1.1
79	26.1	56	13.3	33	0.6
78	25.6	55	12.8	32	0.0

Appendix 2 附录 2

Conversion Table for Common Measures 常用计量单位转换表

Common Measures 常用计量单位	Metric Measures 法制计量单位
1 acre 英亩	100 m^2 米^2
1 bushel 蒲式耳 = 8 gallons 加仑	32 litres 升
1 hectare 公顷 = 100 ares 公亩	10 000 m^2 米^2
1 foot 英尺 = 12 inches 英寸	30. 48 centimeter 厘米
1 furlong 弗隆	201. 17 meters 米
1 gallon 加仑	4 litres 升
1 hand 掌 = 4 inches 英寸	10. 16 centimeters 厘米
1 inch 英寸	2. 54 centimeters 厘米
1 mile 英里 = 8 furlongs 弗隆	1 609 meters 米
1 peck 配克 = 8 quarts 夸脱	8 liter 升
1 pint 品脱 = 1/2 quart 夸脱	0. 5 liter 升
1 pound 磅 = 16 ounces 盎司	454 grams 克
1 quart 夸脱 = 2 pints 品脱	1 liter 升
1 stone 英石 = 14 pounds 磅	6. 356 kilograms 千克
1 ton 吨 = 2,000 pounds	908 kilograms 千克
1 yard 码 = 3 feet 英尺	0. 914 meter 米

Appendix 3 附录 3

Common Horse Breeds

世界常见马品种

Breed 马品种	Origin 产地	Classification 分类	Height(hh)体高 (hands,掌)	Common Coat Colors 常见毛色
Abaga Dark Horse 阿巴嘎黑马	China 中国	Light 轻型马	13～14 掌	Black 黑
Acchetta 阿凯特矮马	Sardinia 撒丁岛	Pony 矮马	13.2～14.2	Bay, black, chestnut 骝、黑、栗
Akhal-Teké 阿哈捷金马	Turkmenistan 土库曼斯坦	Light 轻型马	14.2～15.2	Golden, bay, chestnut 金色、骝、栗
Albino 美国白化马	United States 美国	Light 轻型马	Varies 不定	White only 仅有白色
Alter-Real 阿特瑞尔马	Portugal 葡萄牙	Light 轻型马	13～15.2	Mostly bay or brown 以骝、褐为主
American Buckskin 美国鹿皮马	United States 美国	Light 轻型马	约 14	Buckskin 沙黄
American Cream 美国乳色马	United States 美国	Light 轻型马	12.2～17	Cream-colored 乳黄
American Quarter Horse 美国夸特马	United States 美国	Light 轻型马	15.2～16.1	Any solid color; mostly chestnut 各种单毛色兼有,以栗为主
American Saddlebred 美国骑乘马	United States 美国	Light 轻型马	15～16	Black, bay, brown 黑、骝、褐
American Standarbred 美国标准马	United States 美国	Light 轻型马	14～16	Mostly brown, bay, black, chestnut 多为褐、骝、黑、栗
American Warmblood 美国温血马	United States 美国	Light 轻型马	Varies 不定	Any color 各种毛色兼有
American White 美国白马	United States 美国	Light 轻型马	12.2～17	Snow or milk white 纯白或乳白

（续）

Breed 马品种	Origin 产地	Classification 分类	Height(hh)体高 (hands,掌)	Common Coat Colors 常见毛色
Andalusian 安达卢西亚马	Spain 西班牙	Light 轻型马	15～16.2	mostly Gray 以青为主
Anglo-Arab 盎格鲁阿拉伯马	Europe 欧洲	Light 轻型马	15～16	Bay,chestnut,gray,black 骝、栗、青、黑
Anglo-Argentine 盎格鲁阿根廷马	Argentina 阿根廷	Light 轻型马	15～16	Bay,chestnut,gray,black 骝、栗、青、黑
Appaloosa 阿帕卢萨马	Spain, United States 西班牙,美国	Light 轻型马	14～15.3	Mottled coat pattern 斑纹被毛
Arab 阿拉伯马	Arabia 阿拉伯半岛	Light 轻型马	14.3～16	Bay, brown, chestnut, gray,black 骝、褐、栗、青、黑
Ardennais 阿登马	France, Belgium 法国,比利时	Coldblood 冷血马	15～16	Bay, roan, gray, palomino,chestnut 骝、沙、青、银鬃、栗
Ariège 阿列日马	France 法国	Pony 矮马	13～14	Black 黑
Assateague 安萨狄格矮马	United States 美国	Pony 矮马	约12	All colors 各种毛色兼有
Australian 澳洲矮马	Australia 澳大利亚	Pony 矮马	12～14	Gray 青
Auxios 奥斯瓦马	France 法国	Light 轻型马	15.1～16	Bay,chestnut,roan 骝、栗、沙
Avelignese 艾维林格矮马	Italy 意大利	Pony 矮马	约14	Chestnut,golden 栗、金栗
Baise Horse 百色马	China 中国	Pony 矮马	约11	Bay 骝
Balearic 巴利阿里矮马	Balearic Islands 巴利阿里群岛	Pony 矮马	12～14	Bay,brown 骝、褐

（续）

Breed 马品种	Origin 产地	Classification 分类	Height(hh)体高 (hands,掌)	Common Coat Colors 常见毛色
Bali 巴厘矮马	Indonesia 印度尼西亚	Pony 矮马	12～13	Bay 骝
Barb 柏布马	North Africa 北非	Light 轻型马	14～15	Brown, bay, chestnut, black, gray 褐、骝、栗、黑、青
Barkol Horse 巴里坤马	China 中国	Light 轻型马	约13掌	Mostly bay, chestnut 以骝、栗为主
Baskhkir Curly 巴什基尔卷毛马	Russia 俄罗斯	Light 轻型马	13.1～14	Bay, chestnut, palomino 骝、栗、银鬃
Basque 巴斯克矮马	Spain 西班牙	Pony 矮马	11～14.2	Brown, black, bay, chestnut 褐、黑、骝、栗
Batak 巴特克矮马	Indonesia 印度尼西亚	Pony 矮马	11～13	Any color 各种毛色兼有
Bavarian Warmblood 巴伐利亚温血马	Germany 德国	Warmblood 温血马	15～16	Brown, bay, black, gray, chestnut 褐、骝、黑、青、栗
Beberbeck 波波贝克马	Germany 德国	Light 轻型马	约16	bay, chestnut 骝、栗
Belgian 比利时马	Belgium 比利时	Coldblood 冷血马	17以下	Roan, chestnut, bay, brown 沙、栗、骝、褐
Bhotia 普提亚矮马	India 印度	Pony 矮马	约13	Gray 青
Bosnian 波斯尼亚马	Bosnia-Herzegovin 波斯尼亚-黑塞哥维那	Pony 矮马	12.1～14	Bay, brown, black, gray, chestnut 骝、褐、黑、青、栗
Boulonnais 布洛涅马	France 法国	Coldblood 冷血马	15.3～16.2	Dappled gray, chestnut, bay 斑点青、栗、骝
Breton 布雷顿马	France 法国	Coldblood 冷血马	15～16	Chestnut, bay, gray, roan 栗、骝、青、沙

（续）

Breed 马品种	Origin 产地	Classification 分类	Height(hh)体高 (hands,掌)	Common Coat Colors 常见毛色
Brumby 澳洲野马	Australia 澳大利亚	Light 轻型马	约 15	Any color 各种毛色兼有
Budyonny 布琼尼马	Russia 俄罗斯	Warmblood 温血马	15.1～16	Chestnut,bay,gray 栗、骝、青
Burmese pony 缅甸矮马	Burma 缅甸	Pony 矮马	约 13	Bay,chestnut 骝、栗
Buohai Horse 渤海马	China 中国	Light 轻型马	约 14	Bay,chestnut 骝、栗
Calabrese 卡拉布里亚马	Italy 意大利	Warmblood 温血马	16～16.2	Brown,black,bay,chestnut,gray 褐、黑、骝、栗、青
Camargue 卡马格矮马	France 法国	Pony 矮马	13～14.2	White,light gray 白或淡青
Canadian Cutting Horse 加拿大截牛马	Canada 加拿大	Light 轻型马	15.2～16.1	Any color 各种毛色兼有
Caspian 里海矮马	Iran 伊朗	Pony 矮马	9.2～11.2	Bay,chestnut,gray 骝、栗、青
Chaidamu Horse 柴达木马	China 中国	Light 轻型马	13.2～14 掌	Mostly bay, chestnut, black, gray 以骝、栗、黑、青为主
Chakou Post Horse 岔口驿马	China 中国	Light 轻型马	约 13 掌	Mostly bay 以骝为主
Chickasaw 契卡索马	United States 美国	Light 轻型马	13.2～14.7	Bay, black, chestnut, gray,roan 骝、黑、栗、青、沙
Chincoteague 辛柯狄格矮马	United States 美国	Pony 矮马	约 12	Piebald,skewbald 黑白花、杂色花
Cleveland Bay 克利福兰骝马	England 英格兰	Light 轻型马	16～16.2	Bay,brown 骝、褐
Clydesdale 克莱兹代尔马	Scotland 苏格兰	Coldblood 冷血马	16.2～16.8	Bay,brown,black,roan 骝、褐、黑、沙

(续)

Breed 马品种	Origin 产地	Classification 分类	Height(hh)体高 (hands,掌)	Common Coat Colors 常见毛色
Connemara 康尼马拉矮马	Ireland 爱尔兰	Pony 矮马	13 ~ 14. 2	Gray, black, brown, dun 青、黑、褐、兔褐
Criollo 克里奥罗马	South America 南美	Light 轻型马	14 ~ 15	Any color 各种毛色兼有
Dales 戴尔斯矮马	England 英格兰	Pony 矮马	14. 2 以下	Mostly black 以黑为主
Dali Horse 大理马	China 中国	Light 轻型马	约 12 掌	Mostly Bay, chestnut 以骝、栗为主
Danubian 多瑙河温血马	Bulgaria 保加利亚	Warmblood 温血马	约 15. 2	Black, chestnut 黑、栗
Darashouri 达罗舒里马	Iran 伊朗	Light 轻型马	14. 1 ~ 15	Bay, chestnut, gray 骝、栗、青
Dartmoor 达特姆尔矮马	British Isles 英国	Pony 矮马	11. 2 ~ 11. 3	Bay, brown, black 骝、褐、黑
Datong Horse 大通马	China 中国	Light 轻型马	约 13 掌	Mostly bay, chestnut, black, gray 以骝、栗、黑、青为主
DØle Trotter 多勒快步马	Norway 挪威	Warmblood 温血马	14. 2 ~ 15. 2	Bay, brown, black 骝、褐、黑
DØle Gudbrandsdal 多勒·康伯兰德马	Norway 挪威	Coldblood 冷血马	14. 2 ~ 15. 2	Bay, brown, black 骝、褐、黑
Don 顿河马	Central Asia 中亚	Warmblood 温血马	15. 1 ~ 15. 3	Light bay, chestnut, brown 淡骝、栗、褐
Dutch Draught 荷兰挽马	Holland 荷兰	Coldblood 冷血马	16. 3 以下	Chestnut, bay, gray 栗、骝、青
Dutch Warmblood 荷兰温血马	Holland 荷兰	Warmblood 温血马	约 16	Any color 各种毛色兼有
East Bulgarian 东保加利亚马	Bulgaria 保加利亚	Warmblood 温血马	15 ~ 16	Chestnut, black 栗、黑

（续）

Breed 马品种	Origin 产地	Classification 分类	Height(hh)体高 (hands,掌)	Common Coat Colors 常见毛色
East Friesian 东弗里斯兰马	Germany 德国	Warmblood 温血马	15.2～16.1	Brown, bay, black, gray, chestnut 褐、骝、黑、青、栗
Debao Pony 德保矮马	China 中国	Pony 矮马	9～10掌	Bay, gray, chestnut, black, brown, roan 骝、青、栗、黑、褐、沙
Einsiedler 艾因斯德勒马	Switzerland 瑞士	Light 轻型马	15～16.3	Any color 各种毛色兼有
Erlunchun Horse 鄂伦春马	China 中国	Light 轻型马	约13掌	Mostly gray, bay 以青、骝为主
European Trotter 欧洲快步马	France, USA, Russia 法国，美国，俄罗斯	Light 轻型马	15～16.2	Bay, black, chestnut, gray 骝、黑、栗、青
Exmoor 埃克斯穆尔矮马	England 英格兰	Pony 矮马	11.2～12.3	Bay, brown, dun 骝、褐、兔褐
Falabella 法拉贝拉矮马	Argentina 阿根廷	Pony 矮马	9以下	Any color 各种毛色兼有
Fell Pony 费尔矮马	England 英格兰	Pony 矮马	13～14	Black, brown, bay, gray 黑、褐、骝、青
Finnish 芬兰马	Finland 芬兰	Coldblood 冷血马	14.3～15.2	Chestnut 栗
Frederiksborg 弗雷德里克斯堡马	Denmark 丹麦	Warmblood 温血马	15.1～16.1	Chestnut 栗
Freiburg Saddle Horse 弗赖堡乘用马	Switzerland, Germany 瑞士，德国	Warmblood 温血马	15～16.2	Bay, chestnut, gray, black 骝、栗、青、黑
French Saddle Horse 法兰西乘用马	France 法国	Light 轻型马	15.2～16.3	Bay, chestnut 骝、栗
French Trotter 法国快步马	France 法国	Warmblood 温血马	15.1～16.2	Bay, black, chestnut, gray 骝、黑、栗、青

(续)

Breed 马品种	Origin 产地	Classification 分类	Height(hh)体高 (hands,掌)	Common Coat Colors 常见毛色
Friesian 弗里斯兰马	Netherland 荷兰	Coldblood 冷血马	约15	Black 黑
Furioso-North Star 弗雷索-北极星马	Hungary 匈牙利	Warmblood 温血马	16~16.2	Brown,bay,black 褐、骝、黑
Galiceño 加利西诺矮马	Mexico 墨西哥	Pony 矮马	12~13.2	Bay,black,chestnut 骝、黑、栗
Ganzi Horse 甘孜马	China 中国	Light 轻型马	12~13掌	Gray, chestnut, black, bay 青、栗、黑、骝
Garrano 加里诺矮马	Portugal 葡萄牙	Pony 矮马	10~12	Chestnut 栗
Gayoe 盖欧耶马	Indonesia 印度尼西亚	Pony 矮马	11~13	Bay,chestnut 骝、栗
Gelderland 格尔德兰马	Holland 荷兰	Warmblood 温血马	15.2~16	chestnut,bay,black,gray 栗、骝、黑、青
German Trotter 德国快步马	Germany 德国	Warmblood 温血马	15~16	Bay,chestnut 骝、栗
Gidran Arabian 基特兰阿拉伯马	Hungary 匈牙利	Light 轻型马	16.1~17	Chestnut,bay,black 栗、骝、黑
Gotland 哥特兰马	Sweden 瑞典	Pony 矮马	12~14	Any solid color 各种单毛色兼有
Groningen 格罗宁根马	Holland 荷兰	Warmblood 温血马	15.2~16	Black,brown,bay 黑、褐、骝
Guanzhong Horse 关中马	China 中国	Warmblood 温血马	约15	Chestnut,bay 栗、骝
Guizhou Horse 贵州马	China 中国	Light 轻型马	11~13	Bay,chestnut,gray,black,brown 骝、栗、青、黑、褐
Hackney 哈克尼马	England 英国	Light 轻型马	14.2~16	Black, brown, chestnut, bay 黑、褐、栗、骝

（续）

Breed 马品种	Origin 产地	Classification 分类	Height(hh)体高 (hands,掌)	Common Coat Colors 常见毛色
Hackney Pony 哈克尼矮马	England 英国	Pony 矮马	12.1～14	brown,black,gray,roan 褐、黑、青、沙
Haflinger 哈弗林格矮马	Austria 奥地利	Pony 矮马	约14	Chestnut,palomino 栗、银鬃
Hanoverian 汉诺威马	Germany 德国	Warmblood 温血马	16～17.2	Any solid color 各种单毛色兼有
Hequ Horse 河曲马	China 中国	Light 轻型马	约13.6	Black,bay,gray 黑、骝、青
Highland 高地矮马	Scotland 苏格兰	Pony 矮马	12.2～14.2	Dun, gray, chestnut, bay, black 兔褐、青、栗、骝、黑
Hispano(Spanish Arab) 西班牙阿拉伯马	Spain 西班牙	Light 轻型马	14.3～16	Bay,chestnut,gray 骝、栗、青
Holstein 霍士丹马	Germany 德国	Warmblood 温血马	15.3～16.2	Black,brown,bay,chest-nut 黑、褐、骝、栗
Hucul 胡克尔矮马	Poland 波兰	Pony 矮马	12～13	Dun,bay 兔褐、骝
Iberian horse 伊比利亚马	Iberian Penin-sula 伊比利亚半岛	Light 轻型马	14～15	Gray, bay, black, chest-nut 青、骝、黑、栗
Iomud 约穆德马	Central Asia 中亚	Warmblood 温血马	约15	Gray,bay,chestnut 青、骝、栗
Icelandic 冰岛马	Iceland 冰岛	Pony 矮马	13.2以下	Gray,dun 青、兔褐
Irish Draught 爱尔兰挽马	Ireland 爱尔兰	Coldblood 冷血马	15～17	Brown,bay,chestnut,gray 褐、骝、栗、青
Irish Hunter 爱尔兰亨特马	Ireland 爱尔兰	Light 轻型马	16～17.1	Bay,black,gray,chestnut 骝、黑、青、栗
Italian Heavy Draft 意大利重挽马	Italy 意大利	Coldblood 冷血马	15～17	Bay,chestnut,gray,roan 骝、栗、青、沙

（续）

Breed 马品种	Origin 产地	Classification 分类	Height(hh)体高 (hands,掌)	Common Coat Colors 常见毛色
Jaf 杰弗马	Kurdistan 库尔德斯坦	Light 轻型马	约15	Bay, chestnut, gray 骝、栗、青
Java 爪哇矮马	Indonesia 印度尼西亚	Pony 矮马	约12	Any color 各种毛色兼有
Jianchang Horse 建昌马	China 中国	Light 轻型马	11～12掌	Bay, chestnut, black 骝、栗、黑
Jilin Horse 吉林马	China 中国	Warmblood 温血马	约15	Bay, chestnut, gray 骝、栗、青
Jinjiang Horse 晋江马	China 中国	Light 轻型马	约12掌	Mostly bay 以骝为主
Jinzhou Horse 金州马	China 中国	Warmblood 温血马	约14	Bay, chestnut, black 骝、栗、黑
Jutland 日德兰马	Denmark 丹麦	Coldblood 冷血马	15.2～16	Chestnut, roan, bay, black 栗、沙、骝、黑
Kabardin 卡巴金马	Russia 俄罗斯	Warmblood 温血马	14.1～15.1	Bay, brown, black 骝、褐、黑
Karabair 卡拉巴伊尔马	Uzbekistan 乌兹别克斯坦	Warmblood 温血马	14.2～15	Gray, bay, chestnut 青、骝、栗
Karabakh 卡拉巴赫马	Azerbaijan 阿塞拜疆	Warmblood 温血马	约14	Golden dun, chestnut, bay, gray 沙黄、栗、骝、青
Kathiawari 卡提阿瓦马	India 印度	Pony 矮马	14～15	Any color 各种毛色兼有
Kazakh 哈萨克马	Kazakhstan, China 哈萨克斯坦，中国	Pony 矮马	12.1～13.1	Any color 各种毛色兼有
Kladruber 克拉杜波马	Czechoslovakia 捷克斯洛伐克	Warmblood 温血马	16.2～18	Black or gray 黑、青
Knabstrup 纳普斯特鲁马	Denmark 丹麦	Warmblood 温血马	约15.3	Roan 沙毛

(续)

Breed 马品种	Origin 产地	Classification 分类	Height(hh)体高 (hands,掌)	Common Coat Colors 常见毛色
Konik 考尼克矮马	Poland 波兰	Pony 矮马	12~13.3	Bay,dun,palomino 骝、兔褐、银鬃
Kustanair 库斯塔奈马	Kazakhstan 哈萨克斯坦	Light 轻型马	14~15.1	Any solid color 各种单毛色兼有
Kyrgyz Horse 柯尔克孜马	China 中国	Light 轻型马	13.2~14 掌	Mostly bay, chestnut, black, gray 以骝、栗、黑、青为主
Landais 朗德矮马	France 法国	Pony 矮马	11.1~13	Brown,black,bay,chestnut 褐、黑、骝、栗
Latvian Harness Horse 拉脱维亚挽马	Latvia 拉脱维亚	Warmblood 温血马	15.1~16	Bay,brown,chestnut 骝、褐、栗
Libyan Barb 利比亚柏布马	Libya 利比亚	Light 轻型马	14~15	Brown,black,bay,chestnut,gary 褐、黑、骝、栗、青
Lichuan Horse 利川马	China 中国	Light 轻型马	约12	Gray,bay,chestnut,black 青、骝、栗、黑
Limousin Halfbred 利穆赞半血马	France 法国	Warmblood 温血马	16~17	chestnut or bay 栗或骝
Lipizzaner 利比扎马	Austria 奥地利	Light 轻型马	15~16	Mostly gray 以青为主
Lithuanian Heavy Draft 立陶宛重挽马	Lithuania 立陶宛	Coldblood 冷血马	15~16	Chestnut, black, roan, bay,gray 栗、黑、沙、骝、青
Lokai 洛卡伊马	Uzbekistan 乌兹别克斯坦	Warmblood 温血马	14~14.2	Golden chestnut, bay, gray 金栗、骝、青
Lusitano 卢西塔诺马	Portugal 葡萄牙	Light 轻型马	15~16	Gray,bay,chestnut 青、骝、栗
Macedonian 马其顿矮马	Yugoslavia 南斯拉夫	Pony 矮马	约13	Chestnut,bay,black,gray 栗、骝、黑、青

（续）

Breed 马品种	Origin 产地	Classification 分类	Height(hh)体高 (hands,掌)	Common Coat Colors 常见毛色
Malapolski 曼拉波斯基马	Poland 波兰	Warmblood 温血马	15. 3 ~ 16. 2	Any solid color 各种单毛色兼有
Manipur 曼尼普尔矮马	Manipur, India 印度曼尼普尔	Pony 矮马	11 ~ 13	Bay, brown, gray, chestnut 骝、褐、青、栗
Maremmana 摩雷曼纳马	Italy 意大利	Light 轻型马	15 ~ 15. 3	Bay, chestnut, black 骝、栗、黑
Marwari 马尔瓦尔矮马	India 印度	Pony 矮马	14 ~ 15	Any color 各种毛色兼有
Mecklenburg 梅克伦堡马	Germany 德国	Warmblood 温血马	15. 3 ~ 16	Brown, bay, chestnut, black 褐、骝、栗、黑
Mèrens 梅隆马	France 法国	Pony 矮马	13 ~ 14. 1	Black 黑
Metis Trotter 梅第快步马	Russia 俄罗斯	Warmblood 温血马	15. 1 ~ 15. 3	Gray, black, chestnut 青、黑、栗
Missouri Fox Trotter 密苏里狐步马	United State 美国	Light 轻型马	14 ~ 17	Any color 各种毛色兼有
Mongolian 蒙古马	Mongolia, China 蒙古国，中国	Pony 矮马	12 ~ 14	Brown, black, mouse dun 褐、黑、鼠灰
Morab 摩拉伯马	United State 美国	Light 轻型马	14. 3 ~ 15. 2	Any solid color 各种单毛色兼有
Morgan 摩根马	United State 美国	Light 轻型马	14 ~ 15. 2	Bay, brown, black, chestnut 骝、褐、黑、栗
Murakoz 穆拉克兹马	Hungary 匈牙利	Coldblood 冷血马	约16	Chestnut, bay, brown, black 栗、骝、褐、黑
Murgese 穆尔格斯马	Italy 意大利	Warmblood 温血马	14 ~ 16	Chestnut, brown, black, gray 栗、褐、黑、青

(续)

Breed 马品种	Origin 产地	Classification 分类	Height(hh)体高(hands,掌)	Common Coat Colors 常见毛色
Mustang 北美野马	United States 美国	Light 轻型马	14~15	All colors 各种毛色兼有
New Forest 新福里斯特矮马	England 英国	Pony 矮马	12~14.2	Any colorbut piebald or skewbald 除花毛外各种毛色兼有
Ningqiang Pony 宁强矮马	China 中国	Pony 矮马	约11掌	Mostly bay, chestnut, gray 以骝、栗、青为主
Nonius 诺聂斯马	Hungary 匈牙利	Warmblood 温血马	14.2~16.2	Bay, black, or brown 骝、黑或褐
Noriker 诺里克马	Austria, Germany 奥地利,德国	Coldblood 冷血马	15.2~16.1	Bay, chestnut 骝、栗
North Swedish 北方瑞典马	Sweden 瑞典	Coldblood 冷血马	15.1~15.3	Dun, brown, chestnut, black 兔褐、褐、栗、黑
North Swedish Trotter 北方瑞典快步马	Sweden 瑞典	Warmblood 温血马	15.1~15.3	Dun, brown, chestnut, black 兔褐、褐、栗、黑
Norwegian Fjord Pony 挪威峡湾马	Norway 挪威	Pony 矮马	13~14	Dun 兔褐
Novokirghiz 新吉尔吉斯马	Central Asia 中亚	Warmblood 温血马	14.1~15.1	Any solid color 各种单毛色兼有
Oldenburg 奥尔登堡马	Germany 德国	Warmblood 温血马	16.2~17.2	Bay, brown, black, gray 骝、褐、黑、青
Orlov Trotter 奥尔洛夫快步马	Russia 俄罗斯	Warmblood 温血马	15.1~17	Gray, black, bay 青、黑、骝
Paint/Pinto 美国花马	United States 美国	Light 轻型马	Varies 不定	Piebald 黑白花
Palomino 帕洛米诺马	United States 美国	Light 轻型马	14.2~15.3	Gold coat with white mane and tail 被毛金黄而鬃尾为白

（续）

Breed 马品种	Origin 产地	Classification 分类	Height(hh)体高 (hands,掌)	Common Coat Colors 常见毛色
Paso Fino 帕索菲诺马	Spain, South America 西班牙，南美洲	Light 轻型马	13 ~ 15. 2	All colors 各种毛色兼有
Peneia 皮尼亚矮马	Greece 希腊	Pony 矮马	10. 1 ~ 14	Bay, black, chestnut, gray 骝、黑、栗、青
Percheron 佩尔什马	France 法国	Coldblood 冷血马	15. 2 ~ 17	Gray, black 青、黑
Peruvian Paso 秘鲁帕索马	Peru 秘鲁	Light 轻型马	14 ~ 15. 2	Mostly bay and chestnut 以骝、栗为主
Pindos 品德斯矮马	Greece 希腊	Pony 矮马	12 ~ 13	Dark gray, bay, black, brown 暗青、骝、黑、褐
Plateau Persian 高原波斯马	Iran 伊朗	Light 轻型马	约 15	Gray, bay, or chestnut 青、骝或栗
Pleven 普列文马	Bulgaria 保加利亚	Warmblood 温血马	约 15. 2	Chestnut 栗
Poitevin 普瓦图马	France 法国	Coldblood 冷血马	15 ~ 17	Dun, bay, gray, black, palomino 兔褐、骝、青、黑、银鬃
Pony of the Americas 美国矮马(POA)	United States 美国	Pony 矮马	11. 2 ~ 13	Dappled color 斑毛
Rhineland Heavy Draft 莱茵兰德重挽马	Germany 德国	Coldblood 冷血马	16 ~ 17	Bay, chestnut, or red roan 骝、栗或沙栗
Rocky Mountain Horse 落基山马	United States 美国	Warmblood 温血马	14. 2 ~ 16	Any solid color 各种单毛色兼有
Russian Heavy Draft 俄罗斯重挽马	Ukraine 乌克兰	Coldblood 冷血马	14. 1 ~ 15	Chestnut, bay, or roan 栗、骝、沙
Salerno 萨莱诺马	Italy 意大利	Warmblood 温血马	16 ~ 16. 2	Any solid color 各种单毛色兼有

（续）

Breed 马品种	Origin 产地	Classification 分类	Height(hh)体高 (hands,掌)	Common Coat Colors 常见毛色
Sandalwood 檀香木矮马	Indonesia 印度尼西亚	Pony 矮马	12~13	Any color 各种毛色兼有
Sanhe Horse 三河马	China 中国	Warmblood 温血马	约15	Bay, chestnut 骝、栗
Sardinian 撒丁尼亚矮马	Sardinia 撒丁岛	Pony 矮马	12.1~13	Bay, brown, black, liver chestnut 骝、褐、黑、深栗
Schleswig Heavy Draft 什勒斯威格重挽马	Germany 德国	Coldblood 冷血马	15.1~16.1	Any solid color 各种单毛色兼有
Shagya Arab 沙迦·阿拉伯马	Hungary 匈牙利	Light 轻型马	15~16	Gray 青
Shandan Horse 山丹马	Gansu China 中国甘肃	Light 轻型马	约14	Bay, chestnut 骝、栗
Shetland 设特兰矮马	England 英格兰	Pony 矮马	11.2 以下	Paint, brown, chestnut, gray, black 花毛、褐、栗、青、黑
Shire 夏尔马	England 英格兰	Coldblood 冷血马	约17	Bay, brown 骝、褐
Skyros 斯基罗斯矮马	Greece 希腊	Pony 矮马	9.1~11	Gray, bay, brown, palomino 青、骝、褐、银鬃
Sokolsky 索科尔斯基马	Poland, Russia 波兰,俄罗斯	Warmblood 温血马	15~16	chestnut, brown, gray 栗、褐、青
Soviet Heavy Draft 苏维埃重挽马	Russia 俄罗斯	Coldblood 冷血马	约15	bay, chestnut, roan 骝、栗、沙
Spanish Barb 西班牙柏布马	Spain, United States 西班牙,美国	Light 轻型马	13.3~14.1	Any color 各种毛色兼有
Spiti 斯皮特矮马	India 印度	Pony 矮马	约12	Gray 青

（续）

Breed 马品种	Origin 产地	Classification 分类	Height(hh)体高 (hands,掌)	Common Coat Colors 常见毛色
Spotted Saddle Horse 斑点乘用马	United States 美国	Light 轻型马	14～16	Spotted coloring 斑毛
Suffolk 萨福克马	England 英国	Coldblood 冷血马	15.2～16.2	Chestnut 栗
Sumba 松巴矮马	Indonesia 印度尼西亚	Pony 矮马	12～13	Any color 各种毛色兼有
Sumbawa 松巴哇矮马	Indonesia 印度尼西亚	Pony 矮马	12～13	Any color 各种毛色兼有
Swedish Ardennes 瑞典阿登马	Sweden 瑞典	Coldblood 冷血马	15～16	Black, bay, chestnut, brown 黑、骝、栗、褐
Swedish Warmblood 瑞典温血马	Sweden 瑞典	Warmblood 温血马	16.2～17	Any solid color 各种单毛色兼有
Tengchong Horse 腾冲马	China 中国	Light 轻型马	11～12掌	Mostly bay, chestnut 以骝、栗为主
Tennessee Walking Horse 田纳西走马	United States 美国	Light 轻型马	15～16	All solid colors 各种单毛色兼有
Tersky 捷尔斯基马	Russia 俄罗斯	Warmblood 温血马	14.3～15.1	Gray 青
Thoroughbred 纯血马	England 英格兰	Light 轻型马	14.2～17	Any solid color 各种单毛色兼有
Tibetan Horse 藏马	Tibet, China 中国西藏	Light 轻型马	约12	Any color 各种毛色兼有
Tieling Draft 铁岭挽马	China 中国	Coldblood 冷血马	约15	Bay, black 骝、黑
Timor 帝汶矮马	Indonesia 印度尼西亚	Pony 矮马	9～11	Black, bay, or brown 黑、骝或褐
Toric 托利马	Estonia 爱沙尼亚	Warmblood 温血马	15～15.2	Bay or chestnut 骝或栗

（续）

Breed 马品种	Origin 产地	Classification 分类	Height(hh)体高 (hands,掌)	Common Coat Colors 常见毛色
Trakehner 特雷克纳马	Germany, Poland 德国,波兰	Warmblood 温血马	16 ~ 16. 2	Any solid color 各种单毛色兼有
Turkish Horse 土耳其马	Turkey 土耳其	Pony 矮马	13 ~ 14	Any solid color 各种单毛色兼有
Ukrainian Riding Horse 乌克兰乘用马	Ukaraine 乌克兰	Light 轻型马	15. 1 ~ 16. 1	Chestnut, bay, black 栗、骝、黑
Viatka 维亚特卡马	Russia 俄罗斯	Pony 矮马	13 ~ 14	Bay, gray, mouse dun, palomino 骝、青、鼠灰、银鬃
Vladimir Heavy Draft 弗拉基米尔重挽马	Russia 俄罗斯	Coldblood 冷血马	15. 1 ~ 16. 1	Chestnut, bay, brown, black 栗、骝、褐、黑
Waler 澳洲威尔士马	Australia 澳大利亚	Light 轻型马	14. 2 ~ 16	Any solid color 各种单毛色兼有
Welsh Pony 威尔士矮马	Wales 威尔士	Pony 矮马	12 ~ 13. 2	Any solid color 各种单毛色兼有
Wenshan Horse 文山马	China 中国	Light 轻型马	11 ~ 12 掌	Mostly bay, chestnut, gray 以骝、栗、青为主
Westphalian 威斯特伐利亚马	Germany 德国	Warmblood 温血马	16 ~ 17	Any solid color 各种单毛色兼有
Wielkopolski 韦尔科波斯奇马	Poland 波兰	Warmblood 温血马	15. 1 ~ 16	Any solid color 各种单毛色兼有
Wumeng Horse 乌蒙马	China 中国	Light 轻型马	约 12 掌	Mostly bay, chestnut 以骝、栗为主
Wurttemberg 符腾堡马	Germany 德国	Warmblood 温血马	约 16	Black, bay, chestnut, brown 黑、骝、栗、褐
Xilingol Horse 锡林郭勒马	China 中国	Light 轻型马	约 14 掌	Mostly bay, chestnut, black, gray 以骝、栗、黑、青为主

（续）

Breed 马品种	Origin 产地	Classification 分类	Height(hh)体高 (hands,掌)	Common Coat Colors 常见毛色
Xinihe horse 锡尼河马	China 中国	Warmblood 温血马	14.2～14.6	Bay, chestnut, or black 骝、栗或黑
Yanqi Horse 焉耆马	China 中国	Light 轻型马	约14	Bay, chestnut, black, gray 骝、栗、黑、青
Yili horse 伊犁马	Sinkiang China 中国新疆	Light 轻型马	14～15	Bay, chestnut 骝、栗
Yiwu horse 伊吾马	China 中国	Light 轻型马	14～15	Bay, black 骝、黑
Yongning Horse 永宁马	China 中国	Light 轻型马	约12掌	Mostly chestnut, bay, black, gray 以栗、骝、黑、青为主
Yorkshire Coach Horse 约克夏挽马	Ireland 爱尔兰	Coldblood 冷血马	约16.2	Bay or brown 骝、褐
Yunnan Pony 云南矮马	China 中国	Pony 矮马	11～12	Bay, chestnut, black, gray 骝、栗、黑、青
Yushu Horse 玉树马	China 中国	Light 轻型马	约13掌	Mostly gray, bay 以青、骝为主
Zemaituka 泽迈图卡矮马	Lithuania 立陶宛	Pony 矮马	13～14	Bay, black, mouse dun, palomino 骝、黑、鼠灰、银鬃
Zhongdian Horse 中甸马	China 中国	Light 轻型马	12～13掌	Mostly bay, chestnut, black 以骝、栗、黑为主

Notes 注：1. Horses are classified as coldbloods, warmbloods, light breed, and ponies here.
这里将马分为冷血马、温血马、轻型马和矮马四种。
2. Most domestic breeds can also be classified as ponies according to body heights.
许多国内马品种根据体高也可划分为矮马。
3. Coat colors rank in order of frequency.
毛色以出现频率多寡为序。

Appendix 4　附录 4

National and International Horse Organizations
国内外马业相关机构

Name 机构名称	Website 网址
American Driving Society 美国驾车协会	http://www.americandrivingsociety.org/
American Horse Council 美国马匹理事会	http:// www.horsecouncil.org/
American Jockey Club 美国赛马会	http://www.jockeyclub.com/
American Miniature Horse Association 美国迷尼马协会	http://www.amha.org/
American Vaulting Association 美国马上体操协会	http://www.americanvaulting.org/
Asian Racing Federation 亚洲赛马协会	http://www.asianracing.org/
British Equestrian Federation 英国马术联合会	http://www.bef.co.uk/
British Horse Society 英国马会	http://www.bhs.org.uk/
British Jockey Club 英国赛马会	http://www.thejockeyclub.co.uk/
Chinese Equestrian Association 中国马术协会	http://equestrian.sport.org.cn/
China Horse Industry Association 中国马业协会	http://www.chinahorse.org
Fédération Equestre Internationale(FEI) 国际马术联合会	http://www.horsesport.org/

(续)

Name 机构名称	Website 网址
Federation of International Polo 国际马球联合会	http://www.fippolo.com/
General Stud Book 纯血马登记册	http://www.weatherbys.net/
Hong Kong Jockey Club 中国香港赛马会	http://www.hkjc.com/english/index.asp
International Arabian Horse Association 国际阿拉伯马协会	http://www.arabianhorses.org/
International Racing Bureau 国际赛马管理局	http://www.irbracing.com/
Japan Racing Association 日本中央赛马会	http://www.japanracing.jp/
Macau Jockey Club 中国澳门赛马会	http://www.macauhorse.com/
National Horse Museum 中国马文化博物馆	http://www.china-horse.com/
Professional Rodeo Cowboys Association 职业牛仔竞技协会	http://www.prorodeo.com/
The Farriers' Registration Council 英国蹄铁匠注册委员会	http://www.farrier-reg.gov.uk/
United Stated Equestrian Federation 美国马术联合会	http://www.usef.org/

Appendix 5　附录5

World's Major Horse Races 世界各国及地区主要赛事

Table 1　Flat Races 表1　平地赛

Race Name 赛事名称	Grade 级别	Location 地点	Month 月份	Age/Sex 年龄/性别	Distance 赛程(米)
Australia 澳大利亚					
Australian Cup 澳大利亚杯	G-1	Flemington, Melbourne 费明顿,墨尔本	March 3月	3YO +	2 000
Australian Derby 澳洲德比赛	NG-1	Randwick, Sydney 兰德威克,悉尼	April 4月	3YO	2 400
AJC Australian Oaks 澳洲赛马会欧克斯大赛	NG-1	Randwick, Sydney 兰德威克,悉尼	April 4月	3YO, F	2 400
Caulfield Cup 考菲尔德杯	G-1	Flemington, Melbourne 费明顿,墨尔本	November 11月	3YO +	2 400
Doncaster Handicap 唐克斯特让磅赛	G-1	Randwick, Sydney 兰德威克,悉尼	April 4月	3YO +	1 600
Golden Slipper Stakes 金拖鞋大赛	G-1	Rosehill, Sydney 玫瑰岗,悉尼	March 3月	2YO	1 200
Melbourne Cup 墨尔本杯	G-1	Flemington, Melbourne 费明顿,墨尔本	November 11月	3YO +	3 200
Newmarket Handicap 纽马基特让磅赛	G-1	Flemington, Melbourne 费明顿,墨尔本	March 3月	2YO +	1 200
Stradbroke Handicap 斯特拉布洛克让磅赛	G-1	Eagle Farm, Brisbane 老鹰牧场,布里斯班	June 6月	2YO +	1 400
Victoria Derby 维多利亚德比赛	G-1	Flemington, Melbourne 费明顿,墨尔本	November 11月	3YO	2 500
VRC Oaks 维多利亚欧克斯大赛	G-1	Flemington, Melbourne 费明顿,墨尔本	November 11月	3YO, F	2 500

(续)

Race Name 赛事名称	Grade 级别	Location 地点	Month 月份	Age/Sex 年龄/性别	Distance 赛程(米)
Canada 加拿大					
Breeders' Stakes 育马者大赛	NG-1	Woodbine, Toronto 活拜赛马场,多伦多	August 8月	3YO	2 400
Canadian International Stakes 加拿大国际大赛	G-1	Woodbine, Toronto 活拜赛马场,多伦多	October 10月	3YO +	2 400
Prince of Wales Stakes 威尔士亲王大赛	NG-1	Fort Erie, Ontario 伊利堡,安大略	July 7月	3YO	1 900
The Queen's Plate 女王杯	NG-1	Woodbine, Toronto 活拜赛马场,多伦多	June 6月	3YO	1 600
China-Hong Kong 中国香港					
Hong Kong Cup 香港杯	G-1	Sha Tin Racecourse 沙田马场	December 12月	3YO +	2 000
Hong Kong Mile 香港一哩赛	G-1	Sha Tin Racecourse 沙田马场	December 12月	3YO +	1 600
Hong Kong Sprint 香港短途赛	G-1	Sha Tin Racecourse 沙田马场	December 12月	3YO +	1 000
Hong Kong Vase 香港瓶	G-1	Sha Tin Racecourse 沙田马场	December 12月	3YO +	2 400
China-Macau 中国澳门					
Director's Cup 董事杯	G-1	Taipa Racecourse 氹仔马场	May/June 5/6月	3YO +	1 500
Macau Cup 澳门杯	G-2	Taipa Racecourse 氹仔马场	August 8月	3YO +	1 500
Macau Derby 澳门德比赛	G-1	Taipa Racecourse 氹仔马场	May 5月	4YO	1 800
Macau Gold Cup 澳门金杯	G-1	Taipa Racecourse 氹仔马场	May/June 5/6月	3YO +	1 800
Macau Guineas 澳门几尼赛	G-2	Taipa Racecourse 氹仔马场	March 3月	4YO	1 500

(续)

Race Name 赛事名称	Grade 级别	Location 地点	Month 月份	Age/Sex 年龄/性别	Distance 赛程(米)
China-Mainland 中国大陆					
Silk Road China Cup 丝绸之路·中国杯	NG-1	Xiyu Racecourse 西域赛马场	July 7月	3YO +	2 000
China Breeders' Cup 中国育马者杯	NG-1	Xilinguole Racecourse 锡林郭勒赛马场	October 10月	2YO	1 000
Horse Capital CHIA Cup 中国马会杯·马都赛	NG-1	Yijinhuoluo Racecourse 伊金霍洛赛马场	July 7月	3YO +	1 650
France 法国					
French 1 000 Guineas 法国一千几尼赛	NG-1	Longchamp, Paris 隆尚赛马场,巴黎	May 5月	3YO, F	1 600
French 2 000 Guineas 法国两千几尼赛	G-1	Longchamp, Paris 隆尚赛马场,巴黎	May 5月	3YO, C	1 600
Prix du Jockey Club (Derby) 法国德比赛	G-1	Chantilly Racecourse 尚蒂伊赛马场	June 6月	3YO, C/F	2 100
French Oaks (Prix de Diane) 法国欧克斯奖金赛	G-1	Chantilly Racecourse 尚蒂伊赛马场	June 6月	3YO, F	2 100
Grand Prix de Paris 巴黎大奖赛	G-1	Longchamp, Paris 隆尚赛马场,巴黎	July 7月	3YO	2 400
Prix de l'Arc de Triomphe 凯旋门大奖赛	G-1	Longchamp, Paris 隆尚赛马场,巴黎	October 10月	3YO +	2 400
Germany 德国					
Bayerisches Zuchtrennen 杜美亚大赛	G-1	Munich 慕尼黑赛马场	July/Aug 7/8月	3YO +	2 000
Deutschland-Preis 德国大奖赛	G-1	Düsseldorf 杜塞尔多夫赛马场	June/July 6/7月	3YO +	2 400
German Derby 德国德比赛	G-1	Hamburg 汉堡赛马场	July 7月	3YO, C/F	2 400

（续）

Race Name 赛事名称	Grade 级别	Location 地点	Month 月份	Age/Sex 年龄/性别	Distance 赛程（米）
Grosser Preis von Baden 冯巴登大奖赛	G-1	Baden-Baden 巴登-巴登	September 9月	3YO +	2 400
Preis der Diana(Oaks) 德国欧克斯大赛	G-1	Düsseldorf 杜塞尔多夫赛马场	June 6月	3YO,F	2 200
Preis von Europa 欧洲大奖赛	G-1	Cologne 科隆赛马场	September 9月	3YO +	2 400
Rheinland-Pokal	G-1	Cologne 科隆赛马场	August 8月	3YO +	2 400
Great Britain 英国					
Champion Stakes 冠军赛	G-1	Newmarket 纽马基特	October 10月	3YO +	1 600
Doncaster Cup 唐克斯特杯	G-2	Doncaster 唐克斯特	September 9月	3YO +	3 600
Eclipse Stakes 日食奖金赛	G-1	Sandown Park 沙丘园	July 7月	3YO +	1 600
Epsom Derby 艾普森德比赛	G-1	Epsom Downs 艾普森·唐斯赛马场	June 6月	3YO,C/F	约2 400
International Stakes 国际奖金赛	G-1	York Racecourse 约克赛马场	August 8月	3YO +	约2 000
The Oaks Stakes 欧克斯奖金赛	G-1	Epsom Downs 艾普森·唐斯赛马场	June 6月	3YO + ,F	2 400
One Thousand Guineas Stakes 一千几尼赛	G-1	Newmarket 纽马基特	April/May 4/5月	3YO,F	1 600
St. Leger Stakes 圣莱切奖金赛	G-1	Doncaster 唐克斯特	September 9月	3YO +	约2 900
Two Thousand Guineas Stakes 两千几尼赛	G-1	Newmarket 纽马基特	April/May 4/5月	3YO	2 400

（续）

Race Name 赛事名称	Grade 级别	Location 地点	Month 月份	Age/Sex 年龄/性别	Distance 赛程(米)
Ireland 爱尔兰					
Irish Champion Stakes 爱尔兰冠军赛	G-1	Leopardstown 李奥柏赛马场	September 9月	3YO +	2 000
Irish 1000 Guineas 爱尔兰一千几尼赛	G-1	Curragh 克瑞赛马场	May 5月	3YO,F	1 600
Irish 2000 Guineas 爱尔兰两千几尼赛	G-1	Curragh 克瑞赛马场	May 5月	3YO	1 600
Irish Derby 爱尔兰德比赛	G-1	Curragh 克瑞赛马场	June/July 6/7月	3YO,C/F	2 400
Irish Oaks 爱尔兰欧克斯奖金赛	G-1	Curragh 克瑞赛马场	July 7月	3YO,F	2 400
Italy 意大利					
Gran Premio del Jockey Club 赛马会大奖赛	G-1	Milan 米兰	October 10月	3YO +	2 400
Gran Premio di Milano 米兰大奖赛	G-1	Milan 米兰	June 6月	3YO +	2 400
Italian 1000 Guineas 意大利一千几尼赛	G-2	Rome 罗马	May 5月	3YO,F	1 600
Italian 2000 Guineas 意大利两千几尼赛	G-2	Rome 罗马	April/May 4/5月	3YO,C	1 600
Italian Derby 意大利德比赛	G-1	Rome 罗马	May 5月	3YO,C/F	2 400
Italian Oaks 意大利欧克斯奖金赛	G-2	Milan 米兰	June 6月	3YO,F	2 200
Japan 日本					
Japan Cup 日本杯	G-1	Tokyo Racecourse 东京赛马场	November 11月	3YO +	2 400
Japan Cup Dirt 日本沙地赛	G-1	Tokyo Racecourse 东京赛马场	November 11月	3YO +	2 100

（续）

Race Name 赛事名称	Grade 级别	Location 地点	Month 月份	Age/Sex 年龄/性别	Distance 赛程(米)
Japanese Grand Prix 日本大奖赛	G-1	Nakayama Racecourse 中山赛马场	December 12月	3YO +	2 500
Mile Championship 一哩锦标赛	G-1	Kyoto 京都赛马场	November 11月	3YO +	1 600
Sprinters Stakes 短途大奖赛	G-1	Nakayama Racecourse 中山赛马场	Sept/Oct 9/10月	3YO +	1 200
Takamatsunomiya Kinen 高松宫纪念赛	G-1	Chukyo 中京竞马场	March 3月	4YO +	1 200
Takarazuka Kinen 宝塚纪念赛	G-1	Hanshin 阪神赛马场	June 6月	3YO +	2 200
Tenno Sho 天皇赏	G-1	Kyoto 京都赛马场	April 4月	4YO +	3 200
Tokyo Daishoten 东京大赏典	G-1	Ohi 大井赛马场	December 12月	3YO +	2 000
Yasuda Kinen 安田纪念赛	G-1	Tokyo Racecourse 东京赛马场	June 6月	3YO +	1 600
New Zealand 新西兰					
Auckland Cup 奥克兰杯	G-1	Auckland 奥克兰	March 3月	3YO +	3 200
Easter Handicap 复活节评磅赛	G-1	Auckland 奥克兰	March 3月	3YO +	1 600
New Zealand 1000 Guineas 新西兰一千几尼赛	G-1	Canterbury 坎特伯雷	November 11月	3YO, F	1 600
New Zealand 2000 Guineas 新西兰两千几尼赛	G-1	Canterbury 坎特伯雷	November 11月	3YO	1 600
NZE International Stakes 新西兰国际奖金赛	G-1	Waikato 怀卡托	February 2月	3YO +	2 000
New Zealand Oaks 新西兰欧克斯奖金赛	G-1	Trentham, Wellington 惠灵顿	March 3月	3YO, F	2 400
Wellington Cup 惠灵顿杯	G-1	Wellington 惠灵顿	January 1月	3YO +	3 200

（续）

Race Name 赛事名称	Grade 级别	Location 地点	Month 月份	Age/Sex 年龄/性别	Distance 赛程(米)
South Africa 南非					
Cape Argus Guineas 开普艾格斯几尼赛	G-1	Cape Town 开普敦	January 1月	3YO	1 600
Vodacom Durban July 德班七月评磅赛	G-1	Greyville 格雷维尔赛马场	July 7月	3YO +	2 200
J&B Metropolitan Stakes 约翰内斯堡都市大赛	G-1	Cape Town 开普敦	January 1月	3YO +	2 000
South African Classic 南非经典大赛	G-1	Johannesburg 约翰内斯堡	March 3月	3YO	1 800
South African Derby 南非德比赛	G-2	Johannesburg 约翰内斯堡	April 4月	3YO	2 450
United Arab Emirates(UAE)阿拉伯联合酋长国					
Al Quoz Sprint 阿尔括兹短途赛	G-1	Meydan Racecourse, Dubai 迈丹赛马场,迪拜	March 3月	3YO +	1 200
Dubai Turf 迪拜草地赛	G-1	Meydan Racecourse, Dubai 迈丹赛马场,迪拜	March 3月	4YO +	1 800
Dubai Golden Shaheen 迪拜金隼赛	G-1	Meydan Racecourse, Dubai 迈丹赛马场,迪拜	March 3月	3YO +	1 200
Dubai Sheema Classic 迪拜经典赛	G-1	Meydan Racecourse, Dubai 迈丹赛马场,迪拜	March 3月	4YO +	2 410
Dubai Gold Cup 迪拜金杯赛	G-2	Meydan Racecourse, Dubai 迈丹赛马场,迪拜	March 3月	4YO +	3 200
Dubai World Cup 迪拜世界杯	G-1	Meydan Racecourse, Dubai 迈丹赛马场,迪拜	March 3月	4YO +	2 000
Godolphin Mile 哥德芬一哩赛	G-2	Meydan Racecourse, Dubai 迈丹赛马场,迪拜	March 3月	3YO +	1 600

（续）

Race Name 赛事名称	Grade 级别	Location 地点	Month 月份	Age/Sex 年龄/性别	Distance 赛程(米)
UAE Derby 阿联酋德比赛	G-2	Meydan Racecourse, Dubai 迈丹赛马场*,迪拜	March 3月	3YO	1 900
United States of America 美国					
Belmont Stakes 贝尔蒙特大赛	G-1	Belmont Park 贝尔蒙特公园	June 6月	3YO	2 400
Breeders' Cup Classic 育马者杯古典大赛	G-1	Changes 不定	November 11月	3YO +	2 000
Breeders' Cup Distaff 育马者杯雌驹赛	G-1	Changes 不定	November 11月	3YO + ,F	1 800
Breeders' Cup Juvenile 育马者杯两岁雄驹赛	G-1	Changes 不定	November 11月	2YO,C	1 700
Breeders' Cup Juvenile Fillies 育马者杯两岁雌驹赛	G-1	Changes 不定	November 11月	2YO,F	1 700
Breeders' Cup Mile 育马者杯一哩赛	G-1	Changes 不定	November 11月	3YO	1 600
Breeders' Cup Sprint 育马者杯短途赛	G-1	Changes 不定	November 11月	3YO +	1 200
Breeders' Cup Turf 育马者杯草地赛	G-1	Changes 不定	November 11月	3YO +	2 400
Brooklyn Invitational Stakes 布鲁克林邀请赛	G-2	Belmont Park 贝尔蒙特公园	June 6月	4YO +	2 400
Donn Handicap 唐纳让磅赛	G-1	Gulfstream Park 湾流公园	February 2月	4YO +	1 900
Hollywood Gold Cup 好莱坞金杯赛	G-1	Hollywood Park 好莱坞公园	July 7月	3YO +	2 000
Jockey Club Gold Cup 赛马会金杯赛	G-1	Belmont Park 贝尔蒙特公园	October 10月	3YO +	2 000

* The Meydan Racecourse was built in 2010 upon the same site of the Nad Al Sheba Racecourse.
迈丹赛马场于2010年在诗柏赛马场的原址上重建。

（续）

Race Name 赛事名称	Grade 级别	Location 地点	Month 月份	Age/Sex 年龄/性别	Distance 赛程(米)
Kentucky Derby 肯塔基德比赛	G-1	Churchill Downs 邱吉尔·唐斯	May 5 月	3YO	2 000
Kentucky Oaks 肯塔基欧克斯大赛	G-1	Churchill Downs 邱吉尔·唐斯	May 5 月	3YO,F	1 800
Metropolitian Handicap 大都会让磅赛	G-1	Belmont Park 贝蒙园	Early June 6 月初	3YO +	1 600
Preakness Stakes 普瑞克尼斯大赛	G-1	Pimlico,Maryland 皮姆利克,马里兰	May 5 月	3YO	1 900
Santa Anita Handicap 圣安妮塔让磅赛	G-1	Santa Anita Park 圣安妮塔园	March 3 月	4YO +	2 000
Santa Anita Derby 圣安妮塔德比赛	G-1	Santa Anita Park 圣安妮塔园	April 4 月	3YO	2 000
Suburban Handicap 市郊让磅赛	G-2	Belmont Park 贝蒙园	July 7 月	4YO +	2 000
Travers Stakes 特拉维斯大赛	G-1	Saratoga 沙拉托加	August 8 月	3YO	2 000
Whitney Handicap 惠特尼让磅赛	G-1	Saratoga 沙拉托加	August 8	3YO +	1 800
Woodward Stakes 伍德沃德大赛	G-1	Saratoga 沙拉托加	September 9 月	3YO +	1 800

Notes 注：

1. This list of flat races is sorted by country.
   平地赛以国家名称字母为序。
2. Distance of a race may vary over years, and the distance listed here is the present one.
   许多赛事的赛程都有所变动,表中列出者为目前赛程。
3. Location of a race may vary over years, and the place listed here is the present one.
   许多赛事的地点都有所变动,表中列出者为目前举办地。
4. Status of a race may change over years, and the grade listed here is the present one.
   许多赛事级别都有所变动,表中列出者为目前级别。
5. Abbreviations 缩写词：C- Colt 雄驹；F-filly 雌驹；G-Grade 级别；NG-National Grade 国内级别；m- meter 米

Table 2 Steeplechases

表 2 越野障碍赛

Name of Steeplechase 越野障碍赛名称	Country 国家	Location 地点	Month 月份	Age 年龄	Distance 赛程(米)
Grand Pardubice 帕尔杜比策越野障碍赛	Czech 捷克	Pardubice Racecourse 帕尔杜比策赛马场	October 10 月	6YO +	7 240
Grand Steeple-Chase de Paris 巴黎越野障碍赛	France 法国	Auteuil Racecourse 奥特伊赛马场	May 5 月	5YO +	5 800
Irish Grand National Steeple-chase 爱尔兰越野障碍赛	Ireland 爱尔兰	Fairyhouse Racecourse, Meath 仙屋赛马场,米斯郡	March 3 月	5YO +	5 834
Gran Premio Merano 梅拉诺越野障碍赛	Italy 意大利	Maia Racecourse, Merano 迈亚赛马场,梅拉诺	September 9 月	4YO +	5 000
Nakayama Grand Jump 中山越野障碍赛	Japan 日本	Nakayama Racecourse 中山赛马场	April 4 月	4YO +	4 250
Cheltenham Gold Cup 切尔腾纳姆金杯赛	UK 英国	Cheltenham Racecourse 切尔滕纳姆赛马场	March 3 月	5YO +	5 331
Grand National Steeplechase 全英越野障碍赛	UK 英国	Aintree Racecourse, Liverpool 爱茵垂赛马场,利物浦	April 4 月	5YO +	7 218
King George VI Chase 国王乔治六世越野障碍赛	UK 英国	Kempton Park Racecourse 金顿公园赛马场	January 1 月	4YO +	4 827
Breeders' Cup Grand National 育马杯全美越野障碍赛	USA 美国	Far Hills, New Jersey 远山赛马场,新泽西	October 10 月	4YO +	4 225
Maryland Hunt Cup 马里兰狩猎杯	USA 美国	Worthington Valley, Maryland 沃辛顿谷赛马场,马里兰州	April 4 月	5YO +	6 436
Iroquois Steeplechase 易洛魁越野障碍赛	USA 美国	Percy Warner Park, Tennessee 波西—华纳公园,田纳西州	May 5 月	4YO +	4 827
Carolina Cup 卡罗莱纳杯	USA 美国	Springdale, South Carolina 春谷赛马场,南卡罗莱纳	March 3 月	4YO +	3 420
Colonial Cup 殖民杯	USA 美国	Springdale, South Carolina 春谷赛马场,南卡罗莱纳	November 11 月	4YO +	4 425

Appendix 6 附录 6

Official Results of Olympic Equestrian Events
奥运会马术奖牌统计表
Table 1 Dressage Individual Competition
表 1 盛装舞步个人赛

Medal 奖牌	Country 国家	Rider 骑手	Mount 坐骑	Points 得分
1912 STOCKHOLM 斯德哥尔摩				
Gold 金牌	Sweden 瑞典	Carl Bonde	Emperor	15. 00
Silver 银牌	Sweden 瑞典	Gustaf Boltenstern, Sr.	Neptune	21. 00
Bronze 铜牌	Sweden 瑞典	Hans von Blixen-Finecke, Sr.	Maggie	32. 00
1920 ANTWERP 安特卫普				
Gold 金牌	Sweden 瑞典	Jainne Lundblad	Uno	15. 00
Silver 银牌	Sweden 瑞典	Bertil Sandström	Sabel	21. 00
Bronze 铜牌	Sweden 瑞典	Hans von Rosen	Running Sister	32. 00
1924 PARIS 巴黎				
Gold 金牌	Sweden 瑞典	Ernst Linder	Piccolomini	276. 40
Silver 银牌	Sweden 瑞典	Bertil Sandström	Sabei	275. 80
Bronze 铜牌	France 法国	Xavier Lesage	Plumard	265. 80
1928 AMSTERDAM 阿姆斯特丹				
Gold 金牌	Germany 德国	Carl von Langen-Parow	Draufgänger	237. 42
Silver 银牌	France 法国	Charles Marion	Linon	231. 00
Bronze 铜牌	Sweden 瑞典	Ragnar Ohlson	Gunstling	229. 78
1932 LOS ANGELES 洛杉矶				
Gold 金牌	France 法国	Xavier Lesage	Taine	343. 75
Silver 银牌	France 法国	Charles Marion	Linon	305. 42
Bronze 铜牌	USA 美国	Hilram Tuttle	Olympic	300. 50

（续）

Medal 奖牌	Country 国家	Rider 骑手	Mount 坐骑	Points 得分
1936 BERLIN 柏林				
Gold 金牌	Germany 德国	Heinz Pollay	Kronos	1 760. 00
Silver 银牌	Germany 德国	Friedrich Gerhard	Absinth	1 745. 50
Bronze 铜牌	Austria 奥地利	Alois Podhajsky	Nero	1 721. 50
1948 LONDON 伦敦				
Gold 金牌	SWI 瑞士	Hans Moser	Hummer	492. 50
Silver 银牌	France 法国	Andre Jousseaume	Harpagon	480. 00
Bronze 铜牌	Sweden 瑞典	Gustaf-Adolf Boltenstern	Trumf	477. 50
1952 HELSINKI 赫尔辛基				
Gold 金牌	Sweden 瑞典	Henri Saint Cyr	Master Rufus	561. 00
Silver 银牌	Denmark 丹麦	Lis Hartel	Jubilee	541. 50
Bronze 铜牌	France 法国	André Jousseaume	Harpagon	541. 00
1956 STOCKHOLM 斯德哥尔摩				
Gold 金牌	Sweden 瑞典	Henri Saint Cyr	Juli	860. 00
Silver 银牌	Denmark 丹麦	Lis Hartel	Jubilee	850. 00
Bronze 铜牌	Germany 德国	Liselott Linsenhoff	Adular	832. 00
1960 ROME 罗马				
Gold 金牌	SOV 苏联	Sergei Filatov	Absent	2 144. 00
Silver 银牌	SWI 瑞士	Gustav Fischer	Wald	2 087. 00
Bronze 铜牌	Germany 德国	Josef Neckermann	Asbach	2 082. 00
1964 TOKYO 东京				
Gold 金牌	SWI 瑞士	Henri Chammartin	Woermann	1 504. 00
Silver 银牌	Germany 德国	Harry Boldt	Remus	1 503. 00
Bronze 铜牌	SOV 苏联	Sergei Filatov	Absent	1 486. 00
1968 MEXICO CITY 墨西哥城				
Gold 金牌	SOV 苏联	Ivan Kizimov	Ikhor	1 572. 00
Silver 银牌	Germany 德国	Josef Neckermann	Mariano	1 546. 00

（续）

Medal 奖牌	Country 国家	Rider 骑手	Mount 坐骑	Points 得分
Bronze 铜牌	Germany 德国	Reiner Klimke	Dux	1 537. 00
1972 MUNICH 慕尼黑				
Gold 金牌	Germany 德国	Liselott Linsenhoff	Piaff	1 229. 00
Silver 银牌	SOV 苏联	Yelena Petushkova	Pepel	1 185. 00
Bronze 铜牌	Germany 德国	Josef Neckermann	Venetia	1 177. 00
1976 MONTREAL 蒙特利尔				
Gold 金牌	SWI 瑞士	Christine Stückelberger	Granat	1 486. 00
Silver 银牌	Germany 德国	Harry Boldt	Woyceck	1 435. 00
Bronze 铜牌	Germany 德国	Reiner Klimke	Mehmed	1 395. 00
1980 MOSCOW 莫斯科				
Gold 金牌	Austria 奥地利	Elisabeth Theurer	Mon Cherie	1 370. 00
Silver 银牌	SOV 苏联	Yuri Kovshov	Igrok	1 300. 00
Bronze 铜牌	SOV 苏联	Viktor Ugryumov	Shkval	1 234. 00
1984 LOS ANGELES 洛杉矶				
Gold 金牌	Germany 德国	Reiner Klimke	Ahlerich	1 504. 00
Silver 银牌	Denmark 丹麦	Anne-Grethe Jensen	Marzog	1 442. 00
Bronze 铜牌	SWI 瑞士	Otto Hofer	Limandus	1 364. 00
1988 SEOUL 首尔(汉城)				
Gold 金牌	Germany 德国	Nicole Uphoff	Rembrandt	1 521. 00
Silver 银牌	France 法国	Margit Otto-Crepin	Corlandus	1 462. 00
Bronze 铜牌	SWI 瑞士	Christine Stückelberger	Gauguin de Lully	1 417. 00
1992 BARCELONA 巴塞罗那				
Gold 金牌	Germany 德国	Nicole Uphoff	Rembrandt	1 626. 00
Silver 银牌	Germany 德国	Isabell Werth	Gigolo	1 551. 00
Bronze 铜牌	Germany 德国	Nikolaus “Klaus” Balkenhol	Goldstern	1 515. 00
1996 ATLANTA 亚特兰大				
Gold 金牌	Germany 德国	Isabell Werth	Gigolo	235. 09

(续)

Medal 奖牌	Country 国家	Rider 骑手	Mount 坐骑	Points 得分
Silver 银牌	Holland 荷兰	Anky van Grunsven	Olympic Bonfire	233. 02
Bronze 铜牌	Holland 荷兰	Sven Rothenberger	Weyden	224. 94
2000 SYDNEY 悉尼				
Gold 金牌	Holland 荷兰	Anky van Grunsven	Bonfire	239. 18
Silver 银牌	Germany 德国	Isabell Werth	Gigolo	234. 19
Bronze 铜牌	Germany 德国	Ulla Salzgeber	Rusty	230. 57
2004 ATHENS 雅典				
Gold 金牌	Holland 荷兰	Anky van Grunsven	Salinero	74. 208
Silver 银牌	Germany 德国	Ulla Salzgeber	Rusty	76. 524
Bronze 铜牌	Spain 西班牙	Beatriz Ferrer-Salat	Beauvalais	76. 667
2008 香港 *				
Gold 金牌	Holland 荷兰	Anky Van Grunsven	Salinero	78. 680
Silver 银牌	Germany 德国	Isabell Werth	Satchamo	76. 650
Bronze 铜牌	Germany 德国	Heike Kemmer	Bonaparte	74. 455
2012 London 伦敦				
Gold 金牌	UK 英国	Charlotte Dujardin	Valegro	90. 089
Silver 银牌	Holland 荷兰	Adelinde Cornelissen	Parzival	88. 196
Bronze 铜牌	UK 英国	Laura Bechtolsheimer	Mistral Hojris	84. 339
2016 Rio de Janeiro 里约热内卢				
Gold 金牌	UK 英国	Charlotte Dujardin	Valegro	93. 857
Silver 银牌	Germany 德国	Isabell Werth	Weihegold Old	89. 071
Bronze 铜牌	Germany 德国	Kristina Bröring-Sprehe	Desperados FRH	87. 142

* Note: The equestrian events of the 29th Olympic Games were held in China-Hong Kong.

注:第 29 届奥运会马术项目在中国香港举行。

Table 2 Dressage Team Competition
表2　盛装舞步团体赛

Medal 奖牌	Country 国家	Riders 骑手	Mounts 坐骑	Points 得分
1912 STOCKHOLM 斯德哥尔摩		Not Held 没有举行		
1920 ANTWERP 安特卫普		Not Held 没有举行		
1924 PARIS 巴黎		Not Held 没有举行		
1928 AMSTERDAM 阿姆斯特丹				
Gold 金牌	Germany 德国	Carl Friedrich Freiherr von Langen-Parow	Draufgänger	237. 42
		Hermann Lnkenbach	Gimpel	224. 26
		Eugen Freiherr von Lotzbeck	Caracaila	208. 04
Silver 银牌	Sweden 瑞典	Ragnar Ohlson	Gunstling	229. 78
		Janne Lundblad	Blackmar	226. 70
		Carl Bonde	Ingo	194. 38
Bronze 铜牌	Netherlands 荷兰	Jan van Reede	Hans	220. 70
		Pierre Versteegh	His Excellence	216. 44
		Gerard Le Heux	Valerine	205. 82
1932 LOS ANGELES 洛杉矶				
Gold 金牌	France 法国	Xavier Lesage	Taine	1 031. 25
		Charles Marion	Linon	916. 25
		André Jousseaume	Sorelta	871. 25
Silver 银牌	Sweden 瑞典	Bertil Sandström	Kreta	964. 00
		Thomas Byström	Gulliver	880. 50
		Gustaf-Adolf Boltenstern, Jr.	Ingo	833. 50
Bronze 铜牌	USA 美国	Hiram Turtle	Olympic	901. 50
		Isaac Kitts	American Lady	846. 25
		Alvin Moore	Water Pat	829. 00
1936 BERLIN 柏林				
Gold 金牌	Germany 德国	Heinz Pollay	Kronos	1 760. 00
		Friedrich Gerhard	Absinth	1 745. 50
		Hermann von Oppeln-Bronikowski	Gimpel	1 568. 50

（续）

Medal 奖牌	Country 国家	Riders 骑手	Mounts 坐骑	Points 得分
Silver 银牌	France 法国	Andre Jousseaume	Favorite	1 642. 50
		Gerard de Ballore	Debaucheur	1 634. 00
		Daniel Gillois	Nicolas	1 569. 50
Bronze 铜牌	Sweden 瑞典	Gregor Adlercreutz	Teresina	1 675. 00
		Sven Coliiander	Kal	1 530. 50
		Folke Sandstrom	Pergoia	1 455. 00
1948 LONDON 伦敦				
Gold 金牌	France 法国	Andre Jousseaume	Harpagon	480. 00
		Jean Saint Fort Paillard	Sous Ies Ceps	439. 50
		Maurice Buret	Saint Ouen	349. 50
Silver 银牌	USA 美国	Robert Borg	Klingsor	473. 50
		Earl Thomson	Pancraft	421. 00
		Frank Henry	Reno Overdo	361. 50
Bronze 铜牌	Portugal 葡萄牙	Fernando Pais da Silva	Matamas	411. 00
		Francisco Valadas	Feitico	405. 00
		Luiz Mena e Silva	Fascinante	366. 00
1952 HELSINKI 赫尔辛基				
Gold 金牌	Sweden 瑞典	Henri Saint Cyr	Master Rufus	561. 00
		Gustaf-Adolf Boltenstern, Jr.	Krest	531. 00
		Gehnall Persson	Knaust	505. 50
Silver 银牌	Switzerland 瑞士	Gottfried Trachsel	Krusus	531. 00
		Henri Chammartin	Wohler	529. 50
		Gustav Fischer	Solimon	518. 50
Bronze 铜牌	Germany 德国	Heinz Pollay	Adular	518. 50
		Ida von Nagel	Afrika	503. 00
		Fritz Thiedemann	Chronist	479. 50
1956 STOCKHOLM 斯德哥尔摩				
Gold 金牌	Sweden 瑞典	Henri Saint Cyr	Juli	860. 00
		Gehnall Persson	Knaust	821. 00
		Gustaf-Adolf Boltenstern, Jr.	Krest	794. 00

（续）

Medal 奖牌	Country 国家	Riders 骑手	Mounts 坐骑	Points 得分
Silver 银牌	Germany 德国	Liselott Linsenhoff	Adular	832.00
		Hannelore Weygand	Perkunos	785.00
		Anneliese Küppers	Afrika	729.00
Bronze 铜牌	Switzerland 瑞士	Gottfried Trachsel	Kursus	807.00
		Henri Chammartin	Wohler	789.00
		Gustav Fischer	Vasello	750.00
1960 ROME 罗马		Not Held 没有举行		
1964 TOKYO 东京				
Gold 金牌	Germany 德国	Harry Boldt	Remus	889.00
		Reiner Klimke	Dux	837.00
		Josef Neckermann	Antoinette	832.00
Silver 银牌	Switzerland 瑞士	Henri Chammartin	Woermann	870.00
		Gustav Fischer	Wald	854.00
		Marianne Gossweiler	Stephan	802.00
Bronze 铜牌	SOV 苏联	Sergei Filatov	Absent	847.00
		Ivan Kizimov	Ikhor	758.00
		Ivan Kalita	Moar	706.00
1968 MEXICO CITY 墨西哥城				
Gold 金牌	Germany 德国	Josef Neckermann	Mariano	948.00
		Reiner Klimke	Dux	896.00
		Lisdott Linsenhoff	Piaff	855.00
Silver 银牌	SOV 苏联	Ivan Kizimov	Ikhor	908.00
		Ivan Kalita	Absent	879.00
		Yelena Petushkova	Pepel	870.00
Bronze 铜牌	Switzerland 瑞士	Gustav Fischer	Wald	866.00
		Henri Chammartin	Wolfdietrich	845.00
		Marianne Gossweiler	Stephan	836.00

（续）

Medal 奖牌	Country 国家	Riders 骑手	Mounts 坐骑	Points 得分
1972 MUNICH 慕尼黑				
Gold 金牌	SOV 苏联	Yelena Petushkova	Pepel	1 747.00
		Ivan Kizimov	Ikhor	1 701.00
		Ivan Kalita	Tatif	1 647.00
Silver 银牌	Germany 德国	Liselott Linsenhoff	Piaff	1 763.00
		Josef Neckermann	Venetia	1 706.00
		Karin Schlüter	Lisotto	1 614.00
Bronze 铜牌	Sweden 瑞典	Ulla Håkansson	Ajax	1 649.00
		Ninna Swaab	Casanova	1 622.00
		Maud von Rosen	Lucky Boy	1 578.00
1976 MONTREAL 蒙特利尔				
Gold 金牌	Germany 德国	Harry Boldt	Woyceck	1 863.00
		Reiner Klimke	Mehmed	1 751.00
		Gabriela Griilo	Ultimo	1 541.00
Silver 银牌	Switzerland 瑞士	Christine Stückelberger	Granat	1 869.00
		Ulrich Lehmann	Widin	1 425.00
		Doris Ramseier	Roch	1 390.00
Bronze 铜牌	USA 美国	Hilda Gurney	Keen	1 607.00
		Dorothy Morkis	Monaco	1 559.00
		Edith Master	Dahlwitz	1 481.00
1980 MOSCOW 莫斯科				
Gold 金牌	SOV 苏联	Yuri Kovshov	Igrok	1 588.00
		Viktor Ugryumov	Shkval	1 541.00
		Vita Misevych	Plot	1 254.00
Silver 银牌	Bulgaria 保加利亚	Peter Mandazhiev	Stchibor	1 244.00
		Svetoslav Ivanov	Aleko	1 190.00
		Georgi Gadjev	Vnimatelen	1 146.00

（续）

Medal 奖牌	Country 国家	Riders 骑手	Mounts 坐骑	Points 得分
Bronze 铜牌	Romania 罗马尼亚	Anghelache Donescu	Dor	1 255.00
		Dumitru Veliku	Decebal	1 076.00
		Petre Rosca	Derbist	1 015.00
1984 LOS ANGELES 洛杉矶				
Gold 金牌	Germany 德国	Reiner Klimke	Ahlerich	1 797.00
		Uwe Sauer	Montevideo	1 582.00
		Herbert Krug	Muscadeur	1 576.00
Silver 银牌	Switzerland 瑞士	Otto Hofer	limandus	1 609.00
		Christine Stückelberger	Tansanit	1 606.00
		Amy-Cathérine de Bary	Aintree	1 458.00
Bronze 铜牌	Sweden 瑞典	Ulla Hakanson	Flamingo	1 589.00
		Ingamay Bylund	Aleks	1 582.00
		Louise Nathhorst	Inferno	1 459.00
1988 SEOUL 首尔(汉城)				
Gold 金牌	Germany 德国	Nicole Uphoff	Rembrandt	1 458.00
		Monica Theodorescu	Ganimedes	1 433.00
		Ann-Kathrin Linsenhoff	Coruage	1 411.00
Silver 银牌	Switzerland 瑞士	Christine Stückelberger	Gauguin de Lully	1 430.00
		Otto Hofer	Andiamo	1 392.00
		Daniel Ramseier	Random	1 342.00
Bronze 铜牌	Canada 加拿大	Cynthia Ishoy	Dynasty	1 363.00
		Ashley Nicoll	Reipo	1 308.00
		Gina Smith	Malte	1 298.00
1992 BARCELONA 巴塞罗那				
Gold 金牌	Germany 德国	Nicole Uphoff	Rembrandt	1 768.00
		Isabell Werth	Gigolo	1 762.00
		Nikolaus“Klaus” Balkenhol	Goldstern	1 694.00

（续）

Medal 奖牌	Country 国家	Riders 骑手	Mounts 坐骑	Points 得分
Silver 银牌	Netherlands 荷兰	Anky van Grunsven	Olympic Bonfire	1 631.00
		Ellen Bontje	Olympic Larius	1 577.00
		Tineke Bartels de Vries	Olympic Courage	1 534.00
Bronze 铜牌	USA 美国	Carol Laveil	Gifted	1 629.00
		Charlotte Bredahl	Monsieur	1 507.00
		Robert Dover	Lectron	1 507.00
1996 ATLANTA 亚特兰大				
Gold 金牌	Germany 德国	Isabel Werth	Gigolo	1 915.00
		Monica Theodorescu	Grunox	1 845.00
		Klaus Balkenhol	Goldstern	1 793.00
		Martin Schaudt	Durgo	1 781.00
Silver 银牌	Holand 荷兰	Anky van Grunsven	Olympic Bonfire	1 893.00
		Sven Rothenberger	Weyden	1 854.00
		Tineke Bartels-de Vries	Olympic Barbria	1 690.00
		Gonnelien Rothenberger	Olympic Dondolo	1 673.00
Bronze 铜牌	USA 美国	Michelle Gibson	Peron	1 880.00
		Guenter Seidel	Graf George	1 734.00
		Steffen Peters	Udon	1 695.00
		Robert Dover	Metallic	1 649.00
2000 SYDNEY 悉尼				
Gold 金牌	Germany 德国	Isabell Werth	Gigolo	1 908.00
		Nadine Capellmann	Farbenfroh	1 867.00
		Ulla Salzgeber	Rusty	1 829.00
		Alexandra Simons de Ridder	Chacomo	1 857.00
Silver 银牌	Holand 荷兰	Ellen Bontje	Silvano	1 786.00
		Anky van Grunsven	Bonfire	1 875.00
		Arjen Teeuwissen	Goliath	1 831.00
		Coby van Baalen	Ferro	1 873.00

(续)

Medal 奖牌	Country 国家	Riders 骑手	Mounts 坐骑	Points 得分
Bronze 铜牌	USA 美国	Susan Blinks	Flim Flam	1 725.00
		Robert Dover	Ranier	1 678.00
		Guenter Seidel	Foltaire	1 695.00
		Christine Traurig	Etienne	1 746.00
2004 ATHENS 雅典				
Gold 金牌	Germany 德国	Ulla Salzgeber	Rusty	78.208
		Martin Schaudt	Weltall	73.417
		Hubertus Schmidt	Wansuela Suerte	72.333
		Heike Kemmer	Bonaparte	71.292
Silver 银牌	Spain 西班牙	Beatriz Ferrer-Salat	Beauvalais	74.667
		Rafael Soto	Invasor	72.792
		Juan Antonio Jimenez	Guizo	71.292
		Ignacio Rambla	Oleaja	64.750
Bronze 铜牌	USA 美国	Deborah McDonald	Brentina	73.375
		Robert Dover	Kennedy	71.625
		Günter Seidel	Aragon	69.500
		Lisa Wilcox	Relevant 5	68.792
2008 Hong Kong 香港				
Gold 金牌	Germany 德国	Hieke Kemmer	Bonaparte	72.250
		Nadine Capellmann	Elvis Va	70.083
		Isabell Werth	Satchmo	76.417
Silver 银牌	Holand 荷兰	Hans Peter Minderhoud	Nadine	69.625
		Imke Schellekens-Bartels	Sunrise	70.875
		Anky van Grunsven	Salinero	74.750
Bronze 铜牌	Denmark 丹麦	Anne van Olst	Clearwater	67.375
		Zu Sayn-Wittgenstein	Digby	70.417
		Andreas Helgstrand	Don Schufro	68.833

(续)

Medal 奖牌	Country 国家	Riders 骑手	Mounts 坐骑	Points 得分
2012 LONDON 伦敦				
Gold 金牌	UK 英国	Carl Hester	Uthopia	80. 571
		Laura Bechtolsheimer	Mistral Hojris	77. 794
		Charlotte Dujardin	Valegro	83. 286
Silver 银牌	Germany 德国	Helen Langehanenberg	Diva Royal	77. 571
		Dorothee Schneider	Desperados	76. 254
		Kristina Sprehe	Damon Hill	78. 937
Bronze 铜牌	Holand 荷兰	Ankyvan Grunsven	Salinero	74. 794
		Edward Gal	Undercover	75. 556
		Adelinde Cornelissen	Parzival	81. 968
2016 Rio de Janeiro 里约热内卢				
Gold 金牌	Germany 德国	Sönke Rothenberger	Cosmo	76. 261
		Dorothee Schneider	Showtime FRH	82. 619
		Kristina Bröring-Sprehe	Desperados FRH	81. 401
		Isabell Werth	Weihegold Old	83. 711
Silver 银牌	UK 英国	Spencer Wilton	Super Nova Ⅱ	73. 613
		Fiona Bigwood	Orthilia	74. 342
		Carl Hester	Nip Tuck	76. 485
		Charlotte Dujardin	Valegro	82. 983
Bronze 铜牌	USA 美国	Allison Brock	Rosevelt	73. 824
		Kasey Perry-Glass	Dublet	73. 235
		Steffen Peters	Legolas 92	74. 622
		Laura Graves	Verdades	80. 644

Table 3 Show Jumping Individual Competition
表 3 个人场地障碍赛奖牌榜

Medal 奖牌	Country 国家	Rider 骑手	Mount 坐骑	Time 时间
1900 PARIS 巴黎				
Gold 金牌	Belgium 比利时	Aimé Haegeman	Benton Ⅱ	2:16.0(m)
Silver 银牌	Belgium 比利时	Georges van de Poele	Windsor Squire	2:17.6(m)
Bronze 铜牌	France 法国	Louis de Champsavin	Terpsichore	2:26.0(m)
1912 STOCKHOLM 斯德哥尔摩				
Gold 金牌	Sweden 瑞典	Jean Cariou	Mignon	4.00*; 5.00**
Silver 银牌	Sweden 瑞典	Rabod Wiihelm von Kröcher	Dohna	4.00*; 7.00**
Bronze 铜牌	Sweden 瑞典	Emanuel de Blommaert de Soye	Clonmore	5.00*
1920 ANTWERP 安特卫普				
Gold 金牌	Italy 意大利	Tommaso Lequio di Assaba	Trebecco	2.00
Silver 银牌	Italy 意大利	Aiessandro Valerio	Cento	3.00
Bronze 铜牌	Sweden 瑞典	Carl-Gustaf Lewenhaupt	Mon Coeur	4.00
1924 PARIS 巴黎				
Gold 金牌	SWI 瑞士	Aiphonse Gemuseus	Lucette	6.00
Silver 银牌	Italy 意大利	Tommaso Lequio di Assaba	Trebecco	8.75
Bronze 铜牌	Poland 波兰	Adam Krolikiewicz	Picador	10.00

(续)

Medal 奖牌	Country 国家	Rider 骑手	Mount 坐骑	Time 时间
1928 AMSTERDAM 阿姆斯特丹				
Gold 金牌	Germany 德国	Frantisek Ventura	Eliot	0. 00 *
Silver 银牌	France 法国	Pierre Bertran de Balanda	Papillon	2. 00 *
Bronze 铜牌	Sweden 瑞典	Charley Kuhn	Pepita	4. 00 *
1932 LOS ANGELES 洛杉矶				
Gold 金牌	Japan 日本	Takeichi Nishi	Uranus	8. 00
Silver 银牌	USA 美国	Harry Chamberiin	Show Girl	12. 00
Bronze 铜牌	Sweden 瑞典	Clarence von Rosen, Jr.	Empire	16. 00
1936 BERLIN 柏林				
Gold 金牌	Germany 德国	Kurt Hasse	Tora	4. 00 * ; 4. 00 ** (59. 2 s)
Silver 银牌	Romania 罗马尼亚	Henri Rang	Delfis	4. 00 * ; 4. 00 ** (72. 8 s)
Bronze 铜牌	Hungary 匈牙利	Józse fvon Platthy	Sellö	8. 00 * ; 0. 00 ** (62. 6 s)
1948 LONDON 伦敦				
Gold 金牌	Mexico 墨西哥	Humberto Mariles Cortes	Arete	6. 25 * ; 0. 00 **
Silver 银牌	Mexico 墨西哥	Rubén Uriza	Harvey	8. 00 * ; 0. 00 ** (49. 1 s)
Bronze 铜牌	France 法国	Jean François d'Orgeix	Sucre de Pomme	8. 00 * ; 4. 00 ** (38. 9 s)

（续）

Medal 奖牌	Country 国家	Rider 骑手	Mount 坐骑	Time 时间
1952 HELSINKI 赫尔辛基				
Gold 金牌	France 法国	Pierre Jonquères d'Oriola	All Baba	8. 00 * ; 0. 00 ** (40. 0 s)
Silver 银牌	Chile 智利	Oscar Cristi	Bambi	8. 00 * ; 4. 00 ** (44. 0 s)
Bronze 铜牌	Germany 德国	Fritz Thiedemann	Meteor	8. 00 * ; 8. 00 ** (38. 5 s)
1956 STOCKHOLM 斯德哥尔摩				
Gold 金牌	Germany 德国	Hans Günter Winkler	Halla	4. 00
Silver 银牌	Italy 意大利	Raimondo d'Inzeo	Merano	8. 00
Bronze 铜牌	Italy 意大利	Piero d'Inzeo	Uruguay	11. 00
1960 ROME 罗马				
Gold 金牌	Italy 意大利	Raimondo d'Inzeo	Posiliipo	12. 00
Silver 银牌	Italy 意大利	Piero d'Inzeo	The Rock	16. 00
Bronze 铜牌	UK 英国	David Broome	Sunslave	23. 00
1964 TOKYO 东京				
Gold 金牌	France 法国	Pierre Jonquères d'Oriola	Lutteur	9. 00 *
Silver 银牌	Germany 德国	Hermann Schridde	Dozent	13. 75 *
Bronze 铜牌	UK 英国	Peter Robeson	Firecrest	16. 00 * ;0. 00 ** (61. 0 s)

(续)

Medal 奖牌	Country 国家	Rider 骑手	Mount 坐骑	Time 时间
1968 MEXICO CITY 墨西哥城				
Gold 金牌	USA 美国	William Steinkraus	Snowbound	4.00 *
Silver 银牌	UK 英国	Marion Coakes	Stroller	8.00 *
Bronze 铜牌	UK 英国	David Broome	Mister Softee	12.00 * ;0.00 ** (35.3 s)
1972 MUNICH 慕尼黑				
Gold 金牌	Italy 意大利	Graziano Mancinelli	Ambassador	8.00 * ; 0.00 **
Silver 银牌	UK 英国	Ann Moore	Psaim	8.00 * ; 3.00 **
Bronze 铜牌	USA 美国	Neal Shapiro	Sloopy	8.00 * ; 8.00 **
1976 MONTREAL 蒙特利尔				
Gold 金牌	Germany 德国	Alwin Schockemohle	Warwick Rex	0.00 *
Silver 银牌	Canada 加拿大	Michel Vaillancourt	Branch County	12.00 * ; 4.00 **
Bronze 铜牌	Belgium 比利时	François Mathy	Gai Luron	12.00 * ; 8.00 **
1980 MOSCOW 莫斯科				
Gold 金牌	Poland 波兰	Jan Kowalczyk	Artemoi	8.00 *
Silver 银牌	SOV 苏联	Nikolai Korolkov	Espadron	9.50 *
Bronze 铜牌	Mexico 墨西哥	Joaquin Perez Heras	Alymony	12.00 * ;4.00 ** (43.23s)

（续）

Medal 奖牌	Country 国家	Rider 骑手	Mount 坐骑	Time 时间
1984 LOS ANGELES 洛杉矶				
Gold 金牌	USA 美国	Joe Fargis	Touch of Class	4.00 * ; 0.00 **
Silver 银牌	USA 美国	Conrad Homfeld	Abdullah	4.00 * ; 8.00 **
Bronze 铜牌	SWI 瑞士	Heidi Robbiani	Jessica V	8.00 * ; 0.00 **
1988 SEOUL 首尔（汉城）				
Gold 金牌	France 法国	Pierre Durand	Jappeloup	1.25 *
Silver 银牌	USA 美国	Greg Best	Gem Twist	4.00 * ; 4.00 ** (45.70s)
Bronze 铜牌	Germany 德国	Karsten Huck	Nepomuk	4.00 * ; 4.00 ** (54.75s)
1992 BARCELONA 巴塞罗那				
Gold 金牌	Germany 德国	Ludger Beerbaum	Classic Touch	0.00
Silver 银牌	Holland 荷兰	Piet Raymakers	Ratina Z	0.25
Bronze 铜牌	USA 美国	Norman Dellojoio	Irish	4.75
1996 ATLANTA 亚特兰大				
Gold 金牌	Germany 德国	Ulrich Kirchhoff	Jus de Pommes	1.00 *
Silver 银牌	SWI 瑞士	Willi Melliger	Calvaro	4.00 * ; 0.00 ** (38.07)
Bronze 铜牌	France 法国	Alexandra Ledermann	Rochet M	4.00 * ; 0.00 ** (41.46)

(续)

Medal 奖牌	Country 国家	Rider 骑手	Mount 坐骑	Time 时间
2000 SYDNEY 悉尼				
Gold 金牌	Holand 荷兰	Jeroen Dubbeldam	Sjiem	4.0
Silver 银牌	Holland 荷兰	Albert Voorn	Lando	4.0
Bronze 铜牌	KSA 沙特	Khaled Al Eid	Khashm Al Aan	4.0
2004 ATHENS 雅典				
Gold 金牌	Brazil 巴西	Rodrigo Pessoa	Baloubet du Rouet	8.0*, 4.0** (49.42 s)
Silver 银牌	USA 美国	Chris Kappler	Royal Kaliber	8.0* (ret. 退出)
Bronze 铜牌	GER 德国	Marco Kutscher	Montender 2	9.0*
2008 Hong Kong 香港				
Gold 金牌	Canada 加拿大	Eric Lamaze	Hickstead	0.0*; 0.0** (38.29)
Silver 银牌	Sweden 瑞典	Rolf-Goran Bengtsson	Ninja	0.0*; 4.0** (38.29)
Bronze 铜牌	USA 美国	Beezie Madden	Authentic	4.0*; 0.0** (35.25)
2012 LONDON 伦敦				
Gold 金牌	Switzerland 瑞士	Steve Guerdat	Nino des Buisson-nets	0.0*
Silver 银牌	Holland 荷兰	Gerco Schroder	London	1.0*; 0.0** (49.79 s)
Bronze 铜牌	Ireland 爱尔兰	Cian O'Connor	Blue Loyd 12	1.0*; 4.0** (46.64 s)

(续)

Medal 奖牌	Country 国家	Rider 骑手	Mount 坐骑	Time 时间
2016 Rio de Janeiro 里约热内卢				
Gold 金牌	UK 英国	Nick Skelton	Big Star	0.0 * ; 0.0 ** (42.82)
Silver 银牌	Sweden 瑞典	Peder Fredericson	All In	0.0 * ; 0.0 ** (43.35 s)
Bronze 铜牌	Canada 加拿大	Eric Lamaze	Fine Lady 5	0.0 * ; 4.0 ** (42.09 s)

Note: * *total faults in the first two rounds*; ** *faults in jump-off (time/s in parentheses)*

注：* 表示前两轮总罚分；** 表示决胜赛罚分(括号中为时间/秒)

Table 4 Show Jumping Team Competition
表4 团体场地障碍赛奖牌榜

Medal 奖牌	Country 国家	Riders 骑手	Mounts 坐骑	Faults 罚分
1912 STOCKHOLM 斯德哥尔摩				
Gold 金牌	Sweden 瑞典	Carl-Gustaf Lewenhaupt	Medusa	2.00
		Gustaf Kilman	Gatan	10.00
		Hans von Rosen	Lord Iron	13.00
Silver 银牌	France 法国	Michel d'Astafort	Amazone	5.00
		jean Cariou	Mignon	8.00
		Bernard Meyer	Allons-y	19.00
Bronze 铜牌	Germany 德国	Sigismund Freyer	Ultimus	9.00
		Wilhelm Graf von Hohenau	Pretty Girl	13.00
		Ernst-Hubertus Deloch	Hubertus	18.00
1920 ANTWERP 安特卫普				
Gold 金牌	Sweden 瑞典	Claes Konig	Tresor	2.00
		Daniel Norting	Eros Ⅱ	6.00
		Hans von Rosen	Poor Boy	6.00

（续）

Medal 奖牌	Country 国家	Riders 骑手	Mounts 坐骑	Faults 罚分
Silver 银牌	Belgium 比利时	Henri Laame	Biscuit	2.75
		André Coumans	Lisette	5.25
		Herman de Gaiffier d'Herstroy	Miss	8.25
Bronze 铜牌	Italy 意大利	Ettore Caffaratti	Traditore	1.50
		Alessandro Alvisi	Raggio di Sole	6.25
		Giulio Cacciandra	Fortunello	11.00
1924 PARIS 巴黎				
Gold 金牌	Sweden 瑞典	ÅkeThelning	Loke	12.00
		Axel Ståhle	Cecil	12.25
		Åge Lundstrom	Anveis	18.00
Silver 银牌	Switzerland 瑞士	Alphonse Gemuseus	Lucette	6.00
		Werner Stuber	Girandole	20.00
		Hans Biihler	Sailor Boy	24.00
Bronze 铜牌	Portugal 葡萄牙	Antonio Borges de Almeida	Reginald	12.00
		Hélder de Souza Martins	Avro	19.00
		José Mousinho de Albuquerque	Hetrugo	22.00
1928 AMSTERDAM 阿姆斯特丹				
Gold 金牌	Spain 西班牙	José Navarro Morenés	Zapatazo	0.00
		Marquez de los Trujillos	Zalamero	2.00
		Julio Garcia Fernandez	Revistada	2.00
Silver 银牌	Poland 波兰	Kazimierz Gzowski	Mylord	0.00
		Kazimierz Szosland	Ali	2.00
		Michai Antoniewicz	Readgleadt	6.00
Bronze 铜牌	Sweden 瑞典	Karl Hansen	Gerold	0.00
		Carl Björnstierna	Kornett	2.00
		Ernst Hallberg	Loke	8.00

（续）

Medal 奖牌	Country 国家	Riders 骑手	Mounts 坐骑	Faults 罚分
1932 LOS ANGELES 洛杉矶				
No team medals awarded becaused no team of three riders completed the course. 本届团体障碍赛中参赛团队都没能完成比赛，所以没有颁发奖牌。				
1936 BERLIN 柏林				
Gold 金牌	Germany 德国	Kurt Hasse	Tora	4.00
		Marten von Barnekow	Nordland	20.00
		Heinz Brandt	Alchimist	20.00
Silver 银牌	Holland 荷兰	Johan Jacob Greter	Ernica	12.00
		Jan Adrianus de Bruine	Trixie	15.00
		Henri Louis van Schaik	Santa Bell	24.50
Bronze 铜牌	Portugal 葡萄牙	José Beltrao	Biscuit	12.00
		Marquez de Funchal	Merle Blanc	20.00
		Luis Menae Sitva	Faussette	24.00
1948 LONDON 伦敦				
Gold 金牌	Mexico 墨西哥	Humberto Mariles Cortés	Arete	6.25
		Rubén Uriza	Harvey	8.00
		Alberto Valdes	Chihuchoc	20.00
Silver 银牌	Spain 西班牙	Jaime Garcia Cruz	Bizarro	12.00
		José Navarro Morenés	Quorum	20.00
		Marcelino Gavilán	Forajido	24.50
Bronze 铜牌	Great Britain 英国	Harry Llewellyn	Foxhunter	16.00
		Henry Nicoll	Kilgeddin	16.00
		Arthur Carr	Monty	35.00
1952 HELSINKI 赫尔辛基				
Gold 金牌	Great Britain 英国	Wilfred White	Nizefella	8.00
		Douglas Stewart	Aherlow	16.00
		Henry Llewellyn	Foxhunter	16.75

（续）

Medal 奖牌	Country 国家	Riders 骑手	Mounts 坐骑	Faults 罚分
		Oscar Cristi	Bambi	8.00
Silver 银牌	Chile 智利	Cesar Mendoza	Pfllan	12.00
		Ricardo Echeverria	Undo Peal	25.75
		William Steinkraus	Hollandia	13.25
Bronze 铜牌	USA 美国	Arthur McCashin	Miss Budweiser	16.00
		John Russell	Democrat	23.00
1956 STOCKHOLM 斯德哥尔摩				
		Hans Gunter Winkler	Halla	4.00
Gold 金牌	Germany 德国	Fritz Thiedemann	Meteor	12.00
		Alfons Lutke-Westhues	Ala	24.00
		Raimondo d'Inzeo	Merano	8.00
Silver 银牌	Italy 意大利	Piero d'Inzeo	Uruguay	11.00
		Salvatore Oppes	Pagoro	47.00
		Wilfred White	Nizefella	12.00
Bronze 铜牌	Great Britain 英国	Patricia Smythe	Flanagan	21.00
		Peter Robeson	Scorchin	36.00
1960 ROME 罗马				
		Hans Gunter Winkler	Halla	9.25 * ,4.00 **
Gold 金牌	Germany 德国	Fritz Thiedemann	Meteor	8.00 * ,8.00 **
		Alwin Schockemohle	Ferdi	8.50 * ,8.75 **
		Frank Chapot	Trail Guide	8.00 * ,12.0 **
Silver 银牌	USA 美国	William Steinkraus	Ksar d'Esprit	12.5 * ,9.00 **
		George Morris	Sinjon	8.50 * ,16.0 **
		Raimondo d'Inzeo	Posilllipo	4.00 * ,4.00 **
Bronze 铜牌	Italy 意大利	Piero d'Inzeo	The Rock	24.0 * ,8.00 **
		Antonio Oppes	The Scholar	24.5 * ,16.0 **

（续）

Medal 奖牌	Country 国家	Riders 骑手	Mounts 坐骑	Faults 罚分
1964 TOKYO 东京				
Gold 金牌	Germany 德国	Hermann Schridde	Dozent	12.5 * ,1.25 **
		Kurt Jarasinski	Torro	9.75 * ,12.5 **
		Hans Giinter Winkler	Fidelitas	17.5 * ,15.0 **
Silver 银牌	France 法国	Pierre Jonquères d'Oriola	Lutteur B	9.00 * ,0.00 **
		janou Lefebvre	Kenavo D	16.0 * ,16.0 **
		Guy Lefrant	Monsieur de Littry	20.0 * ,16.75 **
Bronze 铜牌	Italy 意大利	Piero d'Inzeo	Sunbeam	12.0 * ,12.5 **
		Raimondo d'Inzeo	Posillipo	16.0 * ,12.0 **
		Graziano Mancinelli	Rockette	16.0 * ,20.0 **
1968 MEXICO CITY 墨西哥城				
Gold 金牌	Canada 加拿大	James Elder	The Immigrant	9.25,18.00
		James Day	Canadian Club	18.00,18.00
		Thomas Gayford	Big Dee	22.25,17.25
Silver 银牌	France 法国	Janou Lefebvre	Rocket	17.25,12.50
		Marcel Rozier	Quo Vadis	21.50,12.00
		Pierre Jonquères d'Oriola	Nagir	17.75,29.50
Bronze 铜牌	Germany 德国	Alwin Schockemöhle	Donald Rex	13.00,5.75
		Hans Giinter Winkler	Enigk	11.50,16.75
		Hermann Schridde	Dozent	33.75,36.50
1972 MUNICH 慕尼黑				
Gold 金牌	Germany 德国	Fritz Ligges	Robin	4.00,4.00
		Gerhard Wiltfang	Askan	8.00,4.00
		Hartwig Steenken	Simona	4.00,8.00
		Hans Giinter Winkler	Torphy	8.00,8.00

(续)

Medal 奖牌	Country 国家	Riders 骑手	Mounts 坐骑	Faults 罚分
		William Steinkraus	Main Spring	0. 00,4. 00
Silver 银牌	USA 美国	Neal Shapiro	Sloopy	8. 25,0. 00
		Kathryn Kusner	Fleet Apple	20. 0,12. 0
		Frank Chapot	White Lightning	8. 00,28. 0
Bronze 铜牌	Italy 意大利	Vittorio Orlandi	Fukner Feather	4. 00,4. 00
		Raimondo d'Inzeo	Fiorello Ⅱ	8. 00,4. 00
		Graziano Mancinelli	Ambassador	20. 0,8. 00
		Piero d'Inzeo	Easter Light	87. 25,48. 0
1976 MONTREAL 蒙特利尔				
Gold 金牌	France 法国	Hubert Parot	Rivage	8. 00,4. 00
		Marcel Rozier	Bayard de Maupas	8. 00,4. 00
		Marc Roguet	Belle de Mars	8. 00,16. 0
		Michel Roche	Un Espoir	24. 0,8. 00
Silver 银牌	Germany 德国	Alwin Schockemöhle	Warwick Rex	4. 00,8. 00
		Hans Gunter Winkler	Torphy	12. 0,4. 00
		Sönke Sonksen	Kwept	8. 00,12. 0
		Paul Schockemohle	Agent	16. 0,8. 00
Bronze 铜牌	Belgium 比利时	Eric Wauters	Gute Sitte	8. 00,7. 00
		Francois Mathy	Gai Luron	12. 0,8. 00
		Edgar Gupper	Le Champion	12. 0,16. 0
		Stanny van Paeschen	Porsche	16. 0,20. 0
1980 MOSCOW 莫斯科				
Gold 金牌	SOV 苏联	Vyacheslav Chukanov	Gepatit	4. 00,0. 00
		Viktor Pohanovsky	Topky	8. 00,0. 25
		Viktor Asmayev	Reis	4. 00,7. 25
		Nikolai Korolkov	Espadron	8. 00,4. 00

(续)

Medal 奖牌	Country 国家	Riders 骑手	Mounts 坐骑	Faults 罚分
		jan Kowalczyk	Artemor	4.00,8.00
Silver 银牌	Poland 波兰	Wieslaw Hartman	Norton	12.0,12.0
		Marian Kozicki	Bremen	33.5,4.00
		Janusz Bobik	Szampan	16.0,24.0
		Joaquin Perez Heras	Alymony	8.00,4.00
Bronze 铜牌	Mexico 墨西哥	Alberto Valdes Lacarra	Lady Mirka	8.00,12.25
		Gerardo Tazzer Valencia	Caribe	23.25,8.50
		Jesus Gomez Portugal	Massacre	27.25,8.00
1984 LOS ANGELES 洛杉矶				
		Joe Fargis	Touch of Class	0.00,0.00
Gold 金牌	USA 美国	Conrad Homfeld	Abdullah	8.00,0.00
		Leslie Burr	Albany	4.00,8.00
		Melanie Smith	Calypso	0.00,WD 退赛
		Michael Whitaker	Overton Amanda	8.00,0.00
Silver 银牌	Great Britain 英国	John Whitaker	Ryans Son	16.0,4.75
		Steven Smith	Shining Example	19.0,8.00
		Timothy Grubb	Linky	0.00,28.25
		Paul Schockemöhle	Deister	4.00,4.00
Bronze 铜牌	Germany 德国	Peter Luther	Livius	8.00,4.00
		Franke Sloothaak	Farmer	8.00,11.25
		Fritz Ligges	Ramzes	17.0,12.0
1988 SEOUL 首尔(汉城)				
		Ludger Beerbaum	The Freak	0.25,4
Gold 金牌	Germany 德国	Wolfgang Brinkmann	Pedro	9.00,1
		Dirk Hafemeister	Orchidee	4,8
		Franke Sioothaak	Walzerkönig	0,-

（续）

Medal 奖牌	Country 国家	Riders 骑手	Mounts 坐骑	Faults 罚分
		Joe Fargis	Mill Pearl	4,0. 25
Silver 银牌	USA 美国	Greg Best	Gem Twist	4,4
		Lisa Jacquin	For the Moment	4. 25,4
		Anne Kursinski	Starman	8,8
		Pierre Durand	Jappeloup	5,0
Bronze 铜牌	France 法国	Michel Robert	La Fayette	5. 25,4. 75
		Frédéric Cottier	Flambeau C	12. 0,4. 00
		Hubert Bourdy	Morgat	8. 50,8. 00
1992 BARCELONA 巴塞罗那				
		Jos Lansink	Egano	0,0
Gold 金牌	Netherlands 荷兰	Piet Raymakers	Ratina	0,4
		Jan Tops	Top Gun	4,4
		Bert Romp	Waldo	14. 50,24. 75
		Thomas Fruhrmann	Genius	0,0
Silver 银牌	Australia 澳大利亚	Hugo Simon	Apricot D	0,4
		Jörg Munzner	Graf Grande	4. 25,8. 50
		Boris Boor	Love Me Tender	WD 退赛
		Hervé Godignon	Quidam de Revei	4. 75,0
Bronze 铜牌	France 法国	Hubert Bourdy	Razzia du Poncel	8,4
		Eric Navet	Quito de Baussy	4,12
		Michel Robert	Nonix	0,20. 25
1996 ATLANTA 亚特兰大				
		Franke Sioothaak	Joli Coeur	60. 25,0
Gold 金牌	Germany 德国	Ludger Beerbaum	Ratina Z	0,0. 25
		Ulrich Kirchhoff	Jus dePommes	0. 75,0. 75
		Lars Neiberg	For Pleasure	0,12

（续）

Medal 奖牌	Country 国家	Riders 骑手	Mounts 坐骑	Faults 罚分
		Leslie Burr-Howard	Extreme	14,0
Silver 银牌	USA 美国	Peter Leone	Legato	4,0
		Michael Matz	Rhum IV	4,4
		Anne Kursinski	Eros	0,8
		Rodrigo Pessoa	Tomboy	0,0.75
Bronze 铜牌	Brazil 巴西	Luiz Felipe Azevedo	Cassiana	8,4
		Alvaro Miranda Neto	Aspen	0.25,8
		Andre Johannpeter	Calei	4.25,8
2000 SYDNEY 悉尼				
		Ludger Beerbaum	Goldfever	20 * ; 16.25 **
Gold 金牌	Germany 德国	Lars Nieberg	Esprit Frh	8 * ; 0 **
		Marcus Ehning	For Pleasure	0 * ; 7 **
		Otto Becker	Cento	0 * ; 0 **
		Markus Fuchs	Tinka's Boy	0 * ; 8 **
Silver 银牌	Switzerland 瑞士	Beat Maendli	Pozitano	8 * ; 0 **
		Lesley McNaught	Dulf	15 * ; 8.5 **
		Willi Melliger	Calvaro V	0 * ; 0 **
		Rodrigo Pessoa	Baloubet Du Rouet	0 * ; 0 **
Bronze 铜牌	Brazil 巴西	Luiz Felipe De Azevedo	Ralph	8 * ; 0 **
		Alvaro Miranda Neto	Aspen	4 * ; 12 **
		Andre Johannpeter	Calei	8 * ; 16 **
2004 ATHENS 雅典				
		Ludger Beerbaum	Goldfever	0 * ; 0 **
Gold 金牌	Germany 德国	Marco Kutscher	Montender	0 * ; 0 **
		Otto Becker	Cento	5 * ; 4 **
		Christian Ahlmann	Coster	4 * ; 8 **

（续）

Medal 奖牌	Country 国家	Riders 骑手	Mounts 坐骑	Faults 罚分
Silver 银牌	USA 美国	Beezie Madden	Authentic	0*; 0**
		Chris Kappler	Royal Kaliber	0*; 4**
		McLain Ward	Sapphire	8*; 8**
		Peter Wylde	Fein Cera	12*; 12**
Bronze 铜牌	Sweden 瑞典	Rolf-Goran Bengtsson	Mac Kinley	0*; 0**
		Peder Fredericson	Magic Bengtsson	8*; 4**
		Malin Baryard	Butterfly Flip	8*; 4**
		Peter Eriksson	Cardento	4*; 10**
2008 BEIJING 北京				
Gold 金牌	USA 美国	Mclain Ward	Sapphire	0*,4**
		Laura Kraut	Cedric	4*,0**
		Will Simpson	Carlsson Vom Dach	8*,8**
		Beezie Madden	Authentic	11*,4**
Silver 银牌	Canada 加拿大	Jill Henselwood	Special Ed	18*,0**
		Eric Lamaze	Hickstead	0*,4**
		Ian Millar	In Style	4*,0**
Bronze 铜牌	Norway 挪威	Stein Endresen	Le Beau	4*,12**
		Morten Djupvik	Casino	12*,4**
		Geir Gulliksen	Cattani	12*,5**
		Tony Andre Hansen	Camiro	1*,1**
2012 LONDON 伦敦				
Gold 金牌	Great Britian 英国	Scott Brash	Hello Sanctos	4; 0; 4(48.01)
		Peter Charles	Vindicat	8; 5; 0(61.27)
		Ben Maher	Tripple X	0; 4; 0(48.14)
		Nick Skelton	Big Star	0; 0; 0(47.27)

（续）

Medal 奖牌	Country 国家	Riders 骑手	Mounts 坐骑	Faults 罚分
Silver 银牌	Netherlands 荷兰	Marc Houtzager	Tamino	0; 0; 4(52. 40)
		Gerco Schroder	London	4; 4; -
		Maikel Van Der Vleuten	Verdi	0; 0;8(48. 18)
		Jur Vrieling	Bubalo	8; 8; 0(48. 54)
Bronze 铜牌	Saudi Arabia 沙特阿拉伯	Ramzy Al Duhami	Bayard van de Villa There	0; 4
		Abdullah Al Saud	Davos	0; 4
		Kamal Bahamdan	Noblesse des Tess	1; 5
		Abdullah Waleed Sharbatly	Sultan	4; 6
2016 Rio de Janeiro 里约热内卢				
Gold 金牌	France 法国	Philippe Rozier	Rahotep de Toscane	4; 1
		Kevin Staut	Reveur de Hurtebise	0; 0
		Roger-Yves Bost	Sydney Une Prince	1; 1
		Pénélope Leprevost	Flora de Mariposa	0;NS 不计
Silver 银牌	USA 美国	Kent Farrington	Voyeur	0;1
		Lucy Davis	Barron	0; 4
		McLain Ward	HH Azur	0; 0
		Elizabeth Madden	Cortes ‘C’	8; WD 退赛
Bronze 铜牌	Germany 德国	Daniel Deusser	First Class	0; 4; 0(42. 68)
		Meredith Beerbaum	Fibonacci	0;5; 0(55. 35)
		Ludger Beerbaum	Casello	0; 4; 0(44. 58)
		Christian Ahlmann	Taloubet Z	4; 0; NS 不计

Note: * *faults in first round*; ** *faults in second round.*

注：* 表示首轮罚分；** 表示第二轮罚分。

Table 5 Three-Day Event Individual Competion

表5 个人三日赛奖牌榜

Medal 奖牌	Country 国家	Rider 骑手	Mount 坐骑	Points 得分
1912 STOCKHOLM 斯德哥尔摩				
Gold 金牌	Sweden 瑞典	Axel Nordlander	Lady Artist	46. 59
Silver 银牌	Germany 德国	Friedrich von Rochow	Idealist	46. 42
Bronze 铜牌	France 法国	Jean Cariou	Cocotte	46. 32
1920 ANTWERP 安特卫普				
Gold 金牌	Sweden 瑞典	Graf Helmer Mörner	Germania	1775. 00
Silver 银牌	Sweden 瑞典	Åge Lundström	Yrsa	1738. 75
Bronze 铜牌	Italy 意大利	Ettore Caffaratti	Caniche	1733. 75
1924 PARIS 巴黎				
Gold 金牌	Holland 荷兰	Adolph van der Voort van Zijp	Piece	1976. 00
Silver 银牌	Denmark 丹麦	Frode Kirkebjerg	Meteor	1853. 50
Bronze 铜牌	USA 美国	Sloan Doak	Pathinder	1845. 50
1928 AMSTERDAM 阿姆斯特丹				
Gold 金牌	Holland 荷兰	Chailes Pahud de Mortanges	Marcroix	1969. 82
Silver 银牌	Holland 荷兰	Gerard de Kruyff	Va-t-en	1967. 26
Bronze 铜牌	Germany 德国	Bruno Neumann	Ilja	1944. 42

(续)

Medal 奖牌	Country 国家	Rider 骑手	Mount 坐骑	Points 得分
1932 LOS ANGELES 洛杉矶				
Gold 金牌	Holland 荷兰	Charles Pahud de Mortanges	Marcroix	1813. 833
Silver 银牌	USA 美国	Earl Thomson	Jenny Camp	1811. 000
Bronze 铜牌	Sweden 瑞典	Clarence von Rosen	Sunnyside Maid	1809. 416
1936 BERLIN 柏林				
Gold 金牌	Germany 德国	Ludwig Stubbendorff	Nurmi	-37. 70
Silver 银牌	USA 美国	Earl Thomson	Jenny Camp	-99. 90
Bronze 铜牌	Denmark 丹麦	Hans Mathiesen-Lunding	Jason	-102. 20
1948 LONDON 伦敦				
Gold 金牌	France 法国	Bernard Chevallier	Aiglonne	+4. 00
Silver 银牌	USA 美国	Frank Henry	Swing Low	-21. 00
Bronze 铜牌	Sweden 瑞典	Robert Selfelt	Claque	-25. 00
1952 HELSINKI 赫尔辛基				
Gold 金牌	Sweden 瑞典	Hans von Blixen-Finecke, Jr.	Jubal	-28. 33
Silver 银牌	France 法国	Guy Lefrant	Verdun	-54. 50
Bronze 铜牌	Germany 德国	Wilhelm Busing	Hubertus	-55. 50

(续)

Medal 奖牌	Country 国家	Rider 骑手	Mount 坐骑	Points 得分
1956 STOCKHOLM 斯德哥尔摩				
Gold 金牌	Sweden 瑞典	Petrus Kastenman	Iluster	-66.53
Silver 银牌	Germany 德国	August Lütke-Westhues	Trux von Kamax	-84.87
Bronze 铜牌	UK 英国	Francis Weldon	Kilbarry	-85.48
1960 ROME 罗马				
Gold 金牌	AUS 澳大利亚	Lawrence Morgan	Salad Days	+7.15
Silver 银牌	AUS 澳大利亚	Neale Lavis	Mirrabooka	-16.50
Bronze 铜牌	SWI 瑞士	Anton Buihler	Gay Spark	-51.21
1964 TOKYO 东京				
Gold 金牌	Italy 意大利	Mauro Checcoli	Surbean	+64.40
Silver 银牌	ARG 阿根廷	Carlos Moratorio	Chalan	+56.40
Bronze 铜牌	Germany 德国	Fritz Ligges	Donkosak	+49.20
1968 MEXICO CITY 墨西哥城				
Gold 金牌	France 法国	Jean-Jacques Guyon	Pitou	-38.86
Silver 银牌	UK 英国	Derek Allhusen	Lochinvar	-41.61
Bronze 铜牌	USA 美国	Michael Page	Foster	-52.31

（续）

Medal 奖牌	Country 国家	Rider 骑手	Mount 坐骑	Points 得分
1972 MUNICH 慕尼黑				
Gold 金牌	UK 英国	Richard Meade	Laurieston	+57. 73
Silver 银牌	Italy 意大利	Alessandro Argenton	Woodland	+43. 33
Bronze 铜牌	Sweden 瑞典	Jan Jonsson	Sarajevo	+39. 67
1976 MONTREAL 蒙特利尔				
Gold 金牌	USA 美国	Edimund “Tad” Coffin	Bally Cor	-114. 99
Silver 银牌	USA 美国	J. Michael Plumb	Better and Better	-125. 85
Bronze 铜牌	Germany 德国	Karl Schultz	Madrigal	-129. 45
1980 MOSCOW 莫斯科				
Gold 金牌	Italy 意大利	Euro Federico Roman	Rossinan	-108. 60
Silver 银牌	SOV 苏联	Aleksandr Binov	Galzun	-120. 80
Bronze 铜牌	SOV 苏联	Yuri Salnikov	Pintset	-151. 60
1984 LOS ANGELES 洛杉矶				
Gold 金牌	NZE 新西兰	Mark Todd	Charisma	-51. 60
Silver 银牌	USA 美国	Karen Stives	Ben Arthur	-54. 20
Bronze 铜牌	UK 英国	Virginia Holgate	Priceless	-56. 80

(续)

Medal 奖牌	Country 国家	Rider 骑手	Mount 坐骑	Points 得分
1988 SEOUL 首尔(汉城)				
Gold 金牌	NZE 新西兰	Mark Todd	Charisma	-42.60
Silver 银牌	UK 英国	Ian Stark	Sir Wattle	-52.80
Bronze 铜牌	UK 英国	Virginia Holgate Leng	Master Craftsman	-62.00
1992 BARCELONA 巴塞罗那				
Gold 金牌	AUS 澳大利亚	Matthew Ryan	Kibah Tic Toc	-70.00
Silver 银牌	GER 德国	Herbert Blöcker	Feine Dame	-81.30
Bronze 铜牌	NZE 新西兰	Blyth Tait	Messiah	-87.60
1996 ATLANTA 亚特兰大				
Gold 金牌	NZE 新西兰	Blyth Tait	Ready Teddy	-56.80
Silver 银牌	NZE 新西兰	Sally Clark	Squirrel Hill	-60.40
Bronze 铜牌	USA 美国	Kerry Millikin	Out and About	-73.70
2000 SYDNEY 悉尼				
Gold 金牌	USA 美国	David O'Connor	Custom Made	34.00
Silver 银牌	AUS 澳大利亚	Andrew Hoy	Swizzle In	39.80
Bronze 铜牌	NZE 新西兰	Mark Todd	Eyespy Ⅱ	42.00

（续）

Medal 奖牌	Country 国家	Rider 骑手	Mount 坐骑	Points 得分
2004 ATHENS 雅典				
Gold 金牌	UK 英国	Leslie Law	Shear L'eau	44. 4
Silver 银牌	USA 美国	Kimberly Severson	Winsome Andante	45. 2
Bronze 铜牌	UK 英国	Pippa Funnell	Primmore's Pride	46. 6
2008 BEIJING 北京				
Gold 金牌	Germany 德国	Hinrich Romeike	Marius	54. 20
Silver 银牌	USA 美国	Gina Miles	McKinlaigh	56. 10
Bronze 铜牌	GBR 英国	Christina Cook	Miners Frolic	57. 40
2012 LONDON 伦敦				
Gold 金牌	Germany 德国	Michael Jung	Sam	40. 60
Silver 银牌	Sweden 瑞典	Sara Algottson Ostholt	Wega	43. 30
Bronze 铜牌	Germany 德国	Sandra Auffarth	Opgun Louvo	44. 80
2016 Rio de Janeiro 里约热内卢				
Gold 金牌	Germany 德国	Michael Jung	Sam FBW	40. 90
Silver 银牌	France 法国	Astier Nicolas	Piaf de B'Neville	48. 00
Bronze 铜牌	USA 美国	Phillip Dutton	Mighty Nice	51. 80

Table 6 Three-Day Event Team Competion

表6 团体三日赛奖牌榜

Medal 奖牌	Country 国家	Riders 骑手	Mounts 坐骑	Points 得分
1912 STOCKHOLM 斯德哥尔摩				
Gold 金牌	Sweden 瑞典	Axel Nordlander	Lady Aitist	46. 59
		Nils Adlercreutz	Atout	46. 31
		Ernst Casparsson	Irmelin	46. 16
Silver 银牌	Germany 德国	Friedrich von Rochow	Idealist	46. 42
		Richard Graf von Schaesberg-Tannheim	Grundsee	46. 16
		Eduard von Lutcken	Blue Boy	45. 90
Bronze 铜牌	USA 美国	Benjamin Lear	Poppy	45. 91
		John Montgomery	Deceive	45. 88
		Guy Henry, Jr.	Chiswell	45. 54
1920 ANTWERP 安特卫普				
Gold 金牌	Sweden 瑞典	Graf Hehner Mörrner	Germania	1775. 00
		Åge Lundström	Yrsa	1738. 75
		Georgvon Biaun	Diana	1543. 75
Silver 银牌	Italy 意大利	Ettore Caffaratti	Caniche	1733. 75
		Garibaldi Spighl	Otllo	1647. 50
		Giulio Cacciandra	Facetto	1353. 75
Bronze 铜牌	Belgium 比利时	Roger Moeremans d'Emaus	Sweet Girt	1652. 50
		Oswald Lints	Martha	1515. 00
		Jules Bonvaiet	Weppelghem	1392. 50
1924 PARIS 巴黎				
Gold 金牌	Holland 荷兰	Adolph van der Voort van Zijp	Silver Piece	1976. 00
		Charles Pahud de Mortanges	Johnny Walker	1828. 00
		Gerard de Kruyff	Addio	1493. 50
Silver 银牌	Sweden 瑞典	Claës König	Bojar	1730. 00
		Carl Torsten Sylvan	Amita	1678. 00
		Gustaf Hagelin	Varius	1335. 50

（续）

Medal 奖牌	Country 国家	Riders 骑手	Mounts 坐骑	Points 得分
Bronze 铜牌	Italy 意大利	Alberto Lombardi	Pimplo	1572. 00
		Alessandro Alvsi	Capiligio	1536. 00
		Emanuele di Pralormo	Mount Felix	1404. 50
1928 AMSTERDAM 阿姆斯特丹				
Gold 金牌	Holland 荷兰	Charles Pahud de Mortanges	Marcroix	1969. 82
		Gerard de Kruyff	Va-t-en	1967. 26
		Adolph van der Voort van Zijp	Silver Piece	1928. 60
Silver 银牌	Norway 挪威	Bjart Ording	And Over	1912. 98
		Arthur Quist	Hildalgo	1895. 14
		Eugen johansen	Baby	1587. 56
Bronze 铜牌	Poland 波兰	Michal Antoniewicz	Moja Mita	1822. 50
		Józef Trenkwald	Lwi Pazur	1645. 20
		Karol Rommel	Doneuse	1600. 22
1932 LOS ANGELES 洛杉矶				
Gold 金牌	USA 美国	Earl Thomson	Jenny Camp	1811. 000
		Harry Chamberlin	Pleasant Smiles	1687. 833
		Edwin Argo	Honolulu Tomboy	1539. 250
Silver 银牌	Holland 荷兰	Charles Pahud de Mortanges	Marcroix	1813. 833
		Karel Johan Schummelketel	Duiveltje	1614. 500
		Aemout van Lennep	Henk	1260. 750
Bronze 铜牌	No medal awarded. 没有颁发铜牌。			
1936 BERLIN 柏林				
Gold 金牌	Germany 德国	Ludwig Stubbendorff	Nurmi	-37. 70
		Rudolf Lippert	Fasan	-111. 60
		Konrad Freiherr von Wangenheim	Kurfurst	-527. 35
Silver 银牌	Poland 波兰	Henryk Rojcewicz	Arlekinlll	-253. 00
		Zdzislaw Kawecki	Bambino	-300. 70
		Seweryn Kulesza	Tosca	-438. 00

（续）

Medal 奖牌	Country 国家	Riders 骑手	Mounts 坐骑	Points 得分
Bronze 铜牌	UK 英国	Alec Scott	Bob Clive	−117.30
		Edward Howard-Vyse	Blue Steel	−324.00
		Richard Fanshawe	Bowie Knife	−8754.20
1948 LONDON 伦敦				
Gold 金牌	USA 美国	Frank Henry	Swing Low	−21.00
		Charles Anderson	Reno Palisade	−26.50
		Earl Thomson	Reno Rhythm	−114.00
Silver 银牌	Sweden 瑞典	Robert Selfelt	Claque	−25.00
		Olof Stahre	Komet	−70.00
		Sigurd Svensson	Dust	−70.00
Bronze 铜牌	Mexico 墨西哥	Humberto Mariles Cortés	Parral	−61.75
		Raúl Campero	Tarahumara	−120.50
		Joaquin Solano Chagoya	Malinche	−123.00
1952 HELSINKI 赫尔辛基				
Gold 金牌	Sweden 瑞典	Hans von Blixen-Finecke, Jr.	Jubal	−28.33
		Olof Stahre	Komet	−69.41
		Karl Folke Frolen	Fair	−124.20
Silver 银牌	Germany 德国	Wilhelm Büsing	Hubertus	−55.50
		Klaus Wagner	Dachs	−65.66
		Otto Rothe	Trux von Kamax	−114.33
Bronze 铜牌	USA 美国	Charles Hough	Cassivellannus	−70.66
		Walter Staley	Craigwood Park	−168.50
		John Wofford	Benny Grimes	−348.00
1956 STOCKHOLM 斯德哥尔摩				
Gold 金牌	UK 英国	Francis Weldon	Kilbarry	−85.48
		A. Laurence Rook	Wild Venture	−119.64
		Albert Hill	Countryman Ⅲ	−150.36

（续）

Medal 奖牌	Country 国家	Riders 骑手	Mounts 坐骑	Points 得分
Silver 银牌	Germany 德国	August Lütke-Westhues	Trux von Kamax	-84.87
		Otto Rothe	Sissi	-158.04
		Klaus Wagner	Prinze B	-233.00
Bronze 铜牌	Canada 加拿大	John Rumble	Cilroy	462.53
		James Eider	Colleen	-193.69
		Brian Herbinson	Tara	-216.50
1960 ROME 罗马				
Gold 金牌	Australia 澳大利亚	Lawrence Morgan	Salad Days	+7.15
		Neale Lavis	Mirrabooka	-16.50
		J. William Roycroft	Our Solo	-118.83
Silver 银牌	Switzerland 瑞士	Anton Biihler	Gay Spark	-51.21
		Hans Schwarzenbach	Burn Trout	-131.45
		Rudolf Gunthardt	Atbara	-203.36
Bronze 铜牌	France 法国	Jack Le Goff	Image	-72.91
		Guy Lefrant	Nicias	-208.50
		Jean Raymond Le Roy	Gardern	-234.30
1964 TOKYO 东京				
Gold 金牌	Italy 意大利	Mauro Checcoli	Surbean	+64.40
		Paolo Angioni	King	+17.87
		Giuseppe Ravano	Royal Love	+3.53
Silver 银牌	USA 美国	Michael Page	Grasshopper	+47.40
		Kevin Freeman	Gallopade	+17.13
		J. Michael Plumb	Bold Minstrel	+1.33
Bronze 铜牌	Germany 德国	Fritz Ligges	Do nkosak	+49.20
		Horst Karsten	Condora	+36.60
		Gerhard Schulz	Balza X	-29.07

（续）

Medal 奖牌	Country 国家	Riders 骑手	Mounts 坐骑	Points 得分
1968 MEXICO CITY 墨西哥城				
Gold 金牌	UK 英国	Derek Allhusen	Lochinvar	−41.61
		Richard Meade	Comishman V	−64.46
		Reuben Jones	The Poacher	−69.86
Silver 银牌	USA 美国	Michael Page	Foster	−52.31
		James Wofford	Kilkenny	−74.06
		J. Michael Plumb	Plain Sailing	−119.50
Bronze 铜牌	Australia 澳大利亚	Wayne Roycroft	Zhivago	−95.05
		Brien Cobcroft	Depeche	−108.76
		J. William Roycroft	Warrathoola	−127.55
1972 MUNICH 慕尼黑				
Gold 金牌	UK 英国	Richard Meade	Laurieston	+57.73
		Mary Gordon Watson	Cornishman V	+30.27
		Bridget Parker	CornishGold	+7.53
Silver 银牌	USA 美国	Kevin Freeman	Good Mixture	+29.87
		Bruce Davidson	Plain Sailing	+24.47
		J. Michael Plumb	Free and Easy	−43.53
Bronze 铜牌	Germany 德国	Harry Kiugmann	Christopher Rob	+8.00
		Ludwig Goessing	Chikago	−0.40
		Karl Schultz	Pisco	−25.60
1976 MONTREAL 蒙特利尔				
Gold 金牌	USA 美国	Edmund Coffin	Bally Cor	−114.99
		J. Michael Plumb	Better and Better	−125.85
		Bruce Davidson	Irish Cap	−200.16
Silver 银牌	Germany 德国	Karl Schultz	Madrigal	−129.45
		Herbert Blocker	Albrant	−213.15
		Helmut Rethemeier	Pauline	−242.00

（续）

Medal 奖牌	Country 国家	Riders 骑手	Mounts 坐骑	Points 得分
Bronze 铜牌	Australia 澳大利亚	Wayne Roycroft	Laurenson	-178.04
		Mervyn Bennett	Regal Reign	-206.04
		J. William Roycroft	Version	-215.46
1980 MOSCOW 莫斯科				
Gold 金牌	SOV 苏联	Aleksandr Blinov	Galzun	-120.80
		Yuri Salnikov	Pintset	-151.60
		Vaiery Volkov	Tskheti	-184.60
Silver 银牌	Italy 意大利	Euro Federico Roman	Rossinan	-108.60
		Anna Casagrande	Daleye	-266.20
		Mauro Roman	Dourakine 4	-281.40
Bronze 铜牌	Mexico 墨西哥	Manuel Mendivil Yocupicio	Remember	-319.75
		David Barcena Rios	Bombon	-362.50
		Jose Luis Perez Soto	Quelite	-490.60
1984 LOS ANGELES 洛杉矶				
Gold 金牌	USA 美国	Karen Stives	Ben Arthur	-54.20
		Torrance Watkins Fleischmann	Finvarra	-60.40
		J. Michael Plumb	Blue Stone	-71.40
Silver 银牌	UK 英国	Virginia Holgate	Priceless	-56.80
		Lucinda Green	Regal Realm	-63.80
		Ian Stark	Oxford Blue	-68.60
Bronze 铜牌	Germany 德国	Dietmar Hogrefe	Foliant	-74.40
		Bettina Overesch	Peacetime	-79.60
		Claus Erhorn	Fair Lady	-80.00
1988 SEOUL 首尔（汉城）				
Gold 金牌	Germany 德国	Claus Erhorn	Juslyn Thyme	-62.35
		Matthias Baumann	Shamrock	-68.80
		Thies Kaspareit	Sherry	-94.80

（续）

Medal 奖牌	Country 国家	Riders 骑手	Mounts 坐骑	Points 得分
Silver 银牌	UK 英国	Ian Stark	Sir Wattle	-52.80
		Virginia Holgate Leng	Master Craftsman	-62.00
		Karen Straker	Get Smart	-142.00
Bronze 铜牌	New Zealand 新西兰	Mark Todd	Charisma	-42.60
		Judith"Tinks" Pottinger	Volunteer	-65.80
		Andrew Bennie	Grayshott	-162.80
1992 BARCELONA 巴塞罗那				
Gold 金牌	Australia 澳大利亚	Matthew Ryan	Kibah Tic Toe	-70.00
		Andrew Hoy	Kiwi	-89.40
		Gillian Rolton	Peppermint Grove	-129.20
Silver 银牌	New Zealand 新西兰	Blyth Tait	Messiah	-87.60
		Vicky Latta	Chief	-87.80
		Andrew Nicholson	Spinning Rhombus	-115.40
Bronze 铜牌	Germany 德国	Herbert Blöcker	Feine Dame	-81.30
		Ralf Ehrenbrink	Kildare	-108.40
		Cord Mysegaes	Ricardo	-110.40
1996 ATLANTA 亚特兰大				
Gold 金牌	Australia 澳大利亚	Wendy Schaeffer	Sunburst	-61.00
		Phillip Dutton	True Blue Girdwood	-69.40
		Andrew Hoy	Darien Powers	-73.45
Silver 银牌	USA 美国	Wendy Schaeffer	Sunburst	-61.00
		Karen O'Connor	Biko	-105.60
		David O'Connor	Giltedge	-76.00
		Bruce Davidson	Heyday	-79.50
Bronze 铜牌	New Zealand 新西兰	Blyth Tait	Chesterfield	-70.10
		Andrew Nicholson	JageimeisterⅡ	-100.65
		Vaughn Jefferis	Bounce	-97.80

（续）

Medal 奖牌	Country 国家	Riders 骑手	Mounts 坐骑	Points 得分
2000 SYDNEY 悉尼				
Gold 金牌	Australia 澳大利亚	Phillip Dutton	House Doctor	63. 60
		Andrew Hoy	Darien Powers	45. 60
		Stuart Tinney	Jeepster	41. 00
		Matt Ryan	Kibah Sandstone	60. 20
Silver 银牌	GBR 英国	Ian Stark	Jaybee	WD 退赛
		Jeanette Brakewell	Over To You	61. 60
		Pippa Funnell	Supreme Rock	45. 40
		Leslie Law	Shear H2O	54. 00
Bronze 铜牌	USA 美国	Nina Fout	3 Magic Beans	86. 00
		Karen O'Connor	Prince Panache	43. 00
		David O'Connor	Giltedge	46. 80
		Linden Wiesman	Anderoo	DSQ 取消
2004 ATHENS 雅典				
Gold 金牌	France 法国	Nicolas Touzaint	Galan de Sauvagere	33. 4
		Jean Teulere	Espoir de la Mare	46. 4
		Didier Courreges	Débat D'Estruval	60. 6
		Cedric Lyard	Merveille	70. 6
Silver 银牌	GBR 英国	Philippa Funnell	Primmore's Pride	42. 6
		Leslie Law	Shear L'Eau	44. 4
		Mary King	King Solomon Ⅲ	56. 0
		Jeanette Brakewell	Over To You	57. 8
Bronze 铜牌	USA 美国	Kimberly Severson	Winsome Adante	41. 2
		Amy Tryon	Poggio Ⅱ	51. 8
		Darren Chiacchia	Windfall 2	52. 6
		John Williams	Carrick	60. 8

(续)

Medal 奖牌	Country 国家	Riders 骑手	Mounts 坐骑	Points 得分
2008 Hong Kong 香港				
Gold 金牌	GER 德国	Peter Thomsen	The Ghost of Hamish	102.90
		Frank Osthol	Mr. Medicott	57.80
		Andreas Dibowski	Butts Leon	57.20
		Ingrid Klimke	Abraxxas	54.70
		Hinrich Romeike	Marius	54.20
Silver 银牌	AUS 澳大利亚	Shane Rose	All Luck	70.50
		Sonja Johnson	Ringwould Jaguar	58.80
		Lucinda Fredericks	Headley Britannia	59.60
		Clayton Fredericks	Ben Along Time	57.40
		Megan Jones	Irish Jester	55.00
Bronze 铜牌	GBR 英国	Sharon Hunt	Tankers Town	95.10
		Daisy Dick	Spring Along	79.90
		William Fox-Pitt	Parkmore Ed	64.20
		Kristina Cook	Miners Frolic	57.40
		Mary King	Call Again Cavalier	64.10
2012 LONDON 伦敦				
Gold 金牌	Germany 德国	Peter Thomsen	Barny	71.70
		Dirk Schrade	King Artus	50.60
		Ingrid Klimke	Butts Abraxxas	48.30
		Sandra Auffarth	Opgun Luovo	44.80
		Michael Jung	Sam	40.60
Silver 银牌	GBR 英国	Nicola Wilson	Opposition Buzz	55.70
		Mary King	Imperial Cavalier	42.10
		Zara Phillips	High Kingdom	53.20
		Kristina Cook	Miners Frolic	43.00
		William Fox-Pitt	Lionheart	53.30

（续）

Medal 奖牌	Country 国家	Riders 骑手	Mounts 坐骑	Points 得分
Bronze 铜牌	New Zealand 新西兰	Jonelle Richards	Flintstar	71.70
		Jonathon Paget	Clifton Promise	52.90
		Carolina Powell	Lenamore	57.80
		Andrew Nicholson	Nereo	45.00
		Mark Todd	Campino	46.50
2016 Rio de Janeiro 里约热内卢				
Gold 金牌	Germany 德国	Ingrid Klimke	Hale-Bob Old	39.50
		Sandra Auffarth	Opgun Louvo	41.60
		Michael Jung	Sam FBW	40.90
		Julia Krajewski	Samourai du Thot	44.80
Silver 银牌	France 法国	Mathieu Lemoine	Bart L	39.20
		Thibaut Vallette	Qing du Briot	41.00
		Astier Nicolas	Piaf De B'Neville	42.00
		Karim Laghouag	Entebbe	43.40
Bronze 铜牌	Australia 澳大利亚	Chris Burton	Santano Ⅱ	37.60
		Shane Rose	CP Qualified	42.50
		Sam Griffiths	Paulank Brockagh	46.30
		Stuart Tinney	Pluto Mio	56.80

Appendix 7 China Horse Industry Association

The history of the China Horse Industry Association(CHIA) can be traced back to the month of March 1976. In October 2002, the Association acquired its present name by merging the National Horse Breeding Committee and China Stud Book Committee. The first council and the second council were respectively established on 24th Oct 2002 and 22nd Sept 2007, and the chairman is WuChangxin, fellow member of the Chinese Academy of Science.

The third CHIA General Assembly, held in March 2014, was elected, with JiaYouling, the first National Chief Veterinary Officer, as President and Yue GaoFeng as Secretary General. Buhe, the vice-chairman of the 8th and 9th Standing Committee of the National People's Congress, Simayi Maimaiti, the vice-chairman of the 10th Standing Committee of the National People's Congress, Wuyunqimuge, Vice Chairman of the 11th Standing Committee of the National People's Congress, and Wu Changxin, fellow member of the Chinese Academy of Science were invited to be the CHIA Honorary Presidents.

As one of the 4A social organizations of China, CHIA has an independent legislative status. In 2015, CHIA was awarded the title of "National Advanced Social Organization" by the Ministry of Civil Affairs. Besides being a full member of the International Stud Book Committee(ISBC) and an affiliate member of the Asian Racing Federation(ARF), a registering authority member of the World Arabian Horse Organization(WAHO), it is also one of the founders of the International Akhal-Teke Horse Association(IAHA) and an observer member of the European Trotting Union.

Committed to developing China's horse industry and scientifically managing the association's business, CHIA works hard to establish and maintain a healthy and harmonious environment for the industry. Based on the principles of honesty, fairness, and transparency, CHIA will try all means to provide the best services for its members, promote the development of the horse industry and the prosperity of Chinese horse culture, and contribute to the realization of the Chinese dream of national rejuvenation and economic prosperity.

CHIA is committed to the following missions: 1) to enact rules and regulations, assist government in managing horse industry, and promote the development of the industry; 2) to work out standards for horse feeding, diseases prevention, and professional training; 3) to maintain stud books for Chinese equine animals and carry out selective breeding and performance assessment in accordance with related rules; 4) to establish registration for Chinese horse breeds, and promote cooperation between Chinese and foreign registries; 5) to maintain the registration of Chinese Thoroughbreds in accordance with the ISBC regulations, and publish the annual China Stud Book accordingly; 6) to carry out performance assessment(speed, pulling power, and endurance) in accordance with the Animal Husbandry Law of the People's Republic of China(2005) and the Regulations of Breeding Stock Registration(2006); 7) to organize professional training, provide consultation, and promote product development, technology transfer, and professional education; 8) to organize exposition, fair, and formulate industrial planning design and feasibility study under governmental commission to meet the market requirements; 9) to organize and coordinate horse show, auction, import/export, races and events, and improve the Chinese horse culture in general; 10) to carry out drug tests for race horses in accordance with certain rules and by approval of the authority; 11) to publish periodicals, books, photos, and audio-visual products according to related rules and market needs.

During the past years, CHIA has not only initiatedand hosted the China Horse Culture Festival, the World Horse Culture Forum, the 2014 International Akhal-Teke Horse Association Special Conference, the 15th World Equine Veterinary Association Conference, China Grand National Racing, the CHIA Beijing Time, the CHIA Conference on Horses at Tianshan Mountain, the CHIA Professional Training in Guangdong, the CHIA Conference in Shanghai, and the National Forum of Sec-Generals but also proposed to set September 19 as the annual "Horse Lover's Day" in China.

CHIA participated in the formulation of the National Equine Industry Development Plan (2020 – 2025), the National Seed Industry Development for 14th Five-Year Plan (Horse and Donkey), and the horse industry development plan in Xinjiang Autonomous Region and the Inner Mongolia Autonomous Region, and promoted the National People's Political Consultative Conference's "Proposal on Accelerating the Transformation and Upgrading of China's Modern Horse Industry". Aiming to promote the horse industry in China from the top, CHIA put forward the concept of Steed Diplomacy, advocated the founding of the National Horse Guards Parade, made great efforts in establishing the National Stud Farm, and endeavored to formulate and develop China's horse industry and racing events in the long term.

At the InternationalAkhal-Teke Horse Association Special Conference and China Horse Culture Festival hosted by China Horse Industry Association in May 2014, Chinese President Xi Jinping remarked as follows, Horse plays an important part in traditional Chinese culture, and horse culture in China has a long story to tell. At present we need exactly the power, spirit, and stamina of this noble animal for construction, development, and innovation in all areas. For me the horse is not only a symbol of perseverance and persistence but also a symbol of unfailing courage and endurance. Today, China is galloping forward with a faster pace in order to realize the nation's dream of great rejuvenation.

Abiding by the guidelines listed above, CHIA will carry forward the Family-country emotion, "Taking family as the country, country as the home, dream as the horse, share worries for the country", and make greater efforts together with individuals and institutions interested in the development of China's horse industry, and contribute more to the realization of the Chinese dream of national rejuvenation.

Address: China Horse Industry Association (Base of The National Animal Husbandry Station), No. 1 Area, Ma Xin Zhuang, Shunyi District, Beijing, P. R. China.

Tel: +86 – 010 – 60488801Fax: +86 – 010 – 60488665 E – Mail: international@ chinahorse. org

Website: www. chinahorse. org

附录7 中国马业协会

中国马业协会(简称“中国马会”),英文名称为 China Horse Industry Association,源起于 1976 年3月,1999 年9月23日在“全国马匹育种委员会”和“中国纯血马登记管理协会”的基础上登记成立了“中国马业协会”。第一届理事会于2002 年10月24日成立,第二届理事会于2007 年9月22日成立,理事长为中国科学院院士吴常信。

2014年3月29日,选举产生第三届中国马业协会理事会,会长为首任国家首席兽医官贾幼陵,秘书长为岳高峰。特别邀请第八届、第九届全国人大常委会副委员长布赫,第十届全国人大常委会副委员长司马义·艾买提,第十一届全国人大常委会副委员长乌云其木格,中国科学院院士吴常信为中国马会名誉会长。

中国马会具有社会团体法人资格,是4A 级全国性社会组织,2015 年获民政部“全国先进社会组织”称号,是国际纯血马登记管理委员会成员、世界阿哈尔捷金马(汗血马)协会发起成员国、亚洲赛马联盟成员、世界阿拉伯马协会成员、世界运动马繁育联盟成员、欧洲速步赛马联盟观察国成员。

中国马会坚持科学地建设和管理协会,开放地发展和壮大协会;求真务实,勇于创新,营造和维护一个健康、平等、和谐的中国马业大环境;公开、公平和公正地推动各项工作,积极为会员做好服务;推动中国马业进步和中国马文化繁荣,为构建世界马业新格局贡献中国智慧,为实现中华民族伟大复兴的中国梦贡献力量。

中国马会职能及业务范围包括:加强行业自律,制定行规行约,协助政府有关部门完善行业管理,促进中国马业科学健康发展;参与制定马匹和马场的饲养、防疫、专业人员配备等标准,推进标准化管理;依照有关规定,制定和实施马属动物血统登记、种马选择和马匹比赛成绩认定的管理规则;接受农业农村部委托,负责全国良种马登记工作,完善马匹品种登记制度,加强与国外马(驴)品种登记组织的交流与合作;按照国际纯血马登记管理委员会(ISBC)规则,全权负责中国境内纯血马的登记和管理工作;出版发行中国纯血马登记册;根据《中华人民共和国畜牧法》和《优良种畜登记规则》相关规定,组织实施马匹性能测定(速力、挽力、持久力等)的工作;开展行业培训,提供产业咨询,组织专业交流,促进产品开发与技术推广,加强专业化队伍建设;受政府委托承办或根据市场和行业发展需要,编制可行性研究报告和规划设计;组织展览会、展销会;组织或协调有关单位进行评比、拍卖、进出口贸易、赛马、马术竞技比赛及登记管理等工作,弘扬民族文化,提高全民马文化水平;依照有关规定,组织实施对马匹特别是运动马违禁物质检测;出版发行《中国马会》期刊和《中国赛马年鉴》等图书、图片、音视频资料;运营官方媒体。

中国马会成功创办和主办了中国马文化节,世界马文化论坛、2014 世界汗血马协会特别大会、第15届世界马医大会等具备全球马业影响力的重要品牌活动。另有中华民族大赛马、中国马会北京时间、中国马会天山论马、中国马会广东讲习、中国马会上海论道、中国马会海南作为、中国马会好马山东、全国各级马业社团组织会长书记秘书长论坛、“一带一路”国际马展等,倡导确立了每年9月19日为“中国爱马日”。

中国马会参与了《全国马产业发展规划(2020—2025 年)》《国家种业“十四五”发展规划》(马驴)、新疆及内蒙古自治区马产业发展规划的制定,促成了全国政协“关于加快推动我国现代马产业转型升级的建议”。积极推进中国马业的顶层设计,提出了“骏马外交”的时代理念;倡导筹建“国家仪仗马队”;与百余个国家地区、国际组织建立和保持着良好的合作关系。全力推进“国家种马场”立项;探索与推动符合国情的赛马事业和中国马业中长期发展规划。

2014 年 5 月 12 日,国家主席习近平在北京出席由中国马会主办的世界汗血马协会特别大会暨中国马文化节主席会议时指出,马在中华文化中具有重要地位,中国的马文化源远流长。建设国家需要万马奔腾的气势,推动发展需要快马加鞭的劲头,开拓创新需要一马当先的勇气。马是奋斗不止、自强不息的象征,马是吃苦耐劳、勇往直前的代表。

以梦为马是中国马会的初心,司牧安骥是中国马会的使命,知行合一是中国马会的准则,家国情怀是中国马会永恒的情怀。

地址:北京市顺义区北小营镇马辛庄 1 号院 中国马业协会(全国畜牧总站顺义基地)
电话:010 - 60488865 Fax: +86 - 010 - 60488665 邮箱: international@ chinahorse. org
网址:www. chinahorse. org

Appendix 8 附录8

Equine Resources on the Internet
马科学网络资源

Name/Topic 网名/主题	Website 网址
Agriculture Online 农业在线	http://www.agriculture.com/
American Academy of Equine Art 美国马艺术研究院	http://www.aaea.net/
Arabian Horse 阿拉伯马	http://www.arabianhorses.org/
Carriage Association of America 美国马车协会	http://www.caaonline.com/
Equestrian Sports 马术	http://www.equestrian.co.uk/
Equine Art 马艺术	http://www.equineart.com/
Equine Info 马信息	http://www.equineinfo.com/
Equine Research Center 马研究中心	http://www.erc.on.ca/
Equijournal 马杂志	http://www.equijournal.com/
Equiweb 马网	http://www.equiweb.co.uk/
Equiworld 马世界	http://www.equiworld.net/
Equusite 马科学	http://www.equusite.com/
Evolution of Horse 马进化	http://www.talkorigins.org/faqs/horses/
Harness Racing 轻驾车赛	http://www.harnessracing.com/
Horse Books 马书籍	http://www.horsebooksonline.com/
Horse Breeds 马品种	http://www.horsebreed.com/
Horse Carriage 马车	http://www.buggy.com/
Horse Farm Net 马场	http://www.horsefarm.com/
Horse Forum 马论坛	http://www.horseforum.net/
Horse Genetics 马遗传	http://www.vgl.ucdavis.edu/~lvmillon/
Horse Racing 赛马	http://horseracing.about.com/
Horse Racing of USA 美国赛马	http://www.bamhorseracing.com/
Horse Reproduction 马匹繁育	http://www.equine-reproduction.com/

（续）

Name/Topic 网名/主题	Website 网址
Horse Rules 马规章	http://www.horserules.com/
Horseshoes 马蹄铁	http://www.horseshoes.com/
Horseshoeing 蹄铁装钉	http://www.horseshoeing.com/
Horse Racing 赛马	http://www.horseracing.com/
Horse Tack 马具	http://www.horsetack.com/
Horseweb 马网	http://www.horseweb.com/
International Museum of the Horse 国际马博物馆	http://www.imh.org/
Kentucky Equine Research 肯塔基马研究	http://www.ker.com/
Kentucky Horse Park 肯塔基马公园	http://www.kyhorsepark.com/
Merck Veterinary Manual 默克兽医手册	http://www.merckvetmanual.com/
Model Horse 马模型	http://www.breyerhorses.com/
NetVet 网上兽医	http://netvet.wustl.edu/
Olympic Equestrian Events 奥运会马术	http://www.olympic.org/uk/sports/
Pony Club 矮马俱乐部	http://www.pcuk.org/
Saddle 马鞍	http://www.horsesaddleshop.com/
Sidesaddle 侧鞍	http://www.sidesaddle.org/
Thoroughbreds 纯血马	http://www.tbheritage.com/
Thoroughbreds Times 纯血马时代	http://www.thoroughbredtimes.com/
Toy Horse 玩具马	http://www.toyhorse.co.uk/
United States Pony Club 美国矮马俱乐部	http://www.ponyclub.org/
WebPony 矮马	http://www.webpony.com/
Western Horseman 西部骑手	http://www.westernhorseman.com/
Worldwide Farrier Directory 蹄铁匠	http://www.farriers.com/

Bibliography 参考书目

崔堉溪,1987. 养马学[M]. 北京:农业出版社.

李卫平,2002. 爱马人手册[M]. 北京:中国铁道出版社.

宋继忠,2004. 完全实用马术[M]. 北京:中国铁道出版社.

宋继忠,2005. 中英马术俱乐部培训系列教程[M]. 北京:中国铁道出版社.

谢成侠,1959. 中国养马史[M]. 北京:科学出版社.

谢成侠,1987. 中国马驴品种志[M]. 上海:上海科学技术出版社.

王铁权,1992. 中国的矮马[M]. 北京:北京农业大学出版社.

Belknap, Maria Ann, 1997. *Horsewords: The Equine Dictionary*[M]. North Pomfret: Trafalgar Square Publishing.

Bergsten, Vera, 1998. *Illustrated Dictionary of Equine Terms*[M]. Crawford: Alpine Publications.

Boulin-Néel, Lætitia, 2004. *Le Cheval Lusitanien*[M]. Paris: Editions Lariviere.

Brown, Jeremy Houghton, Powell - Smith, et al, 1984. *Horse and Stable Management*[M]. London: Granada Publishing.

Bryant, Jennifer O, 2000. *Olympic Equestrian*[M]. Lexington: Eclipse Press.

Cunha, Tony J, 1991. *Horse Feeding and Nutrition*[M]. San Diego: Academic Press.

Dossenbach, Hans, Monique, 1987. *The Noble Horse*[M]. New York: Portland House.

Evans, James Warren, 1977. *The Horse*[M]. New York: W. H. Freeman and Company.

Goody, Peter C, 1983. *Horse Anatomy: A Pictorial Approach to Equine Structure*[M]. London: J. A. Allen & Co Ltd.

Hughes, Christine E., Oliver, et al, 1987. *Practical Stable Management*[M]. London: Pelham Books.

Laidlaw, Caroline, 2001. *Amazing Horses*[M]. London: Penguin Books.

Marlin, David, Nankervis, et al, 2002. *Equine Exercise Physiology*[M]. Oxford: Blackwell Science.

Orwell, George, 1996. *Animal Farm*[M]. Orlando: Signet Classics.

Parker, Rick, 2003. *Equine Science*[M]. 2nd edition. New York: Thomson Delmar Learning.

Price, Steven D, et al, 1998. *The Whole Horse Catalog*[M]. 3rd edition. New York: Simon & Schuster.

Rossdale, Peter, 1981. *Horse Breeding*[M]. Devon & New York: David & Charles Publishers.

Sewell, Anna, 2007. *Black Beauty*[M]. London: Penguin Classics.

Shaffer, Peter, 1978. *Equus*[M]. London: Penguin Books.

Sponenberg, D. Phillip, Beaver, et al, 1983. *Horse Color*[M]. College Station: Texas A & M University Press.

US Olympic Committee, 2001. *A Basic Guide to Equestrian*[M]. Torrance: Griffin Publishing Group.